AIR POLLUTION HANDBOOK

McGRAW-HILL HANDBOOKS

ABBOTT AND STETKA · National Electrical Code Handbook, 10th ed.
ALIJIAN · Purchasing Handbook
AMERICAN INSTITUTE OF PHYSICS · American Institute of Physics Handbook, 2d ed.
AMERICAN SOCIETY OF MECHANICAL ENGINEERS · ASME Handbooks:
  Engineering Tables
  Metals Engineering—Design
  Metals Engineering—Processes
  Metals Properties
AMERICAN SOCIETY OF TOOL AND MANUFACTURING ENGINEERS:
  Die Design Handbook
  Handbook of Fixture Design
  Manufacturing Planning and Estimating Handbook
  Tool Engineers Handbook, 2d ed.
BEEMAN · Industrial Power Systems Handbook
BERRY, BOLLAY, AND BEERS · Handbook of Meteorology
BLATZ · Radiation Hygiene Handbook
BRADY · Materials Handbook, 9th ed.
BURINGTON · Handbook of Mathematical Tables and Formulas, 3d ed.
BURINGTON AND MAY · Handbook of Probability and Statistics with Tables
CARROLL · Industrial Instrument Servicing Handbook
COCKRELL · Industrial Electronics Handbook
CONDON AND ODISHAW · Handbook of Physics
CONSIDINE · Process Instruments and Controls Handbook
CROCKER · Piping Handbook, 4th ed.
CROFT AND CARR · American Electricians' Handbook, 8th ed.
DAVIS · Handbook of Applied Hydraulics, 2d ed.
DUDLEY · Gear Handbook
ETHERINGTON · Nuclear Engineering Handbook
FACTORY MUTUAL ENGINEERING DIVISION · Handbook of Industrial Loss Prevention
FINK · Television Engineering Handbook
FLÜGGE · Handbook of Engineering Mechanics
FRICK · Petroleum Production Handbook, 2 vols.
GUTHRIE · Petroleum Products Handbook
HARRIS · Handbook of Noise Control
HARRIS AND CREDE · Shock and Vibration Handbook, 3 vols.
HENNEY · Radio Engineering Handbook, 5th ed.
HUNTER · Handbook of Semiconductor Electronics, 2d ed.
HUSKEY AND KORN · Computer Handbook
JASIK · Antenna Engineering Handbook

JURAN · Quality Control Handbook, 2d ed.
KALLEN · Handbook of Instrumentation and Controls
KETCHUM · Structural Engineers' Handbook, 3d ed.
KING · Handbook of Hydraulics, 4th ed.
KNOWLTON · Standard Handbook for Electrical Engineers, 9th ed.
KOELLE · Handbook of Astronautical Engineering
KORN AND KORN · Mathematical Handbook for Scientists and Engineers
KURTZ · The Lineman's Handbook, 3d ed.
LA LONDE AND JANES · Concrete Engineering Handbook
LANDEE, DAVIS, AND ALBRECHT · Electronic Designers' Handbook
LANGE · Handbook of Chemistry, 10th ed.
LAUGHNER AND HARGAN · Handbook of Fastening and Joining of Metal Parts
LE GRAND · The New American Machinist's Handbook
LIDDELL · Handbook of Nonferrous Metallurgy, 2d ed.
MAGILL, HOLDEN, AND ACKLEY · Air Pollution Handbook
MANAS · National Plumbing Code Handbook
MANTELL · Engineering Materials Handbook
MARKS AND BAUMEISTER · Mechanical Engineers' Handbook, 6th ed.
MARKUS · Handbook of Electronic Control Circuits
MARKUS AND ZELUFF · Handbook of Industrial Electronic Circuits
MARKUS AND ZELUFF · Handbook of Industrial Electronic Control Circuits
MAYNARD · Industrial Engineering Handbook
MEITES · Handbook of Analytical Chemistry
MERRITT · Building Construction Handbook
MOODY · Petroleum Exploration Handbook
MORROW · Maintenance Engineering Handbook
PERRY · Chemical Business Handbook
PERRY · Chemical Engineers' Handbook, 3d ed.
SHAND · Glass Engineering Handbook, 2d ed.
STANIAR · Plant Engineering Handbook, 2d ed.
STREETER · Handbook of Fluid Dynamics
STUBBS · Handbook of Heavy Construction
TERMAN · Radio Engineers' Handbook
TRUXAL · Control Engineers' Handbook
URQUHART · Civil Engineering Handbook, 4th ed.
WALKER · NAB Engineering Handbook, 5th ed.
WOODS · Highway Engineering Handbook
YODER, HENEMAN, TURNBULL, AND STONE · Handbook of Personnel Management and Labor Relations

AIR POLLUTION HANDBOOK

Edited by
PAUL L. MAGILL, SENIOR SCIENTIST
FRANCIS R. HOLDEN, PH.D., SENIOR CHEMIST
CHARLES ACKLEY, EDITOR

Stanford Research Institute
Menlo Park, California

Editorial Consultant
FREDERICK G. SAWYER, PH.D.
Director of Industrial and Public Relations
The Ralph M. Parsons Company
Los Angeles, California

McGRAW-HILL BOOK COMPANY
NEW YORK TORONTO LONDON

1956

AIR POLLUTION HANDBOOK

Library of Congress Catalog Card Number: 55-11934

7 8 9 - MP - 9 8

39490

CONTRIBUTORS

Richard D. Cadle, Ph.D., Manager, Atmospheric Chemical Physics Section, Stanford Research Institute, Menlo Park, California.

Jacob Cholak, Ch. E., Associate Professor and Director of Division of Industrial Hygiene Technology, Institute of Industrial Health, University of Cincinnati, Cincinnati, Ohio.

A. M. Clark, Assistant Chief Engineer, Custodis Construction Company, Inc., 157 Chambers Street, New York, New York.

R. C. Corey, Chief, Division of Solid Fuels Technology, United States Bureau of Mines, Pittsburgh, Pennsylvania.

Harry C. Ehrmantraut, Ph.D., Senior Biophysicist, Stanford Research Institute, Menlo Park, California.*

L. L. Falk, Ph.D., Meteorologist, Engineering Department, E. I. du Pont de Nemours & Company, Inc., Wilmington, Delaware.

Sheldon K. Friedlander, Ph.D., Department of Chemical Engineering, Columbia University, New York, New York.

Carl A. Gosline, Jr., Field Group Supervisor, Engineering Department, E. I. du Pont de Nemours & Company, Inc., Wilmington, Delaware.

E. N. Helmers, S.B., S.M., Engineering Department, E. I. du Pont de Nemours & Company, Inc., Wilmington, Delaware.

Russell H. Hendricks, Ph.D., Chemist, Department of Agricultural Research, American Smelting & Refining Company, Salt Lake City, Utah.

H. F. Johnstone, Ph.D., Professor, Head, Division of Chemical Engineering, University of Illinois, Urbana, Illinois.

Morris Katz, M.Sc., Ph.D., Chairman, Canadian Section, Technical Advisory Board of Air Pollution, International Joint Commission, Defence Research Chemical Laboratories, Ottawa, Canada.

K. E. Lunde, Manager, Industrial Air Pollution Research Section, Stanford Research Institute, Menlo Park, California.

Paul L. Magill, Senior Scientist, Department of Chemistry, Stanford Research Institute, Menlo Park, California.

Hans Neuberger, Sc.D., Head, Department of Meteorology, The Pennsylvania State University, State College, Pennsylvania.

Gordon W. Newell, Ph.D., Senior Biochemist, Stanford Research Institute, Menlo Park, California.

Andrew A. Nichol, Senior Ecologist, Stanford Research Institute, Menlo Park, California.

Hans A. Panofsky, Ph.D., Professor, Department of Meteorology, The Pennsylvania State University, State College, Pennsylvania.

* Present affiliation: Spinco Division, Beckman Instruments Corp., Belmont, California.

Max S. Peters, Professor, Division of Chemical Engineering, University of Illinois, Urbana, Illinois.

John J. Phair, M.D., Dr. P.H., Professor, Department of Preventive Medicine and Industrial Health, College of Medicine, University of Cincinnati, Cincinnati, Ohio.

Paul H. Phillips, Ph.D., Professor of Biochemistry, University of Wisconsin, Madison, Wisconsin.

Ilia G. Poppoff, Associate Physicist, Stanford Research Institute, Menlo Park, California.

Walter H. Rupp, B.S., Technical Advisor, Esso Engineers, Esso Research and Engineering Company, Linden, New Jersey.

H. B. Schneider, Vice President, Custodis Construction Co., Inc., 22 West Monroe Street, Chicago, Illinois.

Frank L. Seamans, LL.B., Smith, Buchanan, Ingersoll, Rodewald and Eckert, Attorneys at Law, Pittsburgh, Pennsylvania.

Z. Sekera, Ph.D., Professor, Department of Meteorology, University of California at Los Angeles, Los Angeles, California.

Konrad T. Semrau, Associate Chemical Engineer, Stanford Research Institute, Menlo Park, California.

Leslie Silverman, Sc.D., Associate Professor of Industrial Hygiene Engineering, School of Public Health, Harvard University, Boston, Massachusetts.

Wayne T. Sproull, Ph.D., Head, Electrical Research, Western Precipitation Corporation, Los Angeles, California.

Carsten Steffens, Ph.D., Senior Scientist, Stanford Research Institute, Menlo Park, California.

Moyer D. Thomas, Sc.D., Chemist and Agronomist, Department of Agricultural Research, American Smelting & Refining Company, Salt Lake City, Utah.*

* Present affiliation: Senior Scientist, Department of Chemistry, Stanford Research Institute, Menlo Park, California.

FOREWORD

In dealing with a problem so complex as air pollution, the cooperation of experts in many different disciplines is necessary. Each of the contributors to this Handbook was chosen because he could speak authoritatively on one or more phases of the subject, and each is to be regarded as responsible for the material in his section.

The Handbook is designed to be used section by section. Each section is independent and treats a single aspect of the broad problem. Generally, Sections 1 and 2 state the situation, Sections 3 through 6 pertain to the science of air pollution, Sections 7 through 9 discuss effects, and Sections 10 through 14 cover various corrective techniques.

Thanks are due to the members of the Advisory Board of the Handbook for their help in selecting contributors and in planning the book; to Stanford Research Institute for providing the time of authors, editors, and proofreaders; to Gordon B. Bell for his over-all assistance; and to Merle C. Holmwood, who typed the entire manuscript.

PREFACE

The Air Pollution Handbook brings together in readily available form useful information on the major aspects of the problem of air pollution.

Such a book seems necessary in view of the increasing importance of air pollution, both as a field for scientific investigation and as a factor in our daily lives. The trend toward urbanization and greater industrial concentration has led, among other things, to congestion in residential areas and heavier use of city highways. These, in turn, have resulted in more severe and more widespread contamination of our atmosphere. Where man is, there pollution is. Whereever man's life processes, both individual and societal, pour wastes into a small area, serious contamination results. If the trend continues, moreover, the problem of air pollution is likely to become even more drastic in the future.

The public and industry are aware of the situation, and much has been done in the way of abatement and control. However, much still needs to be done. Those concerned with air pollution have been handicapped by the complexity of the problem. For one thing, "air pollution" means different things to different people. To the householder, it may mean eye irritation and soiled clothing; to the farmer, damaged vegetation and, through ingestion of this vegetation, damaged livestock; to the airline pilot, dangerously reduced visibility; to industry, problems of process control and public relations. "Air pollution" in Los Angeles is not the same as "air pollution" in Pittsburgh, and both are different from the situation called by the same name in Detroit or London.

There are further complications. It is sometimes hard to remember that remedial action on any given problem may require chemists, engineers, meteorologists, plant physiologists, biologists, and others. All of these or any combination of them may contribute to the solution of a particular problem. It is not always easy for these men to understand the concepts, and sometimes even the language, of workers in sciences other than their own. Too, pertinent published information is widely scattered. The field or laboratory worker, on the other hand, is sometimes confronted with an embarrassment of riches: too many techniques to choose from and no way to tell which are the best and most satisfactory for his individual problem. In addition, the application of scientific techniques to some of the aspects of air pollution is still a comparatively recent development.

Public representatives responsible for city planning likewise have their difficulties. What are the questions raised by the admission of new industry to an area? Will a pollution problem result? What factors of terrain and meteorology can be utilized in planning for new industrial zones so as to minimize the danger of such a problem?

The industrialist seeking to improve plant operation and maintain good public relations needs to know the dangers of air pollution. He needs to know what methods are available for the control of stack emission, for example, and which are the best and most economical.

Obviously there is need for a compilation of information on the subject, for an authoritative statement of what we know and where we stand, for a sort of "progress report" on the state of the art. Since the nature of a subject determines the nature of material written about it, such a volume must contain information on widely differing levels: some highly theoretical, some quite practical. Some aspects need research to put to use the valuable basic formulas thus far evolved; on some the engineering work has already been done.

In short, a book is needed which bridges the gaps between the various disciplines involved, and between these disciplines and interested laymen such as city planners and industrialists. A book is needed which provides basic source material on the many facets of air pollution.

Such are the purposes of this book.

—The Editors

CONTENTS

SECTION 1

AIR POLLUTION SOURCES AND THEIR CONTROL

BY WALTER H. RUPP

1.1 INTRODUCTION

It seems fitting and proper to examine the problem of air pollution control from the viewpoint of you—a human being. As an average man, you breathe 22,000 times each day. You take in 35 lb of air each day. The air is your main link to life. It far exceeds your consumption of food and water.

Therefore, you have a general interest in air pollution control. You may have a specific interest if you are an engineer, a smoke-control officer, a student, an executive, a lawyer, a public-relations man, a medical doctor, a botanist, a chemist, or a plant operator.

A little more background before plunging into the job: many billions of tons of air pass over the United States population each day, say a figure of 5 billion tons per day in an air layer extending only 25 ft above the ground.

How then can this huge weight of fresh air become polluted? The answer is that it does not, not all of it. But local air over counties, cities, and plant areas does become polluted because the capacity of the air flowing over such local areas is not sufficient to dilute the contaminants below a certain threshold level. Hence, the fresh air becomes unclean, polluted air. The threshold level where fresh air becomes polluted air can be fixed with certainty for many individual gases and dusts. But rarely can air pollution be traced to one pollutant. The normal condition includes mixtures and combinations of many contaminants. Effects of contaminant intermingling and accumulation are only now under study. It is a fruitful field for investigation and research.

What of the sources of air pollution? Every person contributes to it in some degree. Some say it is the price of industrialization. If so, projection into the future leads to the conclusion that added knowledge and action are required to avoid widespread air pollution in the United States. Studies are proceeding on many fronts. Air over industrialized areas is being assayed for vapors, gases, droplets, spores, pollen grains, dusts, bacteria, and radioactive particles. At the same time, the possible sources of these contaminants are being studied.

The theme of this section is a phase of air pollution control sometimes overlooked, the elimination of primary sources of pollution. Use of improved process technology is the key.

Control of air pollution sources can be practiced in old and new installations. A blend of research, development, and application of improved technology is the nation's traditional method of raising living standards. This method has been used on past air pollution problems and will continue to be used.

1.2 ABATEMENT AND CONTROL OF SOURCES

Source control in the pure sense deals with the elimination, before or during ultimate consumption, of potential air contaminants contained in raw materials. A classic example of this definition is the removal of all sulfur from raw coal and crude petroleum. The sulfur is collected in a pile at the mine or well for by-product use. The purified coal and petroleum are still processed and used as at present. The net result—a decrease in air contamination.

Unfortunately, lack of technical knowledge and economic incentive prohibits full practice of this classical method. Control of air pollution sources must strike a compromise in all but a few isolated cases. Thus, in the coal and petroleum example, the effort to remove unwanted sulfur is now distributed along the line from raw-material sources to ultimate consumer. Some sulfur is removed at the mine and well; more sulfur is removed at the washery and refinery; the remaining sulfur is usually discharged to the atmosphere or otherwise handled by the ultimate consumer in accordance with his local requirements and regulations.

Source control and abatement of formed contaminants are complementary practices in the campaign against air pollution. Source control prevents the emission of contaminants to the atmosphere; abatement renders the emission of contaminants to the atmosphere harmless and inoffensive.

The two control fields are not 100 per cent efficient. They normally overlap at many points. The usual answer is to combine the two methods for the benefit of the whole. In the coal and petroleum example, again, as much sulfur as is feasible can be converted to hydrogen sulfide in some processing operation and used to manufacture sulfuric acid—that is termed source control. The remaining sulfur can be burned and the resulting mixture of sulfur dioxide and flue gases scrubbed and discharged at high velocity through a tall stack—that is termed abatement.

1.3 THE SOURCE-CONTROL PROBLEM

The high living standard of the United States is due in large part to the strong upward trend in industrial production. New ideas, research, invention, development, risk taking, and mass production are all ingredients of progress. In the desire for higher industrial production and efficiency, there were few in the early years who observed that social relationships and amenities were not keeping step with industrial production. This was true for such "public-relations" matters as water pollution, industrial safety, and social-security benefits, as well as air pollution.

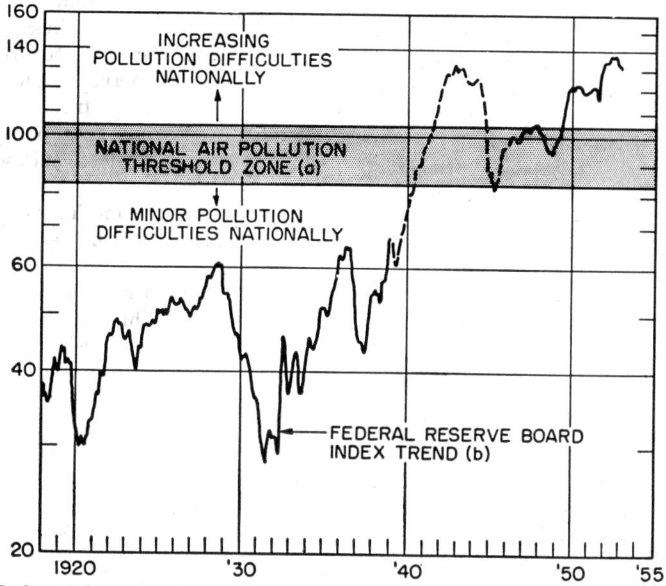

FIG. 1-1. Industrial production and air pollution threshold. (*a*) Threshold zone superimposed by author. (*b*) Long-term trend of Federal Reserve Board Index of Industrial Production, average 1947–1949 = 100. World War II period shown by dashed line pending more extended review of war-production figures. (*Adapted from chart in National City Monthly Letter on Business and Economic Conditions, p.* 5, *The National City Bank of New York, January,* 1954.)

The air pollution problem started at the time when the effort put into industrial production began to rise faster than the effort to control air pollution sources. There is no month or year or perhaps even decade in the past when it can be said, "That is when it started," nor is it necessary to fix the time. Even when no industry existed, there were forms of natural air pollution.

What is important is to take a long-range view into the future, to plan the trend of source control from here on, in all its varied applications. Some background for guidance in future planning seems worthwhile.

When an air pollution threshold zone is reached, greatly expanded efforts are necessary to control pollution. This section considers the control of pollution at the source by using tools such as improved chemical reactions, processes, products, and operating techniques. The alternative, purposely decreasing industrial production and lowering living standards below the threshold zone, does not warrant serious consideration.

Figure 1-1 attempts to illustrate a principle by linking the industrial-production trend with an "air pollution threshold"—both on a national basis. It suggests the necessity for increased effort to reduce air pollution as industrial production rises. The national air pollution threshold is fixed arbitrarily as having been reached sometime during the 1940s when major difficulties occurred at Los Angeles and Donora. St. Louis and Pittsburgh had acted effectively somewhat earlier to improve their local conditions. A number of other difficulties and civic accomplishments could be cited in these and subsequent years as a part of the general outbreak of serious pollution problems attributed to threshold crossing (3). Plainly, the capacity of the atmosphere to disperse the air contaminants from man-made sources had been exceeded. Threshold crossings were only brief and rarely violent, but the implications were clear. Air pollution control had become of national interest. Hundreds of communities increased past efforts or initiated studies of their local conditions.

A sketch of the social-technical effects of this threshold crossing might be of interest. The entire May, 1947, issue of *Industrial and Engineering Chemistry* was devoted to papers on industrial waste as a follow-up of the April, 1947, meeting of the American Chemical Society at Atlantic City. Most of the papers described industrial-waste problems associated with water and stream purity. Some small attention was given to air-purity problems also. In March, 1949, the American Chemical Society discussed "Atmospheric Contamination and Purification" at the San Francisco meeting. A number of other societies, associations, and technical and medical groups held allied discussions in that general period. Something of a climax was reached in May, 1950, when a national meeting was held in Washington, D.C. This was the U.S. Technical Conference on Air Pollution. It had as its chairman Dr. Louis C. McCabe. Subsequently, the meeting's papers and discussions were published in a single-volume book (11).

1.4 CONTAMINANT GROUPS

An inventory of air contaminants is a necessary first step to control. Starting off with a look at the national problem seems to provide a desirable foundation. Air pollution is normally a local problem at the city or county level. The problem is infrequently an interstate one, and even more rarely international. Any implication that source control must follow a national pattern like peace control through armaments should be avoided. Each interested part of the nation should want to survey its own local conditions and plan control accordingly.

Much can be learned from a national inventory of air contaminants. Such a survey would serve to guide thinking and to direct control efforts.

1.4.1 Mass Emission of Contaminants

Figure 1-2 is an estimation of the relative quantities of the major air contaminants over the continental United States.

The data used for preparing this chart are quite scattered, representing averages or totals over an area of 3 million square miles, and are merely indicators at best. An examination of Fig. 1-2 will disclose some interesting magnitudes, however. Masses of three natural contaminants are included for reference. The industrial contaminants can be divided into four major groups.

The first three, smoke, dust, and sulfur oxides, are predominantly by-products of what might be termed "external combustion." These are the leftovers from the heat and power generation that supplies the energy for the industrial revolution.

The fourth group is mostly the by-product of more modern industry. The processing of raw materials to obtain better products results in increased air-contamination problems. The by-products of internal combustion are included to emphasize the complex reactions in gasoline and diesel engines.

A word about carbon dioxide and carbon monoxide might be in order here. These two gases are by far the most common products of fuel combustion (along with water vapor). None of these three gases seems to qualify as an air contaminant in the accepted sense. The naturally pure atmosphere contains carbon dioxide and water vapor as necessary gases for living animal and plant life. Carbon monoxide is normally converted to carbon dioxide by reactions in the atmosphere. Even before conversion, the carbon monoxide concentration outdoors rarely approaches the nuisance threshold on a local or national basis.

The mass emission of carbon dioxide as a result of fuel combustion in the United States during 1953 was estimated to be about 3 billion short tons per year. Carbon

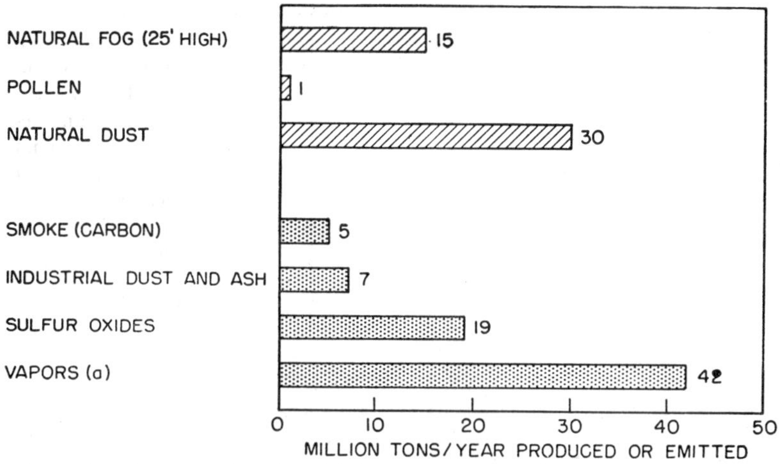

FIG. 1-2. Major air contaminants. (*a*) Includes hydrocarbons, nitrogen dioxide, aldehydes, fluorides, organic acids, ammonia, tars, etc. (*Calculated using data from numerous published sources.*)

monoxide emission approximates about 60 million short tons per year. This item alone about equals the combined total of all other industrial contaminants. Should carbon monoxide ever reach an undesirable concentration level, there are many methods available for converting it to carbon dioxide.

1.4.2 Some Natural Contaminants

Natural contaminants from sea, forest, desert, volcano, and field have been with us a long time. There seems to be little point in dwelling on these contaminants other than to say that they are not in themselves needed to support human life. This is also true of industrial contaminants. The real problem is to eliminate harmful man-made contaminants wherever they occur. While this is being accomplished, it might bring some comfort to recognize the presence of unnecessary natural air contaminants also.

Natural fog is generally not classified as an air pollutant. Fog containing droplets of air contaminants has, however, been included as one of our more prevalent and unwelcome natural pollutants. For comparison, the yearly weight of water contained in fogs to a height of 25 ft is somewhat less than the yearly weight of all sulfur dioxide released to the United States atmosphere.

Pollen dust is another natural pollutant. The total ragweed-pollen emission weighs far less than that of natural dust. Yet, because of its peculiarly irritating properties, about 4 million North Americans are affected regularly each year with varying degrees of hay-fever discomfort.

Natural dust appears to be present in far greater weight than industrial smoke and dust combined. In many cases human activities are also responsible for increased natural dust.

1.4.3 Industrial Contaminants

Changing to an industrial civilization has increased the burden of foreign contaminants in the atmosphere. The present state of industrial progress finds the United

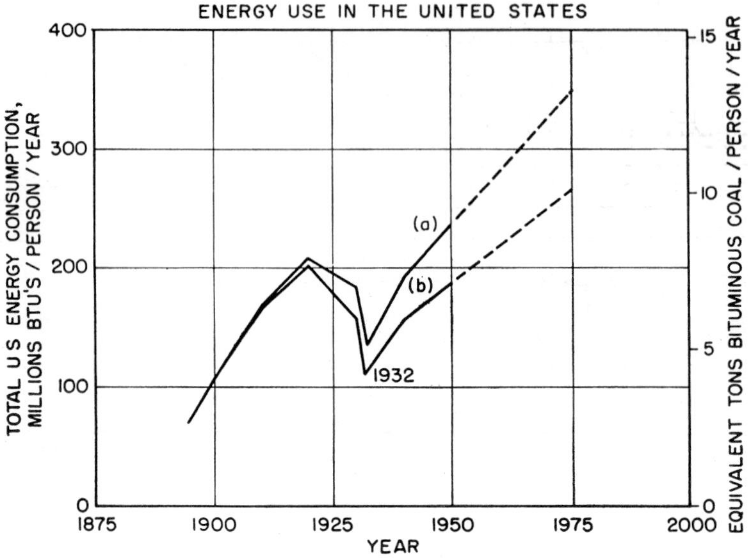

FIG. 1-3. Energy use in the United States. (*a*) Including fuels, chiefly gasoline, for internal-combustion engines. (*b*) Excluding fuels, chiefly gasoline, for internal-combustion engines. (*Calculated using data from the following references: W. S. Paley, "Resources for Freedom," vol. III, A Report to the President by the President's Materials Policy Commission; Petroleum Facts and Figures, 9th and 10th eds., American Petroleum Institute, New York; U.S. Bureau of Mines; U.S. Bureau of the Census.*)

States with a widely diversified list of man-made pollutants. About 100 contaminants have been identified and probably many others could be.

Of the four groups summarized in Fig. 1-2, most past effort has been exerted in reducing the first three (smoke, dust, and sulfur dioxide emissions). Present concentrations of smoke, dust, and sulfur dioxide can be traced chiefly to the increased use of fuel energy. The trends in energy use and efficiency in the United States are shown in Figs. 1-3 and 1-4.

Two trends are included on Fig. 1-3. The top curve shows the total yearly energy consumption per person in the United States and includes the fuels consumed in internal-combustion engines (chiefly gasoline). The lower curve excludes internal-combustion fuels. This figure illustrates the fast-growing consumption of internal-combustion fuels as represented by the increasing gap between the curves. Conversion of raw materials, such as coal, to useful energy has been steadily increasing in efficiency since the 1920s. This, in effect, is another example of source control working to decrease air pollution from external combustion. At the same time, the demand for useful energy from external-combustion sources has been increasing since the early 1930s. Predictions say that the trend of both curves will be generally upward during the next two decades.

The trend in energy efficiency—that is, the percentage of potential energy in fuels converted to useful purposes—is shown in Fig. 1-4.

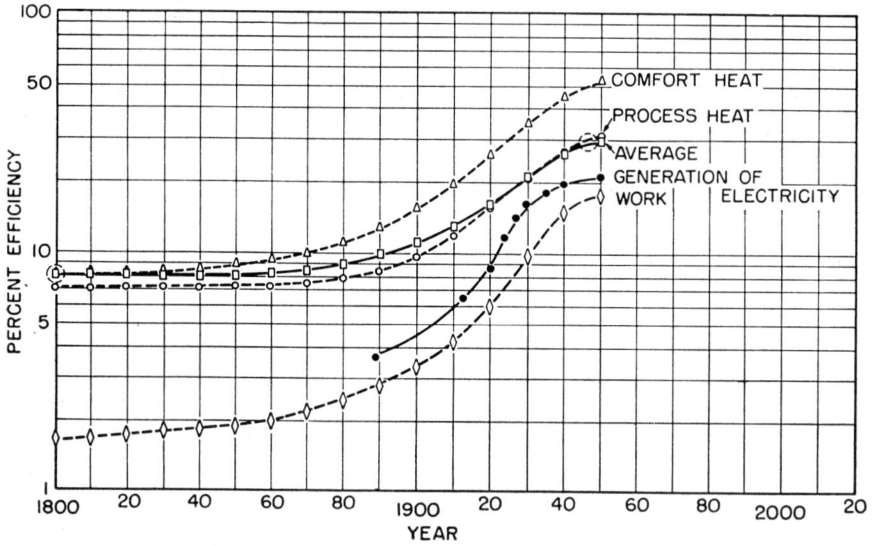

FIG. 1-4. Energy efficiency. (*After P. C. Putnam, "Energy in the Future," p. 90, D. Van Nostrand Company, Inc., New York,* 1953. *Courtesy of United States Atomic Energy Commission.*)

The fourth category (other vapors) is a broad grouping. Many of the pollutants included in the mass shown are of comparatively recent origin or have only recently been suspected. Some might be exonerated when research studies and surveys are completed. Other contaminants now unknown undoubtedly will be added in the future.

Table 1-1 outlines a further breakdown of air contaminants by type.

Table 1-1. Air Contaminants

Group	*Specific Examples*
Fine solids (less than 100μ in diameter); Coarse particulates (greater than 100μ in diameter)	Carbon, fly ash, $CaSiO_3$, ZnO, $PbCl_2$
Sulfur compounds	SO_2, SO_3, hydrogen sulfide, mercaptans
Organic compounds	Aldehydes, hydrocarbons, tars
Nitrogen compounds	NO, NO_2, NH_3
Oxygen compounds	O_3, CO, CO_2
Halogen compounds	HF, HCl
Radioactive compounds	Radioactive gases, aerosols, etc.

The classifications themselves are few in number. Practically every contaminant released anywhere in the United States falls into these groupings. Combinations of contaminants may result in amplified effects. Studies by various investigators have indicated that some contaminants undergo chemical reaction in the atmosphere. The end products sometimes are more noxious than the original contaminants. For example, it is claimed that unsaturated hydrocarbons react with NO_2 in sunlight to form smog (see Chemical Reactions in Contaminated Atmospheres, Art. 3.5 of this handbook).

Reactions in the atmosphere must be recognized and considered in planning constructive source-control programs. The sampling and identification of atmospheric reaction by-products is a new phase of air pollution research currently gaining much attention. The study of these combined effects on human, animal, and plant life comprises an important field for research.

It should not be implied that every last one of the air contaminants falling within these groupings must be eliminated to restore satisfactory air purity. Rather, the harmful contaminants should be identified and reduced selectively for practical results.

1.4.4 Ground-level Pollution Effects

There are several other variables worthy of mentioning in appraising potential air pollution sources. Ground-level concentration of contaminants is now recognized by air pollution technologists as the controlling factor (rather than the older concept of contaminant concentration in a stack, for example). Theoretically derived relationships by Bosanquet and Pearson (2) can be simplified for source-control illustration by the following equation:

$$\text{Maximum ground concentration of contaminant} = \frac{(\text{constant}) \times (\text{mass rate of emission of contaminant})}{(\text{wind velocity}) \times (\text{elevation of emission squared})}$$

Thus, for every chemical contaminant or group of contaminants there is a maximum threshold concentration which can be exceeded only at the risk of nuisance or damage. This threshold varies widely, depending upon chemical type and sensitivity of the person, animal, or plant living at ground level near the contaminant source. The following is a cursory examination of stack dispersion. For a more detailed technical discussion of the problem, see Evaluation of Weather Effects, Sec. 5 of this handbook.

The ground-level concentration of contaminant varies inversely according to wind velocity to the first power and height of emission to the second power. Thus, if the contaminant is released into a stream of increased-velocity wind, the ground-level concentration will be lower because of better diffusion with air. Also, if the contaminant is released at a higher point in the atmosphere, the ground-level concentration will be lower because of better diffusion with air.

Some general conclusions that can be drawn from this formula are of significance. Ground-level concentration of a gaseous contaminant released from a stack is not dependent on the stack concentration but rather is directly dependent on the mass rate of emission of the contaminant without regard to dilution within the stack. If the wind velocity doubles, the ground-level concentration of contaminant will be one-half that at the lower wind velocity. If the stack height is doubled, the ground-level concentration of contaminant will be one-quarter that with the lower stack height.

This is a greatly simplified relationship. The simplicity assumes smooth, flat terrain, single-point source of contaminant, stable and nonturbulent wind currents, no rain, gaseous contaminants, or suspended fine particles, and the resulting maximum ground concentration occurring at a distance of about ten stack heights from the emission source. The first investigators and other subsequent investigators have developed refinements of the general theory to permit wider application under practical conditions.

It might be well to mention a few practical hints here. Freedom in selection of stack layout and process and mechanical design features varies widely from industry to industry. In general, it is beneficial from an air pollution control viewpoint to build several stacks and disperse them over a wide area rather than to concentrate the emission at a single point source. Effective stack height can be gained by discharging the gases at as high a temperature as possible to give them buoyancy. Similarly, effective stack height can be gained by discharging the gases at as high a velocity as possible. In some cases, natural draft is insufficient and forced draft must be used. Each situation is specific but fortunately can be reduced to a sound technical and economic analysis.

1.5 SOURCE-CONTROL METHODS

Existing plants and processes usually can be altered to decrease air pollution by adding new abatement equipment. Additions of electrical precipitators, cyclones, scrubbing towers, high stacks, and similar equipment are only a few general possibilities. An example of process modification is substitution of oxygen for air in gas manufacture and in blast furnaces. The resulting vent gases are lower in volume and contaminant content. Improvements in manufacturing cost and product yield and quality often result from this method. Future trends in this direction will be discussed later in this section.

1.5.1 Practicing Source Control

Control of air pollution sources is usually most easily attained in the development stage of a new project and in the architecture-design phase of a completely new plant. It is here that the best arrangement of new process ideas can be applied without complications of dismantling and revising existing equipment.

The advance selection of new pollution-free processes and practices is largely in the power of planning designers working with management cooperation. This team has become increasingly air pollution conscious. The management approach might be typified by the 1952 speech of Henry du Pont (6) in which he said that (among other attitudes) "our top management policy is to insist, before approval of a project is given, that all pollution problems be solved in advance. . . ." The process-engineering approach presented in 1952 by Gosline (7) before the Manufacturing Chemists' Association suggests that planning for waste disposal should begin in the research and pilot-plant stage and be followed by creation of a waste-disposal flow sheet listing all sources of potential contamination with composition and characteristics. This team approach is widely accepted throughout the United States today.

1.5.2 Confidential Source-control Methods

The processes and practices used in source control are usually covered by patents or licenses available to any interested second party. Thus, the cost of research and development is recovered and an incentive provided to continue technological progress. At times, however, the processes must be considered by a company as being confidential. These processes normally involve those for which considerable cost was required to develop "know-how" without justifying patent coverage. Every effort is made by most companies to draw a line between new air pollution technology for the public good and new process and product improvements for preservation of the company's competitive situation.

Perhaps most new technical information on air pollution is derived from government groups such as the Bureau of Mines and university-industry-sponsored organizations. A survey (1) by the American Society of Mechanical Engineers listed 37 groups at work on air pollution research and technology in 1953.

1.6 THE IMPORTANCE OF ECONOMICS

The job of the engineer is to do at reasonable cost the things that anyone can do with unlimited resources. There are many unknowns in the effects of air pollution on men, animals, plants, and materials. Much research is planned and will be required to tie down these effects. The control phase of air pollution is not quite so complex, technically speaking. Almost every engineering problem could be solved today—over a period of time and at a cost. The real question is where and how to direct control efforts to the greatest public advantage.

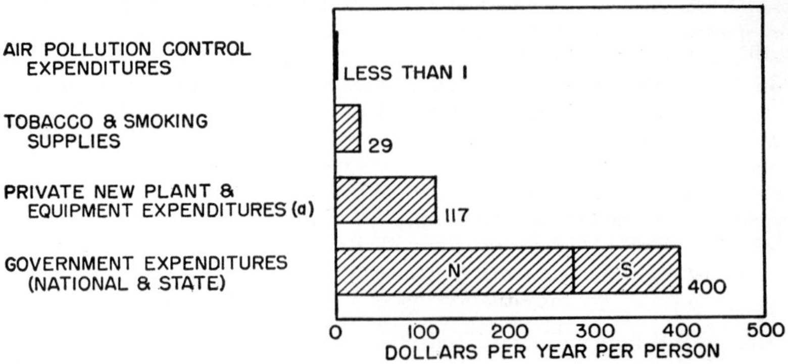

FIG. 1-5. Air pollution control expenditures as compared with other selected expenditures, covering 1950—a "normal" year. (*a*) Excludes agriculture. (*Calculated using data from the U.S. Department of Commerce, including the Bureau of the Census, and other sources.*)

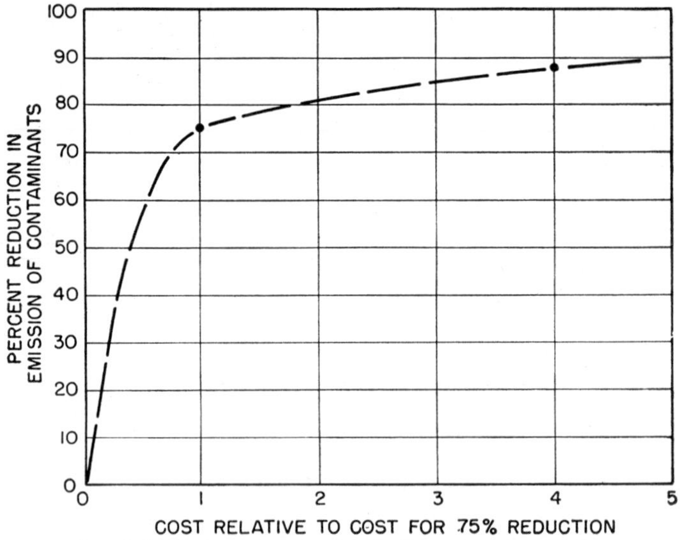

FIG. 1-6. Cost vs. degree of reduction of contamination.

Aside from knowing the specific air contaminants which must be reduced first, the next question is cost of reduction. Figure 1-5 compares air pollution control expenditures, new plant expenditures, and government expenditures per person. The cost of one item of personal enjoyment is also injected for comparison. The author is inclined to believe that the air pollution control costs shown in Fig. 1-5 are a very minimum. In most cases, industry is inclined to list as control costs only those items not showing any tangible monetary return. The public-relations, employee-relations, and "good-neighbor" expenditures are generally listed. Some small percentage of the costs does show by-product credits.

Some studies for one industry indicate that, instead of less than $1 per year per person, a maximum figure might be nearer $5 per year per person. The maximum figure includes many costs leading to reduced air pollution whether justified or not. Thus, the cost of any new process or plant bought as a process improvement and causing a *reduction in air pollution* would be accounted for in the maximum.

Whether or not the foregoing cost concept is used, air pollution control expenditures seem moderate. Some increase in the future is predicted, but attention is required to be certain that value is received.

Many authorities seem to agree on a tentative future target. The required degree of correction might be based on contaminant emissions during the 1945 to 1948 era. Figure 1-6 attempts to blend cost and degree of correction to attain a satisfactory ground concentration of contaminant. If it is possible to speak nationally on a complex relationship such as this one, it appears that a 75 per cent reduction in air contamination from the 1945 to 1948 era is a fair target. It is at least a common goal for uniformity. A range higher than this will at present increase the cost burden without proportional benefits in a majority of cases. This figure is offered as a guide for planning. Each local pollution situation must be studied to arrive at a true specific curve, but the curves will generally be similar and will follow the law of diminishing returns.

1.7 SOURCE CONTROL OF MAJOR EMISSIONS

Principles of source control are illustrated best by specific examples. A cross section of various activities where sources of air pollution are being brought under control has been selected in the hope that continued efforts and new ideas will be stimulated. The principles followed will be along practical and commercial lines rather than theoretical concepts.

The order of consideration is more or less historical:

1. Smoke (carbon)
2. Industrial dust and ash
3. Sulfur oxides
4. Fumes and odors (internal-combustion products, refuse burning, hydrocarbons, etc.)

Following the treatment of these contaminant groups, future trends in source control will be discussed.

1.7.1 Smoke (Carbon)

More effort probably has been spent on the study of the combustion of fuels than on any other technical subject. More people have employed combustion in home and factory than any other chemical reaction. Source control to improve fuel quality has been practical for many years, as have the techniques for improved combustion.

This article draws a distinction between "external fuel combustion" for heat and power generation and "internal combustion" as utilized in engines. A subsequent article in this section will deal with air pollution from internal combustion.

Many excellent handbooks are available on combustion theory, research, and practices. Examples are the handbooks by Johnson and Auth (8) and de Lorenzi (5). Both cover a range of combustion technology. With all the literature and know-how available, it is a wonder that any chimney emits black smoke. However, it is easy to oversimplify and overtheorize on the subject of black smoke. Few people will admit to being to blame for the problem. Essentially, the basic causes lie in the areas of operation, maintenance, design and installation, obsolescence, and overload. The fact remains that all fuels can be burned smokelessly and efficiently.

Black smoke is an economic waste, as well as a nuisance. Emission of smoke is like waving a flag of negligence for all to see. Laws and regulations have been simplified to permit easy detection of smoke. One example of an uncomplicated instrument for checking smoke intensity appears in Fig. 1-7. The Ringelmann scale of smoke measurement is used by most communities interested in smoke prevention. Almost anyone can judge smoke emission with the smoke card and a watch. Many rugged and efficient instruments also have been developed for refining smoke measurements and combustion-efficiency control to give continuously recorded results.

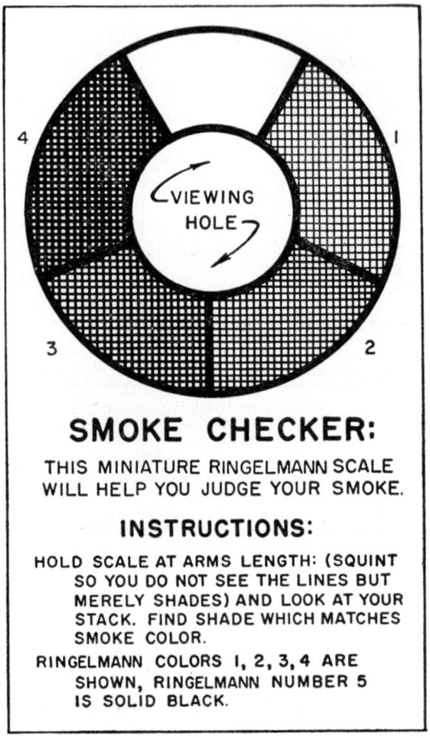
4
1
VIEWING
HOLE
3
2
SMOKE CHECKER:
THIS MINIATURE RINGELMANN SCALE WILL HELP YOU JUDGE YOUR SMOKE.
INSTRUCTIONS:
HOLD SCALE AT ARMS LENGTH: (SQUINT SO YOU DO NOT SEE THE LINES BUT MERELY SHADES) AND LOOK AT YOUR STACK. FIND SHADE WHICH MATCHES SMOKE COLOR.
RINGELMANN COLORS 1, 2, 3, 4 ARE SHOWN, RINGELMANN NUMBER 5 IS SOLID BLACK.

FIG. 1-7. Ringelmann scale. (*Courtesy of Ess Instrument Co., Bergenfield, N.J.*)

An interesting relationship of Ringelmann numbers and efficiency loss is outlined in Fig. 1-8.

Each fuel requires a balance of many technical and economic factors to reach best combustion conditions. Usually more extensive and costly combustion equipment and controls are available than those presently being used for the control of inherent smoking tendencies of some fuels. Carried to an extreme, there is little skill required in burning hydrogen-oxygen perfectly and with no contaminants. On the other hand, much thought and skill are necessary to handle the burning of highly volatile coal without smoke. The Bituminous Coal Research organization, Battelle Institute, and other interested groups have devoted considerable effective effort to the study of the smokeless combustion of coal and the design of suitable combustion devices.

Source control of smoke usually finds a balance between the selection of fuel for burning qualities and improved combustion devices. The background of United States fuel consumption is brought out by Fig. 1-9. In the seven postwar years covered by this chart, United States consumption of fuel for all nonmilitary purposes has risen from 29 thousand trillion Btu to 35 thousand trillion, a gain of more than 20 per cent. Among the individual fuels, natural gas shows the most spectacular growth; its use has virtually doubled. The rise in gasoline consumption is, of course, a result of the huge increase in highway traffic. (Gasoline is shown separately from other oil fuels because its chief service, as an automotive fuel, cannot be performed by any nonoil fuel.) At the right is a comparison of the sources and disposition of all United States oil in 1952, when United States production and consumption were the highest in history.

Progress has been continuous in handling and burning all types of fuel. As new raw-materials sources are developed, substitution and shifting of consumer's preference will always occur. It is often helpful to compare the United States status on fuels with that of Great Britain, where the air pollution prevention crusade had its start. The comparison in Table 1-2 on page 1-14 illustrates the types of fuels used in energy production by the two nations.

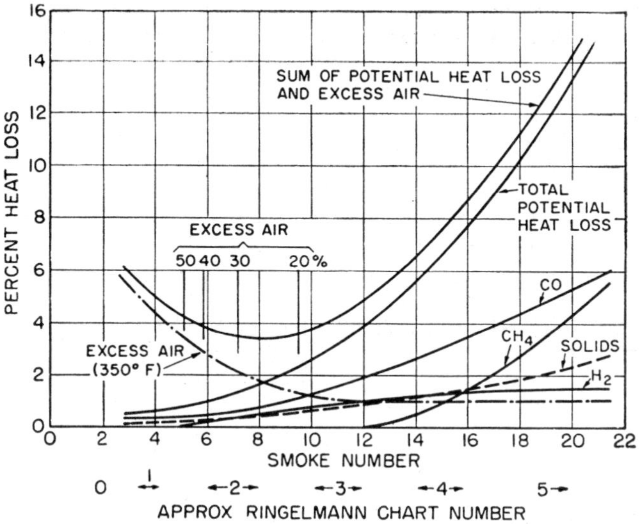

FIG. 1-8. Efficiency loss vs. smoke intensity. (*Fuel Research Technical Paper No.* 53, *Fig.* 14 (1946). *By permission of the Controller of Her Britannic Majesty's Stationery Office.*)

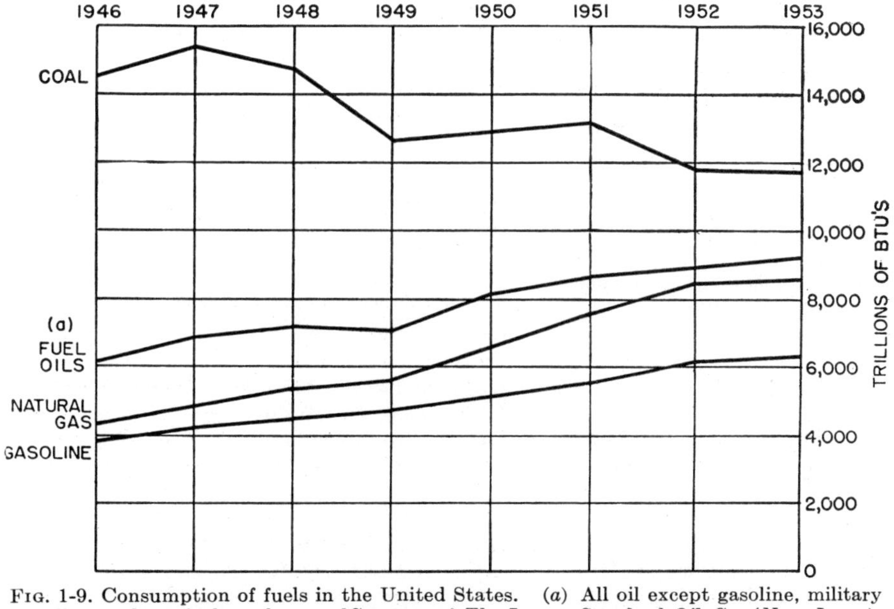

FIG. 1-9. Consumption of fuels in the United States. (*a*) All oil except gasoline, military supplies, and nonfuel products. [*Courtesy of The Lamp, Standard Oil Co.* (*New Jersey*), *June*, 1953.]

Combustion Principles. It is generally agreed that fuel technologists have carried forward the empirical art of combustion to a more advanced stage than reaction theory. The more important chemical-engineering principles for good smoke-free combustion of all types of fuels are the following:*

* After M. L. Smith and K. W. Stinson, "Fuels and Combustion," McGraw-Hill Book Company, Inc., New York, 1952; and A. J. Johnson and G. H. Auth, "Fuels and Combustion Handbook," McGraw-Hill Book Company, Inc., New York, 1951.

1. Proper air-to-fuel ratio
2. Adequate mixing of air and fuel at the proper time (especially combustible gases vaporized from the fuel)
3. Sufficient ignition temperatures on combustible gases
4. Adequate furnace volume to allow time for burning
5. Proper firebox setting height (adequate height above the grate)

Of the variables in combustion, the size of the fuel particles has the greatest effect on flame propagation—the smaller the size, the greater the propagation. Turbulence can be created either by external energy input to the fuel, to the air, or by the use of energy in the products of combustion. Furnaces handling solid fuels such as coal, coke, wood, and refuse use turbulent combustion gases to sweep clean new surfaces for combustion. Fluid-fuel burners handling oil, gas, and powdered coal are characterized by violent turbulence of the fuels as they pass from burner to furnace. Research efforts on combustion have been very rewarding as demonstrated by Table 1-3. The trend of increased useful power per pound of coal consumed is a tribute to the engineering profession. It is also well to pause and imagine the air pollution source reduction from this technological improvement.

Table 1-2. Comparison of United States and British Fuel Consumption[a]

Domestic consumption of energy[b]	Quantity		Ratio to coal	
	United States[c]	Great Britain[d]	United States	Great Britain
Population, million persons	151.2	49.2		
Solid fuels, million short tons:				
Coal (anthracite and bituminous)	498	226	1.0	1.0
Lignite	3.1	Negl.	0.006	
Coke (except petroleum coke)	69	30	0.14	0.13
Manufactured fuels (briquets)	2.4	1.1	0.005	0.005
Wood	No data	0.5	......	0.002
Liquid fuels, million bbl:				
Refined petroleum products, natural gasoline, and benzol	2,110	123	4.2	0.55
Gaseous fuels, billion cu ft[a]:				
Natural gas	5,200	Negl.	10.4	
Manufactured gas	1,500	725	3.0	3.2
Electricity, billion kwhr:				
From all sources	390	56	0.78	0.25

[a] 1950 figures, except gaseous fuels for the United States which are 1949 figures.
[b] Some fuels are raw materials for other energy sources, e.g., coal for coke, all fuels for electricity, etc.
[c] Excludes noncontiguous territories except for coal and coke (little consumption).
[d] Great Britain except for coal and petroleum which include Northern Ireland, i.e., United Kingdom.
From Frederick Brown, ed., "Statistical Yearbook of World Power Conference," No. 6, The Central Office, World Power Conference, London, 1952.

Table 1-3

Year	*Pounds of Coal per Kilowatthour of Electrical Energy*
1900	7.0
1940	1.34
1950	1.25
New plants	0.75

From the President's Materials Policy Commission, The Promise of Technology, in "Resources for Freedom," vol. IV, p. 5, U.S. Government Printing Office, Washington, D.C., June, 1952.

Coal, Coke, and Wood. The major burden of past energy generation has been carried by solid fuels, primarily coal. A wide choice of coals is available, ranging from highly volatile bituminous to comparatively smokeless fuels such as anthracite and coke. Table 1-4 lists the various types of coal being marketed. It is entirely

Table 1-4. Types of Coal Available

Type of Coal	*Smoke-tendency Index, % Black × Hours*[a]
Anthracite	0
Semianthracite	0 (1 test)
Subbituminous B	0 (2 tests)
Subbituminous C	0 (2 tests)
Low-volatile bituminous	0–50
Medium-volatile bituminous	20–60
High-volatile B bituminous, High-volatile C bituminous and subbituminous A	50–130
High-volatile A bituminous	70–180

[a] Range of most values reported. No tests on lignite.
After Rose and Lasseter, *Heating, Piping and Air Conditioning*, **11**, No. 2, 119–122 (1939).

feasible to burn any of the coals listed without smoke emission. The correct burner and proper operating conditions can make all the fuels "smokeless." One major problem lies in modernizing millions of installations sufficiently so that smoke is eliminated. This involves a time factor rather than one of scientific knowledge.

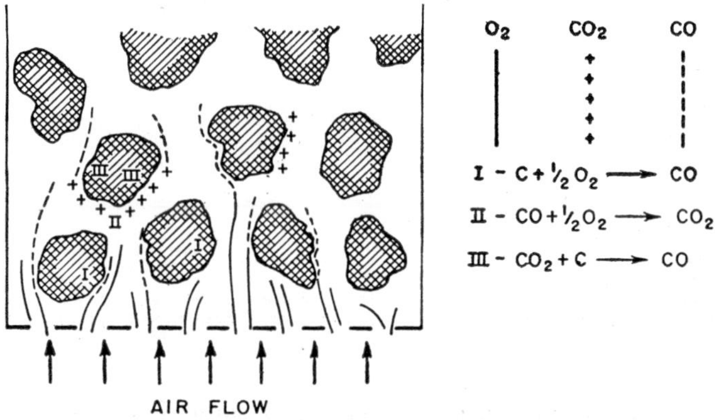

FIG. 1-10. Combustion reactions. Plane through fuel bed showing sequence of reactions. (*Courtesy of L. C. McCabe, ed., "Air Pollution," chap.* 46 *by R. C. Corey, p.* 395, *McGraw-Hill Book Company, Inc., New York,* 1952.)

Complex chemical reactions occur in a fuel bed of coal or coke. A somewhat simplified description of a reaction sequence is shown in Fig. 1-10. Reactions carried out to completion, as shown in this fuel bed, will result in smokeless combustion.

Wood is an important fuel in some sections of the United States, especially the Pacific Northwest. It will be a replaceable natural resource as long as the sun shines. Important differences exist in the analyses of wood and coal, as shown in Table 1-5. Wood usually contains less ash and sulfur than coal. The higher content of volatile matter and moisture (with lower fixed carbon content) places emphasis on proper furnace design to prevent smoke when handling wood. Manufacture of charcoal from wood is one method of pollution source control practiced in limited degree.

The combustion of wood has been studied by a number of authorities. Rosene (14)

Table 1-5. Wood and Coal Analyses

Analysis	Wood—hogged waste, Douglas fir		Coal		
			Bituminous		Anthracite
	Wet[a]	Dry	High vol., Wash.	Low vol., Ark.	Pa.
Proximate analysis, wt %:					
Moisture	36	0[b]	4	3.4	5
Volatile matter	52	82	38	16	4
Fixed carbon	11	17	47	72	82
Ash	0.5	0.8	11	9	9
Ultimate analysis, wt %:					
Hydrogen	8	6	6	4	2.3
Carbon	33	52	69	80	82
Oxygen	58	41	12	5	6
Nitrogen	0.1	0.1	1.5	1.7	0.7
Sulfur	0	0	0.5	1.0	0.5
Ash	0.5	0.8	11	9	9

[a] As received and usually fired.
[b] 6.5 weight per cent moisture if air-dried.
After U.S. Bureau of Mines data reported in L. S. Marks, ed., "Mechanical Engineers' Handbook," 5th ed., McGraw-Hill Book Company, Inc., New York, 1951; and A. J. Johnson and G. H. Auth, "Fuels and Combustion Handbook," McGraw-Hill Book Company, Inc., New York, 1951.

has listed practical methods of wood combustion. In general, the design of furnaces for burning wood differs from that for coal furnaces in the following ways:

1. Larger area of refractory surface in the primary fuel-drying zone (usually two-zone furnaces)
2. Higher proportion of overfire secondary air above the grate to primary air through the grate
3. Larger combustion space, often in the form of a secondary air space, to burn the volatiles

Furnaces for dry-wood waste differ from ones for wet wood in that the fuel-drying zone is less important.

Oil and Gas Fuels. The variety and choice in the selection of oil and gas fuels are as extensive as for solid fuels. Popular oil fuels used in the United States for external combustion are:

Liquefied petroleum gas—LPG (propane, butane, or mixtures)
Gasoline (white)
Kerosene
Heating oils Nos. 1, 2 (distillates)
Fuel oil No. 4 (blend of distillate and residual usually)
Bunker fuel Nos. 5, 6 (residual oils)

The heavy bunker fuel oils, or residuals, are most important from a heat- and power-generation viewpoint. Table 1-6 gives some general specifications of commercial fuel oils.

Methods have been developed to custom-refine crude petroleum to very specific needs. Highly unsaturated hydrocarbons of the olefin and diolefin families, as well as aromatics, are removed from fuels by treating with sulfuric acid, clay, bauxite, hydrogen, sulfur dioxide, or other agents to improve burning qualities. Olefins, diolefins, and aromatics characteristically burn with smoky, sooty flames compared to paraffins

Table 1-6. Fuel-oil Grades[a]

Grade of fuel oil[b]	Flash point, °F	Pour point, °F	Water and sediment, % by vol.	Carbon residue on 10% bottoms	Ash, % by wt	Distillation temperatures, °F: 10% point	Distillation temperatures, °F: 90% point	Distillation temperatures, °F: End point	Saybolt viscosity, sec: Universal at 100°F		Saybolt viscosity, sec: Furol at 122°F		Kinematic viscosity, centistokes: At 100°F		Kinematic viscosity, centistokes: At 122°F		Gravity, °API	Corrosion at 122°F (50°C)[d]
	Min	Max	Max	Max	Max	Max	Max	Max	Max	Min	Max	Min	Max	Min	Max	Min	Min	
No. 1: A distillate oil intended for vaporizing pot-type burners and other burners requiring this grade of fuel	100 or legal	0	Trace	0.15	....	420	...	625	...	...	...	..	2.2	1.4	...	..	35	Pass
No. 2: A distillate oil for general-purpose domestic heating for use in burners not requiring No. 1 fuel oil	100 or legal	20[c]	0.10	0.35	....	[e]	675	...	40	...	...	..	(4.3)	....	...	..	26	
No. 4: An oil for burner installations not equipped with preheating facilities	130 or legal	20	0.50	....	0.10	...	...	...	125	45	...	..	(26.4)	(5.8)				
No. 5: A residual-type oil for burner installations equipped with preheating facilities	130 or legal	..	1.00	....	0.10	...	...	...	...	150	40	..	....	(32.1)	(81)			
No. 6: An oil for use in burners equipped with preheaters permitting a high-viscosity fuel	150	..	2.00[f]	....	....	...	...	...	...	...	300	45	....	....	(638)	(92)		

[a] Recognizing the necessity for low-sulfur fuel oils used in connection with heat-treatment, nonferrous metal, glass and ceramic furnaces, and other special uses, a sulfur requirement may be specified as follows:

Grade of Fuel Oil	*Sulfur, Max %*
No. 1	0.5
No. 2	1.0
No. 4	No limit
No. 5	No limit
No. 6	No limit

Other sulfur limits may be specified only by mutual agreement between the purchaser and the seller.

[b] It is the intent of these classifications that failure to meet any requirement of a given grade does not automatically place an oil in the next lower grade unless in fact it meets all requirements of the lower grade.

[c] Lower or higher pour points may be specified whenever required by conditions of storage or use. However, these specifications shall not require a pour point lower than 0°F under any conditions.

[d] Report as passing when the copper test strip shows no gray or black deposit.

[e] The 10 per cent point may be specified at 440°F maximum for use in other than atomizing burners.

[f] The amount of water by distillation plus the sediment by extraction shall not exceed 2.0 per cent. The amount of sediment by extraction shall not exceed 0.50 per cent. A deduction in quantity shall be made for all water and sediment in excess of 1.0 per cent.

After ASTM Standards on Petroleum Products and Lubricants, p. 184, November, 1953.

and naphthenes. Burning qualities of some pure hydrocarbons are shown in Fig. 1-11. Heavy residual fuel oils are not treated for improvement in burning qualities as are heating oils and kerosene. However, proper selection of burner and firebox design will result in smokeless operation with any oil fuel. Some heavier oil fuels require care in the selection and design of fuel preheating and atomizing equipment.

Gas fuels are available throughout the United States. The distribution factor usually dictates that only one gaseous fuel is marketed in a local area. Thus, natural gas

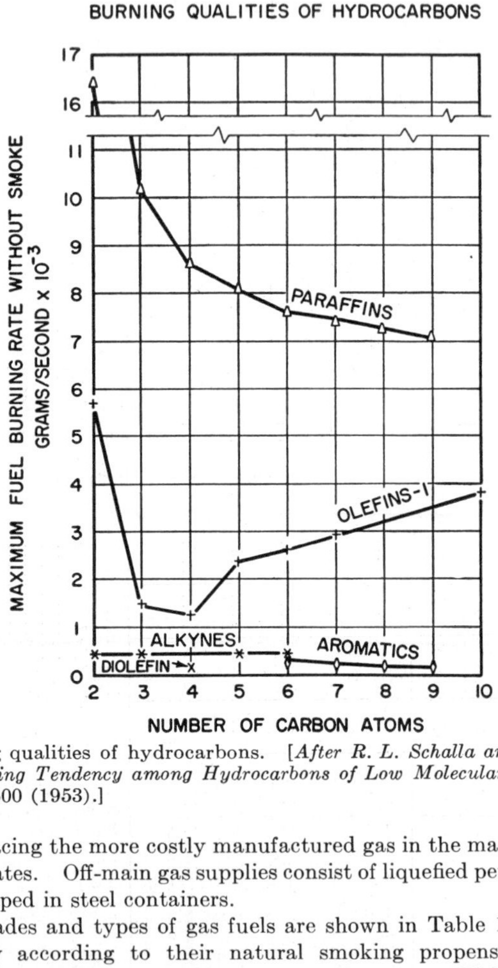

FIG. 1-11. Burning qualities of hydrocarbons. [*After R. L. Schalla and G. E. McDonald, Variation in Smoking Tendency among Hydrocarbons of Low Molecular Weight, Ind. Eng. Chem.*, **45,** 1497–1500 (1953).]

is gradually displacing the more costly manufactured gas in the mains throughout the eastern United States. Off-main gas supplies consist of liquefied petroleum gas (LPG) which can be shipped in steel containers.

The various grades and types of gas fuels are shown in Table 1-7. The order of listing is roughly according to their natural smoking propensities. A properly designed burner can handle any gas fuel without smoke. With the exception of acetylene and hydrogen, the gaseous fuels listed are used for commercial heating purposes. Gas fuels are the cleanest fuels available and have contributed greatly toward reducing air pollution in the United States.

The common asset of oil and gas fuels is their relative ease of mixing with air. Dispersion and diffusion are normally rapid. Oil fuels can be heated to reduce viscosity. This brings them close to the physical properties of gaseous fuels. The ease of flow control of oil and gas fluids can be approached with powdered solid fuels.

An interesting special procedure for decreasing smoke from the combustion of emer-

Table 1-7. Variation in Gas Fuel Composition

Type of gas	Composition, %							
	Hydrogen	Methane	Ethane	Propane	Butane	Illuminants		Other[a]
						Ethylene	Benzene	
Industrial acetylene	..	....	..	..	..	C_2H_2—98	.....	2
Carbureted water, coke	37	8	1	..	..	6.7	(3.6)[b]	43.7
Coke oven	46	32	..	..	..	3.5	0.5	18
Oil, Pacific Coast	49	26	..	..	..	2.7	1.1	21.8
Coal, continuous vert. retort	54	24	..	..	..	1.5	1.3	19.2
Refinery	6	31	20	33[c]	1	0.2	.....	4.8
Reformed, natural	47	27	3	..	..	0.6	0.6	22.4
Comm. butane (from nat. gas)	..	....	..	6	94	[d]	[d]	
Comm. propane (from nat. gas)	..	....	2	97	1	[d]	[d]	
Casinghead	..	37	14	..	..	....	.....	49
Natural, Pittsburgh, Pa.	..	83	16	..	..	....	.....	1
Sewage	2	68	..	..	..	....	.....	30
Water, coke	47	1	..	..	..	....	.....	52
Producer, coke	12	0.5	..	..	..	0.0	0.0	87.5
Blast furnace	2	0.2	..	..	..	....	.....	97.8
Industrial hydrogen	97	0.3	..	..	..	....	.....	2.7

[a] Other gases in analyses: CO, CO_2, N_2, O_2, etc. These gases do not produce smoke.
[b] Propylene, butylene, and other light hydrocarbons.
[c] Usually contains some propylene.
[d] Usually contains olefins (25 to 30 per cent) when made from refinery gas. Rank in listing would be higher in such case, i.e., C_4 second and C_3 third.
Data from various sources.

(*a*) (*b*)

FIG. 1-12. Use of steam to eliminate smoke. (*a*) No steam used. (*b*) With steam—no smoke. (*Courtesy of Esso Petroleum Co., Ltd.*)

gency releases of waste fuels is the mixing of steam or air with the fuel gases before release in a high flare stack. Even highly reactive or smoky gases can be burned in this way. Another advantage is the elimination of costly storage of this excess gas during the emergency. The result is illustrated in Fig. 1-12.

A number of investigators have reported on different approaches to this design problem. Several foolproof commercial designs of burners are on the market. The relationship between steam consumption and unsaturated hydrocarbon content of the flare gas for 2 deg of smoke improvement appears in Table 1-8.

Table 1-8. The Use of Steam for the Suppression of Smoke

Per cent (by weight) unsaturates in gas	Pounds of steam per pound of gas	
	No smoke	No trailing smoke
0	0.15	0.12
5	0.28	0.21
10	0.41	0.30
15	0.55	0.40
20	0.68	0.49
25	0.81	0.59
30	0.95	0.68

After Warren H. Smolen, Smokeless Flare Stacks, *Petroleum Processing*, **6**, No. 9, pp. 978–982 (1951), and subsequent data.

Process Carbon. Perhaps the carbon-black industry is a classical example of progress made in handling smoke or, specifically, fine carbon particles. This industry is also noteworthy for pointing the way early in a trend to improved processing con-

FIG. 1-13*a*. Carbon-black emissions. Channel-black plant in Texas (old type).

sciousness. By developing improved processing techniques, carbon-black yield has increased from 2 lb per thousand cubic feet of gas to 7 lb per thousand cubic feet. Still further improvement is expected.

Figure 1-13 compares an old-type plant with the latest design. This improved technology is a good example of source control in practice.

FIG. 1-13*b*. Carbon-black emissions. Furnace-black plant in Kansas (new type). (*Courtesy of the United Carbon Co.*)

Summary. Smoke prevention is good business. No elaborate tools are necessary to detect the presence of smoke. Great strides have been made in reducing smoke in the United States. An example is illustrated vividly in Fig. 1-14. Many people have labored over the years to cleanse Pittsburgh's atmosphere and the air of many other American cities. The beginnings and problems facing one of the earliest smoke and air pollution control associations make interesting reading as described by Wurts (15).

Contamination of air by smoke has been surveyed in many localities. The technology of improvement is available and freely offered by competent groups. An example of a survey with practical recommendations covering a highly industrialized

FIG. 1-14. Pittsburgh's "evolution." The photograph at the left was taken about 12 noon on a "smoggy" day in November, 1934, by H. W. Graham, J&L's Vice President-Research. In November, 1950, Graham asked John E. Stewart of the Hazelwood Research Laboratory to photograph the same scene (right), on a "bad" day. Stewart's photograph proves that Pittsburgh no longer deserves the nickname "The Smoky City"—thanks to smoke control. (*Courtesy of Men and Steel, Jones and Laughlin Steel Corp., December, 1953.*)

area in West Virginia (Kanawha Valley) is given by Lammers (9). Surveys of this type point the way to cleaner air.

Smoke emission cannot be corrected entirely by substitution of inherently clean-burning fuels. The economic structure for energy generation is far too specific for treatment here. It should be recorded that for one locality the fuels available for combustion fall into the pattern shown in Table 1-9. The real challenge to research

Table 1-9. Pricing Structure for Fuels[a]

Fuel	*Price per Million Btu*
Industrial fuels:	
Bituminous coal (run-of-mine); Bunker fuel oil; Natural gas (N.Y. state)	$0.37–$0.45
Residential fuels:	
Anthracite coal (stove); Coke oven (domestic–f.o.b. N.Y. and N.J. ovens); Heating oil (No. 2); Commercial propane (LPG)	0.59– 0.77
Specialty fuels:	
Diesel oils (Shore); Gasoline (White); Kerosene	0.68– 1.03

[a] Using average f.o.b. price for 1951 in New York, N.Y., area except natural gas. Coal prices are for carload lots; coke price is price per net ton f.o.b. ovens (merchant sales) in New York and New Jersey; oil-product prices are refinery and tank-terminal prices for tank-car lots; and natural-gas price is average value of gas for industrial uses (probably a comparable f.o.b. price).

Data from various sources.

and development groups is to upgrade all smoke-producing fuels. Those fuels which are low-priced might be improved in attractiveness to the consumer; those fuels which are high-priced might be improved in efficiency of use.

The release of black smoke to the atmosphere can be prevented in practically all cases except unforeseeable emergencies. At this time, there is no other field in air pollution control where operator vigilance is more rewarding. Until the day when there is a perfect pairing between device design and fuel consumed, some human care is necessary.

The general trend is toward increased use of cleaner fuels, water power, and atomic energy for power generation. However, large central powerhouses using coal as the energy source are still being planned. This is a logical trend because large new installations can be designed and operated free of air contamination.

1.7.2 Industrial Dusts

Over all, the effect of industrialization has been to increase the solid-dust content of the atmosphere. Sources range from major emissions such as dust in vent gases from combustion and processing operations down to minor ones such as rubber-tire dust and tobacco aerosols, commonly called smoke.

Table 1-10 breaks down some major sources of industrial dust from a national viewpoint.

Table 1-10. Dust Sources

Source	*Examples*
Combustion	Fuel burning (coal, wood, fuel oil, fuels containing additives) Incineration (house and garden trash, sewage) Other (tobacco smoking, forest fires)
Solids handling and processing	Loading and unloading raw materials Mixing and packaging solids Size reduction (crushing and grinding of ores, stone, cement, fertilizer, rock) Industries using solids (metal refining, foundries, petroleum catalytic cracking, roofing and wallboard manufacture) Food processing (grains, spray-dried milk)
Vaporizing operations	Pyrometallurgical operations (zinc and lead oxides from nonferrous metal processing, silica from aluminum industry) Emission of chemical vapors (part of which later become solid crystals)
Earth-moving operations	By construction and mining By agriculture operations By nature By transportation (cars, humans)
Others	Housecleaning, etc. Rubber-tire abrasion

Many investigators have studied the theory and technology of forming and removing dust particles from gases. In general, a particle, 10μ or less in diameter, will be suspended in air and tend to act in accordance with gas laws. Figure 1-15 records the particle-size range for some common materials handled by industry.

The largest sources of industrial dust are combustion processes. Perhaps 75 per cent of the industrial dust (exclusive of soot) comes from fuel combustion. All major fuels—coal, coke, and residual fuel oil—contain some ash or solid, nonburnable contaminants. Gaseous and distillate fuels do not contain ash or solid contaminants, although these fuels can be sources of soot under conditions of improper combustion. The ash content of some major fuels is summarized in Table 1-11.

Much effort has gone into reducing these solid contaminants in fuels. It is good business to do so up to a certain point. The contaminants cost money to handle through the producer's channels. They cause maintenance problems and shutdowns of the consumer's equipment by depositing slag in tubes and building up deposits in fireboxes. A study of the national fuel-consumption trend illustrates that source control has been effective in reducing ash in fuels. This trend is outlined in Fig. 1-16. It is anticipated that this trend will continue as noted on the chart by the dashed line.

Some dust control falls into the class of good housekeeping. Prevention of dust recycle by street cleaning and flushing, by oiling, and by paving of roads and cleared

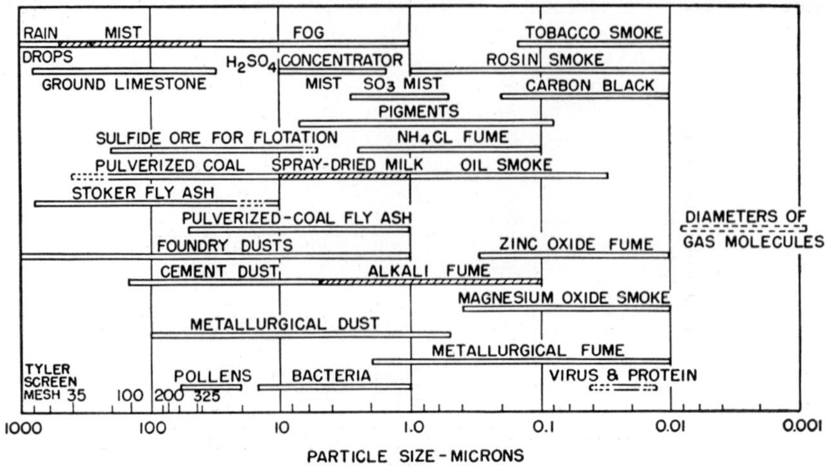

FIG. 1-15. Particle sizes for some common materials. (*Courtesy of L. C. McCabe, ed., "Air Pollution," chap.* 16 *by H. P. Munger, p.* 159, *McGraw-Hill Book Company, Inc., New York*, 1952.)

areas are constructive steps. Little new research is needed to direct these corrective steps.

Coal and Coke. Coal technologists have been developing improved methods of ash removal for some time. The efforts of the coal industry to improve the quality of coal by removing ash are shown in Fig. 1-17.

Table 1-11. Ash Content of Solid and Liquid Fuels

Selected Fuels	*Ash, Wt %*
Anthracite coal, Pa	10.7
Bituminous coal:	
Low volatile, W.Va	3.9
Medium volatile, W.Va	5.4
High volatile A, Pa	9.1
High volatile B, Ky	5.9
High volatile C, Ill	8.6
By-product coke	10.7
Charcoal	3.0
Wood, typical nonresinous, seasoned	1.0
Diesel oil 1-D	0.01 max spec.
Fuel oil No. 2, heating	0
Fuel oil No. 6, bunker	0–0.2

Data from L. S. Marks, ed., "Mechanical Engineer's Handbook," 5th ed., McGraw-Hill Book Company, Inc., New York, 1951; M. L. Smith and K. W. Stinson, "Fuels and Combustion," McGraw-Hill Book Company, Inc., New York, 1952; and private communications.

The inclusion of more ash in the raw mine coal is due chiefly to mechanical loading and depletion of better coal beds. These trends tend to mask the benefits of coal cleaning. Figure 1-17 shows that the refuse removed from raw coal has increased twofold since 1929. This increase is undoubtedly due to technological improvements, as well as higher ash in the raw coal. Mechanical cleaning of coal lagged behind mechanical loading because of construction restrictions during World War II. Mechanical cleaning is increasing again, and greater benefits in cleaner coal will be apparent as more of the coal is washed. Industry and government groups have developed numerous coal-washing processes. Table 1-12 presents methods of cleaning coal in the United States.

The more popular processes and typical performance data are given in Table 1-13. The efficiency of these commercially used methods of ash removal are constantly being improved. Each coal source is different, and empirical assay methods are necessary

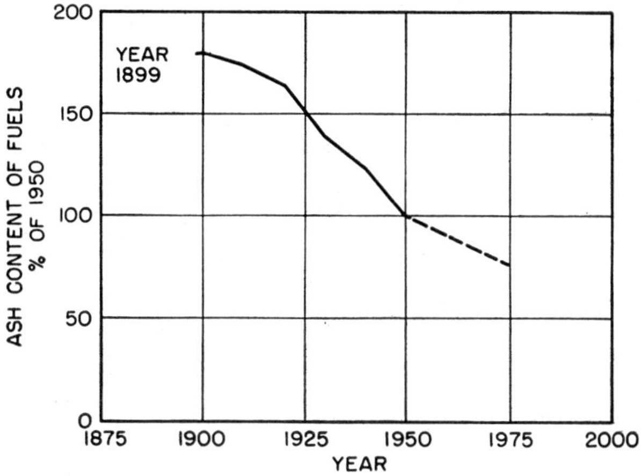

FIG. 1-16. Ash content of United States fuels. (*Calculated using data from various published sources.*)

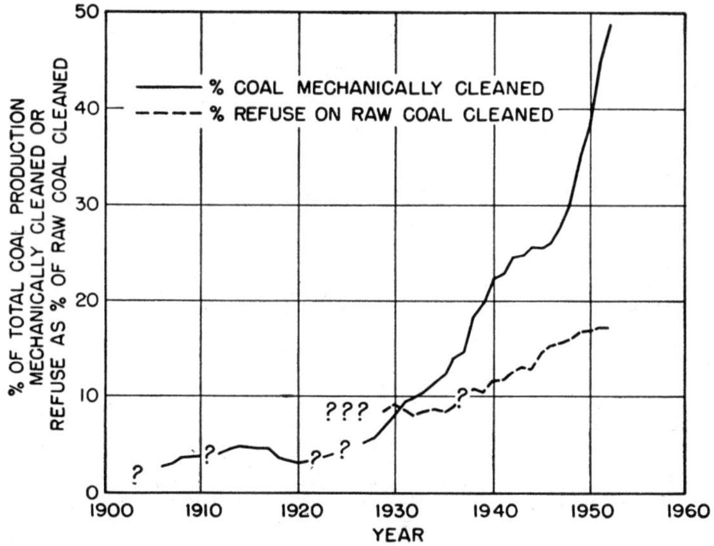

FIG. 1-17. Trends in the cleaning of coal. (*U.S. Bureau of Mines publications.*)

to determine the best process. Competitive market conditions, demands of customers, and economics determine the degree of ash removal for each installation. Some coals marketed are dustproofed for retail markets by coating with calcium chloride or hydrocarbon oils.

Coking of coal (especially fine sizes), briquetting, and other methods of smokeless-fuel manufacture serve to decrease dust emission. Efficiency in use depends on many factors other than the fact that the coal is smokeless. In general, carry-over and fly ash depend on the type of firing (hand, stoker, spreader, pulverizer) and the size and quality of the coal. For example, small fuel sizes on a spreader would increase pollution. Large-size high-volatile coal, on the contrary, would produce little pollution. Low-volatile coals may produce more carry-over and fly ash than coals of highest volatility. The amount of material retained as clinkers or firebox residue depends on

Table 1-12. Methods of Cleaning Coal

Cleaning Methods	*Cleaned by Type, % of Total*[a]	
I. Gravity stratification		
A. Wet processes:		
1. Jigs	42.4	
2. Launders	4.3	
3. Upward-current classifiers	9.7	
4. Heavy mediums	14.1	
a. High-density suspensions		
b. High-density solutions		
5. Tables	2.4	
		92.3[b]
B. Dry processes:		
1. Pneumatic tables, 2. Air jigs, 3. Air-sand	7.7	
		7.7
II. Nongravity processes		
1. Froth flotation	Included with heavy med.	
Total		100

[a] In 1951 when 45 per cent of coal produced was cleaned.
[b] Including combinations of individual wet processes—19.4 per cent.
Classification from National Research Council Committee, Yancey and Geer, "Chemistry of Coal Utilization," vol. 1, chap. 16, John Wiley & Sons, Inc., New York, 1945.
Percentage from U.S. Bureau of Mines Mineral Market Report No. 2102.

Table 1-13. Performance Data for Various Methods of Cleaning Coal

Type equipment	Baum jig	Dense mediums	Menzies cone classifier	Stump air-flow jig
Location	Washington	Pennsylvania	West Virginia	
Coal type	Roslyn bed (high vol. bitu. A)	(Anthracite)	Eagle seam (bituminous)	Seam A
Coal size	3 in screenings (run of mine)	Chestnut	1¼″–⅜″	6 mesh-0
Ash, wt %:				
Feed	15.4	34.7	10.3	20.6
Washed	10.5	10.5	4.4	12.0
Refuse	58.3	75.2	62.4	42.1
Yield of washed coal, wt %	89.7	62.7	89.9	71.5
Equipment details	4 cell—2 compartments	....	2—10′ diam units	2 units in series
Throughput, tons/hr	100	....	182	

Data after Yancey and Geer, *U.S. Bur. Mines, Rept. Invest.* 3371, January, 1938; and Mitchell, ed., "Coal Preparation," 2d ed., American Institute of Mining and Metallurgical Engineers, New York, 1950.

fusion temperature of ash, furnace temperature and construction, rate of burning, and many other factors. Thus it is necessary to integrate the fuel and the furnace for best over-all results.

Progress is being made in developing dustless and smokeless fuels of the briquet and packaged type. These fuels can be manufactured from coal and wood fines too small for other uses. Table 1-14 lists a few of the more popular bulk fuels of this type.

The final line of defense against air pollution from combustion of coal and coke is the installation of dust-removal devices by the consumer. Many reliable devices are available for removing particles in the size range normally contained in combustion flue gases. These devices also serve to remove soot particles with the ash.

In boiler-plant design, several factors should be mentioned which influence fly-ash emission. Although coals may have the same ash content, the type of fly ash formed may differ between coals. Fly ash from strip-mine coal has been found to be lower in conductivity and more difficult to recover than the fly ash from deep-mine coal. The manner in which coal is fed to a furnace will also influence the ash characteristics. Pulverized-coal feeders give smaller size ash than do stoker-coal feeders. Dust loadings also are higher from conventional furnace outlets firing pulverized coal. Thus, the cost of recovery equipment as influenced by these factors is taken into account in studying boiler-plant economics.

Source control has been applied to coal combustion in a number of ways. Aside from the steady trend in reducing the weight of coal burned per unit of useful energy, burners have been developed for removing ash before it enters the furnace. The cyclone burner is an example of this trend. When pulverized coal is fired in a cyclone burner, most of its ash is removed as molten slag from the burner body. Thus, the ash is disposed of without requiring extensive flue-dust abatement and collecting equipment.

Table 1-14. Fuel Briquets and Packaged Fuels

Raw Fuel	*Raw Fuel Used in* 1952, *Per Cent*
Fuel Briquets	
Anthracite:	
Pennsylvania	22
Semi- and other	2
Bituminous:	
Low volatile	58
High volatile	4
Petroleum coke	6
Residual coke from oil gas manufacture; Semicoke (lignite char)	8
	100[a]
Packaged Fuel	
Bituminous:	
Low volatile	91
High volatile	5
Petroleum coke; Semianthracite	4
	100

[a] In addition, briquets (stoker fuel) and logs are made from wood sawdust.
From U.S. Bureau of Mines, Mineral Market Report, M.M.S. No. 2177.

Fly ash recovered from flue gases formed in the combustion of pulverized coal has been studied for further use. Consolidated Edison (12) in New York City has been partially replacing cement with fly ash in concrete mixes. They report that fly ash can be used up to 20 per cent by weight of portland cement. Laboratory tests and construction experience have proved that fly-ash concrete is stronger, more durable, more waterproof, and easier to handle than conventional concrete. It is also less expensive. The savings are estimated to be about 6 per cent of the cost of ordinary poured concrete. This company has used fly ash as a partial replacement for cement since about 1952. Three construction projects thus far have utilized the fly-ash concrete.

This outlet is another example of converting a former liability to an asset. The technological problem was studied for years by government agencies, industrial-research institutes, and companies like Consolidated Edison. Where formerly the fly ash (recovered from mechanical and electrical precipitators) had been loaded in barges, carried to sea, and dumped as waste, it now has value. Economic savings result from elimination of disposal cost and from an improved product.

Fuel Oils. Distillate products do not contain appreciable amounts of dust-forming contaminants except in cases where additives are purposely blended to improve quality. Examples of this are tetraethyllead which is used to improve the octane number of gasoline and various chemicals used to improve the burning of diesel oils.

The United States trend in residual-oil technology has been toward a steady increase in depth of cut on crude. Research has been aimed at upgrading crude into higher-priced products than residual fuel oil. This has resulted in a tendency to concentrate natural petroleum-ash and salt contaminants into the bottoms cut. At the same time, processes for removing salt have come into play. Chief among these is desalting of crude oils at the well or refinery. At the present time, over 50 per cent of the national crude-petroleum volume passes through desalting facilities. A number of effective processes are available to remove water-soluble salts from crude oil. Residual-oil quality is helped by desalting because otherwise the salt tends to collect in this nondistillate fuel. A distinct uptrend in crude desalting capacity has been occurring since about 1946.

Typical crude-oil desalter performance and the effect on residual fuel oil are summarized in Table 1-15 for a low-salt-content crude. Some crude oils contain as much as 100 to 500 lb of salt per thousand barrels of crude.

Table 1-15. Typical Crude-oil Desalting Results

	Crude oil	Residual fuel oil
Source	West Texas	West Texas
Specific gravity, °API	32	12
Viscosity	50 Universal at 100°F	175 Furol at 122°F
Yield on crude, %	100	10[a]
Salt content in, PTB[b]	27	
Salt content out, PTB	1.5	
Salt content, without desalter, PTB	....	270
With desalter, PTB	....	15
Reduction in salt content, %	94.5	94.5

[a] Minimum merchantable fuel-oil yield with conventional, modern refining facilities (no coking).
[b] Salt content (NaCl) expressed as pounds per thousand barrels (PTB).

Upgrading of residual oils by coking processes to form lighter products is on the increase. Desalting of crudes thus becomes more important because coking concentrates salt in the carbonaceous fraction.

Process Dusts. An examination of manufacturing industries and individual plants will show that most emit particulate matter. A major source of dust is combustion. Iron and steel mills, foundries, nonferrous smelters of copper, zinc, and lead, flour and cement mills, oil refineries, and the multitude of large and small dust producers are faced with dust handling as an important part of their processing technology.

Ingenuity fired by incentive is a rewarding combination. Foundry dusts and fumes are particularly difficult to handle because of the extremely fine particle size. Figure 1-18 demonstrates an example of the effectiveness of source control in a foundry. Both crucibles are handling the same raw-charge stock. The one with the slag cover not only prevents the escape of dust and fumes at the source, but also improves the process.

The use of finely ground or fluidized solids is widely practiced in the catalytic cracking of petroleum. These process dusts have 20 to 45 lb/cu ft aerated density and when aerated behave like fluids. The solids are useful in transferring large heat loads from one vessel to another. The preferred particle size for these fluidized solids ranges from 20 to 80μ. Particles smaller than 20μ in diameter tend to escape with the flue gases. Abatement devices such as cyclones, electrical precipitators, and allied apparatus for fine-dust removal are used to reduce air pollution within the process itself.

Summary. The purposeful generation of dusts is often advantageous to many processes and is a final product in flour milling, cement, gypsum, and allied industries. As a general rule, technical knowledge to prevent dust emission to the atmosphere is widely distributed. Modern technology continues to value the processing and engineering properties of dusts and, hence, tends to increase dust sources. It is the major

FIG. 1-18. Nonferrous crucibles, one with cover, other without. (*Courtesy of Allen, Viets, and McCabe, Control of Metallurgical and Mineral Dusts and Fumes in Los Angeles County, Calif., U.S. Bur. Mines, Inform. Circ.* 7627, *April,* 1952.)

job of abatement equipment to carry the burden of controlling dust emission to the atmosphere.

1.7.3 Sulfur Oxides

The effect of industrialization has been to increase the emission of all types of sulfur compounds to the atmosphere. By far the largest mass emission consists of sulfur dioxide (SO_2) generated by the oxidation of raw sulfur compounds in all types of fuel combustion and in ore refining. Major sulfur contaminants consist of colorless and invisible sulfur dioxide gas and hygroscopic sulfur trioxide (SO_3). The sulfur trioxide quickly picks up water vapor from the atmosphere to form sulfuric acid. This results in the familiar plume of sulfuric acid plants and the blue haze of many combustion operations. Some sulfur compounds such as mercaptans are objectionable because of their bad odor. These and other odorous contaminants will be covered subsequently in this section.

Sulfur is almost always present with petroleum, gas, and coal as found in the raw state. In fact, some experts theorize that sulfur must be present to form oil in the first place. Sulfur is one of the most sought-after chemical raw materials for one segment of the chemical industry and somewhat unpopular elsewhere. The total sulfur released to the atmosphere in the United States each year is about double the mass consumed. A breakdown is shown in Fig. 1-19. Progress is being made in source control of sulfur oxide emissions by removing sulfur compounds from the more concentrated sources. The sulfur compounds are then converted to solid or liquid sulfur for use as by-products. The majority of sulfur oxide emissions are in relatively low concentrations.

No discussion on pollution from sulfur-containing gases would be complete without a review of the sulfur-supply situation and especially the efforts at by-product sulfur

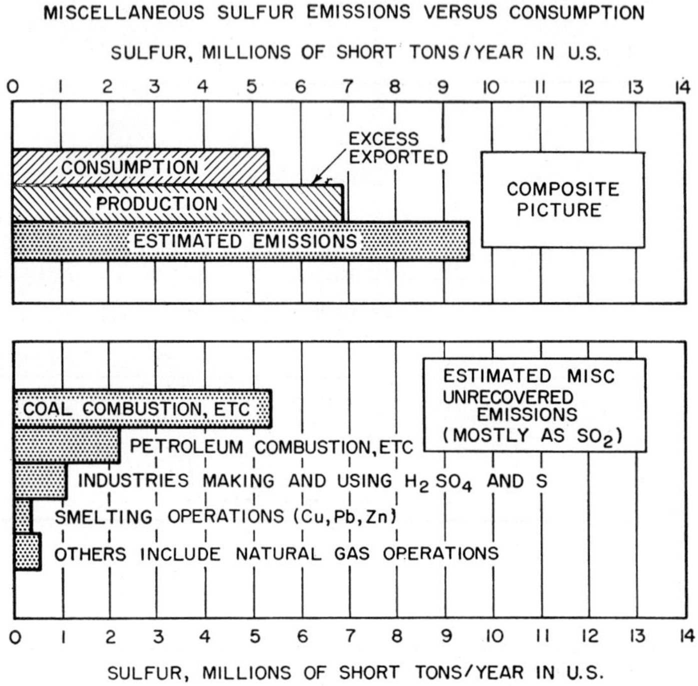

FIG. 1-19. Miscellaneous sulfur emissions vs. consumption. (*Consumption and production data from U.S. Bureau of Mines for* 1951. *The estimated emissions are calculated using data from various published sources.*)

recovery. As Fig. 1-20 shows, increases in the recovery of sulfur by-products are planned. The increase will come chiefly from processing petroleum-refinery and natural gases. There will also be an increase in the recovery of sulfur as sulfuric acid from smelter gases. However, despite these increases (which were spurred by a post-World War II sulfur shortage), the additional recovery will still account for less than 10 per cent of the total unrecovered sulfur emissions from miscellaneous sources. This fact points out why research effort is continuing to emphasize source control by reducing sulfur content before fuels reach consumers.

Table 1-16 lists the sulfur content of the more popular and representative fuels.

Sulfur dioxide ground concentrations can be controlled in many cases by the use of high stacks. Ultimately, the sulfur dioxide will return to the earth's surface in other chemical forms. There are numerous cases, however, where the air reservoir is not inexhaustible in its capacity for diffusing gases such as sulfur dioxide. Some of the alternative source-control methods are discussed later in this section.

A composite of United States fuels shows a steady decrease in the sulfur content of fuels. Figure 1-21 illustrates this trend.

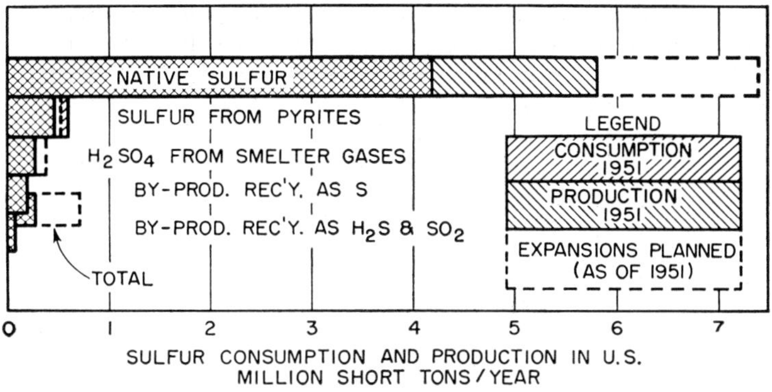

PRODUCT RECOVERY OF SULFUR COMPOUNDS

The up-trend in by-product recovery of sulfuric acid and sulfur is illustrated as follows:

	Sulfur equivalent, short tons/year					
	1943–1947 avg	1948	1949	1950	1951	1952
By-product sulfuric acid (as sulfur)	270	210	187	242	270	283
Elemental sulfur	29	49	64	159	206	281
Other (H_2S, SO_2 as sulfur)	22	29	42	47	67	75

FIG. 1-20. By-product recovery of sulfur compounds. [*U.S. Bur. Mines, Mineral Market Report, M.M.S. No.* 2111; *Anon., Oil and Gas J.* **51,** 199 (1952).]

Table 1-16. Sulfur Content of Selected Fuels

Fuel	*Sulfur, Wt %*
Anthracite coal, Pa	0.7
Bituminous coal	0.3–5.0
Low volatile, Pa	1.8
Medium volatile, Va	1.0
High volatile A, Ky	0.6
High volatile B, Ohio	1.2
High volatile C, Ind	2.3
By-product coke	1.0
Charcoal, willow	0.1
Wood, typical nonresinous, seasoned	0
Fuel oil No. 6 (bunker C)	0.3–3.0
Seaboard	1.2
Mid-continent	0.3
Fuel oil No. 2 (heating)	1.0 max spec.
Diesel oil 1-D	0.5 max spec.
Kerosene, 41-3°API, water white	0.05–0.1
Gasoline, 1951 regular, U.S.A	0.09 (0.01–0.4)
Mid-Atlantic states	0.06 (0.02–0.1)
California	0.23 (0.1–0.4)
Liquefied petroleum gas	0–25 grains/100 cu ft
Manufactured gas	2–10 grains/100 cu ft
Natural gas	0–10 max grains/100 cu ft

Data from various sources.

Coal and Coke. There is always a premium on locating and mining low-sulfur coal. Progress has been made in upgrading coals by purification processes at the mine. Some of the more popular coal-preparation processes were discussed previously under dust problems. The major methods used for coal preparation involve the use of jigs, bearing mediums, and classifiers. Table 1-17 gives a typical example of sulfur reduction by coal washing. The removal of sulfur and ash contaminants is accomplished

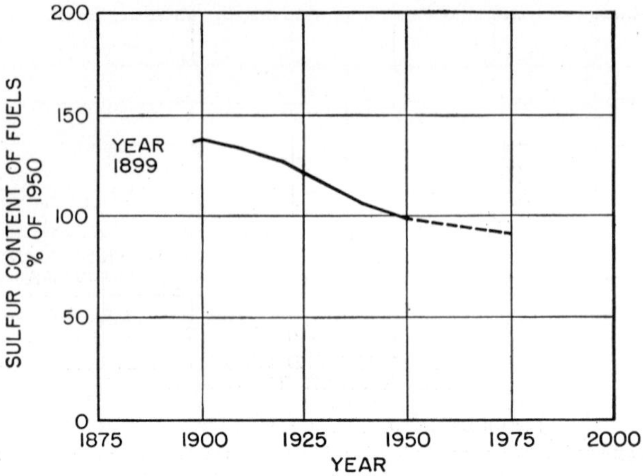

Fig. 1-21. Sulfur content of United States fuels. (*Calculated using data from various published sources.*)

together. The remaining sulfur is trapped in the coal lumps and can be reached only by grinding to a smaller particle size.

The effect on sulfur reduction of converting coal to coke is shown in Table 1-18 for a typical example. Sulfur is transferred from the solid fuel to the vaporized products of carbonization. The sulfur in the gas, light oils, and tar products is in more available form than in coal. At least the problem of removal is transferred from widely scattered home and industrial users to large by-product processors where a higher degree of operating vigilance is possible.

Petroleum. Low-sulfur oil and gas reservoirs also command a premium price on their production. Technological developments have resulted in narrowing the price spread for the sulfur level of crude petroleum by developing upgrading processes appli-

Table 1-17. Sulfur Removal from Coal

Type equipment	Jig
Type coal	Alabama
Size coal, in	0–1
Sulfur in raw coal, wt %	1.72
Sulfur in clean coal, wt %	1.35
Yield of clean coal, wt %	83.0
Reduction in sulfur level, %	21.5
Reduction in sulfur weight contained in clean coal, %	35.0

After Yancey and Parr, *Ind. Eng. Chem.*, **16,** 501–508, 1924.

Table 1-18. Reduction of Sulfur by Coking[a]

	Coal	Coke
Analysis, wt %:		
Sulfur	0.89	0.71
Moisture	4.43	2.43
Ash	7.26	9.31
Volatile matter	30.73	1.14
Fixed carbon	60.22	89.92
Total coke yield, wt %	76.3	
Reduction in sulfur level, %	20	
Reduction in sulfur weight contained in coke, %	39	

[a] Average operation of seven coke plants.

After A. J. Johnson and G. H. Auth, "Fuels and Combustion Handbook," McGraw-Hill Book Company, Inc., New York, 1951. Averages data from Ess, *Iron Steel Engr.*, **25,** C-3–37 (1948).

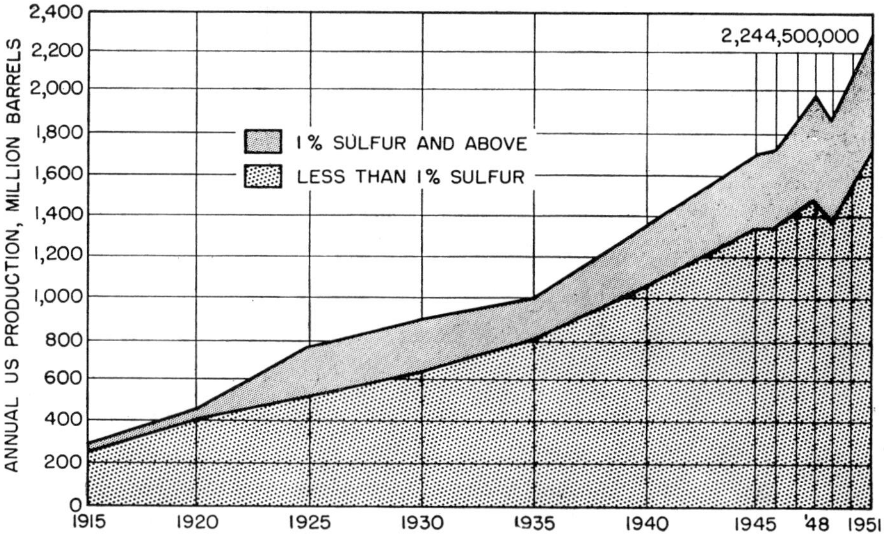

FIG. 1-22. Crude-oil sulfur content. (*Courtesy of Staff Report, Petroleum Processing,* p. 1687, *November,* 1953.)

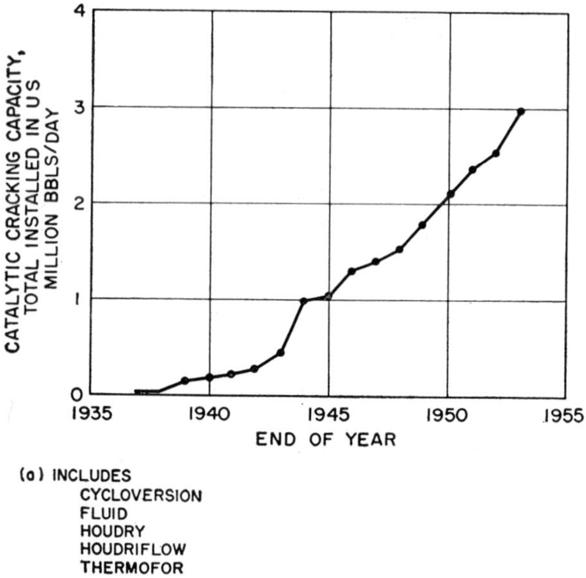

FIG. 1-23. Growth of catalytic cracking. (*Compiled using published data from many sources.*)

cable to many petroleum products. The need for this source control is illustrated by Fig. 1-22. The sulfur content of new crude petroleum discoveries in the United States and throughout the world is increasing.

Improved refining processes have been developed to keep pace with and to anticipate the higher sulfur trend. The increased use of catalytic cracking instead of thermal cracking of oils results in higher yields of distillate products per barrel of crude.

Catalytic cracking capacity has been steadily rising since 1938, as indicated in Fig. 1-23. Catalytic-cracking processes are used for both light- and heavy-distillate feed stocks. High-boiling sulfur compounds contained in the feed oils are converted mostly to hydrogen sulfide gas and to lower-boiling sulfur compounds. The hydrogen sulfide and lower-boiling compounds can be removed economically from the cracked products by many processes.

The sulfur balance for a typical catalytic-cracking operation is shown in Table 1-19.

A trend toward increased coking of residual oils is becoming apparent. Table 1-20 presents a typical sulfur balance for one coking process now under commercial development. Processes for extracting sulfur from oils are gaining in importance. Distillate fuels can be reduced in sulfur content by numerous processes. Some specifications call for an odor-threshold level. This odor specification is often met by converting malodorous sulfur compounds to nonodorous types without removing sulfur from the fuel. Examples of compounds used in these sweetening processes are copper chloride, lead sulfide, doctor, and sodium hypochlorite. Other conversion-type processes can be operated to remove sulfur from oils. The processes listed in Table 1-21 cover the more popular refining methods for removing sulfur from oils.

Table 1-19. Sulfur Balance for Catalytic Cracking of Gas Oil[a]

	Sulfur Distribution, %
Feed (2.5% wt sulfur, 25°API gravity)	100
Raw gas products:	
C_1-C_4 fractions	36[b]
Raw liquid products:	
C_5-430°F gasoline	3[b]
Light gas oil	11[b]
Heavy gas oils to fuel	43
Coke:	
Burned in regenerator	7
Total	100

[a] Typical of fluid catalytic-cracking process.
[b] Sulfur compounds removed from these products.

Table 1-20. Sulfur Balance for Residual Oil Coking[a]

	Sulfur Distribution, %
Feed (4.5% wt sulfur, 4°API gravity)	100
Raw gas products:	
C_1-C_4 fractions	27[b]
Raw liquid products:	
C_5-430°F	3[b]
430–1050°F	37[b]
Coke:	
Product coke	25
Coke burned	8
Total	100

[a] Typical of fluid coking process.
[b] Sulfur compounds removed from these products.

One of the more promising developments for both converting and extracting sulfur compounds from liquids is hydrogen treatment using various catalysts. Table 1-22 lists the available processes in this growing field. Hydrogen treatment of many oils and solids is under active research and development. Most of the sulfur in petroleum oils can be converted to hydrogen sulfide by hydrogen treatment. Quality is improved for most products, and yield is higher for all products compared with older treating methods.

Table 1-23 gives performance data for a typical hydrogen treatment of an industrial distillate gas oil. By converting the sulfur compounds to hydrogen sulfide, an ultimate recovery as solid or liquid sulfur is possible by the primary processor. Thus the sulfur oxides emitted to the atmosphere by the consumers are reduced by source control.

Sulfuric Acid. Sulfuric acid has long been the most widely used treating agent for sulfur removal and other purification steps. Present trends in such industries as oil

Table 1-21. Removal of Sulfur from Distillate Fuels

Process	Solvent	Sulfur compounds removed
	By Extraction	
Caustic washing	Alkali hydroxide solution	Hydrogen sulfide, some mercaptans
Mercapsol, Solutizer, Tannin solutizer, Unisol	Alkali hydroxide solution plus various solubility promoters	Hydrogen sulfide mercaptans
Edeleanu	Sulfur dioxide	Cyclic sulfur compounds
Acid treating	Sulfuric acid	Sulfides, disulfides, sulfates, mercaptans
	By Catalytic Conversion[a]	
Cobalt molybdate	Cobalt molybdate	Sulfides, disulfides, mercaptans
Cycloversion	Bauxite	
Gray	Fuller's earth	

[a] Raw sulfur in feed is partially converted to volatile sulfur compounds removable by distillation, chemical extraction, etc.

Reference: W. L. Nelson, "Petroleum Refinery Engineering," 3d ed., McGraw-Hill Book Company, Inc., New York, 1949.

Table 1-22. Catalytic Processes for Removing Sulfur from Oils with Hydrogen

Re-forming and Hydrogenation

Catforming[a]	Hyperforming[a]
Catalytic reforming[a]	Orthoforming[a]
Cycloversion[a]	Platforming[a]
High-pressure hydrogenation[b]	Thermofor reforming[a]
Houdriforming[a]	Ultraforming[a]
Hydroforming, fixed bed[c]	
Fluid[a]	

Treating[d]

Autofining[a]	Hydropretreating
Diesulforming[a]	Hydrotreating
Gulfining	Sovafining
Hydrofining[a]	Unifining[a]
Hydrodesulfurization[a]	
Hydrobon	

References: [a] *Oil Gas J.*, Mar. 21, 1955.
[b] *Petroleum Refiner*, Process Handbook, April, 1947.
[c] *Oil Gas J.*, Mar. 29, 1951.
[d] *Oil Gas J.*, June 6, 1955.

Table 1-23. Sulfur Balance for Gas Oil Hydrodesulfurization

	Sulfur Distribution, %
Feed (2.9% wt sulfur, 21.3°API gravity)	100
Raw gas products:	
Gas fractions (by difference)	89.9
Raw liquid products:	
Under 430°F	Negl.
430–500°F	0.1
500°F plus	10.0
Total	100.0

After Hoog, Klinkert, and Schaafsma, Shell Hydrodesulfurization Process, American Petroleum Institute Mid-Year Meeting, New York City, May, 1952.

refining point to a declining use of sulfuric acid. However, a number of processes are used commercially for restoring spent sulfuric acid, as given in the following list:

Drum-type concentrator (air-blown)
Flash-film vacuum-type concentrator
Vacuum concentrator
Acid-decomposition process
Alkylation acid-regeneration process
Sludge-conversion process

Sulfuric acid is still the most widely used and economical inorganic acid. It will continue to be widely used in the steel, oil, chemical, and by-product coke industries for removing undesirable sulfur and unsaturated compounds from final products.

Gases. The removal and recovery of hydrogen sulfide from gas streams has been steadily assuming importance. Natural-gas purification, petroleum-cracking operations, and chemical by-product manufacture all require efficient removal and recovery processes.

The following is a list of the hydrogen sulfide processing methods now available:*

Caustic soda	Sodium phenolate
Girbotol	Thylox
Iron oxide	Tripotassium phosphate
Lime	Vacuum carbonate
Seaboard	

A flow plan of a typical combination hydrogen sulfide scrubbing and sulfur-conversion plant is shown in Fig. 1-24. Depending upon the composition of the feed-gas stream,

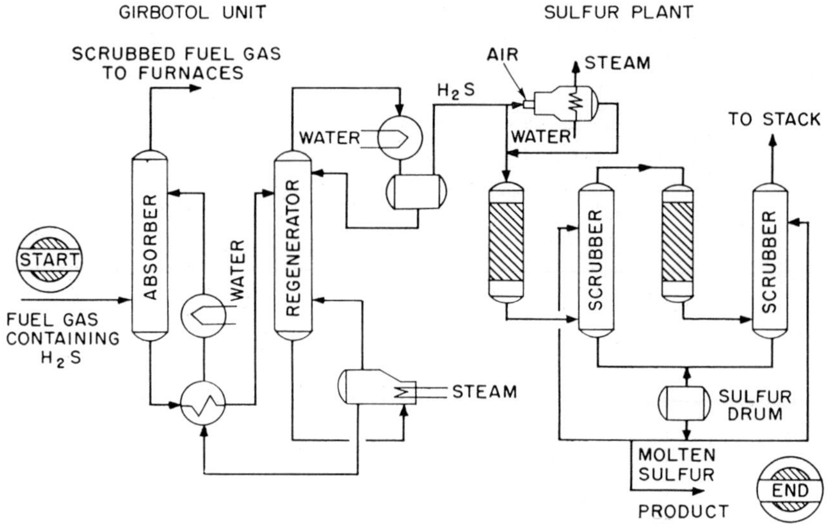

FIG. 1-24. Flow plan of hydrogen sulfide scrubbing and sulfur-conversion plant.

the hydrogen sulfide content of the sulfur-plant inlet gas will usually vary from 60 to 98 per cent hydrogen sulfide. One of the research problems now being investigated is the elimination of the gas-scrubbing step and the manufacture of sulfur directly from a feed gas low in hydrogen sulfide. A flow diagram of a new plant practicing this latter operation is shown in Fig. 1-25.

In many cases, the sulfur is not available in a form so convenient as hydrogen sulfide. Particularly in the combustion of coal and residual fuel oil, the sulfur is burned to sulfur oxides and appears in the hot flue gases. The concentration of sulfur oxides in flue gases emitted to the atmosphere is normally low. Thus the sulfur oxides usually can be readily diffused with air to avoid ground-level fumigation.

In a few places it is sometimes necessary to consider flue-gas scrubbing. Several installations at large British boilerhouses have been made and operated where pollu-

* W. L. Nelson, "Petroleum Refinery Engineering," 3d ed., McGraw-Hill Book Company, Inc., New York, 1949.

tion is critical and the neighborhood is congested. A process flow plan of one type of flue-gas scrubbing process is shown in Fig. 1-26. According to some recent investigations, it is desirable to reheat the flue gases to improve atmospheric diffusion.

An interesting economic comparison of two source-control routes for preventing atmospheric contamination from high-sulfur fuel gases in a critical area is presented in

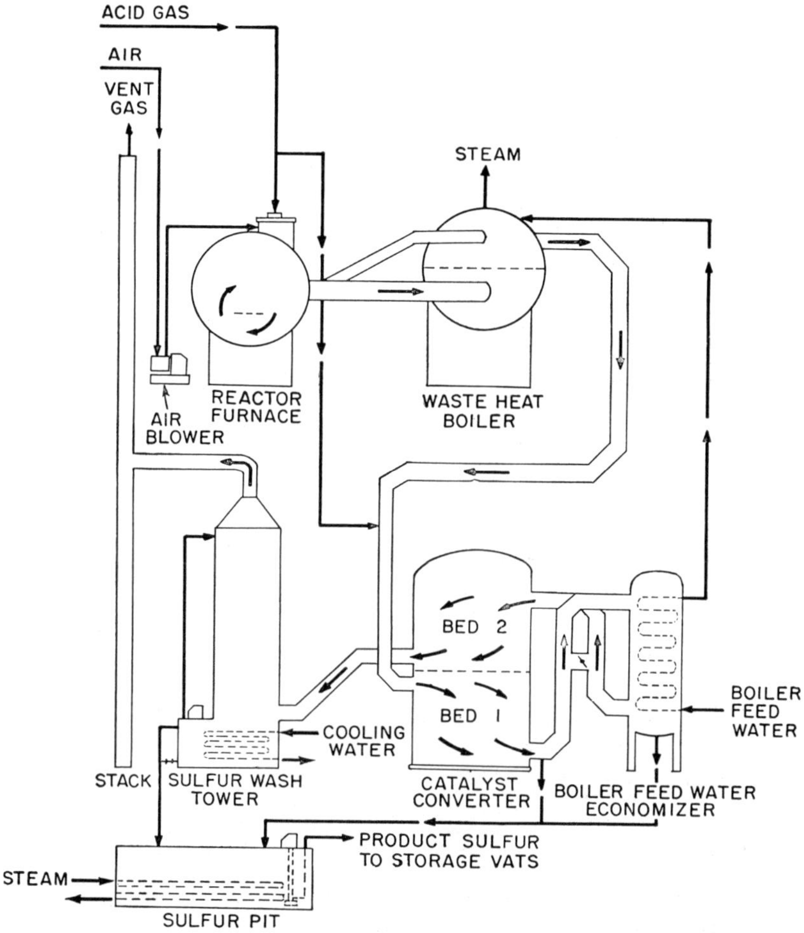

FIG. 1-25. Flow diagram of the Slaughter Field sulfur plant. Hydrogen sulfide content of feed gas is 15.5 per cent (mol.). (*Courtesy of J. F. Mullowney, Petroleum Processing, p.* 1345, *September*, 1953.)

Table 1-24. This single case clearly indicates that the sulfur by-product route is better. Flue-gas scrubbing must also be handled carefully to prevent water pollution.

A similar economic comparison of two source-control routes for preventing atmospheric contamination from sulfur-bearing fuel oil in a critical area is presented in Table 1-25. The data indicate that flue-gas scrubbing is more efficient in liquid-fuel combustion (and probably coal combustion) than removal of sulfur from the fuel.

Ore Refining. Nonferrous smelters handle raw ores containing sulfur compounds. Processing operations convert the sulfur materials to SO_2 and SO_3. Table 1-26 gives an idea of the problem faced by some typical smelters.

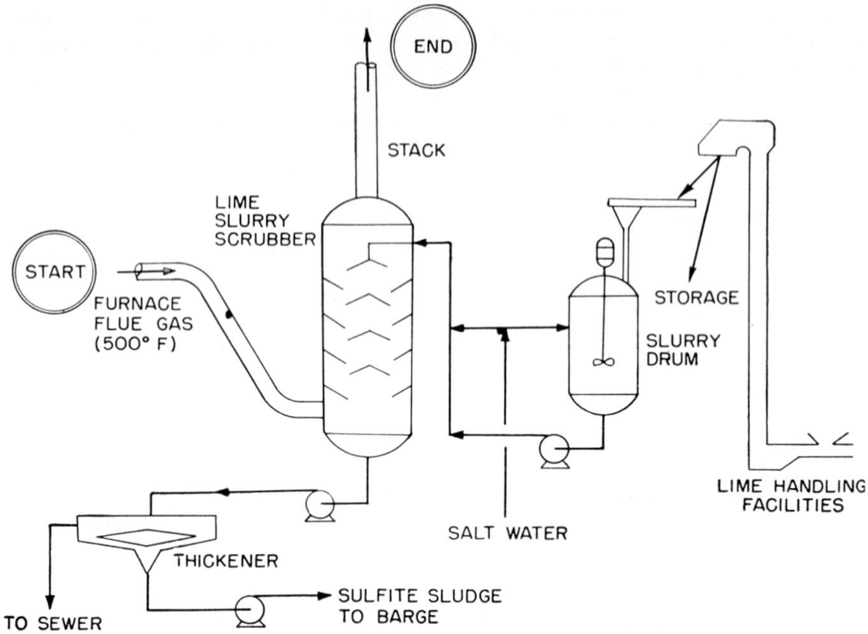

FIG. 1-26. The scrubbing of sulfur oxides from flue gas.

Table 1-24. Costs of Removal of Sulfur

	Girbotol unit plus sulfur plant	Lime-slurry flue-gas scrubber plant
Gas rate, million SCF/stream day:		
Fuel gas	11	
Flue gas	....	200
Sulfur in gases, short tons/stream day	25	25
Total investments, incl. utilities and offsite, $	1,100,000	3,400,000
Operating costs, $/calendar day:		
Total, incl. depreciation and taxes	900	2,700
H_2S fuel value, at $4/short ton H_2S	100	
Produce credit, sulfur at $30/short ton	(680)	
Net cost, $/calendar day	320	2,700
$/year	120,000	1,000,000
Sulfur produced, short tons/calendar day	22.5	

Here again is an example of source control to the extent that ore concentration excludes some sulfur-bearing gangue. In addition to source control, extensive abatement devices such as filters, cyclones, and scrubbers are used to alleviate air contamination. Research and development have made it possible to recover useful by-products from many of these plants. Sulfuric acid and other chemicals are being manufactured from sulfur oxides contained in vent gases.

Blue Haze. A familiar result of sulfur-compound combustion is the formation of a blue haze at or near the top of flue stacks. This haze is caused by the oxidation to sulfur trioxide of 1 to 5 per cent of the sulfur under normal conditions of excess air usage. The sulfur trioxide is hydrated by water vapor in the stack gas or in the atmosphere to form aerosols of sulfuric acid. These aerosols are in such a particle size range that they appear blue against a dark background. Typical conditions resulting in sulfur trioxide production are shown in Fig. 1-27.

Table 1-25. Economic Comparison of Sulfur-removal Systems

Reduction in atmospheric sulfur emission, %	Fuel oil burned, ¢/bbl	
	Fuel-oil desulfurization[a]	Flue-gas scrubbing[b]
0 (4.2% wt sulfur fuel oil).........	0	0
30..............................	50	25–35
98..............................	No reasonable process available	50–75

[a] Based on large-scale and efficient refining processes.
[b] Flue gas scrubbed with calcium carbonates or calcium hydroxides corresponding to a scale of about 20,000 bbl/day of fuel oil burned.

Table 1-26. Typical Sulfur-removal Problems

	Copper				Lead				Zinc			
Metal company.	Noranda Mines, Ltd.		Kennecott Copper Co.		Bunker Hill & Sullivan Min. & Conc. Co.		St. Joseph Lead Co.		Anaconda Copper Min. Co.		New Jersey Zinc Co.	
Location.......	Noranda, Quebec		Bingham, Utah		Kellogg, Idaho		Herculaneum, Mo.		Great Falls, Mont.		General company experience	
Type plant.....	Concentrating, roasting, smelting		Concentrating		Concentrating, roasting, smelting		Concentrating, roasting, smelting		Concentrating, roasting, electrolysis			
Analysis, wt %..	S	Cu	S	Cu	S	Pb	S	Pb	S	Zn	S	Zn
Ore as mined....	29[a]	2.3[a]	1.5	...	....	....	0.9	2	....	..	5–30	
Concentrates....	39[a]	10.5[a]	25–30	...	....	....	17–18[b]	66–75[b]	32	54	32 ± 2	
Roaster feed....	28	6	.....	...	13	52.6	...	..	Same as conc.			
First roast......	None				7.5	52.9	7.5	49	None			
Final calcine of smelter feed..	14	7	.....	...	1.6	51.8	3.2	49	2.3	60	0.5–2.5	

[a] Ore as mined is fed directly to roaster and is not feed for concentrates.
[b] Gravity concentrate — flotation concentrate.
Bray, "Nonferrous Production Metallurgy," John Wiley & Sons, Inc., New York, 1947.
Information courtesy of Kennecott Copper Co., The New Jersey Zinc Co. (of Pennsylvania), St. Joseph Lead Co.

Blue haze can be reduced in a number of ways. Figure 1-28 illustrates the effect of some variables explored in experimental work on a commercial processing furnace. Leaks of flue gas through the brickwork of one furnace contained 60 per cent of the sulfur as SO_3 compared with 1 to 5 per cent of the sulfur as SO_3 in stack flue gas. This high percentage is due to the catalytic effect of the refractory surface and can be eliminated easily by holding proper draft pressure on the firebox.

Some experiments have also been made on the use of firebox additives such as dolomite or hydrated lime for reducing slagging, corrosion, and blue haze. Figure 1-29 shows the results of a commercial experiment in which hydrated lime was injected into

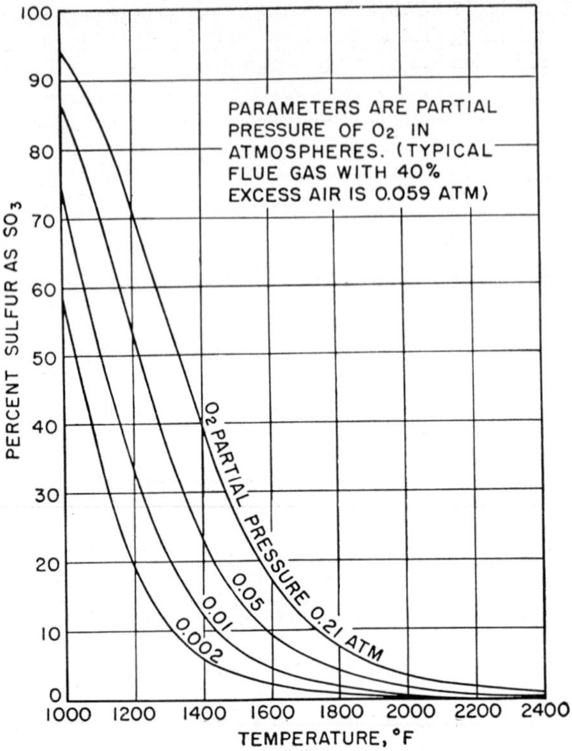

FIG. 1-27. Sulfur trioxide formation. (*From unpublished company source.*)

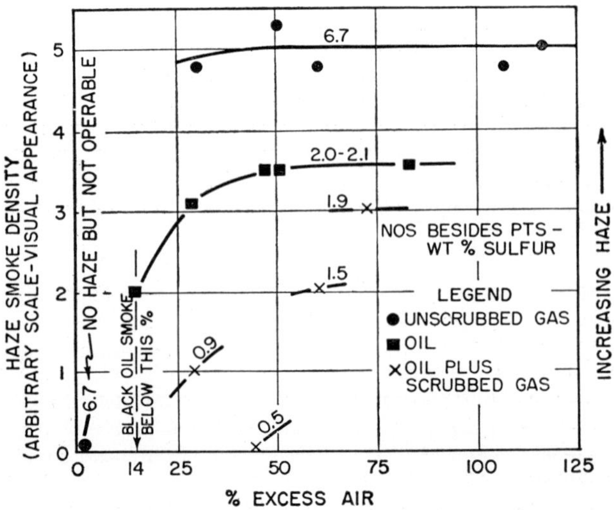

FIG. 1-28. Blue-haze reduction. (*From unpublished company source.*)

the firebox. Some abatement device to reduce the dust emission in flue gas with lime injection might be necessary under some conditions.

Summary. By far the largest mass of sulfur contaminants is in the form of sulfur oxides—sulfur dioxide and sulfur trioxide. These contaminants are often released as large single-point source emissions, particularly from energy generation and fuel- and ore-refining operations. Source control is necessary and reaches a high state of development in dealing with sulfur emissions. The cost of abatement devices usually

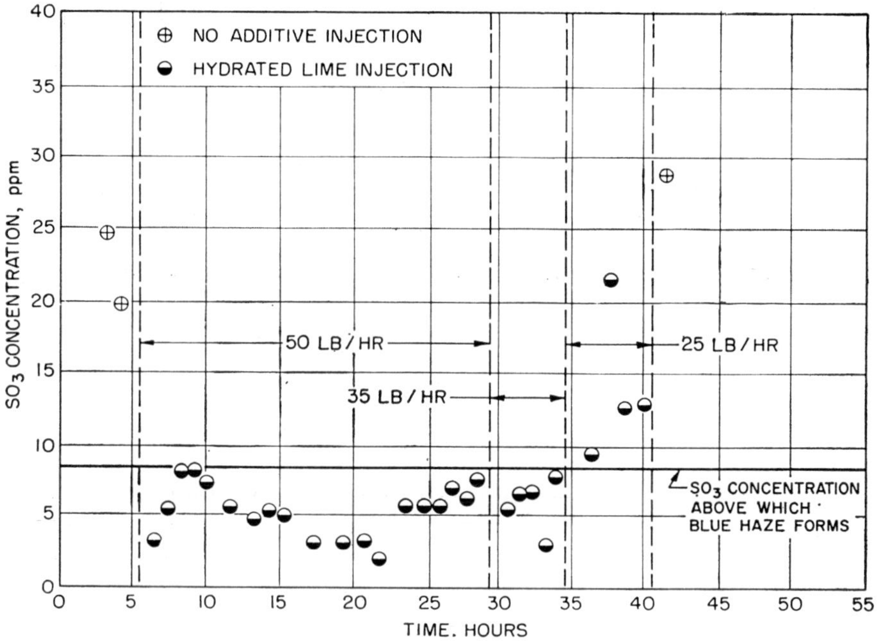

FIG. 1-29. Reduction of blue haze using hydrated lime. (*Courtesy of Esso Standard Oil Company.*)

exceeds the cost of source control without corresponding benefits. Source control is particularly attractive where product improvements in the form of higher quality and higher yields are realized along with air pollution reduction.

There are measures that the consumer of fuels can use to combat air pollution nuisance. High stacks help in preventing nuisance from fumigation at ground level. In some cases, flue-gas scrubbers are useful. But in the main analysis, the problem of sulfur reduction falls on the doorstep of the primary refiner of raw materials.

1.7.4 Fumes and Odors

Industrialization has increased the mass of fumes, odors, and similar contaminants released to the atmosphere, but it can hardly be said that industrial contaminants are more evil-smelling than natural odors. Dead fish on a hot August day, pig pens in a humid valley, an excited skunk—these are some never-to-be-forgotten olfactory experiences of natural origin. The salvation lies in their isolated occurrence and rather limited nuisance area.

So it is with man-made odors. Though large in cumulative mass, most odors come from millions of vents, leaks, and assorted discharges of hundreds of chemical compounds. Odors are the most difficult to define of all air pollution factors. The human

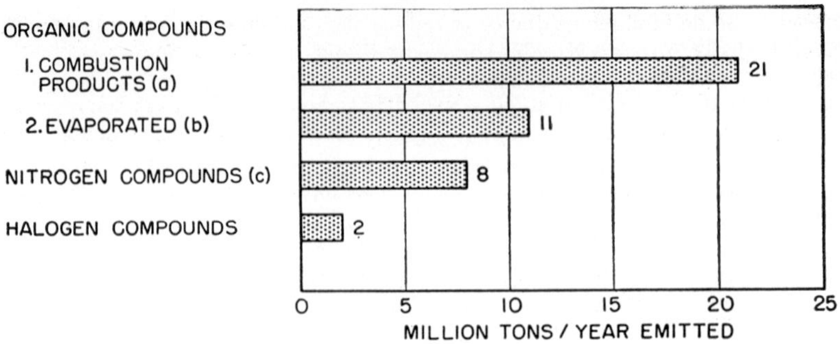

(a) INCLUDES COMPOUNDS RESULTING FROM COMBUSTION OF COAL, OIL, GASOLINE, REFUSE, ETC

(b) INCLUDES EVAPORATION LOSSES OF HYDROCARBONS, NATURAL GAS, ETC, BEFORE COMBUSTION, PLUS MISCELLANEOUS VAPORS SUCH AS SOLVENTS, ETC

(c) EXPRESSED AS NITROGEN DIOXIDE

FIG. 1-30. Major emissions. (*Calculated from numerous published sources.*)

Table 1-27. Combustion Fumes

Source of pollution	Losses to atmosphere, lb per unit of consumption								
	Unit of consumption	Non-ether-soluble aerosols	Ether-soluble aerosols	Oil mist[a]	Total hydro-carbons	Acety-lene	Total alde-hydes[b]	Oxides of nitro-gen[c]	Organic acids[d]
Automobile exhaust:[e]									
Idling	1 bbl gasoline	0.30	f	3.2	29.6	13.3	0.30	0.00	0.09
Cruising	1 bbl gasoline	0.24	0.14	0.00	2.1	0.31	0.14	5.2	0.22
Accel.-decel.	1 bbl gasoline	0.45	0.37	0.97	34.3	0.87	0.30	3.3	0.12
Diesel-bus exhaust,[g] cruising	1 bbl fuel	3.2	f	f	Trace	f	0.46	9.4	1.3
Incinerator,[h] domestic	1 ton rubbish	17.1	29.2	0.00	0.00	0.00	5.1	10.6	27.4
Boiler,[i] fuel-oil-burning	1 bbl oil	f	f	f	Trace	f	Trace	5.2	0.56

[a] At equilibrium conditions, approximately 90 per cent of this mist will exist as a vapor in the atmosphere.
[b] Calculated as formaldehyde.
[c] Calculated as nitrogen dioxide.
[d] Calculated as acetic acid.
[e] Studebaker Champion, 1947, odometer reading 40,000 miles, used; engine in average mechanical condition. Acceleration-deceleration conditions were simulated on a dynamometer by accelerating to 40 mph and decelerating to 20 mph continuously in high gear at full and closed throttle, respectively.
[f] Not measured.
[g] Diesel bus, 210-hp two-stroke cycle used. Cruising conditions were simulated on a dynamometer by operating at 30 mph and 15 bhp.
[h] Standard 3-cu-ft domestic incinerator used. Rubbish, composed of 46 per cent leaves, 2 per cent paper, 15 per cent grass, 14 per cent wood, and 3 per cent rags, was burned at a rate of 29 lb/hr.
[i] Boiler, serving large power plant, consumed 100 bbl of fuel oil (9.2°API) per hour.

Courtesy of Los Angeles County Air Pollution Control District, Los Angeles, Calif., Second Technical and Administrative Report on Air Pollution Control, p. 37, 1950–1951.

Correlations developed from special studies.

nose is still an amazingly accurate tool for discerning contaminants in a few parts per million of air and often even parts per billion. No man-made mechanical or electronic instrument serves so well.

An inventory of United States fume and vapor emissions is almost like counting grains of sand on a mile of beach. No reliable estimates have been made. Figure 1-30 is an effort to try to break down the mass emissions of some more widely prevalent fumes, odors, and vapor-type releases. Once again the products of combustion—energy for transportation—predominate. Internal-combustion engines bestow pleasures and benefits, but they also create some air pollution problems. Table 1-27 presents some comparative data on the products of combustion from automobiles, diesel buses, domestic incinerators, and fuel-oil burning.

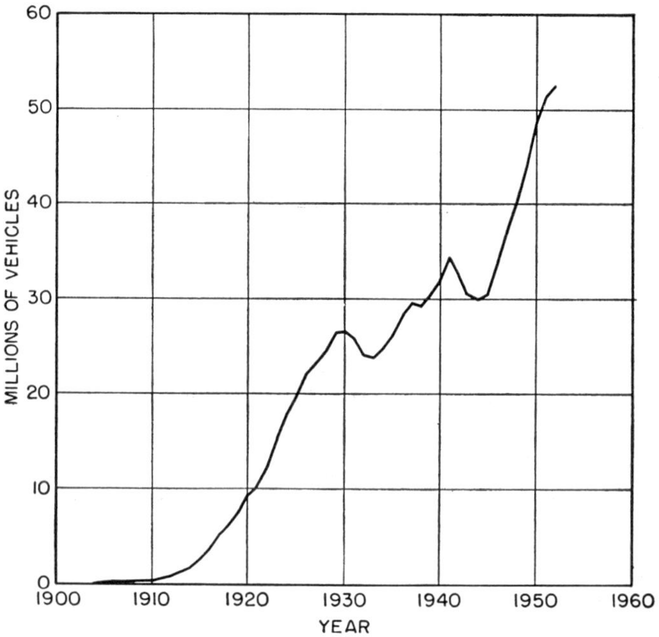

FIG. 1-31. Motor-vehicle registrations in the United States. (*U.S. Bureau of Public Roads*, 1900 *to* 1952.)

One method of design which might be applied to odor-producing plants is the following. In a chemical-processing plant designed to recover fissionable material from used reactor fuel elements, air is drawn into processing areas and away from all areas containing personnel. Wastes are run through a central location where leaks are checked continually and automatically. Perhaps this method is too expensive and elaborate for general industrial plants where less harmful contaminants are handled. The principles seem to be sound—continuous ventilation for personnel, followed by collection and processing of the entire vent gas and air stream for removal of critical contaminants.

Internal Combustion. Many investigators have studied engine and fuel theory, design, performance, and improvement for many years. Recognition of the effect of exhaust-gas contaminants on air pollution (particularly in cities) is a more recent trend. Magill and Littman (10) have summarized the Los Angeles situation, where automobile exhausts are a major source of air pollution.

The rise in automotive use is illustrated in Fig. 1-31. Although each vehicle releases only a small mass of contaminants, the large number increases the total rapidly in large cities. The near-ground-level release of engine-exhaust gases amplifies the problem. Running the exhaust pipe on large trucks to extend above the top of the cab appears to offer one quick and easy method of getting increased mixing with air and reduced ground-level concentration.

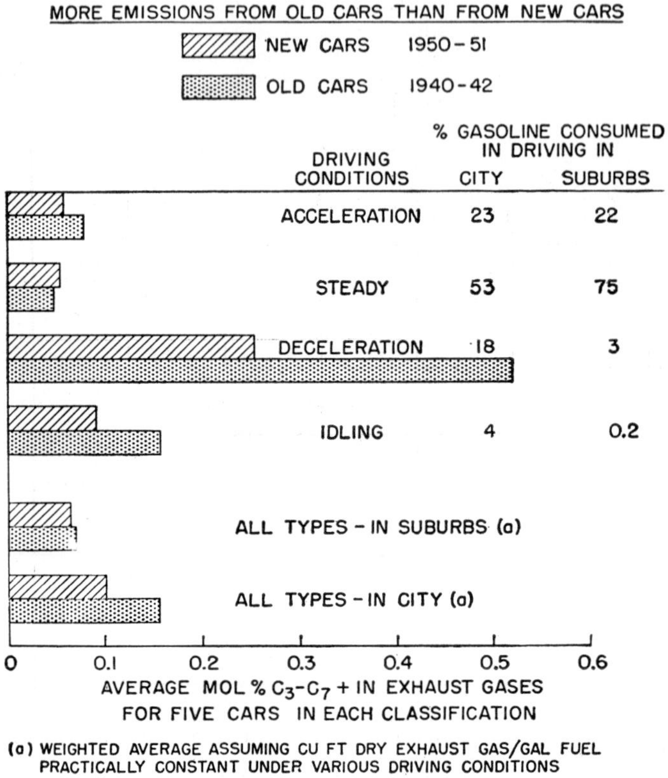

FIG. 1-32. More emissions from old cars than from new cars. (*After P. L. Magill, D. H. Hutchison, and J. M. Stormes, Proc. 2d Nat. Air Pollution Symposium, pp.* 71–83, *Stanford Research Institute, Los Angeles, Calif.*, 1952.)

A more fundamental approach is being taken after studies of automobile exhausts showed the typical effects for old and new cars, shown in Fig. 1-32. The trend in old- and new-car use is illustrated by Fig. 1-33. Deceleration is the chief cause of increased exhaust contamination. Old engines generate more contaminants than new engines, as expected.

Improved engine exhausts from both gasoline- and diesel-powered vehicles are being studied from several angles. Better mixing of air and fuel, improved engine design by control of combustion reactions, use of after-burners in the exhaust tail pipe, and variation in fuel composition, possibly by the use of additives, are some of the present possibilities.

This is a complex problem. To give a brief idea of the chemical-engineering sequences involved, Fig. 1-34 is offered. The automobile-engine designer for this particular case has to plan a process feeding 0.06 cu cm of gasoline and 500 cu cm of air at a temperature of about 110°F for the 20-mph speed (1,000 rpm). The mixture enters the cylinder containing 100 cu cm of exhaust gas from the previous cycle. The temperature rises to about 250°F at the end of the intake stroke. The mixture in the cylinder is then compressed to 180 psia and 600°F. Then the spark ignites the mixture, and the combustion reaction is carried out in about 0.024 sec. The pressure

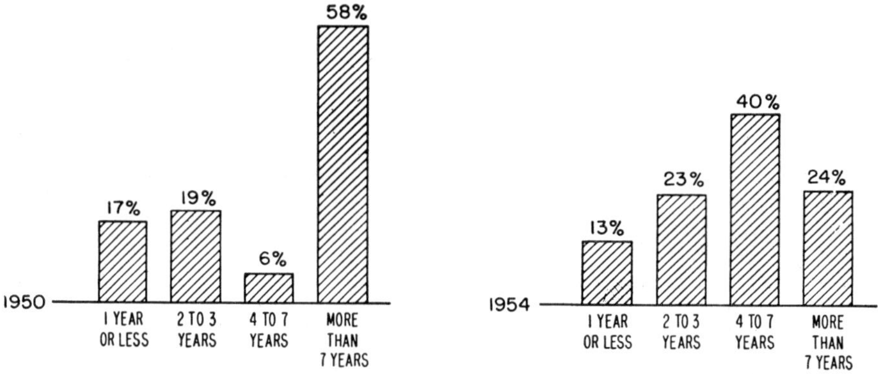

FIG. 1-33. Age of automobiles owned in the United States. *(Courtesy of Federal Reserve Board, Pick-S, and New York Herald Tribune, July 11, 1954.)*

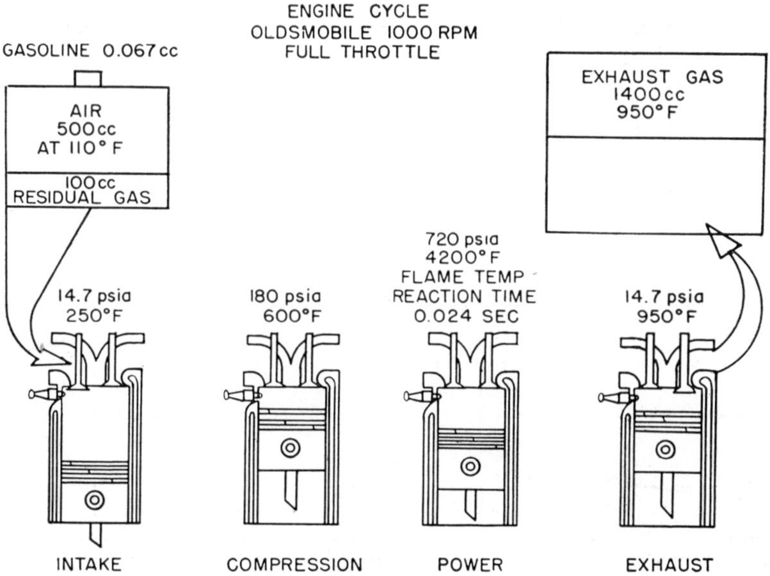

FIG. 1-34. Internal-combustion reactions. *(From unpublished company source.)*

increases up to 720 psia with a moving flame front at about 4200°F. The 1,400 cu cm of exhaust gases must then be removed at a temperature of about 950°F. At about 1,000 rpm there are 500 firing strokes per minute; so the above sequence occurs 500 times per minute. In addition, the process must be prepared to handle a range of gasoline feed rates from idling to 60-mph speed.

Aside from careful and continued maintenance plus following responsible advice regarding fuel and lubricant specifications, there is no present reasonable way of eliminating exhaust-gas contaminants. Gasoline engines operate most efficiently with

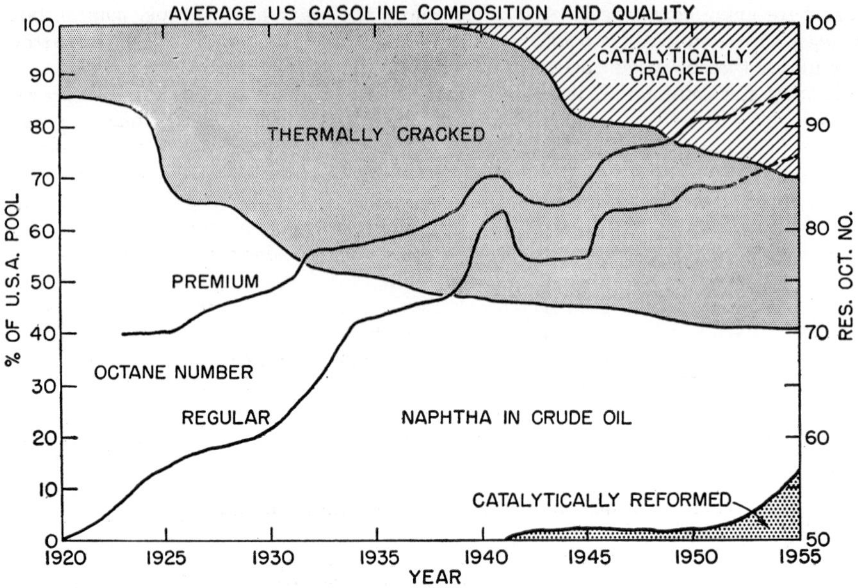

FIG. 1-35. Average United States gasoline composition and quality. (*From unpublished company source.*)

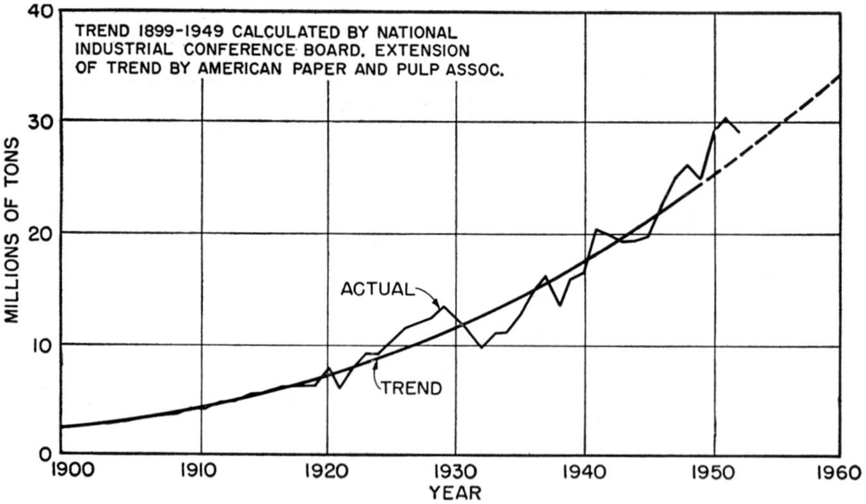

FIG. 1-36. Increase in paper refuse. [*Courtesy of L. T. Stevenson, Paper Trade J.*, **136,** 130–133 (1953).]

but 80 per cent of the theoretical air requirement. This tends to give incomplete combustion.

Nitrogen oxides are found in exhaust engine gases, probably because of fixation of atmospheric nitrogen. Economical source control for these reactions is difficult—leading to the conclusion that abatement measures are more promising.

Gasoline engines and fuels are being improved constantly. In 1953, only 0.7 gal of gasoline was required to do the work that 1.0 gal did in 1925. Source control is at work here also. Figure 1-35 shows the changing trend in motor-fuel composition. Increased percentages of cracked or re-formed gasolines from catalytic processes not only increase useful power per gallon, but also tend to reduce sulfur content.

Refuse Incineration. True source control of waste materials now burned or buried would be so widely distributed and varied as to defeat itself from the start. Substitution of metallic wrappings or noncombustible containers for paper is one example difficult to accept. It would be true source control if every bit of waste material was collected and carefully reprocessed. The United States is too well endowed with resources to justify such a procedure; so rubbish and garbage must be destroyed. The "cut-and-fill" method of burying refuse in dumps is time-honored. In most locations, refuse is burned.

Figure 1-36 shows the increasing quantities of paper refuse. Incineration of refuse is chiefly an abatement problem. Numerous efficient designs are available for burning either sorted or unsorted rubbish and garbage. Some of the incinerators use secondary fuel to help destroy the refuse. The major field of improvement seems to involve elimination of small and inefficient domestic incinerators in favor of municipal plants. Table 1-28 demonstrates that this procedure is effective in reducing noxious emissions.

Table 1-28. Emission from Odorous Domestic Incinerators

Effluent	Incinerator effluent analysis, wt % of feed	
	Domestic	Municipal
Organics	13.7	0.06
Aldehydes (as formaldehyde)	0.2	0.07
Acids (as acetic acid)	0.04	0.03
Sulfur oxides (as SO_2)	0.1	0.1
Nitrogen oxides (as NO_2)	0.025	0.1
Ammonia	0.1	0.02
Solids (carbon)	0.12	1.2

After Stanford Research Institute, The Smog Problem in Los Angeles County, Third Interim Report, Western Oil and Gas Association, Los Angeles, Calif., 1950.

The wide range of feeds to incinerators makes proper design a most complex problem. Corey, Spano, Schwartz, and Perry (4) have reported on experiments covering a systematic investigation of incineration principles.

Evaporation. The sources of practically all evaporation emissions are volatile liquids. Source control can rarely change the characteristics of the liquids without destroying their usefulness. Many evaporation-prevention devices are available to reduce emissions of the most volatile liquids to negligible quantities.

The petroleum industry handles the major percentage of volatile fluids in the United States. Some of the more popular devices used in this industry are the following:

Gastight cone and dome roof tanks (with pressure-vacuum vent)*
Floating roof tanks:†

- Pan-type
- Pontoon
- Double-deck pontoon

* All tankage handling volatile stocks should be painted white or aluminum. Insulation of the roof and sometimes the shell is also desirable for volatile stocks.

† These can be equipped with one, two, or three roof seals, depending upon service. Three seals are used for especially odorous stocks.

Variable-vapor-space tanks:

- Wet-seal lifter
- Dry-seal lifter
- Vapor dome
- Vapor sphere
- Dry-seal gas holder

Pressure tanks:

- Hemispheroid
- Spheroid
- Globe roof
- Sphere

Interconnected tankage (with or without a recovery compressor)

Each storage problem must be considered specifically to arrive at optimum results, as illustrated in Table 1-29. The effectiveness of evaporation-loss control for one large, complex oil-refining organization is shown in Table 1-30.

An interesting new development has been tried for use with existing fixed-roof tankage. Some data are listed in Table 1-31.

A minor trend toward use of liquefied petroleum gas as motor fuel is evident. This

Table 1-29. Effect of Tank Type on Loss

Tank type	Gastight cone roof with white roof paint	Double-seal pontoon-type floating roof	$2\frac{1}{2}$-lb pressure tank	Operating conditions
Case I				Tank capacity, 50,000 bbl Days in service, 365 Tank throughput, 100,800,000 gal/yr Avg tank outage (cone roof), 24 ft Product RVP, 10 lb Avg product temp, 61°F Product value, 9.01¢/gal Pumping rate into tank, 84,000 gal/hr Pumping rate out of tank, 8,400 gal/hr Avg wind velocity, 10 mph Turnovers/month, 4
Breathing loss, gal/yr	50,250	0	0	
Windage loss, gal/yr	0	8,480	0	
Pumping loss, gal/yr	168,350	0	168,350	
Total loss, gal/yr	218,600	8,480	168,350	
Installation cost, $	85,000	107,500	130,000	
Incremental cost over gastight cone roof, $	0	22,500	45,000	
Value of savings, $/yr	0	18,930	4,530	
Increase (decrease) in maintenance over gastight cone roof, $/yr	0	50	(5)	
Net savings, $/yr	0	18,880	4,535	
Payout on incremental cost, months	0	11	119	
Case II				Tank capacity, 50,000 bbl Days in service, 365 Tank throughput, 8,400,000 gal/yr Avg tank outage, 22 ft Product RVP, 10 lb Avg product temp, 61°F Product value, 9.01¢/gal Pumping rate into tank, 84,000 gal/hr Pumping rate out of tank, 8,400 gal/hr Avg wind velocity, 10 mph Turnovers/month, ⅙
Breathing loss, gal/yr	48,990	0	0	
Windage loss, gal/yr	0	8,480	0	
Pumping loss, gal/yr	14,030	0	14,030	
Total loss, gal/yr	63,020	8,480	14,030	
Installation cost, $	85,000	107,500	130,000	
Incremental cost over gastight cone roof, $	0	22,500	45,000	
Value of savings, $/yr	0	4,915	4,415	
Increase (decrease) in maintenance over gastight cone roof, $/yr	0	50	(5)	
Net savings, $/yr	0	4,865	4,420	
Payout on incremental cost, months	0	55	121	

Courtesy of Committee on Loss Prevention, Esso Standard Oil Co., *National Petroleum News* reprint, p. 6, July 13 and 27, 1949.

Table 1-30. Effectiveness of Evaporation-loss Control

	Loss on crude refined, %	
	1940	1st 6 months 1953
Tankage }	1.9	0.38
Sewers and separators }		0.15
Pump and line leaks }	0.6	0.09
Flare gas }		0.08
Other known losses }		0.11
Unknown losses }		0.30
Total	2.5	1.11

Table 1-31. Evaporation Prevention in Cone-roof Tanks Using Microballoons[a]

	Control tank[b]		Foam covering[b]	
Evaporation loss, bbl/day	2.7		0.36	
Reduction in evaporation loss, %	...		86	
Oil properties	Initial	Final	Initial	Final
API	33.5	31.9	35.3	34.9
RVP, psi	3.0	1.25	4.7	4.5
IBP	129	158	128	120
5%	178	240	211	225
20%	336	366	330	349

[a] Microballoon technique of Standard Oil Co. (Ohio). Microballoons are small spherical hollow particles containing sealed-in gas at about atmospheric pressure. A ½- or 1-in.-thick blanket of these particles forms the foam covering. Particles are usually made of plastic and float on oil—similar to a floating roof.

[b] 25,000-bbl working tanks with open vents.

After Ellerbrake and Veatch, Microballoons and Their Use in Reducing Evaporation Losses of Crude Oil, 33d Annual Meeting of American Petroleum Institute, Chicago, Ill., Nov. 9, 1953.

product must be handled in closed pressure cylinders. Hence, evaporation loss to the atmosphere is normally negligible. Some work is being done to use low-volatility kerosene or gas oils for driving motor vehicles through gas-turbine engines. Any appreciable future trend of this nature would decrease evaporation losses at the source.

Odors. The sources of odors are so varied that only generalities can be drawn. Source control of odors can be carried out by the following routes:

1. Change of the composition of process materials or removal of causative impurities
2. Dilution of odorous air with relatively clean air
3. Masking the odor with a less objectionable additive
4. Combustion to odorless or nonobjectionable products
5. Injection of a reactive substance into the gas with sufficient contact before discharge
6. Conversion to a tolerable compound in a reaction scrubber
7. Sorption in a suitable solvent
8. Removal of vapor by condensation
9. Filtration through a bed of granular sorbent
10. Removal of odor-bearing mist or dust by a separator for particulate matter

A majority of these methods involve source-control principles.*

In addition to these odor-control routes involving chemical-engineering techniques, it should also be mentioned that minimizing leakage and using conservation storage devices to reduce evaporation losses are fundamental. The process and economic improvements possible are specific to many industries and beyond the scope here.

One increasing trend is evident, particularly among oil refineries and chemical plants. Waste streams of water are collected and processed at a central point by blowing with hot flue gas or steam. Sometimes spent acid is used to neutralize alkaline wastes. The odorous gases are passed through a firebox. The deodorized and purified water flows to disposal basins.

Odor masking on a commercial scale is a relatively new development with the following possible application routes:†

1. Spraying, vaporizing, or atomizing the selected odorant into air-gas streams
2. Adding directly to a process whenever possible
3. Adding to scrubbing liquors
4. Spreading or floating on contaminated surfaces without dilution

Catalytic combustion of various burnable contaminants has been studied by several investigators for reducing air pollution. Figure 1-37 shows the principles of this

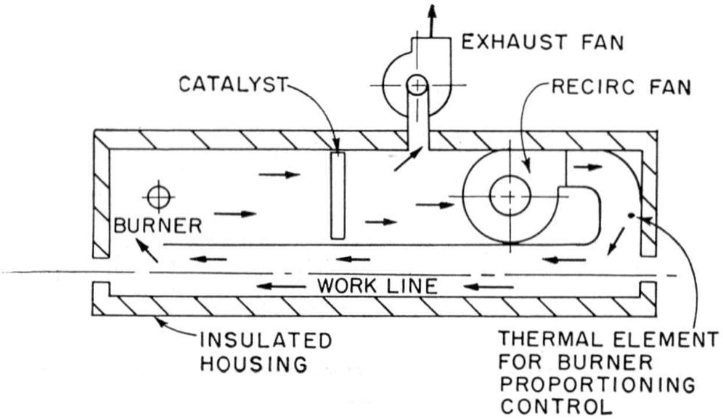

Fig. 1-37. Catalytic combustion of vapor emissions. Sectional view through conveyor-type solvent-evaporation oven showing catalyst controlling solvent emission at source. (*Courtesy of L. C. McCabe, ed., "Air Pollution," chap.* 32 *by R. J. Ruff, pp.* 259–263, *McGraw-Hill Book Company, Inc., New York,* 1952.)

process. The advantages are decomposition of contaminants at low temperature level. A typical application is described as follows:‡

Source of odorous vapors	Paint solvent
Chemical type	Mixture of toluene, xylene, butanol and higher alcohols, ketones and phenols
Catalyst	Platinum alloy and alumina
Inlet temperature to catalyst	400°F (min)
Anticipated catalyst life	4,000–5,000 hr

Summary. Source control of fumes and vapor emissions is complicated by the wide range of specific contaminants encountered. Internal-combustion exhaust emissions, refuse incinerators, and hydrocarbon evaporation to the atmosphere comprise the major contamination masses.

* Courtesy of W. N. Witheridge, Odor Control Practices, chap. 31, p. 251, in L. C. McCabe, "Air Pollution," McGraw-Hill Book Company, Inc., New York, 1952.

† Courtesy of B. K. Tremaine, Masking Industrial Malodors, *Air Repair,* **3,** No. 2, November, 1953.

‡ After L. C. McCabe, Atmospheric Pollution, *Ind. Eng. Chem.,* **45,** No. 3, 109A (1953).

Control of fumes and vapors has been studied for many years from equipment-design viewpoints. Increased attention is being directed to the air pollution technology involved. There are actually tens of millions of sources emitting fumes and vapors in the United States. A combination of control and abatement at the source appears to afford the best results.

1.8 INDUSTRIAL CHECK LIST

Many local surveys have indicated that industrial sources emit the majority of air contaminants. Commercial, residential, and motor-vehicle sources are important but usually secondary. These private sources of pollution are large in total, but small individually, and hence difficult to control.

Of all the sources, combustion of fuels is predominant. Table 1-32 breaks down the fuel uses in the United States. It has been shown that the percentage of cleaner-burning fuels has been increasing. But energy generation has been climbing faster than composite fuel quality, resulting in increased pollution.

Table 1-32. Fuels Used in the United States[a]

1952	Coal		Fuel oil		Natural and manufactured gas
	Bituminous	Anthracite	Distilled	Residual	
Quantity	....	.....	[b]	....	[c]
Trillion btu	10,980	890	2,700	2,800	8,240
Million tons	418	35			
Million bbl	....	.....	463	445	
Billion cu ft	....	.....	....	....	7,845
	Consumption, %				
Consumer					
Retail	16.4	80.2			
Fuel briquets	....	1.3			
Heating oil	....	.....	60.4	17.8	
Residential	....	.....	....	....	22.4
Commercial	....	.....	....	....	7.1
Electric power	24.7	10.7	....	....	11.6
Gas and electric power plants	....	.....	1.8	15.8	
Railroads	9.1	1.8	14.7	9.0	
Oil-company use	....	.....	1.7	12.2	
Petroleum refineries	....	.....	....	....	6.8
Field use	....	.....	....	....	19.5
Coke plants	23.3	0.2			
Steel	1.6	1.1			
Cement	1.9	0.5			
Smelters, mines, and manufacturing plants	....	.....	9.3	35.5	
Other industrial	23.0	4.2	....	....	32.6
Other uses, including military	....	.....	12.1	9.7	
Total	100.0	100.0	100.0	100.0	100.0

[a] Excluding fuel for vessel bunkers.
[b] Including diesel oil.
[c] Manufactured gas is 0.3 volume per cent of total.
References: Calculated from data of U.S. Bureau of Mines and American Gas Association.

Figure 1-38 (pages 1-52 to 1-54) is a check list and relates each industry with the specific contaminant type it normally releases.

A somewhat more generalized treatment of the more important source-control methods is given in Table 1-33 on page 1-55.

SOURCES OF INDUSTRIAL CONTAMINANTS

MATERIALS, INDUSTRIES, ETC. (a)	FINE SOLIDS	COARSE PARTICULATES	COMPOUNDS: SULFUR	COMPOUNDS: ORGANIC	COMPOUNDS: NITROGEN	COMPOUNDS: OXYGEN	COMPOUNDS: HALOGEN	COMPOUNDS: RADIOACTIVE
MINERAL FUELS								
COAL (ANTHRACITE, BITUMINOUS & LIGNITE)	✓	✓	✓	✓	✓	✓		
NATURAL GAS			✓	✓	✓	✓		
PETROLEUM (CRUDE, NAT. GASO., BENZOL)			✓	✓	✓	✓		
NONMETALLIC MINERALS (EX. FUELS)								
ASPHALT AND RELATED BITUMENS	✓	✓	✓	✓		✓		
BARITE	✓	✓	✓					
BORON MINERALS	✓	✓						
CEMENT	✓	✓				✓		
CLAYS	✓	✓				✓		
GYPSUM	✓	✓						
LIME	✓	✓				✓		
PHOSPHATE ROCK	✓	✓						
POTASSIUM SALTS	✓	✓						
PUMICE & PUMICITE	✓	✓						
PYRITES	✓	✓	✓					
SALT	✓	✓					✓	
SAND & GRAVEL	✓	✓						
SLATE	✓	✓						
STONE	✓	✓						
SULFUR	✓	✓	✓					
TALC, PYROPHYLLITE & GROUND SANDSTONE	✓	✓						
OTHER NONMETALLIC MINERALS	✓	✓	✓	✓	✓	✓	✓	✓

(a) INCLUDING PROCESSING TO AND INCLUDING END USE. FUELS CONSIDERED PART OF PROCESSING. MAJOR EMISSION TYPES ARE CHECKED.

FIG. 1-38. Sources of industrial contaminants.

	FINE SOLIDS	COARSE PARTICULATES	COMPOUNDS: SULFUR	ORGANIC	NITROGEN	OXYGEN	HALOGEN	RADIOACTIVE
METALS (b)								
BAUXITE (AND ALUMINUM ITSELF)	✓	✓			✓	✓	✓	
COPPER	✓	✓	✓	✓	✓	✓		
IRON ORE	✓	✓	✓	✓	✓	✓		
LEAD	✓	✓	✓	✓	✓	✓		
MANGANIFEROUS ORE	✓	✓	✓	✓	✓	✓		
TITANIUM	✓	✓	✓			✓		
ZINC	✓	✓	✓	✓	✓	✓		
OTHER	✓	✓	✓			✓	✓	✓
NONDURABLE GOODS INDUSTRIES								
FOOD, BEVERAGE AND KINDRED PRODUCTS	✓	✓		✓	✓			
TOBACCO MANUFACTURES	✓	✓		✓	✓	✓		
TEXTILES AND APPAREL		✓		✓				
PAPER AND ALLIED PRODUCTS	✓	✓	✓	✓	✓	✓		
PRINTING AND PUBLISHING	✓	✓		✓				
CHEMICALS AND ALLIED PRODUCTS								
INDUSTRIAL INORGANIC CHEMICALS	✓	✓	✓		✓	✓	✓	✓
INDUSTRIAL ORGANIC CHEMICALS								
PLASTIC MATERIALS	✓	✓	✓	✓	✓	✓	✓	
SYNTHETIC FIBERS	✓	✓	✓	✓	✓	✓	✓	
ORGANIC CHEMICALS	✓	✓	✓	✓	✓	✓	✓	
ALCOHOLS				✓	✓	✓		
OTHER	✓	✓	✓	✓	✓	✓	✓	✓
DRUGS AND MEDICINES	✓	✓	✓	✓	✓		✓	✓
SOAP AND RELATED PRODUCTS	✓	✓	✓	✓	✓	✓	✓	
PAINTS AND ALLIED PRODUCTS	✓	✓	✓	✓	✓	✓	✓	

(b) ALSO RECOVERY PLANTS

Fig. 1-38. Sources of industrial contaminants. *(Continued)*

	Fine solids	Coarse particulates	Compounds: Sulfur	Organic	Nitrogen	Oxygen	Halogen	Radioactive
VEGETABLE AND ANIMAL OILS				✓	✓	✓		
OTHER, INCLUDING FERTILIZERS, TOILET PREPARATIONS, WOOD CHEMICALS, CARBON BLACK, ETC.	✓	✓	✓	✓	✓	✓	✓	
PETROLEUM AND COAL PRODUCTS	✓	✓	✓	✓	✓	✓	✓	
COKE AND BYPRODUCTS	✓	✓	✓	✓	✓	✓	✓	
RUBBER PRODUCTS	✓	✓	✓	✓	✓	✓	✓	
LEATHER AND LEATHER PRODUCTS			✓	✓	✓	✓	✓	
OTHER NONDURABLE GOODS	✓	✓	✓	✓	✓	✓	✓	✓
DURABLE GOODS INDUSTRIES								
PRIMARY METALS								
IRON AND STEEL	✓	✓	✓	✓	✓	✓	✓	
OTHER PRIMARY METALS	✓	✓	✓	✓	✓	✓	✓	
FABRICATED METAL PRODUCTS	✓	✓	✓	✓	✓	✓	✓	
MACHINERY								
NON-ELECTRICAL, INCLUDING APPLIANCES	✓	✓	✓	✓	✓	✓		
ELECTRICAL, INCLUDING COMMUNICATION	✓	✓	✓	✓	✓	✓		
TRANSPORTATION EQUIPMENT	✓	✓	✓	✓	✓	✓		
INSTRUMENTS AND RELATED PRODUCTS	✓	✓		✓				
LUMBER, FURNITURE, ETC.	✓	✓	✓	✓	✓	✓		
STONE, CLAY AND GLASS PRODUCTS	✓	✓		✓				
MISCELLANEOUS DURABLE GOODS	✓	✓	✓	✓	✓	✓	✓	

	Fine solids	Coarse particulates	Compounds: Sulfur	Organic	Nitrogen	Oxygen	Halogen	Radioactive
GENERAL								
TRANSPORTATION								
RAILROADS	✓	✓	✓	✓	✓	✓	✓	?
OTHER, INCLUDING PETROLEUM DRIVEN			✓	✓	✓	✓		?
MATERIALS HANDLING	✓	✓	✓	✓	✓	✓	✓	
ELECTRIC AND GAS UTILITIES	✓	✓	✓	✓	✓	✓	✓	?
CONSTRUCTION								
FARM	✓	✓		✓				
INDUSTRIAL AND COMMERCIAL	✓	✓		✓				
RESIDENTIAL	✓	✓		✓				
ROAD, ETC.	✓	✓		✓				
HUMAN ACTIVITY								
HEATING	✓	✓	✓	✓	✓	✓	✓	
SMOKING	✓	✓		✓	✓	✓		
INCINERATION, INCLUDING BACKYARD	✓	✓	✓	✓	✓	✓	✓	
SEWAGE			✓	✓	✓	✓	✓	
CLEANING AND LAUNDRY	✓	✓	✓	✓	✓	✓	✓	
PAINTING				✓				
COOKING				✓	✓	✓		

FIG. 1-38. Sources of industrial contaminants. (*Continued*)

Table 1-33. Generalized Source-control Methods

Improve Conditions

1. Better control and observance of operations
2. Better combustion
3. Better maintenance
4. Proper instrumentation
5. Good housekeeping
6. Better materials handling

Change Operations

1. Modified design or process
2. New, modern design[a]
3. Different design or process[a]
4. Consolidation of many units into one unit which is more efficient[a]
5. Different and/or better raw materials
6. Different fuel and/or better combustion equipment
7. Removal of contaminants from feed, products, etc. (treating)
8. Concentrated treating agents
9. Use of pollutants formerly wasted
10. Produce gas or liquid product instead of dust-producing solid (fuels, etc.)
11. Controlled dust removal
12. Pelletizing, etc.
13. Use of additives, odor masking, etc.
14. Closed, indirect cooling, etc., instead of direct cooling, etc.

Use of Abatement-type Measures

1. Enclosed process for vapor recovery
2. Improved process covers and recovery equipment, especially tankage
3. Process handling of effluent streams for contaminant removal (solid, liquid, and vapor)
4. Better integration of abatement equipment with process
5. Selection of fuels, materials, and process equipment compatible with space for abatement equipment
6. Modification of process conditions to permit better operation of abatement equipment
7. Combustion of effluents (catalytic and direct)

Other Methods

1. Fix fine particles (soil stabilization, etc.)
2. Find causes for unusual solids attrition and vapor generation
3. Reduce vaporization caused by radiation, etc.

[a] New plant construction required.

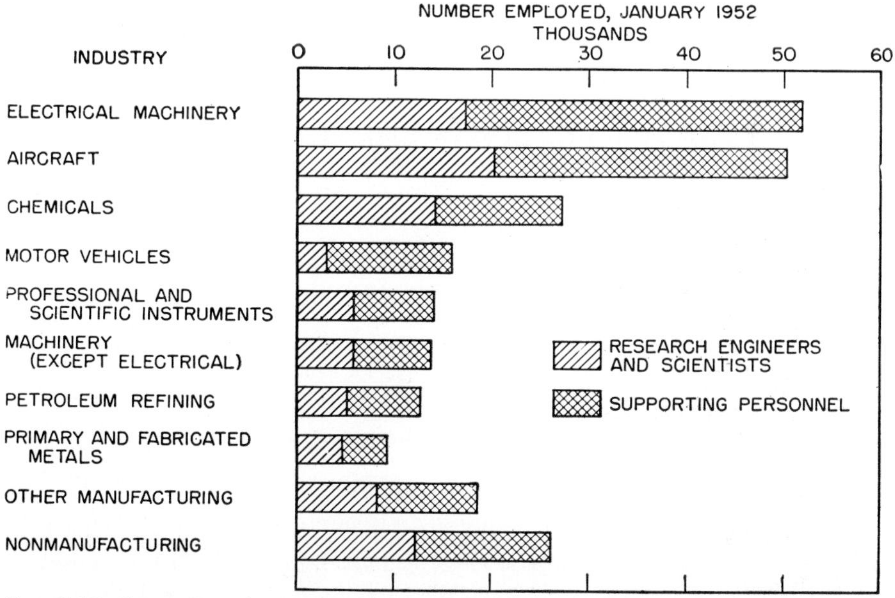

FIG. 1-39. Inventory of research personnel. (*Courtesy of U.S. Department of Labor, Bureau of Labor Statistics, Scientific Research and Development in American Industry, Bull.* 1148, *p.* 5, 1953.)

1.9 FUTURE TRENDS

The series of 1952 reports by the President's Materials Policy Commission (13) provides some uniform ground for looking ahead; however, there is no unanimous opinion on the validity of all predictions for future technology trends.

Major factors in all improved technology—air pollution control is included—are the numbers, incentives, and ideas of United States research personnel. Figure 1-39 on page 1-55 shows the research and supporting personnel employed by various industries. With such personnel available, technological progress should continue.

Planning and looking into the future is practiced by industry and government alike. A check list on the future is included as Fig. 1-40 on page 1-57.

One of many examples of progressive industrial research is illustrated in Fig. 1-41 on page 1-58.

Natural gas is in plentiful supply now. It is a fuel commodity enjoying popularity and prosperity. Along with other endeavors, its use has contributed greatly to reducing air pollution in the United States. But at this growing stage, this industry is also looking ahead for new processes to manufacture gas substitutes. This symbolizes the real spirit of research and development people everywhere.

1.10 ACKNOWLEDGMENTS

The author wishes to express appreciation to Harley S. Anders for valuable contributions and assistance and to the Esso Research and Engineering Co. (formerly the Standard Oil Development Co.) for permission to include unpublished material in Sec. 1.

REFERENCES

1. ASME Air Pollution Committee: Air Pollution Activities, *Mech. Eng.*, **75**, 712–714 (1953).
2. Bosanquet, C. H., and J. L. Pearson: *Trans. Faraday Soc.*, **32**, 1249–1264 (1936).
3. *Chem. Eng. News*, **32**, No. 12, 1108–1113 (1954). Smog Scramble Spans Nation.
4. Corey, R. C., L. A. Spano, C. H. Schwartz, and H. Perry: Experimental Study of Effects of Tangential Overfire Air on the Incineration of Combustible Wastes, *Air Repair*, **3**, No. 2, 109–116 (1953).
5. de Lorenzi, O.: "Combustion Engineering," Combustion Engineering–Superheater, Inc., New York, 1950.
6. du Pont, H. B.: Management Looks at Air Pollution, *Proc. 2d Nat. Air Pollution Symposium*, pp. 58–61, Stanford Research Institute, Los Angeles, Calif., 1952.
7. Gosline, C. A.: Planning Prevents Air Pollution, Air Pollution Abatement Conference, Manufacturing Chemists' Association at New York City, Feb. 25–26, 1952.
8. Johnson, A. J., and G. H. Auth: "Fuels and Combustion Handbook," McGraw-Hill Book Company, Inc., New York, 1951.
9. Lammers, H. B., et al.: "A Study of Smoke and Air Pollution in the Kanawha Valley," Coal Producers' Committee for Smoke Abatement, Cincinnati, Ohio, 1949.
10. Magill, Paul L., and F. E. Littman: Air Pollution in Los Angeles, American Society of Mechanical Engineers Meeting, New York City, Nov. 29 to Dec. 4, 1953.
11. McCabe, Louis C., ed.: "Air Pollution," McGraw-Hill Book Company, Inc., New York, 1952.
12. Pearson, A. S., and T. R. Galloway: Fly Ash Improves Concrete and Lowers Its Cost, *Civil Eng.*, **23**, 592–595 (1953).
13. President's Materials Policy Commission: "Resources for Freedom," vol. IV, "The Promise of Technology," U.S. Government Printing Office, Washington, D.C., 1952.
14. Rosene, John: General Examples of Industrial Furnaces and Fuels, Report to the City of Tacoma, Wash., 1952.
15. Wurts, T. C.: American Industry Lends Generous Support to the Air Pollution Control Association, *Smokeless Air*, No. 85, pp. 116–118, Spring, 1953.

	FINE SOLIDS	COARSE PARTICULATE	SULFUR COMPOUNDS	ORGANIC COMPOUNDS	NITROGEN COMPOUNDS	OXYGEN COMPOUNDS	HALOGEN COMPOUNDS	RADIOACTIVE COMPOUNDS	
NEW ENERGY SOURCES									
1. INDUSTRIAL ATOMIC ENERGY	✓	✓	✓	✓	✓	✓		✓	
2. HEAT PUMPS	✓	✓	✓	✓	✓	✓			
3. SOLAR	✓	✓	✓	✓	✓	✓			
4. WIND	✓	✓	✓	✓	✓	✓			
5. TROPICAL WATER HEAT DIFFUSION	✓	✓	✓	✓	✓	✓			
6. WATER POWER BY CONTROLLED RAIN	✓	✓	✓	✓	✓	✓			
7. TIDAL ENERGY	✓	✓	✓	✓	✓	✓			
IMPROVEMENTS IN ENERGY USE									
1. NEW ENGINES--JET, TURBINE, ROCKET, ATOMIC, ETC.			✓	✓	✓	✓			
2. NEW POWER TRANSMISSION SYSTEMS--TORQUE CONVERTER, AUTOMATIC SHIFTS, ETC			✓	✓	✓	✓			
3. NEW THERMO-CHEMICAL COMBUSTION DEVICES			✓	✓	✓	✓			
NEW PROCESSES ALLIED WITH ENERGY									
1. GASIFICATION OF COAL, LIGNITE, OIL, ETC.	✓	✓	✓	✓	✓	✓			
2. CONTROLLED BIOLOGICAL PHOTOSYNTHESIS--CHLORELLA, ETC.	✓	✓	✓	✓	✓	✓			
3. LIQUID FUELS SYNTHESIS--SHALE OIL, COAL, LIGNITE, ETC.	✓	✓	✓	✓	✓	✓			

(*a*)

	FINE SOLIDS	COARSE PARTICULATE	SULFUR COMPOUNDS	ORGANIC COMPOUNDS	NITROGEN COMPOUNDS	OXYGEN COMPOUNDS	HALOGEN COMPOUNDS	RADIOACTIVE COMPOUNDS	
NEW PROCESSING TRENDS									
1. SUBSTITUTION OF OXYGEN FOR AIR IN OXIDATION REACTIONS, COMBUSTION, ETC.	✓	✓			✓				
2. COAL HYDROGENATION			✓						
3. WIDER USE OF HYDROGEN TREATING AND HYDRO-REACTIONS			✓						
4. WIDER USE OF ELECTRIC FURNACES IN METALLURGY	✓						✓		
5. ATOMIC RADIATION PROCESSING								✓	
6. SULFUR EXTRACTIONS FROM RAW MATERIALS AT WELL AND MINE			✓						
NEW STRUCTURAL TRENDS									
1. WIDER USE OF PLASTICS--REQUIRING LESS ENERGY THAN METALS	✓	✓	✓	✓	✓	✓			
2. ALLOYS FOR OPERATION ABOVE 2000°F UNDER STRESS				✓	✓	✓			
3. WIDER USE OF IMPROVED INSULATION FOR HEAT RECOVERY	✓	✓	✓	✓	✓	✓			
GENERAL									
1. INCREASED USE OF ELECTRICITY FOR RESIDENTIAL HEAT	✓	✓	✓	✓	✓	✓		✓	
2. DECENTRALIZATION OF INDUSTRY	✓	✓	✓	✓	✓	✓	✓		
3. INCREASED SUBURBAN LIVING	✓	✓	✓	✓	✓	✓	✓		

(*b*)

FIG. 1-40. Some probable future source-control trends. A check mark under contaminant group indicates an anticipated change in air pollution trend. (*After "Resources for Freedom," vol. IV, "The Promise of Technology," U.S. Government Printing Office, Washington, D.C., June, 1952; and various other sources.*)

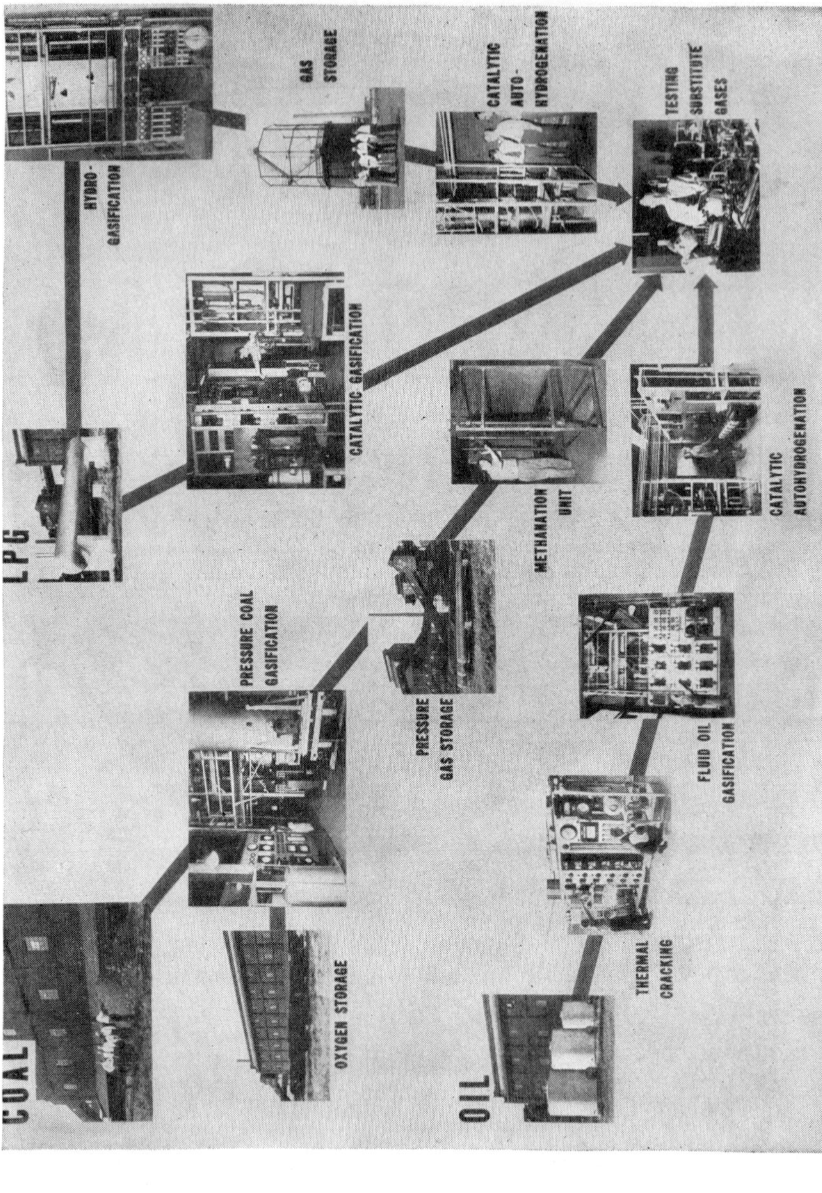

FIG. 1-41. Development of natural gas substitutes. (*Courtesy of Institute of Gas Technology.*)

SECTION 2

CITY PLANNING, INDUSTRIAL-PLANT LOCATION, AND AIR POLLUTION

BY MORRIS KATZ

2.1 INTRODUCTION

The pollution of the air we breathe by waste products resulting from the varied activities of the population and industry has attained serious proportions in many communities. Urban centers, which, a few years ago, were unaware of the extent of the air contamination in their midst because adverse effects were not readily apparent, are now confronted with recurrent smogs, the prevalence of eye irritation among the residents, and other unpleasant experiences. The continued expansion of industry and growth of population, the creation of new industries during and after World War II, and the shift of the population from rural to urban communities to supply the labor demand are factors which have contributed to the present status of the atmospheric pollution problem in America.

Public and technical interest in the control of industrial effluents and radioactive wastes continues to grow in the postwar period. The smog disaster at Donora, Pa., in 1948, and the widespread publicity given the Los Angeles smog problem in the last few years have aroused public concern on a nationwide basis about the menace to public health and welfare. The nuisance and economic aspects of excessive air contamination are matters with which the public has an older acquaintance. In agricultural and forest areas, extensive damage to vegetation or livestock has been caused by liberation of excessive quantities of sulfur dioxide, lead and arsenic, fluorine compounds, and other effluents in smelting and metallurgical refining operations. In many of our large cities and urban areas the objectionable nuisance and economic effects are evident in accelerated corrosion of metals; deterioration of stonework, other building materials, and plant and household equipment; soiling of laundry, textiles, and other fabrics; and increased weathering of paints and finishes. Communities where pollution is high suffer from frequent occurrence of smogs, accompanied by reduced visibility, the prevalence of eye irritation among the residents, and costly traffic dislocation at airports and on the highways.

Such conditions present a challenge to the scientist, the engineer, and air pollution control groups. The successful manner in which the problem is being attacked is attested by the immense and constantly growing literature in this field and by the number of technical conferences and symposia conducted in the past 5 years. Significant contributions to methods of investigation, meteorology, effects on plants and animals, and collection and control of pollutants were made at symposia sponsored by the American Chemical Society at San Francisco in 1949, Stanford Research Institute at Pasadena in 1949 and 1952, the United States Technical Conference convened at the request of President Truman at Washington in May, 1950, and the Twelfth International Congress of Pure and Applied Chemistry at New York in September, 1951. Other notable conferences have been held recently, such as those conducted by the Manufacturing Chemists' Association, the American Society of Mechanical Engineers, the American Industrial Hygiene Association, and the American Meteorological Society.

These technical conferences have been of great service in making available to industry, science, and government the outstanding developments in the diversified fields of physics, chemistry, meteorology, and engineering. Such information provides guidance for the correct approach to the problem of national control and regulation of atmospheric pollution. Although there are gaps in our present knowledge of the subject, sufficient information and experience are available for intelligent use in the planning of industrial-plant locations, zoning of areas in relation to pollution, and organization of air pollution control groups. The primary objective of planning is to ensure that the normal growth of industry and population does not result in raising the level of pollution to the point where it becomes a persistent nuisance and a potential hazard to health.

This section will deal with such questions as the following:

1. City planning, industrial-plant location, and zoning. Some case histories will be analyzed to show the success or failure of certain abatement efforts. Various factors must be considered in the organization of an air pollution control group in order that effective liaison may be maintained between the local agency and representatives of industry and county, state, and Federal government agencies. The problem of voluntary versus compulsory abatement merits careful consideration. Planning for the future involves a consideration of the influence of potential air pollution problems on city planning, industrial-plant location, industrial and residential zoning, and the need for special control regulations.

2. The organization and functions of air pollution control groups.

3. The relation between nature, extent, and distribution of contaminants as revealed by air pollution surveys and the measures necessary to reduce or control pollution.

4. The influence of meteorological factors and topography on dilution, dispersion, and control of effluents.

2.2 INDUSTRIAL-PLANT LOCATION AND ZONING

In planning the location of an industrial plant, the factor of air pollution control must be included in the basic considerations which have to be assessed by modern plant management. No longer is it sufficient to consider only the availability of raw materials, labor force, transportation, water supplies, and markets. Past neglect of air pollution planning has proved to be a costly error for many large undertakings which have been faced with expensive damage claims, litigation, and the necessity for the installation of control equipment after several years' operation. It is usually much simpler and less expensive to provide for control features in the design stage than to be compelled to add these after the plant has been constructed and placed in operation.

If the new plant is to be located in an area which is already industrialized, it is sound practice to undertake a preoperational survey to determine the existing levels of contaminants under prevailing meteorological conditions. The results of such a survey, in conjunction with known operational data on the scale of contemplated emissions from the new sources, would provide information on the extent to which waste products could be safely discharged to the atmosphere without producing too much contamination. An intelligent appraisal of the problem requires a knowledge of the specific effects of the major contaminants to be discharged to the atmosphere in relation to the topography, population, and land use of the area surrounding the site. Certain contaminants are more toxic to vegetation and animals than to people. A rural and agricultural area is more sensitive to sulfur dioxide and fluorides than is an urban community. Hydrogen sulfide has little effect on vegetation but is obnoxious and even dangerous to human life in comparatively low concentrations.

A preliminary air pollution survey is also useful if the site to be selected is located in a suburban or rural area. Frequently, the area under consideration is subjected to exotic pollution, i.e., contamination from distant sources. It is important to establish what these concentration levels may be in relation to the proposed scale of operations. Suspended particulates, fluorides, sulfur dioxide, and other gases may be carried great distances from large industrial communities to predominantly rural areas. The existing conditions may be tolerated by the rural dwellers until a new plant commences activities in the immediate neighborhood.

The ideal site for disposal of airborne wastes is comparatively level terrain in a region where the average wind velocity is of the order of 10 mph or more and where deep temperature inversions are a rare occurrence. An additional advantage is gained if the site is not upwind of valuable farm land or a populated community. The plant-site property should be large enough so that maximum concentrations of effluents, at ground level downwind, occur well within company premises rather than on surrounding private property. Valley sites require more pollution control than level or undulating terrain, especially when the average wind speed is less than 10 mph. Stacks must be tall enough so that the effective height of the plume will permit atmospheric wastes to be carried out of the valley rather than to be trapped below the level of the surrounding hills.

A large power plant burning 5,000 to 10,000 tons of pulverized fuel per day may discharge 300 to 600 tons of sulfur dioxide daily and a large amount of fly ash if the coal contains about 3 per cent of sulfur. With the removal of fly ash by the operation of electrostatic precipitators or other dust collectors of 95 per cent efficiency, the plant may still release upward of 25 to 50 tons of ash per day. About 30 per cent of this ash may be deposited in the neighborhood of the plant. If the effective discharge height is 400 ft, most of this deposition will occur within about 1,600 to 4,000 ft of the plant in a wind of 10 mph. Let us suppose that 75 per cent of the dust is deposited within a radius of 1 mile. The monthly dustfall rate over this area of about 3 sq miles will then be approximately 54 to 108 tons/sq mile. The sector in the prevailing wind direction will have a greater rate of deposition, and other sectors will be correspondingly lower, depending upon the frequency distribution of wind direction. Proper zoning of an operation of this magnitude would require that the industrial site be located on about 2,500 to 5,000 acres in order to minimize the hazard from both dust and sulfur dioxide damage on adjacent private property. Under favorable conditions of topography and micrometeorology, the effective dispersion of such a large amount of waste sulfur dioxide would require stacks 400 to 600 ft tall.

Modern control measures for smelters and steel, aluminum, and other large plants include not only the recovery of a substantial portion of the waste sulfur, fluorides, and dust from metallurgical operations, but also the ownership of sufficient land to prevent the occurrence of excessive ground concentrations or damage to valuable farm or forest property. Such land ownership provides a method of industrial zoning by private management. This pattern can be assisted greatly by municipal or county planning for industrial-plant location. This involves zoning restrictions so planned that residential areas and certain heavy industries will not be located too close to each other. Although the location of new industries will result inevitably in the growth of new communities or the expansion of existing towns, such growth could be planned and zoned to prevent the encroachment of residences in the immediate vicinity of factories, especially in the prevailing downwind direction.

Certain chemical plants may discharge vapors and gases which react in low concentrations in the atmosphere to produce eye-irritating and other unpleasant effects by direct or photochemical reactions. Thus a synthetic-rubber plant, producing or utilizing butadiene and styrene, located adjacent to a chlorine or hydrochloric acid plant, may create a pollution problem as a result of joint operations where none would exist if these plants were situated at a reasonable distance from each other. Halogen compounds and unsaturated cyclic hydrocarbons, such as styrene, can react in extremely low concentrations in the air in sunlight to produce lachrymators. Such conditions have actually occurred at Baton Rouge, La., in the Sarnia–Port Huron area of the Great Lakes, along the Houston Ship Channel, and elsewhere. These incidents provide an added reason for industrial-plant zoning to prevent the haphazard location of industrial activities and to segregate certain types of industries.

In the zoning of land for industrial use it is apparent that due consideration must be given to all factors which govern the diffusion and dispersion of atmospheric contaminants, including topography of the area, frequency and speed of prevailing winds, and stability of the air.

2.3 PLANNING FOR THE FUTURE

Many cities are now paying the price of past failure to plan for future orderly development. Smoke, dust, odors, and poor zoning practices have destroyed fine residential sections and created blighted areas in the central parts of some cities. The resultant unattractive living and business conditions tend to drive people away to the suburbs. The deterioration in property values soon makes it uneconomic to maintain buildings in good repair or to modernize them, and the whole neighborhood degenerates into a slum area. The newer built-up sections undergo the same process of decay in time as factories and other business enterprises creep into locations where they are not desirable.

It is impossible to assess accurately the huge economic losses incurred by such conditions. In the United States, surveys within recent years indicate that the losses from smoke damage alone amount to about $1.5 billion annually (19). There can be no successful solution to the problem of eliminating blighted areas in our large cities until air pollution is properly controlled. However, in many instances intelligent zoning and pollution control measures must be carried out on a regional rather than a local basis. This requires cooperation between counties and states as well as between adjoining municipalities. Thus in the Philadelphia and Delaware River Valley industrial area, there is a tristate situation with regard to pollution control involving Pennsylvania, New Jersey, and Delaware. A similar situation exists with respect to large cities and factories bordering New York and New Jersey and along some points of the international boundary between Canada and the United States.

Extensive reconstruction and development programs have been undertaken by a number of large cities within the last few years to remove blighted areas, plan for future orderly growth, and make living more attractive within city limits. This involves the expenditure of large amounts of money from both public and private sources for slum clearance, housing developments, transportation facilities, and air pollution control equipment. Successful results are being achieved by this process of planning and rebuilding, especially with the cooperation of community leaders representing industry, labor, and public-spirited bodies. A few examples of what is taking place in some of the larger cities are given below.

2.3.1 Detroit

An old section in the downtown heart of Detroit near the waterfront has been undergoing redevelopment on an extensive scale under the City Planning Commission. New office and other buildings are being erected for municipal, county, and private business purposes as blighted and slum areas are torn down and the land is cleared. Land use in this area is restricted to activities which will not give rise to air pollution. Public parks, recreational and educational facilities, and improvements in highway transportation and parking are features of this new development. Other blighted areas are slated to be cleared out and rebuilt in the future.

According to its chief, B. Linsky, the Bureau of Smoke Inspection and Abatement cooperates effectively with the agencies responsible for city planning and zoning by furnishing technical assistance related to air pollution problems and their control. Enforcement of zoning ordinances, which specify the types of industrial operations that can be carried on within certain districts, is in the hands of the Building Inspection Bureau of the Department of Buildings and Safety Engineering. Questions concerning land use by acid plants, smelters, cement plants, oil refineries, and other plants which may cause undesirable pollution are referred to the Smoke Abatement Bureau in order to determine the special zoning restrictions required to protect residential areas and housing developments.

The Detroit Metropolitan Area Regional Planning Commission is designing a master plan of land use for an area which embraces 3½ counties. The situation is complicated by the fact that a great deal of the heavy industry in the region lies along the waterfront, down river from the heart of Detroit but upwind from the city in the prevailing southwesterly wind direction. It is planned to isolate these areas, containing steel mills, blast furnaces, foundries, refineries, and cement and chemical plants, by buffer zones from residential sections.

Information and data from the Smoke Abatement Bureau are useful to the Zoning Appeals Board of Detroit in cases involving air pollution. This Board has the authority to permit the extension of land use for industrial operations provided that such factories will not emit waste products which might be considered detrimental to the surrounding neighborhood. The Smoke Bureau renders technical assistance and advice to many smaller municipalities surrounding Detroit, such as Wyandotte, Highland Park, Ecorse, and Dearborn. There are over 100 politically independent communities in the Detroit metropolitan region. Since 1949, over $14 million has been expended in Detroit alone for air pollution control equipment. This does not include the costs to industry of abatement measures and devices incorporated in modernization and expansion of plant facilities.

The current international study on air pollution in the Detroit River area includes such factors as the abatement of vessel smoke on the river and the effects of pollutants on health, vegetation, soiling, visibility, and corrosion on an area-wide basis. The results of this study will be useful to municipalities and industry in planning future control activities and industrial expansion. Vessel owners have expended several million dollars on smoke abatement since 1950 in the conversion of ships from coal to oil firing and in the installation of mechanical underfeed stokers, air-steam overfire jets, smoke indicators, and other devices to eliminate or abate the smoke nuisance.

2.3.2 Philadelphia

The Greater Philadelphia area is undergoing a large-scale redevelopment as one of the most rapidly growing regions in the United States. There are over 2 million residents living at present within city limits and several million more in the immediately surrounding neighborhoods. City planning and expansion are geared to a model air pollution control law, passed in 1954, which replaces the one of 1948. In the center of the city an old railroad embankment has been removed and modern office buildings, department stores, and shops, with off-street parking facilities, are being constructed in an area formerly blighted. This Triangle District, extending from the Schuylkill River to the former Broad Street Station of the Pennsylvania Railroad, is being rebuilt according to a plan which includes high-income housing and light industry as well as a building center which, on a more modest scale, will rival Rockefeller Center in New York. The Dock Street market area is to be converted to warehouses or light industry, and the market will be moved to a new location on the bank of the Delaware River, several miles south. Industry of Philadelphia, in its battle against smoke, fumes, dust, and odors, has spent over $40 million in the past few years.

Amendments to the new pollution law which were recommended by the Chamber of Commerce of Greater Philadelphia on behalf of industry were accepted by the city and incorporated into this law. The city's air pollution board gives due consideration to the advice of a technical advisory committee of the Chamber, representing major segments of industry and commerce, in regard to any contemplated changes in regulations. This liaison between city officials and the Chamber's technical committee is on a permanent basis.

A zoning ordinance which has been in existence since 1933 is now being revised in order to extend protection to residential areas both old and new from encroachment by industry and commerce. However, success in the fight against slums and blighted areas depends not only on the strict enforcement of the housing code and air pollution law but also on the voluntary cooperation of industry and the public. The city's comprehensive program of urban renewal places great emphasis also on the rehabilitation and conservation of homes and neighborhoods.

2.3.3 Pittsburgh

The fruitful effort of Pittsburgh and Allegheny County in the abatement and control of air pollution is one of the principal factors which have led to the reconstruction of the city's downtown Golden Triangle district, at a cost which will ultimately amount to about $1 billion. Action by city-planning bodies to arrest the decay of the central section of the city prior to 1943 was largely abortive. This section was afflicted with smoke and polluted air from steel mills and other industries and from railroad locomotives. The situation appeared almost hopeless until the Mellon banking and industrial interests organized the Allegheny Conference on Community Development, which prepared comprehensive redevelopment plans that were backed by state authority in 1945. The state of Pennsylvania established a large park area, Point Park, at the apex of the triangle created by the joining of the Monongahela and Allegheny Rivers to form the Ohio. The district adjoining Point Park was declared a blighted area under the terms of the state law and, in 1950, the work of wrecking tenement and other buildings and reconstructing this whole downtown area was commenced. The Redevelopment Authority, with the mayor of Pittsburgh as chairman, guaranteed the financial backers of this plan that smoke and flood control would be adequate to prevent future deterioration.

Within 4 years, the former "Smoky City" has been transformed into a clean, modern metropolis. New skyscraper office buildings have arisen in the Gateway Center of the city, flanking Point Park. Large underground and aboveground parking garages have been built. There is a new landscaped park above the underground garage. Included in this project are a large hotel and an apartment house. The Penn-Lincoln Parkway will move traffic over this area from east to west via a new bridge at the point of the triangle and a tunnel under the Monongahela River. A large housing project, using Federal assistance, will remove slums and rebuild streets in the Hill District, a height of land above the Golden Triangle. This area will contain a public auditorium and arena, symphony hall and theater, television center, and modern apartments, all integrated into a properly zoned development.

Industry has benefited under this reconstruction program by making it possible to acquire condemned slum areas and blighted land for plant expansion. Sections of the steel industry, which had considered the possibility of moving out of Pittsburgh, prior to 1949, have been modernized, and new open-hearth furnaces have been built to increase capacity by over 2 million tons annually. The redevelopment plan has been extended to parts of the metropolitan district outside the Golden Triangle. Blighted land is first acquired by the Redevelopment Authority and subsequently sold to industrial, charitable, or public interests for land use and construction in conformity with the over-all plan for the city's future development.

This metropolitan area of over 2 million people has demonstrated that even the most difficult problems relating to pollution control and city regeneration can be solved if there is effective cooperation between the public, industry, and local and state governments.

2.3.4 Other Urban Communities

Although planning for future orderly development is recognized by nearly every large city of North America as essential, success depends in large measure on the nature of zoning laws and the manner in which they are enforced. Cleveland is a good example of a city which has controlled its air pollution problem effectively and has also applied to its advantage sound community planning with adequate zoning protection. The key to this city's planning for the future is a yearly inventory of real property, which was started in 1931. Since that time there has been available a detailed collection of maps and data covering every aspect of community change and growth, such as number and location of homes, stores, offices and industrial establishments, shopping centers and apartments, sale of land, remodeling and wrecking of buildings, and many other facts. This inventory has served to indicate the trend in growth of neighboring communities, the location of new business and industrial enterprises, and the revisions necessary in city planning and zoning.

In Los Angeles the worst blighted housing areas parallel fairly closely the location of the principal industrial districts, according to studies by the City Planning Commission. These are the districts which show the highest incidence of air pollution and smog. Control of air pollution is therefore essential to prevent further deterioration of housing areas. Installation of pollution-control equipment and devices by industry in the 5 years since 1949 has cost over $21 million, in addition to unknown expenditures for the removal of dust, fumes, and gaseous wastes in the construction of new plants. The huge oil industry in the area, with a capacity which will soon reach about 1 million barrels per day, has made the greatest major contribution to control by reducing hydrocarbon and sulfur dioxide losses. However, many more millions will be expended before the troublesome smog problem is finally eliminated.

The Southern California Air Pollution Foundation is the most recent organization to concern itself with alleviation of Los Angeles smog. It is a nonprofit organization, incorporated in November, 1953, and is supported by a considerable group of industrial, business, and university people under a board of 18 trustees. The Foundation evaluates existing research and gives financial support to new research projects in existing laboratories. It does not propose to spend time and money building its own laboratories but will assist in the coordination of activities of other industrial, university, and government groups investigating smog problems. Findings and information gained by the Foundation will be made available to constituted authorities to bring about the control of smog. Contributions toward an annual budget of about $1 million are made by corporations, organizations, and individuals, as a public service.

It is recognized that zoning of industry in relation to air pollution is a matter of extreme importance in the Los Angeles area. However, more extensive meteorological data are required for the proper solution of this problem. Satisfactory zoning can be achieved only if there is full cooperation and coordination of efforts by all planning commissions within the county. It was urged, in the report of a special committee on air pollution made to Governor Knight of California in 1953, that this cooperative action should not be limited to Los Angeles County but should also be extended to adjoining counties, with the establishment of additional air pollution control districts.

The city of Houston, Tex., is one of the few industrial centers which has yet to pass a zoning ordinance. In the metropolitan area that includes all of Harris County is an industrial strip of about 25 miles along the Houston Ship Channel. The whole Gulf Coast area experienced a rapid expansion of industry during World War II, and many of the larger plants were established along the Ship Channel. As steel, synthetic-rubber, oil-refinery, fertilizer, cement, heavy chemical, and petrochemical plants came into existence, air pollution problems were intensified. The terrain is quite flat, and meteorological conditions are generally favorable for the dispersion of air contaminants, with prevailing winds from the southeast to south for about 10 months of the year. Atmospheric inversions seldom last for more than a few hours at a time. Residential communities have grown up on the north and south banks of the Houston Ship Channel as corporate entities, but these towns are surrounded by the city of Houston. There are today seven such corporate subdivisions within city limits, some with their own zoning regulations.

In 1948 the Houston Chamber of Commerce organized an industrial-pollution committee to deal with stream pollution and excessive contamination of the ship channel by industrial and sanitary wastes. The scope of this committee was later expanded to deal with air pollution abatement. A survey carried out in June, 1952, indicated that since 1947 industrial firms had expended about $8.6 million on remedial measures for air pollution and planned to spend an additional amount of nearly $4.5 million in the near future. The Harris County Health Unit maintains a section on stream and air pollution control. Each industrial establishment is expected to reduce its individual pollution to some degree by taking appropriate measures. Most of the plants along the ship channel are now engaged in this voluntary system of controlling atmospheric wastes. Most of the damage to vegetation and property which has already occurred is on the north side of the channel. It has been stated that greater quantities of chemicals are discharged to the air along the ship channel than in any other industrial region of North America.

2.3.5 Basic Requirements

Some conception of the basic requirements for sound city planning and zoning may be gained by a glance at conditions which arise in the absence of any plan or design for community development. Small businesses and industries invade residential areas; these begin to deteriorate, lose property values, and decay. The expansion of industrial plants into areas settled by workers eventually drives these workers to seek living accommodation in other sections of town many miles from their work. On the other hand, homes may grow up inside factory districts. Instead of compactly organized neighborhoods with appropriate business and shopping centers, ribbon streets or narrow, intermixed communities may stretch out for miles. Uncontrolled smoke, dust, fumes, and odors accelerate the deterioration of such hapazard and disorderly developments and create blighted areas.

It is almost an axiom, therefore, that city planning must be continuous, flexible, and adapted to the local requirements. It must be based on correct and sufficient data to anticipate the growth trends and changes in living habits which the city will undergo in the near future. Provision must be made for transportation, traffic, and parking facilities as well as accommodation for industry, commerce, homes, and apartments, so that the city can grow without sections being blighted and ultimately having to be reconstructed at enormous cost to the taxpayer.

To implement a plan conceived by a competent commission or authority, a zoning law is required to define the permissible areas for all functions of the community, such as light and heavy industry, business or commerce, parks and playgrounds, warehouses, office buildings, apartments and dwellings. Many cities with good zoning regulations have nullified them by granting exemptions whenever sufficient pressure has been applied to permit land use not in conformity with the regulations. Each city requires its own zoning ordinance, and it must fit the local conditions. As land usage changes, the zoning regulations must be modified to meet new conditions. Rural or suburban areas bordering a city must also be subjected to zoning to promote orderly growth.

The effects of air pollution on future zoning are being taken into consideration by some city-planning commissions. Thus, in Chicago and Detroit, proposals are under study to zone industry on the basis of the nature and extent of the emission from each plant rather than by the type of manufacturing process or operation. This provides an incentive for improved pollution control in the design of a new plant, because a more desirable site may be available if the plant can meet the stack-emission and other requirements in the preferred zone. As more information is gained about the effects of specific contaminants and the levels of ground concentration which may be permissible with respect to each type of waste product, zoning ordinances will contain performance provisions on specified permissible contamination limits rather than indefinite terms such as "detrimental, noxious, or offensive." The regulations would be more helpful to industry if positive performance standards for various classes of operations were required instead of being expressed in terms of what may not be done. However, the drafting of such performance zoning regulations requires a high order of technical skill and knowledge on the part of air pollution control agencies or planning bodies. At present, factors such as stack height, velocity and temperature of stack gases, wind velocity, and prevailing meteorological conditions in a given neighborhood are neglected in the setting up of control regulations based on permissible stack concentrations or, as in Los Angeles County, emission standards are based on process weight per hour (mass rate of emission).

With the continued growth of cities and industrial communities, air pollution control is assuming increasing importance in planning and zoning. Meteorological and topographical factors must be taken into account in the location of industrial and residential areas. It is essential to protect industry from the unreasonable encroachment of residential developments, and vice versa. In future planning and zoning, land use for certain types of industrial operations will depend on the extent to which such plants can control objectionable contamination. Area zoning will include more extended use of land for public parks, gardens, and wooded tracts to prevent overcrowding of sources of emission.

2.4 ORGANIZATION AND FUNCTIONS OF AIR POLLUTION CONTROL GROUPS

Many cities provide for local administration of air pollution control by a bureau or department. In the case of a large city, where sufficient funds are available, the organization may be a full department of air pollution control headed by a commissioner or director of recognized professional education, training, and experience in the fields of public health, sanitation, industrial hygiene, and air pollution. The commissioner has the power and jurisdiction to regulate and control the emission to the atmosphere of objectionable or harmful substances, including smoke, soot, fly ash, dust, fumes, gases, vapors, and odors. He is responsible for the enforcement of all laws, rules, and regulations with respect to such emissions and usually has the power to compel the attendance of witnesses and to take their testimony under oath.

The adoption and promulgation of rules, regulations, and amendments which may be deemed necessary for the proper control or elimination of pollution are usually delegated to a board of air pollution control which consists of about five or more members. In the City of New York, the chairman of this board is the Commissioner of Air Pollution Control. The Commissioner of Health and the Commissioner of Housing and Buildings are also members of this board, in addition to two experts appointed by the mayor. The total staff engaged in air pollution control in New York City in 1953 was about 75, with a budget of about $375,000.

The present department and board of air pollution control of New York City were established in 1952. At that time the existing rules and regulations were readopted, but it is anticipated that changes will be made in the air pollution code as continued studies make such changes desirable. Among current problems to be solved are the smoke, fly ash, and odors from 10,000 apartment-house incinerators burning 400,000 tons of refuse annually. In addition 12 municipal incinerators contribute to smoke conditions, there is contamination from 1,400,000 registered automobiles using 1,100,000,000 gallons of gas a year, besides the trucks and buses, and numerous industrial sources.

The city of Cleveland maintains a Division of Air Pollution Control under a commissioner who directs the functions of three bureaus. The division is one of the branches of the Department of Public Health and Welfare. The budget approved for 1953 was $191,000, which provides for normal operation of the division. The personnel consists of 41 persons, divided among three bureaus and a laboratory, which serves all three bureaus.

The Bureau of Smoke Abatement is headed by a chief. He is charged with the responsibility of enforcing the smoke-abatement code. Under his direction, two engineers, sixteen smoke inspectors, one chief inspector, an accountant, and three stenographers carry on the work of the Bureau.

The Bureau of Industrial Nuisances at the present time is staffed by a total of five persons. The scope of their work is that of abating industrial nuisances originating from processes of industry other than those of combustion devices—namely, agricultural fertilizer plants, chemical plants, steel mills.

The Bureau of Industrial Hygiene deals with the health of the workers within the plant. This Bureau normally has a chief and two assistants. The laboratory is completely equipped to do air pollution sampling and is staffed by three chemists.

The air pollution code does not, at present, amply cover industrial-nuisance work. Expansion of the air pollution ordinance to cover more emissions than are now provided for is unlikely in the near future because the establishment of limitations that are fair to all concerned is difficult without further study. The Division of Air Pollution Control is responsible for the enforcement of the air pollution ordinance. There is also a Citizens Committee on Air Purification, originally appointed by the mayor, which is vitally interested in the effectiveness of the program and assists in maintaining good public relations.

The Division functions entirely within the boundaries of the city of Cleveland, but has the full cooperation of neighboring municipalities, and sometimes gives them assistance in solving and abating air pollution problems. The Division does not feel that it will be necessary to intensify its program beyond the city borders while such cooperation exists.

It is the opinion of this city that air pollution control is a local problem and that policy and enforcement should be determined by the municipality immediately concerned. The need for state or Federal regulation does not exist now and would only add another source of appeal if legal prosecution were to be forced.

There are three groups working on air pollution control problems in Chicago. The City Department of Air Pollution Control was created in December, 1952, out of the older Smoke Inspection and Abatement Department. Its function is to enforce the provisions of the regulations in the local ordinance primarily designed to reduce pollution by smoke.

The Cleaner Air Committee of the Chicago Association of Commerce is a voluntary organization which operates through subcommittees, each of which represents a specialized section of industry, namely, the Railroad Subcommittees, the Gray Iron Subcommittee, the Electric Furnace Subcommittee. These subcommittees handle their individual problems in air pollution control. The Midwestern Air Pollution Prevention Association is organized to promote research and education in the field of air pollution control; its interests are primarily concentrated in Greater Chicago and its environs. The two latter organizations are not connected formally with the city enforcement agency.

Studies are being made to include an adequate advisory board of engineers to revise the rules and regulations, taking into account the present state of knowledge with respect to control equipment available to industry. Aside from occasional consultations, no consistent attempt has been made to coordinate the functions of whatever agencies exist outside Chicago proper with Chicago's department and efforts.

The city of Baltimore has operated a Division of Smoke Control for over 20 years in the Department of Public Works. In 1952, an organization to study atmospheric pollution, known as the Division of Air Pollution Control, was set up in the Bureau of Industrial Hygiene of the City Health Department. The head of this Division is also the director of the local Bureau of Industrial Hygiene.

At the present time it is considered that the Baltimore Nuisance Ordinance of the city code of 1927 is a sufficient regulation for the control of air pollution. Until such time as more knowledge is gained on the over-all problem in Baltimore, it is felt that legislation on pollution control, particularly of a punitive type, should be held in abeyance.

The Baltimore Bureau of Industrial Hygiene has secured the assistance of the Division of Occupational Health of the U.S. Public Health Service and maintains effective liaison with the Division of Industrial Health and Air Pollution Control of the State Health Department, as well as the local Division of Smoke Control.

The city of Pittsburgh relies on a smoke bureau consisting of a superintendent, an assistant, twelve inspectors, one chemist, and the necessary office force to administer its smoke ordinance under the Department of Public Health. Apart from smoke limitations and prohibitions on the use or consumption of certain fuels for hand-firing or surface-burning types of equipment, there is no specific ordinance in connection with air pollution. Contaminants such as soot, cinders, acids, fumes, or gases are prohibited under a public-nuisance clause of the ordinance.

In order to keep the public informed and interested in the activities of the Pittsburgh Smoke Bureau, a representative group of civic-minded persons has been organized for a number of years into a committee called the United Smoke Council. Interest in air pollution control is stimulated by this Council through radio addresses, lectures, newspaper articles, and other mediums conducive to good public relations.

There is no formal liaison between the activities of the Pittsburgh Smoke Bureau and the air pollution control groups in surrounding areas such as Allegheny County; however, the city of Pittsburgh was consulted and rendered advisory services during the planning period which preceded the adoption of the Allegheny County pollution control ordinance.

The Allegheny County Bureau of Smoke Control operates under an ordinance which was made effective in 1949. This Bureau functions under a director and had a total staff of 12 in 1952, with an annual budget of \$55,000. The heavy concentration of steel plants, railroads, and other industrial and public activities in the area made it advisable to extend pollution control beyond the corporate limits of the city of Pittsburgh into the whole of Allegheny County. In 1947, an Advisory Committee was formed under Dr. Weidlein, president of the Allegheny Conference on Community Development and president of the Mellon Institute, to study the pollution problems of the county and recommend an ordinance which would control and abate this pollution effectively. After a satisfactory ordinance was evolved and adopted, industry was given a period of 5 years in which to carry on research and development work in order to solve problems connected with the control of metallurgical dust and fumes for which no remedy was available at that time. The general provisions of the Allegheny County ordinance became effective in 1954. In the meantime, the railroads, which had practically no diesel locomotives in operation 5 years ago, have converted to diesels almost entirely. The blast and open-hearth furnaces of the steel industry are being equipped with electrostatic precipitators and other forms of dust-collection equipment. All power and heating plants have been equipped with fly-ash collecting apparatus, and control measures have been applied to cement and coke plants. The experience of Allegheny County with large-scale cooperation by industry through the activities of the Advisory Committee has been a highly successful one.

In 1953 the city of Philadelphia adopted a new air pollution control ordinance after finding that existing atmospheric contamination is detrimental to the health, comfort, living conditions, welfare, and safety of its inhabitants. The new ordinance provides for greater control and more effective regulation of pollution nuisances. Administration and enforcement are assigned to the Department of Public Health, Air Pollution Control Division. All applications, plans, and specifications for the construction, conversion, or alteration of any installation or device that may result in atmospheric contamination or in an air pollution nuisance are transmitted by the Department of Licenses and Inspections to the Department of Public Health for approval.

In addition, an Air Pollution Control Board has been set up with powers to make regulations to prevent and control air pollution, as well as to hear and review objections to any action or decision of the Department of Public Health with respect to the creation or emission of air pollution nuisances. In the case of control measures requiring a major installation to correct a violation of the ordinance, the control board is authorized to allow a reasonable time for completion of such works.

In Detroit, the Bureau of Smoke Abatement is the air pollution control group under the office of the Commissioner of the Department of Buildings and Safety Engineering. The Bureau has the responsibility for enforcing regulations governing the abatement of smoke and air pollution; determining contributing factors; investigating, testing, and analyzing atmospheric pollution; examining plans and equipment for proper installation; making annual inspections of all fuel-burning equipment, except in residences; and promoting public interest in smoke abatement and control of air pollution. The staff engaged in these activities consists of a chief smoke inspector, five senior assistant engineers and industrial hygienists, seventeen inspectors, and four clerical employees. Two industrial hygienists, and air and stack sampling equipment of the Smoke Abatement Bureau, are assigned to the Division of Industrial Hygiene of the Department of Health for direct supervision (in order to avoid duplicating laboratory facilities). Although the Bureau is responsible for air pollution control, this does not relieve the Fire Department and the Board of Health of their statutory responsibility over airborne health hazards in outside atmospheres.

The local air pollution control agencies of neighboring municipalities and townships have informal working relationships with Detroit at the present time. An area-wide "conference" or "council" of a more formal character, but without legal responsibility, is being planned. The Detroit Smoke Abatement Bureau is represented in the Metropolitan Regional Planning Committee on industrial land use. Detroit has little contact with the state agencies concerned with air pollution control, except as they work together in the area-wide study of the International Joint Commission. Various departments of the city, such as the Department of Buildings and Safety Engineering and the Department of Health, cooperate with the Federal government in the International Joint Commission study. The Bureau also draws upon the U.S. Bureau of Mines and the U.S. Public Health Service for technical information on air pollution problems. The Chief of the Smoke Abatement Bureau has been designated as the liaison agent for industry and the International Joint Commission study.

The air pollution group which has been investigating conditions in the Greater Detroit–Windsor area, along the Detroit River, for the International Joint Commission of the United States and Canada, pursuant to the terms of a joint reference adopted in 1949, consists of a Technical Advisory Board and assisting field staff. The Board contains four United States members under a United States chairman and four Canadian members under a Canadian chairman. These members have been appointed by the International Joint Commission and are drawn from various Federal departments, the province of Ontario, and the city of Detroit.

Under the terms of the reference, the Board is empowered to investigate and report on major sources of pollution which may affect the public health, general welfare, and property interests on either side of the international boundary, but remedial or control measures may be recommended only for smoke and fly ash from Great Lakes vessels and others plying the Detroit River.

The Board holds meetings at regular intervals to plan and coordinate its investigations, which are organized on a joint basis. The studies deal with vessel smoke and its control; the determination of levels of pollution from dustfall, suspended particulate matter, and gases such as sulfur dioxide, aldehydes, and oxidants on an area-wide basis; identification and determination of particle-size distribution of particulate pollution and of their effects on public health and vegetation; and the determination of the influence of meteorological factors on diffusion and dissemination of pollutants.

The field staff of the Board is insufficient to perform all the tasks required; consequently, active cooperation and assistance are rendered by the Detroit Bureau of Smoke Abatement, the Detroit Division of Industrial Hygiene, and the Cincinnati Laboratories of the U.S. Public Health Service. The Detroit Department of Health has a major stake in the health and morbidity studies of the International Joint Commission on the United States side in furnishing a staff, accommodations, and supervision. The Canadian Department of National Health and Welfare, in cooperation with the Division of Industrial Hygiene of the Ontario Department of Health, under which the health study is organized, furnishes the direction, supervision, and field enumerators for this phase of the over-all study on the Canadian side. There is close cooperation between the Technical Advisory Board, its Windsor Laboratory and field staff, and the Windsor Air Pollution Control Department. The studies of the effects of air pollutants on vegetation are organized under a subcommittee of the Technical Advisory Board consisting of representatives from the Department of Botany of the University of Michigan, the Detroit Department of Parks and Recreation, the Canadian Department of Agriculture, the city of Windsor, and other municipalities along the Detroit River.

Representatives of major industries in the area have been informed of the objectives and progress made in the environmental studies through meetings of a Joint Industry Conference. Although control of pollution from land-based sources is primarily the responsibility of the municipalities concerned, this is not so for vessel smoke. Since 1952, the Technical Advisory Board, with the cooperation of vessel owners and operators, has instituted a scheme of voluntary control involving permissible smoke-emission objectives for vessels equipped with various types of fuel-burning equipment. Ultimately, it is hoped that all hand-fired coal-burning vessels will be equipped with automatic stokers of either the underfeed- or spreader-stoker type. When this major conversion has taken place, it will be reasonable to expect that all vessels on the Detroit River will be able to reduce their smoke emission in conformity with the model ASME code or equivalent standards. A smoke code involving legal control will require international sanction under some kind of formal agreement between Canada and the United States, because there is no local legal jurisdiction over this traffic.

Communities along the Detroit River are keenly interested in this international effort to abate and control the soot, fly ash, poor visibility, and detrimental soiling effects caused by excessive vessel smoke emission. However, the data on other major pollutants and findings of the International Joint Commission will be useful in guiding these municipalities in their planning for future air pollution control work.

The Los Angeles County Air Pollution Control District is organized in a flexible manner in order to cope with the complex problems involved in its comprehensive control program. The organization under a competent director and assistant director consists of four divisions: (1) office management, (2) inspection, (3) engineering, and (4) engineering testing, development, and control. In 1951 there were 82 staff positions in this control district, consisting of 61 technical and 21 nontechnical positions. Enforcement of the county air pollution law through a system of permits places a heavy burden on the engineering division for processing applications for permits involving almost every conceivable type of industrial processing and storage equipment which may give rise to air pollution. Policing of the entire 4,083 sq miles of Los Angeles County is the responsibility of the inspection division. Regulations approved in November, 1949, require permits for the operation and installation of orchard heaters, of which there were over 1 million. A 3-year plan of replacement has been in operation to remove obsolete heaters.

One of the most important administrative phases of the work of this district is the task of keeping the public informed of the problems involved and the progress being made in the development of the air pollution control program. This is accomplished by disseminating information through radio, television, and the newspapers. Continued support is received and cooperation maintained with many civic organizations and officials throughout Los Angeles County. All these efforts are directed at maintaining public confidence in the control program.

Appeals from the ruling of an air pollution control officer may be made before a hearing board composed of two lawyers and an engineer. This board may grant relief from the general limitations set by the control regulations for the period needed to study possible methods of controlling the offending emissions or the time required to construct and install remedial equipment. An appeal from the decision of the hearing board may be taken to the superior court in order to test the findings in the case.

Rules and regulations are set up by the Los Angeles Air Pollution Control Board, consisting of a professor of mechanical engineering, a professor of chemistry, a meteorologist, an industrialist, and a realtor. Legal advice is furnished by the county counsel and his deputy.

To assist in the research investigations of the Air Pollution Control District, two full-time and five part-time consultants are retained. Problems of the control district which require specialized knowledge, facilities, and development of techniques for their solution are assigned to universities and outside laboratories on a contract basis. Outstanding contributions to air pollution control have been made for this district by the California Institute of Technology and the Universities of California, Southern California, and California at Los Angeles. The over-all annual cost of the work of this control district is nearly half a million dollars, which represents a per capita cost of about 11 cents for the more than 4 million people of Los Angeles County.

Since June, 1947, Stanford Research Institute has carried on research of high quality to determine the impurities in Los Angeles smog, including the isolation and identification of materials responsible for smog formation, poor visibility, eye irritation, and other factors. More than 30 SRI scientists and a number of outside consultants have contributed to this work, which is sponsored by the Western Oil and Gas Association and the American Petroleum Institute. In addition to presenting information and new data on this complex smog problem, the SRI team of scientists has made important contributions in the field of analytical chemistry relating to the measurement of aerosols and in the development of continuous automatic analytical techniques, in particular the determination of "oxidants" or ozone and gaseous and solid fluorides.

Recently, the Southern California Air Pollution Foundation, a nonprofit organization, was formed by community leaders in the fields of industry, business, education, and civic affairs to investigate the nature, causes, and effects of air pollution in the Southern California area, including the Los Angeles district. The investigations of this organization are intended to aid governmental and other agencies in abating air pollution and will not duplicate research work in progress at California universities and at the Los Angeles Air Pollution Control District. This Foundation, in addition to undertaking a broad program of research, will correlate the mass of information already compiled by public and private groups.

The work of air pollution control groups in other cities of the Far West has been summarized recently by Sawyer (52). At San Francisco the Department of Public Health is the municipal body responsible for air pollution control, under a local ordinance dating back to 1925. The city of Oakland, Calif., has a number of local ordinances governing emissions from smelting of ores and other practices which might give rise to detrimental pollution as well as a public-nuisance law. The control bureau is under the local board of health. Responsibility for the control of cinders and fly ash is shared with the fire-prevention bureau. Data are being collected to determine if there is need for modification of local legislation. Berkeley, Calif., has no local air pollution regulations and depends upon the general-nuisance law of the state for control action under the city health department in cooperation with state health authorities.

Portland, Ore., has ordinances regulating emissions and also an Air Pollution Committee which advises concerning new legislation and can act as an appeal board. The responsible municipal group for control is the Bureau of Buildings. Close relations are maintained between the Department of Industrial Hygiene of the State Board of Health and the Sanitary Division of the City Health Bureau. In 1951, Oregon passed a state law which authorizes its sanitary authority to develop a comprehensive program for the prevention and control of air pollution, to conduct investigations relating to its causes, to collect and disseminate information relating to control measures, and to promulgate and enforce the rules and regulations of the sanitary authority. The latter authority also has the power to abate nuisances created by air pollution.

The cities of Seattle and Spokane, Wash., have no local air pollution regulations at present. However, studies are in progress to determine the type of legislation which is needed. The general-nuisance laws of the state of Washington may be resorted to under the powers of the State Pollution Control Commission. Both Washington and Oregon have recently had extensive experience with fluorine damage to vegetation and livestock. Tacoma, Wash., has established a Division of Air Pollution Control under the Department of Public Welfare in close cooperation with the Health Department. Regulation of emissions is governed by local ordinances and a permit system.

Although a number of counties in the United States have organized air pollution control districts, and a considerable number of states have enacted legislation of one kind or another, the only Federal government action taken has been with respect to the city of Washington, D.C., which is under its jurisdiction. A similar situation with regard to the lack of Federal legislation exists in Canada. In Great Britain, emission of atmospheric wastes for heavy industry is regulated at the national level under the provisions of the British Alkali, etc., Works Act through the Ministry of Health. Chief inspectors for England, Wales, and Scotland operate directly under the Ministry. The industrial areas are divided into a number of districts in which the district inspector is the responsible control official. The British Ministry of Health also cooperates directly with local authorities who are assisted in an advisory capacity by a statutory committee and various regional committees. Smoke prevention on a national scale is organized under the British Ministry of Fuel and Power which operates a central fuel-efficiency division and regional offices in various districts. There is a liaison between fuel industries and their associations and various industrial firms and voluntary organizations such as the National Smoke Abatement Society. Directly under the government is a third group known as the Department of Scientific and Industrial Research which carries on investigations of atmospheric pollution and operates a fuel-research station.

In the United States, Federal interest in air pollution is mounting not only as an aftermath of the Donora disaster of 1948, but also as a result of the increase in the number of air pollution problems involving interstate and international action. The United States Technical Conference on Air Pollution, organized by an interdepartmental committee at the request of President Truman in May, 1950, was attended by over 750 outstanding experts representing private industry, the universities, and local, state, and Federal governments from the United States and other countries. About 95 papers were presented before panel meetings, covering analytical methods and instrumentation; chemical, meteorological, agricultural, and public health aspects of the problem; equipment for control; and legislation. The Conference proposed at its conclusion that the Federal government undertake to assist in the solution of air pollution problems (3). Since then several attempts have been made to pass legislation dealing specifically with Federal assistance in research and control of air pollution. An amendment to the Housing Act of 1954 which would provide for air pollution prevention was discussed in extensive hearings before a Senate committee. This amendment would authorize Federal aid to business enterprises in financing the purchase, installation, or construction of any smoke-abatement or air pollution prevention devices. Special tax benefits to permit the rapid amortization of such equipment or remedial works were proposed. In addition, the bill would authorize the appropriation of Federal funds up to $5 million for a program of technical research and studies concerned with (1) the causes of excessive smoke and air pollution, (2) remedial measures and equipment designed to prevent or abate excessive emissions of deleterious waste products, and (3) guidance and assistance to local communities in smoke abatement and air pollution prevention and control.

2.5 COMPULSORY VERSUS VOLUNTARY POLLUTION ABATEMENT

There is no doubt that legislation to control atmospheric pollution is on the increase in municipalities, counties, and states. Recurrent smog episodes, such as those of Donora, Los Angeles, and London, and the economic consequences of sulfur dioxide and fluoride emissions, as well as the nuisance effects of air pollution, have created a widespread public interest in the problem. However, public clamor for control measures has led, in some instances, to the enactment of hasty and ill-conceived legislation, with the result that private industry has had to undertake uneconomic and costly remedial measures or abandon plans for new plant construction in certain areas. Smog experience in the Los Angeles area and elsewhere indicates that all human activities bear a proportionate share in contributing to the over-all pollution, including domestic fuel burning, incineration, operation of vehicles for public and private transportation, industrial processes, and other operations. As more knowledge is acquired regarding the toxicity of specific contaminants and advances in science and engineering make it possible to reduce stack emissions of undesirable products, there is increasing public pressure to incorporate such knowledge into more restrictive control legislation.

So many municipalities have adopted smoke or air pollution control ordinances that compulsory abatement on a city-wide basis is an established fact. In a few areas, such as the Gulf Coast region of Texas (outside Harris County) and the San

Francisco Bay area of California, control of smog is still on a voluntary basis. The strongest advocates of legislative control are city and state government officials and some sections of the public. Voluntary control has had considerable success in the Houston, Tex., area where the local chamber of commerce and a community council air pollution committee have been actively engaged in gaining the support of major industries to undertake remedial measures aimed at pollution control. The San Francisco Bay Area Council is an active proponent of voluntary control and abatement. This bay area comprises nine counties, only one of which has been organized into a pollution control district in accordance with the California state law of 1947. The Council's activities are aimed at forestalling the development of an acute smog problem by inducing the major industrial operators and others to take corrective steps to reduce source emissions in the immediate future.

Even when pollution control is placed on a compulsory basis by the enactment of suitable legislation, voluntary cooperation is still required in regional problems involving city-county relations and in those cases where smog transcends boundaries between counties, states, and countries. Water and air pollution problems arising between the United States and Canada are referred to the International Joint Commission for investigation and report, including the recommendation of remedial measures and their cost allocation. The references to this body are made under the terms of the Boundary Waters Treaty of 1909. If voluntary control provides an effective remedy, there is no need for legal sanctions, which require the approval of the United States Congress and the Canadian Parliament.

Most large plants and industrial organizations have the necessary engineering and technical skill or can employ consultants to solve their pollution problems and reduce their emissions to acceptable limits without resorting to compulsory measures. Legal control is necessary to compel abatement by recalcitrants and certain sections of the public. It is doubtful if voluntary cooperation could be effective in abatement of pollution from domestic fuel burning, backyard and apartment-house incinerators, orchard heaters, trucks, buses, and private automobiles. Thousands of small plants need advice and guidance in the solution of waste problems. Such advice could be rendered by municipal control officers, as small plants usually do not have the staff required for this type of work.

Successful administration of control regulations requires the cooperation of both industry and the public through education and good public relations. Smoke abatement in the city of Pittsburgh was brought about not merely by passing and trying to enforce ordinances, a procedure which had failed in the past, but by obtaining the effective backing of civic-minded groups representing industry, commerce, and the public. In enforcement of its latest ordinance, the Pittsburgh Bureau of Smoke Prevention has adopted the policy that its inspectors are not policemen but combustion engineers who must be able to help and advise the owner of a plant with a smoke problem. Very little legal action has been necessary. Although over the past few years there have been about 1,675 violations a year, only about 16 of these reach the magistrate's court annually, because of the educational and cooperative, rather than punitive, approach to its program employed by the Pittsburgh smoke-control bureau (2).

2.5.1 The Permit System

One of the main objections voiced by industry to stringent control of pollution is directed against the operation of the permit system. This system is operated in a number of industrial communities under local regulations which not only prohibit the emission of specified contaminants as nuisances or as being in excess of certain standards, but also require that authorized construction permits be obtained prior to the erection or alteration of equipment which may release contaminants to the air. Usually the ordinances stipulate that operating permits for such equipment are also required. In addition, the legislation may direct the enforcement agency to make periodic inspections of the licensed equipment to determine whether it is being operated in accordance with the provisions of the local ordinance.

Hansen has reviewed the arguments for and against the operation of this system (3). Although the permit system has been used in conjunction with smoke-control ordinances, there is no reason to assume that regulations controlling air pollution or smog should be drafted along similar lines or that the license system is essential. The abatement of smoke, soot, and fly ash by operation of proper combustion equipment and control devices or by substitution of other fuels for soft coal is now well understood by combustion engineers. It is a comparatively simple process in contrast to the complexities of eliminating or reducing the varied waste products from metallurgical, refining, chemical, and other plants which employ widely divergent processes and manufacturing techniques. In many industrial operations, satisfactory control of pollution must await the results of further progress in research.

The permit system implies that the local enforcement agency is better qualified to decide what type of process or control equipment should be used than the industrial management of a plant whose experts have designed, invented, or tested the operation through the pilot-plant stage. Thus the Los Angeles County regulations require the prior submission and approval of plans covering any article, machine, equipment, or other contrivance the use of which may cause either the issuance or the elimination, reduction, or control of air contaminants. The operation of such a system places a heavy burden on industry while increasing the administrative costs and responsibilities of the local air pollution control department. Processing of applications for permits would require a highly competent engineering staff, and it is doubtful if a single agency could obtain the specially qualified personnel necessary to pass judgment on the effectiveness of plans and specifications covering the many different types of processes and equipment submitted for approval. The industrial growth of Los Angeles County has been so rapid that the Engineering Division of the Control District has been unable to process the increased number of applications for permits. In order to avoid delays in construction, it has been necessary to assign engineering final inspections to the Inspection Division, which is charged with the responsibility of policing the entire 4,083 sq miles of Los Angeles County.

The Air Pollution Abatement Committee of the Manufacturing Chemists' Association, Inc., has recently stated its conception of the basis for a fair and workable law controlling air pollution. The following quotations concern the stand taken by this committee on the responsibility of industrial operators and the proposals for a system of registration of points of emission rather than mandatory permits or licenses.

2.5.2 Responsibility of Industrial Operators

"Regardless of the mechanism by which air pollution control is administered, the ultimate responsibility for limiting the release of pollutants to tolerable quantities must rest with the operators of the units from which the pollution comes. However, in some instances, as in California, there has been reluctance to give industry latitude in discharging this responsibility. A reflection of this attitude is the permit system, which in effect licenses limited pollution, entailing approval of complete design plans, operation under temporary variance permits, issuance of final permits, and other time- and money-consuming features that are burdensome to both administrators and operators, and wholly unnecessary.

"One recent article by public officials pleads that 'No administrative agency is so well staffed that it can ferret out each new pollution source by field observations alone,' and indicates the need for a system that will 'put the administrative agency on official note of each new pollution source.' This is a reasonable position to take, and a plan discussed in a section to follow indicates how this can be done with minimum effort and yet leave the responsibility for proper control of emissions where it belongs, squarely on the shoulders of the operator."

2.5.3 Registration of Points of Emission

"Intelligent promotion of remedial measures for air pollution control must be based on recognition of specific undesirable effects and their causes. In order to act effectively, the state bureau must be informed about waste emissions to the air throughout the state. This can be accomplished most efficiently by requiring that each point of emission be registered with the bureau. The data called for should include (*a*) location, (*b*) size of outlet, (*c*) height of outlet, (*d*) rate of emission, (*e*) composition of effluent.

"Registration should be made on a state-wide basis to aid the state bureau in defining localized air pollution zones and to facilitate correlating undesirable effects with the sources of pollution that are responsible. There should be certain exemptions from registration, such as single non-recurring emissions, accidental discharges, safety vents, small units, etc., the details of which are probably best left to the state bureau.

"As a precaution against obsolescence of registered data, it should be incumbent upon each operator to re-register as often as necessary to keep the registered data representative of current operating performance.

"Expected emissions from new installations should be registered as far in advance of operation as feasible, e.g., upon completion of design. This requirement is logical as a curb on excessive emissions before they occur—prevention is better than cure—by insuring due consideration of air pollution control in planning new installations. If the expected emissions appear excessive to the state bureau, the bureau should issue a warning to this effect. Such a warning, on record also with the local air pollution control commission, would be a strong deterrent to proceeding without being certain of adequate provision for air pollution control. If the operator remained convinced that his emissions would not be harmful, however, coercion should not be applicable until actual proof of harmful effect is established. Registration of actual emissions should be required as soon as representative operating data can be obtained.

"This plan is superior in every respect to a permit or licensing system. It provides the state bureau with the information needed without the trouble of reviewing plans, processing applications, and issuing permits, the time and talent for which can be much more fruitfully spent in seeking solutions to problems revealed by the registered information. At the same time the operator has maximum freedom of action in meeting the problem, and is not called upon to reveal extraneous information about raw materials, processes, equipment, or products that are often secret."

2.6 CASE HISTORIES OF ABATEMENT EFFORTS

2.6.1 Smelters

In the early days of smelter operations, uncontrolled heap roasting in the open air was a common practice and resulted in enormous damage to vegetation from sulfur dioxide or to livestock from arsenic, lead, and other impurities. Later on, the use of furnace roasting and low stacks resulted in only minor alleviation of the problem, and in many instances the output of sulfur dioxide was much too high for effective dispersion by prevailing meteorological and topographical conditions. Even at the present time the loss of elemental sulfur to the atmosphere from zinc plants, lead, copper, and nickel smelters is greater than the total of native sulfur production. In the United States, within recent years, about 700,000 to 950,000 short tons of by-product sulfuric acid is produced annually at copper and zinc smelters, but this represents only about 10 to 15 per cent of the total acid manufactured annually and only about 6 per cent of the sulfur wasted from these sources (31).

Swain (63) has given an excellent historical review of the successful solution of air-contamination problems at a number of smelters, located at Ducktown, Tenn., Anaconda, Mont., Salt Lake City, Utah, and Trail, B.C. One of the earliest cases of sulfur dioxide damage to vegetation in the United States involved two copper smelters near Ducktown, not far from the Georgia border. Operations here, during the first decade of this century, resulted in complete denudation of valuable timberlands, accompanied by soil erosion, over a considerable area in the vicinity. Later on tall stacks were installed, but because of the high waste-gas output and mountainous terrain, this resulted in carrying a trail of injury for 30 miles across the forests of northern Georgia. Matters came to a crisis when the state of Georgia brought suit against Tennessee to compel the smelting companies to cease operations. The Federal government investigated the nature and extent of the injury, and it was apparent that smelter operations could be continued only if the sulfur output were drastically reduced.

The problem was solved by the erection and successful operation of a large lead-chamber process to convert the waste sulfur to sulfuric acid, with a capacity of about 1,000 tons of acid daily, when in full production. Most of this acid is being utilized by the phosphate plants of the South for the production of fertilizer. In some years the income accruing from this by-product recovery of the sulfur dioxide formerly wasted actually exceeds the value of the copper refined.

At Anaconda, prior to 1905, the emissions from low stacks of a smelter plant operating on arsenical copper ore killed all vegetation, and caused heavy losses of livestock, over the surrounding agricultural and forest area. A new smelter, one of the largest of its kind in the world, was then erected with four stacks, each 200 ft in height. When these measures failed to reduce the injury to livestock, a new stack, 300 ft high and 30 ft in diameter, was erected on a spur of the neighboring mountains, 700 ft above the roasting furnaces and 1,100 ft above the floor of Deer Lodge Valley. However, reliance on the collection of arsenical waste fume was placed on settling chambers and flue ducts of 6 million cu ft capacity. These were ineffective because of the small size of the arsenic trioxide particles, most of which were in the submicron range. Successful solution of the problem was finally achieved in 1910 by the installation of an enormous Cottrell electrical-precipitation system, and injury to livestock disappeared for all time.

Four smelters were operated at one time in the fertile valley of the Jordan River near Salt Lake City, because of favorable factors such as proximity to ore bodies, transportation, and water supplies. The marshland along the south shore of Great Salt Lake served as a convenient location for the disposal of tailings and water-borne wastes. However, the proximity of these operations to sensitive crops on farm and pasture lands soon resulted in damage claims and litigation. The claims for injury involved the effects of sulfur dioxide on vegetation and lead poisoning of livestock. The measures taken to relieve this situation have been recounted elsewhere in this section, under meteorological control in the Salt Lake Valley (Sec. 2.7.5).

While some of the smelters originally operating in this valley were consigned to the scrap heap, smelting operations are still in progress at this date. Such operations in this highly sensitive agricultural area are possible only because of the application of the fruits of long-term scientific research on the effects of sulfur dioxide on vegetation (65,67), the use of tall stacks and high temperatures to assist in the dispersion of waste gas, the collection of metallurgical fumes and dusts by electrical precipitation, and other high-efficiency methods. At the Garfield smelter, about 20 per cent of the sulfur released during smelting is recovered as sulfuric acid. During the growing season a close check is kept on the concentrations of gas present during ground fumigations, and careful records are made of effects on crops, if any. The contributions to our general knowledge of the sulfur dioxide problem made by investigators such as O'Gara, Fleming, Thomas, Hill, and Swain have been of outstanding importance.

Among other contributions of research to the broad problem of smelter-smoke injury may be mentioned the classical investigation of the Selby Smelter Commission, 1913–1915, into conditions with respect to the Selby lead smelter located on San Francisco Bay (27). The report of this Commission represented at that time a milestone in progress on the behavior of sensitive plants to sulfur dioxide fumigations under controlled environmental conditions and methods of conducting pollution surveys in the field. The later experimental work in the Salt Lake Valley by experts of the American Smelting and Refining Co., up to about 1925, may be looked upon as a continuation of this work on the subject. The investigations of the American Smelting and Refining Co. were resumed in 1925 by Hill, Thomas, and coworkers. Thomas (65) has recently given a concise account of this outstanding research, which for the next 25 years or more played a dominant role in the control of smelter-smoke pollution in the Salt Lake Valley. Automatic analyzers were developed for sulfur dioxide (Thomas Autometer) and for carbon dioxide in air, which have been used in extensive laboratory and field studies on the effects of gas absorption by various species of plants on photosynthesis, respiration, yield, and chemical composition, not only in the Salt Lake Valley but elsewhere (66,67).

The international case which involved the lead and zinc smelter of the Consolidated Mining and Smelting Company of Canada at Trail, B.C., and agricultural and forest interests in northern Stevens County, Washington, is also a noteworthy example of the successful solution of an extremely difficult air pollution problem. The history of this case will be reviewed later in this section. As in the Salt Lake district, the research into the effects of sulfur dioxide on vegetation was of great value in establishing the extent of the damage caused and the permissible levels of concentrations in ground fumigations which could be tolerated by vegetation and, indirectly, the amount of waste gas which could be safely discharged at the source, depending upon the meteorological and other conditions. The published record of these investigations under the auspices of the National Research Council of Canada (28) has been regarded by Swain (63) as "a noteworthy contribution to the literature of the smoke problem in its wide scope and the high quality of the work done."

For the smelting industry, as a whole, great strides have been made in the control of smelter-smoke damage by scientific investigation of sulfur dioxide injury to plant life and of tolerance levels of gas and by application of remedial measures. The latter include the use of high stacks, 400 to 600 ft or more, and high temperatures for the discharge of waste gases, precipitation or collection of metallurgical dusts and fumes, continuous automatic measurement of ground concentrations, and application of meteorological control, wherever possible, by accurate forecasting of critical weather conditions during the growing season. It is also possible and economically profitable, in many cases, to install recovery plants for the conversion of excess sulfur dioxide to liquid sulfur dioxide, sulfuric acid, fertilizers, or elemental sulfur.

2.6.2 Industrial Communities

Great Britain. The evils of excessive pollution by coal smoke have been apparent in England for centuries. As long ago as 1273 the burning of coal in London was prohibited by Act of Parliament. However, subsequent economic pressure caused an evident relaxation of this drastic law, because there is mention of the poisonous use of coal in the reign of Queen Elizabeth I, and a description of the evils of London smoke is given in John Evelyn's book "Fumifugium" in 1661. By the end of the nineteenth century, the atmospheric pollution problem created by the industrial revolution had grown to major proportions, not only in London, but in many other urban communities throughout England. Extensive damage to farm crops, forest and park land, stonework of buildings, and other materials was clearly apparent not only from the products of the incomplete combustion of coal, but also from sulfur dioxide, hydrochloric acid, nitric acid, oxides of nitrogen, and other fumes and gases from smelters and chemical plants.

Various attempts were made at smoke abatement during the first half of the nineteenth century, but it was not until 1875 that the Public Health Act provided effective means of dealing with smoke nuisances from any furnace used in manufacturing operations. Exemptions were granted to private dwellings, mines, smelters, and certain specified trade premises. In 1881 and 1882 the Alkali, etc., Works Regulation Acts were passed to deal with pollution of the air by hydrochloric acid from alkali works and chemical waste products from other sources such as smelters, cement plants, sulfuric and nitric acid plants, fertilizer and tar works. The regulations under these acts were amended and consolidated by an Act passed in 1906. This Act, with various Works Orders established in the period 1928 to 1950, governs pollution from heavy industry in England and Scotland today (43).

The Public Health Smoke Abatement Act of 1926, and that of 1936, consolidated legislation with respect to emission of smoke in quantity sufficient to constitute a nuisance. Local authorities have the power to institute regulations limiting the duration of smoke emission and to merge into groups for smoke-abatement purposes. The first case in history where sulfur-emission control in a coal-burning plant was set as a prerequisite to its erection was the Battersea Station of the London Power Company in 1927. The government had sanctioned the erection of this plant without sulfur dioxide restrictions until a protest was made by municipal authorities. The trend of modern requirements for power plants in urban districts in England is the installation of scrubbing processes for the reduction of sulfur dioxide in the stack gas to yield an exit concentration of about 50 ppm.

The operation of the Alkali, etc., Works Act under the British Ministry of Health has been effective in controlling pollution by heavy industry. The Act is administered by chief inspectors for England and Scotland, with an appropriate staff. They are empowered to enter premises of industry, make stack-gas analyses to determine whether wastes emitted are in conformity with standards set under the Act, investigate complaints, and give evidence. The annual reports of the chief inspectors are submitted to Parliament by the Minister of Health. Some of the emission restrictions under this Act are as follows:

1. Hydrochloric acid. Every alkali plant is required to condense waste acid gas to the extent of 95 per cent and also to prevent the escape of more than 0.2 grain of hydrochloric acid per cubic foot of air, smoke, or chimney gas.

2. Sulfuric anhydride. It is specified that in the manufacture of sulfuric acid the total acidity of waste gas in each cubic foot of residual gases, before admixture with air or other gases, should not exceed 4.0 grains of SO_3.

3. The Minister may, by order, subject to ratification by Parliament, require owners of cement and smelting works to adopt the best practicable means for preventing the discharge into the atmosphere of any noxious or offensive gas, including fume.

Although considerable progress has been made in Great Britain in the abatement of smoke from industrial sources by the installation of more efficient fuel-burning equipment, there has been much less success in the the control of the emission of fly ash, sulfur dioxide, and other waste products from industrial chimneys. Little progress has been made in the control of air pollution or smoke from domestic sources. Nearly all the fuel burned in domestic furnaces and fireplaces consists of bituminous coal, accounting for about 18 per cent of the total annual consumption of solid fuel. However, the efficiency of combustion is so low that, on the average, about twice as much smoke per ton of coal is produced when burned domestically as when consumed by industry. The effect of domestic pollution is increased still further by the low height of the chimneys in comparison with industrial stacks and also by the fact that it occurs almost entirely during the winter months when atmospheric conditions favor the increased prevalence of fogs and temperature inversions. It has been estimated that the smoke, ash, and sulfur dioxide pollution produced annually in Great Britain from the combustion of fuels amounts to over 8 million tons. This causes far more damage to vegetation and property than pollution from any other source (43).

Probably the worst smog disaster on record occurred during a 4-day fog from December 4 to 9, 1952, in London. The smog filled the Thames Valley to a width of 20 miles and extended to a greater distance along the valley in each direction. The depth of the smoke and fog layer varied between about 150 and 400 ft, with high landmarks and hilltops protruding into the haze above this layer. About 4,000 people died in the Greater London area, most of the fatalities occurring among the population from age forty-five and upward (48,68).

From measurements of suspended smoke and sulfur dioxide which have provided daily results for a number of years at 12 test sites in the Greater London area, it was found that in the central city area the highest daily average concentrations of smoke and sulfur dioxide were between 5 and 10 times the averages of a normal December, while in the outer city area the highest daily average concentrations were between 2 and 5 times the normal for smoke and between 2 and 7 times that for sulfur dioxide. On the assumption that the concentration of carbon monoxide was of the order of 34 ppm in London on a typical winter day, a smog increase ratio of 6 would indicate the possibility that values of 0.02 per cent may have been reached over several days. The maximum allowable exposure for an 8-hr working period is about 0.01 per cent. Other contaminants normally present in the atmosphere would probably increase during the smog in the same ratio as smoke and sulfur dioxide.

The time lag between the arrival of smog and the beginning of increased mortality was less than one day. During the period of several days when the over-all level of pollution was on the increase, the number of deaths rose correspondingly, and when the pollution decreased the death rate declined also (68).

St. Louis. The efforts of the city of St. Louis to reduce the nuisance of excessive atmospheric contamination provide an excellent example of what can be achieved by enlisting the cooperation of the whole population. This city had suffered for many years from a particularly smoky atmosphere which not only caused damage and nuisance effects, but also during inversion periods was responsible for such poor visibility as to cause extensive and costly traffic dislocation. A great deal of the trouble was due to the widespread use of bituminous coals containing about 35 per cent of volatile matter, as much as 5 per cent of sulfur, and a relatively high percentage of ash. In 1937 the city passed an ordinance which specified that no coal used for combustion should contain more than 12 per cent of ash or 2 per cent of sulfur on a dry basis, and if the size was less than 2 in. in diameter it had to be washed prior to delivery. However, it was realized that people could hardly be expected to burn smoky fuel without making smoke. Consequently, the above ordinance was improved in 1940. The amended measure proposed that high-volatile bituminous coal be burned only in mechanical stokers for either domestic or industrial purposes and that all hand-fired fuels be of a smokeless type. The latter was defined for coal as one containing not over 23 per cent volatile matter on a dry basis. Furthermore, plans for all new fuel-burning appliances and alterations to existing equipment had to be submitted to the Smoke Commissioner's office for approval (9). The city was given the authority to buy and sell smokeless fuel in an emergency.

These ordinances represented the first attempt by a city to enact air pollution legislation of the "preventive" type in contrast to the "punitive" type where the onus is placed on the public to adhere to arbitrary regulations without being provided with the necessary means of compliance, e.g., requiring home owners to refrain from making smoke with inherently smoky fuel. Other communities, such as Detroit, Cincinnati, Pittsburgh, and Milwaukee County, have now adopted similar ordinances. The application of these regulations under competent administration resulted in a marked improvement in the atmosphere of St. Louis. According to visual observations of the U.S. Weather Bureau, smoke conditions have been reduced by about 75 per cent, or sufficient to eliminate effectively the smoke nuisance of former years. Comparative data on sulfur dioxide conditions in St. Louis lead to the conclusion that average sulfur dioxide concentrations were 83 per cent lower in the winter and 73 per cent lower in the summer of 1950 than for the comparable periods of 1936–1937 (54). These improvements have been reflected to a corresponding degree in the condition of vegetation, metals, paints, and stonework.

Pittsburgh and Allegheny County. The air pollution problems of Pittsburgh and Allegheny County are particularly acute because of the large concentration of heavy industry, the population of nearly 1.5 million people, the availability of inexpensive, unlimited deposits of high-volatile bituminous coal, the large numbers of municipalities exercising police powers in the county, and the unique topography, which facilitates the retention of air pollutants in large masses of still air. Early Pittsburgh ordinances, passed in 1895 and 1911, limiting the emission of smoke were invalidated by the courts on the grounds that compliance with the ordinances would impose a hardship on users of coal. The ordinance passed in 1917 was much more lenient than the one presently in effect (18). This latter was modeled to a great extent on the St. Louis ordinance of 1940 and was based on the principle that mechanical equipment must be used to burn high-volatile coal smokelessly and that where no such equipment is available, smokeless fuel must be used. Regulations also exist regarding the emission of soot, cinders, noxious acids, fumes, gases, and fly ash. The sale, transportation, and consumption of any solid fuel, for hand firing, containing more than 20 per cent volatile matter on a dry basis is also prohibited. Under the terms of the ordinance all new installations, whether coal, gas, or oil, must be approved by the Bureau of Smoke Prevention, and annual inspections of heating equipment are carried out. The increasing number of diesel locomotives, which in 1953 constituted about 71 per cent of all locomotives, has solved the smoke problem of the railroads.

The smoke-control ordinance recently enacted by Allegheny County is a little less stringent in some respects than the one applying to Pittsburgh. For example, the county ordinance specifically exempts bessemer converters, except that it imposes an obligation to join a 5-year research program. The city ordinance, on the other hand, specifies that it is not permissible to emit "such quantities of soot, cinders, noxious acid, fumes, or gases in such place or manner as to cause injury, detriment, or nuisance to any person or the public," and also that fly-ash emission is subject to an ASME test for quantity and particle size (18).

The application of the Pittsburgh ordinance has been so successful that visibility observations made by the U.S. Weather Bureau show that between 1946 and 1953 heavy smoke has been reduced by 94.4 per cent and total smoke by 69.4 per cent (11). Although the smoke provisions of the ordinance are now working effectively, the provisions regarding fly-ash and air pollution generally are still in the experimental stage (2).

Both industry and the public have cooperated wholeheartedly in air pollution abatement measures for Allegheny County. The assistance of an advisory committee of 17 persons, under the chairmanship of Dr. Weidlein of the Mellon Institute, representing the steel, coal, railroad, power, and other industries, as well as labor and public organizations, has contributed to the success of control efforts. During the 5-year period since 1949, industry has been engaged in a major research effort to design and equip plants with the most modern units for the collection of fly ash from power plants, dust and fume from blast furnaces and cement plants, coal dust from coking operations, and many other miscellaneous abatement measures. Fires from burning mine-refuse piles have been extinguished, and spontaneous combustion has been prevented.

Los Angeles County. The state of California passed an act in 1947 which acknowledges that serious pollution conditions exist within certain parts of the state and empowers county supervisors to set up an air pollution control district in every county, after due notice is given and other formalities, including the holding of public hearings, are completed. This act forms the basis of authority for the organization of the Los Angeles County Air Pollution Control District, which was set up soon afterward. Its rules and regulations, which are based in part on the Health and Safety Code of the state, set up a permit system for both the erection and operation of any equipment or installation which might cause the emission of air contaminants. The rules also cover the operation and installation of air pollution control equipment. Owing to the complex nature of the Los Angeles air pollution problem, extensive investigations were undertaken on the nature and extent of atmospheric contamination. An exhaustive survey was made of all known sources of emission.

In the earlier control work of this district the main emphasis was placed on reduction in the emission of smoke, dust, metallurgical fumes, and sulfur dioxide. The major sources of contamination in this area include:

1. Products of combustion from gaseous and liquid fuels (no coal burned), industrial, municipal, and domestic incinerators, and exhaust gases of motor vehicles
2. Dust and fumes of mineral and metal industries, grain and feed processes, chemical paint and soap industries
3. Sulfur dioxide from petroleum refining, chemical processes, and combustion operations
4. Unsaturated hydrocarbons from petroleum refining, storage, hauling, and marketing of petroleum and its products, and motor-exhaust gases

The control regulations forbid the discharge into the atmosphere, from any source of particulate matter in excess of 0.4 grain/cu ft at a temperature of 60°F and pressure of 14.7 psi. After research and experimentation into the capacity of the metallurgical and foundry industry to control the discharge of dust or fumes, a regulation was adopted forbidding the discharge of such contaminants in accordance with a sliding scale based on the process weight per hour of solid material. A top limit of 40.0 lb/hr was set for large industries processing materials at the rate of 60,000 lb/hr or more. The sliding scale of permissible dust emission requires stricter compliance by large industries than by small ones, on the assumption that use of air pollution control equipment of high efficiency is more practical and more economical for the larger industries. The regulations also forbid the discharge of sulfur compounds in excess of 0.2 per cent by volume in the stack gas (42).

A comparison of the 1948 and 1951 daily emissions from major sources in Los Angeles County (38) indicates that smoke has been reduced by 31 per cent, dust and fumes by 46 per cent, and sulfur dioxide by 50 per cent, through the installation of control equipment. These abatement efforts have not, however, eliminated the troublesome smog problem. A determined attempt is being made to control the emission of hydrocarbons from the petroleum industry, the smoke from rubbish burning in backyard incinerators, and the fumes from automobile exhausts. As shown in Tables 2-1 and 2-2, the total known emissions of aerosols, acid gases, and aldehydes to the atmosphere amounted to about 895 tons per day in 1951. On the other hand, the estimated losses to the air of total hydrocarbons for 1951 reached a level of 2,130 tons per day. A considerable fraction of these emissions is due to the activities of the public involving vehicular transportation and domestic incineration.

Control devices in operation and under construction in the minerals, metals, and other industries will, when they are completed, reduce the total amount of pollutants from these sources below the 1940 level. The oil industry and the Control District

Table 2-1. Known Emissions to Los Angeles County Atmosphere—Aerosols, Acid Gases, and Aldehydes

Tons per day

Source	Aerosols			Oxides of nitrogen (as NO_2)			Sulfur dioxide			Aldehydes (as formaldehyde)			Organic acids (as acetic acid)		
	1940	1948	1951	1940	1948	1951	1940	1948	1951	1940	1948	1951	1940	1948	1951
Chemical, paint, roofing, rubber, and soap industries	6	13	11	N[a]	N	N	10	21	25	N	N	N	N	N	N
Food and fertilizer industries	7	12	8	0	0	0	0	0	0	0	0	0	0	0	0
Fuel-oil burning	3	5	3	50	85	85	100	170	170	N	N	N	3	7	7
Incineration—domestic	60	90	90	15	21	21	N	N	N	8	10	10	42	55	55
Incineration—commercial, industrial, municipal, and dumps	85	115	34	20	27	7	N	N	N	10	14	4	50	69	18
Metals industries	12	28	16	0	0	0	1	2	2	0	0	0	0	0	0
Mineral industries	30	51	26	0	0	0	N	N	N	0	0	0	0	0	0
Petroleum refining	0	8	5	0	0	0	260	380	80	0	0	0	0	0	0
Transportation—gasoline	14	23	25	71	120	132	13	20	21	7	10	10	4	7	7
Transportation—diesel	3	5	7	4	7	10	1	2	2	0	1	1	1	2	3
Totals	220	350	225	160	260	255	385	595	300	25	35	25	100	140	90

[a] Negligible.

Reference: Los Angeles County Air Pollution Control District, Los Angeles, Calif., First and Second Technical and Administrative Reports on Air Pollution Control in Los Angeles County, 1948–1949 and 1950–1951.

are investigating intensively the control of hydrocarbon pollutants from refinery operations and from evaporation of crude oil and gasoline in storage. Most of the harmful materials can be controlled, and substantial reductions in emissions and losses due to evaporation have been made since 1951.

Plans have been formulated and tests are being made to reduce the amount of aerosols and remove the partially oxidized hydrocarbons and other smog-forming components of automobile exhausts. The oxides of nitrogen formed in most combustion processes are also being investigated because these compounds catalyze the oxidation of unsaturated hydrocarbons in sunlight. However, gains made in industrial-pollution control may be jeopardized by the continued growth of the population and the consequent increase in domestic-incineration waste products and automobile-exhaust gases. Known corrective measures for domestic incineration would entail controlled collection or disposal facilities for all combustible refuse. The problem of transportation sources is much more difficult to solve and would require the design and installation of control devices to remove or oxidize completely the harmful components of exhaust gases; or it might require revolutionary technological changes in internal-combustion engines and fuels.

International Windsor-Detroit Problem. In 1949, upon receipt of complaints from the cities of Windsor and Detroit and from citizens in the area that the atmosphere was being contaminated by excessive amounts of smoke, soot, fly ash, and other impurities in quantities detrimental to the public health and welfare and to property interests on either side of the international boundary, the problem was referred to the International Joint Commission. The terms of reference were broad in scope and involved an investigation of the major sources of pollution, including smoke from Great Lakes vessels plying the Detroit River as well as the varied effluents from industrial operations. The questions to be answered involved the determination of the possible effects of air contaminants on public health, safety, vegetation, and

Table 2-2. Known Emissions to Los Angeles County Atmosphere—Hydrocarbons

Estimated tons per day

Source	Total hydrocarbons			Olefinic unsaturation: Total			Olefinic unsaturation: C_4, C_5, C_6 per molecule			More than C_3 per molecule			Acetylene		
	1940	1948	1951	1940	1948	1951	1940	1948	1951	1940	1948	1951	1940	1948	1951
Chemical, paint, roofing, rubber and soap industries	13	25	N[a]	N	N	7	N	N	4	13	25	28	N	N	N
Petroleum marketing	67	115	120	10	17	18	8	16	17	57	95	102	0	0	0
Petroleum production....	1,700	270	270	N	N	N	N	N	N	120	100	100	N	N	N
Petroleum refining..	500	830	830	80	135	135	27	47	48	370	610	610	N	N	N
Transportation—gasoline....	500	850	875	80	133	140	25	42	46	320	540	580	50	85	90
Totals....	2,780	2,090	2,130	170	285	300	60	105	115	880	1,370	1,420	50	85	90

[a] Negligible.

Reference: Los Angeles County Air Pollution Control District, Los Angeles, Calif., First and Second Technical and Administrative Reports on Air Pollution Control in Los Angeles County, 1948–1949 and 1950–1951.

property and identification of the major sources responsible, if any. Furthermore, if vessels were found to be creating an excessive amount of pollution, the most practicable remedial measures were to be determined, as well as the cost of such measures and by whom such cost should be borne.

The area under investigation lies on both sides of the Detroit River for a distance of about 30 miles and extends about 15 miles inland from the river. The estimated population on the Michigan side is about 2,775,000 and on the Canadian side about 160,000. This region contains the third largest concentration of industry in North America. Although the automotive industry occupies a dominant position in both Detroit and Windsor, there are large manufacturing operations which produce steel and pig iron, heavy chemicals, paints and varnishes, pharmaceuticals, electrical appliances, stoves and furnaces, brass and iron-foundry products, machine and cutting tools, synthetic resins and plastics, solvents and organic chemicals, rubber products and adhesives, caustic soda, calcium chloride and chlorine, salt and soda ash, phosphoric acid and phosphate, pulp and paper, and refined-petroleum products.

During the navigation season the Detroit River probably carries the largest vessel traffic in the world. The vessel fleets of the Great Lakes make about 30,000 passages of the Detroit River in transporting an enormous volume of cargo, including iron ore, coal, and grain.

The annual domestic and industrial consumption of coal and solid fuels is estimated at about 15 million tons on the United States side and 650,000 tons on the Canadian side of the international boundary. In addition there is the solid fuel consumed by vessel traffic on the Detroit River. This yields a total consumption of about 16 million tons annually. On the basis of an average sulfur content of 1.5 per cent and the release of about 90 per cent of this sulfur to the atmosphere during combustion, it is estimated that approximately 430,000 tons of sulfur dioxide is discharged to the atmosphere annually from solid fuels alone. This total is augmented further by the sulfur dioxide released in the combustion of liquid and gaseous fuels.

Many other contaminants are released to the air as solids, liquids, gases, and vapors in substantial quantities from public and industrial operations. These include (1) smoke, soot, fly ash, organic and tarry matter from products of combustion; (2) carbon monoxide, hydrocarbons, aldehydes, peroxides, and organic acids from internal-combustion engines; (3) sulfur compounds, including sulfur trioxide, sulfuric acid, and hydrogen sulfide; (4) fluorine and chlorine compounds, oxides of nitrogen, nitric acid, and ammonia; (5) metallic fumes, oxides, and dusts from metallurgical, chemical, and mechanical operations.

The environmental studies have been designed to correlate with meteorological investigations and to delineate areas of high, intermediate, and low pollution. Some of the findings with respect to the amount and composition of dustfall and suspended particulates and the concentration levels of various gaseous contaminants have been presented already in the tables and elsewhere (12,34). Sulfur dioxide pollution varies in intensity and frequency according to a pattern which shows diurnal as well as seasonal trends. The daily cycle of aerosol pollution also exhibits a diurnal variation, with a well-defined peak in the early morning falling to a minimum in the early afternoon and rising during the night. This follows the cycle of temperature lapse rate and turbulence. There is a correlation between intensity of sulfur dioxide and high concentrations of suspended particulate matter.

Smog visitations of several hours' to several days' duration have been noted during temperature-inversion periods that are most frequent during the months of April, May, July, October, November, and December. On many occasions the greatly reduced visibility, accompanied by higher than usual contamination, has been noted for a distance of 15 miles inland from the Detroit River. A number of such visitations caused eye irritation to residents of the area. Crop damage and injury to shrubs, ornamental plants, and trees have been observed.

One of the foremost problems in this area has been the excessive emission of smoke from vessels plying the Detroit River. Thousands of observations have been made by trained observers, using the Ringelmann chart method, in each navigation season since 1949. There has been noted a marked improvement in smoke performance from 1951 to date. In 1950, about 77.5 per cent of the vessels on the Detroit River were found to be emitting excessive amounts of smoke for 51 per cent of the time. By 1953, these figures had been reduced to 33.6 per cent of the vessels and 27.5 per cent of the time.

These results have been achieved through the cooperation of the Lake Carriers and Dominion Marine Associations in a program of improvement of firing methods and boiler equipment, as well as by the institution of a voluntary system of Smoke Emission Objectives, which are revised frequently. These objectives are made more stringent as improved methods and equipment are developed to reduce smoke. Owners of vessels are notified by the Technical Advisory Board if their vessels fail to conform to these objectives, which are designed to meet the best practical performance which may be expected from the various broad classes of boiler equipment presently installed on Great Lakes vessels.

Owners are being encouraged to convert hand-fired fuel-burning equipment to automatic stokers. It is believed that with proper equipment the burning of solid fuel will give just as good results as liquid fuel in regard to smoke emission. The model ASME Smoke Code has been adopted for both oil-burning vessels and those equipped with underfeed stokers using solid fuel.

A comprehensive health study is in progress in the Greater Detroit–Windsor area to determine the chronic effects on public health of long-continued exposure to pollutants. The acute effects of smog disasters are well known and have provided dramatic warnings of the danger from the accumulation of foreign material in the air under conditions when natural dispersion and dilution processes fail. The chronic effects on health have never been evaluated conclusively to date on any known population group (49). During the past 100 years many attempts have been made to apply mortality statistics in order to show the relationship between fog, smoke, and smog and mortality rates from various diseases (50,51). Mills (45,46) and others (25) have reported correlations between dustfall and respiratory disease in some cities of the United States. Such statistics can be readily refuted in the absence of data on the nature of the contamination and factors related to the population itself, because the statistics are collected for other purposes.

The methods used in the Detroit-Windsor health study have been reported by Clayton (12) and Peart and Josie (49). These have consisted of the following steps:

1. A review of existing health data in the Detroit River area for comparison with other cities.

2. A sample morbidity survey to measure and compare the intensity and types of sickness and symptoms in well-defined areas of high and low pollution. The survey has been designed to match sample populations based on socioeconomic factors in localized areas which have different pollution levels.

3. Control samples of households have been selected in a population area which is similar to the high- and low-pollution areas of Windsor and Detroit, except that the level of pollution is less than that of the low-pollution areas.

4. Continuous measurements of air pollution levels by the most modern techniques, observations of meteorological factors, and records of ill health in the various study areas are being made concurrently on a "round-the-clock" basis.

2.7 METEOROLOGY AND POLLUTION CONTROL

2.7.1 Diurnal Variation of Turbulence

The air near the surface of the earth usually moves in an irregular manner, the motion consisting of a series of gusts and lulls accompanied by more or less rapid changes in wind direction. This eddying motion or turbulence serves effectively to carry away and disperse combustion products and objectionable wastes under ordinary atmospheric conditions. In clear summer weather, over level land, the gustiness exhibits a diurnal variation during a 24-hr day, the amplitude of the oscillations rising to a maximum about noon and decreasing to a minimum at night. This phenomenon is accompanied by a diurnal variation in the vertical gradient of temperature with height near the ground. Soon after sunrise the air near the ground becomes warmer than that at higher levels because the incoming solar radiation raises the temperature of the ground. As time goes on the rate of decrease of temperature with height, or the lapse rate, increases to a maximum about midday or early afternoon. Thereafter the lapse rate decreases as the sun descends in the sky, until an approximately isothermal condition of the atmosphere near the ground is reached before sunset. After sunset the surface of the earth cools more rapidly than the air above it. This, in turn, cools the air near the surface, and this air layer becomes cooler than the air above. This process produces a temperature inversion in which the temperature increases with height. The inversion persists throughout the night in varying intensity, until shortly after sunrise when the whole process is repeated (62).

2.7.2 Favorable and Unfavorable Meteorological Factors

During the period of the day when gustiness and lapse rate are increasing, the air is in a state of instability because of disturbances of mechanical and thermal origin. The capacity of the atmosphere for dilution and dispersion of contaminants is then at its highest. Particles of smoke, aerosol, and gaseous wastes will be diluted rapidly to harmless proportions. At sunset and during the night, the contaminants over large cities and industrial areas tend to collect in the inversion layer, extending from several hundred to a thousand feet or more above ground level. In the Detroit-Windsor area, the Sudbury, Ontario, nickel-smelting district, and over many large cites, such meteorological conditions give rise to regular early morning fumigations of short duration and varying intensity (30,32).

In cloudy, overcast weather, the gustiness or turbulence is fairly uniform both day and night. The extent of dispersion of atmospheric pollution will then depend mainly on the wind velocity and topographical features of the terrain. During periods of low wind velocity or calms in such weather, the level of contaminants increases. At last a balance is reached which depends upon the scale of industrial and domestic activities and the ease with which airborne pollutants can be dispersed downwind and crosswind under the cloud layer. Under inversion conditions the ceiling may be much lower, and higher concentrations of pollutants will be built up, giving rise to serious conditions if the inversion period is prolonged over several days' duration.

It is therefore apparent that meteorological factors, when favorable for dispersion, will result in the rapid mixing and dilution to harmless proportions of relatively large quantities of gaseous and aerosol waste products. A minimum of control measures will suffice during periods of high turbulence in clear weather. Such controls, to keep a city atmosphere reasonably clean, include the use of proper combustion equipment to prevent release of smoke, the collection of dust and fly ash of particle size greater than 10 to 20μ by means which are available without excessive cost, and the elimination of acid gases such as sulfur dioxide, oxides of nitrogen, hydrogen sulfide, and by-products from refinery, chemical, and metallurgical operations by means of stacks of sufficient height for proper dispersal. For the greater part of the year, such simple measures would be effective over many industrial areas.

2.7.3 Natural Air-cleaning Processes

Natural air-cleaning processes include the removal of smoke, dust, and gaseous waste products by dew, rain, and snow; the deposition of aerosols by convection settling and impaction on land and other irregular surfaces; and the absorption of gases by vegetation, soil, and water. Turbulent mixing does not remove smoke from the air but merely dilutes it and is, therefore, not a cleaning process in the strict sense. However, when smoke-laden air currents traveling at sufficiently high velocity are deflected by stationary surfaces, even the finer smoke particles may be deposited in much the same manner as the action of a cyclone, centrifugal separator, or jet impactor. There is no doubt that appreciable quantities of smoke are removed from the air by water drops precipitated as rain, by cumulus clouds because of the continuous movement of ice and water particles within them, and by fog droplets as they sink to ground level. In high concentrations, smoke and other aerosols tend to coagulate to form chains and larger particles which deposit by settling according to gravity and Stokes' law. All these factors tend to cleanse the air of impurities which otherwise would accumulate and in time would render life impossible over large sections of the earth.

A study of the wind and weather conditions in a particular industrial area or city is of the utmost importance in the assessment of data from air pollution surveys, formulation of abatement and control measures, industrial-site selection and planning of new towns, or expansion of existing communities. For this purpose, each location may be considered unique, and only the simplest general measures are applicable everywhere. Much is now known concerning the diffusion of matter in the atmosphere; various theories of turbulence and formulas may be employed with some assurance in determining the extent of dilution of smoke and gas from a given source, under different meteorological conditions (see Secs. 5 and 9). These theories and equations, involving emission temperatures and rates, stack parameters such as height, diameter, and exit velocity, and meteorological factors involving wind speed, direction, and atmospheric stability, have been employed successfully in many stack-design and pollution-control problems (7,20,64).

2.7.4 Basic Principles of Control in Stack Design

The basic principles of control in stack meteorology may be stated briefly as follows:

1. The ground-level concentration of an effluent downwind from a stack or point source is directly proportional to the mass rate of emission at the source. The dilution of stack gas with excess air does not have, therefore, any appreciable influence on the ground concentration. This is due to the fact that the extent of dilution of stack gas that is practicable is usually insignificant in comparison with later dilution of the smoke plume in the atmosphere by eddy diffusion. Stack dilution is a significant factor only if it appreciably increases the velocity of the exit gas, thereby increasing the effective stack height. However, the effective stack height depends also upon the temperature and density of the discharge gas, as well as its velocity. Stack dilution, therefore, may lower the temperature of the exit gas, thus lowering the height of the plume above the top of the stack.

2. The effective stack height is the sum of the actual height of the stack above ground level plus the total plume rise due to the velocity of the exit gas and the difference in density between this gas and the atmosphere. The rise of the plume is influenced considerably by the wind speed and atmospheric stability conditions. An increase in wind velocity reduces the height of the plume by a factor which involves the cube of the wind speed (8,61).

3. The downwind ground concentration is inversely proportional to the wind velocity.

4. The downwind ground concentration is inversely proportional to the distance from the stack raised to an exponent which varies from 1.5 to 2, depending upon the theory employed and the meteorological conditions. In general, dilution with distance is much less under inversion conditions than during average or unstable atmospheric conditions.

5. The maximum concentration at ground level downwind is inversely proportional to the square of the effective stack height. This is an important principle which points the way to a solution of many fume-damage problems. For example, if by doubling the effective stack height the maximum ground concentration can be reduced to one-quarter of previous values, great relief can often be expected in cases of damage to susceptible species of vegetation.

6. The distance of the maximum downwind concentration from the source depends upon atmospheric conditions. Under unstable or average turbulence conditions, this distance is about 10 to 20 stack heights from the source. In very turbulent air this point may be as close as 5 stack heights. On the other hand, as the atmosphere approaches stability the point of maximum concentration will move to greater distances from the source. Under stable inversion conditions, the smoke plume assumes the form of a narrow ribbon which may travel a great distance before reaching ground level.

2.7.5 Meteorological Control

Some of the foregoing concepts were applied many years ago on an empirical basis in meteorological control practice in the smelting industry. Thus emission of considerable quantities of sulfur dioxide at relatively high temperatures from tall stacks no longer constitutes a hazard to vegetation at Murray, Utah, and Selby, Calif. A solution was thus furnished for smoke problems which had existed earlier when the gases were released from lower stacks and at lower temperatures (64). It had been estimated in investigations of the American Smelting and Refining Co. at Salt Lake City, 30 years ago, that each degree Fahrenheit of smoke temperature in exit gas above that of the surrounding environment is equivalent to about 2.5 ft of additional stack height (64).

Salt Lake Valley. Probably the earliest recorded case of application of the principles of meteorological control was that of two smelters near Salt Lake City, Utah. Injury to animals as well as to vegetation from such operations led to a situation wherein the smelters were threatened with a permanent injunction. In 1921, Judge T. D. Johnson of the Federal Court accepted the recommendations of an appointed commissioner who had carried out an extended study of plant operations and field conditions. Operations at each plant were to be permitted without limitation on the amount of ore roasted, provided that gases were discharged from stacks at least 450 ft high and that additional buoyancy be given such gases by maintaining, between sunrise and sunset, a temperature difference of at least 75°C between the stack gases and the atmosphere, during the growing season. It was also stipulated that suspended solids and mists were to be removed by Cottrell precipitation or bag filtration. According to Swain (63) there has been no injury to plant and animal life from these sources since these regulations were put into effect.

Correlation of field data on weather conditions, gas concentrations, and sulfur dioxide injury to vegetation in the Murray smelter area indicated that the following factors were necessary before plant damage could occur:

1. A temperature above 5°C
2. Daylight
3. A relative humidity above 70 per cent
4. A prevailing wind carrying gas-laden air for 3 hr or more

These facts led to a "sea-captain plan" of smelter operation during the growing season. Whenever these four weather conditions coincided, the operator of the company weather station notified the plant superintendent, and the chief source of sulfur dioxide, the roasters, were shut down. Roaster operations were not resumed until there was a change in wind direction, a drop in relative humidity, or a great decrease in light intensity because of approaching darkness. This remedy was remarkably effective, but roaster operations were not sufficiently flexible to permit its use over extended periods. The plan was discontinued for economic reasons (63), but the principle of this pioneering work was destined to be applied more successfully in the international case of the Trail smelter.

Trail, B.C. Case History. The Canadian Smelter at Trail, B.C., is the largest lead-zinc smelter in the British Commonwealth, with a daily metal output of approximately 1,000 tons. It is situated in a relatively narrow, mountainous valley on the west bank of the Columbia River, about 11 miles by river from the international boundary. The first complaints of damage to farm and forest lands in northern Stevens County, Washington, arose in 1925 soon after enlargement of the smelter and erection of stacks about 410 ft high. During the next 5 years the sulfur dioxide emission increased from a previous high of 330 tons per day to about 600 to 650 tons, owing to increased production of metals. In August, 1928, following further complaints of widespread damage in the United States, the problem was referred to the International Joint Commission. After investigation and lengthy hearings, the Commission assessed damage claims against the company at $350,000 up to the end of 1931. In that year sulfur-recovery works, consisting of sulfuric acid and fertilizer plants, came into operation.

However, 3 years later, upon receipt of further complaints of alleged damage, a convention was signed between Canada and the United States, under which it was

Table 2-3. Maximum Permissible Sulfur Emission at Trail, B.C.

Time	Turbulence bad (1) Wind not favorable	Turbulence bad (2) Wind favorable	Turbulence fair (3) Wind not favorable	Turbulence fair (4) Wind favorable	Turbulence good (5) Wind not favorable	Turbulence good (6) Wind favorable	Turbulence excellent (7) Wind not favorable and favorable
Growing Season							
Midnight to 3 A.M.	2	6	6	9	9	11	11
3 A.M. to 3 hr after sunrise	0	2	4	4	4	6	6
3 hr after sunrise to 3 hr before sunset	2	6	6	9	9	11	11
3 hr before sunset to sunset	2	5	5	7	7	9	9
Sunset to midnight	3	7	6	9	9	11	11
Nongrowing Season							
Midnight to 3 A.M.	2	8	6	11	9	11	11
3 A.M. to 3 hr after sunrise	0	4	4	6	4	6	6
3 hr after sunrise to 3 hr before sunset	2	8	6	11	9	11	11
3 hr before sunset to sunset	2	7	5	9	7	9	9
Sunset to midnight	3	9	6	11	9	11	11

agreed to submit the issues to an international tribunal for final settlement. This tribunal appointed two technical consultants, R. S. Dean and R. E. Swain, to assess the scientific evidence on either side. In 1938, after extended hearings, a further award of $78,000 in damages was made to cover injury to vegetation during the period since 1932. The tribunal retained jurisdiction until Oct. 1, 1940. Its technical consultants, with an assisting staff, were asked to conduct an investigation, mainly meteorological, in order to formulate a regime which would assure freedom from injury to vegetation in the future. In the meantime, sulfur recovery had been increased by the company through the enlargement of sulfuric acid and fertilizer plant capacity. By 1940, sulfur emission had been reduced to only one-third of the maximum reached in 1930.

The investigation of the scientific aspects of this case for the Canadian side was undertaken by the National Research Council of Canada in 1929. The work, which was presented as evidence at the various hearings, was published in book form in 1939 (28). These investigations on maximum permissible levels of sulfur dioxide concentration for the normal development of susceptible species of plants and the meteorological investigations of the technical consultants (16) and of Hewson (26) laid the groundwork for the operating regime. The mass rate of emission of sulfur dioxide is controlled by regulating the operation of Dwight and Lloyd roasting furnaces in accordance with meteorological conditions. Among these are wind direction and velocity, atmospheric turbulence, time of day or night, and the recorded ground concentrations of a Thomas sulfur dioxide analyzer (16) located at Columbia Gardens, in the river valley, about 5 miles southeast of the smelter. The operating regime is shown in Table 2-3, which indicates the maximum permissible sulfur emission expressed in tons per hour.

Operating Regime. General restrictions and provisions are as follows:

1. If the Columbia Gardens recorder indicates 0.3 ppm or more of sulfur dioxide for two consecutive 20-min periods during the growing season, and if the wind direction is not favorable, emission shall be reduced by 4 tons of sulfur per hour or shut down completely when the turbulence is bad, until the recorder shows 0.2 ppm or less of sulfur dioxide for three consecutive 20-min periods.

If the Columbia Gardens recorder indicates 0.5 ppm or more of sulfur dioxide for three consecutive 20-min periods during the nongrowing season and if the wind direction is not favorable, emission shall be reduced by 4 tons of sulfur per hour or shut down completely when the turbulence is bad, until the recorder shows 0.2 ppm or less of sulfur dioxide for three consecutive 20-min periods.

2. In case of rain or snow, the emission of sulfur shall be reduced by 2 tons/hr. This regulation shall be put into effect immediately when precipitation can be observed from the smelter and shall be continued in effect for 20 min after such precipitation has ceased.

3. If the slag re-treatment furnace is not in operation, the emission of sulfur shall be reduced by 2 tons/hr.

4. If the instrumental reading shows turbulence excellent, good, or fair, but visual observations made by trained observers clearly indicate that there is poor diffusion, the emission of sulfur shall be reduced to the figures given in column 1 if the wind is not favorable or column 2 if the wind is favorable (see Table 2-3).

5. When more than one of the restricting conditions provided for in 1, 2, 3, and 4 occur simultaneously, the highest reduction shall apply.

6. If, during the nongrowing season, the instrumental reading shows turbulence fair and wind not favorable but visual observations by trained observers clearly indicate that there is excellent diffusion, the maximum permissible emission of sulfur may be increased to the figures in column 5. The general restrictions under 1, 2, 3, and 5, however, shall be applicable.

Whenever the smelter shall avail itself of the foregoing provisions, the circumstances shall be fully recorded and a copy of such record shall be sent to the two governments within one month.

7. Nothing shall relieve the smelter from the duty of reducing the maximum sulfur emission below the amount permissible according to the tables and the preceding general restrictions and provisions, as the circumstances may require for the prudent operation of the plant.

Remedial Works. The Trail smelter case represents an outstanding example of the successful scientific solution of a difficult international problem in air pollution. In the late 1920s the Consolidated Mining and Smelting Company sent experts abroad to study every known process for sulfur dioxide recovery, synthesis of ammonia, and manufacture of sulfuric acid. The conversion of an airborne waste to marketable by-products eventually involved an investment of over $15 million in plant construction and equipment. At the present time, the recovery plants at Trail consist of six acid plants with a total capacity of 1,300 tons of 100 per cent sulfuric acid per day, an ammonia synthesis plant of 240 tons per day capacity, and plants for the production of ammonium sulfate, ammonium nitrate, and ammonium phosphate fertilizers.

The total current loss to the atmosphere from all operations at Trail is less than 9 per cent of the sulfur charged (36).

Control of Radioactive Wastes. The application of a practical method of meteorological control to the nuclear reactor at the Brookhaven National Laboratory is described by Smith (58). This control plan includes forecasting of dispersion as well as analysis of current weather conditions in order to provide safety. A series of templates are used to show the contours of radiation dosage at ground level for a particular combination of wind speed and gustiness, based on the results of oil-fog tests, from a tower, at various heights up to 400 ft. The forecasting technique is based primarily on empirical relationships between synoptic and micrometeorological variables as determined by observation and climatological studies. One of the main features of the scheme is the classification and prediction of horizontal gustiness in four basic types, as a measure of turbulence.

Assessment of Meteorological Control. It is evident that the successful use of meteorological control in plant operations depends upon the nature of such operations. In the Trail case, the roasting of the ore was sufficiently flexible so that, during periods favorable for dispersion of waste sulfur dioxide gas, a sufficient stock of calcined ore or metal oxide could be accumulated and stored to prevent shutdown of the smelter when the scale of roasting was reduced during unfavorable weather conditions. Continuous processes cannot be reduced or shut down according to meteorological conditions without serious economic loss. If applied to plants in city areas on a large scale, such control would result in considerable unemployment as well as economic loss to industry. Gosline (21) contends that this type of control and forecasting are last resorts and that forecasts are likely to be wrong about 15 per cent of the time. It is preferable, therefore, to collect most of the pollutant at the source, if possible, or alternatively to find better methods of dispersing it in the atmosphere.

2.8 TOPOGRAPHY AND INDUSTRIAL-PLANT LOCATION

2.8.1 General Considerations

The atmosphere differs markedly in its ability to disperse contaminants effectively at different locations because of variations in topographical features and microclimate. Even on level land, such microfeatures as the character of the surface (i.e., smooth, rough, bare of cover, soil, grass, crop plants, trees, differences in elevation, etc.) affect the gustiness or local vertical distribution of temperature and wind. Close to the ground there exist much greater diurnal variations of humidity and temperature than can be found a few feet higher.

Each site to be considered for location of an industrial plant is therefore unique in this respect. Similar considerations apply to the larger areas covered by communities and cities. Locations at land-water boundaries, in undulating or hilly terrain, in valleys, in valleys in mountainous areas, and in locations of various other types present different problems in air pollution control. In coastal or lakeside areas, the land and sea type of air circulation or breeze is an important feature in dispersion. The daytime on-shore breeze, which results when the air over land becomes warm more rapidly than that over water, may extend only a short distance or many miles inland. The wind pattern may be reversed at night in clear weather to yield the land breeze. This would blow toward the water and thus disperse contaminants which tend otherwise to accumulate below an inversion layer at night.

2.8.2 Valleys

Valleys are poor locations for large industrial plants and communities if air pollution is the dominant consideration. The air pollution disasters of the Meuse Valley,

Belgium (39), and at Donora, Pa. (53), were attended by considerable illness and loss of life. They took place in valleys where, during prolonged 4- to 5-day inversion periods, the contaminants discharged from industrial operations accumulated in an almost stagnant air layer. In both instances the valley sides were only several hundred feet in height, but this proved sufficient to restrain adequate circulation of air. The situation becomes much more critical if the plant is located at the bottom of a relatively narrow valley with mountains rising fairly steeply on either side. The circulation of air in the valley, particularly during the summer period, is then due mainly to the differential heating and cooling of the valley sides. Solar heat in the daytime warms the air in contact with the valley walls, causing a breeze upward along the slopes. At night, cold air forms at higher elevations and moves down the slopes into the bottom of the valley, giving rise to a mountain breeze. The resulting conditions cause the accumulation of pollutants in a shallow layer above the valley floor during the night.

2.8.3 Topographical Effects at Trail, B.C.

The unfavorable topographical effects on dispersion of smoke in a mountainous valley are best illustrated in the Trail, B.C., smelter location. This smelter is situated on the west bank of the Columbia River about 11 miles by river channel north of the international boundary between Canada and the United States. From about 15 miles north of Trail to 35 miles downstream from the boundary, this river flows through a valley ranging in width from about ½ to 3 or 4 miles, bounded by mountain ranges that rise to heights of several thousand feet on each side. The slopes above the river are usually a succession of terraces of varying width, but at some points the mountains rise steeply from the channel and at others there are often steep gravelly slopes. The river flows south to within 1 mile above Trail and then turns sharply to the southeast for about 6 miles. Near Columbia Gardens it turns south to the boundary and then swings southwesterly for about 40 miles in northern Stevens County, Washington. The Pend Oreille River enters the Columbia River from the east at a narrow tributary valley almost at the international boundary. Other tributary valleys enter the Columbia River Valley from the east and west, at various points south of the boundary. The deep valley of the Columbia provides a channel for movement of gas-laden air in this rugged mountain terrain (16).

Diurnal Fumigations. Investigations in the Trail smelter case revealed the remarkable phenomenon of the simultaneous occurrence of sulfur dioxide fumigations at recorder stations at different points along the valley for a distance of about 40 miles. This could not have occurred by the flow of gas-laden air along the valley through the medium of surface wind movement. It must have occurred rather by the simultaneous descent of a continuous layer of contaminated air which had formed at upper levels above the valley floor some hours previously under isothermal or inversion conditions. This type of fumigation was found to be diurnal in character and occurred most frequently during the summer growing season (16).

Mechanism of Diurnal Fumigations. The meteorological conditions and mechanism for this phenomenon were elucidated by Hewson (26). Sulfur dioxide sampling by airplane indicated that in calm weather, especially in the early morning, the effluent gases from the stacks rise about 400 ft above the tops of the two main stacks, soon level out and spread horizontally along the axis of the prevailing wind movement. An atmospheric stratum several hundred feet in thickness, containing sulfur dioxide, slowly spreads over a large area along the valley, especially during the night. With the rising of the sun, the radiational heating, which is fairly uniform along the valley floor but is more intense on the western mountain slopes than on the shaded eastern slopes, leads to a disturbance of the thermal balance. The gas-laden air which had accumulated in the upper atmospheric levels at about 1,000 ft above the bottom of the valley is then brought to ground level almost simultaneously at all points in the valley, as the turbulence increases. Usually the concentration of sulfur dioxide in this type of fumigation rises rapidly to a maximum and then drops off exponentially to traces within 2 or 3 hr. The maximum frequently occurs between 9 and 10 o'clock in the morning. A fundamental feature of the program of smoke control at this smelter is to anticipate this type of diurnal fumigation from an analysis of the turbulence and other meteorological conditions and to reduce the emission of sulfur dioxide from about 3 A.M. to about 8 to 11 A.M.

2.8.4 Topographical Effects at Los Angeles, Calif.

Los Angeles County suffers from a smog problem of unusual complexity. Both meteorological and topographical factors combine to prevent the normal diffusion and dispersion of waste products released to the air. This region has a population of 3 to 5 million inhabitants and is by far the largest subtropical heavily industrialized urban area in the world. Although smog was known in Los Angeles long before World War II, the rapid expansion of industry and the consequent increase in population during and immediately after this war accentuated the air pollution problem. The more noticeable smog visitations occur on about 60 days of the year and are characterized by greatly reduced visibility and the prevalence of severe eye irritation among the residents. The annual average of temperature inversion was estimated at 262 days during the 3-year period 1943 to 1945.

Topography. The topography of this county is a large basin fronting the Pacific Ocean to the west, with a ring of high mountains on the other three sides. An inversion layer of stratified air extends over this basin at levels which usually range between 1,000 and 3,000 ft. As a rule, the inversion layer has an upward slope both to the west and east. Because it rests against the mountain range on the east side of the basin, the flow of air eastward, out of the area, is prevented. The sea breeze in the daytime and the land breeze at night merely shift the inversion layer up and down but do not disperse it.

Inversion Mechanism. This temperature-inversion layer is brought about by the presence of a semipermanent high-pressure area over the North Pacific Ocean, extending from the west coast of the United States to beyond the Hawaiian Islands. This high-pressure area is present during the major portion of the summer and fall months, although its position may vary in longitude and latitude. It is an enormous elliptical column of air, moving in a clockwise direction on its axis and inclined slightly to the southwest. The air paths at intermediate levels in this column are steeply inclined to the earth's surface, the lower edge of this layer being adjacent to the California coast and its upper edge being suspended in mid-Pacific to the north of the Hawaiian Islands. The air moving toward California around the northern side of this pressure area is descending, moving through levels of increased pressure, and is therefore being warmed by compression. In consequence, the air arriving over Los Angeles County from the ocean, at levels above the surface air layer, is already much warmer than the surface air. This temperature difference is accentuated when air over the ocean surface, which has been cooled from below by contact with relatively cold ocean water offshore, is swept inland into the basin (6).

The inversion layer is like a canopy over the Los Angeles basin, preventing both vertical and lateral dispersion of contaminants. The natural haze, composed of salt from ocean spray and dust particles from soil and vegetation, is augmented by smoke, fumes, and gases from industrial and domestic activities, including vehicular traffic. The larger aerosol particles are removed by settling, but the fine aerosols and gaseous pollutants tend to build up at the top of the atmospheric layer which lies just below the base of the inversion stratum. The worst conditions occur when the man-made pollution accumulates over several days in calm weather and the base of the inversion layer is forced to its lowest level, thus bringing the polluted air to the ground (59).

2.8.5 Topographical Considerations in Industrial-site Location

The Trail and Los Angeles areas are probably extreme examples of the adverse influences of topography and weather factors on location of industry in relation to pollution. There are many other areas in which the rugged features of the terrain and the meteorological conditions combine to limit the extent of ordinary diffusion and dilution of man-made contaminants. In the planning and selection of industrial-plant sites, the usual practice has been to consider such factors as availability of raw materials, transportation, water supplies, and markets. The disposal of liquid wastes in rivers, streams, lakes, and coastal waters in such a manner as to avoid excessive contamination has become recognized as a problem which must be considered from the viewpoint of public health and welfare. Less attention is usually paid to the possibility of abnormal air pollution.

In the last few years, however, the importance of potential air pollution in industrial-site selection and planning is commanding more attention. Due consideration must be given the microfeatures of terrain and climate which, if favorable, ensure the rapid and efficient disposal of atmospheric wastes in an inexpensive manner. It is important to know prevailing wind direction, wind-speed pattern, degree of turbulence, persistence of unstable and inversion conditions, variations in elevation and obstacles of the terrain, and the probable influence of other industrial plants and buildings in the vicinity. The eddies produced by nearby buildings and man-made obstacles may result in erratic dispersion of gas. This must be taken into account in the design of stacks of sufficient height and flow velocity to carry away the waste products safely. Such data may be obtained by air pollution surveys and micrometeorological investigations. For pollution control it is important to observe the phenomena in the lower atmosphere, below 500 ft.

A new technique of increasing importance is the testing of smoke dispersion from small-scale models in wind tunnels. This has been employed successfully in the design of stacks for power plants and other buildings where the diffusion pattern is controlled by the eddies produced by the geometry of proposed and existing nearby structures (15). Recently a wind tunnel has been built at New York University which is designed specifically to study stack-dispersion problems (14).

2.9 EFFLUENTS AND AIR POLLUTION CONTROL

The number, nature, and concentration of pollutants which may be found in the air of cities and industrial communities are as variable and complex as the diversity and scale of activities of the population. Apart from the smoke, sulfur dioxide, and ash which contaminate the atmosphere from the combustion of various fuels, each industrial source contributes in varying degree some specific contaminants peculiar to the raw materials used and products manufactured. In addition, there are sources of pollution from transportation and traffic by railways, vessels, trucks, and private passenger vehicles. Recent comprehensive studies of the nature of atmospheric pollution in cities and industrial communities in America and Great Britain have indicated the area distribution, concentration levels, and wide variety of contaminants found in the atmosphere, including diurnal and seasonal fluctuations (10,17,34,38). Systematic observations are of great value in assessing the effectiveness of abatement programs, town planning, and zoning. The observations should be on a more or less continuous basis, at an adequate number of selected stations, and in correlation with data on weather conditions.

Over most large communities, smoke, ash, and sulfur dioxide constitute the major fraction, by far, of the total amount of pollution. It is probable that the total emission of sulfur dioxide over the world from all sources is much greater than the combined total of smoke and ash. In Great Britain, from the annual combustion of 180 million tons of coal, it is estimated that, on the average, the pollution emitted in millions of tons per annum amounts to 2.4 for smoke, 0.6 for ash, and 5.2 for sulfur dioxide (43). Annual sulfur losses to the atmosphere from world utilization of crude oil and coal have been calculated to be over 5 and 25 million tons, respectively (31).

2.9.1 Deposited Matter

The lower atmosphere varies markedly in its content of gaseous and suspended impurities from place to place. It is cleanest over the ocean. If this standard is taken as unity, then the average pollution of rural air would be 10 times greater; pollution over small towns would be about 35 times as much; and over cities, pollution would be 150 times greater than that of ocean air (37). In fact, under unfavorable meteorological conditions and with large industrial sources of pollution, the city levels of pollution may be several thousand times greater than the values found over the oceans. The larger particles of smoke and dust eventually settle out and can be measured in dustfall containers. The contributed wind-blown dust of natural origin, such as sand and soil particles, is not likely to exceed the smoke and dust from chimneys in city areas by more than about 10 per cent, except during severe dust storms or when smoke is brought in from distant forest fires.

The dustfall distribution in cities is a valuable indication of the amount of fly ash and dust deposited from stack emissions. The results are usually expressed in tons per square mile per month and will serve to indicate sources and areas of high and low contamination. Most of the particles collected in this manner are larger than 20 to 40μ in size. It is natural to expect, therefore, that such data will reflect the scale of industrial and domestic activity based on combustion of solid fuels and processes involving coarse dust emission. A comparison over a period of time at selected stations will indicate the success or failure of pollution-abatement efforts directed toward control of excessive emission. The monthly dustfall for a number of cities in North America and Great Britain is shown in Table 2-4. Values of 50 to 100 tons/sq mile/month or more are found usually in the larger, more heavily polluted cities which burn large quantities of coal.

Table 2-4. Mean Dustfall for Various Cities
(Tons per square mile per month)

City	Year	Monthly mean	City	Year	Monthly mean
Detroit	1946	51.7[a]	Rochester	1942	26.4[c]
	1950	59.8[a]	Pittsburgh	1951	45.7[d] (insoluble matter only)
	1951	62.1[a]			
	1952	64.1[a]			
	1953	72.1[a]	Toronto	....	31.7–54.2[e]
New York	1953	67.5[b]	Birmingham	1939–1944	27.8[f]
Chicago	1947	61.2[c]	Glasgow (East)	1939–1944	29.6[f]
Cincinnati	1946	34.0[c]	Leeds (Park Square)	1939–1943	35.9[f]
Los Angeles	1948	33.3[c]	Manchester (Philips Park)	1939–1944	42.9[f]

[a] Detroit Department of Buildings and Safety Engineering.
[b] New York State Chamber of Commerce.
[c] McCabe et al., *Ind. Eng. Chem.*, **41,** 2486 (1949).
[d] City of Pittsburgh, Department of Public Health Report, 1951.
[e] E. A. Allcut, *Eng. J.* (*Can*)., **30,** 154 (1947).
[f] The Investigation of Atmospheric Pollution. 26*th Rept. Dept. Sci. Ind. Research* (*Brit.*), 1949.

The solid impurities collected in dustfall cans or jars consist of water-soluble and insoluble components, tarry and organic combustible material, and ash. The soluble fraction in heavily polluted areas is high in sulfate, and the solution is strongly acidic or low in pH value. The distribution of these components of deposited matter from samples collected in the Greater Windsor, Ontario, area is shown in Table 2-5. Each dustfall station is roughly indicative of conditions over an area of about ¼-mile radius.

Table 2-5. Mean Values for Dustfall in Greater Windsor Area
(Tons per square mile per month)

Area	Dustfall	Components of deposited material					pH (water-soluble fraction)
	Total solids	Water insoluble	Water soluble	Tar	Other combustible matter	Ash	
Test Period: August to December, 1951							
Industrial	92.2	69.2	21.1	1.48	21.0	45.9	5.42
Industrial-residential	53.9	45.0	11.0	0.62	13.0	28.7	6.31
Residential-semirural	35.9	26.9	8.8	0.61	11.0	16.1	6.81
Test Period: May to September, 1952. Nonheating Season							
Industrial-residential	36.1	31.4	4.3	1.23	10.3	21.3	6.9
Residential-semirural	18.8	16.7	2.1	0.81	3.6	11.3	7.3
Test Period: October, 1952, to February, 1953. Heating Season							
Industrial-residential	92.1	69.1	22.9	1.70	13.5	52.0	6.1
Residential-semirural	53.0	40.8	12.3	2.59	8.2	30.0	6.8

2.9.2 Smoke and Suspended Matter

Smoke and suspended particulate matter consist of small particles which are generally less than 2 or 3μ in size. The particles have a low settling velocity, are readily transported by wind currents, and in the smaller sizes behave almost like gas molecules. They are generally collected by filtration, impaction, or thermal or electrostatic precipitation methods in air pollution surveys of communities. The size range and other properties of particles generated in many industrial operations are indicated in Table 2-6. The blast furnaces of the iron and steel industry, the nonferrous metallurgical industry, the manufacture of sulfuric acid, and foundry and smelting operations are some of the sources of suspended particulate matter, in addition to carbon smoke from combustion of coal and oil. True smokes and many metal oxide fumes such as those of lead, zinc, arsenic, cadmium, and beryllium contain particles which are mainly in the submicron range.

Daily Cycle. The concentration of suspended particulate matter is a valuable index of pollution of the air in cities and centers of industry. The variation in concentration is closely related to meteorological conditions and human habits. Observations in many cities indicate a daily cycle with a diurnal rhythm, one maximum occurring in the morning from about 8 to 9 o'clock and the other in the late afternoon or evening. This diurnal variation reflects the influence of solar heating and cooling, turbulence, and other factors (15,35,56). It has been observed in such widely sepa-

Table 2-6. Properties of Some Typical Aerosols

Type of disperse system	Size range of particles, diameter, μ	Electro-magnetic, spectrum, μ	Terminal velocity, cm/sec, due to gravity settling in air, 20°C and 1 atm
Raindrops	5,000–500		Turbulent motion $v_t = k_1\sqrt{pD}$ for particles down to about 1,000μ
Natural mist	500–40		p = particle density
Natural fog	40–1		D = particle diameter, μ
		Dusts	
Foundry sand	2,000–200	↑ Infrared, 420–0.7μ ↓	Intermediate region for particles between 1,000 to 100 μ
Ground limestone, fertilizers	800–30		$v_i = k_2 p^{2/3} D$
Sand tailings from flotation	400–20		
Pulverized coal	400–10		
Ground sulfide ore for flotation	200–4		
Foundry shake-out dust	200–1		Stokes' law for streamline region, applicable to particles from 100 to 1μ
Cement	150–10		
Fly ash	80–3		$v_s = KpD^2/n = K_3pD^2$
Silica dust (in silicosis)	10–0.5		
Pigments	8–1		
Pollens	60–20		n = coefficient of viscosity of air, poises
Plant spores	30–10		
Bacteria	15–1		
		Fumes and Mists	
Metallurgical fumes	100–0.1		Cunningham's correction for particles in range 1 to 0.10μ
H_2SO_4 concentrator mist	10–1		$v_c = v_s\left(1 + \frac{1.72l}{D}\right)$
Alkali fume	2–0.1		
SO_3 mist	3–0.5		l = mean free path of gas molecules, μ
NH_4Cl fume	2–0.1		$v_c = v_s\left(1 + \frac{0.172}{D}\right)$ in air
Zinc oxide fume	0.3–0.03		
		Smokes	
Oil smoke	1.0–0.03		Velocity due to Brownian motion exceeds velocity of gravity settling for particles less than 0.1μ
Rosin smoke	1.0–0.01		Einstein equation:
Tobacco smoke	0.15–0.01		$x = k_4\sqrt{t/D}$
Carbon smoke	0.2–0.01		x = average displacement in cm of spherical particle in air in time, t sec
Normal impurities in quiet atmosphere	1.0–0.01		

	Spheres	Irregular shape
k_1 =	24	16
k_2 =	0.41	0.26
k_3 =	0.0030	0.002
k_4 =	0.00068	

References: C. E. Miller, Pointers on Selecting Equipment for Industrial Gas Cleaning, *Chem. Met. Eng.*, **45**, 132 (1938); and L. Silverman, What Process Wastes Cause Air Pollution? *Chem. Eng.*, **58**, No. 5, 132–135 (1951).

rated areas as Glasgow, Leicester, and other cities in Great Britain and New York, western Pennsylvania, and the Detroit-Windsor region.

The concentration of suspended particulate matter, expressed as milligrams per cubic meter, varies from below 0.10 to as high as 1 or 2 in cities. The range of concentration for a number of cities, industrial and residential areas, is indicated in Table 2-7. During periods of smog when the atmospheric pollution load increases, the amount of suspended particulate matter may rise to two or three times the daily mean or higher (48,53). In a smog disaster in London, England, which occurred in December, 1952, and caused an estimated death rate of more than 4,000 above normal, the smoke or fine particulate matter in the air rose to over 4 mg/cu m during a 6-day period.

Table 2-7. Concentration of Suspended Particulate Matter in the Atmosphere of a Number of Communities

(Milligrams per cubic meter)

Community	Min	Max	Mean
Baltimore—Downtown area	0.21	1.72	0.87
Industrial area	0.05	1.46	0.38
Cincinnati—Industrial area	0.01	1.98	0.42
Residential area	0.01	1.30	0.28
Donora	0.00	2.50	0.74
Los Angeles, Detroit, San Francisco, Washington	0.25	0.47	0.51
Windsor—Areas of high pollution[a]	0.036	0.47	0.21
Areas of moderate pollution[a]	0.001	0.56	0.15
Areas of low pollution[a]	0.003	0.21	0.08

[a] October to December, 1951.

References: J. Cholak, The Nature of Atmospheric Pollution in a Number of Industrial Communities, *Proc. 2d Nat. Air Pollution Symposium*, p. 6, Stanford Research Institute, Los Angeles, Calif., 1952; and Morris Katz, Part II—Investigation of Environmental Contaminants by Continuous Observations and Area Sampling, *Am. Ind. Hyg. Assoc. Quart.*, **13,** No. 4, 211–225 (1952).

Chemical Composition. Aerosol contaminants in city air termed "suspended matter" are extremely complex in chemical composition. More than 20 metallic elements have been found by chemical and spectrographic analysis, in addition to carbon or soot and tarry organic material. The average composition is indicated in Table 2-8. The most abundant metallic elements are silicon, calcium, aluminum, and iron. Relatively high quantities of magnesium, lead, copper, zinc, and manganese may also be found. The concentrations of these elements will depend upon the nature of the principal industries and the effectiveness of measures taken to control particulate emissions (10). The distribution of lead in the air of cities has been correlated with density of vehicular traffic because lead compounds are widely used as antiknock substances in gasoline (10,24).

Small particles of suspended matter, less than about 5μ in size, are considered to be more of a nuisance and of greater importance to public health and welfare. Such contaminants discolor and soil clothing, painted surfaces, and buildings and seriously reduce visibility because of their light-scattering properties. Particles smaller than about 2 to 3μ are inhaled readily into the lungs, whereas those above 5μ are excluded for the most part by being trapped in the nasal passages. The effective control of aerosol emissions of fine particle size at the source represents a serious economic and technical problem in many industries. For example, the extreme toxicity of small amounts of beryllium dust has resulted in adverse health effects in several communities. Investigation has revealed that the toxicity of beryllium is several hundred times or more greater than that of metals such as lead, cadmium, or arsenic. This constitutes a major problem in control for the beryllium industry (60).

Table 2-8. Average Composition of Suspended Matter in the Atmosphere of a Number of Communities
(Micrograms per cubic meter)

Element	Baltimore	Cincinnati	Donora	Los Angeles	Windsor
Silicon	....	....	...	2.5–6.0	6.21
Calcium	....	16.0	....	2.5	3.26
Aluminum	4.0	4.4	6.0	2.5–6.0	2.75
Iron	15.1	12.5	14.0	2.5–6.0	2.49
Magnesium	....	7.0	....	1.1–2.5	0.81
Lead	1.0	2.7	4.5	0.8–6.0	0.614
Manganese	0.3	0.3	0.6	Trace–0.24	0.281
Copper	0.4	0.86	1.0	0.1–0.24	0.247
Zinc	....	2.0	40.0	....	0.197
Titanium	0.5	1.0	....	Trace–0.24	0.067
Tin	0.2	0.2	0.2	....	0.046
Molybdenum	....	....	....	....	0.034
Barium	....	....	....	0.1–0.24	0.027
Nickel	....	....	....	....	0.026
Vanadium	0.2	0.2	....	....	0.017
Chromium	....	....	....	....	0.011
Cadmium	0.2	0.22	2.0	....	0.008
Beryllium	0.0005	0.0095	....	....	0.006

References: J. Cholak, The Nature of Atmospheric Pollution in a Number of Industrial Communities, *Proc. 2d Nat. Air Pollution Symposium*, p. 6, Stanford Research Institute, Los Angeles, Calif., 1952; and Morris Katz, Sources of Pollution, *Proc. 2d Nat. Air Pollution Symposium*, pp. 95–105, Stanford Research Institute, Los Angeles, Calif., 1952.

Table 2-9. Composition of Suspended Matter in the Atmosphere of Cincinnati and Windsor
(Micrograms per cubic meter)

Element	Cincinnati			Windsor		
	Min	Max	Mean	Min	Max	Mean
Silicon	......	......	....	1.05	20.84	6.21
Calcium	0.4	65.0	16.0	0.85	13.88	3.26
Aluminum	0.1	25.0	4.4	0.71	6.94	2.75
Iron	0.1	110.0	12.5	0.59	8.77	2.49
Magnesium	5.0	10.0	7.0	0.20	2.64	0.81
Lead	0.1	11.5	2.7	0.125	1.805	0.614
Manganese	Trace	2.0	0.3	0.064	0.910	0.281
Copper	0.01	32.0	0.86	0.013	0.830	0.247
Zinc	0.1	14.0	2.0	0.00	0.89	0.197
Titanium	0.1	20.0	1.0	0.005	0.18	0.067
Tin	0.0	12.0	0.2	0.110	0.110	0.046
Molybdenum	......	......	....	Trace	0.087	0.034
Barium	......	......	....	0.00	0.18	0.027
Nickel	......	......	....	0.006	0.069	0.026
Vanadium	0.2	0.4	0.2	0.00	0.28	0.017
Chromium	......	......	....	0.00	0.042	0.011
Cadmium	0.1	0.42	0.22	0.001	0.046	0.008
Beryllium	0.0001	0.06[a]	0.0095[a]	0.00	0.073	0.006

[a] Dubious value.

References: J. Cholak, The Nature of Atmospheric Pollution in a Number of Industrial Communities, *Proc. 2d Nat. Air Pollution Symposium*, p. 6, Stanford Research Institute, Los Angeles, Calif., 1952; and Morris Katz, Part II—Investigation of Environmental Contaminants by Continuous Observations and Area Sampling, *Am. Ind. Hyg. Assoc. Quart.*, **13**, No. 4, 211–225 (1952).

2.9.3 Gases

Sulfur Dioxide. Sulfur is a component of nearly all fuels based on wood, coal, or petroleum. During combustion the major fraction of the sulfur is converted to sulfur dioxide, which escapes in the stack gas. Other large-scale sources of sulfur dioxide are from smelting and roasting of sulfide ores, refining of metals, combustion of hydrogen sulfide in natural-gas and oil-refinery operations, and various chemical processes. As sulfur is essential to the production of sulfuric acid, fertilizers, pulp and paper, explosives, paints and pigments, certain textiles, rubber, and many other products necessary to our economy, control of atmospheric wastes containing sulfur can be extremely profitable in some circumstances.

The current shortage of sulfur from natural salt-dome and other deposits, which in the United States have an estimated life of only about 20 to 25 years, has made the control and recovery of atmospheric wastes containing sulfur a matter of major importance. The sulfur losses to the atmosphere are greater than the total production of native sulfur, as shown in Table 2-10.

Table 2-10. World Native Sulfur Production and Estimated Sulfur Emitted to Atmosphere Annually
(Metric tons)

Year	Sulfur production			Sources of sulfur loss					
	U.S. crude exports	U.S. consumption	Canadian consumption	Zinc plants	Lead smelters	Copper smelters	Nickel smelters	Crude oil	Coal
1937	644,005	1,800,000	....	1,300,000	269,000	4,690,000	776,000	4,078,000	23,250,000
1938	....	....	....	1,252,000	272,000	4,076,000	782,000	3,976,000	22,000,000
1939	627,814	1,595,000	....	1,320,000	277,000	4,348,000	922,000	4,171,000	24,650,000
1940	....	....	....	1,297,000	283,000	4,961,000	1,008,000	4,284,000	26,980,000
1941	729,464	2,239,000	253,004	1,399,000	276,000	5,270,000	1,060,000	4,441,000	27,800,000
1942	....	....	....	1,440,000	273,000	5,581,000	1,192,000	4,186,000	28,250,000
1943	654,393	2,532,000	281,280	1,462,000	252,000	5,512,000	1,238,000	4,513,000	25,700,000
1944	....	....	....	1,300,000	218,000	5,158,000	1,199,000	5,182,000	
1945	918,691	2,961,000	266,013	1,019,000	200,000	4,400,000	999,000	5,189,000	
1946	1,189,072	2,907,000	....	1,123,000	187,000	3,670,000	749,000	5,500,000	
1947	1,299,060	3,540,000							

Sulfur dioxide is therefore one of the principal constituents of air pollution. Some idea of its distribution and frequency of occurrence is helpful to the control engineer. The gas, which is readily absorbed by vegetation and soil and water surfaces, accelerates corrosion of wire, metals, and many other materials. The concentrations which have been found in many industrial areas and in the air of cities are shown in Table 2-11. Concentrations greater than 0.5 ppm by volume become injurious to the leaves of susceptible green plants if exposure is prolonged for more than several hours under optimum relative humidity, light, temperature, and other growth conditions (29,67).

High average and maximum concentrations of sulfur dioxide have been reported in many of the larger cities of the United States and elsewhere. The annual consumption of coal in such cities amounts to many millions of tons. The sulfur content of the fuel varies from about 1 per cent for good-quality anthracite to over 4 per cent for some grades of bituminous (high-volatile) coal. Thus the annual consumption for Pittsburgh is about 20 million tons, Greater Detroit 15 million, St. Louis 5.5 million, and Birmingham 11 million.

The average concentration varies with the amount and sulfur content of the fuel used. In certain areas there is also a distinct seasonal variation, the average concentrations rising during the cold months of the heating season. Since the industrial consumption of fuel does not vary greatly throughout the year, this variation is due mainly to domestic use and demand for space heating. In recent years there has been a progressive decrease in the average atmospheric concentrations of several large cities because of the increased use of lower-sulfur coal and natural gas or oil for space heating. Air pollution control measures in St. Louis prohibit the use of high-sulfur

Table 2-11. Concentration of Gaseous Pollutants in the Atmosphere of a Number of Communities

Community	Oxides of nitrogen, ppm NO_2		Aldehydes, ppm formaldehyde		Chloride, ppm		Ammonia, ppm	
	Range	Mean	Range	Mean	Range	Mean	Range	Mean
Baltimore.........	0.05–0.79	0.26	0.025–0.12	0.067	0.001–0.11	0.037	0.005–0.053	0.018
Cincinnati........	0.00–0.55	0.23	0.00 –0.27	0.070	0.00 –0.19	0.033	0.00 –0.089	0.023
Donora...........	0.00–0.63	0.15	....	.....	0.00 –0.3	0.071		
Los Angeles.......	0.00–0.50	0.10	0.00 –1.0	0.18	....	.....	0.15	
Windsor..........	0.03–3.48	0.89	0.00 –0.85	0.20	0.00 –0.618	0.095	0.02 –3.070	0.216

Community	Sulfur dioxide, ppm		Fluoride, ppm HF		Oxidant	
	Range	Mean	Range	Mean	Range	Mean
Baltimore—Industrial........	0.01–0.46	0.074	0.00–0.08	0.018		
Rural.....................	0.00–0.11	0.023	0.00–0.021	0.008		
Cincinnati—Industrial........	0.00–0.46	0.064	0.00–0.025	0.005		
Residential................	0.00–0.27	0.044	0.00–0.025	0.006		
Donora.....................	0.00–0.50	0.15	0.00–0.03	0.006		
Los Angeles.................	Trace–0.60	0.05[a] 0.20[b]	0.00–0.025	0.008	0–0.56	0.07[a] 0.35[b]
Windsor–Industrial[c]..........	0.00–1.454	0.093				
Semirural.................	0.00–0.876	0.006				

[a] Mean for days of good visibility.
[b] Mean for days of reduced visibility.
[c] 1951–1952 values.

coal. The trend toward increased consumption of natural gas, instead of coal, for space heating and some industrial operations is likely to accelerate in the near future and will result in a considerable reduction in sulfur dioxide pollution and smoke.

Hydrogen Sulfide and Organic Sulfides. As a rule, these gases are not liberated in appreciable quantities by industrial operations in city communities. They cause odor nuisances when present in the air at concentrations 10 to 1,000 times smaller than the lowest concentration of sulfur dioxide detectable by smell. Mercaptans and hydrogen sulfide are evolved in the manufacture of coke, distillation of tar, petroleum and natural-gas refining, manufacture of viscose rayon, and in certain chemical processes. Such effluents are either burned to sulfur dioxide before liberation to the air or else are absorbed in "purifiers" containing iron oxide. In Great Britain the iron sulfide, or "spent oxide," from such purification measures is used for the production of sulfuric acid on a considerable scale (31). The recovery of hydrogen sulfide by the Girbotol and other processes is becoming standard practice in the petroleum industry of the United States and represents a source of elemental sulfur of increasing importance (31).

Hydrogen Fluoride and Chloride. Fluorides may be present in the atmosphere in gaseous or solid forms as effluents from the aluminum industry, from the manufacture of phosphate fertilizer, from brick plants, or from pottery and ferroenamel works. Small amounts are also liberated in the combustion of coal and from fluxing agents used in making foundry iron and in miscellaneous cupola and blast-furnace operations. Hydrogen fluoride and fluorine are important air contaminants in extremely low concentrations, 0.001 to 0.10 ppm by volume. The gas accumulates in the leaves of plants and can be absorbed in forage to such an extent as to cause a disease called fluorosis in animals. This disease causes mottling of teeth and affects the bone structure. Gladiolus, prune, apricot, and peach, among other sensitive plants and flowers, are extremely susceptible to hydrogen fluoride in concentrations as low as 0.02 to 0.05 ppm. This has created a problem of special concern to agriculture in the vicinity of large aluminum-refining operations. Fluorides, especially gaseous compounds, rapidly attack glass by etching the surface.

The concentrations which have been found in the air of a number of communities are shown in Table 2-11 from data presented by Cholak (10). The values are lower than those of virtually all other pollutants which may occur in the atmosphere from industrial activities. However, the high degree of toxicity of fluorine compounds renders the control of such emissions imperative for industries making aluminum and phosphate fertilizer.

Hydrochloric acid and chlorine caused widespread damage to vegetation and property, in Great Britain, in the early days of the alkali industry when the by-products of the Leblanc soda process were allowed to escape into the atmosphere (43). The first Alkali, etc., Works Regulation Act of Great Britain, passed in 1863, specified a 95 per cent recovery of such emissions, and an amendment to this Act in 1874 limited the discharge of hydrochloric acid to 0.2 grain per cubic foot of flue gas, before dilution with other stack gas or air. However, recovery of such by-products is relatively easy and profitable so that the escape of significant quantities seldom occurs, except under accidental conditions. The modern alkali industry is based on the electrolysis of common salt, and by-products are usually carefully controlled. Average concentrations of chlorides reported for a number of cities, calculated as chlorine in ppm, vary from about 0.016 to 0.095 (10,35).

Oxides of Nitrogen and Ammonia. It is probable that oxides of nitrogen are the second most abundant atmospheric contaminant in many communities, ranking next to sulfur dioxide. These oxides are important by-products of the chemical industry in the manufacture of nitric acid, sulfuric acid by the chamber process, nitration of organic compounds, manufacture of nylon intermediates, and in many high-temperature processes from the air and raw materials employed. Significant amounts are liberated in the exhaust gases of trucks and passenger automobiles. Magill and Hutchison (41) have estimated, from studies of various contaminants released to the atmosphere from the exhaust gases of gasoline-burning vehicles, that automobiles in the Los Angeles area consume a daily average of 12,000 tons of gasoline and emit about 230 tons of nitrogen oxides. In the combustion of natural gas by domestic gas appliances and furnaces, the products of combustion may contain 15 to 50 ppm of oxides of nitrogen. The most favorable conditions for the formation of these oxides in the combustion chamber of furnaces are high temperature followed by rapid cooling of the products of combustion. Industrial furnaces and internal-combustion engines operating on gaseous fuel may release up to 500 ppm or more of nitrogen oxides in the waste gases (24).

Oxides of nitrogen, as such, are more dangerous to public health than to vegetation. Both nitric oxide and peroxide are readily absorbed by the blood, after inhalation into the lungs, and combine with the hemoglobin to form an addition complex similar to that caused by the action of carbon monoxide. In this respect, these oxides are more injurious than carbon monoxide in equivalent concentrations. It is standard practice in the chemical industry to absorb and recover significant quantities of oxides of nitrogen.

Ammonia and ammonium salts are not important air contaminants. Ammonia is an important raw material in the fertilizer and organic-chemical industries and in the manufacture of nitric acid by the oxidation process. Its recovery is a matter of fundamental importance in the economical operation of such processes and in the manufacture of gas from coal.

Aldehydes. Lower aldehydes may be present in the atmosphere of some industrial communities in higher concentrations than sulfur dioxide. This is true for Los Angeles and Cincinnati, as shown in Table 2-11. Relatively high aldehyde levels have also been found in the industrial section of Windsor, Ontario, near the Detroit River. Formaldehyde is irritating to the eyes of some persons in concentrations as low as about 0.25 ppm by volume, although several ppm are required to affect the majority of persons exposed (38). In the Los Angeles district pronounced eye irritation is experienced by large numbers of persons whenever intense smog conditions prevail. However, the aldehyde concentrations alone are not high enough to account for this effect.

Cholak (10) has shown that aldehydes in the air vary not only with the nature of the activity of an area but also with the density of motor traffic. The incomplete oxidation of motor fuel and lubricating oils leads to the formation of aldehydes and organic acids. The products of combustion of natural gas may also contain aldehydes. It is probable that atmospheric contamination by aldehyde is due more to the exhaust and blow-by gases of motor vehicles than to the waste products of industrial and heating-plant sources.

Smog Gases. Although the term "smog" was coined to denote smoke and fog conditions in urban and industrial areas, the composition and effects of pollution by smog gases are best known from studies on the Los Angeles problem. Severe economic losses to vegetation, the occurrence of eye irritation among the residents, and rapid deterioration of rubber are attributed to the action of organic ozonides, peroxides, and acids. Organic acids, aldehydes, and saturated hydrocarbons in the concentrations found in the Los Angeles smog atmosphere caused no injury characteristic of the smog damage or silver-leaf symptoms to sensitive plants, such as endive, romaine lettuce, and spinach. However, a mixture of ozonides and other peroxidic compounds, acids and aldehydes formed by the interaction of ozone and olefins in low concentrations, did produce smog symptoms identical with those found on crops in the field. When plants were exposed to the oxidation products of unsaturated hydrocarbons reacting with nitrogen oxides in sunlight, similar damage was produced on the leaves. It was also found that, when ozone or oxides of nitrogen were combined with certain olefins in sunlight, the products were irritating to the eyes at the concentrations to be expected in the Los Angeles atmosphere. In experimental fumigations on crop plants, typical smog damage was produced by the oxidation products of ozone and straight-chain olefins of C_5 and C_6 atoms. Higher olefins of the C_7, C_8, and C_9 series were also effective in producing silver-leaf damage on sensitive plants when converted to the corresponding ozonides and other products but were not so active as 1-pentene and 1-hexene. The products of ozonization of the fraction of cracked gasoline boiling between 59 and 69°C (containing C_5 and C_6 olefins) also produced typical damage on vegetation.

Similar results were obtained when plants were treated with the vapors of cracked gasoline or 1-hexene, nitrogen dioxide, and sunlight or ultraviolet light. Nitrogen dioxide alone, with or without ultraviolet light, or olefins alone in the presence of sunlight, did not produce these effects (23,38,65).

The composition and concentration of the impurities in the Los Angeles atmosphere as determined by studies of the Los Angeles County Air Pollution Control District (38) are shown in Table 2-12. These analyses of the atmosphere, studies

Table 2-12. Concentrations of Pollutants in the Los Angeles Atmosphere
(Maximum values as measured over downtown Los Angeles on various days)

Concentrations, ppm by vol			Concentrations, mg/cu m		
Pollutant	Day of good visibility	Day of reduced visibility	Pollutant	Day of good visibility	Day of reduced visibility
Acrolein	....	Present	Aluminum	0.003	0.008
Lower aldehydes	0.07	0.4	Calcium	0.006	0.007
Carbon monoxide	3.5	23.	Carbon	0.035	0.132
Formaldehyde	0.04	0.09	Iron	0.003	0.010
Hydrocarbons	0.2	1.1	Lead	0.002	0.042
Oxidant	0.1	0.5	Ether-soluble aerosols	0.012	0.120
Oxides of nitrogen	0.08	0.4	Silicon	0.007	0.028
Ozone	0.06	0.3	Sulfuric acid	0.000	0.110
Sulfur dioxide	0.05	0.3			

Reference: Los Angeles County Air Pollution Control District, Los Angeles, Calif., First and Second Technical and Administrative Reports on Air Pollution Control in Los Angeles County, 1948–1949 and 1950–1951.

on the extent of emission of contaminants from all known sources, and the foregoing effects of smog lend considerable support to the theory that the harmful constituents are formed by reactions which occur in the atmosphere between certain impurities after they are released. This poses a difficult problem in pollution control.

2.9.4 Theory of Smog Formation

Several theories have been propounded to account for the formation of smog in the Los Angeles atmosphere by interaction of impurities with molecular oxygen, atomic oxygen, and ozone. The most comprehensive one is that of Haagen-Smit (23). It was found that during smogs there was a considerable increase in the concentrations of ozone and oxidant material. The concentrations reached as high as 30 to 40 ppm or more, much higher than most other areas in the world where tests have been made (5,23). The ozone is not found at night but only in daylight hours, beginning to form simultaneously throughout the Los Angeles basin in smoggy air shortly after dawn (40).

These facts indicate the photochemical formation of ozone or oxidant from impurities by the action of sunlight. The atmospheric studies on smoke, dusts, fumes, sulfur dioxide, oxides of nitrogen, and organic pollutants such as hydrocarbons and aldehydes have been used by Haagen-Smit to establish the fundamental principles underlying the theory of the formation of smog through photochemical reactions. Thus sulfur dioxide, nitrogen dioxide, and aldehydes may absorb ultraviolet radiation in the wavelengths present at ground level and react in their excited states with molecular oxygen to produce atomic oxygen. Although the amount of atomic oxygen produced photochemically by aldehydes and sulfur dioxide is limited by the concentrations of these impurities, because the reactions are irreversible, this is not so for nitrogen dioxide. In the latter case, the absorption of ultraviolet light leads to the rupture of a bond to form atomic oxygen and nitric oxide. Reaction of the products with molecular oxygen leads to the formation of ozone and the regeneration of the nitrogen dioxide. The nitrogen dioxide is therefore available for repetition of the process, unless converted to nitric acid or used up in organic substitution reactions. Even low concentrations of nitrogen dioxide could produce relatively large amounts of atomic oxygen to form ozone or to react with organic pollutants to yield compounds which could cause eye irritation, crop damage, and reduced visibility. The ozone formed during smog could account for the accelerated rubber cracking, while the oxidation of sulfur dioxide to the trioxide, with subsequent formation of sulfuric acid aerosol, as well as existing smoke, submicron dusts, and fumes would still further reduce the visibility.

2.10 DEPENDENCE OF CONTROL MEASURES ON NATURE OF CONTAMINANTS

It is obvious from the discussion in the foregoing article on the diversified nature of atmospheric contaminants that successful control measures in a particular community cannot be instituted without a comprehensive knowledge of the amounts of particular pollutants emitted from all major sources in relation to the concentrations found in the air. If, however, the only requirement is to control visible smoke and fly-ash emissions, there is little necessity for more than a simple approach involving application of the Ringelmann chart method and observations on dustfall rates to determine the efficacy of control measures involving fuel specifications, improved firing equipment, and dust collection at the source. Most of the earlier efforts at air pollution control were directed toward the reduction in emission of smoke and fly ash in the urban communities and sulfur dioxide and metallurgical fumes in smelter areas. With the increasing scope of our knowledge of the air pollution problem, and advances in engineering equipment and technology, it has become possible to control at the source a wide variety of dusts, fumes, mists, vapors, and gases from most process operations. The extent of such air pollution prevention measures is now limited only by such factors as financial considerations, public demand, public relations, and legal requirements.

Barkley of the U.S. Bureau of Mines (4) has recently reviewed the history of air pollution prevention in the United States. Since the turn of the century many cities and a number of states have passed laws aimed at smoke abatement. Within recent years such regulations have become more comprehensive and, in some instances, specify the emissions to be prohibited and standards of measurement for various dusts, sulfur dioxide, and other stack wastes (in addition to smoke). Among these more comprehensive regulations may be mentioned those of Los Angeles and Allegheny Counties. Model smoke-abatement ordinances have been proposed by committees representing the American Society of Mechanical Engineers, the Air Pollution Control Association, the American Society of Heating and Ventilating Engineers, and others. In Canada recently, a model ordinance was proposed by a committee representing the Engineering Institute of Canada (13). The ASME published in 1949 an information bulletin containing 15 example sections of a smoke-regulation ordinance (47). These sections have been used by more than 100 cities as the basis for regulations in existing and new smoke ordinances.

Although a number of states have enacted legislation concerning air pollution, such as Massachusetts, Rhode Island, Pennsylvania, New York, Ohio, Wisconsin, Iowa, Louisiana, Oregon, and California, there has been no comparable Federal legislation, to date, respecting the nation as a whole. State laws are usually enacted to give authority to cities or counties to control air pollution. The prevention of excessive air pollution is usually considered to be a local affair, but pollutants cannot be confined by arbitrary boundaries. Since impurities may be carried far afield from industrial and community sources, the problem of control can transcend local boundaries and become a matter for interstate, Federal, and even international action, as in the case of water pollution.

Progress in the control of air pollution can be assessed to some extent by consideration of a number of case histories of abatement efforts. In spite of all that has been done in the last 50 years of smoke abatement and air pollution prevention activities in cities and industrial areas, the basic problem has not been solved, and over-all progress is considerably less than that accomplished in the fields of water sanitation and regulations regarding food and drugs. Many efforts have been limited to investigation and measurement, while it has not been found possible to eliminate or reduce pollution drastically at the source because of political and economic factors, public apathy, the magnitude of the problem, lack of funds for research into methods of control, and the fact that recovery of industrial wastes in many cases merely adds to the cost of the manufactured products. The latter factor introduces a hardship for industries which are required to install control equipment in one area while competitors are free to discharge to the air similar wastes in another part of the country.

REFERENCES

1. Allcut, E. A.: The Smoke Problem, *Eng. J.* (*Can.*), **30,** 154–160, 164 (1947).
2. Alpern, Anne X.: Problems in Strict Enforcement of Air Pollution Control, in Louis C. McCabe, ed., "Air Pollution," p. 722, McGraw-Hill Book Company, Inc., New York, 1952.
3. "Air Pollution," Proceedings of the United States Technical Conference on Air Pollution, Louis C. McCabe, ed., McGraw-Hill Book Company, Inc., New York, 1952.
4. Barkley, J. F.: Review of Air Pollution Prevention in the United States, presented at the annual meeting of the American Society of Mechanical Engineers, New York, Nov. 26 to Dec. 1, 1950.
5. Bartel, A. W., and J. W. Temple: Ozone in Los Angeles and Surrounding Areas, *Ind. Eng. Chem.*, **44,** 857–861 (1952).
6. Beer, C. G. P., and L. B. Leopold: Meteorological Factors Influencing Air Pollution in the Los Angeles Area, *Trans. Am. Geophys. Union*, **28,** No. 2, 173–192 (1947).
7. Bosanquet, C. H., and J. L. Pearson: The Spread of Smoke and Gases from Chimneys, *Trans. Faraday Soc.*, **32,** 1249–1264 (1936).
8. Bosanquet, C. H., W. F. Carey, and E. M. Halton: Dust Deposition from Chimney Stacks—Appendix VII (Note G), The Institute of Mechanical Engineers, London, 1949.
9. Carter, J. H.: Atmospheric Pollution. Enforcement Principles and Standards, *Meteorol. Monographs*, **1,** No. 4, 1–3 (1951).
10. Cholak, J.: The Nature of Atmospheric Pollution in a Number of Industrial Communities, *Proc. 2d Nat. Air Pollution Symposium*, p. 6, Stanford Research Institute, Los Angeles, Calif., 1952.
11. City of Pittsburgh, Department of Public Health Report, 1951–1953.
12. Clayton, G. D.: Epidemiologic Approach, *Proc. 2d Nat. Air Pollution Symposium*, p. 110, Stanford Research Institute, Los Angeles, Calif., 1952.
13. Committee on Atmospheric Pollution in Canada, Report, *Eng. J.* (*Can.*), **34,** 448–452 (1951).
14. Davidson, W. F.: Studies of Stack Discharge under Varying Conditions, *Combustion*, **23,** 49–51 (1951).
15. Davidson, W. F.: A Study of Atmospheric Pollution, *Monthly Weather Rev.*, **70,** 225–234 (1942).
16. Dean, R. S., and R. E. Swain: Report Submitted to the Trail Smelter Arbitral Tribunal, *U.S. Bur. Mines, Bull.* 453, 1944.
17. Department of Scientific and Industrial Research (Brit.): Atmospheric Pollution in Leicester—A Scientific Survey, *Tech. Paper No.* 1, 1945.
18. German, R. H.: Problems of Compliance with Air Pollution Ordinances in Allegheny Co., Pa., "Air Pollution," Louis C. McCabe, ed., p. 695, McGraw-Hill Book Company, Inc., New York, 1952.
19. Gibson, W. B.: The Economics of Air Pollution, *Proc. 1st Nat. Air Pollution Symposium*, p. 109, Pasadena, Calif., 1949.

20. Gosline, C. A.: Dispersion from Short Stacks, *Chem. Eng. Prog.*, **48,** 165–172 (1952).
21. Gosline, C. A.: Meteorology Applied to Air Pollution Abatement, *Meteorol. Monographs*, **1,** No. 4, 4–5 (1951).
22. Great Britain: The Investigation of Atmospheric Pollution, *Dept. Sci. Ind. Research (Brit.)*, *26th Rept.*, 1949.
23. Haagen-Smit, A. J.: Chemistry and Physiology of Los Angeles Smog, *Ind. Eng. Chem.*, **44,** 1342 (1952).
24. Hall, E. L.: Products of Combustion of Gaseous Fuels, *Proc.* 2d *Nat. Air Pollution Symposium*, p. 84, Stanford Research Institute, Los Angeles, Calif., 1952.
25. Heimann, H., W. F. Reindollar, H. P. Brinton, and R. Sitgreaves: Health and Air Pollution, *Arch. Ind. Hyg. and Occupational Med.*, **3,** 399 (1951).
26. Hewson, E. W.: Atmospheric Pollution by Heavy Industry, *Ind. Eng. Chem.*, **36,** 195–201 (1944).
27. Holmes, J. A., E. C. Franklin, and R. A. Gould: Report of the Selby Smelter Commission, *U.S. Bureau of Mines, Bull.* 98, 1915.
28. Katz, Morris, and F. E. Lathe: Effect of Sulphur Dioxide on Vegetation, *Nat. Research Council Can.*, No. 815, 1939.
29. Katz, Morris: Sulfur Dioxide in the Atmosphere and Its Relation to Plant Life, *Ind. Eng. Chem.*, **41,** 2450–2465 (1949).
30. Katz, Morris: The Distribution and Dispersion of Contaminants in the Atmosphere, *Proc. 12th Intern. Cong. Pure and Appl. Chem.*, New York, 1951. *Defence Research Chem. Lab. Report No.* 67, Ottawa, Can., 1951.
31. Katz, Morris, and R. J. Cole: Recovery of Sulfur Compounds from Atmospheric Contaminants, *Ind. Eng. Chem.*, **42,** 2258–2269 (1950).
32. Katz, Morris: The Atmospheric Pollution Problem in Canada, *Chemistry in Can.*, **4,** No. 8, 25–32 (1952).
33. Katz, Morris: Sulfur from Industrial Wastes, *Chemistry in Can.*, **4,** No. 9, 27–34 (1952).
34. Katz, Morris: Sources of Pollution, *Proc. 2d Nat. Air Pollution Symposium*, pp. 95–105, Stanford Research Institute, Los Angeles, Calif., 1952.
35. Katz, Morris: Part II—Investigation of Environmental Contaminants by Continuous Observations and Area Sampling, *Am. Ind. Hyg. Assoc. Quart.*, **13,** No. 4, 211–225 (1952).
36. King, R. A.: Economic Utilization of Sulfur Dioxide from Metallurgical Gases, *Ind. Eng. Chem.*, **42,** 2241–2258 (1950).
37. Landsberg, H. E.: Climatology and Its Part in Pollution, *Meteorol. Monographs*, **1,** No. 4, 7–8 (1951).
38. Los Angeles County Air Pollution Control District, Los Angeles, Calif., First and Second Technical and Administrative Reports on Air Pollution Control in Los Angeles County, 1948–1949 and 1950–1951.
39. Mage, J., and G. Batta: Results of the Investigation into the Cause of the Deaths Which Occurred in the Meuse Valley during the Fogs of December, 1930, *Chimie & industrie*, **27,** 961–975 (1932).
40. Magill, P. L., and R. W. Benoliel: Air Pollution in Los Angeles County, *Ind. Eng. Chem.*, **44,** 1347–1355 (1952).
41. Magill, P. L., D. H. Hutchison, and J. M. Stormes: *Proc. 2d Nat. Air Pollution Symposium*, pp. 71–83, Stanford Research Institute, Los Angeles, Calif., 1952.
42. McCabe, L. C., P. P. Mader, H. E. McMahon, W. J. Hamming, and A. L. Chaney: Industrial Dusts and Fumes in the Los Angeles Area, *Ind. Eng. Chem.*, **41,** 2486–2493 (1949).
43. Meetham, A. R.: "Atmospheric Pollution," Pergamon Press Ltd., London, 1952.
44. Miller, C. E.: Pointers on Selecting Equipment for Industrial Gas Cleaning, *Chem. Met. Eng.*, **45,** 132 (1938).
45. Mills, C. A.: Urban Air Pollution and Respiratory Diseases, *Am. J. Hyg.*, **37,** 131–141 (1943).
46. Mills, C. A.: Relationship of Smoke to Pneumonia and Other Infections of Respiratory System, *Cincinnati J. Med.*, **27,** 624–633 (1946).
47. Model Smoke Law Committee: Example Sections for a Smoke Regulation Ordinance, Fuels Division, American Society of Mechanical Engineers, May, 1949.
48. National Smoke Abatement Society: The London Fog, *Smokeless Air*, No. 85, pp. 100–111, spring, 1953, Chandos House, Buckingham Gate, London, S.W.
49. Peart, A. F. W., and G. H. Josie: Planning a Study to Determine the Effects of Air Pollution on Health, *Arch. Ind. Hyg. and Occupational Med.*, **7,** 326–338 (1953).
50. Russel, W. T.: The Influence of Fog on Mortality from Respiratory Diseases, *Lancet*, pp. 335–339, Aug. 16, 1924.
51. Russel, W. T.: The Relative Influence of Fog and Low Temperature on the Mortality from Respiratory Disease, *Lancet*, pp. 1128–1130, Nov. 27, 1926.

52. Sawyer, F. G.: Air Pollution, *Ind. Eng. Chem.*, **43**, 2687–2693 (1951).
53. Schrenk, H. H., H. Heimann, G. D. Clayton, W. M. Gafafer, and H. Wexler: Air Pollution in Donora, Pa., *U.S. Public Health Service, Public Health Bull.* 306, 1949.
54. Schueneman, Jean J.: Atmospheric Concentrations of Sulfur Dioxide in St. Louis—1950, St. Louis, Mo., Division of Health, Industrial Hygiene Section, Report, 1950.
55. Setterstrom, Carl, P. W. Zimmerman, and W. Crocker: Effect of Low Concentrations of SO_2 on Yield of Alfalfa and Cruciferae, *Contribs. Boyce Thompson Inst.*, **9**, 179–198 (1938).
56. Shaw, N., and J. S. Owens: "The Smoke Problem of Great Cities," Constable Co., Ltd., London, 1925.
57. Silverman, L.: What Process Wastes Cause Air Pollution? *Chem. Eng.*, **58**, No. 5, 132–135 (1951).
58. Smith, M. E.: The Forecasting of Micrometeorological Variables, *Meteorol. Monographs*, **1**, No. 4, 50–55 (1951).
59. Stanford Research Institute: The Smog Problem in Los Angeles County, First Interim Report, Western Oil and Gas Association, Los Angeles, Calif., 1948; Second Interim Report, 1949; Third Interim Report, 1950.
60. Stokinger, H. E., and S. Laskin: Air Pollution and the Particle Size Toxicity Problem—I and II, *Nucleonics*, **5**, No. 6, 50 (1949); **6**, No. 3, 15 (1950).
61. Sutton, O. G.: The Dispersion of Hot Gases in the Atmosphere, *J. Meteorol.*, **7**, No. 5, 307–312 (1950).
62. Sutton, O. G.: "Atmospheric Turbulence," Methuen & Co., Ltd., London, 1949.
63. Swain, R. E.: Smoke and Fume Investigation, *Ind. Eng. Chem.*, **41**, 2384–2388 (1949).
64. Thomas, M. D., G. R. Hill, and J. N. Abersold: Dispersion of Gases from Tall Stacks, *Ind. Eng. Chem*, **41**, 2409–2417 (1949).
65. Thomas, M. D.: Gas Damage to Plants, *Ann. Rev. Plant Physiol.*, **2**, 293–322 (1951).
66. Thomas, M. D.: The Present Status of the Development of Instrumentation for the Study of Air Pollution, *Proc. 2d Nat. Air Pollution Symposium*, p. 16, Stanford Research Institute, Los Angeles, Calif., 1952.
67. Thomas, M. D., and G. R. Hill: Relation of SO_2 in the Atmosphere to Photosynthesis and Respiration of Alfalfa, *Plant Physiol.*, **12**, 309–383 (1937).
68. Wilkins, E. T.: Air Pollution in a London Smog, *Mech. Eng.*, **5**, 426–428 (1954).

SECTION 3

CHEMISTRY OF CONTAMINATED ATMOSPHERES

BY R. D. CADLE AND P. L. MAGILL

3.1 INTRODUCTION

The solution of many air pollution problems requires considerable knowledge of the chemistry both of the natural atmosphere and of contaminated atmospheres. During recent years there has been growing awareness of this fact. For example, the physical chemistry of nucleation is of considerable significance to problems resulting from the decrease in visibility produced by contaminants. London fog and the "smogs" which are combinations of smoke and fog probably, at least in part, result from contaminants which serve as nuclei for fog formation. Air fields in Alaska are plagued with "ice fogs" which form at temperatures below −40°F. These ice fogs may likewise result from water vapor and nuclei poured into the atmosphere by nearby heating plants. Los Angeles smog is an example of a contaminated atmosphere in which chemical reactions undoubtedly occur on a large scale (6). These reactions probably contribute to the unpleasant properties of the smog (7). The reactions apparently involve both the naturally occurring and the man-made constituents of the atmosphere.

The purpose of this section is to discuss the chemistry of contaminated atmospheres and also the chemistry of the natural atmosphere as it is related to air pollution problems. Thus emphasis is placed on the chemistry of the troposphere, the layer of air in which man lives. The troposphere extends from the surface of the earth to the lower edge of the stratosphere at about 8 miles above the earth's surface. The stratosphere extends from about 8 to about 60 miles above the surface of the earth. Above the stratosphere lies the ionosphere, which is characterized by the presence of large concentrations of ions and activated molecules produced by the action of ultraviolet radiation on the air molecules. The chemistry of the complex reactions which occur in the ionosphere is beyond the scope of this section.

3.2 THE EARTH'S NATURAL ATMOSPHERE

A polluted atmosphere is generally considered to be an unnatural atmosphere. While it is true that there are numerous natural materials in the earth's atmosphere that are at times quite objectionable, such as pollens and dust, such materials are not usually referred to as air pollutants. To understand what man has added to the atmosphere, it is necessary to know the composition of the atmosphere provided by nature, and something of its history.

The average composition of the earth's atmosphere near the surface is shown in Table 3-1. In addition to the constituents which occur in more or less constant concentration in the atmosphere, there are a number of decidedly variable constituents of natural origin. Some of these are listed in Table 3-2.

Table 3-1. Average Composition of the Atmosphere

Gas	Composition by volume, ppm	Composition by weight, ppm	Total mass, $\times 10^{20}$g
N_2	780,900	755,100	38.648
O_2	209,500	231,500	11.841
A	9,300	12,800	0.6555
CO_2	300	460	0.0233
Ne	18	12.5	0.000636
He	5.2	0.72	0.000037
CH_4	2.2	1.2	0.000062
Kr	1	2.9	0.000146
N_2O	1	1.5	0.000077
H_2	0.5	0.03	0.000002
Xe	0.08	0.36	0.000018

Table 3-2. Variable Constituents of the Earth's Atmosphere of Natural Origin

Water
Meteoric dust
Sodium chloride
Airborne soil
Nitrogen dioxide (electrical discharge)
Sulfur dioxide, Hydrogen chloride, Hydrogen fluoride } Volcanic origin
Hydrogen sulfide (sulfide bacteria, volcanic origin)
Ozone (formed photochemically and by electrical discharge)
Pollen
Bacteria
Spores
Condensation nuclei

The atmosphere weighs approximately 4.5×10^{15} metric tons. It presses down upon the earth with a pressure of 2,016 lb/sq ft. At the surface of the earth it has a density of 0.0013 g/cu m, and it decreases in density as the distance from the earth is increased. The atmosphere gradually thins out into space. Above about 600 km the atoms and molecules describe free elliptical orbits in the earth's gravitational field (42).

The proportions of the different components of the atmosphere remain essentially constant from the earth's surface to a height of about 10 miles. Above this height gravitational separation probably begins, although it does not become important until above about 80 miles.

The earth's atmosphere has not always had its present composition, although the composition has remained relatively unchanged throughout most of geological time (42). Reconstruction of the early history of the atmosphere depends upon the theory for the formation of the earth which one accepts. It is generally agreed that the predecessors of our present atmosphere consisted mainly of carbon dioxide, nitrogen, and water. It may have been similar in composition to the gases emitted by present-day volcanoes, which do not contain any oxygen. In some manner, life on earth was started and the chlorophyll-containing plants converted some of the carbon dioxide and water into oxygen. Whether these phenomena occurred simultaneously or consecutively is probably immaterial to us at present. What is important is that changes have occurred and changes are still occurring, and the probable nature of these changes helps us understand the chemistry of our atmosphere.

The hydrosphere and biosphere must be considered in any discussion of the atmosphere. There is a constant shift of substances between these three spheres. The gases as we know them exist in the atmosphere; the hydrosphere contains all the waters of the oceans, the lakes, and the streams; and the biosphere comprises all living matter. Of the three, the biosphere has the smallest weight. They may be arranged in the following order of relative weights (54):

Biosphere	1
Atmosphere	300
Hydrosphere	69,100

There is also a constant exchange between the earth's crust and the atmosphere.

The common constituent gases of the atmosphere have been rather intensively studied. Much is known about the reactions in which such gases have taken part since the earth's formation, approximately 3 billion years ago.

3.2.1 Oxygen

The atmosphere is constantly acquiring oxygen from chlorophyll-bearing plants, which take up carbon dioxide and evolve oxygen. Some of the oxygen in the atmosphere dissolves in the hydrosphere. At the same time, oxygen is being removed from the atmosphere by the oxidation of the various reducing materials in the earth's crust such as sulfur, iron oxides, and manganese salts. Goldschmidt (22) has calculated the amount of oxygen lost from the atmosphere by the oxidation of FeO to Fe_2O_3, MnO to MnO_2, and S to SO_3. He based his calculations on the analysis of the ratio of ferric to ferrous iron in various types of rocks (Table 3-3) and on the weight of igneous rocks which had been eroded from the earth during geologic time. Depending upon certain assumptions made in the calculation, the amount of "fossil" oxygen is between about 0.56 and 0.26 kg/sq cm of the earth's surface. The amount of oxygen in the hydrosphere is equivalent to 0.002 kg/sq cm, and in the atmosphere to 0.230 kg/sq cm. Thus the total free and fossil oxygen is between 0.79 kg/sq cm and 0.49 kg/sq cm. In other units, between 2.5×10^{15} and 4.0×10^{15} metric tons of oxygen has been produced by plants in the 3 billion years of the earth's existence, a

Table 3-3. Ratio of Fe_2O_3 to FeO in Various Rocks

	Igneous rocks	Shales	Terrigenous mud	Sandstones
% Fe_2O_3	3.08	4.03	5.07	1.08
% FeO	3.80	2.46	2.30	0.30
Molecular quotient, $2Fe_2O_3/FeO$	0.73	1.47	1.98	3.24

large part of this at present being combined oxygen in sedimentary rocks. Goldschmidt also analyzed the oxygen cycle throughout geological time and found that the amount of oxygen which could have been produced by chlorophyll-bearing plants was consistent with these figures.

It should be pointed out that there are at least two other hypotheses to explain the occurrence of oxygen in the atmosphere. One is that free oxygen was formed by thermal decomposition of water vapor while the earth and its atmosphere were very hot. However, it is difficult to get around the objection that the hydrogen and oxygen would later recombine on contact with the still hot surface of the earth. Another hypothesis is that photochemical decomposition of water vapor occurred, at a later and cooler stage, in the upper atmosphere. However, it is unlikely that there was ever sufficient water vapor at high altitudes to account for all the oxygen which was produced.

3.2.2 Carbon Dioxide

All living matter on the earth is involved in the reactions of the carbon dioxide in the atmosphere. Carbon dioxide is ejected into the atmosphere by volcanoes, by decay of organic matter, and by the combustion of fuels. It is also in equilibrium with the dissolved carbon dioxide in the waters of the earth. In addition, much is stored in the limestones and dolomites; carbon which was once atmospheric carbon dioxide is stored in coal and petroleum. Goldschmidt has estimated the following values for the weight of carbon dioxide per square centimeter of the earth's surface (22):

Limestone and dolomite	6,560 g/sq cm
Coal, bitumen, humus, and the biosphere	0.67–3.1 g/sq cm
Hydrosphere	20 g/sq cm
Atmosphere	0.4 g/sq cm

The annual production of carbon dioxide by respiration and decay is approximately 0.040 g/sq cm/year. Over the entire earth's surface of 5.1×10^{18} cm there is thus produced 2×10^{11} metric tons. The entire atmosphere contains approximately 2×10^{12} metric tons. In other words, decay and respiration are adding carbon dioxide to the atmosphere at the rate of approximately 10 per cent per year.

The combustion of fuels also is adding to the carbon dioxide, but in a relatively insignificant manner. In 1951 the entire world had an energy consumption equivalent to 2.5×10^9 metric tons of coal (65). Approximately 20 per cent of this energy was hydroelectric power in Europe and somewhat less in the United States. If it is assumed that 15 per cent of the total world's energy is hydroelectric power and that coal contains 90 per cent carbon, then the combustion of coal to carbon dioxide would add only 7×10^9 metric tons of carbon dioxide per year or one-thirtieth the amount produced by decay and respiration.

Geological evidence suggests that relatively little change has occurred in the concentration of carbon dioxide from Archeozoic times to the present. However, there must have been ceaseless minor fluctuations. Such data as exist show that the carbon dioxide content of the atmosphere has increased about 30 ppm in the last 50 years. The fact that it has not increased more is probably because of its consumption by plants and because the waters of the earth act as a huge reservoir that contains 50 times as much dissolved as is mixed in our gaseous atmosphere.

3.2.3 Nitrogen

The amount of nitrogen that undergoes change is small compared with that of oxygen or carbon dioxide. Nitrogen is removed from air by both organic and inorganic processes. Organic processes involve nitrogen-fixing microorganisms and some blue-green algae (42). The inorganic processes produce oxides of nitrogen by electrical discharges and photochemical reactions in the upper atmosphere. The amount of organically fixed nitrogen far exceeds that fixed by inorganic processes. Hutchinson (27) has estimated biological fixation at 0.07 mg of nitrogen per square centimeter of the earth's surface per year and nonbiological fixation at not more than 0.0035 mg/sq cm/year. Much of this nitrogen is eventually returned to the atmosphere by the decay of organic matter. Adel (1), who has suggested the presence of nitrous oxide in the atmosphere, concludes that it is supplied to the atmosphere as part of the earth's nitrogen cycle.

The presence of nitrous oxide has been confirmed by Migeotte (43), using infrared spectroscopy, and by Slobod and Krogh (61), who condensed it from the atmosphere and determined its concentration with a mass spectrometer. Hutchinson has estimated that the average content of the combined nitrogen in sediments is 510 g/ton. Multiplying this by Goldschmidt's figure of 170 kg/sq cm for the total amount of sediments formed during geologic time gives 86.7 g/sq cm for fossil nitrogen.

Nitrogen is also a constituent of all igneous rocks, which contain an average of 0.04 cu cm of nitrogen per gram or approximately 0.005 per cent by weight. It is evidently present in combined form, since it may be released from the rocks as ammonia by means of fusion with soda ash. Ammonia is also a constituent of volcanic gases.

3.2.4 Methane

This gas was found by spectrographic identification of its absorption spectrum in sunlight. The spectrum is modified by its passage through the earth's atmosphere. Its concentration seems to be uniform over the earth's surface. It is presumed to have a constant source which may be either decay of biological products or escape from the earth's deposits. The relative extent of these two processes was determined by measuring the concentration of radioactive C^{14} in the methane of air. Methane from biological sources contains 0.095×10^{-12} g of C^{14} per gram of carbon, while mineral methane is inactive. Apparently the methane of the atmosphere is mainly of biologic origin.

3.2.5 Ozone

Large gaps exist in our knowledge of atmospheric ozone near the ground. Much has been written concerning its formation and existence in the stratosphere. Within the stratosphere it may exist at concentrations between 600 and 800 pphm by volume. The concentrations at ground level are highly variable both at one locality and from place to place. Little is known concerning the effect of weather conditions on its concentration, particularly with reference to its vertical distribution. In most areas of the world where it has been measured, its concentration at ground level ranges between 0 and 5 pphm. However, there is considerable evidence that the concentration near the ground is often considerably greater. Wilson et al. have reported concentrations of ozone greater than 20 pphm in Alaska (69). Bartel and Temple (3), using chemical analytical techniques, studied the concentration of ozone in southern California. They found concentrations of ozone as high as 20 to 50 pphm in the Los Angeles area, on Mt. Wilson, and on the desert. These results were confirmed by members of the staff of the Stanford Research Institute who also found unusually high concentrations of ozone on Catalina Island and in northern California at Palo Alto (62). Further confirmation was obtained by the Stanford Research Institute by adsorption of gases from the Los Angeles atmosphere on silica gel at liquid oxygen temperature (−183°C), followed by spectroscopic examination of the gases desorbed when the gel was allowed to warm to −80°C (62). How much of this ozone in the

California atmosphere is of natural origin and how much is produced by photochemical reactions involving atmospheric impurities is not known at present.

3.3 AEROSOLS

In our everyday life, visual perception plays an important role and dominates the other senses with which we are equipped. It is therefore not surprising that the average person associates the term atmospheric pollution with clouds of dust and palls of smoke. The word "particulate" is frequently used to describe the particles in aerosols. However, gas is also particulate; so when we speak of aerosols we mean at least groups of molecules. In this article, aerosols are set apart from the gases such as nitrogen, oxygen, argon, carbon dioxide, and ozone as being present in larger groups than individual molecules of gases. Dust particles are larger than fume or smoke particles, and although they may be suspended in the atmosphere, they settle according to Stokes' law. The smaller fume or smoke particles, on the other hand, may be either liquid or solid and exhibit Brownian motion.

3.3.1 General Properties of Aerosols

When a liquid or solid is dispersed to form a cloud or a smoke, many of its properties are greatly altered. For example, the surface per unit weight (specific surface) is greatly increased; the nature of the surface may greatly influence the behavior of the material. Typical diameters of a number of aerosols are shown in Table 3-4. It is assumed that the particles are spherical in shape and that their specific gravities are not changed by dispersion. These assumptions are not strictly correct, nor are the typical diameters of the particles truly representative in all cases. Nonetheless, the figures give a general indication of the size of such particles.

Table 3-4. Typical Diameters of Aerosol Particles

Aerosol	*Diameter, μ*
Tobacco smoke	0.25
NH_4Cl smoke	0.1
H_2SO_4 mist	0.8–5.5
ZnO smoke	0.05
Coal-mine air	10
Flour-mill air	15–20
Cement mill (kiln exhaust)	10
Grain-elevator air	15
Fog	50
Talc dust	10
Pigments	1–4

As a direct result of the increase in surface area, the properties that depend upon the activity of the surface molecules are greatly enhanced. Substances that may be ignited only with difficulty in air in their massive state will explode if ignited in the form of dust; examples are sugar, coal, starch, iron, and lead. Adsorption plays a much more important part in disperse systems than in other heterogeneous systems which are not so highly dispersed. Since adsorption is a specific phenomenon depending upon the nature of the adsorbing and adsorbed substances, the particles of the dispersed phase may adsorb molecules of the dispersion medium or of a particular constituent of the dispersion medium.

The presence of suspended solids or liquids in a gas renders it more sensitive to thermal radiation than a pure gas. Cloud or smoke particles suspended in a gas diminish the thermal transparency and absorb heat radiation. Particles become warmed and communicate their heat by conduction to the gas immediately surrounding them, which gas may be quite transparent to the radiant heat.

On the other hand, if the radiation is intercepted before it reaches the suspended particles, the particles may cool quickly by radiating to a colder surface and assume a lower temperature than that of the gas. If the gas contains sufficient water vapor to saturate it at the temperature of these cooling particles, some of it will condense upon them. In this way ground fogs and mists are sometimes formed (19).

If particles suspended in a gas are smaller in diameter than the mean free path of the molecules, which is about 0.1μ at atmospheric pressure and room temperature, they will be driven to and fro incessantly by the irregular impacts of molecules of the air. If a particle is substantially larger than 0.1μ and is suspended in a thermally and electrically uniform gas, the molecular bombardment will be uniform in all directions and will produce no displacement of the particle. If, however, a thermal gradient is set up in the gas, for example, by introducing a hot object into it, the gas molecules in the neighborhood of the hot object become warm and move more energetically than those more remote. Any such particle, therefore, suspended in the gas in the immediate vicinity of the object will be repelled from it. If the particles are small enough to exhibit Brownian motion, they will pursue a zigzag path away from the hot object. The object thus becomes surrounded by a region of warm, dust-free air.

A portion of the particles in practically all aerosols is charged. The particles may become charged in a variety of ways, depending largely on the manner of dispersion. For example, when a liquid is dispersed in a gas, the increase in specific surface is generally associated with the electrification of the droplets. When water vapor, as steam, expands suddenly into the air through a narrow orifice, the dense cloud of condensate is found to be electrically charged. Particles may acquire their charges by direct association with ions produced by ionization of the dispersion medium. Ionization can be produced by an electric discharge, such as the corona discharge of the electrostatic precipitator. It can also be produced by nuclear radiation.

According to Whytlaw-Gray and Patterson (68), the electrical character of a smoke depends very largely on the method of its formation. Smokes formed by electric arc and violent chemical reactions are highly electrified, and their individual particles appear to carry a considerable number of unit charges. On the other hand, smokes volatilized at lower temperatures usually contain very few charged particles initially, but these particles rapidly acquire charges as the system ages. Whytlaw-Gray and Patterson were not able to show that the electrification ordinarily acquired by smokes influences the rate of coagulation. There is even considerable doubt that charging all the particles to one sign would greatly affect the rate of coagulation.

The rate at which coagulation occurs is often an important property of aerosols. Apparently, every collision between particles results in coagulation. Equations have been developed by Smoluchowski and by Whytlaw-Gray for calculating the number of collisions in unit time which occur in a monodisperse aerosol.* If the effect of settling is neglected, the particle concentration in such an aerosol varies inversely with time according to the equation

$$\frac{1}{n} - \frac{1}{n_0} = Ht$$

where n_0 = initial particle concentration
H = coagulation constant

The differential equation of this process is

$$-\frac{dn}{dt} = Hn^2$$

According to the theory of Smoluchowski, H is equal to $4kT/3\eta$, where k is Boltzman's constant, η is the viscosity of the dispersing medium, and T is the absolute

* That is, an aerosol in which all the particles are the same size.

temperature. H is equal to 3.0×10^{-10} cu cm/sec in air at $T = 293°K$. Thus the rate of coagulation is independent of particle size.

This equation is applicable only to particles that are large compared to the mean free path l. For smaller particles, the Cunningham correction, which takes into account "slippage" of the particles between the molecules, must be applied. The equation then becomes

$$-\frac{dn}{dt} = H\left(1 + \frac{0.9l}{r}\right)n^2$$

While these equations were developed for monodisperse aerosols, experimental work by Whytlaw-Gray, and more recently by Gillespie and Langstroth (20), has shown that they apply fairly well to a number of aerosols which have a wide range of particle size. However, there is considerable variation in the experimental value of H (Table 3-5).

Table 3-5. Coagulation Constants (H) for a Number of Aerosols

Dispersed Material	$H \times 10^8$, cu cm/min
Theoretical value	1.8
Ammonium chloride	3.3–4.7
Magnesium oxide	8.5–10.9
Copal resin	9.3
Carbon	14.1
Zinc oxide	18.9–19.6
Silica powder	26.0–28.0

Data from T. Gillespie and G. O. Langstroth, Coagulation and Deposition in Still Aerosols of Various Solids, *Can. J. Chem.*, **30,** 1003–1011 (1952).

The rate of coagulation at ordinary concentrations is quite low. For example, the concentration of particles above 0.5μ in diameter in Los Angeles smog is about 2,000 particles per cubic centimeter of air. If we rewrite the equation for the rate of coagulation in terms of the per cent coagulation per hour, we obtain

$$-100\frac{dn}{n} = 1.08 \times 10^{-4}n$$

Thus the per cent coagulation per hour in such a contaminated atmosphere would be $1.08 \times 10^{-4} \times 2{,}000$, or about 0.2 per cent. If we use the coagulation constant for carbon (Table 3-5) instead of the theoretical constant, the calculated coagulation rate is still only 1.6 per cent per hour.

3.3.2 Sources of Dusts

The atmosphere would contain dust even in the absence of human activity. Meteoric dust from outer space arrives at the earth at the rate of about 1,000 tons/year. However, this amount is exceedingly small compared to other sources. Volcanic eruptions and dust storms can contribute large amounts of dust to the earth's atmosphere. It has been estimated that the volume of fine ash thrown into the atmosphere by a strong volcanic eruption amounts to as much as 100 billion cu yd (33). A single dustfall in 1901 (26) deposited 2 million tons of dust on the African desert and the continent of Europe. In February, 1903, 10 million tons of red dust from northwest Africa was deposited over England (58). Dust from the American dust bowl was clearly visible in the city of Washington, D.C.

The dust in the atmosphere has a wide range of chemical composition. The composition of dust from natural sources is as varied as are the various portions of the earth's crust.

Human activities, such as the manufacture of various products, introduce organic and inorganic particles into the atmosphere. These products include steel, rubber, lime, and a tremendous variety of other items. In fact, almost all operations involving the burning of coal introduce some dust (that is, soot and fly ash) into the atmosphere. Some of the sources are not so obvious. Large quantities of rubber (50 tons/day in Los Angeles) are worn from tires on city streets. Much of this is possibly injected into the atmosphere in dust form.

3.3.3 Condensation Nuclei

Vapors in general must be greatly supersaturated before condensation will occur in the absence of foreign particles. A supersaturation of about 4.2 times is required for water vapor (35). However, if the vapor contains particles which can act as nuclei for the condensation, little supersaturation is required. Condensation nuclei are important because of the way in which they affect our everyday life. They are largely responsible for fogs and rain. Without them, condensation would occur mainly on the walls of buildings and other exposed surfaces of the earth.

Natural sources appear to be the main producers of nuclei. Such sources include volcanic eruptions, ocean spray, and natural combustion (such as some forest fires). Man-made sources include many industrial operations, modern transportation facilities (locomotives, boats, automobiles, and airplanes), and domestic combustion processes. Enormous quantities of nuclei are often produced by single sources. An average grass fire extending over 1 acre produces some 20,000 billion-billion (2×10^{22}) nuclei. If these nuclei were distributed through a column of air having a cross section of 1 acre and a height of 10,000 ft, there would still be a concentration of about 2 billion particles per cubic centimeter (48). Coste and Wright (12) showed that a flame of commercial coal gas lit for only 15 sec increased the number of nuclei in a chamber from 109,000 to 860,000 per cubic centimeter. Amelung and Landsberg (2) found that the concentration of nuclei in a ventilated kitchen containing a large operating gas range exceeded 500,000 nuclei per cubic centimeter while the outside air contained only about 25,000 nuclei per cubic centimeter.

Most nuclei range in size from about 0.001 to 0.1μ. Their weight has been estimated at 10^{-14} to 10^{-18} g or an equivalent aggregate of about 10^6 molecules (48). It is of interest that when water vapor is sufficiently supersaturated to be self-nucleating, that is, when condensation occurs in the absence of foreign nuclei, the nuclei consist of aggregates containing only about 80 water molecules (35).

The number of nuclei in the atmosphere usually exceeds the number of dust particles by a factor of several thousand (13). This has been attributed to the fact that the dust particles, being larger, are subject to considerable sedimentation, while the nuclei remain suspended because of convection currents and Brownian movement (48). However, it is probably also true that much larger numbers of nuclei than of dust particles are produced by many dispersion processes.

The existence and concentration of nuclei can be determined by drawing a sample of air into a chamber, saturating the air with water vapor, and adiabatically expanding the air. The expansion produces cooling and supersaturation of the air. Condensation of water occurs on any nuclei which may be present. The concentration of the resulting droplets and thus of the nuclei may readily be determined. If the air is freed from condensation nuclei, for example, by filtration through a cotton plug, no condensation occurs until the expansion almost reaches 1.25, corresponding to a supersaturation of about 4.2 times. The well-known Aitken nuclei counter and the C.T.R. Wilson cloud chamber are devices which can be used for making nuclei counts.

The vapor pressure of a liquid is affected by the curvature of the surface of the liquid. At a given temperature, the vapor pressure at a convex liquid surface of sufficiently small radius of curvature is greater, while that at a concave surface is less, than that at a plane surface.

The vapor pressures p above a plane surface and p' above a curved surface, whose radius of curvature is r, are related to each other and to r by the expression

$$\log \frac{p'}{p} = \frac{2\sigma}{r\rho RT}$$

where σ = surface tension of liquid
ρ = density of liquid
R = gas constant
T = absolute temperature

This equation applies in general to the vapor pressure over any convex surface. When the surface is concave, the equation is identical except that the second term is preceded by a negative sign. For most liquids, including water, the effect of curvature is unappreciable until the radius of the drop is 0.01μ or less. Droplets of water much smaller than this evaporate rapidly, even in a saturated or slightly supersaturated atmosphere. Conversely, such droplets could not act as condensation nuclei unless the atmosphere were sufficiently supersaturated that the partial pressure of the water vapor in the air exceeded the vapor pressure of the droplets. As was mentioned previously, self-nucleation of water vapor in air occurs when the supersaturation exceeds about 4.2 times. Aggregates of water molecules containing about 80 molecules are formed by a series of bimolecular collisions (35). These have a sufficiently low vapor pressure that they can serve as condensation nuclei at that degree of supersaturation.

For charged liquid surfaces

$$\log \frac{p'}{p} = \frac{1}{RT\rho}\left(\frac{2\sigma}{r} - \frac{e^2}{8\pi r^4}\right)$$

where e = quantity of charge per square centimeter of surface
p = vapor pressure above the plane, uncharged surface

Thus the presence of a charge tends to decrease the vapor pressure.

Foreign nuclei, that is, nuclei other than aggregates of water molecules, act by lowering the vapor pressure of the water which becomes associated with them. Such nuclei may be:

1. Inert particles possessing plane surfaces, the vapor pressure at such a surface being equal to that of the bulk liquid and, therefore, lower than the pressure of supersaturated vapor.

2. Inert particles possessing porous surfaces. Once a film of vapor molecules has formed by adsorption, vapor will condense upon the concave surface, and as the radius of curvature of the pores is further diminished it will condense readily upon the surfaces, even from vapor at partial pressures far below its saturation point.

3. Inert particles carrying an electrical charge.

4. Substances that have a strong chemical affinity for the vapor. For example, hygroscopic substances such as sulfur trioxide combine with water vapor to form droplets of relatively low vapor pressure upon which more vapor easily condenses even from unsaturated atmospheres. Such droplets have a permanently lowered vapor pressure. They therefore remain stable even when the air has become unsaturated. The stability of a city fog, even in a highly unsaturated atmosphere, has been attributed to the presence of various substances such as sulfuric acid dissolved in it.

Looking at the problem from a somewhat different point of view, foreign nuclei can be considered to act by catalyzing the formation of aggregates of water molecules which act as nuclei, the foreign particles serving to decrease the work of formation of the aggregates (48). The work of formation W of nuclei during a self-nucleation process can be calculated from the equation

$$W = \frac{16}{3} \frac{\pi\sigma^3 V_B}{(kT \ln p'/p)^2}$$

where σ = surface tension
V_B = volume per molecule of the droplet
k = Boltzman's constant
p'/p = degree of supersaturation

Note that when $p'/p = 1$, that is, when there is no supersaturation, W is infinite and self-nucleation cannot occur. When foreign nuclei are present, W must be multiplied by a factor which takes into account the contact angle θ between the growing aggregate of water molecules and the surface of the foreign particle. When this surface is flat, the factor is

$$\frac{(2 + \cos\theta)(1 - \cos\theta)^2}{4}$$

When θ is 180°, the condition for a droplet just touching the surface, the factor is unity and the limiting equation for self-nucleation is recovered. As θ decreases to zero, the factor decreases to zero. Thus with decreasing θ the amount of supersaturation required to obtain mist or fog likewise decreases.

3.3.4 Biological Aerosols

Aerosols of biological origin are probably not important in direct chemical reactions that may occur in the atmosphere. However, the study of aerobiology has provided information that may be important as it applies to the transport of other materials in the atmosphere.

Biological aerosols have been found great distances from their sources. Living spores of various fungi have been collected with an airplane above the Caribbean Sea 600 miles from their nearest source, while allergens have been identified at least 1,500 miles from their probable origin. In spite of the usually low concentrations at the source, marine bacteria have been collected 80 miles inland from the nearest seacoast (29).

Marine bacteria are removed from the sea only when the surface is stirred sufficiently to produce spray. Since on the average the number of marine bacteria seldom exceeds 500 per milliliter of sea water, the population of marine bacteria in the air must be sparse everywhere except perhaps at times of rough sea or in the vicinity of coastal breakers (29).

Occasionally sea water becomes highly contaminated with microorganisms that are carried into the atmosphere. One such instance is reported by Woodcock (70) in which *Dinoflagellate gymnodium* was present in sufficient quantities to cause the sea to appear red. Ocean spray under these conditions produced severe respiratory irritation among persons who came in contact with the spray.

Proctor (53) has reported microorganisms at various levels in the air up to 20,000 ft. Spore-forming bacteria were predominant, and 29 species of bacteria were identified. He found the ratio of dust particles to microorganisms (bacteria and molds) to be about 100:1.

Living spores of several common molds were caught in a spore trap released from the balloon *Explorer* II at 72,500 ft and set to close at 36,000 ft (56).

Many observers have reported great irregularities in concentrations collected at different levels of the atmosphere, often finding heavier concentrations at higher altitudes than at lower levels. There are several recorded instances of such biologic stratification. For example, Durham (14) noted that abrupt spore ceilings are often marked by a cloud layer or a visible haze. He noted one instance in which a detached cloud of fungus spores consisting almost entirely of Alternaria was encountered about 1,000 ft above a visible 4,000-ft haze line that marked the ceiling for ragweed and a ground cloud of other spores.

3.4 GASEOUS IMPURITIES

Gaseous impurities are emitted into the air in tremendous quantities. For example, approximately 700 to 1,000 tons of volatile hydrocarbons are emitted daily into the atmosphere of Los Angeles from automobile exhausts (38). The concentrations of impurities can range from practically zero to about 100 per cent at the source. Furthermore, the variety of gaseous impurities is enormous. However, there are a few gases and classes of gases which seem to deserve special attention because of their prevalence in the atmospheres of many cities or near many industrial operations. Some of these will be discussed.

3.4.1 Sulfur Dioxide

Sulfur dioxide occurs in the contaminated atmospheres of cities at concentrations up to several parts per million (by volume) (11). It produces acidity in rain water and fogs and is a major source of corrosion of buildings and metal objects. Johnstone (30) has pointed out that the combustion of fuels, smelting operations, refineries, and chemical plants are the chief sources of sulfur dioxide in industrial cities. A power station burning 5,000 tons of coal per day may discharge 500 tons of sulfur dioxide into the air.

3.4.2 Hydrogen Sulfide

Hydrogen sulfide is usually an important contributor to air pollution problems only under rather localized conditions. This probably results in part from the fact that hydrogen sulfide is easily burned to sulfur dioxide. Thus it is not generally released into the atmosphere following combustion. However, because of its great toxicity, it is a hazard in the oil-refining industry, tanneries and other industries where animal matter is handled, the manufacture of sulfur dyes, the manufacture of artificial silk by the viscose process, and the rubber industry (28). Even when present in the air in concentrations below the level of physiological significance, it discolors lead paints (30).

3.4.3 Oxides of Nitrogen

Oxides of nitrogen are produced by the combustion of organic matter and are thus introduced into the atmosphere from automobile exhausts, furnace stacks, incinerators, and many other similar sources. The oxides of nitrogen may be formed from nitrogen compounds in the organic material, but they are also formed by nitrogen fixation. Concentrations as high as 0.5 ppm have been found in the Los Angeles atmosphere and as high as 0.8 ppm in Baltimore (11). The oxides include nitric oxide (NO), nitrogen dioxide (NO_2), nitrogen pentoxide (N_2O_5), and the hydrated nitrogen pentoxide, nitric acid (HNO_3). The nitrogen dioxide is in equilibrium with its dimer, nitrogen tetroxide.

Nitrogen dioxide has received considerable attention as an air pollutant because it is a hazard in numerous industries. Its insidious nature as a poison was emphasized by the Cleveland Clinic fire which occurred in May, 1929. Many deaths occurred following that fire as a result of the inhalation of the nitrogen dioxide produced from burning X-ray film.

3.4.4 Ammonia

Ammonia is another nitrogen compound which is frequently present in the air. It has been found in the atmospheres of Los Angeles and of Charleston, W. Va., at concentrations as great as 0.2 ppm, and in the atmospheres of other cities at somewhat lower concentrations (11,63). The relatively high concentration at Charleston has been attributed to the fact that ammonia is a by-product and raw material in the extensive organic-chemical industry of the area.

Ammonia is a product of many combustion processes, including domestic incineration and the operation of automobiles (39). It is also discharged from certain refinery operations and is always detectable in the air near stockyards.

3.4.5 Hydrogen Fluoride

Fluorides in general, including both hydrogen fluoride and various solid fluorides dispersed as aerosols, have received considerable attention because relatively small amounts produce fluorosis in cattle. The contaminated atmospheres of cities contain lower concentrations of fluorides than of most other contaminants which are commonly determined (11). The maximum concentration found in most cities is about 0.025 ppm. The highest maximum concentration reported has been 0.08 ppm in an industrial area of Baltimore (11).

Fluorides are emitted into the atmosphere by aluminum plants, steel plants, and phosphate-fertilizer plants. Fluorides in the atmosphere have also been attributed to the burning of coal (11). Hydrogen fluoride itself is used in the chemical industry and in refining processes.

3.4.6 Hydrocarbons

Recently, hydrocarbons have received considerable attention as air pollutants because they may participate in reactions in the atmosphere which produce objectionable intermediate compounds and products. Such intermediates and products may contribute to the eye irritation and plant damage resulting from contaminated city air (7,24). The concentrations of hydrocarbons in the atmospheres of most cities have not been measured. This is probably largely due to the difficulty of making such determinations. The concentration of hydrocarbons in Los Angeles smog has been estimated to be about 2 ppm, of which about 1.6 ppm is paraffins and 0.4 ppm is olefins (6). Shepherd et al. (59) chilled volatile contaminants from the Los Angeles atmosphere and by means of the mass spectrometer identified a number of hydrocarbons, including acetylene. Magill, Hutchison, and Stormes (38) investigated the hydrocarbon constituents of automobile-exhaust gases. They estimated that 700 to 1,000 tons of volatile hydrocarbons is emitted daily into the Los Angeles atmosphere from automobile exhausts. Hydrocarbons are also emitted into the atmospheres of cities by oil refineries and by the evaporation of gasoline at service stations.

Local high concentrations of hydrocarbons in the air are common in garages, service stations, dry-cleaning establishments, oil refineries, and many other industries. In general, aliphatic hydrocarbons are not considered to be hazardous if the concentration is less than 20 per cent of the lower inflammable limit, although Patty (51) suggests that this concentration (2,000 ppm in the case of gasoline) may produce unpleasant effects on the human system.

Benzene and other aromatic hydrocarbons are used extensively for solvent and extraction purposes. They often occur locally in the air at relatively high concentrations where they can be quite a health hazard. The concentration of aromatic hydrocarbons in city air is probably generally less than 0.1 ppm, although some aromatic hydrocarbons are present in automobile-exhaust gases (38).

3.4.7 Aldehydes and Ketones

Aldehydes and ketones are introduced into city air in automobile-exhaust gases, incinerator smoke, and stack gases from the combustion of various organic substances.

Aldehydes and ketones are evidently also produced by the oxidation of hydrocarbons after they are admitted into the atmosphere. Such oxidations are discussed later in this section.

Concentrations of aldehydes and ketones have been determined in the air of several cities. Maximum concentrations ranged from 0.12 ppm* (Baltimore) to 1.0 ppm* (Los Angeles). The variations may represent differences in analytical techniques rather than significant differences between the cities. Studies at the Stanford Research Institute revealed that replacement of the usual sodium bisulfite absorption train by an efficient cold-trap system greatly increased the amount of aldehydes and ketones collected from a given volume of Los Angeles air. In the Los Angeles atmosphere, at least, less than half of these substances are formaldehyde, the remainder being ketones and higher aldehydes.

3.4.8 Organic Acids

Most of the processes which emit aldehydes and ketones into the atmosphere probably also emit organic acids. Cadle and Johnston (6) have reported the presence of formic acid in Los Angeles smog. The Los Angeles County Air Pollution Control District (37) has reported finding acids varying from 2 to 12 carbon atoms in chain length, some of which contained hydroxyl and carbonyl groups.

3.4.9 Organic Halides

Organic halides have been identified in Los Angeles smog by several workers, and they are probably present in the air of other cities. The concentrations are probably generally below 0.1 ppm (63,59). Organic halides have been the subject of considerable speculation as possible eye irritants in contaminated city air. Some of these compounds produce eye irritation at very low concentrations. The threshold lachrymatory concentrations for cyanogen chloride, chloroacetophenone, and bromobenzyl cyanide are 0.05, 0.05, and 0.02 ppm, respectively. Unfortunately, the individual compounds are very difficult to identify at such low concentrations.

3.4.10 Carbon Monoxide

Carbon monoxide is commonly found in city air at concentrations up to about 55 ppm (11). Concentrations much higher than this occasionally occur in the open atmosphere or locally in garages, tunnels, behind automobiles, etc. Carbon monoxide is produced by the incomplete combustion of organic material, and automobiles are notorious for their production of this gas.

3.4.11 Carbon Dioxide

Since carbon dioxide exists naturally at a concentration of about 300 ppm in the atmosphere, it is not usually considered to be a pollutant. Nevertheless, huge quantities of carbon dioxide are emitted each day into city air from the combustion of coal, oil, and gasoline. Concentrations of carbon dioxide in the air of industrial areas are at times as high as 600 ppm (11). Fortunately, at least 0.5 per cent (5,000 ppm) of carbon dioxide in air is required before the respiration of man is appreciably affected.

3.5 CHEMICAL REACTIONS IN CONTAMINATED ATMOSPHERES

A variety of chemical reactions occurs in the contaminated atmospheres of cities. The products of such reactions have been blamed for many of the unpleasant proper-

* Parts per million by volume calculated as formaldehyde.

ties of polluted atmospheres. For example, it has been suggested that the products of the oxidation of hydrocarbons in the presence of nitrogen dioxide and sunlight are responsible for the eye irritation, decrease in visibility, plant damage, and cracking of rubber goods which are caused by air pollution in Los Angeles County (24).

Cadle and Johnston have discussed chemical reactions which may occur in Los Angeles smog (6). The following discussion of chemical reactions in contaminated atmospheres is based in part on their paper.

The substances which can react chemically in the atmosphere fall into two groups: major natural constituents of the atmosphere, which are present in high concentrations, and contaminants, usually present at low concentrations. These groups differ in concentration by a factor of 10^4 to 10^6. This large difference in concentration is convenient for purposes of classification of reactions and calls for careful interpretation of the kinetic and photochemical data of the literature. For example, when nitrogen dioxide is irradiated by near-ultraviolet radiation, molecular oxygen and nitric oxide are formed by the following reactions (49):

$$NO_2 + h\nu = NO + O \tag{3-1}$$
$$NO_2 + O = NO + O_2 \tag{3-2}$$

However, in an atmosphere containing a fraction of a part per million of nitrogen dioxide and about 21 per cent of oxygen, the following reaction (where M is any molecule) will be faster than reaction (3-2):

$$O + O_2 + M = O_3 + M \tag{3-3}$$

Suppose that a reaction of the following type occurs:

$$nA + mB = \text{products}$$

and suppose that this reaction obeys the following rate law:

$$-\frac{d[A]}{dt} = k[A]^n[B]^m$$

where $-d[A]/dt$ = rate of consumption of A
$[A]$ and $[B]$ = concentrations of A and B, respectively
k = a constant

If $n + m$ is 2 or larger, that is, if the reaction is kinetically of second or higher order, relatively small decreases in the concentration of A and B would greatly decrease the rate of reaction. The rates of most gas-phase reactions are studied with reactants present at partial pressures between 1 mm of mercury and 1 atm. If a reaction is second or higher order, then a conveniently measured rate under these conditions would imply a completely negligible rate if the reactants were present at fractions of a part per million. The well-known reaction of nitric oxide with oxygen to form nitrogen dioxide illustrates this effect. The equation is

$$2NO + O_2 = 2NO_2 \tag{3-4}$$

This reaction follows the rate law (72)

$$-\frac{d[NO]}{dt} = k[NO]^2[O_2]$$

Since the concentration of oxygen in air is essentially constant, the reaction is kinetically of second order in the atmosphere and the time required for one-half of the nitric oxide to disappear is given by the equation

$$t_{0.5} = \frac{1}{k[O_2][NO]\ \text{initial}}$$

When the concentration of nitric oxide is 10,000 ppm (1 per cent), the half-life is about 36 sec, but when the concentration of nitric oxide is 0.1 ppm, the half-life is about 1,000 hr (72).

The half-life for a first-order reaction does not change with concentration of the reactant. In this case reaction will occur in the atmosphere at a conveniently measurable rate if the energy of activation is below about 25 kcal.*

The velocity constant k for a reaction is usually a function of the temperature. The relationship between velocity constant and temperature can generally be indicated by an equation of the form

$$k = Ae^{-E/RT}$$

where A = a constant having the same dimensions as k
E = activation energy for the reaction
R = gas constant
T = absolute temperature

Values for A and E for a number of reactions which may occur in contaminated atmospheres are presented in Table 3-6.

Table 3-6. Some Constants for Atmospheric Reactions

Reaction	A^a	E, kcal	Ref.
$2NO + O_2 = 2NO_2$	8.0×10^9	0 or negative	17
$NO + O_3 = NO_2 + O_2$	8.0×10^{11}	2.5	31
$2NO_2 + O_3 = N_2O_5 + O_2$	5.9×10^{12}	7.0	32
$O_3 + H_2C = CH_2 \rightarrow$ products	3.5×10^5	0	9
$O_3 + H_3C(CH_2)_3CH{=}CH_2 \rightarrow$ products	5.6×10^6	0	9
$O_3 + HC{\equiv}CH \rightarrow$ products	3.0×10^7	4.8	10
$CH_3 + CH_3CHO \rightarrow CH_4 + CH_3CO$	7.3×10^{11}	7.5	66
$CH_3 + CH_3 \rightarrow C_2H_6$	4.5×10^{13}	0	23
$CH_3 + NO \rightarrow$ products	8.0×10^{11}	0	52
$CH_3 + O_2 \rightarrow CH_3O_2$ (?)	10^{10}	0	49
$H + C_2H_6 \rightarrow H_2 + C_2H_5$	3.4×10^{12}	6.8	4

[a] All the values have the dimensions $cm^3\ mole^{-1}\ sec^{-1}$ except for the reaction of nitric oxide with oxygen which is third-order and has the dimensions $cm^6\ mole^{-2}\ sec^{-1}$.

Photochemically active radiation in sunlight near the earth's surface varies in wavelength from about 2,900 to 7,000 A. Substances which absorb radiation in this region can serve as primary photochemical reactants or as photosensitizers, which function by transferring the absorbed energy to potentially reactant molecules. If a substance is present in small concentrations, it must have a high specific absorption in the 2,900 to 7,000 A wavelength region if it is to initiate photochemical reactions of any importance. If a substance is present in large concentrations, it can absorb relatively weakly and still initiate reactions of considerable importance to the properties of polluted atmospheres. Many reactions can be postulated which might occur among the constituents of contaminated atmospheres. If the kinetics of a postulated reaction has been studied, the extent to which the reaction would occur can usually be predicted. If the reaction has not been investigated, considerable information concerning the likelihood that the reaction would occur to an appreciable extent can often be obtained by thermodynamic considerations. The free energies and heats of formation of a large number of substances are known or can be estimated (5,36,50). These values can be used to calculate the change in free energy and in heat content accompanying postulated reactions. The free energies and heats of formation of a

* Unimolecular first-order reactions generally become second-order at low pressures. However, in the atmosphere the relatively high partial pressures of oxygen and nitrogen would prevent such first-order reactions from becoming second-order at low partial pressures of the reactants (21).

number of substances which may occur in contaminated atmospheres are shown in Table 3-7.

Table 3-7. Some Heats and Free Energies of Formation
(Values in kcal at 25°C. All substances listed in gaseous state.)

Substance	$\Delta H°$	$\Delta F°$	Substance	$\Delta H°$	$\Delta F°$
$O(^3P_2)$	59.1	55.02	$N(^3S'_{1/2})$	85.1	81.0
$O(D_2)$	104.3		N_2O	19.65	24.93
$O_2\left(\frac{1}{\Sigma}+\right)$	37.4		NO	21.6	20.66
			NO_2	8.0	12.27
O_3	34.5	39.4	N_2O_4	3.06	23.44
$H(^2S_{1/2})$	51.9	48.35	N_2O_5	−0.6	26.8
HO	−5.93	−4.80	CH_4	−18.24	−12.20
H_2O	−57.8	−54.64	C_2H_4	11.0	12.3
H_2O_2	−33.6	−24.73	C_2H_6	−20.96	−10.7
S	66.3	56.6	HCHO	−28.7	−24.9
S_8	−5.3	−7.87	HCO_2H	−88.65	−85.1
SO_2	−70.9	−71.74	CO	−26.84	−33.01
SO_3	−93.9	−88.0	CO_2	−94.95	−94.0

References: F. R. Bichowsky and F. P. Rossini, "The Thermochemistry of Chemical Substances," Reinhold Publishing Corporation, New York, 1936; W. M. Lattimer, "Oxidation Potentials," Prentice-Hall, Inc., New York, 1938; and G. S. Parks and H. M. Huffman, "The Free Energies of Some Organic Compounds," Chemical Catalog Company, Inc., New York, 1932.

The changes in heat content (derived from the data of Table 3-7) accompanying a number of reactions which might produce oxygen atoms in the atmosphere are shown in Table 3-8. Oxygen atoms reacting with molecular oxygen would produce ozone. Table 3-8 also shows the maximum wavelengths which would supply the energy required by the endothermic reactions. It was assumed in calculating these wavelengths that each reaction as written absorbs one quantum of the radiation. Almost innumerable other postulated reactions could be included in such a table.

Table 3-8. Some Reactions and Their Energy Deficiencies

Reaction	ΔH, kcal	λ, A
$O_2 \rightarrow O + O$	+118.2	2,420
$O + O_2 + M \rightarrow O_3 + M$	−24.6	Dark
$SO_2 + O_2 \rightarrow SO_3 + O$	+46.1	6,197
$H_2S + O_2 \rightarrow H_2O + S + O$	+72.4	3,919
$O_2 \rightarrow O_2^*$	+37.4	7,639
$NO + O_2^* \rightarrow NO_2 + O$	+8.1	35,271
$NO + O_2 \rightarrow NO_2 + O$	+45.5	6,279
$NO_2 + O_2 \rightarrow NO + O_3$	+48.1	5,940
$CH_4 + O_2 \rightarrow CH_3OH + O$	+28.9	9,885
$C_2H_6 + O_2 \rightarrow C_2H_4 + H_2O + O$	+33.3	8,579
$HCHO + O_2 \rightarrow HCO_2H + O$	−0.8	Dark
$CO + O_2 \rightarrow CO_2 + O$	−8.5	Dark

The thermal reaction of oxygen to form ozone can be used to illustrate the application of such data. Table 3-7 shows that this reaction would occur with an increase in free energy of 39.4 kcal/mole. The equilibrium constant K for the reaction can be calculated from the equation

$$\Delta F° = RT \ln K$$

K is found to be about 10^{-29}. Then

$$\frac{[O_3]}{[O_2]^{3/2}} = 10^{-29}$$

where $[O_3]$ and $[O_2]$ are the concentrations of ozone and oxygen expressed in atmospheres. Substituting 0.2 for $[O_2]$, the equilibrium concentration of ozone from the thermal reaction is found to be about 10^{-30} atm or 10^{-24} ppm. Thus the thermal reaction would produce a negligible concentration of ozone in the atmosphere.

With the exception of the decomposition of oxygen, all the reactions listed in Table 3-8 could, from an energetic standpoint alone, proceed to the right with the light available at the earth's surface. Some could even go in the dark. This does not mean that they actually do. First, the energy required for the initial intermolecular process (the activation energy) is not taken into account. Second, in the case of those reactions requiring light, the ability of one of the reactants to absorb solar radiation is not indicated. It is for this reason that only a few reactions were included in Table 3-8—the list could be expanded indefinitely.

3.5.1 Reactions of Oxides of Nitrogen

Oxides of nitrogen undoubtedly undergo many reactions in the atmosphere, both in sunlight and in the dark. Also, it is possible that they act as photosensitizers. Thus they probably play an important role in the chemistry of contaminated atmospheres.

The reaction of nitric oxide with oxygen to form nitrogen dioxide was discussed previously. For nitric oxide at 1 ppm the half-life is 100 hr, and at 0.1 ppm the half-life is 1,000 hr. After a few days there would be considerable amounts of both nitric oxide and nitrogen dioxide. Given this situation, the following reaction is known to go very rapidly (67):

$$NO + NO_2 + H_2O = 2HNO_2 \tag{3-5}$$

If the concentrations of nitric oxide and nitrogen dioxide are equal, and about 2 per cent of water vapor is present, at equilibrium the concentration of nitrous acid would be 19 per cent of the concentration of nitric oxide. The following reactions attain equilibrium almost instantly, even at 0.1 ppm (72):

$$2NO_2 = N_2O_4 \tag{3-6}$$

$$NO + NO_2 = N_2O_3 \tag{3-7}$$

However, at these low concentrations the equilibriums lie far to the left. Nitrogen dioxide reacts with water vapor and oxygen to give nitric acid vapor:

$$H_2O + 2NO_2 + \tfrac{1}{2}O_2 = 2HNO_3 \tag{3-8}$$

At equilibrium about 90 per cent of the nitrogen oxides would be in the form of nitric acid. The rate of reaction (3-8) is much slower than the gas-phase reaction:

$$H_2O + 3NO_2 = 2HNO_3 + NO \tag{3-9}$$

Reaction (3-8) is probably made up of reactions (3-4) and (3-9), with (3-4) being the slower rate-determining step. When nitrogen dioxide and nitric oxide are present in equal amounts, reaction (3-9) at equilibrium gives nitric acid as only about 1.5 per cent of the nitrogen dioxide (72). Thus for equal amounts of nitric oxide and nitrogen dioxide, the concentration of nitrous acid is more than ten times as great as that of nitric acid at equilibrium, and these equilibriums are attained rather rapidly. To summarize, if 1 ppm of nitric oxide is added to otherwise uncontaminated air containing about 2 per cent of water vapor, after 4 days there will be about 0.4 ppm nitric oxide, 0.4 ppm nitrogen dioxide, and about 0.1 ppm nitrous acid; after several weeks there will be 0.9 ppm nitric acid and 0.1 ppm nitrogen dioxide as the equilibrium mixture.

If the nitric oxide is introduced into an atmosphere containing small amounts of ozone, the situation is quite different. The reaction

$$NO + O_3 = NO_2 + O_2 \qquad (3\text{-}10)$$

has a half-life of about 1.8 sec if both ozone and nitric oxide are present at 1 ppm. The half-life is 18 sec if both are present at 0.1 ppm (31). If nitric oxide is initially present in excess of ozone, the ozone would quickly be used up and there would remain a mixture of nitric oxide, nitrogen dioxide, and nitrous acid vapor. The presence of an excess of nitric oxide in the atmosphere would explain the fact that the atmospheres of a number of cities appear to be virtually free of ozone. If ozone is initially in excess, as it often appears to be in Los Angeles and other cities (3,62), reaction (3-10) is followed by

$$2NO_2 + O_3 = N_2O_5 + O_2 \qquad (3\text{-}11)$$

$$N_2O_5 + H_2O = 2HNO_3 \qquad (3\text{-}12)$$

If ozone and nitrogen dioxide are present at 1 ppm, the half-life of reaction (3-11) is about 8 min (32). The rate of reaction (3-12) has not been measured, but it is known to be extremely fast. The equilibrium for both reactions (3-11) and (3-12) lies far to the right. These considerations suggest that, in the presence of an excess of ozone, a large proportion of the "oxides of nitrogen" in the atmosphere would actually be present as nitric acid vapor. Analyses of the Los Angeles atmosphere have indicated that often about two-thirds of the "oxides of nitrogen" are actually nitric acid vapor or nitrates in aerosol form.

Nitrogen dioxide reacts rapidly with moist solid sodium chloride to produce nitrosyl chloride (72):

$$2NO_2 + NaCl = NOCl + NaNO_3 \qquad (3\text{-}13)$$

Nitric acid reacts with ammonia and amines to form the corresponding nitrates. Nitric oxide reacts rapidly with organic free radicals to stop free-radical chain reactions. Also, nitrogen dioxide reacts rapidly with organic free radicals to produce stable nitroparaffins (15). However, in the dark and at room temperature, nitric oxide, nitrogen dioxide, and nitric acid vapor do not react rapidly with hydrocarbon vapors at concentrations usually existing in the contaminated atmospheres of cities.

The reactions just discussed are all "dark reactions," that is, reactions which occur even in the absence of light. However, numerous photochemical reactions involving oxides of nitrogen may occur in contaminated atmospheres. Nitric oxide absorbs visible and near-ultraviolet light so weakly that it is unlikely that this absorption is responsible for appreciable chemical reaction in contaminated atmospheres. The absorption spectrum of nitric acid vapor is not very well known. There may be moderate or weak absorption in the near-ultraviolet, and such radiation decomposes liquid nitric acid. However, this decomposition may result from absorption by nitrogen dioxide. The following dissociation, analogous to the thermal decomposition, may occur:

$$HNO_3 + h\nu = HO + NO_2\ (?) \qquad (3\text{-}14)$$

Nitrosyl chloride, which might be formed in the atmosphere by reaction (3-13), absorbs near-ultraviolet radiation very strongly to form nitric oxide and chlorine:

$$NOCl + h\nu = NO + Cl \qquad (3\text{-}15)$$

Nitrogen dioxide absorbs strongly in the visible and near-ultraviolet regions. Its photochemistry has been thoroughly investigated. The photolysis of nitrogen dioxide to form nitric oxide and oxygen atoms and the combination of the oxygen atoms with molecular oxygen to form ozone were discussed previously [Eqs. (3-1) and (3-3)].

However, the nitric acid and ozone formed according to Eqs. (3-1) and (3-3), respectively, would react very rapidly to form nitrogen dioxide and oxygen [Eq. (3-10)]. Thus a low steady-state concentration of ozone would quickly be attained. Cadle and Johnston (6) estimated that for 1 ppm of nitrogen dioxide in sunlight the steady-state ozone concentration would be 0.1 ppm and for 0.1 ppm of nitrogen dioxide the steady-state ozone concentration would be about 0.03 ppm. Thus these reactions cannot explain the ozone content of the atmospheres of some cities, where the concentration of ozone is often several tenths of a ppm (3). This result has been qualitatively confirmed at Stanford Research Institute by irradiating mixtures of nitrogen dioxide and air and measuring the ozone produced (62). Haagen-Smit (25) has reported that adding various organic compounds to such mixtures greatly increased the amount of ozone which was formed.

Generally, in the Los Angeles area, ozone is found at high concentrations in the air only when the sun is shining brightly. Recent work by the Stanford Research Institute in the Los Angeles area has involved drawing air into flasks early in the morning, before appreciable ozone has accumulated, and irradiating the flasks with sunlight. Ozone* is formed within the flasks if ozone is later found in the open air. Experiments involving irradiation of the flasks with light which has passed through various filters suggest that light in the wavelength region 3,600 to 4,000 A is most effective for producing ozone in the contaminated air of the Los Angeles area.

3.5.2 Reactions of Sulfur Dioxide

Sulfur dioxide absorbs solar radiation between 2,900 and about 4,000 A, with the maximum absorption probably occurring between 2,900 and 3,000 A. It reacts very slowly with oxygen in sunlight to form sulfur trioxide:

$$2SO_2 + O_2 + h\nu = 2SO_3 \tag{3-16}$$

Gerhard (18) found that the reaction is first-order with respect to sulfur dioxide and that the reaction amounts to about 0.1 to 0.2 per cent per hour in air in intense natural sunlight. He also found that nitrogen dioxide in concentrations of 5 to 20 per cent of the sulfur dioxide concentration had no significant effect on the reaction rate. Sodium chloride nuclei and variation of the relative humidity in the range 30 to 90 per cent also had no effect on the reaction rate. It is of interest to speculate concerning the effect of chlorine on the reaction of sulfur dioxide with oxygen. Chlorine absorbs sunlight strongly, with a maximum absorption coefficient at about 3,400 A. It is a powerful photoactivator. Apparently the primary effect of light is to decompose the molecular chlorine:

$$Cl_2 + h\nu = Cl + Cl \tag{3-17}$$

When an oxidizable substance is exposed to radiation in the presence of chlorine and oxygen, a sensitized oxidation almost always occurs (49). Chlorine is probably formed in the atmosphere by the reaction of nitrogen dioxide with salt [Eqs. (3-13) and (3-15)] and by the reaction of ozone with salt (71):

$$2H^+ + 2Cl^- + O_3 = O_2 + H_2O + Cl_2 \tag{3-18}$$

However, the concentration of salt in the air of most cities seldom exceeds about 0.1 mg/cu m (11,47), and it is doubtful whether enough chlorine would often be produced to affect appreciably the oxidation of sulfur dioxide.

The rate of reaction of sulfur dioxide with ozone, even when both are present at relatively high concentrations, is also very slow. Recent experiments by Cadle (8) showed that, when the concentration of ozone was about 1 per cent, the concentration

* That is, a gas which produces rubber cracking and which liberates iodine from potassium iodide.

of sulfur dioxide 1 to 10 per cent, and the concentration of water vapor 0 to 1 per cent, less than 0.1 per cent of the sulfur dioxide reacted during 24 hr.

3.5.3 Reaction of Ozone with Unsaturated Hydrocarbons

Fairly rapid reaction occurs between ozone and various olefins at concentrations at which they may occur in contaminated atmospheres (6,9). The reaction mechanisms are not precisely known, and a variety of products is obtained. However, the initial rates follow a second-order rate law:

$$-\frac{d[O_3]}{dt} = k[O_3][\text{olefin}]$$

The first stage of the reaction appears to be a bimolecular addition reaction which forms an unstable complex. This seems to be followed by dissociation of the complex into unknown fragments which interact further with the olefin. The subsequent reactions appear to be of the free-radical chain type, although this point has not been established. Final products include formaldehyde, higher aldehydes, and polymers of unknown composition. The reactions may be tentatively summarized as

$$O_3 + \text{olefin} = \text{olefin-}O_3 \text{ complex} \tag{3-19}$$

$$\text{Olefin-}O_3 \text{ complex} = \text{decomposition fragments, including free radicals} \tag{3-20}$$

At high concentrations of reactants a visible aerosol forms, but at concentrations which would probably exist in the smog-laden air of cities no aerosol has been experimentally demonstrated (7).

The reaction of ozone with acetylene has also been investigated (10). This reaction also follows a second-order rate law, but the velocity constant k is so small that the reaction would be very slow in city air. The velocity constants for the reactions of ozone with benzene and with paraffin hydrocarbons are so small that they could not be determined by the methods employed in these investigations.

Initial second-order rate constants and half-lives for the reaction of ozone with various substances are shown in Table 3-9.

Table 3-9. Rate Constant and Half-life for Initial Rate of Reaction of Ozone with Various Substances

Substance reacting with ozone	Half-life if reactants and ozone are 0.2 ppm, min	Half-life if reactants and ozone are 1 ppm, min	Rate constant $ppm^{-1}\ min^{-1}$ at 25°C
Ethylene	1,100	220	0.0045
1-Hexene	330	66	0.015
Cyclohexene	57	12	0.087
Gasoline	380	76	0.013
Acetylene	500,000	24,000	0.00010
Nitric oxide	0.16	0.03	32.0
Nitrogen dioxide	65	13	0.077

It is seen that the reaction of nitric oxide with ozone is 2,000 times faster than that of 1-hexene and ozone. Nitrogen dioxide reacts with ozone about five times faster than 1-hexene. Thus these last two reactions would proceed at the same rate if, for example, the concentrations of nitrogen dioxide and of 1-hexene were 0.2 and 1 ppm, respectively.

3.5.4 Photochemical Reactions of Aldehydes, Ketones, and Olefins

Simple aldehydes and ketones absorb radiation which begins between 3,000 and 4,000 A and extends toward shorter wavelengths, usually terminating between 2,300 and 2,500 A. The absorption coefficients in the solar region are relatively low. The absorption regions for diketones are displaced toward longer wavelengths to an extent which depends upon the proximity of the groups within the molecule (49). Thus glyoxal and diacetyl absorb up to about 5,000 A.

The most important primary photochemical reactions of aldehydes and ketones are to break the molecule into two free radicals. For example, acetaldehyde decomposes into methyl and formyl radicals:

$$CH_3CHO + h\nu = CH_3 + HCO \qquad (3\text{-}21)$$

Acetone decomposes into methyl and acetyl radicals:

$$(CH_3)_2CO + h\nu = CH_3 + CH_3CO \qquad (3\text{-}22)$$

The free acetyl radical slowly decomposes into free methyl radicals and carbon monoxide:

$$CH_3CO = CH_3 + CO \qquad (3\text{-}23)$$

The free radical HCO is remarkably stable (64). However, it would very slowly decompose:

$$HCO = H + CO \qquad (3\text{-}24)$$

$$H + O_2 = HO_2 \qquad (3\text{-}25)$$

It might react directly with oxygen:

$$HCO + O_2 = HO_2 + CO\ (?) \qquad (3\text{-}26)$$

Reactions of free radicals such as those formed by the photolysis of aldehydes and ketones are discussed in the next article. However, it should be pointed out here that large numbers of secondary reactions follow the primary photolysis. Most of the studies of the reaction products have been made in the absence of air. Under these conditions formaldehyde undergoes a chain reaction giving, among other products, carbon monoxide, carbon dioxide, and water. Among the products of the photochemical decomposition of acetaldehyde in the absence of air are methane, carbon dioxide, glyoxal, diacetyl, formaldehyde, and unidentified material of high molecular weight. Acetone forms carbon monoxide, methane, ethane, diacetyl, and material of high molecular weight. In the presence of air, the radicals tend to react rapidly with oxygen. Thus somewhat different compounds are produced. In the presence of air, acetone appears to give, among other products, acetic acid and dimethyl peroxide (16,55,40). Acetaldehyde apparently gives diacetyl peroxide as a major product (44). Irradiation with sunlight of mixtures of diacetyl and air has been found to produce ozone, although not enough to account for the ozone concentrations sometimes found in contaminated atmospheres (25). The reactions involved are not known, but the following are not unreasonable:

$$CH_3COCOCH_3 = 2CH_3CO \qquad (3\text{-}27)$$

$$CH_3CO + O_2 = CH_3COO_2\ (?) \qquad (3\text{-}28)$$

$$CH_3COO_2 + O_2 = CH_3COO + O_3\ (?) \qquad (3\text{-}29)$$

Mono-olefins do not absorb solar radiation in the lower atmosphere. However, conjugated diolefins absorb near-ultraviolet fairly strongly to produce electronically excited molecules which react with oxygen (15):

$$\text{Diolefin} + h\nu = \text{excited diolefin} \quad (3\text{-}30)$$
$$\text{Excited diolefin} + O_2 = \text{free radical} + HO_2 \quad (3\text{-}31)$$

Oxygen absorbs weakly the many atmospheric bands in the visible region to produce excited oxygen molecules:

$$O_2 + H\nu = O_2^* \quad (3\text{-}32)$$

By analogy to reaction (3-31), one could write down

$$O_2^* + RH = R + HO_2\ (?) \quad (3\text{-}33)$$

There appear to be no data for or against reaction (3-33).

3.5.5 Reactions of Organic Free Radicals in the Atmosphere

A number of the reactions which may occur in contaminated atmospheres [reactions (3-20) to (3-23), (3-27) to (3-29), (3-31), (3-33)] yield organic free radicals as products. Other reactions [(3-1), (3-14), (3-15), (3-17), (3-24) to (3-26), (3-31), (3-33)] produce atoms or inorganic free radicals which in some cases would attack any organic molecule, saturated, unsaturated, or aromatic, to produce organic free radicals. The following are expected rapid reactions of atoms or inorganic free radicals to form organic free radicals (64):

$$O + RH = OH + R \quad (3\text{-}34)$$
$$OH + RH = H_2O + R \quad (3\text{-}35)$$
$$Cl + RH = HCl + R \quad (3\text{-}36)$$
$$H + RH = H_2 + R \quad (3\text{-}37)$$
$$HO_2 + RH = H_2O_2 + R \quad \text{or } 2OH + R \quad (3\text{-}38)$$

Reaction (3-38) is known to be very fast, but the products are uncertain and vary with the nature of the reactants. The other cases may vary with the reactant also; for example, oxygen atoms may simply add to an olefin, hydroxyl atoms may split out an alcohol instead of water, etc. The point to emphasize here is that virtually every photochemical reaction expected to occur in the contaminated air of cities may result in the production of an organic free radical.

Many organic free radicals, such as those produced by extracting a hydrogen atom from a hydrocarbon molecule, react rapidly with oxygen to form peroxy free radicals:

$$R + O_2 = RO_2 \quad (3\text{-}39)$$

The peroxy free radicals can react in turn with other organic substances to form organic peroxides and new free radicals:

$$RO_2 + R'H = RO_2H + R' \quad (3\text{-}40)$$

Other reactions expected of organic free radicals include polymerization of olefins (34):

$$R' + R{-}CH{=}CH_2 = R{-}CH(R'){-}CH_2{-} \quad (3\text{-}41)$$

Such oxidations and polymerizations are chain reactions, new radicals being generated by reactions (3-40) and (3-41) to maintain the chains.

In the literature there is some speculation, but no proof, concerning the production of ozone from peroxy free radicals (15):

$$RO_2 + O_2 = RO + O_3\ (?) \quad (3\text{-}42)$$

Reaction (3-29) would be a special case of this reaction. If reaction (3-42) has any tendency to go, atmospheric conditions [with large amounts of oxygen and thus relatively small competition from reaction (3-40)] are highly favorable.

Peroxides formed by reaction (3-40) decompose to form aldehydes, ketones, and water, or they may undergo degradation reactions to give lower aldehydes, hydroxyl radicals, and organic free radicals (15).

Chain reactions involving free radicals terminate when the chain-maintaining radicals react with some substance without forming new free radicals. Terminating steps for free radical chains in contaminated atmospheres include

$$R + NO = \text{unknown products} \tag{3-43}$$

$$R + NO_2 = RNO_2 \tag{3-44}$$

$$R + R' = RR' \tag{3-45}$$

Adsorption and destruction of the radicals occur on surfaces, probably including the surfaces of particulate matter in the atmosphere. Also, one would expect that organic free radicals would be adsorbed and destroyed on the surfaces of most systems used for sampling and analysis of the atmosphere.

Most of the free radicals expected in the atmosphere, such as HO, HO_2, HCO, R, and RO_2, are remarkably stable even though they are highly reactive. These free radicals are capable of removing a hydrogen atom from almost any organic substance upon contact, and thus they are expected to be irritating and harmful to living tissues.

3.5.6 Reactions Involving Particulate Material

Reactions involving particulate material suspended in the atmosphere are of at least three types. One type consists of heterogenous gas-phase reactions, that is, reactions which occur at least in part on the surfaces of particles. The particles may adsorb reactant gases and catalyze the reaction. For example, the decomposition of ozone is catalyzed by solid surfaces. Particles adsorb and destroy free radicals, thereby decreasing the rate of free radical chain reactions. The adsorption of gas on solid surfaces might change the absorption spectrum of the gas for sunlight; for example, oxygen adsorbed on carbon particles might absorb sunlight much more strongly than would free oxygen.

The reactions just described produce relatively little change in the particulate material. A second type of reaction involves both the gases and the particles and produces gross chemical changes in the latter. For example, the rate of oxidation of gaseous sulfur dioxide by oxygen or ozone is very slow. However, contaminated atmospheres may contain relatively large amounts of essentially aqueous droplets (57). Such droplets would dissolve sulfur dioxide from the atmosphere. The sulfurous acid thus formed would quickly be oxidized to sulfuric acid. Such droplets have been found to contain nitrates, sulfates, and organic material. Thus throughout the lower atmosphere reactions typical of aqueous solutions may occur inside such droplets. Reactions of ozone or nitrogen dioxide with salt particles [reactions (3-18) and (3-13)] and of ammonia with sulfuric acid droplets are other examples of such reactions.

A third type involves reactions between suspended particles. An example would be reactions between sulfuric acid droplets and limestone dust. Actual contact between the particles is necessary for such reactions to occur.* Since the rate of agglomeration in city air is very slow, as was pointed out earlier, these reactions would usually be slow except near the sources.

Though at present relatively little is known concerning reactions involving dusts and aerosols, they should not be overlooked as possible chemical or photochemical reactants or catalysts.

REFERENCES

1. Adel, Arthur: Selected Topics in the Infrared Spectroscopy of the Solar System, chap. 10 in G. P. Kuiper, ed. "Atmospheres of the Earth and Planets," University of Chicago Press, Chicago, 1949.

* Assuming negligible vapor pressure of the reactants.

2. Amelung, W., and H. Landsberg: Kerzählungen in Freiluft and Zimmerluft, *Bioklimatische Beiblätter*, **1**, 49–53 (1934).
3. Bartel, A. W., and J. W. Temple: Ozone in Los Angeles, *Ind. Eng. Chem.*, **44**, 857–861 (1952).
4. Berlie, M. R., and D. J. LeRoy: The Reaction of Atomic Hydrogen with Ethane, *Discussions Faraday Soc.*, No. 14, pp. 50–54, 1953.
5. Bichowsky, F. R., and F. P. Rossini: "The Thermochemistry of Chemical Substances," Reinhold Publishing Corporation, New York, 1936.
6. Cadle, R. D., and H. S. Johnston: Chemical Reactions in Los Angeles Smog, *Proc. 2d Nat. Air Pollution Symposium*, Stanford Research Institute, Los Angeles, Calif., 1952.
7. Cadle, R. D., and P. L. Magill: Study of Eye Irritation Caused by Los Angeles Smog, *Arch. Ind. Hyg. and Occupational Med.*, **4**, 74–84 (1951).
8. Cadle, R. D.: Stanford Research Institute, unpublished data.
9. Cadle, R. D., and Conrad Schadt: Kinetics of the Gas Phase Reaction of Olefins with Ozone, *J. Am. Chem. Soc.*, **74**, 6002–6004 (1952).
10. Cadle, R. D., and Conrad Schadt: Kinetics of the Gas Phase Reaction between Acetylene and Ozone, *J. Chem. Phys.* **21**, 163 (1953).
11. Cholak, J.: The Nature of Atmospheric Pollution in a Number of Industrial Communities, *Proc. 2d Nat. Air Pollution Symposium*, Stanford Research Institute, Los Angeles, Calif., 1952.
12. Coste, J. H., and H. L. Wright: The Nature of the Nucleus in Hygroscopic Droplets, *Phil. Mag.*, **20**, 209–234 (1935).
13. Dörffel, K., H. Lettau, and M. Rötschke: Luftkörper Alterung als Austausch-problem auf Grund von Staub und Kerngehaltsmessungen, *Meteorol. Z.*, **54**, 16–23 (1937).
14. Durham, O. C.: Airborne Fungus Spores as Allergens, Aerobiology, *Pub. Am. Assoc. Advance. Sci.*, No. 17, pp. 32–47, 1942.
15. Faraday Society: "Oxidation," Gurney & Jackson, London, 1945.
16. Frankenburg, P. E., and W. A. Noyes, Jr.: Photochemical Studies. XLVII. Liquid Acetone-oxygen and Liquid Acetone-heptane-oxygen, *J. Am. Chem. Soc.*, **75**, 2847–2850 (1953).
17. Frost, A. A., and R. G. Pearson: "Kinetics and Mechanism," John Wiley & Sons, Inc., New York, 1953.
18. Gerhard, E. R.: The Photochemical Oxidation of Sulfur Dioxide to Sulfur Trioxide and Its Effect on Fog Formation, Engineering Experiment Station University of Illinois, Urbana, Ill., Oct. 10, 1953.
19. Gibbs, W. E.: "Clouds and Smokes," P. Blakiston's Son & Company, Philadelphia, 1924.
20. Gillespie, T., and G. O. Langstroth: Coagulation and Deposition in Still Aerosols of Various Solids, *Can. J. Chem.*, **30**, 1003–1011 (1952).
21. Glasstone, Samuel, K. J. Laidler, and Henry Eyring: "The Theory of Rate Processes," McGraw-Hill Book Company, Inc., New York, 1941.
22. Goldschmidt, V. M.: Grundlagen der Quantitativen Geochemie, *Fortschr. Mineral., Krist. Petrog.*, **17**, 112–156 (1933).
23. Gomer, R., and G. B. Kistiakowski: Rate Constant of Ethane Formation from Methyl Radicals, *J. Chem. Phys.*, **19**, 85–91 (1951).
24. Haagen-Smit, A. J.: Chemistry and Physiology of Los Angeles Smog, *Ind. Eng. Chem.*, **44**, 1342–1351 (1952).
25. Haagen-Smit, A. J., C. E. Bradley, and M. M. Fox: Ozone Formation in Photochemical Oxidation of Organic Substances, *Ind. Eng. Chem.*, **45**, 2086–2089 (1953).
26. Hellmann, G., and W. Meinardus: Der grosse Staubfall vom 9–12 März 1901 in Nordafrika, Süd- und Mitteleuropa, *Meteorol. Z.*, **19**, 180–184 (1902).
27. Hutchinson, G. E.: Nitrogen in the Biogeochemistry of the Atmosphere, *Am. Scientist*, **32**, 178–195 (1944).
28. Jacobs, M. B.: "The Analytical Chemistry of Industrial Poisons, Hazards, and Solvents," 2d ed., Interscience Publishers, Inc., New York, 1949.
29. Jacobs, W. C.: "Aerobiology. Compendium of Meteorology," T. F. Malone, ed., American Meteorological Society, Boston, Mass., 1951.
30. Johnstone, H. F.: Properties and Behavior of Air Contaminants, in Louis C. McCabe, ed., "Air Pollution," McGraw-Hill Book Company, Inc., New York, 1952.
31. Johnston, H. S., and H. J. Crosby: Rapid Gas-phase Reaction between Nitric Oxide and Ozone, *J. Chem. Phys.*, **19**, 799 (1951).
32. Johnston, H. S., and D. M. Yost: The Kinetics of the Rapid Gas Reaction between Ozone and Nitrogen Dioxide, *J. Chem. Phys.*, **17**, 386–392 (1949).
33. Köppen, W., and R. Geiger: "Handbuch der Klimatologie," vol. 1, part B, Gebruder Borntraeger, Berlin, 1936.

34. Laidler, K. J.: "Chemical Kinetics," McGraw-Hill Book Company, Inc., New York, 1950.
35. LaMer, V. K.: Nucleation in Phase Transitions, *Ind. Eng. Chem.*, **44,** 1270–1277 (1952).
36. Lattimer, W. M.: "Oxidation Potentials," Prentice-Hall, Inc., New York, 1938.
37. Los Angeles County Air Pollution Control District, Los Angeles, Calif., Second Technical and Administrative Report on Air Pollution Control in Los Angeles County, 1950–1951.
38. Magill, P. L., D. H. Hutchison, and J. M. Stormes: Hydrocarbon Constituents of Exhaust Gases, *Proc. 2d Nat. Air Pollution Symposium,* Stanford Research Institute, Los Angeles, Calif., 1952.
39. Magill, P. L., and R. W. Benoliel: Air Pollution in Los Angeles County. Contribution of Combustion Products, *Ind. Eng. Chem.*, **44,** 1347–1351 (1952).
40. Marcotte, F. B., and W. A. Noyes, Jr.: Photochemical Studies. XLV. The Reactions of Methyl and Acetyl Radicals with Oxygen, *J. Am. Chem. Soc.*, **74,** 783–786 (1952).
41. Marcotte, F. B., and W. A. Noyes, Jr.: The Reaction of Radicals from Acetone with Oxygen, *Discussions Faraday Soc.*, No. 10, pp. 236–241, 1951.
42. Mason, Brian: "Principles of Geochemistry," John Wiley & Sons, Inc., New York, 1952.
43. Migeotte, M. V.: On the Presence of CH_4, N_2O, and NH_3 in the Earth's Atmosphere, chap. 10 in G. P. Kuiper, ed., "The Atmospheres of the Earth and Planets," University of Chicago Press, Chicago, 1949.
44. Mignolet, J.: Effect of Small Amounts of Oxygen on the Photolysis of Acetaldehyde, *Bull. soc. roy. sci. Liège,* **10,** 343–348 (1941).
45. Mills, C. A.: The Donora Episode, *Science,* **111,** 67–68 (1950).
46. Mills, C. A.: The Donora Smog Disaster, *Hygeia,* **27,** 684–685, 722–724 (1949).
47. Moyerman, R. M., and K. E. Shuler: The Concentration of Contaminant Alkali Salts in Ground Level Air, *Science,* **118,** 612–614 (1953).
48. Neuberger, Hans: Condensation Nuclei, *Mech. Eng.*, **70,** No. 3, 221–225 (1948).
49. Noyes, W. A., Jr., and P. A. Leighton: "The Photochemistry of Gases," Reinhold Publishing Corporation, New York, 1941.
50. Parks, G. S., and H. M. Huffman: "The Free Energies of Some Organic Compounds," Chemical Catalog Company, Inc., New York, 1932.
51. Patty, F. A.: "Industrial Hygiene and Toxicology," Interscience Publishers, Inc., New York, 1949.
52. Powell, R. E.: Reaction Kinetics, *Ann. Rev. Phys. Chem.*, **3,** 313 (1953).
53. Proctor, B. E.: The Microbiology of the Upper Air. I, *Proc. Am. Acad. Arts Sci.*, **69,** 315–340 (1934).
54. Rankama, K., and Th. G. Sahama: "Geochemistry," University of Chicago Press, Chicago, 1950.
55. Rice, F. O., and C. E. Schildknecht: The Photochemical Oxidation of Acetone, *J. Am. Chem. Soc.*, **60,** 3044–3047 (1938).
56. Rogers, L. A., and F. C. Meier: The Collection of Microorganisms above 36,000 Feet, *Nat. Geog. Soc., Contrib. Techn. Papers, Stratosphere Ser.*, **2,** 146–151 (1936).
57. Rubin, Sylvan: Liquid Particles in Atmospheric Haze, *J. Atm. & Terrest. Phys.*, **2,** 130–140 (1952).
58. Shaw, Wm. N.: "Manual of Meteorology," 2d ed., Cambridge University Press, New York, 1936.
59. Shepherd, Martin, S. M. Rock, R. Howard, and J. Stormes: Isolation, Identification, and Estimation of Gaseous Pollutants of Air, *Anal. Chem.*, **23,** 1431–1447 (1951).
60. Shilen, Joseph, J. F. Mellor, Jr., A. M. Stang, J. S. Sharrah, and C. D. Robson: The Donora Smog Disaster, Pennsylvania Department of Health, Bureau of Industrial Hygiene, 1950.
61. Slobod, R. L., and M. E. Krogh: Nitrous Oxide as a Constituent of the Atmosphere, *J. Am. Chem. Soc.*, **72,** 1175–1177 (1950).
62. Stanford Research Institute: Unpublished data.
63. Stanford Research Institute: The Smog Problem in Los Angeles County, Third Interim Report, Western Oil and Gas Association, Los Angeles, Calif., 1950.
64. Steacie, E. W. R.: "Atomic and Free Radical Reactions," Reinhold Publishing Corporation, New York, 1946.
65. "United Nations Statistical Yearbook," p. 267, New York, 1952.
66. Volman, D. H., and R. K. Brinton: Reactions of Free Radicals with Aldehydes, *J. Chem. Phys.*, **20,** 1764–1768 (1952).
67. Wayne, L. G., and D. M. Yost: Kinetics of the Rapid Gas-phase Reaction between NO, NO_2, and H_2O, *J. Chem. Phys.*, **19,** 41–47 (1951).

68. Whytlaw-Gray, R., and H. S. Patterson: "Smoke," Edward Arnold & Co., London, 1934.
69. Wilson, W. S., W. B. Guenther, R. D. Lowrey, and J. C. Cain: Surface Ozone at College, Alaska, for 1950, *Trans. Am. Geophys. Union*, **33**, 361–364 (1952).
70. Woodcock, Alfred H.: Note Concerning Human Respiratory Irritation Associated with High Concentrations of Plankton and Mass Mortality of Marine Organisms, *J. Marine Research*, **7**, 56–62 (1948).
71. Yeatts, L. B., Jr., and Henry Taube: The Kinetics of the Reaction of Ozone and Chloride Ion in Acid Aqueous Solution, *J. Am. Chem. Soc.*, **71**, 4100–4105 (1949).
72. Yost, D. M., and H. Russell: "Systematic Inorganic Chemistry," Prentice-Hall, Inc., New York, 1944.

SECTION 4

PHYSICS OF THE ATMOSPHERE

BY H. NEUBERGER, H. PANOFSKY, AND Z. SEKERA

4.1 ATMOSPHERIC OPTICS

4.1.1 Introduction

Meteorological optics is concerned with the propagation of electromagnetic waves through the atmosphere and with changes in them connected with or occurring during their propagation. In optics, however, only electromagnetic waves with a special range of wavelengths, from about 10^{-6} to 10^{-3} cm, are considered. The central part of this range (4×10^{-5} to 8×10^{-5} cm) corresponds to the electromagnetic waves visible to the human eye, and the electromagnetic waves in this and the two adjacent regions (ultraviolet and infrared) are also called "light," if a distinction from other types of electromagnetic waves is desired.

The laws governing propagation and other phenomena must be discussed differently, depending on whether the atmosphere may be considered as a medium with a continuous mass distribution or with the mass concentrated in separate and discrete centers (air molecules, particulate matter, dust, and other particles). Since in air pollution problems the emphasis is on the presence of such centers (pollutants) in the atmosphere, the second case will form the main part of the present discussion.

4.1.2 Characteristics of Electromagnetic Waves

An electromagnetic wave is defined as periodic oscillations in the intensity of the electric and of the magnetic field which travel in a vacuum with a velocity of $c = 3 \times 10^{10}$ cm/sec. In an uncharged, nonconducting, nonmagnetic medium the velocity of propagation is reduced by the factor $1/n$, where n is the refractive index, related to the dielectric constant ϵ of the medium by the relationship $n = \sqrt{\epsilon}$.

The oscillations may be considered to be harmonic; consequently the intensities of the electric field (represented by the electric vector **E**) and of the magnetic field (represented by the magnetic vector **H**) contain a common factor

$$\exp\,[i(\phi - \omega t)]^* \tag{4-1}$$

where ϕ is a function of space coordinates called the phase and

$$\omega = 2\pi\nu = \frac{2\pi v}{\lambda} = kv \tag{4-2}$$

where ν = frequency
λ = wavelength
v = velocity of propagation
k = wave number

The relation

$$\phi(x,y,z) = \text{const} \tag{4-3}$$

represents in the space coordinates a surface of constant phase. The wave is called a plane wave if (4-3) represents a plane and a spherical wave if (4-3) represents a sphere.

From the electromagnetic theory, especially from Maxwell's equations, it follows that the electric and magnetic vectors oscillate in the plane normal to the direction of propagation of the waves, and the magnetic vector is at any instant proportional and perpendicular to the electric vector to the right, when looking in the direction of propagation. For this reason it is sufficient to restrict the attention to only one of these vectors, usually to the electric vector **E**. The type of oscillation of the electric vector **E** in the plane normal to the direction of propagation determines the polarization of the electromagnetic wave.

If the electric vector oscillates in a very random fashion, so that its mean position during a given time interval cannot be determined, the corresponding electromagnetic wave is called neutral or unpolarized. On the other hand, if the electric vector oscillates in a regular way, the electromagnetic wave is said to be polarized. If the end point of the electric vector during one oscillation sweeps out an ellipse, the wave is called elliptically polarized. This ellipse may degenerate into a straight line, in which case the wave is linearly polarized; if the ellipse degenerates into a circle, the case is

* For reasons of simplicity the complex notation is used, and it is understood that only the real or the imaginary part of the complex expression is to be used.

called circular polarization. The direction of the maximum magnitude of the electric vector and the direction of propagation of the wave determine the plane of polarization. For a linearly polarized wave the electric vector and the direction of propagation of the wave determine the plane of polarization. For a linearly polarized wave the electric vector oscillates only in this plane; in the case of circular polarization, the electric vector has no maximum and the plane of polarization is undetermined.

As a consequence of the linear form of the differential equations for electromagnetic waves, the electric vectors **E**, corresponding to different waves from several sources, can be added or subtracted. The superposition of the electromagnetic waves is one of the fundamental principles in the discussion and study of electromagnetic waves.

4.1.3 Intensity, Flux, and Stokes Polarization Parameters

The human eye and other photometric elements which are used for measuring light intensity are unable to describe the character of the above-mentioned electric or magnetic vectors **E** and **H** directly. They respond only to the energy of electromagnetic waves, which is carried by the waves in the direction of their propagation. If the electromagnetic waves (the radiation) propagate from a point source, the amount of their (radiant) energy—which passes through a cone with vertex at the point source and corresponding to a solid angle $d\omega$, per unit time and per unit frequency interval $d\nu$—is proportional to the solid angle. The proportionality factor is called the specific intensity, and it represents the amount of radiant energy passing through a cone (pencil of radiation) of unit solid angle per unit time and unit frequency interval.

In general, the specific intensity varies from place to place, and it also depends upon the orientation of the axis of the cone, as well as upon the wavelength or frequency of the radiation. In the special case, in which the space variation does not exist, the radiation is called homogeneous; when the radiation is the same for all directions, it is called isotropic.

In electromagnetic theory the relationship between the specific intensity and the electric and magnetic vectors is established by means of Poynting's vector. For sufficiently large distance R from the source, the electric and magnetic vectors can be written in the form

$$\mathbf{E} = (kR)^{-1}\mathbf{A}e^{i(kR-\omega t)} \qquad \mathbf{H} = (kR)^{-1}\mathbf{B}e^{i(kR-\omega t)} \tag{4-4}$$

where the amplitude vectors **A** and **B** depend upon the parameters defining the direction of the radiation. The specific intensity in a direction given, for example, by a unit vector **i** is related to the amplitude vectors by the relationship

$$I_\nu = \tfrac{1}{2}k^{-2}\,\mathrm{Re}\,\{\mathbf{A} \times \mathbf{B}^*\}\mathbf{i} \tag{4-5}$$

where the * denotes the complex conjugate and ν the frequency.

So far only radiant energy from a point source has been considered. For an emitting surface $d\sigma$, with its outward normal deviated from the direction of the pencil of radiation by the angle θ, the amount of radiant energy dE emitted into the solid angle $d\omega$, per unit time and unit frequency interval, is furthermore proportional to $d\sigma \cos\theta$,

$$dE_\nu = I_\nu\, d\omega\, d\sigma \cos\theta \tag{4-6}$$

where I_ν represents the specific intensity in the direction given by θ. The total amount of radiant energy passing through $d\sigma$ in all directions can be obtained by integrating (4-6) over all solid angles. This amount of radiant energy is proportional to the area $d\sigma$, and the proportionality factor

$$F_\nu = \int_{4\pi} I_\nu \cos\theta\, d\omega \tag{4-7}$$

is called net flux. It represents the rate of flow of radiant energy across $d\sigma$ per unit area and per unit frequency interval. For a system of polar coordinates with the axis in the direction of the outward normal to $d\sigma$,

$$d\omega = \sin\theta \, d\theta \, d\phi$$

where ϕ represents the azimuth angle from an arbitrary direction. The expression for the net flux then assumes the form

$$F_\nu = \int_0^{2\pi} \int_0^{\pi} I_\nu(\theta,\phi) \cos\theta \sin\theta \, d\theta \, d\phi \tag{4-8}$$

In the special case of parallel radiation, when the point source may be considered to be at an infinite distance, the specific intensity loses its meaning as a measure of radiant energy, since $d\omega \to 0$. The net flux, however, remains an appropriate measure of the radiant energy passing per unit time through a unit area normal to the direction of the parallel radiation.

In the preceding discussion, all quantities were related to a given frequency ν and thus considered as monochromatic. Furthermore all quantities were defined as per unit frequency interval. If the monochromatic values are multiplied by the elementary frequency interval $d\nu$ and integrated from 0 to ∞, the so-called integrated quantities are obtained, i.e., the integrated intensity and integrated net flux, defined by

$$I = \int_0^{\infty} I_\nu \, d\nu \quad \text{and} \quad F = \int_0^{\infty} F_\nu \, d\nu \quad \text{respectively} \tag{4-9}$$

The characteristics of the electric or magnetic vector, which lead to different types of polarization, do not appear in the foregoing consideration of the flow of the radiant energy. For the detection or measurement of the polarization, special instrumental arrangements must be made. For a complete analysis of the state of polarization, two optical systems are necessary: (1) an analyzer, which in an ideal case will transmit electric or magnetic oscillations in only one direction in a plane normal to the direction of the propagation (to a high accuracy realized by the Nicol or Glan-Thompson prism or by special types of polaroid), and (2) retardation plates, which in an ideal case will transmit electric or magnetic oscillations in two perpendicular directions and will introduce a known phase difference between the oscillations along these two directions, the fast and slow axes (realized by different kinds of mica or quartz plates). When the specific intensity of a nonparallel or the net flux of a parallel radiation is observed or measured after passage through an analyzer, two positions of the analyzer (measured, for example, by the angle ψ from an arbitrary direction) can be found where a maximum and a minimum appear, if the radiation is polarized. The state of polarization can be described by the maximum and minimum intensity, and by the angle ψ_0, where the maximum appears, defining the position of the plane of polarization (the plane through the direction of radiation and the direction of maximum intensity). Such a determination is not unique since it is not possible to distinguish between a partial linear polarization (neutral + linear polarization) and an elliptical polarization. If a retardation plate of a known phase difference (retardation) δ is inserted in front of the analyzer and oriented in such a way that the fast axis coincides, for example, with the direction $\psi = 0$, then the intensity behind the analyzer $I(\psi,\delta)$ varies with ψ according to the relationship

$$I(\psi,\delta) = \tfrac{1}{2}[I + Q\cos 2\psi + (U\cos\delta - V\sin\delta)\sin 2\psi] \tag{4-10}$$

The quantities I, Q, U, and V are the so-called Stokes polarization parameters (57). They are all of the same dimensions, i.e., those of specific intensity or of net flux (for parallel radiation), and have a simple physical meaning: if I_e and I_r denote the intensities for $\psi = 0°$ and $90°$, respectively, then

$$I = I_e + I_r \qquad Q = I_e - I_r \tag{4-11}$$

The first parameter, as the sum of two intensities in two perpendicular but otherwise quite arbitrary directions, measures the energy flow as discussed above and is called the total intensity. The second parameter measures their difference. If $\psi = \psi_0$ represents the directions of the maximum intensity (or the plane of polarization), then

$$U = Q \tan 2\psi_0 \tag{4-12}$$

and thus U is a parameter determining the position of the plane of polarization.

The last parameter V measures the ellipticity of polarization. If a and b stand for the lengths of the major and minor axes of the ellipse described by the end of the electric vector and $\tan \beta = b/a$, then

$$\sin 2\beta = \frac{V}{\sqrt{Q^2 + U^2 + V^2}} \tag{4-13}$$

It is easy to see that for neutral (unpolarized) radiation $Q = U = V = 0$; for linearly polarized radiation $V = 0$; for circular polarization $Q = U = 0$, $V \neq 0$.

The basic advantage of the Stokes parameters is that they are additive for several incoherent streams of radiations, i.e., for streams that have no permanent phase relations among themselves. In general the mixture of streams contains some part which is neutral (unpolarized). In such a case of partial polarization, because of the additiveness of these parameters and of the foregoing characteristics of neutral radiation, $I > \sqrt{Q^2 + U^2 + V^2}$. The degree of polarization P is then conveniently introduced as

$$P = \frac{\sqrt{Q^2 + U^2 + V^2}}{I} \tag{4-14}$$

which compares the intensity of all polarized components in the mixtures with the total intensity of all components. If there is no polarized component in the mixture, $P = 0$; if there is no neutral component present (case of total polarization), $P = 1$.

For only linearly polarized components, $V = 0$, and if $\psi = 0$ coincides with the direction of maximum intensity, $U = 0$; then with respect to (4-11)

$$P = \frac{I_{\max} - I_{\min}}{I_{\max} + I_{\min}} \tag{4-15}$$

has the form introduced originally by Rubenson.

4.1.4 Problem of Radiative Transfer in the Atmosphere

The basic problem of atmospheric optics and its applications is that of radiative transfer in the atmosphere, i.e., the study of the changes which a radiation will undergo during its passage through or within the atmosphere.

The changes in radiation are only the consequence of the interaction of the radiation with the matter in the atmosphere. The matter in the atmosphere may be considered to be uniformly distributed and the air considered as a uniform medium with the density varying locally, mostly in the vertical direction. Because of the density variation, the refractive index is not constant and its variation causes bending of the light rays, the so-called astronomical or terrestrial refraction.

Much more important for air pollution problems are the changes in radiation due to its interaction with the discrete particles of matter, which are suspended in the air. There are many kinds of such particles, from the molecules of air constituents to large dust particles or large raindrops. When these particles are struck by an electromagnetic wave of a given wavelength, two different physical processes will, in general, occur. The energy of the radiation received can be transformed into another type of energy such as heat, energy of chemical reaction, or radiation of a different wavelength. In such a case the physical process of the energy transformation is called absorption. Alternatively, the energy received is not transformed, but is reradiated from the particle in the same wavelength. It is reradiated in all directions, but usually with different intensities in different directions. This process is called scattering, and the particle a scattering center. In the visible range, there exist a very few spectral bands with real absorption; the most important absorption bands are in the ultraviolet and infrared regions. On the other hand, the scattering process is not so selective with respect to the wavelength as is the absorption. Its magnitude does vary with wavelength but continuously and thus is present for all wavelengths. In both cases the incident radiation is weakened or attenuated by the amount of radiant energy absorbed or scattered in different directions by the particles. After traversing the length ds in the direction of the propagation, through the absorbing or scattering medium, the intensity I_ν of the incident radiation of frequency ν changes by the amount

$$dI_\nu = -k_\nu \rho I_\nu \, ds \tag{4-16}$$

where ρ is the density of the medium. The proportionality factor k_ν is called mass absorption coefficient $k_\nu^{(a)}$ or mass scattering coefficient $k_\nu^{(s)}$, as the case may be. If both processes are present, $k_\nu = k_\nu^{(a)} + k_\nu^{(s)}$, and this is called the mass attenuation coefficient.

The amount of radiant energy in a pencil of radiation determined by the solid angle $d\omega$ and having the intensity I_ν, which is absorbed or scattered during the passage through the mass element dm of the medium, is equal to

$$k_\nu I_\nu \, dm \, d\omega \tag{4-17}$$

The amount of radiant energy which is scattered from the mass element dm into a solid angle $d\omega'$ deviated from the direction of the incident radiation by the angle θ is equal to

$$I_\nu^{(s)} \, d\omega' = k_\nu^{(s)} I_\nu \, dm \, d\omega \frac{d\omega'}{4\pi} P(\cos\theta) \tag{4-18}$$

In (4-18) the phase function $P(\cos\theta)$ specifies the angular distribution of the intensity of the scattered radiation. If the absorption of the medium is negligible ($k_\nu^{(a)} = 0$), then it is a case of perfect scattering; it follows from the principle of conservation of energy that the total energy of the radiation scattered in all directions from dm must be equal to the energy loss of the incident radiation. Consequently, the expression (4-18) integrated over all solid angles must be equal to (4-17) with $k_\nu \equiv k_\nu^{(s)}$.

$$\int I_\nu^{(s)} \, d\omega' = k_\nu^{(s)} I_\nu \, dm \, d\omega \frac{1}{4\pi} \int P(\cos\theta) \, d\omega' \tag{4-19}$$

Hence the phase function has to satisfy the condition

$$\frac{1}{4\pi} \int P(\cos\theta) \, d\omega' = 1 \tag{4-20}$$

i.e., it has to be normalized to unity. The equation (4-19) gives the method of computing the mass scattering coefficient from the known angular distribution of the specific intensity of scattered radiation.

In the most general case the interaction of the mass dm with the incident radiation may also result in an increase of the intensity I_ν. The gain of the radiant energy can be put equal to

$$j_\nu \, dm \, d\omega \tag{4-21}$$

where the proportionality factor j_ν is called the mass emission coefficient. The emission of the excited dissociated atoms and molecules of the air constituents (N_2, O_2, etc.) occurs in the upper levels of the atmosphere (>80 km) and is very weak, thus being negligible with respect to the sun's radiation during daylight hours. Another type of emission is the heat radiation according to Kirchhoff's law, which is effective in the far infrared region and has very little importance in optical problems. In a scattering medium, however, a virtual emission exists, because of the rescattering of scattered radiation. The radiation from all scattering centers around the mass element *dm* illuminates the particles in the element *dm*, and some part of this radiation is scattered again by the particles in *dm* into the direction of I_ν. The amount of energy thus rescattered into the direction of I_ν is given by (4-18), if I_ν is replaced by

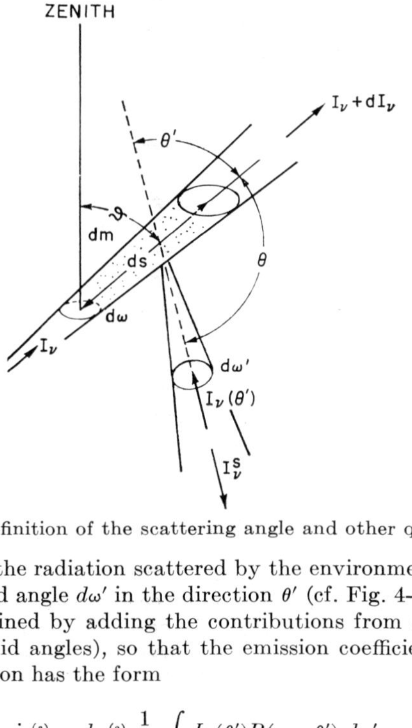

FIG. 4-1. Definition of the scattering angle and other quantities.

the intensity $I_\nu(\theta')$ of the radiation scattered by the environment of *dm* and incident on *dm* through the solid angle $d\omega'$ in the direction θ' (cf. Fig. 4-1). The total gain of radiant energy is obtained by adding the contributions from all directions (i.e., by integrating over all solid angles), so that the emission coefficient $j_\nu^{(s)}$ corresponding to this apparent emission has the form

$$j_\nu^{(s)} = k_\nu^{(s)} \frac{1}{4\pi} \int I_\nu(\theta') P(\cos\theta') \, d\omega' \tag{4-22}$$

The net change of the radiant energy of the incident radiation after passing through *dm* is equal to the sum of the losses, expressed by (4-17) or (4-18), and of the gains, given by (4-21) with (4-22) substituted. After dividing out the common factors, the following equation results:

$$\frac{dI_\nu(\mu,\phi)}{\rho \, ds} = -k_\nu I_\nu(\mu,\phi) + \frac{1}{4\pi} k_\nu^{(s)} \int_{-1}^{+1} \int_0^{2\pi} P(\mu,\phi;\mu',\phi') I_\nu(\mu',\phi') \, d\mu' \, d\phi' \tag{4-23}$$

where the direction of incident or scattered radiation is defined by the zenith angle ϑ and the azimuth ϕ and the parameter $\mu = \cos\vartheta$ is used. Furthermore, $P(\mu,\phi;\mu',\phi')$

denotes the phase function corresponding to the scattering angle between the directions given by (μ,ϕ) and (μ',ϕ'), respectively. Equation (4-23) represents the equation of radiative transfer through a scattering and absorbing medium, if the true emissions previously mentioned are neglected. This equation has the form of an integro-differential equation for the specific intensity I_ν. This equation can be solved, provided that the law of scattering is known, which then permits the determination of the phase function and the mass scattering coefficient.

The foregoing relations refer only to the total specific intensity and are thus valid for unpolarized light. Unfortunately, the scattered radiation is usually polarized. Therefore, for a correct solution of the problem of radiative transfer in the atmosphere, it is necessary to introduce the polarization by considering the other Stokes parameters. This can be done by replacing I_ν in (4-23) by a vector $\mathbf{I}_\nu$, having the four Stokes parameters as components. The equation of radiative transfer will then represent a set of four equations for the four components, in which the phase function will be replaced by a system of 16 functions, defining the linear relationship between the Stokes parameters for the scattered and for the incident radiation. These functions can be considered as the elements of a matrix, called the phase matrix, consisting of four columns and four rows. The equation of radiative transfer has then the same form as (4-23), only the intensity I_ν is replaced by the one column matrix $\mathbf{I}_\nu$ with the elements I_ν, Q_ν, U_ν, and V_ν, and the phase function $P(\mu,\phi;\mu',\phi')$ is replaced by the phase matrix $\mathbf{P}(\mu,\phi;\mu',\phi')$. The solution of the equation of radiative transfer is, in general, very complicated and is discussed for a few special cases by Chandrasekhar (8). Essential simplifications can be achieved by assuming the earth's atmosphere to be plane parallel and of infinite horizontal extent with a stratified density distribution. In such a case the length ds along the direction of the intensity I_ν can be expressed by means of corresponding change in the vertical coordinate dz and the zenith distance of the direction

$$ds = \frac{dz}{\mu} \tag{4-24}$$

Furthermore, since ρ and k_ν are functions of z only, it is convenient to express the vertical distances by means of a parameter called the the optical thickness, defined as

$$\tau = \int_z^\infty k_\nu \rho \, dz \tag{4-25}$$

measured from the top of the atmosphere down. This quantity can be computed from the known dependence of the attenuation coefficient on height. For the case of perfect scattering ($k_\nu \equiv k_\nu^{(s)}$) the optical thickness is the only parameter which enters in the problem as a dependent variable.

In problems dealing with the transfer of solar radiation through the atmosphere, it is also convenient to separate parallel solar radiation from the scattered or diffuse radiation from the atmosphere. If πF_ν is the net flux of the solar radiation incident on the plane-parallel atmosphere in the direction specified by $(-\mu_0,\phi_0)$, then this net flux will be reduced by attenuation according to (4-16), and thus at the level of the optical thickness τ its reduced values will be $\pi F_\nu \exp(-\tau/\mu_0)$. The particles in the mass dm will be illuminated by this reduced flux, and the apparent emission has to be increased by the part of this radiation scattered into the direction (μ,ϕ). In the equation of radiative transfer another term has to be added on the left-hand side, namely,

$$\tfrac{1}{4} k_\nu^{(s)} F_\nu e^{-\tau/\mu_0} P(\mu,\phi;-\mu_0,\phi_0)$$

In the case with polarization and for perfect scattering, the equation of radiative transfer in a plane-parallel atmosphere can be written in the form

$$\mu \frac{d\mathbf{I}(\tau,\mu,\phi)}{d\tau} = \mathbf{I}(\tau,\mu,\phi) - \frac{1}{4} e^{-\tau/\mu_0}\mathbf{P}(\mu,\phi;-\mu_0,\phi_0)\cdot\mathbf{F} - \frac{1}{4\pi}\int_0^{2\pi}\int_{-1}^{+1} \mathbf{P}(\mu,\phi;\mu',\phi')\cdot\mathbf{I}(\tau,\mu',\phi')\,d\mu'\,d\phi' \quad (4\text{-}26)$$

The subscript ν is omitted, since the frequency dependence is included in the value of the optical thickness τ.

Equation (4-26) can be used for the determination of the intensity and the Stokes polarization parameters for the diffuse sky radiation in a plane-parallel atmosphere provided that the law of scattering is known. The law of scattering has been derived from the electromagnetic theory for spherical and ellipsoidal particles. Even for the simplest form, the sphere, the theory of scattering is quite complicated.

4.1.5 Scattering of Light by a Single Spherical Particle

The theory of scattering by a spherical particle is derived, in the original formulation by G. Mie (38) and P. Debye (11), from the following boundary-value problem:

A plane-parallel electromagnetic wave of wavelength λ is incident on a dielectric sphere of radius a and of refractive index m. The oscillations of the incident wave produce forced oscillations in the interior of the sphere and also on the exterior of the sphere. The oscillations of the exterior propagate as a spherical wave from the center of the sphere. The electric vector of the electromagnetic field outside the sphere can be considered as the sum of the electric vectors corresponding to the incident wave and to the scattered wave. The mathematical form of the solution of Maxwell's equations for the electromagnetic field in the interior and on the exterior can be determined from the boundary conditions on the sphere, requiring the continuity of the tangential components of the electric and magnetic vectors. The form of the solution for the interior and exterior of the sphere is expressed as a series in the same spherical wave functions as the incident wave with two sets of coefficients, uniquely determined by the boundary conditions.

From the solutions for the electric vector of the scattered radiation only, it is not difficult to derive the expressions for a set of Stokes parameters. It is more convenient in this case to choose the intensities $I_\perp^{(s)}$ and $I_\parallel^{(s)}$ normal and parallel to the plane of scattering than to choose the parameters I and Q used before. The plane of scattering is determined by the incident and the scattered rays. If similar parameters are introduced for the incident wave, $F_\perp^{(i)}$ and $F_\parallel^{(i)}$ being the components of the net flux normal and parallel to the plane of scattering, then the relationship between the parameters of the wave scattered (into a unit solid angle) and of the incident wave can be written in the form of a system of linear equations:

$$\begin{aligned} I_\perp^{(s)} &= F_1 F_\perp^{(i)} & \qquad U^{(s)} &= F_3 F_u^{(i)} + F_4 F_v^{(i)} \\ I_\parallel^{(s)} &= F_2 F_\parallel^{(i)} & \qquad V^{(s)} &= -F_4 F_u^{(i)} + F_3 F_v^{(i)} \end{aligned} \quad (4\text{-}27)$$

or in a matrix form

$$\begin{pmatrix} I_\perp^{(s)} \\ I_\parallel^{(s)} \\ U^{(s)} \\ V^{(s)} \end{pmatrix} = \begin{pmatrix} F_1 & 0 & 0 & 0 \\ 0 & F_2 & 0 & 0 \\ 0 & 0 & F_3 & F_4 \\ 0 & 0 & -F_4 & F_3 \end{pmatrix} \begin{pmatrix} F_\perp^{(i)} \\ F_\parallel^{(i)} \\ F_u^{(i)} \\ F_v^{(i)} \end{pmatrix} \quad (4\text{-}28)$$

The functions $F_i(\alpha,m;\theta)$, $(i = 1,2,3,4)$, in the scattering matrix of (4-28) are functions of the nondimensional parameter $\alpha = 2\pi a/\lambda$, the refractive index of the sphere m, and the scattering angle θ between the directions of the incident and of the scattered ray. Their explicit form can be written in terms of two infinite series,

$$
\begin{aligned}
S_L(\alpha,m,\theta) &= k^{-1}\sum_1^\infty [a_n(\alpha,m)p_n(\theta) + b_n(\alpha,m)t_n(\theta)]\\
S_R(\alpha,m,\theta) &= k^{-1}\sum_1^\infty [a_n(\alpha,m)t_n(\theta) + b_n(\alpha,m)p_n(\theta)]
\end{aligned}
\tag{4-29}
$$

where a_n, b_n are complex coefficients and

$$
p_n(\theta) = \frac{dP_n(x)}{dx} \qquad t_n(\theta) = n(n+1)P_n(x) - x\frac{dP_n(x)}{dx}
$$

where $x = \cos\theta$, with $P(x)$ denoting the Legendre polynomial of nth degree.
In terms of these series,

$$
\begin{cases} F_1 = S_R S_R^* & F_3 = \text{Re}\,\{S_L S_R^*\} \\ F_2 = S_L S_L^* & F_4 = \text{Im}\,\{S_L S_R^*\} \end{cases}
\tag{4-30}
$$

(The * denotes the complex conjugate quantities.)

The computation of the functions F_i is rather laborious and was done by several authors. The tables usually contain the values of F_1 and F_2, and the first rather extensive tables were published by the National Bureau of Standards (40). The functions F_3 and F_4 have been computed rather recently (51) by means of auxiliary functions published by the National Bureau of Standards (40). The most extensive tables for $\alpha = 0.1 - 30.0$ by steps in 0.1 and for $m = 1.33$, 1.40, 1.44, 1.485, 1.50 have been computed under the direction of R. Penndorf and B. Goldberg (18) and are being prepared for publication. The tables of the functions F_i usually contain another important quantity known as total scattering coefficient or scattering cross section $K(\alpha,m)$, defined as the total energy scattered per second per cross-sectional area of particle, illuminated by a unit flux of neutral light. The total energy scattered per unit time is according to (4-19) equal to $\int(I_\perp^{(s)} + I_\parallel^{(s)})\,d\omega'$; after substitution from (4-27) for a neutral incident wave of unit flux ($F_\perp^{(i)} = F_\parallel^{(i)} = \frac{1}{2}$, $F_u^{(i)} = F_v^{(i)} = 0$),

$$
\int_{4\pi} (I_\perp^{(s)} + I_\parallel^{(s)})\,d\omega' = \frac{\lambda^2}{4\pi}\int_0^\pi (F_1 + F_2)\sin\theta\,d\theta = K(\alpha,m)\pi a^2 \tag{4-31}
$$

so that

$$
K(\alpha,m) = \frac{1}{\alpha^2}\int_0^\pi (F_1+F_2)\sin\theta\,d\theta = \frac{2}{\alpha^2}\sum_1^\infty n(n+1)[|a_n|^2 + |b_n|^2] \tag{4-32}
$$

The coefficients a_n and b_n can be easily developed in the series of powers of α, with the first term in the series for a_n, b_n being of the order α^{2n+1} and α^{2n+3}, respectively. In the series (4-29) for the special case of a very small particle ($\alpha \to 0$), only the terms with a_1 can be then considered, leading to

$$
S_L = k^{-1}\frac{m^2-1}{m^2+2}\alpha^3\cos\theta \qquad S_R = k^{-1}\frac{m^2-1}{m^2+2}\alpha^3 \tag{4-33}
$$

With these values substituted in (4-30), the relation (4-28) reduces to the form

$$
\begin{pmatrix} I_\perp^{(s)} \\ I_\parallel^{(s)} \\ U^{(s)} \\ V^{(s)} \end{pmatrix} = \frac{16\pi^4}{\lambda^4}a^6\left(\frac{m^2-1}{m^2+2}\right)^2 \begin{pmatrix} 1 & 0 & 0 & 0 \\ 0 & \cos^2\theta & 0 & 0 \\ 0 & 0 & \cos\theta & 0 \\ 0 & 0 & 0 & \cos\theta \end{pmatrix} \begin{pmatrix} F_\perp^{(i)} \\ F_\parallel^{(i)} \\ F_u^{(i)} \\ F_v^{(i)} \end{pmatrix} \tag{4-34}
$$

and the expression for the total energy scattered per unit time by the particle, given in (4-31), assumes the form

$$\int (I_{\perp}^{(s)} + I_{\parallel}^{(s)})\, d\omega' = \frac{128\pi^5}{3\lambda^4} a^6 \left(\frac{m^2 - 1}{m^2 + 2}\right)^2 \tag{4-35}$$

The scattered intensity in (4-34) has the same form as the intensity of radiation from an oscillating dipole* with the polarizability

$$\bar{\alpha} = a^3 \frac{m^2 - 1}{m^2 + 2} = a^3 \frac{\epsilon - 1}{\epsilon + 2} \tag{4-36}$$

Lord Rayleigh in 1871 arrived at the same result, assuming only that the volume of the scattering center is very small compared to the wavelength and $\epsilon \to 1$ without any specification of its shape. Consequently, in (4-34) the volume $dV = \frac{4}{3}\pi a^3$ should be introduced, and the result will be valid for any shape of the scattering volume.

The circumstance that the scattering from a very small volume becomes independent of the shape of the volume can be considered as a consequence of another fact. The polarizability $\bar{\alpha}$ is a molecular or atomic constant, which can be related to macroscopically measurable quantities such as the refractive index n of the gas, a quantity different from the refractive index m used above. If N denotes the number of molecules in a unit volume, the total energy scattered by the unit volume is given by the expression in (4-35) multiplied by N, and hence according to the definition of the mass scattering coefficient in (4-19),

$$k_{\nu}^{(s)} = \frac{128\pi^5}{3\lambda^4} \bar{\alpha}^2 \frac{N}{\rho} \qquad (\rho = \text{density}) \tag{4-37}$$

When the Lorentz-Lorenz law for $\bar{\alpha}$ is used, i.e.,

$$\bar{\alpha} = \frac{3}{4\pi N} \frac{n^2 - 1}{n^2 + 2}$$

with n denoting the macroscopic refractive index, then the expression for $k_{\nu}^{(s)}$ can be written in the form

$$k_{\nu}^{(s)} = \frac{24\pi^3}{\lambda^4} \left(\frac{n^2 - 1}{n^2 + 2}\right)^2 \frac{1}{N\rho} = \frac{8\pi^3}{3\lambda^4} \frac{(n^2 - 1)^2}{N\rho} = \frac{32\pi^3 (n - 1)^2}{3\lambda^4 N\rho} \tag{4-38}$$

where the last two expressions represent the approximations to the fact that for a gas n^2 (or n) is very closely equal to 1. The expression in (4-38) thus represents the mass scattering coefficient of a gas of refractive index n and density ρ for the case of molecular scattering. The size or the radius of the molecules does not appear in (4-38), which proves that the result is independent of the shape of the scattering centers. The mass scattering coefficient of molecular scattering for a mixture of gases can be easily computed from the fact that the expression in (4-35) is additive for different constituents of the mixture.

It should be mentioned that the matrix in (4-34) is not equivalent to the phase matrix $\mathbf{P}(\mu,\phi;\mu',\phi')$ in the equation (4-26). The matrix in (4-34) should be first normalized to unity, which can be done by multiplying by the factor $\frac{3}{2}$. Furthermore, the scattering angle should be expressed by the zenith distances and the azimuth of the directions of the incident and of the scattered radiation. Since the plane of scattering does not have a fixed orientation, it is convenient to introduce the intensities I_{ℓ} and I_r, parallel and perpendicular, respectively, to the local vertical plane (containing the point of observation and the zenith) rather than $I_{\perp}^{(s)}$ and $I_{\parallel}^{(s)}$ parallel

* An elementary discussion of the oscillations of a dipole can be found in Ref. 32.

and perpendicular to the scattering plane. The corresponding transformation of the matrix in (4-34) is rather complicated and is given by Chandrasekhar (8).

The characteristics of the scattered radiation for the Rayleigh limit ($\alpha \to 0$) are quite simple. For an unpolarized incident wave $F_{\parallel}^{(i)} = F_{\perp}^{(i)} = \frac{1}{2}F_0$, (4-34) yields for the total intensity of the scattered radiation per unit solid angle the expression

$$I^{(s)} = I_{\perp}^{(s)} + I_{\parallel}^{(s)} = \frac{3}{16\pi} k_\nu^{(s)} \rho [1 + \cos^2 \theta] F_0 \tag{4-39}$$

and for the degree of polarization the expression

$$P = \frac{\sin^2 \theta}{1 + \cos^2 \theta} \tag{4-40}$$

The distribution of the scattered intensity is symmetrical with respect to the direction $\theta = 0$ and $\theta = 90°$, with the maximum in the forward direction equal to the maximum in the backward direction and with the minimum at right angles. The scattered radiation in the forward and backward direction is unpolarized, whereas it is totally linearly polarized at right angles ($\theta = 90°$) and partially linearly polarized in between. The polarization is positive, the plane of polarization being normal to the plane of scattering.

For moderate values of α the computation of the coefficients a_n and b_n cannot be simplified, and the characteristics of scattering in such a case can be derived from the tabulated values of the intensities $I_{\perp}^{(s)}$ and $I_{\parallel}^{(s)}$. For small values of m (≤ 2.0) the asymmetry in favor of forward scattering appears; for very large values of m (opaque particles) the asymmetry is in favor of backward scattering. With increasing values of α, the total intensity of scattered radiation shows great variations with small change in θ. In a polar diagram with the center in the particle, the curve of the intensity has the appearance of a butterfly wing (cf. Fig. 4-2). The intensities $I_{\perp}^{(s)}$ and $I_{\parallel}^{(s)}$ separately have a similar character, with very rapid changes for small increments in θ or α. The secondary maxima and minima in both intensities are not distributed regularly, so that their ratio or the degree of polarization shows similar fluctuations, the degree of polarization changing from positive to negative values several times in the interval $0° \leq \theta \leq 180°$. The total scattering cross section $K(\alpha,m)$ has a more regular character. For small α it increases rapidly along a fourth-order parabola (consequence of the λ^{-4} law of $k_\nu^{(s)}$ for Rayleigh's limit), then reaches a maximum (for $m = 1.33$ at $\alpha = 6.6$, for $m = 1.50$ at $\alpha = 4.3$), and finally decreases in several waves asymptotically toward $K(\alpha,m) = 2$. Over this general trend are superposed numerous relatively small fluctuations (cf. Fig. 4-3). The reason for these violent fluctuations in the intensities and smaller ones in the curve for $K(\alpha,m)$ is that with larger α (larger size) the radiation generated by the incident wave has in addition to the dipole oscillations the character of oscillating multipoles of the order 2^n. The variety of the induced oscillations increases very rapidly with increasing size.

For very large α the Mie-Debye theory should offer results identical with those derived from the laws of geometrical optics, where an individual ray can be followed through its reflection, refraction, and diffraction and the intensity of the external field is obtained by addition of the separate fields with the consideration of their eventual interference. The solution of the problem in terms of infinite series (4-29) is not suitable for the investigation of scattering from a dielectric sphere for very large α. The use of the solution of corresponding integral equations for this purpose is more promising, and several approximations have been recently derived by D. S. Saxon.* The quality and the regions of validity of these approximations are being studied.

The theory of scattering by a dielectric sphere has been extended in two directions.

* Scientific Report No. 9 (1954), Contract AF 19(122)-239, Department of Meteorology, University of California at Los Angeles.

The spherical shape has been replaced by an ellipsoid by Gans (14), but only the dipole oscillations were considered. The effect of the shape has been studied for microwaves (2) and seems to be important only for large refractive indices. Furthermore, the sphere has been replaced by two concentric spheres of different refractive

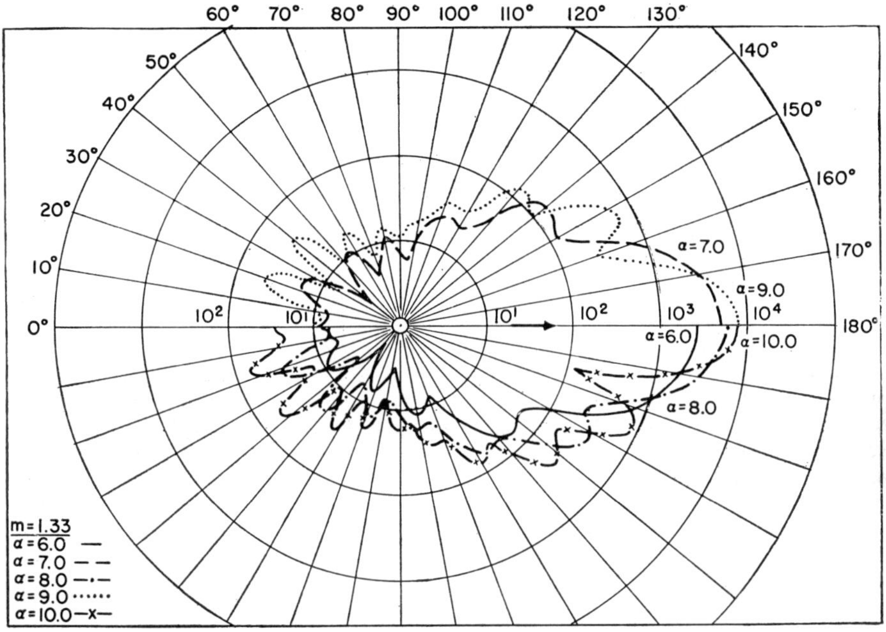

FIG. 4-2. Distribution of the total intensity scattered by a single spherical particle for different scattering angles and different values of α. (Refractive index of the particle $m = 1.33$.)

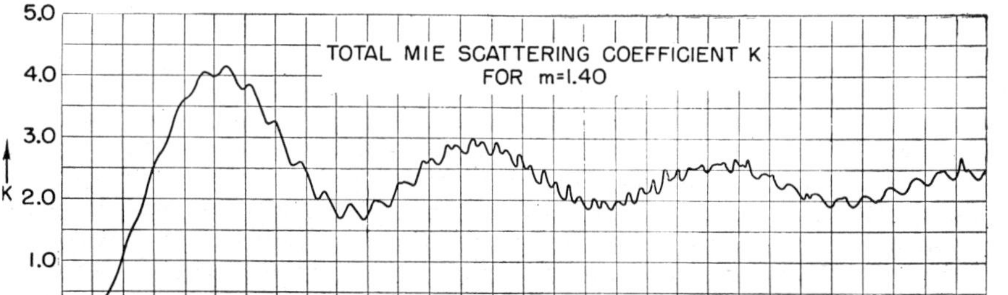

FIG. 4-3. Total scattering coefficient $l(\alpha)$ as a function of the nondimensional parameter α for the refractive index $m = 1.40$.

indices (1). But even with a very thin coating, a particle scatters as if composed entirely of the coating material. The mathematical difficulties in treating these extensions, however, are very large, so that their introduction does not seem to be of great practical value.

4.1.6 Scattering of Light by the Atmospheric Aerosol

The discussion in the previous section refers to a single particle. The atmosphere, however, contains a very large number of particles of different sizes and of different physical properties, in addition to the much larger number of the molecules of its gaseous constituents. The entire system, the particles or the particulate matter and the air molecules, is usually called an aerosol. In deriving the law of scattering for such an aerosol it is usually assumed that all particles may be considered as incoherent point sources of the scattered radiation. This assumption is justified by the random distribution and by the quite irregular motion of individual molecules and other particles which prevent the establishment of any phase relations between the individual scattered waves. Under this assumption the scattering properties of an aerosol can be computed by simple addition of additive physical quantities. Since the Stokes parameters are additive, it is thus sufficient to add the Stokes parameters of scattered waves radiated from each scattering center in a given volume of the aerosol in order to get the law of scattering of the corresponding volume. Since for very small particles, such as molecules, the result is independent of the size, the addition for all such particles is very simple and has been done in the discussion of Rayleigh scattering. In this case the scattering matrix is dependent on the scattering angle only, and all Stokes parameters are obtained by multiplication of a factor containing the mass scattering coefficient given in (4-38).

For particles of a radius comparable to the wavelength (Mie particles), the computation of the volume scattering coefficient is relatively simple. This coefficient can be derived in a similar way as the mass coefficient by putting in (4-17) $k_\nu^{(s)}\,dm = \beta_\nu^{(s)}\,dV$, so that $\beta_\nu^{(s)}$ represents the total energy scattered by a unit volume, illuminated by a neutral radiation of unit intensity per unit solid angle or of unit net flux of parallel radiation. The total energy scattered by the unit volume is given as the sum of the expressions (4-31) corresponding to each individual particle, since the integrand in (4-31) is additive for incoherent point sources. If $N(a,m)$ denotes the number of particles in the particular unit volume of radius a and of refractive index m, then according to (4-31),

$$\beta_\nu^{(s)} = \pi \sum_{a,m} K(\alpha,m)N(a,m)a^2 \tag{4-41}$$

or for a continuous size distribution,

$$\beta_\nu^{(s)} = \pi \sum_{m} \int_0^\infty K(\alpha,m)N(a,m)a^2\,da \tag{4-42}$$

where the sum in (4-41) is to be performed over the discrete values of a and m and in (4-42) $N(a,m)$ denotes the number of particles of the size in the interval $\langle a, a + da\rangle$.

Since the Stokes parameters are additive and the functions of the matrix (4-28) $F_i(\alpha,m;\theta)$, ($i = 1,2,3,4$), are dependent on α and m, the addition of the matrices for each particle has to be performed, which is equivalent to the addition of corresponding terms in all the matrices. In other words, for an aerosol, the Stokes parameters for a scattered radiation will be given by exactly the same equations as (4-28), only the functions $F_i(\alpha,m;\theta)$ have to be replaced by the sums

$$\sum_{a,m} F_i(\alpha,m;\theta)N(a,m) \tag{4-43}$$

or

$$\sum_{m} \int_0^\infty F_i(\alpha,m;\theta)N(a,m)\,da \tag{4-44}$$

As is evident from the tables of the scattering function F_i, they are very rapidly varying functions of α. It is thus necessary to perform the numerical integration using very small intervals da, if an accurate result is desired. On the other hand, if $N(a,m)$ is a very slowly varying function of a, during the summation or integration the fast variations of F_i for small increments in θ are smoothed, so that an aerosol will have much smoother variation of the intensity or polarization distribution with scattering angle than a single particle. This fact can be illustrated by Fig. 4-4, where the distribution of the intensity as a function of the scattering angle is plotted for a normal distribution in and around the same mean ($\alpha = 3.0$), but with different standard deviations. From this fact it is then quite obvious that it is impossible

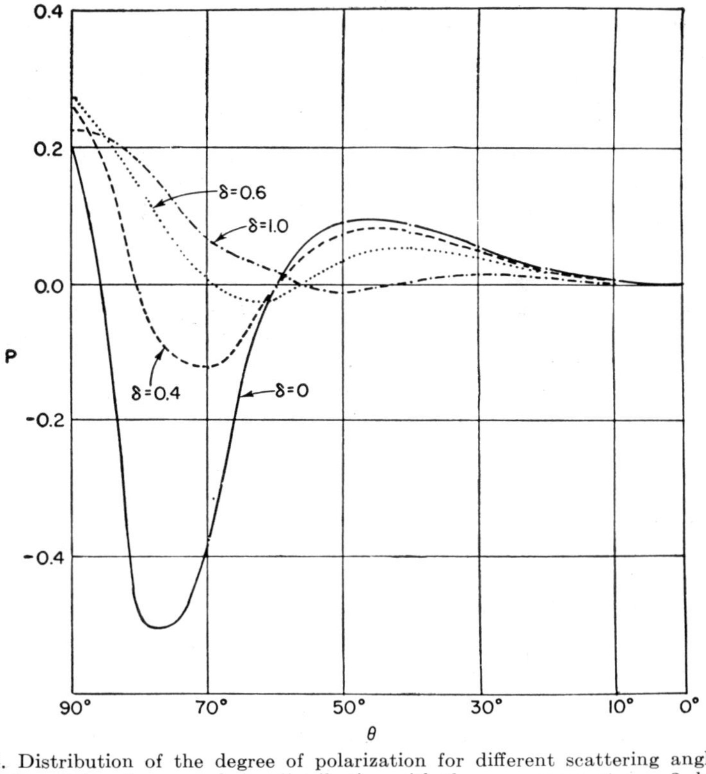

FIG. 4-4. Distribution of the degree of polarization for different scattering angles of a mixture of particles of a normal size distribution with the same mean at $\alpha = 3$, but with different standard deviations σ.

and also improper to express the scattering properties of an aerosol by introducing a mean size.

Since the scattering angle θ enters in the function F_i implicitly, it is necessary to perform the summation in (4-43) and the integration in (4-44) for each scattering angle separately. To avoid this difficulty, it would be desirable to separate the dependence on θ from the dependence on α and m. This can be done by developing the F_i's as a series of functions depending only on the angular variable θ. Since in the series S_L and S_R in (4-29) this separation is done by means of the Legendre polynomials and their derivatives, it is convenient to develop the functions F_i in series of Legendre polynomials in the scattering angle θ (52). The use of Legendre polynomials introduces another great advantage, in that it allows one to replace the scattering angle by the zenith distances of the incident and scattered waves and their azimuth differences, and thus to perform the transformation which, as was mentioned previously in the case of Rayleigh scattering, is necessary for the solution of the equation of radiative transfer in (4-26). For a complete discussion of the scattering properties of an atmospheric aerosol it is necessary to know the actual size distribution of Mie particles in the atmosphere, i.e., the quantity $N(a,m)$ at the ground and its variations with height. According to the most reliable direct measurements, made by Junge (33), under mean conditions, $N(a)$* can be considered proportional to a negative power of a,

$$N(a) \sim a^{-\beta} \qquad 3 \leq \beta \leq 4 \tag{4-45}$$

provided that the air has remained over a continent for a sufficiently long time. Close to the coast, the air contains small droplets or salt nuclei from sea spray in addition to a particle distribution such as given by (4-45). The distribution of particles in a coastal region thus becomes more complicated and more variable. For a large amount of spray particles, their superposition over the curve defined by (4-45) may produce a secondary maximum or a small cusp at their maximum frequency. In such a case the distribution function $N(a)$ can be approximated by two power laws with different exponents β. [See D. J. Moore and B. J. Mason, *Quart. J. Roy. Meteor. Soc.* **80,** 583 (1954).] For smaller amounts the size distribution is better approximated by an exponential law (15),

$$N(a) \sim e^{-\gamma a} \qquad 1.3 \leq \gamma \leq 4.1 \tag{4-46}$$

for $a \geq 0.05\mu$. The change in the distribution of Mie particles with the distance inland from the coast can be also traced indirectly in the variations of the observed atmospheric transmissions in short wavelengths (12).

Several measurements of the vertical distribution of Mie particles indicate an exponential decrease at different rates up to 4 or 5 km, with a slower decrease, a constant value, or even a slight increase above this level (47). This distribution corresponds to a rather normal condition. On other occasions, however, the size distribution with height may show large irregularities, as observed by Waldram (63). The distribution in very high altitudes is, of course, quite unknown. Haze layers observed during high-altitude flights (44) indicate the presence of large particles in considerable amounts even in the stratosphere.

The varying content of Mie particles and eventual irregularities in their size distribution make any quantitative evaluation of the scattering properties of the real atmosphere rather uncertain. Only the normal conditions, represented by distributions such as in (4-45) or (4-46), are accessible to a mathematical analysis. The constant in (4-45) and (4-46) can be conveniently expressed in terms of the total number of all particles of radii $a \geq a_0$. The introduction of a lower limit for the size is justified by the observational fact that the frequency curves indicate a maximum and a decrease toward smaller radii. If the α corresponding to the radius of the maximum frequency is sufficiently small (≤ 0.1), the smaller particles scatter in exactly the same way as the molecules. Their total number is, however, quite negligible with respect to the number of air molecules (10^{19} molecules/cu cm in the lower troposphere) so that their effect can be neglected and their presence can be disregarded. The total number of Mie particles can be obtained by multiplying (4-45) or (4-46) by da and by integrating from a_0 to infinity. After elimination of the undetermined constants in (4-45) and (4-46), the substitution in (4-42) and (4-44) yields the following expressions for the volume scattering coefficient and for the Stokes parameters of the atmospheric aerosol (Mie particles) for the negative power law:

$$\beta_{\nu^{(s)}}(m) = \frac{(\beta - 1)\lambda^2}{4\pi} \bar{N}(m)\alpha_0{}^{\beta-1} \int_{\alpha_0}^{\infty} K(\alpha,m)\alpha^{-\beta+2}\, d\alpha \tag{4-47}$$

$$\bar{F}_i(m,\theta) = (\beta - 1)\bar{N}(m)\alpha_0{}^{\beta-1} \int_{\alpha_0}^{\infty} F_i(\alpha,m;\theta)\alpha^{-\beta}\, d\alpha$$

* Since the existing measurements give the total number of particles without any distinction in refractive index, m is omitted in the argument.

for the exponential law:

$$\beta_\nu^{(s)}(m) = \frac{\sigma\lambda^2}{4\pi}\bar{N}(m)\int_{\alpha_0}^{\infty} K(\alpha,m)e^{\sigma(\alpha_0-\alpha)}\alpha^\lambda\, d\alpha$$

$$\bar{F}_i(m,\theta) = \sigma\bar{N}(m)\int_{\alpha_0}^{\infty} F_i(\alpha,m;\theta)e^{\sigma(\alpha_0-\alpha)}\, d\alpha \qquad \left(\sigma = \frac{\gamma\lambda}{2\pi}\right) \tag{4-48}$$

where $\bar{N}(m)$ denotes the total number (of all sizes) of Mie particles of the refractive index m in a unit volume and $\alpha_0 = 2\pi a_0/\lambda$. Both expressions in either case are still dependent on the wavelength and on the refractive index. Since it is believed that the majority of existing particles are hygroscopic and therefore surrounded by a thin water layer, $m = 1.33$ is used as a mean value. As long as there is no precise information available about the chemical composition of the Mie particles, the preceding value may be considered as a good approximation. The expressions (4-47) and (4-48) are valid at the ground ($z = 0$). For $z > 0$ they are to be multiplied by the factor $e^{-z/h}$, where h can be considered independent of the particle size and represents the height where the total number of particles has decreased to $e^{-1} = 36.7$ per cent of the value at $z = 0$. Above the level $z = z_0$, where the rate of the decrease $1/h$ changes to $1/h'$, the factor $e^{-z/h}$ should be replaced by $\exp\left[-\frac{z}{h'} + z_0\left(\frac{1}{h'} - \frac{1}{h}\right)\right]$. Given the values of the volume scattering coefficient $\beta_\nu^{(s)}$ and the Stokes parameters for the atmospheric aerosol, it is theoretically possible to compute the additional intensities (and the remaining Stokes parameters) due to the presence of Mie particles in the atmosphere, at least at normal conditions. Varying the models as to the size distribution of these particles, it would be possible to determine the effect of different amounts, types, and vertical distributions of aerosol particles. There is no doubt that such a computation is mathematically very complicated and that it can be performed only with the use of modern fast computing machines.

4.1.7 Scattering Properties of a Turbid Atmosphere, Derived from Actual Observations

In spite of the fact that as far back as 1926 Milch had suggested including the Mie particle in the determination of the degree of polarization, the computation as just formulated has not yet been performed. An attempt for an approximate solution was done by Möller (5), who assumed a size distribution of the form given in (4-45), and who included the effect of secondary scattering (scattering process occurring only twice), but did not consider the polarization of the scattered light.

The only means so far available for getting at least an idea about scattering by large aerosol particles is to compare the results of the actual measurements with the theoretical values for a pure molecular atmosphere. In the Rayleigh limit, the phase matrix has a simple form and the equations of radiative transfer can be solved exactly in terms of special X and Y functions, to be determined by successive iteration of simple integral equations, as shown by Chandrasekhar (8). This solution is exact in the sense that the polarization of the scattered light is properly included and no restriction is made with respect to the repetition of scattering processes (multiple scattering). For different optical thicknesses τ, which according to (4-26) is the only dependent parameter, the intensities of the diffuse sky radiation parallel and normal to the sun's vertical have been computed by Chandrasekhar's method (53). From these values the intensity, the degree of polarization, and the position of the plane of polarization can be easily computed for any direction of observation. It is also possible to include the effect of the ground reflection according to Lambert's law (reflected radiation assumed to be neutral and isotropic). For a given composition and vertical distribution of density, the normal optical thickness τ of a molecular atmosphere can be computed as a function of the wavelength and of the height (13). Therefore, for a given elevation above sea level and for a given wavelength, the distribution of the intensity and of the polarization of diffuse sky radiation in a molecular atmosphere can be considered as known (9), and the deviation of the measured value from the theoretical values can be used as a measure of the scattering by large aerosol particles.

There are several quantities which may be used for such a study. Quantities which show the greatest deviation from the molecular theory are obviously preferable for this purpose. Preference should also be given to quantities which can be easily measured and easily interpreted. One element which shows a very great deviation is the intensity of the sky radiation close to the sun (aureole). In the molecular atmosphere, when crossing toward the solar disk, the intensity should suddenly increase by a factor of 10^8, while in the actual atmosphere, when approaching the solar disk, the intensity increases gradually, with the rate dependent on the turbidity of the atmosphere. The measuring technique in this case is quite difficult (increase of the intensity by several orders of magnitude, undesirable solar glare due to reflection, and diffraction from the edges of baffles). The intensity measurement far from the immediate neighborhood of the sun is not difficult, but for a proper interpretation of the results, a detailed knowledge of the extraterrestrial sun radiation is necessary. This difficulty disappears when measuring the degree of polarization, which, as a ratio of two intensities, is independent of the extraterrestrial flux of sun radiation. Since the quantities mentioned vary in a different way with the turbidity, it is preferable to measure simultaneously all these quantities, i.e., the intensity and the polarization of skylight and the optical thicknesses of the atmosphere obtained from the attenuation of the direct solar radiation. All the measurements should be performed in a very narrow spectral range, since the molecular scattering is highly dependent on the wavelength. Furthermore, the measurements should be done when the conditions are not too far from those of a molecular atmosphere, i.e., diffuse sky radiation should be measured when the sky is completely clear. Any cloudiness causes unnecessary complications in the interpretation.

The uncertainty regarding the source intensity in the measurement of the scattered light intensity can be avoided by the introduction of a searchlight beam as a source. From the consideration of primary scattering only, several workers (28) tried to derive the angular distribution of the light scattered by the atmospheric aerosol from the measurement of the scattered intensity at different scattering angles. The disadvantage of this method is usually the low energy level of the scattered light; furthermore, the secondary and higher-order scattering was not included in the interpretation of the results.

The measurement of the intensity and of the polarization of sky radiation has been used so far rather for a general determination of the turbidity of the atmosphere than for a quantitative determination as suggested in the preceding article. The measurement of the total intensity in a broad solid angle around the sun has been and is being used at the Smithsonian Astrophysical Observatories, for the determination of conditions favorable for the measurement of the solar constant. If the intensity exceeds a certain limit, the turbidity is considered too high and the corresponding value of the solar constant is considered unreliable.

The polarization of the sky radiation has been studied mainly in two of its aspects. It is apparent from the distribution of the degree of polarization along the sun's vertical in Fig. 4-5 that the degree of polarization has its maximum at 90° from the sun and vanishes at two points, the so-called neutral points. For a high sun elevation above the horizon, one of these is situated above the sun and is called the Babinet point, the other is below the sun and is called the Brewster point. For low sun elevations, when the Brewster point sets at the horizon, another point, called the Arago point, rises on the antisolar side of the horizon. The entire attention of previous investigations has been directed toward these two main features, i.e., the maximum polarization and the position of the neutral points. The measurements under different conditions of turbidity have shown a definite relationship among the maximum polarization, the position of the neutral points, and the turbidity of the atmosphere. The maximum polarization decreases with increasing turbidity, the decrease being more pronounced for longer wavelengths. The relation between turbidity and the

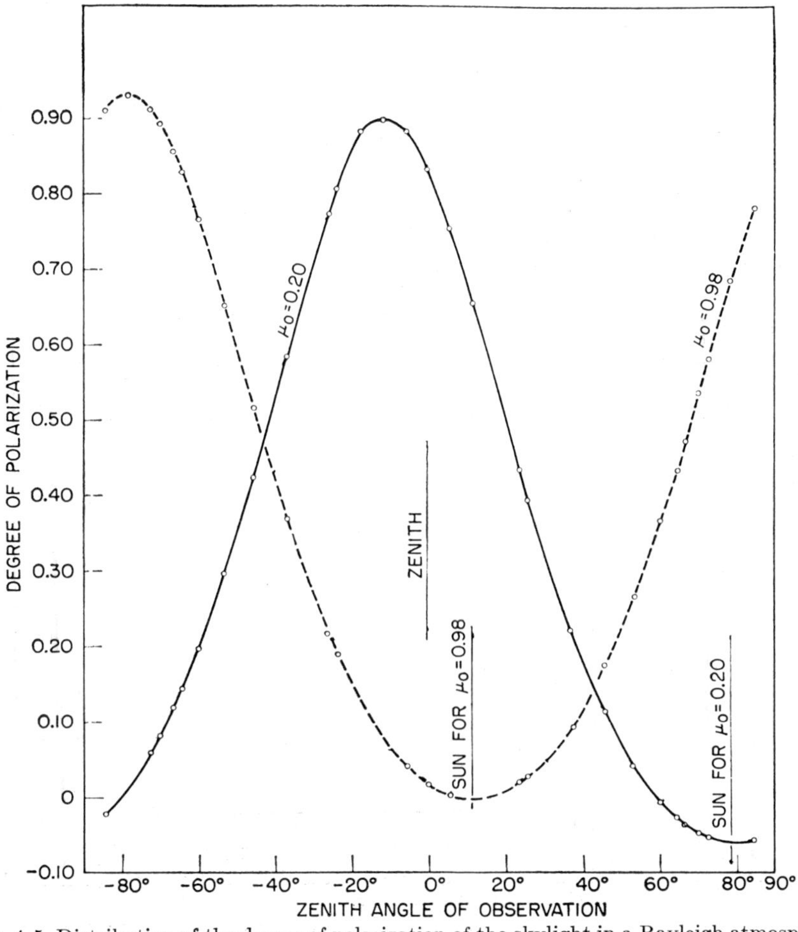

FIG. 4-5. Distribution of the degree of polarization of the skylight in a Rayleigh atmosphere along the sun's vertical for two different sun elevations.

position of the different neutral points is not so simple. In the presence of large quantities of volcanic dust in the atmosphere, the distances of the Babinet point from the sun and of the Arago points from the antisolar point increase quite remarkably (49). Under normal conditions, the distance at the Arago point from the antisolar point increases with an increase in turbidity, again at a much larger rate for longer wavelengths (54). The behavior of the Brewster point with varying turbidity has not been established because of the difficulty in visual observation (at lower altitudes the Brewster point was visible only on very few occasions).

Many observational difficulties have been completely eliminated by a photoelectric polarimeter developed recently, which makes it possible to measure the position of the Brewster point as easily as the other two. Investigations of the relation between the position of the Brewster point and turbidity are in progress. Also an attempt is being made to obtain a more quantitative evaluation of the turbidity condition from the deviation of the measured polarization from the theoretical value for a molecular

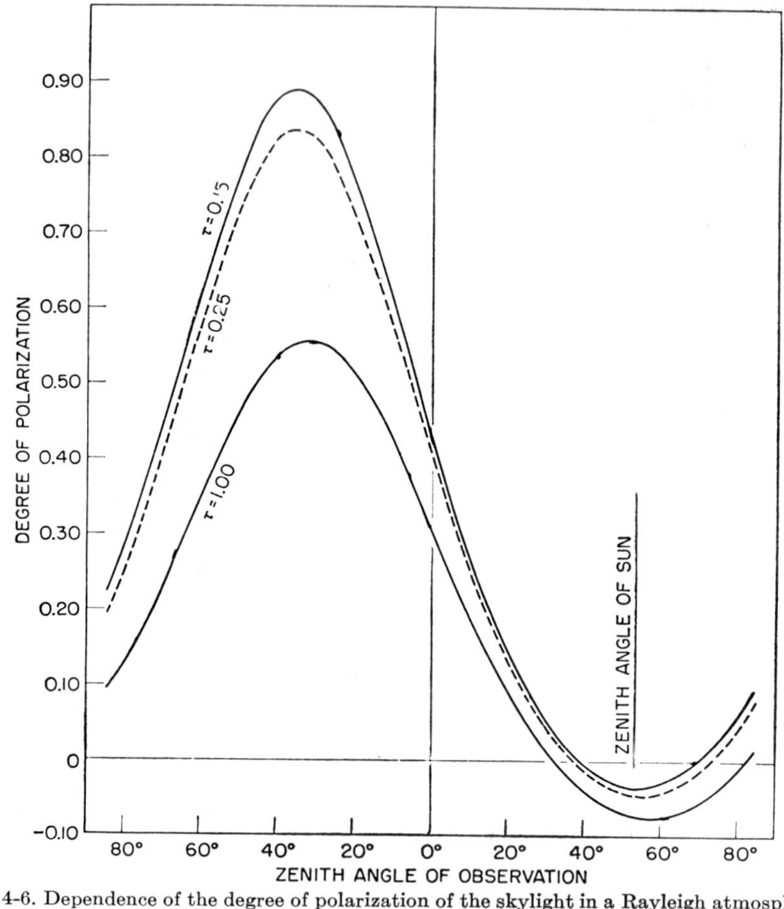

FIG. 4-6. Dependence of the degree of polarization of the skylight in a Rayleigh atmosphere along the sun's vertical on the optical thickness τ.

atmosphere. Such a quantitative evaluation of the turbidity of the actual atmosphere can be done in the following way:

The scattering phenomena in the atmospheric aerosol can be separated into those arising from molecular scattering, represented by the mass scattering coefficient in (4-38) and by the scattering matrix in (4-34), and those attributable to large particle scattering, represented by the volume scattering coefficient and by the scattering matrix in (4-28) with the elements $\bar{F}_i$ computed from (4-47) or (4-48) according to the law of assumed size distribution. After expressing the matrix in terms of quite general directions of incident and scattered radiation (defined by the corresponding pairs of parameters $\mu,\phi;\mu',\phi'$), it can be seen that this matrix contains terms corresponding exactly to the matrix for Rayleigh scattering in (4-34). Therefore the matrix $\mathbf{P}(\mu,\phi;\mu',\phi')$ can be decomposed into two parts

$$\mathbf{P} = \mathbf{P}_R + \mathbf{P}' \tag{4-49}$$

with $\mathbf{P}_R$ corresponding to Rayleigh scattering and the remaining part $\mathbf{P}'$ representing the asymmetric part of the large particle scattering. The same splitting can be done with the intensity $\mathbf{I}$ by writing

$$\mathbf{I} = \mathbf{I}_R + \mathbf{I}' \tag{4-50}$$

where $\mathbf{I}_R$ represents the intensity for molecular scattering using the total optical thickness, resulting from the molecular and large particle scattering, namely,

$$\tau = \int_0^\infty k_\nu^{(s)}\rho\, dz + \int_0^\infty \beta_\nu^{(s)}\, dz \tag{4-51}$$

If this value is substituted in (4-26), the normalization factor will contain the quantity

$$T_\nu = \frac{\beta_\nu^{(s)}}{k_\nu^{(s)}\rho + \beta_\nu^{(s)}}$$

which varies with height or τ but is always smaller than 1. Since $\mathbf{I}'$ can be expected to be smaller than $\mathbf{I}$, (4-26) can be solved by successive approximation. In the equation for $\mathbf{I}'$, $\mathbf{I}$ in the integrand can be replaced by $\mathbf{I}_R$ and $\mathbf{P}$ can be replaced by $\mathbf{P}_R$.

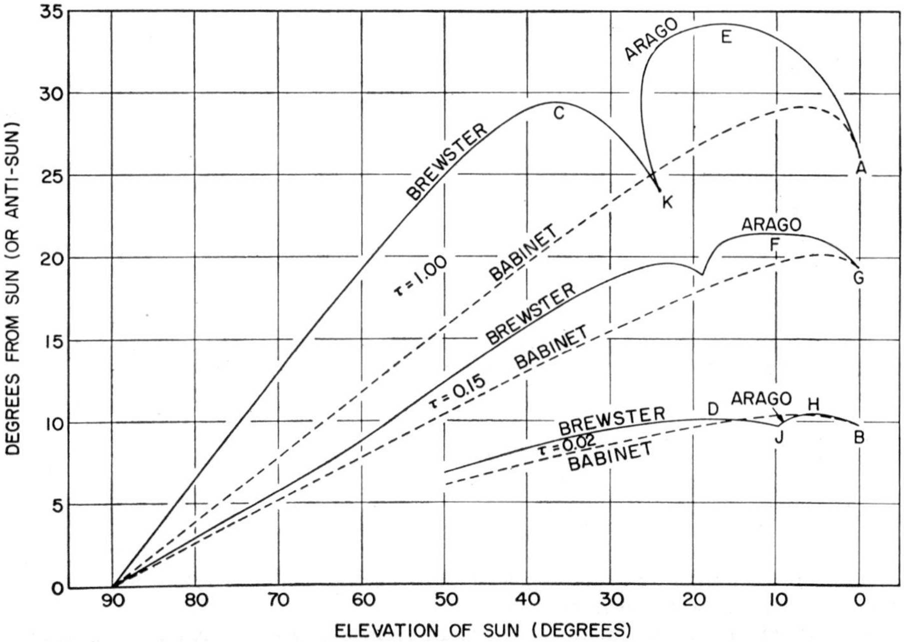

FIG. 4-7. Variation of the position of neutral points of the skylight polarization with the sun's elevation for different optical thicknesses τ, for a Rayleigh atmosphere.

If T_ν depends on τ in a simple way, the equation for $\mathbf{I}'$ can be solved and thus the deviation of the observed polarization from that of the molecular atmosphere, given by the components of $\mathbf{I}'$, can be quantitatively evaluated for any given size distribution of the Mie particles. In the expression (4-47) or (4-48) there is still one undetermined constant $\bar{N}$ which can be uniquely determined from the measured optical thickness in a given frequency (preferably in the red part), when substituted for τ in (4-51).

The theoretical computation of the deviation in polarization of a turbid atmosphere from that of a molecular atmosphere can be done for different size distributions, and the comparison with the actually measured deviations will then enable us to find a size distribution similar to that actually present in the atmosphere.

Unfortunately, the foregoing method cannot be applied immediately for the simple reason that the scattering functions for spherical particles are not yet tabulated in a form which would enable a simple computation of the elements of the scattering matrix **P′**. It is expected that the preparation of such tables will be started in the very near future.

4.1.8 Scattering by Very Large Spherical Particles and Classical Determination of the Size from a Corona or from a Rainbow

When the scattering particles are quite large ($\alpha \geq 50$, which for $\lambda = 0.5\mu$ means $a \geq 4\mu$), then the character of the scattered light approaches that given by geometrical optics (refraction and reflection) and by diffraction theory in a forward direction. The numerous secondary maxima or minima around the mean maximum in the forward direction, already indicated in Fig. 4-2 for $\alpha = 10$, gradually change into a typical diffraction pattern. The scattering angles for the secondary maxima or minima can be found from the equations of the type*

$$\alpha \sin \theta = \text{const} \qquad \text{or} \qquad \sin \theta = \text{const}\,\frac{\lambda}{a} \tag{4-52}$$

their positions are shifted with increasing α toward the forward directions, in an identical way with the Mie scattering ($\alpha < 50$). The character of scattered light depends also primarily upon the size distribution of scattering particles. For a = constant, the shift of secondary extremes with wavelength has as a consequence the appearance of a corona, a system of concentric, differently colored rings around the source (sun or moon). It is quite evident that the purity of the colors is decreased by the increasing spread of the distribution from a constant size, and thus the determination of the size from (4-52) by measuring the scattering angles of maxima appearing in a monochromatic light is limited to the special cases of monodisperse aerosols. The same limitation applies to the rainbow, which appears in the presence of a large number of water droplets of the same size. The rainbow appears as a consequence of the internal reflection of light incident on larger droplets ($a \geq 5 \times 10^{-3}$ cm) in a form of circular arcs of different colors seen by the observer at the angle of view between 40° (blue end) and 42° (red end) from the forward direction of the sun radiation (scattering angle 140° and 138°, respectively). The scattering angles determining the maxima and minima as well as the degree of polarization, as a function of the parameter α, can be found from the theory of scattering and thus can be used for the determination of the particle size from the measurements of the corresponding scattering angles for a known wavelength.

For a nonuniform size distribution the maxima and minima corresponding to different sizes overlap, and as a result of a complete mixture of colors, white, colorless impressions remain, quite characteristic of an overcast sky with a uniform thin cloud layer.

When discussing the problem of scattering by large particles, it should be kept in mind that the larger particles, even in small numbers, are more efficient scatterers than the smaller ones. This fact can be seen from the form of the volume scattering coefficient, where the efficiency of particles can be judged from the factor $K(\alpha,m)\pi a^2$. If $K(\alpha,m)$ is considered to have the same values for particles of $a = 1\mu$ and $a = 10\mu$,

* Cf. the more detailed discussion in Ref. 32.

10 particles of the bigger size contribute by the amount $K(\alpha,m)10\pi 10^2$, equivalent to 10^3 particles of $a = 1\mu$. Therefore, even in the presence of a moderate number of large particles, the efficiency in attenuation is considerable and the characteristics of the scattered light are shifted to those of large particles, the effect of smaller ones becoming unimportant. This fact corroborates the observed low polarization of the overcast sky. A detailed quantitative analysis must be postponed until a more appropriate form of the scattering functions for large particles can be obtained, which will be suitable also for the study of multiple scattering and mutual coherence of scattered waves in the case of larger concentrations of scatterers in small volumes. These two problems seem to be more important for larger particles than for Mie particles discussed before.

4.1.9 Summary

In the preceding paragraphs an attempt is made to present the basic concept of radiative transfer in the atmosphere, with the emphasis on the role of scattering particles of different sizes and different optical properties in this process. Because of its high level of energy (a very advantageous factor for the measurements), the sky radiation has been chosen as the object of the study and the principal ideas on methods of particle-size determination from the skylight measurements are indicated. These methods of determination are more complicated than the classical methods (based on the study of a corona or a rainbow) and are applicable only in the case of monodisperse aerosols. On the other hand, the methods discussed are in many respects still in a stage of development.

The quantitative analysis of the size distribution of particles of a given optical characteristic (refractive index) can be determined only if the distribution law is expressible in a simple mathematical form. Such a model distribution usually involves a few constants which have to be determined from a comparison between theory and actual measurement. For this reason the observations should be made so as to include the simultaneous measurement of as many quantities as possible, e.g., the measurement of sky brightness should be combined with the polarization measurement and with the measurement of the attenuation in direct solar radiation. Since all these are dependent on the wavelength, the measurements should be made in several narrow spectral ranges, extending from the ultraviolet to the infrared part of the spectrum.

Once the approximate distribution of different aerosol particles is known, there is no basic difficulty in the evaluation of the attenuation in the atmosphere and of the effect of sky radiation in other problems of atmospheric optics, such as the problems of the air light in the theory of horizontal or slant visibility and of the illumination of any subject within the atmosphere.

4.2 ATMOSPHERIC ELECTRICITY

The following discussion is confined, essentially, to the atmospheric-electric phenomena occurring in the lower atmosphere, the scene of air pollution. Topics such as thunderstorm electricity, ionospheric phenomena, and the problem of maintenance of the earth's electric field are omitted. As regards the methods and instruments used in measuring atmospheric-electric quantities, the reader is referred to the pertinent literature (7,30,43,61).

Both electrostatic units (esu) in the centimeter-gram-second (cgs) system and practical units (such as volts) are used in the subsequent discussions. Pertinent conversion factors are given in Table 4-1.

Table 4-1. Conversion of Practical and Electrostatic Units

Electric quantity	Practical units (absolute)	Electrostatic units
Charge	1 coulomb 3.34×10^{-10} coulomb	3×10^{9} $cm^{3/2}g^{1/2}/sec$ 1 $cm^{3/2}g^{1/2}/sec$
Potential	1 volt 300 volts	3.34×10^{-3} $cm^{1/2}g^{1/2}/sec$ 1 $cm^{1/2}g^{1/2}/sec$
Capacitance	1 μf 1.11×10^{-6} μf	8.99×10^{5} cm 1 cm
Resistance	1 ohm 8.99×10^{11} ohms	1.11×10^{-12} sec/cm 1 sec/cm
Resistivity	1 ohm-cm 8.99×10^{11} ohm-cm	1.11×10^{-12} sec 1 sec
Conductance	1 mho 1.11×10^{-12} mho	8.99×10^{11} cm/sec 1 cm/sec
Conductivity	1 mho/cm 1.11×10^{-12} mho/cm	8.99×10^{11} sec 1 sec
Current	1 amp 3.34×10^{-10} amp	3×10^{9} $cm^{3/2}g^{1/2}/sec^2$ 1 $cm^{3/2}g^{1/2}/sec^2$

4.2.1 Atmospheric Ions and Their Sources

The air is generally considered a relatively good electric insulator. Nevertheless, when a charged condenser is exposed to air, it will slowly lose its charge. This discharge is due to the electric conductivity of the air produced by carriers of electrostatic charges, the so-called ions. Ions are essentially particles, i.e., molecules or molecule complexes, of the atmospheric constituents that have lost or gained a negative charge, an electron. When an electron has been lost, the residual ion is positively charged; the free electron, almost immediately, attaches itself to another, sometimes neutral particle, in which case it produces a negatively charged ion. Thus, positive and negative ions are produced in pairs. The removal of an electron from a particle requires a certain external energy, the sources of which can be enumerated as follows:

1. Radioactive substances that occur in minute quantities almost everywhere in the rocks and soils of the earth. These substances, such as radium, thorium, etc., radiate α, β, and γ rays. The relatively "soft" α rays ionize only the first few centimeters of air, the somewhat "harder" β rays produce ions in the air to a height of a few meters above the ground, while the γ rays cause ionization to a height up to 1 km. This source furnishes approximately one-third of the ions in the air near the ground, whereas it is practically absent over the oceans, because sea water has only about one-thousandth of the radioactive content of soil (16,17,31,43,62).

2. Radioactive gases (radon, thoron, etc.), resulting from the disintegration of the corresponding substances, escape from the pores in the ground into the adjacent air where they cause, roughly, half of the total ionization observed. Again, this source is negligible over the oceans.

3. Cosmic rays cause about one-sixth of the ionization of air over land and are practically the sole source of ions in air over the oceans.

4. Occasional sources of ionization are represented by lightning or other electric discharges, the friction electricity produced by blowing sand or drifting snow, splashing water from rain showers or waterfalls, short-wave ultraviolet light,* X rays, and combustion processes (16,19,21).

* The sun's ultraviolet light of sufficiently short wavelengths is found only in the upper atmosphere.

The number of ion pairs produced per cubic centimeter of air (near the ground) in 1 sec is of the order of, roughly, 20, but varies over a wide range. The ion productivity of sources 1 and 2 depends on the types of rock or soil that constitute the surface. For example, igneous rocks contain more radioactive matter than do sedimentary rocks, acidic rocks more than do basic rocks.

For a given concentration of radioactive substances in the ground, the emanation of radioactive gases varies with the porosity of the soil; also, when the ground is dry, the rate of emanation is greater than when the ground is wet or frozen. Snow or ice on the ground reduces the amount of radioactive gases that can escape into the air. A drop in air pressure facilitates and a pressure rise impairs emanation from the ground. Similarly, heating of the ground by solar radiation drives radioactive gases from the soil, whereas cooling of the ground by eradiation at night diminishes the rate of escape of these gases. The concentration of radioactive gases in the lowermost layer of the air depends not only on the rate of emanation from the ground, but also on the rate of dissipation, which is governed essentially by the thermal stratification and the motion of the air (see Arts. 4.3 and 4.4).

There are two major types of ions: small ions are formed when an electron is removed from, or becomes attached to, a gaseous molecule or small cluster of molecules; large ions* are electrically charged condensation nuclei (34). Small ions may become large ions by attaching themselves to uncharged condensation nuclei. In addition to these major types of ions, intermediate ions of sizes between the small and the large ions are occasionally observed (7,16,62); electrostatically charged large haze or dust particles and fog or cloud droplets represent the ultralarge ions. The ultralarge ions commonly carry multiple charges, varying from a few to a few hundred elementary charges,† either positive or negative (19). Small, intermediate, and large ions are usually assumed to carry single elementary charges, although the existence of multiple charges on large ions has been observed (29,62).

4.2.2 Ion Mobility

When an ion is placed in an electric field such as that of the earth, it is accelerated in the direction of the electrode that has the opposite charge from that of the ion itself. However, the viscous friction between the ion and the gas, in which the ion is suspended, will almost instantaneously establish an unaccelerated motion. The velocity of this motion is, aside from the density of the gas and of the ion, a function of the mass of the ion and is proportional to the strength of the electric field. The ratio of the ion velocity (centimeters per second) to the electric field intensity (volts per centimeter) is called ion mobility and has the dimensional unit square centimeter per second volt.

The ion mobility is inversely proportional to the density of the air (16); however, the effect of the change in density of the air near the ground can be neglected since it is negligible in comparison to variations in mobility produced by impurities in the air and, particularly, by water vapor. Increasing concentrations of impurities and increasing humidity generally diminish ion mobility (16,62) essentially because of the fact that impurities and water vapor associate with the ions and thereby increase their mass. This association is selective in that under certain conditions the positive ions are more affected than are the negative ions, or vice versa. On the average, as shown in Table 4-2, the mobility of negative ions is greater than that of positive ions, in which also the effect of moisture is evident (43). In view of the fact that the electric conductivity of air is directly proportional to the ion mobility, it is obvious that only the small ions need be considered in this respect. Nevertheless, the bal-

* Also called Langevin ions after their discoverer.
† An elementary charge is the absolute value of the charge of an electron amounting to 4.8×10^{-10} esu.

ance and transformation of small ions is greatly affected by the presence of large ions, as will be shown in the following article. (Intermediate and ultralarge ions will not be considered.)

Table 4-2. Average Ion Mobility near the Ground

Types of ions	Condition of air	Mobility, sq cm/sec volt
Small:		
Positive	Dry	1.4
	Moist	1.1
Negative	Dry	1.9
	Moist	1.2
Intermediate	.....	5×10^{-2}
Large	.....	4×10^{-4}
Ultralarge	.....	5×10^{-7}

4.2.3 Concentration of Ions and Condensation Nuclei

As regards the concentrations of small ions, large ions, and condensation nuclei from which the large ions are derived, great variations with time and place are observed. Therefore, average values are quite useless; nevertheless, some very rough estimates are cited for orientation with respect to the general orders of magnitudes involved. In Table 4-3, the ion concentrations are totals for both positive and negative ions. Near the ground the concentration of positive ions (small and large) is slightly greater (about 10 to 20 per cent) than that of negative ions, particularly under stable* atmospheric conditions. This is, in part, due to the fact that the negatively charged earth's surface attracts the positive and repels the negative ions.

Table 4-3. Orders of Magnitudes of Ion and Condensation-nuclei Concentrations per Cubic Millimeter

Type	Land	Ocean
Small ions	1	1
Large ions	8	<1
Condensation nuclei	20	1

The small-ion concentration is practically the same over land and ocean, although the small-ion production over the ocean is only about 2 ion pairs per cubic centimeter second while that over land is about 15 to 20 ion pairs per cubic centimeter second. The reason for this small difference in concentration of small ions between land and ocean lies in the much greater concentration of condensation nuclei over land; the attachment of small ions to large ions of the opposite charge and to uncharged condensation nuclei represents an important process of depletion of small ions, as will be discussed below. In this connection it should be noted that only charged condensation nuclei need be considered as large ions (16,62). Charged small dust particles can be neglected for atmospheric-electric considerations, because their concentration is smaller by two or more orders of magnitude than that of condensation nuclei (34,42).

The much larger concentrations of condensation nuclei and of large ions over land than over the ocean reflect the efficacy of man-made and natural combustion processes as nuclei sources (42). Over land, the concentration of small ions has been found to be, roughly, inversely proportional to the concentration of large ions. In turn, the large-ion concentration increases more slowly than the concentration of

* See Art. 4.3.

condensation nuclei, so that the ratio of charged to uncharged nuclei is smaller for larger nuclei concentrations (34). In general, the nuclei concentrations measured have shown the greatest variations, the small-ion concentrations the least. A great many nuclei measurements* made at different localities have been summarized by H. Landsberg (34) and are reproduced in Table 4-4. Differences similar to those shown in Table 4-4, although of smaller magnitude, can be assumed to hold for large-ion concentrations (62).

Table 4-4. Condensation-nuclei Concentrations at Various Types of Localities

Locality	Avg per cu mm	Abs max per cu mm
Oceans	1	40
Islands	10	109
Mountains: <1 km	6	155
1–2 km	2	37
>2 km	1	27
Country: seashore	10	150
Inland	10	336
Towns	34	400
Cities	147	4,000

After H. Landsberg, Atmospheric Condensation Nuclei, *Ergeb. kosmischen Physik*, **3**, 155–252 (1938).

The diurnal and seasonal variations of ion and nuclei concentrations are not clear-cut. Although the production of small ions shows, according to G. R. Wait and W. D. Parkinson (62), a maximum during the early morning and a minimum in the early afternoon, the small-ion concentration does not seem to have a simple diurnal variation (16). Roughly, the concentration of small ions is smallest in winter, largest in summer, while that of large ions is opposite (16,62). The temporal variations are complicated by the following chain of relationships: small-ion concentration depends on the rate of ionization which, other things being equal, depends on the diurnal and seasonal variations of the state of the ground, those of the soil temperature, radiation, etc. Moreover, the concentration of small ions varies inversely with that of large ions, which, in turn, depends on the concentration of condensation nuclei. The latter is dependent not only on the variation of the nuclei sources (e.g., domestic heating, industrial activity, etc.) but also on the periodic and aperiodic fluctuations of eddy diffusion which are functions of atmospheric stability and motion.† Only at places far from human habitation, traffic,‡ and industrial activity can a simple relationship to meteorological parameters be expected.

4.2.4 Ion Balance

The number of ions in a unit volume fixed in space is determined by (1) the intensity of ionization, (2) the number of ions neutralized by coalition of some ions with others of the opposite charge, and (3) the transport of ions into the volume minus the transport of ions out of the volume. The transport is produced by the wind, eddy diffusion, and the electrode effect of the electrically charged earth's surface.

The transport of ions by the wind will change the local ion concentration if in the upwind direction a horizontal gradient of the ion concentration exists. Moreover, since the wind also carries condensation nuclei, which facilitate the transformation of small into large ions and hence the neutralization of small ions as will be discussed

* It should be noted that all determinations of condensation-nuclei concentrations made with the Aitken nuclei counter (34,43) are probably too small, according to recent investigations into the method of measurement (26).

† See Arts. 4.3 and 4.4.

‡ In the wooded countryside a few hundred yards downwind from the Pennsylvania Turnpike, a condensation-nuclei concentration of 270,000 per cubic centimeter was measured (43).

below, the existence of a gradient of nuclei upwind is also of importance in the change of local ion concentration. The transport by eddy diffusion is essentially a function of thermal stability, vertical wind gradient, and vertical gradient of ion concentration. Finally, the normal electrode effect tends to cause downward drift of positive ions and upward drift of negative ions because ordinarily the earth's surface is negatively charged. Under disturbed (thunderstorm) conditions the electric field and therefore the direction of ion drift may be reversed. The rate of drift is proportional to the intensity of the electric field and to the ion mobility.

The transport of ions is difficult to appraise quantitatively. For this reason, only the intensity of ionization and the ion neutralization are usually considered and an otherwise steady state is assumed. The assumption is probably a fair approximation, because significant inhomogeneities in the horizontal direction occur mostly near sources of atmospheric pollution, and the electrode effect and eddy diffusion partly compensate each other (16). Nevertheless, this compensation is, particularly in unpolluted air, not perfect because the number of positive ions often exceeds that of negative ions near the ground, as mentioned before.

Assuming all ions to be singly charged, we can enumerate the various possibilities of ion coalition:

1. Positive with negative small ions. This combination neutralizes small ions, i.e., each such combination destroys one pair of ions.
2. Positive small ions with negative large ions. This causes destruction of the ions involved and the reconversion of large ions into uncharged condensation nuclei.
3. Negative small ions with positive large ions. This process is similar to 2.
4. Positive small ions with uncharged nuclei. This causes the transformation of a small ion into a large one, thereby depleting the concentration of small ions and uncharged nuclei, but increasing that of large ions.
5. Negative small ions with uncharged nuclei. This process is similar to 4.
6. Positive with negative large ions. This causes the neutralization and thereby destruction of large ions by pairs, while forming uncharged nuclei. However, the total concentration of nuclei (uncharged and charged) is depleted; the size of each nucleus is enlarged by this coagulation, and subsequent occurrence of processes 4 or 5 produces large ions of smaller mobility.
7. Positive large ions with uncharged nuclei. This coalition merely produces large ions of the same charge but smaller mobility and diminishes the total concentration of condensation nuclei.
8. Negative large ions with uncharged nuclei. This process is the same as 7.

In the case of large ions with multiple charges, coalition with other ions of fewer or more charges of the opposite sign will, aside from changing the mobility, retain the sign of ion charge or reverse it, respectively. Multiple charges will be excluded from our considerations.

Let us designate the concentration of positive and negative small ions by n^+ and n^-, respectively, the large-ion concentration correspondingly by N^+ and N^-, that of uncharged condensation nuclei by N^0, and the concentration to total number of nuclei per unit volume by Z, so that

$$Z = N^0 + N^+ + N^- \tag{4-53}$$

If we now consider the simplest case of aerosol-free air, $Z = 0$, which contains only n^+ and n^-, and if the rate of ionization, i.e., production of ion pairs per unit volume and time, is q, then the rate of recombination, possibility 1 above, of small ions is proportional to the product of the respective concentrations. Then the change in small-ion concentration is

$$\frac{dn}{dt} = q - \alpha n^+ n^- \tag{4-54}$$

where t is the time and α is the recombination coefficient; the latter's value is commonly taken as 1.6×10^{-6} cu cm/sec, although somewhat smaller and larger values have been observed (62). There is a slight dependence of α on the temperature and pressure near the surface of the earth (16). If equilibrium exists between ion formation and ion destruction so that $dn/dt = 0$, then

$$q = \alpha n^+ n^- \tag{4-55}$$

In the case in which the concentration of positive ions is the same as that of negative ions so that $n^+ = n^- = n$, Eq. (4-55) becomes

$$q = \alpha n^2 \tag{4-56}$$

The original assumption that condensation nuclei are essentially absent makes this equation applicable only to steady equilibrium conditions at higher altitudes in the free atmosphere. Near the ground some of the other coalition processes enumerated above must be taken into consideration. In all these processes, the combinations are proportional to the products of the respective concentrations of particles involved; the proportionality factors are called combination coefficients and are designated by the notations given in Table 4-5, in which their median values are also given (62).

Table 4-5. Combination Coefficients

Combination between	Notation	Median value, cu cm/sec
Positive small—negative small ions	α	1.6×10^{-6}
Positive small—negative large ions	$\eta_{\substack{+\\-}}$	2.4×10^{-6}
Negative small—positive large ions	$\eta_{\substack{-\\+}}$	4.5×10^{-6}
Positive small—uncharged nuclei	η_{+}	0.6×10^{-6}
Negative small—uncharged nuclei	η_{-}	1.1×10^{-6}
Positive large—negative large ions	γ	$\sim 10^{-9}$

These coefficients are not constant, but depend on the mobility of the small ions; for this reason the coefficients of the negative small ions are larger than those for the corresponding positive small ions. On the other hand, the coefficients of combination between small ions and charged or uncharged nuclei also become larger as the mobility of the nuclei becomes smaller, such as when these larger particles grow by conglomeration or absorption of moisture from the air (62).

Now, if we assume equilibrium conditions to prevail, the sum of the ion-destruction and ion-transformation processes 1 to 5, which involve small ions, is equal to the ion-production rate q. Since q is the number of pairs of ions produced, it is also the number of positive small ions q_+ as well as the number of negative small ions q_- formed. Thus the balance of positive small ions is represented by the equation

$$q_+ = \alpha n^+ n^- + \eta_{\substack{+\\-}} n^+ N^- + \eta_{\substack{+\\0}} n^+ N^0 = n^+(\alpha n^- + \eta_{\substack{+\\-}} N^- + \eta_{\substack{+\\0}} N^0) \tag{4-57}$$

By reversing the subscript and superscript signs, we obtain a similar equation to represent the balance of negative small ions. Over land, where the values of N^+, N^-, and N^0 are large as compared to n^+ and n^-, the α term representing the recombination of small ions is negligible, so that Eq. (4-57) can be written

$$q_+ = n^+(\eta_{\substack{+\\-}} N^- + \eta_{\substack{+\\0}} N^0) \tag{4-58}$$

Over the oceans or at higher altitudes, where nuclei concentrations are small as compared to the small-ion concentrations, the η terms are negligible, and Eq. (4-57) becomes identical with (4-55).

Under the assumption that processes 7 and 8 do not change the large-ion concentrations and that the large ions are formed solely by processes 4 and 5 and destroyed by processes 2, 3, and 6, the balance equation for positive large ions is

$$\eta_{\substack{+\\0}} n^+ N^0 = \eta_{\substack{-\\+}} n^- N^+ + \gamma N^+ N^- = N^+(\eta_{\substack{-\\+}} n^- + \gamma N^-) \qquad (4\text{-}59)$$

A similar equation for the negative large ions is obtained when the subscript and superscript signs are reversed. The β term in Eq. (4-59) is ordinarily small in comparison with the η terms, because γ is, according to Table 4-5, roughly a thousand times smaller than the values of η, and the large ions usually do not exceed the small

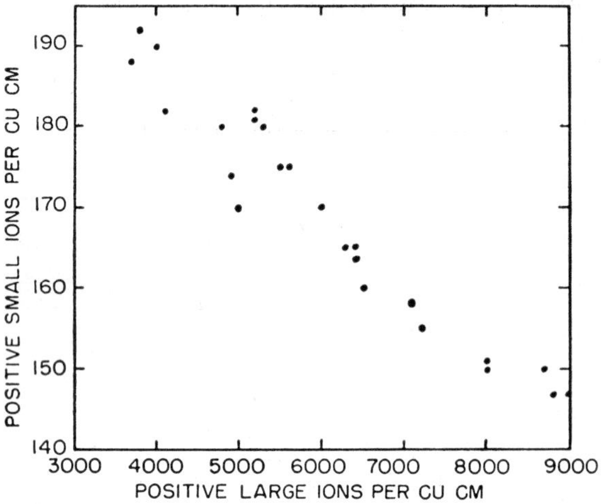

Fig. 4-8. Correlation between concentration of positive small and large ions (October, 1932, to March, 1933, Washington, D.C.).

ions by a factor of more than 100. Therefore, we can often neglect the effect of large-ion recombination, and Eq. (4-59) becomes

$$\eta_{\substack{+\\0}} n^+ N^0 = \eta_{\substack{-\\+}} n^- N^+ \qquad (4\text{-}60)$$

A similar equation holds for the negative large ions. This approximation is not valid in highly polluted air, in which the nuclei concentration may be extremely large. Then, however, we are also not justified in assuming constancy of the various ion and nuclei concentrations, and the balance equations do not apply.

Let us assume, as a further approximation, that Eq. (4-60) is identical with its corresponding equation for negative large ions, i.e., that the production of positive large ions is equal to that of negative large ions. Then, substitution into Eq. (4-58) yields

$$q_+ \approx 2\eta_{\substack{+\\-}} n^+ N^- \qquad (4\text{-}61)$$

Therefore, when q_+ is constant, the concentration of positive small ions is inversely proportional to that of negative large ions. Since the production of negative and positive large ions was assumed to be equal, we also find that the concentration of positive small ions is inversely proportional to that of positive large ions. In Fig. 4-8 the mean hourly values of ion concentrations presented by O. H. Gish (16) have been replotted to show the correlation between positive small and large ions. According to this figure, the concentration of positive small ions appears to be an inverse power function of that of the positive large ions. Because the production of large ions depends on the availability of condensation nuclei and thus on the pollution of the air, the concentration of large ions or the reciprocal of the concentration of small ions can be taken as an index of air pollution.

4.2.5 Mean Life of Ions

From the equations of the balance between ion formation and ion destruction we may determine the mean time interval between these processes, i.e., the mean life L (sec) of the ions. The mean life is the ratio of the existing ion concentration to the ion-production rate. Thus, omitting the charge signs, we find the mean life L_n for small ions to be

$$L_n = \frac{n}{q} \qquad (4\text{-}62)$$

In clean air, where $N \ll n$, q can be substituted from Eq. (4-56), so that

$$L_n = \frac{1}{\alpha n} \qquad (4\text{-}63)$$

Hence, the mean life of small ions in clean air is inversely proportional to the small-ion concentration. Over the ocean (see Tables 4-3 and 4-5), the mean life is between 5 and 10 min. In polluted air, as over land, the approximate equation (4-61) can be substituted into (4-62); hence

$$L_n = \frac{1}{2\eta N} \qquad (4\text{-}64)$$

where η can be taken as either $\eta_{\pm}$ or $\eta_{\mp}$, or their average. The mean life of small ions over land is inversely proportional to the large-ion concentration; it is of the order of a few seconds to, perhaps, a minute. Because the presence of condensation nuclei over land shortens the life of the small ions, the small-ion concentrations over land and ocean are practically the same (Table 4-3), although the ionization is much greater over land.

The approximate mean life L_N of the large ions is obtained by substitution of Eq. (4-60) into the denominator of (4-62) with N in the numerator:

$$L_N = \frac{1}{\eta n} \qquad (4\text{-}65)$$

The mean life of large ions is inversely proportional to the concentration of small ions. In very polluted air, Eq. (4-59) must be substituted into the denominator of Eq. (4-62) (with N in the numerator):

$$L_n = \frac{1}{\eta n + \gamma N} \qquad (4\text{-}66)$$

so that the mean life of large ions also diminishes when the large-ion concentration increases. The mean life of large ions is of the order of 10 min or more.

4.2.6 Conductivity

An object having the electrostatic charge Q and a capacitance C has the electrostatic potential P; the three quantities are related by

$$Q = CP \qquad (4\text{-}67)$$

Owing to the electric conductivity Λ of the air, the charged object will lose its charge at a rate that is proportional to the magnitude of the charge:

$$-\frac{dQ}{dt} = aQ \tag{4-68}$$

or with (4-67)

$$-\frac{dP}{dt} = aP \tag{4-69}$$

where the proportionality factor a is the so-called dissipation coefficient which, for objects ventilated by air speeds of ≥ 2m/sec, is related to the conductivity by the expression (43)

$$a = 4\pi\Lambda \tag{4-70}$$

so that Eqs. (4-68) and (4-69), respectively, become

$$-\frac{dQ}{dt} = 4\pi\Lambda Q \tag{4-71}$$

$$-\frac{dP}{dt} = 4\pi\Lambda P \tag{4-72}$$

A charged object will be discharged by the neutralizing effect of ions of the opposite charge colliding with it. The conductivity Λ^+ due to positive small and large ions having the mobilities k^+ and K^+, respectively, is

$$\Lambda^+ = \epsilon(n^+k^+ + N^+K^+) \tag{4-73}$$

where ϵ is the elementary charge and all ions are assumed to be singly charged. The conductivity Λ^- due to negative ions is, correspondingly,

$$\Lambda^- = \epsilon(n^-k^- + N^-K^-) \tag{4-74}$$

According to Table 4-2, the mobility of large ions is more than three orders of magnitude smaller than that of small ions, so that the large-ion contribution to the electric conductivity of the air is negligible and the large-ion terms in Eqs. (4-73) and (4-74) can be omitted. The conductivities, then, are

$$\Lambda^+ = \epsilon n^+k^+ \qquad \text{and} \qquad \Lambda^- = \epsilon n^-k^- \tag{4-75}$$

Substituting (4-75) into (4-72), we obtain

$$\Lambda^+ = \epsilon n^+k^+ = \frac{1}{4\pi P^-}\frac{dP^-}{dt} \qquad \text{and} \qquad \Lambda^- = \epsilon n^-k^- = \frac{1}{4\pi P^+}\frac{dP^+}{dt} \tag{4-76}$$

Thus the conductivity is proportional to the rate of the relative change in potential of a charged conductor.

Near the earth's surface, $k^+ < k^-$ (Table 4-2), but $n^+ > n^-$; ordinarily, the second inequality overcompensates for the first one, so that $\Lambda^+ > \Lambda^-$, and a negatively charged body loses its charge somewhat more rapidly than does a positively charged one. A rough average value of the conductivity is 2×10^{-4} esu, which, however, is subject to large variations in time and space (7). Since the conductivity is proportional to the small-ion concentration which, in turn, is inversely proportional to the large-ion concentration, the conductivity of the air varies inversely as the degree of air pollution (16,17). Repeated measurements over the oceans have revealed a steady decrease in conductivity for the last 40 years, which fact has been ascribed to the advancing industrialization of the world (62).

4.2.7 Space Charge

The excess in concentration of ions of one sign over that of ions of the opposite sign is called space charge s (16,43), which thus represents the net electric charge per unit volume:

$$s = \epsilon(n^+ + N^+ - n^- - N^-) \tag{4-77}$$

where ϵ is again the elementary charge and n and N are the concentrations of small and large ions, respectively. Because, near the earth's surface, the concentration of positive ions is ordinarily somewhat greater than that of negative ions, the space charge is usually positive. However, negative space charges are not infrequently observed, particularly under shower and thunderstorm conditions. The space charge can be directly determined without measurement of individual ion concentrations (30,43).

Determinations of the space charge near the earth's surface at various places show great variations, ranging in magnitude from a few to more than a thousand excess ions of one sign per cubic centimeter. Rapid fluctuations may be observed when gusty winds sweep up sand, dust, or snow from the ground locally and thereby produce additional intermittent ionization in the surface layer of air.

4.2.8 Potential Gradient

The earth, being a charged conductor, produces an electric field around it. Under undisturbed (fair-weather) conditions, the earth's surface is negatively charged, while the air around it is positively charged. Thus, at any point in the air, there exists an

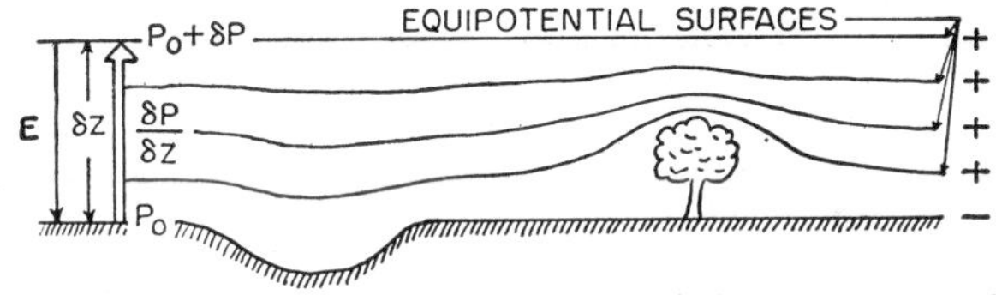

FIG. 4-9. Schematic diagram of the normal electric field above the earth's surface.

electric potential relative to the earth's surface. All points having the same potential form an equipotential surface (see Fig. 4-9) that is subject to deformations by the topography of the underlying earth's surface. By convention, the potential of the earth's surface, P_0, is set at zero. The vector of the electric field strength $\mathbf{E}$ is normal to the potential surfaces and is directed to the negatively charged earth's surface, while the gradient of electric potential points in the opposite direction, so that

$$-\mathbf{E} = \frac{\delta P}{\delta z} \tag{4-78}$$

where δP is the difference in potential of two potential surfaces, the normal distance of which is δz; thus the potential gradient $\delta P/\delta z$ is the difference in potential per unit height, provided that we assume horizontally homogeneous conditions. Over depressions in the surface, the potential gradient is diminished, as can be seen from Fig. 4-9, whereas over protruding objects or convex contours the potential gradient is increased.

If we again assume horizontally homogeneous conditions so that the potential gradient does not change in the horizontal, the space charge s is related to the change in potential gradient with height by Poisson's theorem:

$$-\frac{\delta^2 P}{\delta z^2} = 4\pi s \tag{4-79}$$

A decrease of potential gradient with height is associated with a positive space charge, as normally prevails near the earth's surface.

The magnitude of the potential gradient is, on the average, of the order of 130 volts/m at most places (17). A few annual averages for several stations are cited in Table 4-6, after H. Benndorf and V. F. Hess (43).

Table 4-6. Annual Average Potential Gradients

Station	Latitude	Longitude	Volts/m
Davos	46.8N	9.8E	65
Oceans (Carnegie cruises)	......	......	125
Buenos Aires	34.5S	58.6W	135
Washington, D.C	39.0N	77.1W	180
Potsdam	52.4N	13.1E	200
Kew	51.5N	0.3W	315

After H. Benndorf and V. F. Hess in H. Neuberger, "Introduction to Physical Meteorology," The Pennsylvania State College, Mineral Industries Extension Services, 1951.

The variations of the potential gradient near the earth's surface are, to the greatest extent, due to local variations in the conductivity of the air. The potential gradient is inversely proportional to the conductivity, just as the drop in voltage over a given length of wire is inversely proportional to the conductivity of the wire, or directly proportional to its resistivity. Since the conductivity is an inverse function of the degree of air pollution, the potential gradient is greater in polluted than in clean air. This is evident from Table 4-6, when the potential gradient at Davos, an Alpine resort in Switzerland, is compared with that at Kew located in an industrial region of England.

The diurnal and seasonal variations of the potential gradient at most places are essentially the result of the combined effects of variations in the activity of pollution sources and in meteorological diffusion processes. For example, during conditions of strong, turbulent wind and/or thermally unstable air (see Arts. 4.3 and 4.4), the concentration of air pollutants and thereby the potential gradient are diminished. By contrast, when the concentration of pollutants is high during stable atmospheric conditions, the potential gradient is large. This explains, at least in part, the fact that the potential gradient generally is at a maximum during winter when the concentration of pollutants is high near the ground, whereas it is at a minimum during summer when convective activity is high and pollutants are transported aloft. For example, the average potential gradient at Kew is, roughly, 500 volts/m in January, 240 volts/m in July (26). The diurnal variation generally shows maxima during forenoon and evening, minima in the early morning and early afternoon. The amplitude of the diurnal variation is usually larger in winter than in summer and is of the order of 50 per cent of the mean daily value of the potential gradient.

During rain showers, snowfalls, thunderstorms, etc., very large fluctuations are often observed, and the potential gradient may reverse its sign repeatedly within a few minutes. The magnitude of the potential gradient under such disturbed conditions may reach several thousand volts per meter. It may be mentioned here that the potential gradient is also affected by the electric conditions in the ionosphere which, however, become apparent only at places at which air pollution is absent.

4.2.9 Summary of Relationship between Air Pollution and Atmospheric-electric Phenomena

Polluted air is generally characterized by large concentrations of condensation nuclei. Small ions, which are chiefly responsible for the electric conductivity of the air, readily attach themselves to condensation nuclei to form large ions. Because of their small mobility, large ions do not contribute to the conductivity of the air. Therefore, in polluted air, the small-ion concentration and, consequently, the conductivity are diminished. In turn, because of the inverse relationship between conductivity and potential gradient, the latter is a direct function of the degree of air pollution.

4.3 THERMODYNAMICS

In this article, only the fundamentals of thermodynamics are discussed as far as they are essential for the understanding of the thermal aspects of the atmosphere in air pollution problems. For more extensive treatment, the reader is referred to pertinent textbooks (4,22,25).

All quantities are given in the metric system (cgs) and temperatures in degrees Kelvin, unless otherwise specified.

Differentials are denoted by d if they represent property changes of an individual parcel of air. By contrast, δ denotes the difference in property between two different points in space (25). The rectangular space coordinates x, y, z are used, where x and y lie in the horizontal plane, z in the vertical, counted positive upward.

4.3.1 Composition of Air

For all practical purposes, dry air can be considered a mixture of the gases shown in Table 4-7, which gives their approximate percentages by volume and by mass as well as their molecular weights. These gases, together with traces of hydrogen, helium, neon, and others, are present in practically invariant proportions to more than 10 miles altitude. In addition, the atmosphere, particularly in the lowest few miles, contains variable amounts of carbon dioxide (up to about ½ per cent by volume), water vapor (up to 4 per cent by volume), and aerosol. The aerosol, the collective term for particulate matter suspended in the air, comprises water droplets, ice crystals, dust, and condensation nuclei (34). The latter play a significant role in the transition of water vapor into the liquid or solid phases and in air pollution (42).

Table 4-7. Composition of Dry Air

Element	Vol %	Mass %	Mol wt
Nitrogen	78.1	75.5	28.0
Oxygen	21.0	23.2	32.0
Argon	0.9	1.3	39.9

4.3.2 Physical Parameters

At a given level, the pressure is the weight, on a unit surface, of the column of air above that level, where the weight (or force) of the column is its mass multiplied by the acceleration due to gravity. This acceleration, denoted by g, can be considered constant with altitude for the lower regions of the atmosphere. For practical computations the standard value at 45° latitude, $g = 981$ cm/sec², is usually adequate. It must be noted, however, that this value varies with latitude, from 978 at the equator to 983 at the poles. The unit of pressure is 1 g cm/sec² cm² = 1 dyne/cm²; the pressure at sea level is of the order of 10^6 dynes/cm² = 1 bar; the unit used in meteorological practice is 1,000 dynes/cm² = 1 mb. The standard barometric pressure at sea level is 1,013.3 mb = 29.92 in. Hg ≈ 14.7 lb (force)/in.² For thermodynamic purposes, the standard reference pressure of 1,000 mb = 29.53 in. Hg has been conventionally adopted.

The mass M of a quantum of air is usually considered constant, while the temperature T, volume V, and pressure p are considered variable. Since the mass and volume of air usually cannot be determined, the volume per unit mass, i.e., the specific volume $v = V/M$ or the reciprocal of v, which is the density ρ, is usually employed.

4.3.3 Thermodynamics of Dry Air

The equation of state for ideal gases is

$$pV = \frac{MTR^*}{m} \tag{4-80}$$

$$pv = \frac{TR^*}{m} \tag{4-81}$$

where R^* is the universal gas constant (referring to 1 g of gas) and m is the molecular weight of the gas. Since $v = 1/\rho$, we can also write

$$p = \frac{\rho TR^*}{m} \tag{4-82}$$

This gas law is also applicable to a mixture of ideal gases, such as dry air; according to Dalton's law, the total pressure p of a gas mixture is equal to the sum of partial pressures p_i of the constituents having the molecular weights m_i and the masses M_i, so that according to Eq. (4-80)

$$pV = V\sum p_i = TR^*\sum\left(\frac{M_i}{m_i}\right) = TR^*M\sum\left(\frac{M_i}{M}\frac{1}{m_i}\right) = TR^*M\left(\frac{1}{m}\right) \tag{4-83}$$

where m is the fictitious mean molecular weight of the mixture. We see from the last equality of the equations (4-83) that the mean molecular weight of a gas mixture is the harmonic mean of the molecular weights of the individual gases. Applied to dry air, we find the mean molecular weight to be $m \cong 29$, and the ratio

$$\frac{R^*}{m} = R = 2.87 \times 10^6 \text{ cm}^2/\text{sec}^2 \text{ deg} \tag{4-84}$$

where R is the specific gas constant for dry air. Thus, Eq. (4-82) can be written for dry air in the form

$$p = R\rho T = \frac{RT}{v} \tag{4-85}$$

The law of conservation of energy as expressed by the first law of thermodynamics takes the following form for an ideal gas:

$$dQ = c_v\,dT + p\,dv \tag{4-86}$$

where dQ is the heat added to (+) or subtracted from (−) a unit mass of air (cm/sec² g = ergs/g); c_v is the specific heat at constant volume, thus, for dry air $c_v = 7.1 \times 10^6$ ergs/g deg; dT is the ensuing change in absolute temperature, p the pressure, and dv the change in specific volume. Thus, an addition of heat is equal to the sum of the increase in internal energy ($c_v\,dT$) of the air parcel involved and the work ($p\,dv$) done by the expanding air parcel against the pressure of the air surrounding it. We can substitute from Eq. (4-85) after differentiation and obtain

$$dQ = (c_v + R)\,dT - \frac{RT\,dp}{p} \tag{4-87}$$

From this equation the specific heat $c_p = (dQ/dT)_p$ at constant pressure (thus, $dp = 0$) is obtained:

$$c_p = c_v + R \approx 10^7 \quad \text{ergs/g deg} \tag{4-88}$$

or

$$R = c_p - c_v \tag{4-89}$$

so that (4-87) becomes

$$dQ = c_p\,dT - \frac{RT\,dp}{p} \tag{4-90}$$

In case of vertical motion of air parcels, the effect of pressure change on the temperature of the parcel is so dominant that additions or removals of heat by conduction and radiation processes can be neglected.* The parcel can thus be considered as being neither heated nor cooled by external sources. Any process for which $dQ = 0$ is called an adiabatic process. Equation (4-90) thus becomes the adiabatic equation:

$$\frac{dT}{T} = \frac{R}{c_p}\frac{dp}{p} = 0.287\,\frac{dp}{p} \tag{4-91}$$

which, when integrated between the limits indicated by subscripts 1 and 2, yields Poisson's equation:

$$\frac{T_2}{T_1} = \left(\frac{p_2}{p_1}\right)^{R/c_p} = \left(\frac{p_2}{p_1}\right)^{0.287} \tag{4-92}$$

This equation expresses the relationship between adiabatic pressure and temperature changes. It also permits the definition of the potential temperature of dry air, θ, as the temperature that a parcel of air would assume if it were adiabatically moved from a level with a pressure p and a temperature T to the level of standard pressure $P = 1{,}000$ mb. Thus:

$$\theta = T\left(\frac{P}{p}\right)^{R/c_p} = T\left(\frac{1{,}000}{p\ \text{mb}}\right)^{0.287} \tag{4-93}$$

By definition, the potential temperature is invariant for adiabatic changes of dry air.

4.3.4 Thermodynamics of Water Vapor and Moist Air

Water vapor is of particular meteorological interest; like the other atmospheric gases, it exerts a partial pressure, generally denoted by e. However, unlike the other gases, practically all the water vapor is found in the lowest few miles of the atmosphere, and its small contribution to the total atmospheric pressure is subject to large variations. The partial pressure of water vapor, at which the vapor is in equilibrium with the liquid (or solid) phase, is called saturation pressure e_s; it is a function only of the temperature and is independent of the presence of other gases. The ratio of existing vapor pressure to saturation pressure is called the relative humidity f:

$$f = \frac{e}{e_s} \tag{4-94}$$

The relative humidity can be directly measured by means of a hair hygrometer. The variation of e_s with temperature is roughly shown in Table 4-8. At temperatures below 0°C, the saturation pressure over water is approximately 1 per cent per degree below 0°C higher than that over ice.

Since e_s increases with rise in temperature, the relative humidity decreases or increases when a mass of air is heated or cooled, respectively, without the addition or removal of water vapor. When, under such conditions, the temperature is lowered

* In case of an air parcel moving along a sloping surface, heat changes by conduction and radiation may not be negligible.

without change in pressure, the ratio e/e_s approaches unity; the temperature at which the relative humidity thus reaches 100 per cent (saturation) is called the dew point τ. The dew point can be directly measured by observing the temperature of an internally cooled vessel at which dew forms on its polished surface. Another measurable quantity is the wet-bulb temperature T_w, which is the lowest temperature attainable by evaporating water from the ventilated bulb of a thermometer, without changing the

Table 4-8. Saturation Pressure over Water at Various Temperatures
(In millibars)

T°C.	−20	−15	−10	−5	0	5	10	15	20	25	30
e_s.	1.2	1.9	2.9	4.2	6.1	8.7	12.3	17.0	23.4	31.7	42.4

properties of the air.* If one of the quantities f, τ, or T_w is measured together with T and p, the other two can be determined from tables, or graphically, as will be shown below.

The water-vapor content of the air can also be expressed in terms of the vapor density ρ_w, which is also called absolute humidity, and is usually given in units of grams of vapor per cubic meter of air. Since the molecular weight of water is $m_w = 18$, whereas the "molecular weight of dry air" is $m_w \approx 29$, the density of water vapor is only about 62 per cent of that of dry air, and moist air (the mixture of dry air and water vapor) is less dense than is dry air at the same pressure and temperature. The ratio of vapor density ρ_w to the density of dry air ρ_d is called the mixing ratio w. If m_d is now the mean molecular weight of dry air and $(p - e)$ the pressure of dry air (p the pressure of moist air, e that of the water vapor), we find from Eq. (4-82)

$$(p - e)m_d = \rho_d T R^* \qquad \text{and} \qquad e m_w = \rho_w T R^* \tag{4-95}$$

where the temperatures of the dry air and the water vapor are, of course, the same. Therefore:

$$w = \frac{\rho_w}{\rho_d} = \frac{m_w}{m_d}\frac{e}{p - e} \approx 0.62\,\frac{e}{p - e} \tag{4-96}$$

Another expression is the specific humidity q, which is the ratio of vapor density to the density of the moist air ρ. With Eqs. (4-82) and (4-95) we obtain, by considering that $\rho = \rho_d + \rho_w$,

$$q = \frac{\rho_w}{\rho} = \frac{m_w}{m_d}\frac{e}{p - e(1 - m_w/m_d)} \approx 0.62\,\frac{e}{p - 0.38e} \tag{4-97}$$

Since e is always small as compared to p, the approximation

$$w \approx q \approx 0.62\,\frac{e}{p} \tag{4-98}$$

is generally used; the quantities w and q are frequently expressed in 0/00, i.e., grams of water vapor per kilogram of air.

The saturation mixing ratio w_s is the mixing ratio at saturation and is, according to (4-98),

$$w_s \approx 0.62\,\frac{e_s}{p} \tag{4-99}$$

It is, thus, only a function of temperature (because of e_s) and of pressure. We also see immediately that

$$f = \frac{e}{e_s} \approx \frac{w}{w_s} \approx \frac{q}{q_s} \tag{4-100}$$

* For theoretical treatment of the complicated wet-bulb process, see Refs. 4, 22, 25.

The effect of the water vapor on the density of the air can be accounted for by substituting a slightly higher temperature, the virtual temperature T_v, for the actual temperature and considering the air to be dry. In other words, the virtual temperature is defined as the temperature of dry air having the same density and pressure as that of the actual moist air. With this definition and Eqs. (4-82), (4-95), and (4-98) we find that

$$T_v = T\left(1 + 0.38\frac{e}{p}\right) \cong T\frac{1 + 0.38e}{p} = T(1 + 0.61w) \tag{4-101}$$

Since w rarely exceeds 2 per cent, the difference between the virtual and actual temperature is usually less than 2°, even under conditions of saturation.

The gas equation for moist air can be written

$$p = R\rho T_v \tag{4-102}$$

where the virtual temperature [Eq. (4-101)] takes care of the water-vapor effect on the density of the air.

It can be shown that the adiabatic relationship between pressure and temperature changes of moist, but unsaturated, air is practically the same as that for dry air. If we again use $(p - e)$ as the pressure of dry air, Eq. (4-90) becomes (for 1 g of dry air)

$$dQ = c_p\,dT - \frac{RT\,d(p - e)}{p - e} \tag{4-103}$$

Let 1 g of air be mixed with w g of water vapor so that the total mass of moist air is $(1 + w)$ g. The mixing ratio w will be constant, as long as no condensation occurs, i.e., as long as the air remains unsaturated. By differentiating Eq. (4-96), we find, because $dw = 0$,

$$\frac{d(p - e)}{p - e} = \frac{de}{e} = \frac{dp}{p} \tag{4-104}$$

so that by substitution into (4-103) we obtain the first law of thermodynamics for 1 g of dry air of the same form as Eq. (4-90). We apply a similar equation to w g of unsaturated water vapor:

$$w\,dQ = wc_{wp}\,dT - w\left(R\frac{m}{m_w}\right)T\frac{de}{e} \tag{4-105}$$

Where $c_{wp} \approx 1.9 \times 10^7$ ergs/g deg is the specific heat of water vapor at constant pressure, (Rm/m_w) is, according to Eq. (4-84), the specific gas constant of water vapor, and m/m_w is the ratio of the molecular weight of dry air to that of water vapor. Substituting from Eq. (4-104) into (4-105), we obtain

$$w\,dQ = wc_{wp}\,dT - w\left(R\frac{m}{m_w}\right)T\frac{dp}{p} \tag{4-106}$$

By addition of Eqs. (4-90) and (4-106), we find this law applied to $(1 + w)$ g of moist air:

$$(1 + w)\,dQ = (c_p + wc_{wp})\,dT - \frac{1 + wm}{m_w}\frac{RT\,dp}{p} \tag{4-107}$$

and for the adiabatic conditions $dQ = 0$:

$$\frac{dT}{T} = \left(\frac{1 + wm/m_w}{c_p + wc_{wp}}\right)R\frac{dp}{p} \tag{4-108}$$

Equation (4-108) has the same form as Eq. (4-91) except that the term in parentheses appears instead of $1/c_p$. We may define the reciprocal of this term as the specific heat of moist air at constant pressure c_p'; extracting c_p from the term, we have

$$c_p' = c_p \frac{1 + wc_{wp}/c_p}{1 + wm/m_w} \approx c_p \frac{1 + 1.9w}{1 + 1.6w} \approx c_p \tag{4-109}$$

since w is usually less than 10^{-2}. Thus Eq. (4-108) is, for all practical purposes, identical with Eq. (4-91), and the adiabatic relation for dry air can be applied to moist, unsaturated air.

4.3.5 Thermodynamics of Saturated Air

Atmospheric water has several special characteristics that in many aspects make a rigorous thermodynamic analysis extraordinarily complicated or even impossible. Water is the only constituent of air that naturally occurs in all three phases. Near the point of transition to the liquid or the solid phase, water vapor no longer behaves as an ideal gas. Nevertheless, the laws for ideal gases are applied, because other inadequately fulfilled assumptions obviate the possibility of attaining a high degree of accuracy in pertinent thermodynamic computations.

One of the assumptions usually made in order to simplify thermodynamic considerations in meteorology is that condensation is a discontinuous process that abruptly starts at a relative humidity of 100 per cent. In reality, depending on the properties of the aerosol present, condensation may commence long before saturation, or not until a slight degree of supersaturation is reached (42). Another simplifying assumption is made, namely, that fog or cloud droplets in the air do not freeze when the temperature falls below 0°C. This assumption is, however, well justified, because it has been observed that water droplets remain in a supercooled liquid state at temperatures far below the melting point of ice. Also, when condensation occurs below 0°C, its products frequently—if not always—are, at first, liquid. Nevertheless, water droplets do freeze at temperatures that are lower, the smaller the droplets (23,27). However, at the temperature at which freezing and sublimation occur, the difference in the thermodynamic aspects between the solid and liquid state is very small.

In the transition from the vapor to the liquid state, the latent heat of condensation L is liberated and warms the air in which the condensation takes place. The value of L at 0°C is 2.5×10^{10} ergs/g of water condensed and can be considered constant for most practical purposes, because it diminishes by only about 0.9 per cent for a rise of 10°C in temperature. During sublimation of water vapor into ice, latent heat of sublimation $L_i = 2.8 \times 10^{10}$ ergs/g is released. Again, because supercooled water droplets form more frequently at temperatures below 0°C than do ice crystals, L is more generally used than L_i.

In the case of adiabatic changes of saturated air, we must consider the effect of the latent heat of condensation.

When a parcel of saturated air rises, the ensuing pressure and temperature decrease causes condensation to take place. The removal of water vapor that accompanies the formation of liquid droplets reduces the saturation mixing ratio w_s by the amount $-dw_s$ which is equal to the amount of water condensed per gram of dry air. Therefore, the amount of heat released is $-L\,dw_s$; for the sake of simplicity, we consider the effect of this heat on the air only, neglecting that on the water content.* Thus, in Eq. (4-90), the heat is $dQ = -L\,dw_s$; since this heat comes from the air itself and not from an external source, the process is still considered quasi-adiabatic, although $dQ \neq 0$. This wet-adiabatic or pseudoadiabatic process is described by the equation

* For a more rigorous discussion of adiabatic changes of saturated air, including the separate consideration of the retention of the condensates, on the one hand, and their precipitation, on the other, see Refs. 4, 22, 25.

$$dQ = -L\,dw_s = c_p\,dT - \frac{RT\,dp}{p} \tag{4-110}$$

or

$$\frac{dT}{T} = \frac{R}{c_p}\frac{dp}{p} - \frac{L}{Tc_p}\,dw_s \tag{4-111}$$

Here, the effects of the water vapor have been neglected in accordance with the preceding discussion. It is furthermore assumed that the total water content (liquid and gaseous) of the air remains constant.

The term dw_s can be determined by logarithmic differentiation of Eq. (4-99):

$$\frac{dw_s}{w_s} = \frac{de_s}{e_s} - \frac{dp}{p} \tag{4-112}$$

therefore,

$$dw_s = w_s\left(\frac{de_s}{e_s} - \frac{dp}{p}\right) \tag{4-113}$$

Substitution into Eq. (4-111) yields

$$\frac{dT}{T} = \frac{R}{c_p}\frac{dp}{p} - \frac{Lw_s}{Tc_p}\left(\frac{de_s}{e_s} - \frac{dp}{p}\right) = \frac{R}{c_p}\frac{dp}{p}\left(1 + \frac{Lw_s}{RT}\right) - \frac{Lw_s}{Tc_p}\frac{de_s}{e_s} \tag{4-114}$$

or

$$\frac{dT}{T}\left(1 + \frac{Lw_s}{c_pe_s}\frac{de_s}{dT}\right) = \frac{R}{c_p}\frac{dp}{p}\left(1 + \frac{Lw_s}{RT}\right) \tag{4-115}$$

thus

$$\frac{dT}{T} = \frac{R}{c_p}\frac{dp}{p}\left[\frac{1 + (Lw_s/RT)}{1 + (Lw_s/c_pe_s)(de_s/dT)}\right] \tag{4-116}$$

or substituting for w_s for Eq. (4-99) and introducing the numerical values of the constants:

$$\frac{dT}{T} = \frac{R}{c_p}\frac{dp}{p}\left(\frac{1 + 5.4 \times 10^3\,\dfrac{e_s}{pT}}{1 + 1.55 \times 10^3\,\dfrac{1}{p}\dfrac{de_s}{dT}}\right) = \frac{R}{c_p}\frac{dp}{p}\,\gamma' \tag{4-117}$$

This adiabatic equation for saturated air is similar to that for dry air [Eq. (4.91)] except for a factor γ' which is a function of the pressure and temperature, because e_s and de_s/dT are functions of only the temperature. The change in saturation vapor pressure with change in temperature de_s/dT can be evaluated from saturation vapor-pressure tables such as Table 4-8. The factor γ' is smaller than unity; for example, if $p = 1{,}000$ mb, $t = 273°$K, $e_s = 6.1$ mb, we find, roughly, $de_s/dT \approx 0.5$ mb/°K and $\gamma' = 0.6$. This means that, for a given relative pressure change, the relative temperature change during an adiabatic process is smaller when the air is saturated than when it is dry. The value of γ' approaches unity for very low temperatures, because the amount of water vapor available for condensation, and, therefore, the amount of heat released, decreases with drop in temperature.

4.3.6 Statics

In order to determine the change of pressure with altitude, we consider in Fig. 4-10 a section of an air column having a cross section $\delta x\,\delta y$ and a height δz between the levels z and $(z + \delta z)$. The pressures at the bottom and top of this column section are p and $(p - \delta p)$, respectively, the negative sign indicating that the pressure decreases with increase in altitude.

If δz is small so that the density of the column ρ can be considered constant over this height interval, the mass M of air is density times volume, which reduces to $M = \rho\,\delta z$, because $\delta x\,\delta y = 1$. The weight or force F of this mass exerted on the base of the column section is the product of the mass and the acceleration of gravity g, namely, $F = \rho g\ \delta z$. The force per unit area, i.e., the pressure, due to this mass is the difference of pressure between the top and the bottom of the column section:

$$(p - \delta p) - p = \rho g(z + \delta z - z) \tag{4-118}$$

or

$$-\delta p = \rho g\ \delta z \tag{4-119}$$

In Eq. (4-119), often referred to as the hydrostatic equation, the density can be substituted from Eq. (4-102):

$$g\ \delta z = -RT_v \frac{\delta p}{p} \tag{4-120}$$

This so-called barometric-height formula in differential form postulates an absence of vertical motion. According to (4-120), for a given pressure p, the pressure decrease $-\delta p$ over the height interval δz is greater for lower temperatures; in other words,

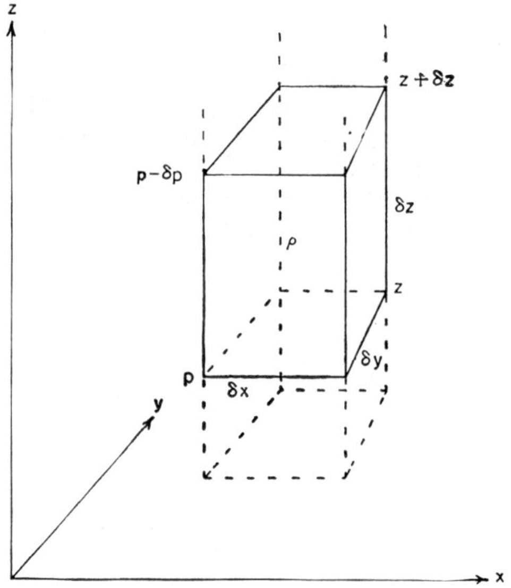

FIG. 4-10. Pressure change with height.

the pressure decreases with increase in height more rapidly in cold than in warm air. For example, over a height of 100 m the pressure decreases from its sea-level value by 1.4 per cent when the mean temperature of the layer is −30°C, but only by 1.1 per cent when the mean temperature is +30°C.

Integration of (4-120) between the height limits z_1 and z_2 and the corresponding pressure limits p_1 and p_2 yields (for constant g and R)

$$\ln \frac{p_1}{p_2} = \frac{g}{r} \int_{z_1}^{z_2} \frac{1}{T_v}\ \delta z \tag{4-121}$$

where T_v is a function of height and g is assumed to be independent of height. If the temperature is practically invariant with height (isothermal stratification), integration of (4-120) results in

$$\ln \frac{p_1}{p_2} = \frac{g}{RT_v} (z_2 - z_1) \tag{4-122}$$

where T_v is the virtual temperature of the air within the layer. Thus, for a given constant temperature, the natural logarithm of pressure decreases linearly with increase in height. In the lower atmosphere, often a linear decrease of temperature with height increase exists such that $T_2 = T_1 - \gamma(z_2 - z_1)$, where γ is the temperature lapse rate, i.e., $-\delta T/\delta z$. In this case, integration of (4-120) leads to

$$\frac{p_1}{p_2} = \left(\frac{T_1}{T_2}\right)^{g/R\gamma} \tag{4-123}$$

When the temperature occasionally increases with increase in height, the lapse rate γ is negative, a condition referred to as inversion.

In practice, height is determined from temperature and pressure measurements with Eq. (4-122) by taking sufficiently small sections from a curve of temperature versus pressure so that the temperature in every curve section can be considered constant at the mean temperature of that section. Subsequent addition of all height increments thus determined furnishes the total height. (For use of graphical methods see Refs. 4 and 22.)

4.3.7 Adiabatic Lapse Rates and the Adiabatic Chart

Let a discrete parcel of dry air be adiabatically displaced through a small vertical distance dz in such a manner that the surrounding air is not disturbed. Let, further, the displacement proceed so slowly that the parcel has everywhere the same pressure p as that of the surrounding air p', so that $dp = \delta p'$ for $dz = \delta z'$, and let it be assumed that the pressure of the surrounding air does not change during the displacement. The dry-adiabatic relative pressure change dp/p in Eq. (4-91) can then be substituted from Eq. (4-120) which now assumes the form

$$\frac{dp}{p} = -\frac{g}{R}\frac{dz}{T'} \tag{4-124}$$

where T' is the temperature of the air surrounding the parcel. Combining this equation with (4-91), we find

$$-\frac{dT}{dz} = \frac{gT}{c_p T'} \tag{4-125}$$

Since for small displacements the ratio of parcel temperature T to the temperature T' of its surroundings is nearly unity, we can adopt the approximation

$$-\frac{dT}{dz} = \frac{g}{c_p} \approx 1°\text{C}/100 \text{ m} \approx 5\tfrac{1}{2}°\text{F}/1{,}000 \text{ ft} = \Gamma \tag{4-126}$$

Γ is the dry-adiabatic lapse rate, according to which the temperature of a parcel of dry air decreases (increases) by 1°C for an adiabatic upward displacement (downward displacement) of the parcel by 100 m. Considering the approximation of Eq. (4-109), the same lapse rate is also valid for a parcel of moist air as long as it remains unsaturated.

At sea level, or near it, the potential temperature defined in Eq. (4-93) is practically equal to the actual temperature in degrees Kelvin. According to its definition, the potential temperature θ at a given height z above sea level, where the actual temperature is T, will be, approximately, $\theta = T + \Gamma z$, provided that z does not exceed a few hundred meters (34).

When a saturated air parcel is displaced upward, the wet-adiabatic equation (4-117) applies. This equation with substitution from (4-124) and again neglecting the small difference between the temperature of the parcel and that of the surrounding air now reads

$$-\frac{dT}{dz} = \gamma' \frac{g}{c_p} \approx \Gamma'°\text{C}/100 \text{ m} \qquad 0 < \Gamma' < 1 \tag{4-127}$$

where the wet-adiabatic lapse rate Γ' is the same function of pressure and temperature as is γ'. Table 4-9 shows various values of Γ' at different temperatures and at pressures of 1,000 and 500 mb.

Table 4-9. Saturation-Adiabatic Lapse Rates

Temp, °C	30	20	10	0	−10	−20	−30
1,000 mb	0.4	0.4	0.5	0.6	0.7	0.9	0.9
500 mb	0.3	0.3	0.4	0.5	0.6	0.8	0.9

In practice, dry-adiabatic and pseudoadiabatic temperature changes are determined by means of thermodynamic diagrams, which contain these changes in the form of two families of curves on a grid having a linear temperature scale as abscissa and a function of pressure ($\ln p$ or $p^{0.287}$) as ordinate.* In addition, a family of lines of equal saturation mixing ratios are generally shown on such charts.

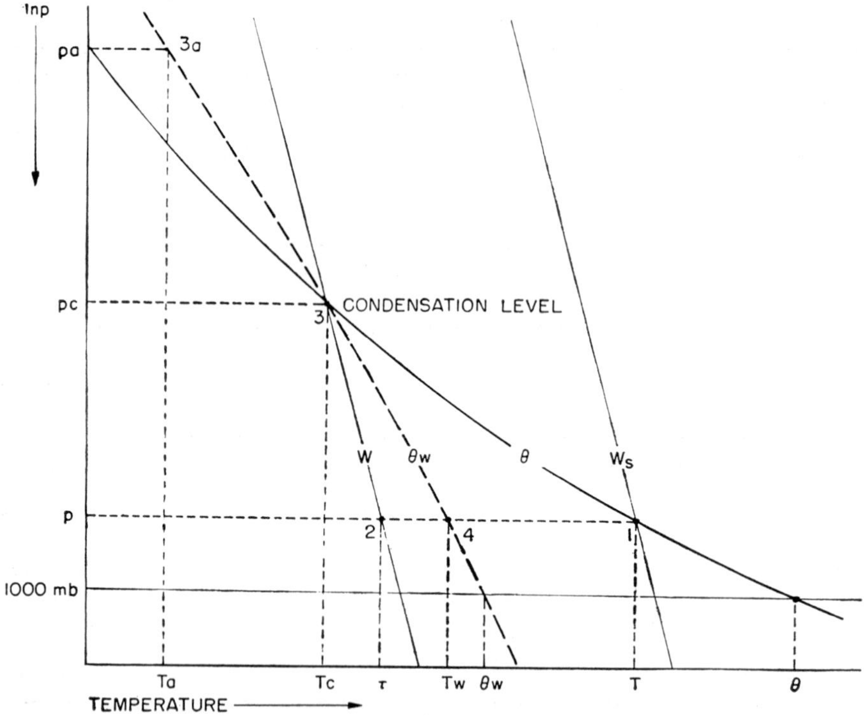

FIG. 4-11. Schematic meteorological diagram.

In Fig. 4-11, which represents a schematic diagram of a pseudoadiabatic chart, let an air parcel have the pressure p, the temperature T, and the mixing ratio w. These properties are represented on the chart as point 1 at the intersection of pressure p and temperature T and point 2 at the intersection of p and w. The slanting heavy line through point 1 marked θ is the dry adiabat along which the parcel moves as long as it remains unsaturated; where the θ line intersects the line p = 1,000 mb, the potential temperature is found on the T scale. The thin line through point 1 represents the saturation mixing ratio w_s [Eq. (4-99)] for the pressure and temper-

* Other thermodynamic diagrams with different grids and curves are also in use (4,22).

ature of the air parcel. The ratio w/w_s is the relative humidity according to Eq. (4-100).* By definition, the existing mixing ratio w is the saturation mixing ratio at the dew point so that the temperature at point 2 represents the dew point τ of the air parcel.

If the parcel having the characteristic points 1 and 2 is lifted adiabatically, point 1 moves upward along the θ line, while point 2 moves along the w line, because the mixing ratio, i.e., the parcel's water-vapor content, does not change. At point 3, where the pressure is p_c and the temperature is T_c, points 1 and 2 merge; this means that the temperature is equal to the dew point and the air has become saturated. The level of point 3, at which cloud formation starts, is called the condensation level, p_c is the adiabatic condensation pressure, and T_c the adiabatic condensation temperature. If the parcel is lifted farther, its now single characteristic point 3 will follow the pseudoadiabat, the heavy broken line, and will, e.g., have a temperature T_a at the pressure p_a.

By analogy to the definition of potential temperature θ, we may define as wet-bulb-potential temperature θ_w the temperature which is obtained at the intersection of the saturation adiabat, that passes through the condensation level, with the pressure of 1,000 mb. It is obvious that θ_w is invariant for both dry- and wet-adiabatic processes, since an air parcel of a given characteristic has only one condensation level through which only one wet adiabat passes. The name wet-bulb-potential temperature is derived from the fact that the temperature at point 4, the intersection of the θ_w line with the p line, is the wet-bulb temperature T_w of the air parcel.

4.3.8 Lapse Rate and Stability

The hydrostatic equation (4-119) is valid on a large scale, i.e., when we consider air over many square miles and periods of several hours. In other words, hydrostatic equilibrium prevails in the atmosphere on a large scale. However, individual smaller portions (parcels) of the atmosphere may be vertically displaced from their equilibrium level by turbulent motion† or by obstacles in the path of the air flow. When an air parcel is thus moved upward or downward, its temperature will change according to the dry-adiabatic Γ or pseudoadiabatic Γ' rate, depending on whether the parcel is unsaturated or saturated, respectively. In order to simplify the subsequent discussion, only Γ will be used, with the provision that Γ' is to be substituted in case of saturated air, unless otherwise specified.

It is unfortunate that, in meteorology, the term lapse rate is used for both the local vertical gradient of temperature $-\delta T/\delta z = \gamma$ [see Eq. (4-123)] and the individual change in temperature of an air parcel with change of its height, the adiabatic lapse rate, which, more rigorously, should be written $-dT/dz = \Gamma$. In other words, γ refers to the vertical distribution of temperature over a given place, whereas Γ refers to the temperature change that an individual parcel undergoes when it changes its level. Whereas Γ is constant and Γ' varies roughly between 0.3 and 0.9°C/100 m (Table 4-9), γ varies over a wide range, including negative values (inversions) and positive values greater than 1°C/100 m, called superadiabatic lapse rates. For example, when the ground is heated by solar radiation, the lowermost layer of air, up to perhaps a meter or more above the ground, may have uniform density throughout. In this case, partially differentiating Eq. (4-85) and then substituting into Eq. (4-119), we obtain, with ρ = const, thus $\delta\rho = 0$:

$$-R\delta(\rho T) = -R\rho\,\delta T = g\rho\,\delta z \tag{4-128}$$

or

$$-\frac{\delta T}{\delta z} = \frac{g}{R} \approx 3.4°\text{C}/100\text{ m} \tag{4-129}$$

* Usually, the relative humidity is measured and, together with the known w_s, furnishes w.

† For discussion of scale and turbulence, see Art. 4.4, The Atmosphere in Motion.

If this lapse rate is exceeded, i.e., if the density increases with height, the stratification of air overturns without external perturbation, which is called autoconvection. Under all other conditions, vertical motions of air parcels, convection, must be initiated by some external disturbance.

On the average, the lapse rate γ is about 0.6°C/100 m, as shown in Fig. 4-12 by the heavy line AB. If, e.g., a parcel P of the air layer having the indicated lapse rate is now displaced upward, its temperature is decreased adiabatically along the thin line $\Gamma\Gamma$; it can be seen that everywhere above its original position the parcel will have a lower temperature than that of the surrounding air (indicated by the line PB). This means that the parcel is denser than the surrounding air and will return to its original position upon removal of the lifting force. If the parcel were lowered from its original level at P, its temperature would be increased adiabatically along

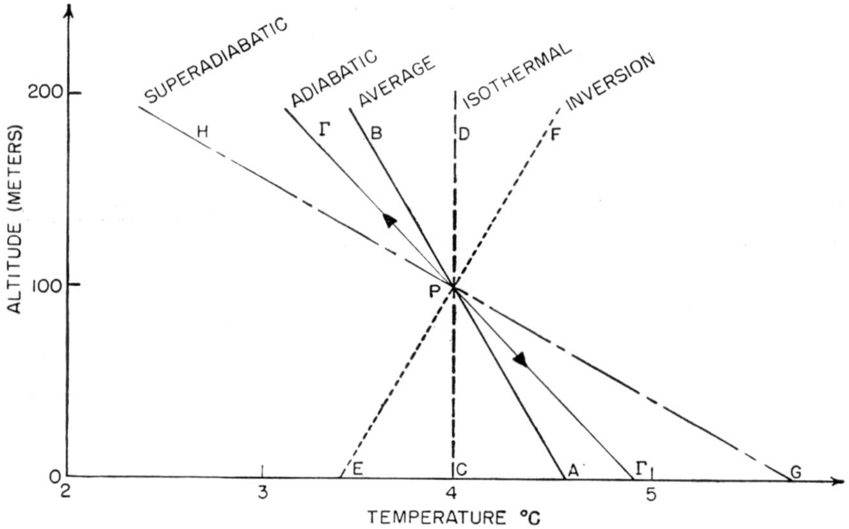

FIG. 4-12. Displacement of air parcel in an air layer having various lapse rates.

the line $\Gamma\Gamma$ and would be higher everywhere than that of the surrounding air (indicated by the line PA). In this case, the parcel, being less dense than the surrounding air, would again tend to return to its original position. This tendency of resisting vertical displacement (upward or downward) is characteristic of stability.

From Fig. 4-12, it is evident that the corresponding temperature and, therefore, density differences between parcel and environment would occur if the lapse rate of the air layer were isothermal, $\gamma = 0$, i.e., if the same temperature prevailed at every level as indicated by the dashed line CD, or if the lapse rate were an inversion, $\gamma < 0$, i.e., if an increase in temperature with height existed as indicated by the dotted line EF. In all these cases of stability the existing lapse rate of the layer is less than adiabatic, $\gamma < \Gamma$.

If the existing lapse rate were the same as the adiabatic rate, $\gamma = \Gamma$, as indicated by the thin line $\Gamma\Gamma$ in Fig. 4-12, a parcel of air, when lifted or lowered adiabatically, would everywhere have the same temperature as the surrounding air and would thus be in equilibrium with it at any level. The parcel would have the same density and, therefore, would not tend to return to its original position. This condition is called neutral, or indifferent equilibrium.

Now, if the existing lapse rate of a layer were greater than adiabatic, i.e., superadiabatic, $\gamma > \Gamma$ as indicated by the dot-dashed line GH, a parcel when lifted would follow the adiabat $\Gamma\Gamma$ and therefore be everywhere warmer and thus less dense than the air surrounding it. This would cause the parcel, after a small initial upward displacement, to continue to move upward by its own buoyancy. Conversely, if displaced downward, the parcel would be colder, thus denser, than the environment and would therefore continue downward after a small initial displacement. This tendency of assisting vertical displacement and of moving farther away from the original position is characteristic of instability.

We may summarize the lapse-rate values γ, characteristic of various stability conditions, as follows:

$$\left.\begin{array}{ll} \gamma < \Gamma & \text{stable} \\ \gamma = \Gamma & \text{indifferent} \\ \gamma > \Gamma & \text{unstable} \end{array}\right\} \qquad (\Gamma' \text{ for } \Gamma \text{ in case of saturated air}) \tag{4-130}$$

The stability conditions can also be expressed in terms of the vertical gradient of potential temperature, if Eq. (4-93) is logarithmically differentiated partially with respect to height z,

$$\frac{1}{\theta}\frac{\delta\theta}{\delta z} = \frac{1}{T}\frac{\delta T}{\delta z} - \frac{R}{c_p}\frac{1}{p}\frac{\delta p}{\delta z} \tag{4-131}$$

Substituting from Eq. (4-120) we obtain

$$\frac{1}{\theta}\frac{\delta\theta}{\delta z} = \frac{1}{T}\left(\frac{\delta T}{\delta z} + \frac{g}{c_p}\right) \tag{4-132}$$

Since $-\delta T/\delta z = \gamma$ and $g/c_p = \Gamma$ [Eq. (4-126)], we can write

$$\frac{\delta\theta}{\delta z} = \frac{\theta}{T}(\Gamma - \gamma) \tag{4-133}$$

Considering that θ/T is always positive, we have from the criteria (4-130) concerning γ

$$\begin{array}{lll} \dfrac{\delta\theta}{\delta z} > 0 & \text{stable} & (\theta \text{ increases with height}) \\ \dfrac{\delta\theta}{\delta z} = 0 & \text{indifferent} & (\theta \text{ constant with height}) \\ \dfrac{\delta\theta}{\delta z} < 0 & \text{unstable} & (\theta \text{ decreases with height}) \end{array} \tag{4-134}$$

In the case of stable or unstable stratification of an air layer or column, a parcel displaced upward will, because of negative or positive buoyancy, be subject to an acceleration. The force causing the vertical acceleration d^2z/dt^2 (z = height, t = time) of the parcel's mass ρV is the difference between the weight of the air displaced by the parcel and the weight of the parcel, where the weight is the respective mass times the acceleration of gravity g. Thus, if we distinguish the properties of the air column from those of the parcel by a prime, we have, since the volume of the displaced air is equal to the volume of the parcel, and therefore cancels out,

$$(\rho' - \rho)g = \rho\frac{d^2z}{dt^2} \tag{4-135}$$

Substituting ρ' and ρ from Eq. (4-85) and considering that the pressures of the parcel and the environment are the same, $p' = p$, we obtain

$$\frac{T - T'}{T'}g = \frac{d^2z}{dt^2} \tag{4-136}$$

According to this equation, the acceleration of the parcel will be positive (upward) or in the direction of the original displacement if the temperature of the parcel T is greater than that of the environment T'. If $T < T'$, the resulting acceleration is negative (downward); if $T = T''$, the acceleration will be zero.

Let the temperature of the parcel at its original level be T_0 (which is, of course, also the temperature of the environment at that level); then, at the height z, to which the parcel has been adiabatically lifted, the temperature of the parcel is $T = T_0 - \Gamma z$, the temperature of the environment is $T'' = T_0 - \gamma z$, where γ is the lapse rate existing in the air column. Substituting these temperatures in the numerator of Eq. (4-136), we find

$$\frac{\gamma - \Gamma}{T'} zg = \frac{d^2z}{dt^2} \qquad (4\text{-}137)$$

We see that the acceleration is, other things equal, greater (1) for greater original displacement z, (2) for lower environmental temperature T', and (3) for greater difference between existing and adiabatic lapse rates, $\gamma - \Gamma$. Also, for the stable condition $\gamma < \Gamma$, the acceleration of the parcel is negative, i.e., in the opposite direction

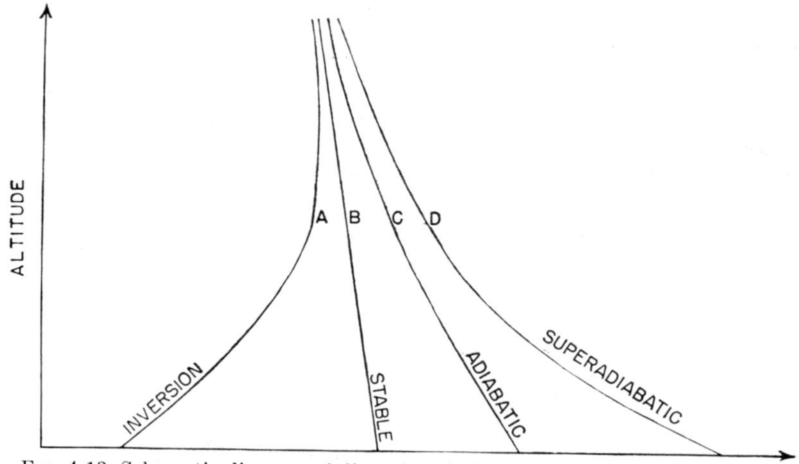

FIG. 4-13. Schematic diagram of diurnal variation of lapse rate on a clear day.

from the direction of original displacement; for the unstable conditions of $\gamma > \Gamma$, the acceleration is positive, i.e., in the direction of original displacement.

It was stated above that the lapse rate γ varies over a wide range, particularly in the lowest few hundred meters of the atmosphere which are of particular interest in problems of air pollution. These variations may occur at a given locality over a period of the order of an hour or even minutes. The causes of these variations are manifold.

A stable stratification of the air near the ground can be produced by a process that either cools the air at a low level or warms the air at a high level or by the simultaneous occurrence of both. Thereby, the air at the higher level can become potentially, or actually in case of inversions, warmer than the air at the lower level. Conversely, when the air is warmed near the ground or cooled aloft or both, unstable conditions are established. The air near the ground may be warmed or cooled, respectively, by the earth's surface being warmer or colder than the air flowing over it. In turn, the surface is heated by solar radiation in daytime and cooled by eradiation of heat at night. These processes are favored under cloudless skies and are suppressed under overcast skies. Therefore, inversions are more likely to occur on clear nights, unstable lapse rates on clear days.

In general, the lapse rates in the lowest air layers are essentially determined by the heating and cooling of the air in the immediate vicinity of the surface or in contact with it. Because the temperature variations of the surface are greater for land than for water surfaces, the corresponding lapse-rate variations are greater over land than over water. Moreover, the diurnal alternation of heating and cooling of the surface under clear conditions causes a corresponding diurnal variation of lapse rates, as shown schematically in Fig. 4-13 (35,48).

4.4 THE ATMOSPHERE IN MOTION

The motion of the atmosphere is both vertical and horizontal; yet the wind-measuring instruments commonly in use measure only the horizontal components of this motion. This practice does not imply that vertical motion is unimportant, for vertical motion produces rain, shakes up airplanes, and is the cause of vertical heat flux and diffusion. The neglect of the vertical motion among the physical properties of the atmosphere normally measured is due to the difficulty of measuring this motion. This difficulty, in turn, stems from the fundamentally different statistical properties of the vertical and horizontal wind components. The vertical motion in the lower layers is limited by the proximity of the ground, whereas the horizontal wind components are not. The influence of the ground decreases with height but increases with increasing scale. Thus we find that 24-hr average vertical velocities at a height of 10,000 ft are of the order of 1 cm/sec, whereas 2-min average vertical velocities at the same level often exceed 100 cm/sec. On the other hand, the order of magnitude of horizontal velocities averaged over 1 min is about the same as that averaged over 1 day, namely, 1,000 cm/sec.

4.4.1 Division of Scale

The magnitude of the wind components fluctuates. The more sensitive the recording instrument, the more rapid the fluctuations appear to be. On the other end of the scale, fluctuations occur with periods of years, centuries, and, probably, millions of years. The high-frequency fluctuations are of little practical interest to the forecasting meteorologist or to a pilot attempting to estimate the time required to reach his destination. Yet, the same high-frequency fluctuations are essential in the diffusion of air pollutants and the air "roughness" that makes airplane flights unsafe or, at least, uncomfortable.

Because the different scales of wind and its fluctuations are important from different points of view, an arbitrary division is effected between the "macrometeorological" scale and the "micrometeorological" scale. These will be simply referred to as "small-" and "large-" scale motions. The division between these two scales can be best explained in terms of the procedure used on a weather map. A weather map normally covers an area as large as the United States, or larger. On these charts, wind vectors are plotted which represent 1- or 2-min averages. These wind vectors are thus affected by variations of many scales, from a few minutes on up. In order to eliminate the small-scale variations, the wind vectors (and similarly other meteorological parameters) are analyzed by drawing smooth lines parallel to the wind through observations 100 km or more apart. The winds represented by these lines are then rather slowly varying functions. They represent averages over 100 km or more, or perhaps time averages over several hours. The field of motion represented by the smooth lines will be called "large-scale" or "mean" motion. All deviations from this large weather-map scale will be called "small-scale."

Thus, we write the total wind speed:

$$V = \bar{V} + V' \tag{4-138}$$

where $\bar{V}$ is the large-scale or mean motion and V' is the small-scale motion, also called "turbulence." In subsequent notations, a bar will denote large-scale, a prime small-scale motion.

To return to the discussion in the introduction: in the case of the small-scale motions, the fluctuations of vertical and horizontal motions are of the same order of magnitude; on the other hand, large-scale vertical velocities tend to be a small fraction of large-scale horizontal velocities.

4.4.2 Properties of Large-scale Horizontal Motions

Of the different categories of motion discussed here, most information is available for large-scale horizontal motion. These motions are near zero at the surface, increase rapidly in the lowest few meters and then more slowly to a maximum averaging about 75 mph at 10 km altitude in winter; in summer, the winds are about half as fast.

The increase of speed with height in the lowest levels is approximately logarithmic, and at higher levels it is more nearly linear. The variation is, however, not at all regular. Limited regions of extremely high speed, such as 175 mph, are common, especially near 10 km in regions of strong horizontal temperature contrasts. Such air currents are relatively narrow and are called jet streams.

The direction of the winds near the surface is quite variable, except in the subtropics and tropics, where it normally has a component from the east. Above 3 km, over almost the whole earth, the wind directions have strong components from the west, but the air is equally likely to move northward or southward. Above 20 km, the normal west winds shift to east winds in the summer.

One of the reasons for the division of scale previously mentioned is the fact that the large-scale horizontal motions can be related quite well to the other atmospheric variables, such as pressure and temperature, with the aid of physical equations. By contrast, a discussion of small-scale motion must largely be based on statistical and empirical relationships.

The relation between wind and pressure follows from Newton's second law applied in the horizontal plane. Two forces* are important: the first, the Coriolis force, is due to the tendency of air to move in a straight line (with respect to space), while the earth rotates underneath the air. The horizontal component of the Coriolis force is always at right angles to the motion, to the right of the motion in the Northern Hemisphere, to the left in the Southern Hemisphere. Its magnitude is $2\omega\bar{V} \sin \phi$, where ω is the angular velocity of the earth, $\bar{V}$ the horizontal wind speed, and ϕ the latitude.

The other force is due to horizontal pressure gradient. Its magnitude is given by $(1/\bar{\rho})(\delta\bar{p}/\delta n)$, where ρ is the air density and $\delta\bar{p}/\delta n$ represents the variation of pressure at right angles to the lines of constant pressure (isobars). The direction of this gradient force is from high to low pressure.

In many cases, the accelerations are small and momentum increase due to transfer by small-scale motion can be neglected at heights above 3,000 ft. Then the two forces will be in balance, and the large-scale wind speed will be given by the "geostrophic wind equation":

$$V = \frac{1}{2\omega\bar{\rho} \sin \phi} \frac{\delta\bar{p}}{\delta n} \tag{4-139}$$

Since, in the Northern Hemisphere, for example, the Coriolis force is to the right of the motion, this balance is possible only when the wind is parallel to the isobars, with high pressure to the right of the wind vector.

* All forces are expressed in terms of force per unit mass.

Because the horizontal component of the Coriolis force diminishes with decrease in latitude and becomes zero at the equator, the geostrophic equilibrium becomes a poor approximation of the actual wind in a region within 20° of the equator. Furthermore, in the case of strongly curved air trajectories, the two important forces are no longer in balance; their resultant produces a centripetal acceleration. A correction can be made for this acceleration by introducing the centrifugal force $\bar{V}^2/R$ in the balance equation, where R is the radius of curvature of the trajectories.

With the geostrophic wind relation, the wind field above 3,000 ft can be determined quite well without any wind measurements, if only the pressure field is known.

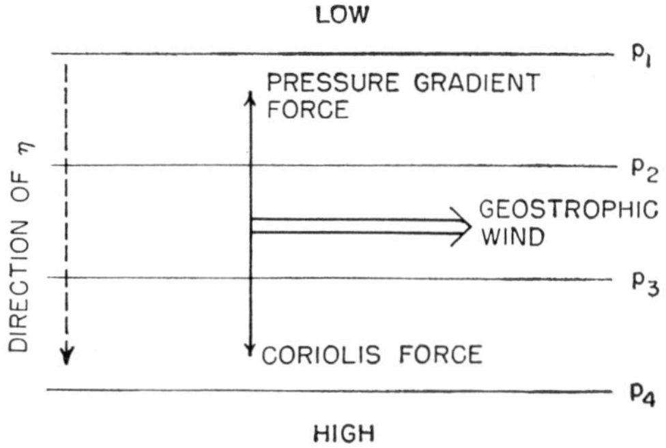

FIG. 4-14. Geostrophic wind.

Since the wind depends on the pressure gradient, and the vertical distribution of pressure is a function of temperature, the vertical wind variation above 3,000 ft also depends on temperature. If the geostrophic wind equation is satisfied, the difference vector between the winds at two heights is parallel to the mean isotherms for the layer, with highest temperatures to the right of the difference vector in the Northern Hemisphere.

Figure 4-15 shows the relation between isotherms and wind-vector difference in the Northern Hemisphere; in the Southern Hemisphere, the temperature gradient would be reversed.

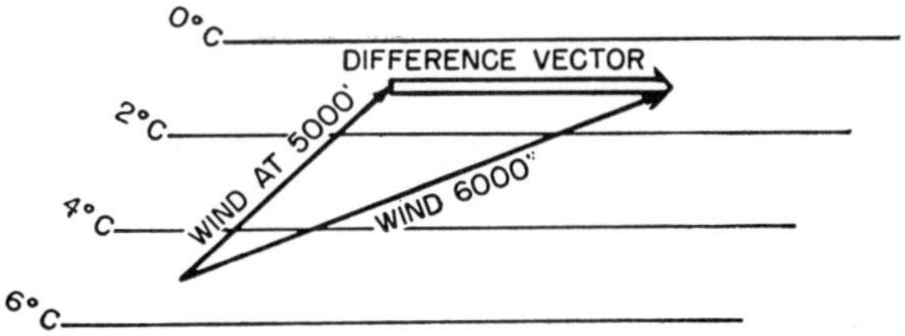

FIG. 4-15. Relation between isotherms and the difference in wind vector between bottom and top of layer.

Since in the lower 10 km the temperature decreases northward and above that level decreases southward, this relation explains the maximum speed of west winds near 10 km.

Below 3,000 ft, a "frictional" force must be added to the pressure gradient and Coriolis forces. Unlike true friction, this force is not produced by molecular motion but by vertical transport of large-scale momentum produced by small-scale vertical motion. But just as is the case with true friction, this pseudofrictional force produced by turbulence is generally in a direction opposite to that of the wind (at least, close to the ground). For this reason, the force is referred to as "friction" and the layer of atmosphere within the lowest 3,000 ft (approximately) as the "friction layer."

The effect of friction is, first, to diminish the wind speed. In consequence, it also decreases the Coriolis force which is proportional to the speed, so that the equilibrium at right angles to the wind speed is upset, and the wind is turned toward lower pressure. Finally, equilibrium is possible with the wind speed less than geostrophic and with an angle between wind and isobars between 10 and 45°, depending on the magnitude of friction. Figure 4-16 illustrates this equilibrium. Over the ocean, the angle between wind near the surface and isobars is generally between 15 and 20°, over land 30° or higher, depending on the roughness of the ground. Largest angles are found in mountainous regions, where the topography produces pronounced channeling effects.

The greatest angles and slowest winds are found close to the ground; with increasing elevation, the wind speed increases and the wind becomes more parallel to the isobars, as shown in Fig. 4-17.

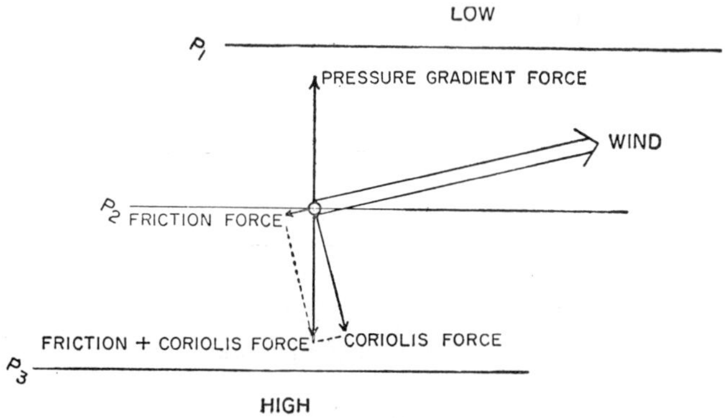

FIG. 4-16. Schematic diagram of force balance in case of wind in the friction layer.

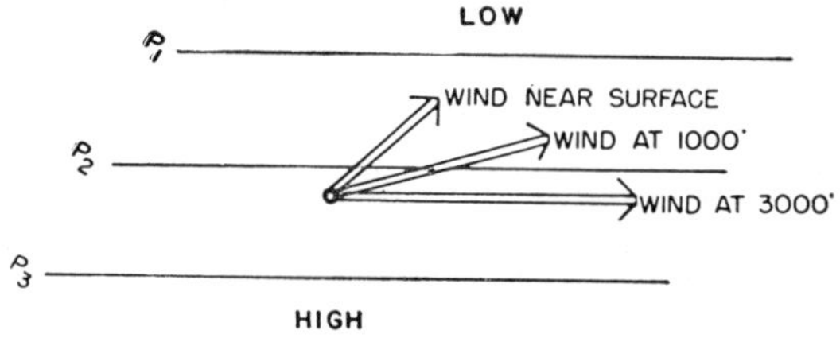

FIG. 4-17. Schematic illustration of the variation of wind with height in the friction layer.

In the lowest layer of the atmosphere (up to perhaps 50 ft) the turning of the wind is negligible. Also, the vertical variation of the wind can be described successfully by relatively simple equations. Under neutral stability conditions, the wind follows well the Prandtl logarithmic law (58):

$$\bar{V} = \frac{1}{k}\sqrt{\frac{\tau_0}{\bar{\rho}}}\ln\frac{z}{z_0} \qquad (4\text{-}140)$$

where k is von Kármán's constant, usually taken as 0.4 by meteorologists, and τ_0 is the wind stress. The density and the stress vary only little in the lowest 50 ft. The wind stress usually is between 1 and 10 dynes/sq cm over land, somewhat smaller over the oceans. The quantity z_0 is the "roughness length" which, as its name implies, has the dimension of a length. Numerically, the roughness length is between one-tenth and one-thirtieth of the size of the roughness elements. Thus, the roughness length varies from about 0.1 cm for short grass to 100 cm for clumps of trees 10 m high. Normally, the stress is not known and must be evaluated from the wind observations at one level. Often, the roughness length must also be determined from wind observations, in which case a minimum of two wind observations at different heights is required.

According to the logarithmic law, the wind vanishes at $z = z_0$, or an extremely short distance above the origin of the vertical coordinate. If the ground is covered by cereal grasses or other tall plants, the wind actually vanishes some distance above the ground. For this reason, the logarithmic law has been generalized to

$$\bar{V} = \frac{1}{k}\sqrt{\frac{\tau_0}{\bar{\rho}}}\ln\left(\frac{z-d}{z_0}\right) \tag{4-141}$$

where d is called the displacement length, which has the same order of magnitude as the height of the plants (58).

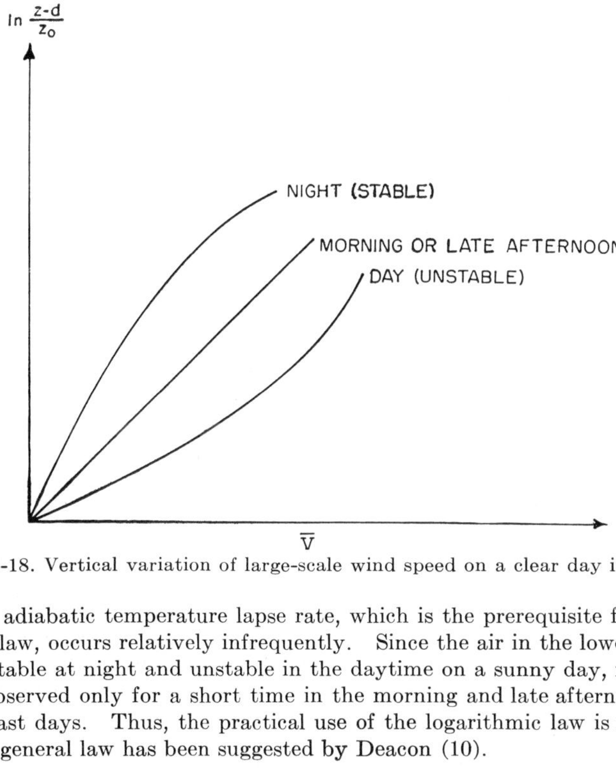

FIG. 4-18. Vertical variation of large-scale wind speed on a clear day in the lowest layers.

An adiabatic temperature lapse rate, which is the prerequisite for the logarithmic wind law, occurs relatively infrequently. Since the air in the lowest 3,000 ft is usually stable at night and unstable in the daytime on a sunny day, neutral lapse rates are observed only for a short time in the morning and late afternoon and on windy, overcast days. Thus, the practical use of the logarithmic law is rather limited. A more general law has been suggested by Deacon (10).

$$\bar{V} = \sqrt{\frac{\tau_0}{\bar{\rho}}}\frac{1}{k(1-\beta)}\left[\left(\frac{z-d}{z_0}\right)^{1-\beta} - 1\right] \tag{4-142}$$

Here β is a parameter which is less than unity when the respective atmospheric layer is stable, greater than unity when the layer is unstable. For $\beta = 1$, Deacon's law reduces to the logarithmic law. The values of β vary from about 0.80 to 1.20. The quantity d is often omitted in Deacon's law. Figure 4-18 shows typical wind distributions described by Deacon's law at night, in the morning, and at midday.

A knowledge of the wind distribution in the lowest layers is required in solving most diffusion problems. Unfortunately, theoretical solutions have not yet been found by the use of either Deacon's or Prandtl's law. Instead, simple power laws of the form

$$\frac{\bar{V}}{\bar{V}_1} = \left(\frac{z}{z_1}\right)^\alpha \tag{4-143}$$

have been used to simplify the mathematical calculations. Here $\bar{V}_1$ is the wind speed at a particular height z_1. Such laws seem to agree quite well with actual wind-speed profiles within a limited height range, for example, between 2 and 10 m. The exponent α varies strongly with thermal stability: the greater the stability, the larger the exponent. For an adiabatic temperature distribution, the value $\alpha = 1/7$ has been widely used (58).

4.4.3 Properties of Large-scale Vertical Motions

Unlike large-scale horizontal motions, large-scale vertical motions can be estimated only from their thermodynamic and kinematic effects. They cannot be measured directly because they average only about 1 cm/sec, whereas the superimposed small-scale vertical motions are 100 times larger and tend to mask the small average values of the large-scale vertical motions. Thus, vertical motions have not been measured as a routine matter but have been the subject of extensive specialized research (39,45). According to these studies, the large-scale vertical velocities have the following properties (in the temperate latitudes of the Northern Hemisphere at levels above 10,000 ft):

1. Regions of ascending motion coincide with regions of cloudiness, regions of descending motion with extensive clear areas.
2. Northward moving air usually has an upward component, southward moving air a downward component. In other words, warm air from the south tends to rise, and cold air from the north tends to sink.

No simple theoretical relation exists which associates large-scale vertical velocities with the temperature or pressure fields.

4.4.4 Properties of Small-scale Motions

Small-scale motions, especially at right angles to the large-scale flow, are of great practical importance because they produce, in those directions, all the transport of momentum, smoke, water vapor, heat, radioactive particles, and polluting gases. However, the small-scale motions have not yet been related theoretically to the better-known large-scale meteorological variables. Such relations form the topic of much current research. Our knowledge of the properties of small-scale motions is based on statistical summaries at relatively isolated spots, mostly close to the ground.

The kinetic energy of the small-scale motions will be referred to as "turbulent energy"; the kinetic energy of the horizontal and vertical components of small-scale motions will be called horizontal and vertical turbulent energy, respectively. We may summarize some of the tentative statistical results regarding the properties of turbulence.

1. Turbulent energy is greatest in unstable air, smallest in stable air. As a consequence, turbulent energy is small near the ground at night, large in the daytime (55).
2. Vertical turbulent energy is somewhat smaller than horizontal turbulent energy. Furthermore, vertical turbulent energy decreases relative to horizontal energy as the scale is increased, that is, when averages over larger periods or over greater distances are considered (37).
3. When the atmosphere is stable, the turbulent variations are almost entirely a high-frequency phenomenon, with fluctuations of the order of a few seconds or fractions of seconds. When the atmosphere is unstable, longer periods (of the order of

minutes) become important. This long-period turbulence is produced by surface heating (convection) rather than by mechanical means, such as wind blowing over rough ground.

4. The scale of vertical turbulence increases with height. In other words, the higher the level at which the turbulent eddies are found, the less effective is the damping influence of the ground. Therefore, fluctuations of longer periods become increasingly important at higher levels (46).

5. The turbulent energy increases with increasing wind speed as well as with increasing vertical variation of the wind. Since large variations of wind with height are generally associated with high wind speeds, the observations do not permit a decision on whether the absolute value of the wind or the wind variation is responsible for the turbulence. Theory suggests that the variation of the wind speed is the important factor (24).

6. The turbulent energy is greater over rough ground than over smooth ground.

4.4.5 Effect of Atmospheric Motions on Air Pollution

Large-scale motions have two effects on air pollution:

1. They transport the pollutants with the mean speed of the wind.
2. The greater the wind speed, the less concentrated are the pollutants, provided the pollution source is a continuous one.

In fact, for many problems the concentration of pollutants can be considered inversely proportional to the mean wind speed, as illustrated in Fig. 4-19. Large-scale vertical motion is negligible, unless the slope of the ground is considerable.

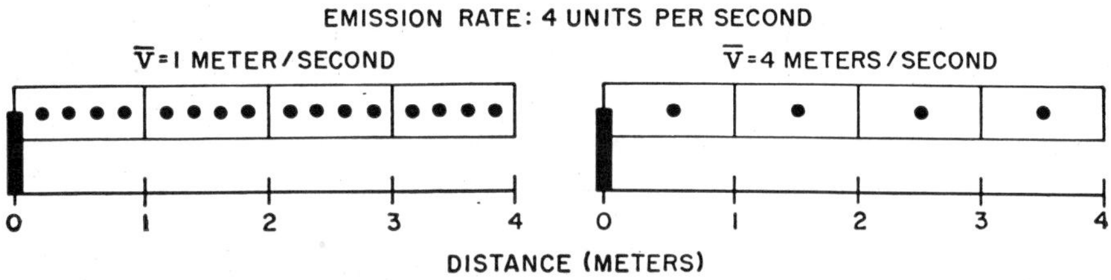

Fig. 4-19. Schematic illustration of effect of wind speed on concentration of pollutants from continuous and constant source.

Small-scale turbulent motions spread out the pollutants at right angles to the mean wind. Small-scale motions in the mean wind direction can be neglected, unless the source of the pollutant is instantaneous. In that case, the turbulence along the wind direction provides for the dispersion in that direction.

As was mentioned above, vertical air currents are likely to develop in unstable air, but are suppressed in stable air. The vertical air currents tend to diffuse effluents from pollution sources and to dilute the concentration of polluting matter; stable conditions, on the other hand, are conducive to accumulation of pollutants near the ground with consequently high concentrations. However, this does not imply that concentrations of pollutants experienced at the ground are necessarily high under stable conditions and low under unstable conditions. Depending on the distance from, and the height of, the source of pollutants, the opposite may be the case (55) as shown schematically for a point A in Fig. 4-20. The greatest ground concentrations from elevated sources are observed when an inversion condition with its high concentration aloft changes to an unstable condition. Then, vertical (both up and down) currents develop and bring the accumulated pollution down to the ground for a brief period (36).

In addition to the thermal influences of the ground on the stability conditions of the air, the wind has an effect on the lapse rate, and vice versa. Turbulence of the air flow, due to roughness of the ground, is largely suppressed in very stable air. On the other hand, strong turbulence (as occurs with strong winds blowing over very rough terrain) transports air upward and downward. Since such transport proceeds adiabatically, vertical mixing will convert any lapse rate (from inversions to superadiabatic lapse rates) into an adiabatic lapse rate. This is schematically shown in Fig. 4-21: an originally stable inversion layer between A and B after vertical mixing assumes an indifferent equilibrium $A'B'$, and an inversion develops above the mixing zone. For this reason, inversions are found to be generally associated with light winds or calm conditions; they may persist through part of the daytime under overcast skies, particularly over ground covered with snow or ice. Strong winds, both

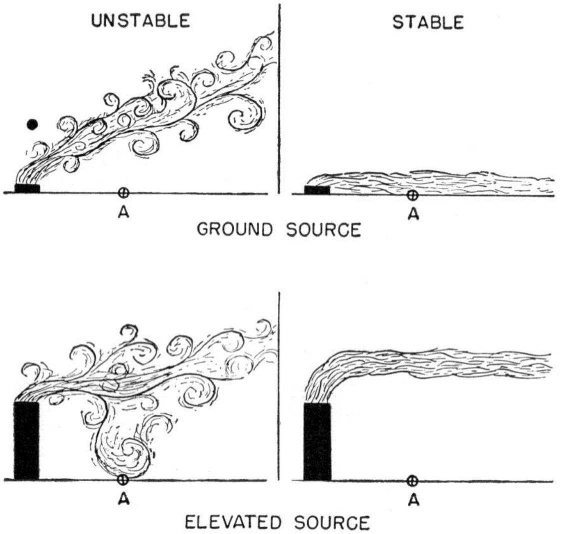

FIG. 4-20. Schematic diagram depicting the behavior of smoke plumes from ground and elevated sources under stable and unstable conditions.

day and night, tend to keep the lapse rate nearly adiabatic, whereas light winds or calm conditions on sunny days permit superadiabatic lapse rates to develop.

For these reasons, in appraising the effect of meteorological conditions on air pollution, thermal properties of the air and its motion must be simultaneously considered.

Other factors important in the qualitative appraisal of air pollution are connected with topography. For example, the "roughness" of the ground increases the vertical dispersion (other things being equal); larger features, such as hills and valleys, have a pronounced channeling effect.

Quantitative theories of air pollution are based on idealized models of topography and atmosphere. The ground is usually assumed to be a flat surface with uniform roughness. Such theories are often based on the diffusion equation for a steady state:

$$\bar{u}\,\frac{\delta\bar{\chi}}{\delta x} = \frac{\delta}{\delta x} K_x \frac{\delta\bar{\chi}}{\delta x} + \frac{\delta}{\delta y} K_y \frac{\delta\bar{\chi}}{\delta y} + \frac{\delta}{\delta z} K_z \frac{\delta\bar{\chi}}{\delta z} \qquad (4\text{-}144)$$

where x, y, z are the rectangular coordinates, with x in the mean wind direction, y at right angles to it in a horizontal plane, and z in the vertical, respectively; K_x, K_y, and K_z are the diffusion coefficients in these directions. They are defined as the ratio of flux of pollutants to their gradient in the respective direction. $\bar{u}$ is the large-scale horizontal wind speed and $\bar{\chi}$ the mean concentration as observed over a period of 3 min or more. The equation implies that the pollutant travels with the air and neither falls because of its weight nor rises because of its low density.

Solutions of the complete equation have been given only under the untenable assumptions of constant diffusion coefficients (50). In the case of an infinite line source near the ground at right angles to the mean wind, only vertical small-scale motions are important as diffusing agents. For this case, Deacon (10) has given a solution, making use of power laws for the vertical variation of wind [Eq. (4-143)] and the vertical variation of the diffusion coefficient K_z (Deacon's wind law implies that K_z is proportional to z^β). Furthermore, the diffusion coefficient for momentum (also

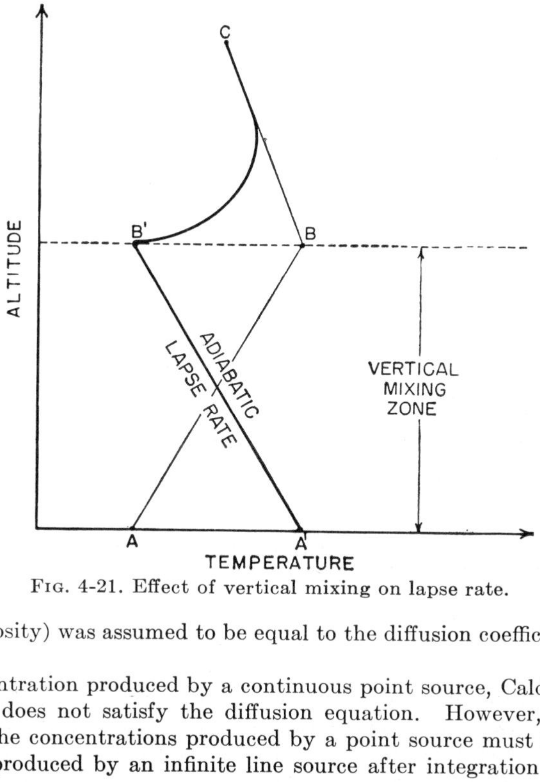

FIG. 4-21. Effect of vertical mixing on lapse rate.

called eddy viscosity) was assumed to be equal to the diffusion coefficient for particulate matter, K_z.

For the concentration produced by a continuous point source, Calder (6) found an expression that does not satisfy the diffusion equation. However, it satisfies the condition that the concentrations produced by a point source must behave like the concentrations produced by an infinite line source after integration from $y = -\infty$ to $y = +\infty$.

The solution for line sources seems reliable for adiabatic or superadiabatic lapse rates; the solution for point sources has been found to be satisfactory under adiabatic conditions.

For the concentration of pollutants produced by elevated sources such as smokestacks, the equations used most frequently are those derived by Sutton (59). These expressions do not satisfy the diffusion equation, but are based on Taylor's theory of diffusion by continuous movement (60). This theory requires the knowledge of the Lagrangian autocorrelation coefficient of small-scale velocities as a function of time, for which Sutton made reasonable assumptions.

With the source at the origin of coordinates, the concentration at the ground due to a source at height h becomes

$$\bar{\chi} = \frac{2Q}{\pi C_y C_z \bar{u} x^{2-n}} \exp\left[-x^{n-2}\left(\frac{y^2}{C_y^2} + \frac{z^2}{C_z^2}\right)\right] \tag{4-145}$$

where Q is the strength of the source in dimensions of mass per unit time, n is a nondimensional constant, and C_y^2 is given by

$$C_y^2 = \frac{4\epsilon^n}{(1-n)(2-n)\bar{u}^n}\left[\frac{\overline{v'^2}}{(\bar{u})^2}\right]^{1-n} \tag{4-146}$$

where ϵ is the macroviscosity, defined by $z_0\sqrt{\tau_0/\bar{\rho}}$, and v' is the small-scale horizontal velocity at right angles to the mean wind $\bar{u}$. A similar expression holds for C_z, with the vertical velocity w' substituted for v'.

Sutton's diffusion equation holds well when the temperature decreases with height at the adiabatic rate or more rapidly. (For discussion of the stable case, see Ref. 3.) Sutton recommends for neutral stability a numerical value of $n = 0.25$. Actually, n is best determined from concentration measurements at the site where the equations are to be applied; some experiments indicate considerably different values of n (56). Since no concentration is found at the ground immediately at the stack, the ground concentration first increases to a maximum with downwind distance from the stack. At greater distances, the diffusion produces again smaller concentrations.

According to Sutton's equation, the ground concentration due to a continuous point source at height h reaches a maximum at a horizontal downwind distance from the source, given by

$$x_{\max} = \left(\frac{h^2}{C_z^2}\right)^{1/(2-n)} \tag{4-147}$$

The height h is often assumed to be equal to the stack height. Actually, it is the height at which the smoke levels off, a level which may be much higher than the stack height. The theory of the relation of h to stack height is in quite an unsatisfactory state.

For given atmospheric conditions, however, the distance from the stack, at which maximum ground concentrations occur, is almost proportional to the stack height. The concentration at this point is proportional to the inverse square of the stack height and is given by

$$\bar{\chi}_{\max} = \frac{2Q}{\epsilon\pi\bar{u}h^2}\frac{C_z}{C_y} \tag{4-148}$$

According to Sutton, these equations are valid for 3-min average concentrations of pollutant. However, average concentrations observed over periods of about an hour are considerably lower than those expected from Sutton's equations. This is due to the fact that the wind direction oscillates in such a way that for periods of several minutes no smoke at all may reach an observer downwind from the source. Lowry (36) suggests a formula for the maximum average concentration over long periods at the ground:

$$\bar{\chi}_{\max} = \frac{2Q}{\epsilon\pi h^2}\frac{a_m}{\bar{u}} \tag{4-149}$$

where a_m is the relative frequency of the most frequent wind direction (where frequencies are recorded in 3° intervals). The distance of the point of maximum concentration from the stack is, according to Lowry, given by the empirical formula

$$x_{\max} = h \csc \sigma \tag{4-150}$$

where σ is the standard deviation of the horizontal wind direction in degrees over a period of 10 to 15 min.

Again, the formulas are inapplicable when the atmosphere is stable; in that case, the vertical diffusion is so weak that no noticeable concentrations are produced at the ground by a stack of height h for distances of at least $50h$.

It should be pointed out that all theories of concentrations are based on ideal conditions, such as uniform roughness, uniform rates of emission, no systematic upward or downward motion of the pollutant, and absence of topographical irregularities.

Nevertheless, the theories give an idea of the orders of magnitude of the concentrations to be expected from particular sources.

REFERENCES

1. Aden, A. L., and M. Kerker: Scattering of Electromagnetic Waves from Two Concentric Spheres, *J. Appl. Phys.*, **22**, 1242–1246 (1951).
2. Atlas, D., M. Kerker, and W. Hitschfeld: Scattering and Attenuation by Non-spherical Atmospheric Particles, *J. Atm. Terrest. Phys.*, **3**, No. 2, 108–119 (1953).
3. Barad, M. L.: Diffusion of Stack Gases in Very Stable Atmospheres, *Meteorol. Monographs*, **1**, No. 4, 9–14 (1951).
4. Beers, N. R.: Meteorological Thermodynamics and Atmospheric Statics, in F. A. Berry, E. Bollay, and N. R. Beers, eds., "Handbook of Meteorology," sec. V, pp. 313–409, McGraw-Hill Book Company, Inc., New York, 1945.
5. Bullrich, K. E., de Bary, and F. Moller: Die Farbe des Himmels. I, II, *Geofis. pura e Appl.*, **23**, 69–110 (1952); **26**, 141 (1953).
6. Calder, K. L.: To be published in *Journal of Meteorology*.
7. Chalmers, J. A.: "Atmospheric Electricity," Oxford University Press, New York, 1949.
8. Chandrasekhar, S.: "Radiative Transfer," Oxford University Press, New York, 1950.
9. Coulson, K. L.: Polarization of Light in the Sun's Vertical for a Rayleigh Atmosphere, *Univ. Calif.* (*Los Angeles*), *Dept. Meteorol., Sci. Rept. No.* 4, *Contract AF* 19(122)–239, 1952; and Neutral Points of Skylight Polarization in a Rayleigh Atmosphere, *Univ. Calif.* (*Los Angeles*), *Dept. Meteorol., Sci. Rept. No.* 7, *Contract AF* 19(122)–239, 1954.
10. Deacon, E. L.: Vertical Diffusion in the Lowest Layer of the Atmosphere, *Quart. J. Roy. Meteorol. Soc.*, **75**, No. 323, 89–103 (1949).
11. Debye, P.: Der Lichtdruck auf Kugeln von beliebigem Material, *Ann. Physik*, **30**, 57–136 (1909).
12. Deirmendjian, D.: Apparent Extraordinary Atmospheric Transmission of Ultraviolet Solar Radiation, submitted to the *Journal of the Optical Society of America* for publication, 1954.
13. Deirmendjian, D.: The Optical Thickness of the Earth's Molecular Atmosphere, submitted to the *Journal of Geophysical Research*, 1954.
14. Gans, R.: Uber die Form ultramikroskopischer Goldteilchen, *Ann. Physik*, **37**, 881–900 (1912).
15. Gilbert, J.: Condensation Nuclei of the Los Angeles Region, Univ. Calif. (Los Angeles), Dept. Meteorol., unpublished research report, 1954.
16. Gish, O. H.: Atmospheric Electricity, in J. A. Fleming, ed., "Physics of the Earth," vol. VIII, "Terrestrial Magnetism and Electricity," chap. IV, pp. 149–230, Dover Publications, New York, 1949.
17. Gish, O. H.: Universal Aspects of Atmospheric Electricity, in "Compendium of Meteorology," pp. 101–119, American Meteorological Society, Boston, Mass., 1951.
18. Goldberg, B.: New Computation of the Mie Scattering Functions for Spherical Particles, *J. Opt. Soc. Am.*, **43**, 1221–1222 (1953).
19. Gunn, Ross: Precipitation Electricity, in "Compendium of Meteorology," pp. 128–135, American Meteorological Society, Boston, Mass., 1951.
20. Guttler, A.: Die Mie-sche Theorie der Beugung durch dielektrische Kugeln mit absorbierenden Kern und ihre Bedeutung fur Probleme der interstellaren Materie und des atmospharischen Aerosols, *Ann. Physik*, **11**, 65–98 (1952).
21. Hagenguth, J. H.: The Lightning Discharge, in "Compendium of Meteorology," pp. 136–143, American Meteorological Society, Boston, Mass., 1951.
22. Haurwitz, B.: "Dynamic Meteorology," McGraw-Hill Book Company, Inc., New York, 1941.
23. Heverly, J. R.: Supercooling and Crystallization, *Trans. Am. Geophys. Union*, **30, No. 2,** 205–210 (1949).

24. Holland, J. Z., and R. F. Myers: A Contribution to the Climatology of Turbulence, Abstract: *Bull. Am. Meteorol. Soc.*, **33,** No. 2, 78 (1952).
25. Holmboe, J., G. E. Forsythe, and W. Gustin: "Dynamic Meteorology," John Wiley & Sons, Inc., New York, 1945.
26. Hosler, C. L., M. D. Burkhart, and H. Neuberger: On the Effect of Time Lapse between Sampling and Expansion in the Aitken Nuclei Counter, *Bull. Am. Meteorol. Soc.*, **33,** No. 6, 251–254 (1952).
27. Hosler, C. L., and C. R. Hosler: An Investigation of the Spontaneous Freezing Point of Small Quantities of Water, *The Penn. State College, Mineral Inds. Expt. Sta., Div. Meteorol. Sci. Rept.* No. 1 [Air Force Cambridge Research Center, Contract AF 19(604)–140], August, 1952. (PB 108,100.)
28. Hulburt, E. O.: Optics of Atmospheric Haze, *J. Opt. Soc. Am.*, **31,** 467–476 (1941).
29. Israel, H.: Bemerkungen zu meinen bisherigen Kernzahlungen und zur Frage der Ionenladung, *Gerlands Beitr. Geophys.*, **40,** 29–43 (1933).
30. Israel, H.: Instruments and Methods for the Measurement of Atmospheric Electricity, in "Compendium of Meteorology," pp. 144–154, American Meteorological Society, Boston, Mass., 1951.
31. Israel, H.: Radioactivity of the Atmosphere, in "Compendium of Meteorology," pp. 155–161, American Meteorological Society, Boston, Mass., 1951.
32. Johnson, J. C.: "Physical Meteorology," Technology Press of Massachusetts Institute of Technology and John Wiley & Sons, Inc., New York, 1954.
33. Junge, Ch.: Gesetzmassigkeiten in der Grossenverteilung atmospharischer Aerosole uber dem Kontinent, *Ber. deut. Wetterdienst.*, U.S. Zone, No. 35, 261–277 (1952).
34. Landsberg, H.: Atmospheric Condensation Nuclei, *Ergeb. kosmischen Physik*, **3,** 155–252 (1938). (In English.)
35. Latimer, W. M., General Meteorological Principles, in "Handbook on Aerosols," chap. 2, pp. 15–39, U.S. Atomic Energy Commission, Washington, D.C., 1950.
36. Lowry, P. H.: Microclimatic Factors in Smoke Pollution from Tall Stacks, *Meteorol. Monographs*, **1,** No. 4, 24–29 (1951).
37. McCormick, R. A.: The Partition of Eddy Energy at 300 Feet above the Surface During Unstable Conditions at Upton, N.Y., Abstract: *Bull. Am. Meteorol. Soc.*, **33,** No. 9, 397 (1952).
38. Mie, G.: Beitrage zur Optik truber Medium, *Ann. Physik*, **25,** 377–445 (1908).
39. Miller, J. E.: Studies of Large Scale Vertical Motions of the Atmosphere, New York University, Meteorological Papers, vol. 1, No. 1, July, 1948.
40. National Bureau of Standards Tables of Scattering Functions for Spherical Particles, Applied Mathematics Series, No. 4, 1949.
41. Neuberger, H.: Arago's Neutral Point: A Neglected Tool in Meteorological Research, *Bull. Am. Meteorol. Soc.*, **31,** 119–125 (1950).
42. Neuberger, H.: Condensation Nuclei; Their Significance in Atmospheric Pollution, *Mech. Eng.*, **70,** 221–225 (1948).
43. Neuberger, H.: "Introduction to Physical Meteorology," The Pennsylvania State College, Mineral Industries Extension Services, 1951.
44. Packer, D. M., and C. Lock: The Brightness and Polarization of the Daylight Sky at Altitudes of 18,000 to 38,000 Feet above Sea Level, NRL Report No. 3731, 1950; *J. Opt. Soc. Am.*, **41,** 473–478 (1951).
45. Panofsky, H. A.: Large-scale Vertical Velocity and Divergence, in "Compendium of Meteorology," pp. 639–646, American Meteorological Society, Boston, Mass., 1951.
46. Panofsky, H. A.: The Variation of the Turbulence Spectrum with Height under Superadiabatic Conditions, *Quart. J. Roy. Meteorol. Soc.*, **79,** No. 339, 150–153 (1953).
47. Penndorf, R.: The Vertical Distribution of the Mie Particles in the Troposphere, Geophysical Research Papers, G.R.D., No. 25, pp. 1–12, 1954; *J. Meteorol.*, **11,** 245–247 (1954).
48. Petterssen, S.: "Weather Analysis and Forecasting," McGraw-Hill Book Company, Inc., New York, 1940.
49. Reeger, E., and H. Siedentopf: Die Streufunktion des atmospharischen Dunstes nach Scheinwerfermessungen, *Optik*, **1,** 15–41 (1946).
50. Roberts, O. F. T.: The Theoretical Scattering of Smoke in a Turbulent Atmosphere, *Proc. Roy. Soc. (London)*, **104 A,** 640 (1923).
51. Sekera, Z.: Additional Scattering Functions for Spherical Particles, *Univ. Calif. (Los Angeles), Dept. Meteorol., Sci. Rept. No.* 2, *Contract AF* 19(122)–239, 1952.
52. Sekera, Z.: Legendre Series for the Scattering Functions for Spherical Particles, *Univ. Calif., (Los Angeles), Dept. Meteorol., Sci. Rept. No.* 5, *Contract AF* 19(122)–239, 1952.
53. Sekera, Z.: Tables Relating to Rayleigh Scattering of Light in the Atmosphere, *Univ. Calif., (Los Angeles), Dept. Meteorol., Sci. Rept. No.* 3, *Contract AF* 19(122)–239, 1952.

Z. Sekera and E. V. Ashburn, Tables Relating to Rayleigh Scattering of Light in the Atmosphere, NAVORD Report No. 3061, China Lake, Calif., 1953.
54. Sekera, Z.: Polarization of Skylight, in "Compendium of Meteorology," p. 83, American Meteorological Society, Boston, Mass., 1951.
55. Smith, M. E.: The Forecasting of Micrometeorological Variables, *Meteorol. Monographs*, **1,** No. 4, 50–55 (1951).
56. Smith, M. E., and I. A. Singer: Personal communication.
57. Stokes, G. G.: On the Composition and Resolution of Streams of Polarized Light from Different Sources, in "Mathematical and Physical Papers," vol. 3, p. 233, Cambridge University Press, New York, 1901. From *Trans. Cambridge Phil. Soc.*, **9,** 399 (1852).
58. Sutton, O. G.: "Atmospheric Turbulence," John Wiley & Sons, Inc., New York, 1949.
59. Sutton, O. G.: "Micrometeorology," McGraw-Hill Book Company, Inc., New York, 1953.
60. Taylor, G. I.: Diffusion by Continuous Movements, *Proc. London Math. Soc.*, **20,** 196 (1922).
61. Torreson, O. W.: Instruments Used in Observations of Atmospheric Electricity, in J. A. Fleming, ed., "Physics of the Earth," vol. VIII, "Terrestrial Magnetism and Electricity," chap. V, pp. 231–269, Dover Publications, New York, 1949.
62. Wait, G. R., and W. D. Parkinson: Ions in the Atmosphere, in "Compendium of Meteorology," pp. 120–127, American Meteorological Society, Boston, Mass., 1951.
63. Waldram, J. M.: Measurement of the Photometric Properties of the Upper Atmosphere, *Quart. J. Roy. Meteorol. Soc.*, **71,** 309–310, 319–336 (1945).

Section 5

EVALUATION OF WEATHER EFFECTS

BY C. A. GOSLINE, L. L. FALK, AND E. N. HELMERS

5.1 INTRODUCTION

5.1.1 Air Pollution and the Weather

For a given situation, severity of air pollution is determined largely by the weather. Even though the total discharge of contaminants to the atmosphere in a given area remains constant from day to day, the degree of air pollution will vary widely. The observation that smoke is dissipated much more quickly and thoroughly on some days than on others is the direct result of differences in the weather. Quantitative means for evaluating these weather effects are necessary to ensure satisfactory and economically optimum solutions to atmospheric pollution problems. This section is designed to present reliable methods for judging the rate of dispersion and its consequences upon the territory exposed.

5.1.2 Applications of Meteorology to Air Pollution Control Engineering

Undoubtedly, weather affects the degree of air pollution; but since weather cannot be controlled, are there any practical applications of meteorology to air pollution control engineering? Granted, the basic method of control is reduction of air contaminants at the source by change in process or collection, but the cost of control depends, to a large extent, on the required degree of reduction. The required reduction, in turn, is a function of the ability of the atmosphere to disperse the contaminant. To assure the most economical solution, the cost of increased collection efficiency or process alteration must be weighed against the cost of increased stack height or meteorological control of emission rates.

The more important applications of meteorology to air pollution control may be summarized as follows:

Determination of Allowable Emission Rates. Atmospheric concentrations of a contaminant downwind from a source of constant discharge vary with weather conditions. Consequently, a knowledge of regional climatology is necessary in order to define the frequency of different levels of contamination. Knowing this frequency and the allowable level of pollution caused by the particular substance in question, the required reduction in emission is determined.

Stack Design. The application of dispersion theory and a knowledge of local weather conditions are necessary to determine the required stack height for a given emission. Meteorology and aerodynamics can also be used in the design of stack nozzles, in the determination of optimum stack velocities and stack-gas temperatures, and in the determination of stack location relative to plant structures and topography.

Planning and Interpreting Air Pollution Surveys. The results of an air pollution survey lose much of their value unless they are correlated with weather conditions at the time of the survey. The location of atmospheric samplers and sampling technique should be based on a thorough consideration of meteorological factors. By proper consideration of weather factors in survey planning and the collection of adequate weather data during the survey, the value of the information obtained is greatly increased and the time period required for the survey is kept to a minimum.

Plant-site Selection. Water, transportation, raw materials, and markets have long been important factors and are often given appreciable study in the comparison and selection of plant sites. In recent years the disposal of liquid wastes has become increasingly critical, and the ability of diluting waters to absorb wastes is usually considered. Because of varying climate and other local factors, the atmosphere in different locations differs in its ability to receive atmospheric wastes without causing a nuisance, and for this reason, climatology also has become an important factor in site selection.

Meteorological Control. An industrial operation may be such that the discharge of a potential atmospheric pollutant can be restricted to periods of weather favorable for dispersion. When weather conditions are unfavorable, production is cut back and discharge of the offending material is reduced. Under extremely unfavorable conditions, the emission of some materials may be stopped. To be successful, such a program requires accurate forecasting of atmospheric conditions and flexibility of operations.

5.1.3 How Air Pollution Affects the Weather

The effects of air pollution on the weather may be of less concern to the air pollution control engineer than the converse, but some effects should at least be recognized. Air pollution affects the weather of a community in several ways: visibility may be reduced, fog frequency and duration may be increased, and the incoming solar radiation may be decreased, particularly in the ultraviolet end of the spectrum.

Neuberger has discussed the significance of condensation nuclei in atmospheric pollution (69) and also the effect of air pollution on the persistence of fog (70). Condensation nuclei are very small hygroscopic particles such as chloride salts and sulfuric acid. Some important sources of condensation nuclei are cosmic dust, ash from volcanic eruptions, wind-blown soil particles, forest fires, salt from evaporated ocean spray, and industrial pollution. These nuclei play an important role in the atmosphere by providing surfaces necessary for condensation. Because of the prevalence of condensation nuclei in an area of industrial pollution, the frequency and duration of fog and possibly rain (54) may be increased. In relatively pure country air a relative humidity of 100 per cent may exist for some time without appreciable reduction in visibility, but in an industrial area, the visibility may be considerably reduced at relative humidities well below saturation.

Visibility is not an adequate index of air pollution because of its dependence on humidity. The frequency of fog is not a reliable index because local effects such as topography and proximity of bodies of water are also important factors in fog formation.

An appreciable depletion of solar radiation in industrial cities due to atmospheric pollution has been reported. Hand (42) compares the total solar and sky radiation received on a horizontal surface in downtown Boston with that received in nearby suburban Blue Hill. The mean radiation at Boston for a 4-year period, October through March, was 82 per cent of the mean radiation for Blue Hill. The greatest depletion of radiation reported (more than 90 per cent) was the result of a dense smoke pall over the city. The loss of ultraviolet radiation is much greater than the loss of total radiation would indicate. A decrease of a few per cent in the total solar radiation resulting from smoke almost completely robs the radiation of its ultraviolet component.

5.2 FUNDAMENTALS OF METEOROLOGY

5.2.1 General

References. The objective of the following discussion is to describe some of the basic concepts with which the air pollution control engineer should be familiar and to indicate sources of weather information and meteorological services. For more detailed information on meteorology, a general text or handbook should be consulted. For a general treatment, see "General Meteorology" by Byers (16) or "Descriptive Meteorology" by Willett (101). The "Handbook of Meteorology" by Berry, Bollay, and Beers (7) contains excellent reference material. Petterssen's "Weather Analysis and Forecasting" (72) is recommended for those interested primarily in weather forecasting. Theoretical meteorology, the application of laws of thermodynamics and hydrodynamics to meteorology, is dealt with in texts on dynamic meteorology: notably, "Physical and Dynamical Meteorology" by Brunt (13) and "Dynamic Meteorology" by Haurwitz (43). "The Compendium of Meteorology" (61) is an excellent survey of the present state of the art. For current articles, including some articles on atmospheric pollution, consult the publications of the American Meteorological Society—*The Bulletin of the American Meteorological Society, The Journal of Meteorology, Meteorological Abstracts and Bibliography,* and *Meteorological Monographs.*

Types of Meteorology. Meteorology includes the study of weather phenomena that vary in scale from the microscopic to the astronomical. Whereas the general circulation of the earth's atmosphere may concern one meteorologist, the distribution of temperature, humidity, and wind velocity in a layer of air only a few centimeters thick covering a few square feet of area may occupy another. The forecaster is primarily concerned with the movement of large-scale high- and low-pressure centers across the countryside, but he cannot neglect the important local effects of topography and water surfaces in making his forecast. The air pollution control engineer must also study the local, relatively small-scale weather of the atmospheric layer through which stack-emission products are dispersed.

The tremendous differences in scale are recognized in meteorological nomenclature by the use of the distinctive terms, meteorology (or macrometeorology) and micrometeorology. As the terms imply, macrometeorology is the study of large-scale weather events while micrometeorology is the study of the small. Willett (102), in describing atmospheric circulations, makes a distinction between primary, secondary, and tertiary circulations. Primary circulation is the general hemispheric wind system affecting the transfer of heat from tropical to polar latitudes. The secondary circulations are the high- and low-pressure circulation cells that appear on weather maps. Tertiary circulations are the relatively small-scale localized disturbances such as land and sea breezes, foehn winds, showers, thunderstorms, tornadoes, etc. The air pollution control engineer is interested in those aspects of the secondary and tertiary circulations which affect atmospheric dispersion of waste products in the atmosphere.

5.2.2 Concepts of Macrometeorology

Air Masses. Modern meteorology is based on the concept of air masses and fronts. An air mass is a vast body of air having properties that vary only slightly in the horizontal over wide areas. The unequal solar heating in tropical and polar latitudes creates a thermal unbalance in the atmosphere. This unbalance causes huge cold air masses to move south from the polar regions and warm air masses to migrate north from the tropics. In general, each high-pressure area represents an individual air mass. To give some idea of the scale of an air mass, the entire United States may be under the influence of only three or four different air masses at any one time. A given air mass is identified by characteristic temperatures, humidities, cloud forms, diurnal variations and vertical gradients of the different air-mass properties, etc.

According to the usual classification, an air mass is designated by geographic origin as either continental (c) or maritime (m) and as either tropical (T) or polar (P). In relation to the surface over which it is passing, it is either warmer (w) or colder (k). Thus a cPk air mass is of continental polar origin and is colder than the surface over which it is traveling. The w and k classification has a special significance from the standpoint of air pollution. A k air mass is heated from below, and the resulting instability in the lower layers promotes good atmospheric dispersion. A w air mass is cooled from below, and the resulting stability causes poor atmospheric dispersion.

Stable air masses that have a tendency to stagnate result in poor dispersion, whereas unstable ones result in much more favorable conditions. Landsberg (54) states that cities of the eastern United States, when dominated by fresh maritime polar air masses, with gusty winds and steep lapse rates, show only one-seventh the pollution of relatively stagnant, modified continental polar air masses. A knowledge of which air masses predominate in a region can thus be of assistance in plant-site selection.

Fronts. The transition in air-mass characteristics from one air mass to another is often quite abrupt. These relatively narrow zones of transition are called fronts. Thus, a front is the boundary between two air masses and may be represented by a line on a weather map. A front is designated as warm or cold, depending on the direction of movement. If a cold air mass is displacing a warm one, the transition is called a cold front. The progress of a front can be followed across the country on a time series of weather maps. In the latitudes of the United States, air masses, fronts, and high- and low-pressure areas move, in general, from west to east.

Fronts are associated usually with low-pressure areas and pressure "troughs." Frontal zones and low-pressure areas are often regions of bad weather—i.e., cloudiness, precipitation, thunderstorm and shower activity, and high winds. Fronts are of importance from the standpoint of air pollution because they mean a "change of air" and often a cleansing of the air by precipitation.

High- and Low-pressure Areas. A weather map is simply a map on which simultaneous weather observations from weather stations over a wide area are plotted. After the map is plotted, it is analyzed by drawing isobars (lines of equal barometric pressure) and fronts and by indicating areas of precipitation, air-mass types, etc. The pressure pattern formed by the isobars is important because it indicates at a glance the direction and speed of the major air currents and the areas of good and bad weather.

Winds blow nearly parallel to the isobars with a speed inversely proportional to the isobar spacing. The direction of the winds is such in the Northern Hemisphere that an observer with his back to the wind has high pressure to his right and low pressure to his left. (In meteorological terminology, a north wind is a wind blowing from the north.) Isobars may be considered to approximate streamlines.

High-pressure areas are usually regions of light winds, clear skies, and good weather. Areas of low pressure tend to be regions of high winds, cloudy skies, precipitation, and bad weather. High-pressure areas are also more likely to be areas of atmospheric stability than low-pressure areas. Because pollutants are diluted or removed from the atmosphere by turbulent mixing, dispersion by the wind, or washing by precipitation, high-pressure areas are usually associated with higher degrees of atmospheric pollution than low-pressure areas.

Mean-pressure Maps. One method of evaluating potential regional weather and air pollution characteristics is to examine monthly or seasonal normal pressure charts. Willett (102) has illustrated this principle by correlating regional air pollution characteristics with monthly mean-pressure charts over North America at the 10,000-ft level. Close isobar spacing in the vicinity of a low-pressure area creates a region of relatively strong air movement and, consequently, an area in which a prolonged period of severe atmospheric pollution or smog condition is not likely to occur. On the other hand, widely spaced isobars near high-pressure areas are more likely to lead to serious pollution problems.

Figures 5-1 and 5-2 show the average pressure pattern at 1,500 m over the United States for the months January and July. The closer isobar spacing in January indicates higher average wind speeds and more favorable conditions for diffusion of atmospheric pollution than exists in July. Also, note the contrast in January between the closely spaced isobars with cyclonic curvature in New England and the widely spaced isobars with anticyclonic curvature in the Southwest.

Willett also points out that, although monthly normal charts indicate the characteristics of the prevailing weather pattern as they affect atmospheric pollution, they do not determine the probable frequency or duration of the exceptional occurrence of extremely severe smog conditions. The frequency of weather conditions favorable for pollution is best determined by a statistical study of regional Weather Bureau

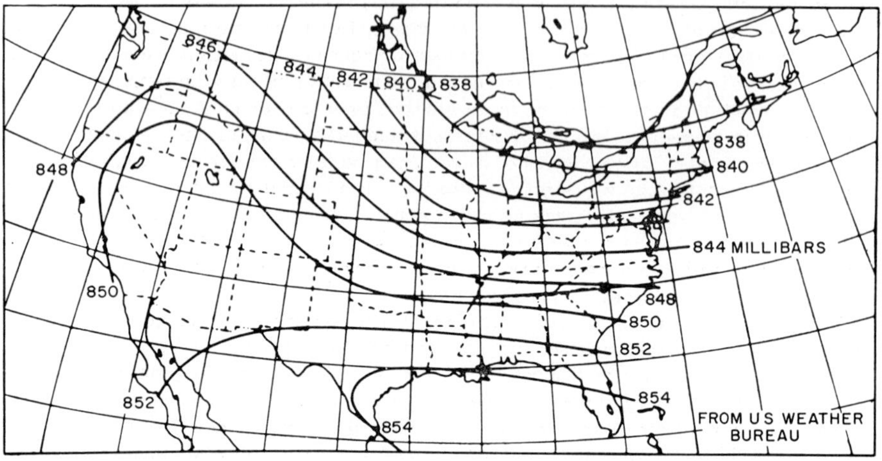

Fig. 5-1. Average pressure pattern at 1,500 m for the United States—January. *(From U.S. Weather Bureau.)*

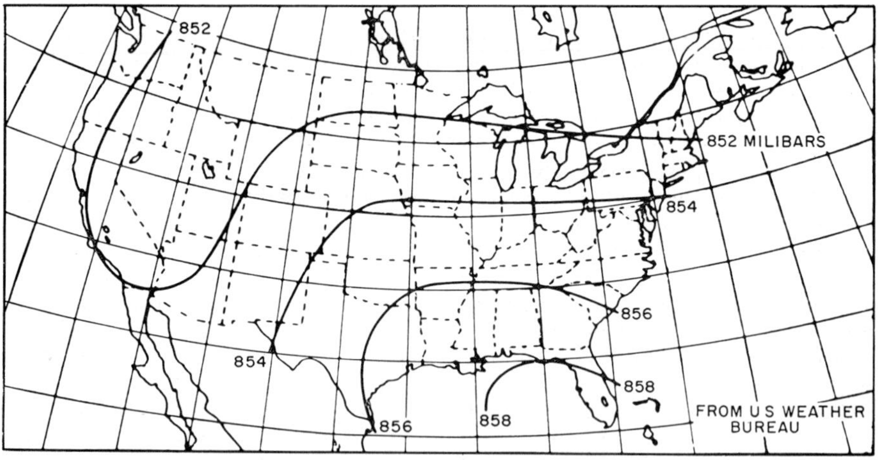

Fig. 5-2. Average pressure pattern at 1,500 m for the United States—July. *(From U.S. Weather Bureau.)*

records. A technique for developing a regional air pollution climatology will be discussed in a following article.

Tertiary Circulations. Some relatively small-scale atmospheric movements often are of primary importance in air pollution problems and warrant special consideration in plant-site selection. Of these small-scale phenomena, the circulation characteristic of valleys is of particular importance. From the viewpoint of air pollution, the bottom of a valley or a low-lying area is an undesirable location for an industrial plant discharging critical quantities of atmospheric pollutants. Under adverse weather conditions the pollutants become trapped in the valley air, and if the discharge continues, the pollutant concentration increases to an objectionable level.

When the prevailing wind speed in a valley area is low, the valley circulation is primarily the result of local cooling or heating. The so-called "drainage" winds (mountain breeze, gravity wind, katabatic wind) result when nocturnal radiation causes cold air to form on the highlands and slopes and then move into the valleys and the low points of the terrain. This gravitational drainage of cold, dense air down even the gentlest slopes is a common occurrence on clear nights with light winds. The accumulation of cold air in valleys produces a very unfavorable dispersion condition. A pollution reservoir is produced because the temperature stratification restricts vertical motion and the valley walls limit the horizontal distribution.

The valley breeze, caused by solar heating of slopes, is the daytime counterpart of the drainage winds. Air is heated in contact with the slopes; this results in an updraft along the valley walls. The resulting instability and vertical currents tend to dissipate atmospheric pollution. Gleeson (34) has written an excellent theoretical treatment of cross-valley winds. These winds are best developed during the summer season when solar heating is a maximum.

Another important effect of local heating is the formation of cumulus clouds and, on occasion, convective showers and thunderstorms. Solar heating of the earth's surface tends to dissipate atmospheric pollution because of the resulting thermal instability and improved dispersion characteristics of the atmosphere.

Land and sea breezes may be important aspects of the circulation in coastal or lakeside areas. Both types of circulation result from the unequal heating and cooling of the air over the land and water. The sea breeze results during the daytime when the air over the land surface is heated more than the air over the water. The resulting circulation is directed onshore. A well-developed sea breeze may have a depth of over 1,000 ft and extend inland for a distance of as much as 30 miles. On many days the sea breeze will not be well developed and may extend only a mile or so inland and have a depth of only a few hundred feet.

On clear nights with light winds, the thermal gradient from land to water is reversed and the sea breeze dies away to be followed by a light land breeze. The land breeze may result in certain advantages from the standpoint of air pollution because pollution that would otherwise tend to accumulate under conditions favorable for a land breeze is blown offshore.

Aside from the circulations produced by local heating and cooling, large-scale atmospheric eddies may be a predominant feature of the local circulation. Eddies on the downwind side of topographic obstructions such as mountains, hills, or other features may result in high-level pollution being carried to the ground in a local area. A large-scale example of this forced-type circulation is the foehn or chinook winds of warm, dry air experienced on the lee side of a mountain range. On a smaller scale, eddies produced by buildings and other man-made obstructions can be responsible for erratic dispersion of stack gases. The complexities of eddy motion in the atmosphere are extremely difficult to generalize, but their existence should not be overlooked in the location and design of a stack. Direct observation in the field and study of a model in a wind tunnel are probably the most satisfactory means of determining the effects of eddies.

5.2.3 Concepts of Micrometeorology

Microclimatology is defined (95) as the detailed study of the climate over a small area, as determined by location and environment. Places only a few hundred feet apart may have radically different climates because of differences in elevation, soil, soil cover, etc. Since many atmospheric-pollution problems are confined to an area within a few miles of a plant or industrial center, a knowledge of the microclimate of that area is essential for a complete understanding of the problem. Many times, the closest U.S. Weather Bureau station is over 5 or 10 miles away from the area of interest, and, without study, it is improper to assume that the microclimate of this area is represented by the climatological records of the Weather Bureau station.

One difficulty often encountered in using records from Weather Bureau stations is that the required type of information is not available. The primary objective of observations made by the Weather Bureau is to determine representative air-mass characteristics as free as possible from the variations that result from the microclimate. On the other hand, the air pollution engineer is interested in just those variations. In view of this difficulty, the Meteorology Panel, U.S. Technical Conference on Air Pollution (44), has recommended that more observations of micrometeorological significance be taken at Weather Bureau stations and that emphasis be placed on obtaining regular observations of temperature and wind gradients in the lowest 500 ft of the atmosphere.

Some aspects of microclimatology and the reasons for climate variation in short distances were discussed under Tertiary Circulations. Of these effects, probably the most important from the standpoint of air pollution are cold-air drainage and topographic effects on the wind. The nature of the earth's surface (i.e., rough, smooth, bare of vegetation, grass, short crops, forest cover, etc.) is of particular importance in determining the local vertical distribution of wind and temperature. For these reasons, microclimatological observations made concurrently with measurements of air-contaminant concentrations are essential for the interpretation of survey results.

Microclimate is also of particular importance in the investigation of fume damage to vegetation. The climate of the first few yards of the atmosphere, or for that matter the first few inches, is a special branch of meteorology (33). The temperature, humidity, and other weather elements near the ground experience much greater diurnal and special variations than these same elements a few feet higher. Weather-instrument shelters are usually above the region of these extreme variations. However, the climate near the ground is the environment of plants, and when fume damage to vegetation is being considered—a problem in which humidity may be a factor—the climate of this shallow layer should be considered.

5.2.4 Meteorological Measurements

The hourly observations taken at Weather Bureau stations are the basis for most of the weather forecasts and climatological data. Instructions for making weather observations and reports and for installation and operation of instruments are outlined in the Bureau's *Circular N* (49). Observations generally consist of ceiling (height of cloud layer above the ground) in feet; sky conditions; visibility in miles; weather conditions (including precipitation, squalls, etc.); obstruction to vision (fog, haze, etc.); temperature; dew point; wind direction and speed; barometric pressure; pressure-change tendency; amount, type, and direction of movement of clouds; and miscellaneous information (thunderstorms, line squalls, etc.).

In addition to the surface observations, a number of stations also make upper-air observations. The two most common types of upper-air observations are pilot-balloon observations (pibals) and radiosonde observations (raobs). Pilot-balloon observations are made by following the rise of a helium-inflated balloon with a theodolite to determine wind speed and direction at various altitudes. A radiosonde is a lightweight instrument including a radio transmitter, which, when attached to a balloon and released, transmits a record of barometric pressure, temperature, and relative humidity as it ascends. The picture of atmospheric stability obtained in this manner is of considerable value to the air pollution control engineer.

The Weather Bureau observations just mentioned are designed primarily "to promote the safety and efficiency of air navigation," and as pointed out in the discussion of micrometeorology, the data are often not sufficient for the air pollution problem at hand. For example, radiosonde observations are taken only two times each day at rather widely scattered locations, and the temperature lapse rate in the lowest layers that is of primary concern in atmospheric diffusion is not sufficiently defined. Also, a more detailed characterization of wind gustiness is usually required than can be obtained from hourly observations. As a result, an atmospheric pollution survey or the study of specific problems usually requires special weather observations at the time of the survey or study.

Two methods for making observations of wind, temperature, and humidity gradients in the lowest few hundred feet of the atmosphere are of particular utility in air pollution studies:

1. Mounting instruments on towers or masts
2. Suspending instruments at various altitudes from captive kite balloons

For long-term measurements or when already available, the use of a mast or tower is usually preferred; but in many instances the expense of construction of a special tower is not warranted and the use of kite balloons provides an excellent alternative. Munger (66,67,68) has described the use of kite balloons for securing air samples, measuring wind and temperature, and firing smoke grenades at various heights. Firing smoke grenades is a useful method for studying air-flow patterns over buildings and rough terrain. The wiresonde (2,3,57,65,74) is a captive-balloon system that can be used to measure temperature and humidity gradients in the lowest few thousand feet of the atmosphere.

5.2.5 Meteorological Services

A statistical study of Weather Bureau climatological data is frequently of considerable value in the comparison of alternative plant sites, in developing air pollution climatology, and in interpreting the results of air pollution surveys. Weather data are currently being punched on IBM cards, and a library for these records has been established in Asheville, N.C. The Weather Bureau Tabulation Unit in Asheville will prepare statistical summaries of data at cost to private industry. This service can be of material help in making air pollution studies. The Climatological and Hydrologic Services Division of the Weather Bureau in Washington, D.C., can also be of considerable assistance in obtaining climatological data.

Since meteorology is a rather specialized science, the service of a specialist is desirable in the conduct of a survey or the study of a particular pollution problem. Some of the larger industrial concerns hire full-time meteorologists to do this type of work, but in the case of smaller plants, this procedure is usually not practical. To fill this need, there are qualified private meteorological consultants who can be contacted through the American Meteorological Society.

5.3 ATMOSPHERIC DIFFUSION

Just as a river or stream is able to absorb a certain amount of pollution without the production of undesirable conditions, the atmosphere can absorb also a certain amount of contamination without bad effects. The self-purification of a stream is primarily the result of biological action and dilution. Dilution of air contaminants in the atmosphere is also of prime importance in the prevention of undesirable levels of pollution. In addition to dilution, several self-purification mechanisms are at work in the atmosphere such as sedimentation of particulate matter, washing action of precipitation, photochemical reactions, and absorption by vegetation, soil, etc.

The dilution of air contaminants is a direct result of atmospheric turbulence and molecular diffusion. The rate of turbulent diffusion is so many thousands of times greater than the rate of molecular diffusion that the latter effect can be neglected in atmospheric-diffusion problems. Atmospheric turbulence, and, hence, atmospheric diffusion, varies widely with weather conditions. Because the rate of atmospheric dilution determines the level of pollution in a given location, methods for estimating the degree of atmospheric diffusion are of paramount importance.

5.3.1 Estimation of Atmospheric Diffusion

Three general methods are available for estimating the effectiveness of atmospheric diffusion in dispersing gaseous and particulate contaminants. These are air pollution surveys, theoretical and empirical methods, and wind-tunnel techniques.

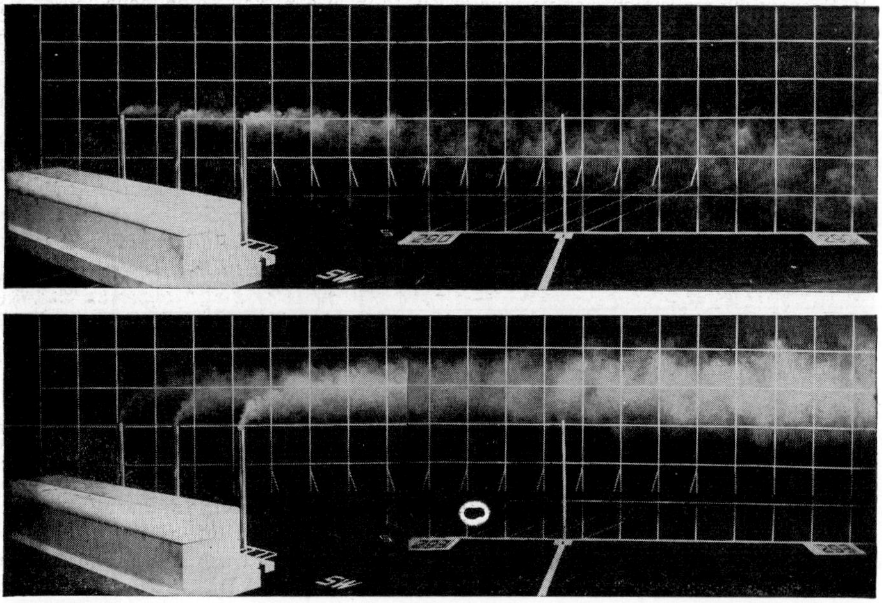

FIG. 5-3. Wind-tunnel test showing the effect of stack velocity on smoke diffusion from a powerhouse. Gas–wind-speed ratio in upper picture is 1, in lower picture, 4. (*Consolidated Edison wind tunnel at New York University.*)

Air Pollution Surveys. An air pollution survey in which complaints are studied and samples of air contaminants taken and analyzed under a variety of weather conditions is the most direct method of estimating diffusion. The use of smoke for the study of flow over rough terrain has been mentioned previously. An understanding of atmospheric diffusion theory and the effects of weather conditions can be of considerable help in the planning of such a survey. The planning and interpretation of surveys will be discussed in more detail later.

Theoretical and Empirical Methods. It is not always possible to solve the problem at hand by means of a survey. An estimate is often required before the construction of a plant or new facility, and one of the other two methods must be used. A number of theoretical and empirical methods for estimating turbulent diffusion have been proposed in recent years. Many of these involve formulas for calculating concentrations of contaminants downwind from an emission source. The application of these methods to practical problems is often extremely difficult because of local topographic and microclimatic conditions, and considerable experience is required for the proper interpretation of results. Nevertheless, these theories are of appreciable value in the solution of many problems. A considerable amount of research is currently in progress on the theoretical and experimental aspects of such methods, and continued refinement of the technique is to be expected.

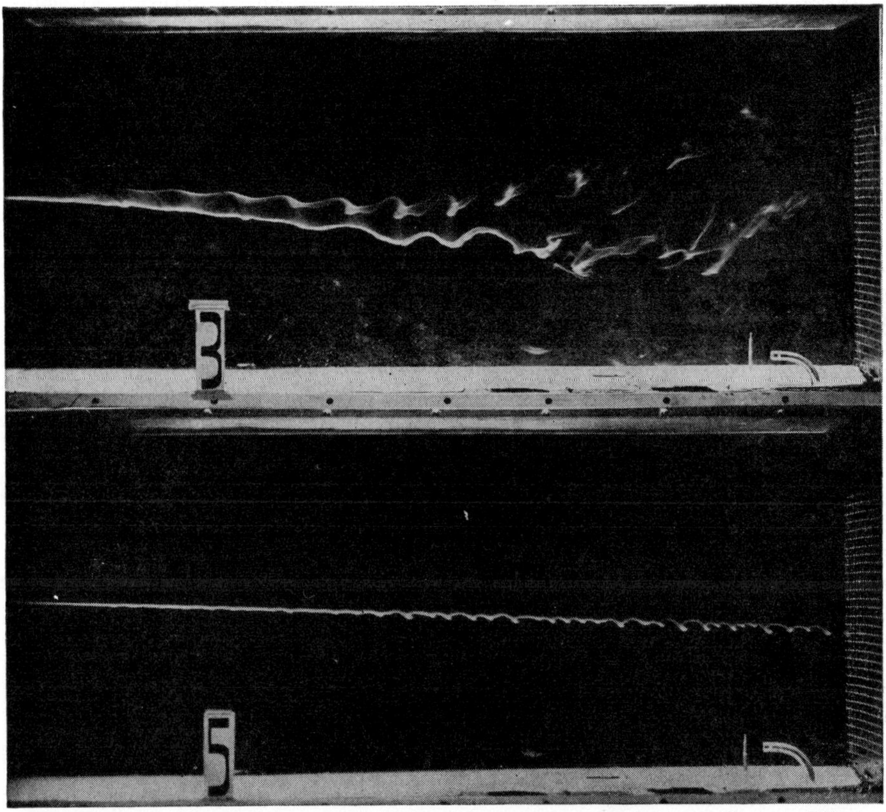

Fig. 5-4. Wind-tunnel test showing the effect of vertical temperature gradient on diffusion of a smoke plume. Plate 5, gradient isothermal; plate 3, decreasing 11°C/ft. (*New York University.*)

Wind-tunnel Techniques. A relatively new approach to the problem is the use of a model in a wind tunnel. This technique has been used with particular success in the design of powerhouse stacks and in other instances where the eddies produced by stack geometry and nearby structures control the dispersion pattern (23,63,85,99). In these instances, the use of existing theoretical formulas is of little help since most of the formulas are for atmospheric diffusion over level terrain.

Recent wind-tunnel tests of particular interest have been conducted for Consolidated Edison of New York in a wind tunnel at New York University that was designed especially for problems in stack diffusion (23). Considerable information of general value has been obtained concerning the effects of the ratio of stack velocity to wind velocity, stack geometry, different types of stack nozzles for increasing stack velocity, and geometry of nearby structures and of the powerhouse itself. It has been shown that fairing or redesign of the powerhouse can improve the dispersion pattern. A wind-tunnel test provides an extremely useful tool for studying particular configurations before and after actual construction. Several investigators have studied the effect of stack velocities on dispersion in wind tunnels (14,23). Figure 5-3 illustrates the type of results obtained by the wind-tunnel technique.

The fundamental problem in all model studies is to assure adequate similitude between model and prototype. Similitude is extremely difficult if not impossible to achieve in many problems of atmospheric diffusion. For gently rolling country, complete similitude cannot be obtained because of the extremely high velocities required to obtain the appropriate Reynolds number. Fortunately, the separation patterns for abrupt angular features such as buildings and more rugged topographic features are essentially independent of viscous effects (Reynolds criterion), and adequate similarity can be obtained at practical wind-tunnel velocities (77).

In many problems, particularly those involving considerable horizontal distances, the vertical atmospheric temperature gradient controls the dispersion pattern. Since the vertical temperature gradient in a wind tunnel is essentially isothermal, conventional tunnels are limited in application to those problems in which geometry controls. A new tunnel at New York University makes it possible to control the temperature gradient. The effect of tunnel temperature gradient on the smoke plume is illustrated by Fig. 5-4. Experiments in this pilot tunnel show that the character of an oil-fog smoke plume can be changed at will by varying the vertical temperature gradient. Much work still is required to establish valid relationships between model and prototype conditions, but this development promises to extend the applicability of wind-tunnel research to a much wider range of atmospheric-diffusion problems.

5.3.2 Atmospheric Turbulence and Turbulence Factors

Nature of Turbulence. Everyone has observed the gusty character of the wind. Sometimes the wind is more variable than at other times, but it is never completely constant in direction or speed. Gusts are a manifestation of atmospheric turbulence, and one method of estimating the magnitude of turbulence is to obtain a quantitative measure of gustiness.

Turbulence theory is extremely complex and cannot be covered in this brief discussion. For an excellent review of the theory of atmospheric turbulence, see Sutton (89,93). Considerable study of the subject is now in progress, but much work remains to be done. A large number of direct and indirect methods have been used to measure and estimate the magnitude of atmospheric turbulence. One of the great needs for both routine and research investigations is standardized instruments for the direct measurement of turbulence (44). In detailed studies of turbulent flow, two types of basic information are required:

1. Instantaneous velocities at fixed locations in the field of flow
2. Movement of individual parcels of fluid

Temporal variation in velocity indicates the intensity of turbulence, and geometric definition of the eddy motion indicates the scale of turbulence. The product of intensity and scale is proportional to the diffusivity, or eddy viscosity, of the atmosphere (77).

Measurement of the intricate variations in wind velocity at a point poses quite a problem in itself, and a number of different instruments have been used. Vertical and horizontal eddy velocities may be measured, using a bidirectional vane which moves simultaneously in both directions and records by a pen and drum. A directional hot-wire anemometer may be used to determine the third component. Gill (45) recently designed an instrument called the "w'-meter" to measure vertical wind components, using a novel float mechanism to produce a mechanical deflection proportional to the square root of the vertical component of the wind force and, therefore, proportional to the vertical component of the wind speed. Other instruments frequently used are Dines pressure-tube anemometers which are very sensitive to gusts, sensitive windmill anemometers and vanes, cup anemometers, and bridled cup anemometers.

For determining the travel of individual gusts a number of ingenious schemes have been tried, including the use of no-lift balloons, soap bubbles, and smoke puffs. These methods often involve photographic techniques.

Because of the random nature of turbulence and the impossibility of completely defining atmospheric flow, it is necessary in practical work to use a number of parameters to indicate the degree of turbulence quantitatively. Since the vertical transport of momentum, heat, and water vapor in the atmosphere are all functions of turbulence, it is apparent that there are a number of possibilities. Some of the parameters useful in studies involving atmospheric pollution problems are:

1. Atmospheric-temperature lapse rate
2. Wind variability (gustiness)
   a. Variability of wind speed
   b. Variability of wind direction
3. Wind-velocity profile
4. Richardson's number

This list by no means exhausts the possibilities, but it contains those parameters which can be used most readily.

Atmospheric-temperature Lapse Rate. Atmospheric turbulence and gustiness vary widely with variations in the vertical temperature gradient in the atmosphere. This relationship between temperature gradient and atmospheric stability is a fundamental concept of meteorology and is important in the field of air pollution. The decrease in temperature with height is called the "temperature lapse rate" γ and is defined by the equation

$$\gamma = -\frac{dT}{dz} \tag{5-1}$$

(See Art. 5.9, Nomenclature.)

The normal daytime condition is for the temperature to decrease with height, but the lapse rate varies widely, and often the temperature increases with elevation. If a layer of the atmosphere is completely mixed, the resulting lapse rate is 5.4°F/1,000 ft (0.98°C/100 m) because of the decrease in barometric pressure with height. This gradient is referred to as the dry-adiabatic lapse rate. On a clear day with pronounced solar heating of the earth, the lapse rate often becomes superadiabatic in the lower layers and the temperature decreases at a rate greater than the dry adiabatic. Under these conditions turbulence is at a maximum, and the atmosphere is said to be in unstable equilibrium because cold, heavy air overlies relatively warm, light air. On the other hand, on a clear night with light winds, the earth is cooled by nocturnal long-wave radiation, and the air next to the earth is cooled by the ground. Under these conditions, the lapse rate may become isothermal or even negative (indicating an increase in temperature with height). A layer with negative lapse rate is called a "temperature inversion." The stratification is stable and turbulence is at a minimum under these conditions since warm air overlies relatively cold air.

Figure 5-5 summarizes the different types of lapse rates, and Fig. 5-6 illustrates several typical atmospheric soundings. It should be pointed out that, although surface heating and cooling usually are of primary importance in determining the lapse rate in the surface layers, other mechanisms are also at work. Turbulence induced by rough topography during high winds tends to produce a dry-adiabatic lapse rate in the lower layers. Subsidence (slow settling or sinking) in high-pressure areas tends to result in a stable lapse rate and inversions aloft (this effect is particularly marked in the eastern part of high-pressure areas in temperate latitudes, i.e., the persistent Pacific high off the coast of Southern California and continental polar (cP) highs moving across the United States), and differential advection of warm and cold air at different elevations produces changes in lapse rate.

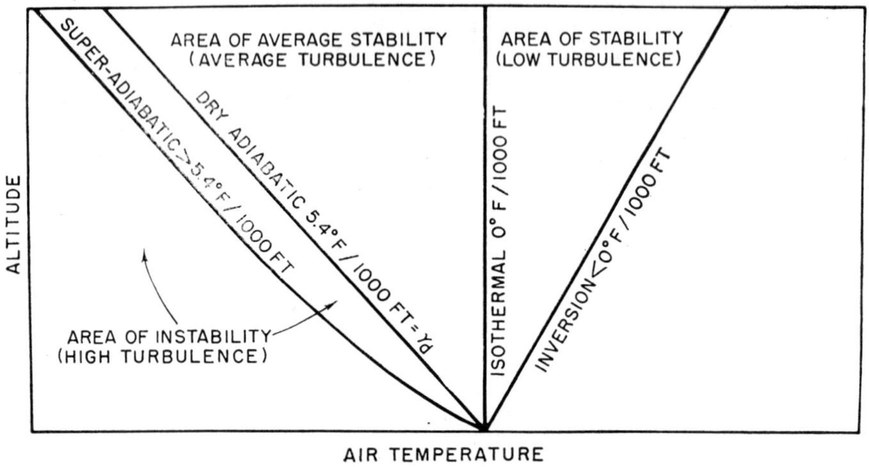

FIG. 5-5. Atmospheric-temperature lapse-rate types.

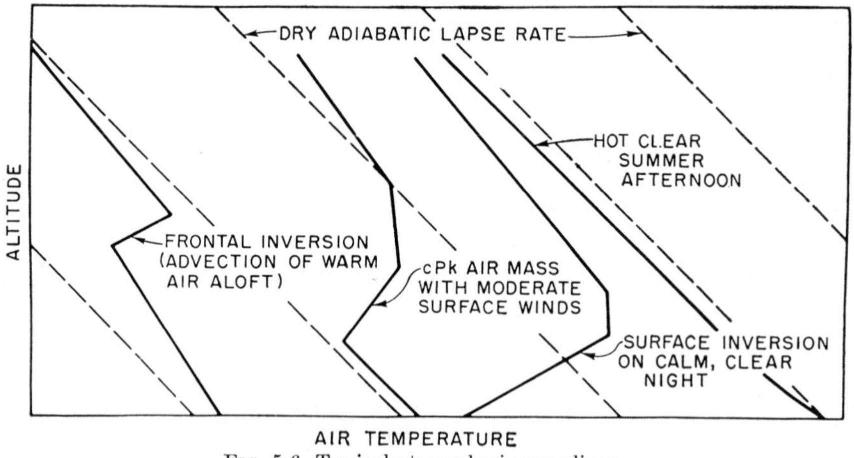

FIG. 5-6. Typical atmospheric soundings.

From the standpoint of air pollution, both stable surface layers and low-level inversions are undesirable because they minimize the rate of dilution of contaminants in the atmosphere. Even though the surface layer may be unstable, a low-level inversion will act as a lid or barrier to vertical mixing, and contaminants will accumulate in the surface layer below the inversion. Stable atmospheric conditions tend to be more frequent and persist longest in the fall, but inversions and stable lapse rates are prevalent at all seasons of the year.

Inversions are often referred to as representing exceptional conditions of atmospheric diffusion. While a good deal has yet to be learned about dispersion during inversions, they must not be considered as uncommon occurrences. In certain locations, inversions that begin at ground level and extend up to a few hundred feet may occur up to 40 to 50 per cent or more of the total hours in the year. In the southeastern United States, such inversions occur on two-thirds to three-fourths of the nights in the year.

Wind Variability (Gustiness). A convenient concept of the turbulent wind velocity at a fixed point is a constant vector $\bar{u}$ in the x direction, representing the mean wind and three mutually perpendicular vectors in the x, y, and z directions with time variable lengths u', v', and w', representing component fluctuations of the mean wind. The vector sum of the four vectors represents the instantaneous wind velocity at the point in question. The u' vector represents the fluctuation or "eddy velocity" in the direction of the mean wind; v', the fluctuation across the mean wind; and w', the vertical fluctuation. By definition, the time average of each of the eddy-velocity components equals zero. A well-damped anemometer will register the mean wind speed and a damped wind vane, the mean direction.

Using this concept, a numerical measure of "gustiness" g' is the root mean square of the ratio of the fluctuations to the mean wind:

$$g_{x'} = \sqrt{\frac{\overline{u'^2}}{\bar{u}^2}} \tag{5-2}$$

$$g_{y'} = \sqrt{\frac{\overline{v'^2}}{\bar{u}^2}} \tag{5-3}$$

$$g_{z'} = \sqrt{\frac{\overline{w'^2}}{\bar{u}^2}} \tag{5-4}$$

Gustiness has been found to be essentially independent of the wind speed (82), but it does depend on atmospheric stability. At low levels, turbulence is nonisotropic (i.e., $g_{x'} \neq g_{y'} \neq g_{z'}$), but some authorities (8) have concluded that turbulence is isotropic at heights in excess of about 80 ft.

A very useful criterion for turbulence based on wind-direction variability has been proposed by Lowry (58). Lowry classified gustiness into four types according to the variability of wind direction. The character of the wind-vane trace for the four types is shown in Fig. 5-7. Representative hourly distributions of 10-sec mean directions taken at 3° intervals corresponding to the Lowry types are shown in Fig. 5-8. The mode of the frequency distribution a_m for type A is less than 10 per cent; for type B, between 10 and 20 per cent; for type C, between 20 and 35 per cent; and for type D, greater than 35 per cent. The same four types may also be used to estimate the standard deviation of the wind direction. The approximate modes and standard deviations for the four types are shown in Table 5-1.

Table 5-1. Relationship of Stack Height, Gustiness, and Distance to Maximum Ground Concentration of Effluent

Gustiness type	Approx. mode (1-hr interval), a_m	Approx. std. deviation (10-min. interval), 0° of direction	Approx. distance from stack to max ground-level conc. stack heights
A	0.075	20°	3
B	0.15	12°	5
C	0.25	5°	12
D	0.50 (Range 0.15–0.60)	<1°	>50

After P. H. Lowry, Microclimate Factors in Smoke Pollution from Tall Stacks, *Meteorol. Monographs*, **1**, No. 4, 24–29 (1951).

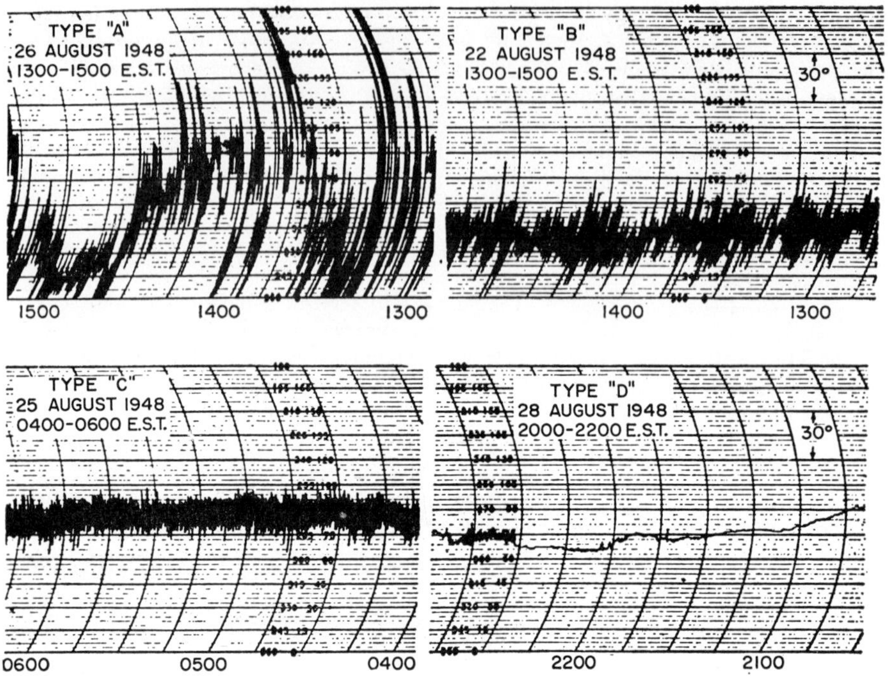

FIG. 5-7. Four types of gustiness shown by the variability of wind direction as indicated by a wind vane at the top of the experimental stack at Brookhaven, according to Smith. *A*, great instability; *B*, moderate instability; *C*, moderate stability; *D*, great stability.

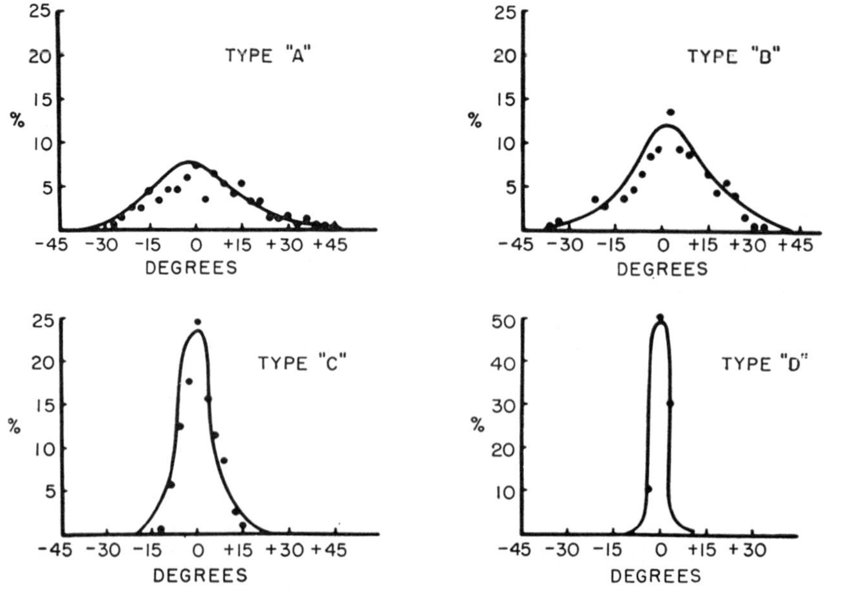

FIG. 5-8. Gustiness types. Representative hourly distributions of 10-sec mean wind direction. (*After Lowry.*)

Other investigators (29) have shown that the wind-direction variability (as represented by the standard deviation of wind direction at 10-sec intervals about the 10-min mean wind direction at 100 ft) shows a semilogarithmic relationship to the temperature difference between 6 and 200 ft. For inversion conditions, the standard deviation of wind direction was in the range of 1 to 5°; for isothermal to dry adiabatic, 5 to 8°; and for superadiabatic, 8 to 30°.

Lapse rates exhibit daily cycles that vary in duration with the time of year. Figure 5-9 shows this relation for Potsdam (33). The solid lines are average lapse-rate values, inversions being negative. The dotted lines show sunrise and sunset times. Note that periods of large lapse rate (strong turbulence) occur longer in summer

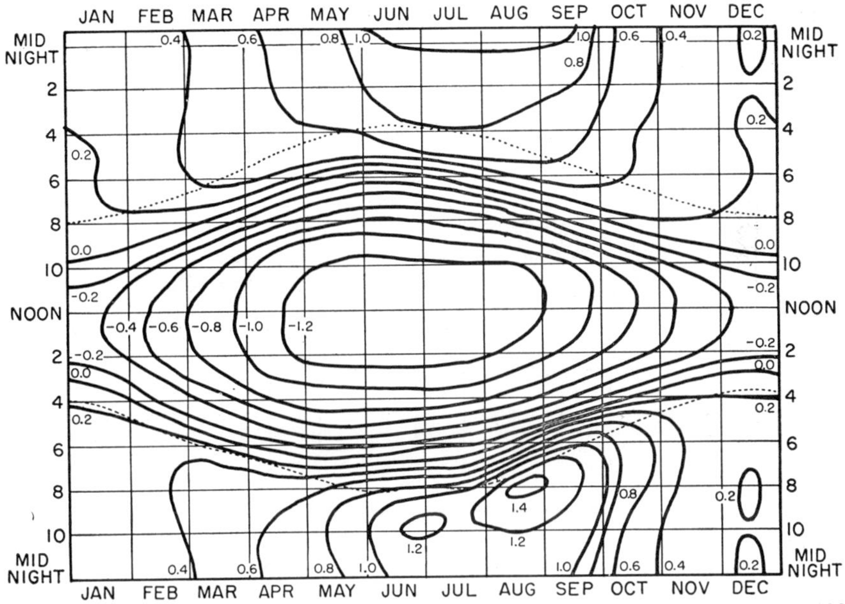

Fig. 5-9. Difference of the air temperature at 2 and 34 m height in Potsdam 1893, 1904. (*From Rudolf Geiger, "The Climate near the Ground," Harvard University Press, Cambridge, Mass.*, 1950; *reprinted by permission of the publishers.*)

because of stronger and longer periods of insolation. Inversions occur more frequently and last longer in winter on account of the longer nights and consequently greater time for heat loss by radiation.

A similar cycle of gustiness as measured by average wind variability is shown in Fig. 5-10 for a southeastern United States location. Strong gustiness is predominant at midday and during the afternoon in summer. The low gustiness associated with inversions occurs for longer periods during winter nights than during summer.

Wind-velocity Profile. The relationship between turbulence, atmospheric stability, and gustiness has been indicated. Amospheric stability is also reflected in the vertical wind-speed profile. In an unstable atmosphere, conditions are favorable for the vertical eddy transfer of momentum, and the variation in wind-speed profile with height is much less than under stable conditions. Thus, measuring the wind speed provides a quantitative means of measuring the degree of turbulence. By application of Taylor's statistical correlation theory of turbulence (94), Sutton (89) shows that for an aerodynamically smooth surface the following relationship applies:

$$\bar{u} = \bar{u}_1 \left(\frac{z}{z_1}\right)^{n/(2-n)} \tag{5-5}$$

where $\bar{u}_1$ is the mean wind speed at the fixed reference height z_1 and n, the turbulence parameter, lies between 0 and 1. The value n is small for unstable atmospheric conditions and tends toward unity with stable conditions. Approximate values for n under different conditions of stability are 0.20, large lapse rate; 0.25, isothermal or small lapse rate; 0.33, moderate inversion; 0.50, marked inversion (90).

Richardson's Number. A criterion, involving both the lapse rate and the wind profile, used to express whether turbulence will subside or increase is Richardson's number (75):

$$Ri = \frac{g/T[-(\partial\bar{T}/\partial z) + \gamma d]}{(\partial\bar{u}/\partial z)^2} \tag{5-6}$$

This number is based on the reasoning that turbulence will subside if the work required to displace air from its equilibrium position in a stable atmosphere is greater than the work done by the eddy stresses. A critical value for the Richardson number

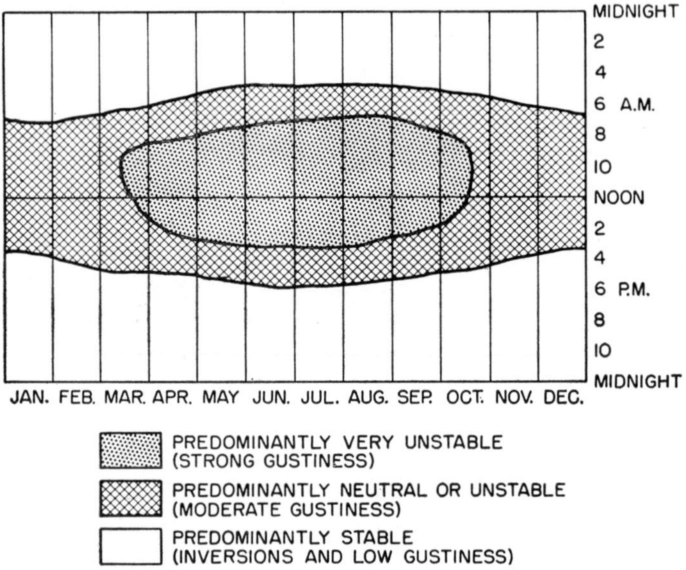

FIG. 5-10. Annual and daily cycle of atmospheric stability for southeastern United States location (based on 1-year record of wind gustiness).

has not been definitely determined, but this concept is sometimes useful in correlations involving atmospheric diffusion.

Other Effects of Turbulence. In addition to the foregoing, a few of the many other observable and variable effects of turbulence are:

1. Shearing stress produced by wind at a horizontal surface (84)
2. Angle between the surface wind and the geostrophic wind (wind direction parallel to the isobars on a weather map)
3. Phase relationship between diurnal temperature and humidity variations at different heights
4. Temperature and humidity fluctuations at a given point
5. Vertical gradient in humidity

5.3.3 Diffusion Theories

A considerable amount of theoretical work has been done to develop formulas to predict atmospheric diffusion. The most useful formulas for this purpose have been developed in England by Bosanquet and Pearson (9) and by Sutton (90,91). In all cases, the formulas contain turbulence parameters or eddy-diffusion coefficients that are functions of atmospheric stability and turbulence. The selection of suitable parameters is of basic importance in application of the theory.

The average values recommended by the authors of the respective formulas may be used. Better still are values determined by diffusion tests made in the field at the particular area under study.

The Sutton formulas are based on Taylor's statistical correlation theory of turbulence and solutions of the classical diffusion equation by Roberts (76). The velocity gradient is considered to be the prime factor in forecasting diffusion in the lower atmosphere. The generalized eddy-diffusion coefficients that appear in Sutton's equation are theoretically functions of the velocity gradient factor n and the gustiness g'.

$$C_y^2 = \frac{4\nu^n}{(1-n)(2-n)\bar{u}_1^n} g_{y'}2(1-n) \tag{5-7}$$

$$C_z^2 = \frac{4\nu^n}{(1-n)(2-n)\bar{u}_1^n} g_{z'}2(1-n) \tag{5-8}$$

The coefficient C_y indicates the magnitude of the horizontal component of diffusion and C_z, the vertical. Under conditions of isotropic turbulence, $C_y = C_z$. The coefficients given by Sutton (90) for different lapse rates and elevations are shown in Table 5-2. Note that above about 80 ft turbulence is considered to be isotropic. Diffusion tests and observations of smoke plumes clearly have indicated that this is not the case under stable atmospheric conditions.

Table 5-2. Generalized Eddy-diffusion Coefficients[a]

Height of source above ground, ft	Large lapse, $n = 0.20$		Zero or small lapse, $n = 0.25$		Moderate inversion, $n = 0.33$		Large inversion, $n = 0.50$	
	C_y	C_z	C_y	C_z	C_y	C_z	C_y	C_z
0	0.42	0.24	0.24	0.14	0.15	0.09	0.12	0.07
33	0.42	0.24	0.24	0.14	0.15	0.09	0.12	0.07
82	0.24		0.14		0.090		0.070	
100	0.23		0.13		0.085		0.065	
150	0.21		0.12		0.075		0.060	
200	0.19		0.11		0.070		0.055	
250	0.18		0.10		0.065		0.050	
300	0.16		0.09		0.055		0.045	
350	0.13		0.07		0.045		0.035	

[a] C has dimensions ft $^{n/2}$.
After O. G. Sutton, The Problem of Diffusion in the Lower Atmosphere, *Quart. J. Roy. Meteorol. Soc.*, **73**, 257–281 (1947).

Another important error in the assumption made by Sutton is that the surface of the earth is aerodynamically smooth. Almost invariably the earth's surface is aerodynamically rough; therefore the Sutton equations tend to overestimate contaminant concentrations downwind from the source of pollution.

The Bosanquet and Pearson equations also are based on classical diffusion theory and dimensional analysis. The diffusion coefficient p has been estimated by several means. These include the tangential wind force per unit area and normal variation in wind speed with height, smoke-plume photographs, and the release of small smoke clouds near the ground. Their other diffusion coefficient q is an index of the horizontal diffusion and has been estimated by the horizontal scatter of small balloons.

The values of p and q commonly used in conjunction with the Bosanquet and Pearson equations are given in Table 5-3.

Table 5-3. Bosanquet and Pearson Turbulence Parameters

Turbulence	p	q	p/q
Low	0.02	0.04	0.50
Average	0.05	0.08	0.63
Moderate	0.10	0.16	0.63

More recently (29), values of p and q have been determined by dispersion tests and the values obtained correlated with wind gustiness. A good correlation was obtained between values of p, in the range of 0.02 to 0.5, and the standard deviations

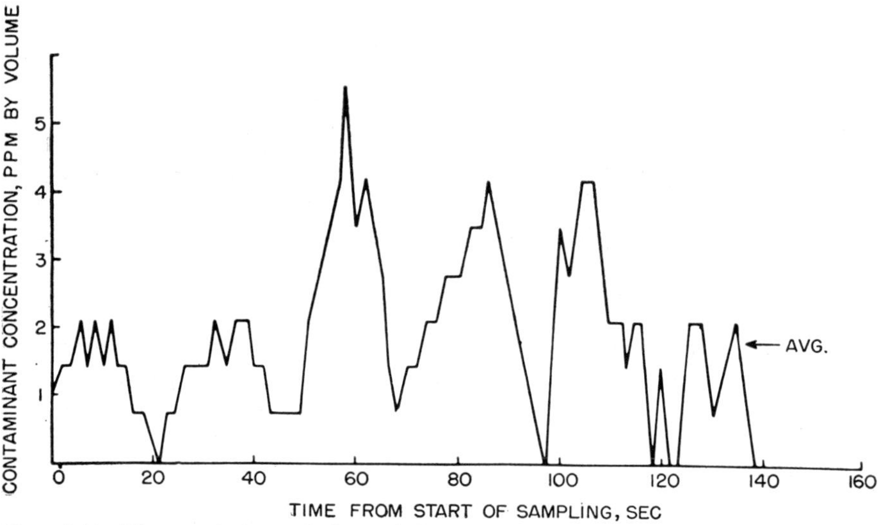

FIG. 5-11. Time variations of downwind contaminant concentration under moderate atmospheric turbulence. (*After Gosline.*)

of the wind speed divided by the mean wind speed. Further, a good correlation was obtained between values of q, in the range 0.06 to 0.65, and the standard deviations of wind direction. In each instance, the standard deviations represented the deviations of wind at 10-sec intervals about the 10-min mean wind at 100 ft.

The expression of atmospheric concentrations of a contaminant downwind from a source of constant emission is complicated because of the extreme variation with time. Figure 5-11 shows the temporal variation in concentration of a contaminant downwind from a 100-ft stack. Gosline (35) found in these tests that only 8 to 30 per cent of the time was there a measurable concentration at a location directly downwind from the source. The average frequency of exposure was 18 per cent. Peak concentrations of only a few seconds' duration corresponded to the passage of a single eddy. Peak of maximum short-time concentrations must be distinguished from the long-period average concentrations. The sampling period should always be indicated along with recorded concentrations and information concerning whether or not an attempt was made to take the sample directly downwind from the source. Gosline (35) states that, close to the stack, 1- to 3-min average maximum concentrations are 6 to 13 times the 30-min or longer concentrations and that the extreme concentrations for a few seconds are 3 to 4 times the 1- to 3-min concentrations. Thus, the 1- to 10-sec maximum may be 50 times the 30-min or longer average.

The foregoing concept is basic and extremely important in the design and interpretation of air pollution surveys, especially when contaminant concentrations downwind from an isolated source are being investigated. If samples of only a few seconds' or minutes' duration are taken, many of them would be expected to indicate little or no pollution, and a wide variation in concentration would be expected in the remaining. In such a case, the maximum observed concentrations would have special significance because they would represent maximum eddy concentrations. Less variation would be expected in the analyses of longer-period downwind samples. An appreciation of the variable concentrations to be expected is also important from the standpoint of estimating whether or not concentrations above tolerable levels are to be expected. In most instances the tolerable concentration of a particular material depends on the duration of exposure.

According to their author, the Sutton equations yield time-mean values based on sampling periods of at least 3 min. However, some recent verifications (87) of the Sutton equations indicate that the values obtained are more representative of the order of magnitude of the maximum concentrations that might be observed. The Bosanquet and Pearson equations are more representative of average concentrations for about 30 min. An exception to this statement results under conditions of "fumigation" when ground-level concentrations several times those predicted by the theoretical equations may result. This phenomenon is described in the following article.

5.4 STACK METEOROLOGY

The term stack meteorology has been used for the study of factors affecting dispersion of pollution from stacks and chimneys. Stacks are expensive items of construction, and there is need for a sound basis for determining the stack heights required for atmospheric disposal of contaminants without creating a pollution problem. Theoretical and empirical methods have been devised for estimating dispersion from point, line, and area sources—continuous and instantaneous (e.g., smoke bombs)—located both at ground level and aloft. Formulas are available for estimating ground-level concentrations of contamination downwind from a source and also for estimating concentrations in the plume itself. The most frequently used formulas in stack meteorology are those giving average ground-level concentrations downwind from an elevated source of constant discharge rate.

Considerable care must be exercised in the application of the diffusion theory because many of the formulas are only now being verified by field tests. The assumptions on which the formulas are based should be recognized to avoid erroneous conclusions resulting from their application.

5.4.1 Plume Types

The appearance of a smoke plume changes radically with wind and atmospheric stability. Figures 5-12 to 5-14 illustrate different configurations of a smoke plume. Church and Gosline (21) classified the different plume types by appearance as looping, coning, and fanning.

The trail of a looping plume, as shown in Figs. 5-12 and 5-13, is extremely broken and erratic, with the trail swinging up and down and from side to side—with an individual puff sometimes reaching the ground relatively close to the stack. Ground-level concentrations of a contaminant at a particular location downwind consequently are highly variable. Looping occurs with unstable (superadiabatic) lapse rates and is indicative of rapid atmospheric diffusion. Coning occurs with neutral stability and wind speeds of 20 mph or more. Diffusion is rapid under these conditions.

FIG. 5-12. A "looping" smoke plume. (*Brookhaven National Laboratory.*)

FIG. 5-13. A "looping" smoke plume. (*Brookhaven National Laboratory.*)

With extreme inversion conditions, diffusion in the vertical is practically nonexistent. Because of slight shifts in wind direction, the plume fans out downwind like a meandering stream in a thin layer. A fanning plume is shown in Fig. 5-14. Under these conditions, over flat terrain, the plume may be visible for many miles. The contaminant frequently cannot be detected at the ground, but if the plume is intercepted by a building or hill, high concentrations result because of the slow diffusion within the plume. The height of the plume above the ground is of great importance in determining its contact with the ground. Assumption of isotropic turbulence under these conditions, e.g., $C_y = C_z$ in the Sutton equations or $q = p\sqrt{2}$ in the Bosanquet and Pearson equations, obviously does not conform to observations.

FIG. 5-14. A "fanning" smoke plume. (*Brookhaven National Laboratory.*)

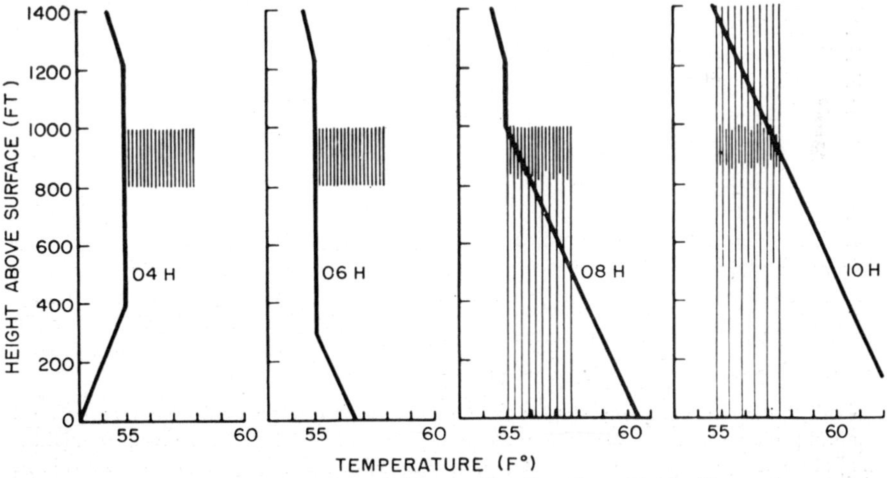

FIG. 5-15. The mechanism of "fumigation." (*According to Hewson.*)

Smooth flow or fanning occurs most frequently at night and highly turbulent flow or looping, in the daytime. The transition from smooth to turbulent flow is abrupt. It usually takes place within an hour or two after sunrise. Since the heating of the atmosphere proceeds from the earth's surface upward, the plumes from short stacks may be looping while the plumes from taller stacks are fanning.

The morning transition from smooth to looping flow is of appreciable significance in air pollution because it results in the phenomenon referred to as "fumigation."

At the breakup, the high concentrations in the fanning plume are rapidly mixed downwind through underlying layers to the surface. The result is a short period of exceptionally high ground-level concentrations. The duration of the high level of surface pollution is about 15 min. The mechanism of fumigation as illustrated by Hewson (46) is shown in Fig. 5-15. The smoke plume is concentrated in a shallow layer aloft until heating from the ground causes instability to reach the level of the plume. Then a high concentration of the pollutant is carried to ground level. On the basis of measurements made at Brookhaven, L. I., Lowry (58) has reported that the average fumigation concentration during the 15-min period is about twenty times the maximum predicted by the Sutton theory.

5.4.2 Influencing Factors

The factors that should be considered in every problem of stack meteorology may be classified under three headings—process factors, source or stack factors, and meteorological factors.

Process Factors

1. Emission rate (including average, maximum, and minimum)
2. Temperature of emission products
3. Form of emission product—dust, fume, mist, spray, or gas
4. Concentrations of gaseous and particulate matter
5. Particle-size distribution and terminal velocities
6. Agglomerating characteristics
7. Chemical properties and possibilities of interaction between constituents
8. Toxicological or nuisance properties

Source Factors

1. Stack height
2. Stack diameter and configuration of exit
3. Stack velocity
4. Relationship of stack to surrounding structures and terrain

Meteorological Factors

1. Wind speed and direction
2. Temperature and humidity
3. Atmospheric stability as indicated by lapse rate, gustiness, etc.
4. Topographic effects

The effect of many of these factors is indicated in a quantitative way in the following article, but the effect of others, such as topography, can only be estimated.

5.4.3 Working Formulas

Gas from Continuous Point Source. The Bosanquet and Pearson (9) equation for the ground-level concentration C downwind a distance x and crosswind a distance y from an elevated source of effective height H emitting gas at a constant rate Q is as follows:

$$C \text{ (ppm)} = \frac{Q10^6}{\sqrt{2\pi}\, pqux^2}\, e^{-\frac{H}{px} - \frac{y^2}{2(qx)^2}} \tag{5-9}$$

The corresponding Sutton (90) equation is

$$C \text{ (ppm)} = \frac{2Q10^6}{\pi C_y C_z u x^{2-n}}\, e^{-\frac{1}{x^{2-n}}\left(\frac{y^2}{C_y^2} + \frac{H^2}{C_z^2}\right)} \tag{5-10}$$

A plot showing the results obtained by application of Eq. (5-9) is shown in Fig. 5-16. The ground-level concentration increases to a maximum at some distance downwind from the stack and then gradually decreases with distance. The ground-level concentration falls off quite rapidly in the y direction to each side of the direct downwind line. The same equations can be used to indicate downwind concentrations of aerosols, provided that their terminal velocities are insignificant in comparison with the eddy velocities in the atmosphere. In calculations involving particulate matter, Q (cfs) may be replaced by M (kg/sec) so that the calculation yields C (mg/cu ft) instead of C (ppm by volume).

The Bosanquet-Pearson and Sutton equations are more easily compared by examination of the maximum ground-level concentrations as determined from each. The

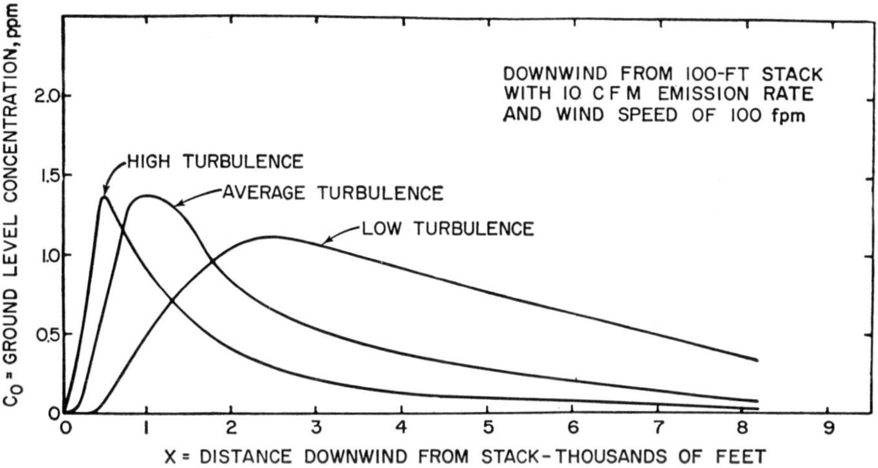

FIG. 5-16. Effect of atmospheric stability on diffusion from a stack.

maximum concentration will occur directly downwind in each case; Bosanquet and Pearson will yield

$$C_{\max}\ (\text{ppm}) = \frac{4Q10^6}{\sqrt{2\pi}\, e^2 u H^2}\,\frac{p}{q} = 2.15\,\frac{Q10^5}{uH^2}\,\frac{p}{q} \tag{5-11}$$

at a distance

$$x_{\max} = \frac{H}{2p} \tag{5-12}$$

Figure 5-17 presents a graphical solution for Eqs. (5-11) and (5-12). The corresponding Sutton equations are

$$C_{\max}\ (\text{ppm}) = \frac{2Q10^6}{e\pi u H^2}\,\frac{C_z}{C_y} = 2.35\,\frac{Q10^5}{uH^2}\,\frac{C_z}{C_y} \tag{5-13}$$

at a distance

$$x_{\max} = \left(\frac{H^2}{C_z^2}\right)^{1/(2-n)} \approx \frac{H}{C_z} \tag{5-14}$$

Assuming that the ratios of the diffusion constants in Eqs. (5-11) and (5-13) are equivalent, the two expressions for maximum ground-level concentration are the same except for a 9 per cent difference in the numerical constants. However, Sutton takes the ratio C_z/C_y equal to unity for heights above about 80 ft (see Table 5-2) while the ratio p/q is apparently in the range 0.35 to 0.7, depending on conditions of atmospheric stability and terrain. Thus, Eq. (5-11) yields values of $C_{\max}$ of less than half of those determined by Eq. (5-13).

In calculating ground-level concentrations, it is difficult to understand why the ratio C_z/C_y should ever be taken as unity. Even though turbulence is isotropic at heights above 80 ft, the plume must be dispersed through the nonisotropic surface layers to produce detectible amounts at ground level. As noted previously, the Sutton assumption of $C_z = C_y$ is not realistic for stable atmospheric conditions even at higher elevations.

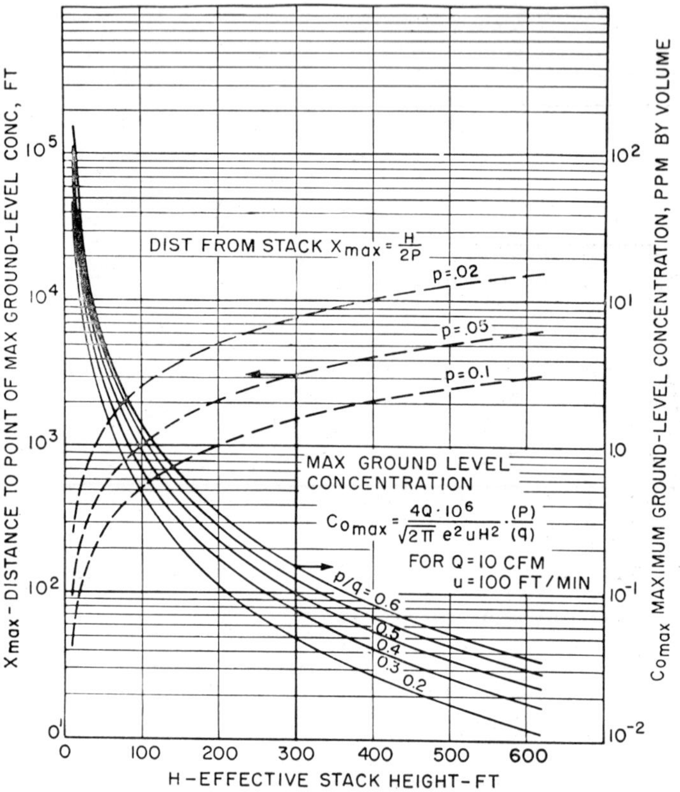

FIG. 5-17. Graphical solution of Bosanquet and Pearson equations for maximum ground-level concentration and distance from stack.

The expression for distance to the maximum ground-level concentration is also of the same form in both instances. Since p equals about 0.05 and C_z equals about 0.13 for average turbulence, the two expressions yield distances of the same order of magnitude. Note that for average turbulence the maximum ground-level concentration is about 8 to 10 stack heights from the source. The over-all range in distance to the maximum ground-level concentration is 5 to 20 stack heights.

A convenient way to express the preceding equations is in terms of the dimensionless ratios R and S where

$$R = \frac{x}{x_{\max}} \tag{5-15}$$

and

$$S = \frac{C}{C_{\max}} \tag{5-16}$$

From the Bosanquet and Pearson equations (5-9), (5-11), and (5-12):

$$S_{y=0} = R^{-2}e^{2(R-1)/R} \tag{5-17}$$

$$\frac{y}{y_{0.1\,\max}} = R\left[\frac{1}{\ln 0.1}\left(\ln S + 2\ln R + \frac{2}{R} - 2\right)\right]^{1/2} \tag{5-18}$$

From the Sutton equations (5-10), (5-13), and (5-14) similar expressions are obtained:

$$S_{y=0} = R^{n-2}e^{1-R^{n-2}} \tag{5-19}$$

$$\frac{y}{y_{0.1\,\max}} = \left\{\frac{R^{2-n}}{\ln 0.1}\left[\ln S + (2-n)\ln R + \frac{1}{R^{2-n}} - 1\right]\right\}^{1/2} \tag{5-20}$$

The value $y_{0.1\,\max}$ is the distance from $C_{\max}$ to the value $S = 0.1$.

Figure 5-18 represents solutions for Eqs. (5-17) and (5-19). Once the location and magnitude of $C_{\max}$ are determined from either Eqs. (5-11) and (5-12) or Eqs. (5-13)

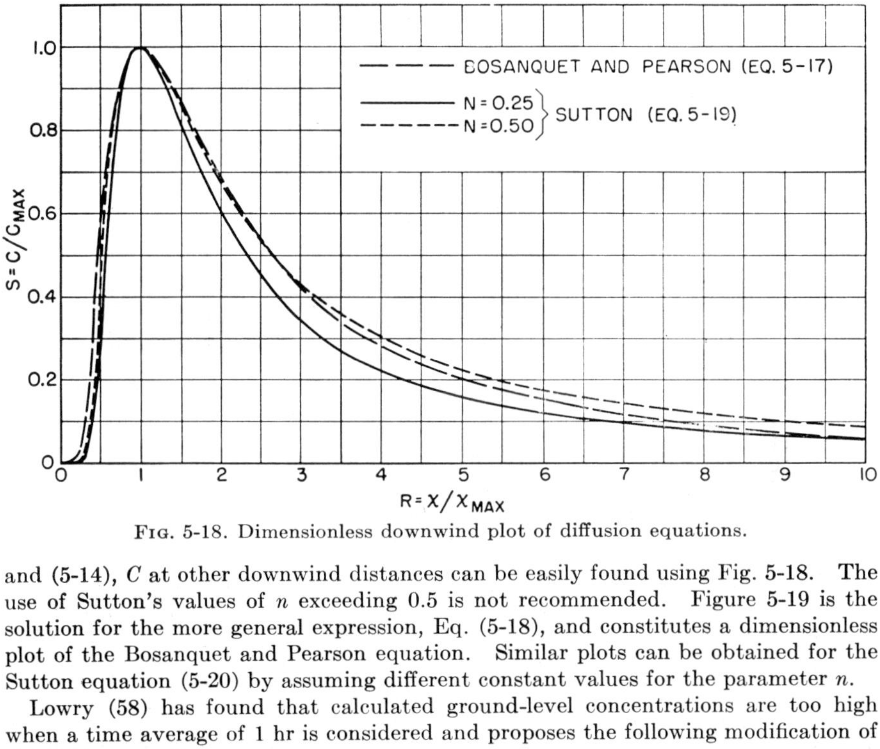

FIG. 5-18. Dimensionless downwind plot of diffusion equations.

and (5-14), C at other downwind distances can be easily found using Fig. 5-18. The use of Sutton's values of n exceeding 0.5 is not recommended. Figure 5-19 is the solution for the more general expression, Eq. (5-18), and constitutes a dimensionless plot of the Bosanquet and Pearson equation. Similar plots can be obtained for the Sutton equation (5-20) by assuming different constant values for the parameter n.

Lowry (58) has found that calculated ground-level concentrations are too high when a time average of 1 hr is considered and proposes the following modification of the Sutton equation:

$$C_{\max}\ (\text{ppm}) = \frac{2Q10^6}{e\pi H^2}\frac{a_m}{u} \tag{5-21}$$

in which the factor a_m takes the place of C_z/C_y in Eq. (5-13). The factor a_m represents the frequency of the most frequent wind direction and probably has a different physical significance than the Sutton ratio C_z/C_y. An approximate relationship between gustiness type and the factor a_m is given in Table 5-1. Lowry also proposes an empirical formula for the distance to the maximum ground-level concentration:

$$x_{\max} = H \csc \sigma \tag{5-22}$$

Examination of Eqs. (5-9) to (5-14) reveals the following basic principles of stack meteorology:

1. Average concentrations of a contaminant downwind from a stack are directly proportional to the discharge rate. An increase in discharge rate by a factor increases ground-level concentrations at all points by the same factor. The discharge rate Q should be calculated at atmospheric temperature for use in the formulas.

2. In general, increasing the dilution of the stack gases by addition of excess air in the stack does not affect ground-level concentrations appreciably. Practical stack dilutions are usually insignificant in comparison to later atmospheric dilution by plume diffusion. Addition of diluting air will increase the effective stack height, however, by increasing the stack exit velocity. This effect may be important at low wind speeds. On the other hand, if the stack temperature is decreased appreciably by the dilution, the effective stack height may be reduced. Stack dilution will have appreciable effect on the concentration in the plume close to the stack.

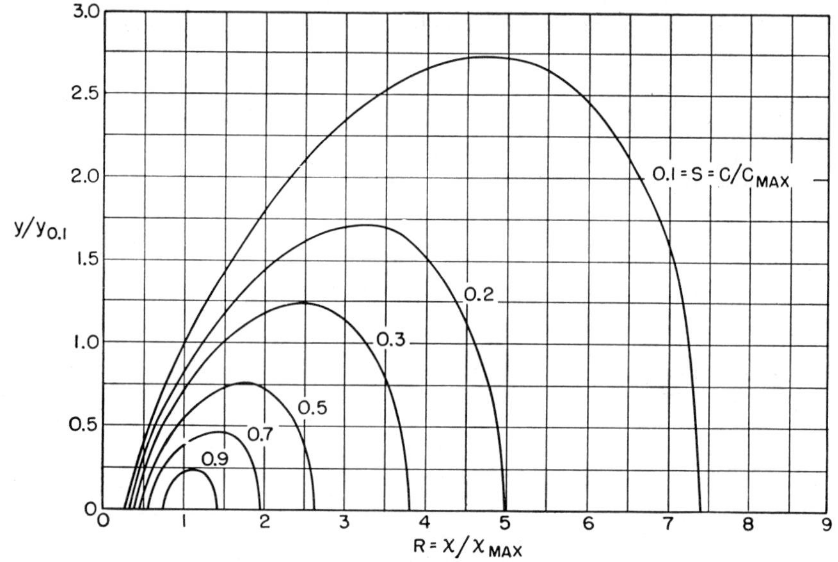

FIG. 5-19. Dimensionless plot of Bosanquet and Pearson equation (5-18).

3. For equivalent effective stack heights, average downwind concentrations are inversely proportional to the wind speed. An increase in wind speed by a factor of 2 reduced ground-level concentrations by one-half. Knowing the wind direction and speed probabilities permits the estimation of frequencies of different levels of pollution at various locations in the neighborhood of the stack.

4. Average downwind concentrations are inversely proportional to the square of the effective stack height. Doubling the stack height reduces maximum ground-level concentrations to one-fourth previous values.

5. The effective stack height is the sum of three terms—the actual stack height, the rise caused by the velocity of the stack gases, and the rise attributable to the density difference between the stack gases and the atmosphere: $H = h_s + h_v + h_t$. (See page 5-33 for methods of calculating effective stack heights.)

6. The stack equations are based on smooth, level terrain. The effect of topography may be estimated to some extent by correcting the effective stack height to account for differences in grade elevation.

7. The location of the maximum ground-level concentration is a function of atmos-

pheric stability. During unstable atmospheric conditions, the maximum occurs relatively close to the stack. As the atmosphere becomes increasingly stable, the location of the maximum ground-level concentration moves farther away from the stack.

Verification of Theoretical Formulas. The verification of the theoretical diffusion formulas of Sutton, of Bosanquet and Pearson, and of others by actual field test is considerably more difficult than one might suppose. The fairly elaborate sampling and meteorological equipment required and the difficulties in the statistical handling of data are two of the primary reasons that better knowledge is still lacking on the accuracy and range of applicability of the theoretical formulations.

A number of independent workers (27,29,35,87,96) have attempted to check the validity of the theory with varying success. A direct comparison of the conclusions of the different investigators is somewhat difficult because of variations in the sampling technique. The values of atmospheric concentration of a contaminant obtained in the field are invariably influenced by such factors as location of sampling stations, duration of sampling period, and the statistical treatment of the data obtained.

In general, two methods of verification may be used: (1) correlation of measured concentrations with concentrations calculated, using the formulas containing the appropriate theoretical diffusion coefficients, (2) calculation of the diffusion coefficients by substitution of the test data into the theoretical formulas and correlation of the coefficients so obtained with meteorological parameters and with the theoretical coefficients. The former method has been used most often, but the latter has certain advantages in that it establishes coefficients that may be used for the solution of practical problems in the area under study.

The conclusions reached in five independent verifications of the theoretical equations may be summarized as follows:

1. Based on ground-level measurements of nitrogen dioxide downwind from an 80-ft stack under different weather conditions, Gosline (35) concluded that average concentrations for 30 min or longer compared within about twofold of the results predicted by Bosanquet and Pearson, by Sutton, and by Church (20). Examination of the individual data indicates little choice among the methods, although the Sutton equation may produce large errors for points close to the stack. The expected averages were calculated from $p = 0.05$, $q = 0.08$ for the Bosanquet equation, and $n = 0.20$ (unstable), $n = 0.25$ (neutral) for the Sutton equation.

2. Thomas et al. (96) concluded that agreement was good between diffusion of sulfur dioxide from tall smelter stacks as determined by automatic recorders and the theoretical equations of Bosanquet and Pearson and of Sutton, assuming $C_z = p = 0.05$, $C_y = q = 0.07$ and Sutton's $n = 0$. Note that Sutton's coefficients were not chosen in accordance with Sutton's recommended values (see Table 5-2).

3. Smith (87) compared the results of Sutton's formulas with actual hourly mean ground-level concentrations of oil fog obtained from diffusion tests made over flat, relatively smooth terrain. His results and conclusions were:

   a. For unstable lapse conditions, Sutton results are high by a factor of about 10. The maximum concentration occurred in the range of 5 to 20 stack heights downwind.

   b. For neutral conditions, Sutton results are high by a factor of about 4. The distance to maximum concentration occurred in the range of 20 to 50 stack heights.

   c. For inversion conditions, there is no agreement between observation and theory. No oil fog was detected at ground level as far as 350 stack heights.

   d. Sutton's predictions would be expected to compare with mean concentrations in individual parcels of effluent carried to the ground under lapse and neutral conditions.

4. Eisenbud and Harris (27) found from tests in several localities that the measured concentrations were less than concentrations as predicted by the Sutton equation by as much as a factor of 5. This observation is based on 1-hr samples taken only during daylight hours.

5. Falk (29) has calculated the Bosanquet and Pearson parameters from dispersion tests made over gently rolling, partly wooded terrain. Preliminary conclusions are:

a. The wind-velocity variability can be used to determine dispersion parameters.
b. Equation (5-11) underestimates the maximum ground-level concentration but agrees within a factor of 2 with the measured concentration if the results are multiplied by a factor of 1.8. The ratio p/q is in the range of 0.35 to 0.7.
c. The distance to the maximum ground-level concentration was found to be about one-third greater than would be indicated by Eq. (5-12).
d. At distances greater than $x_{\max}$, the ground-level concentration decreases at a greater rate than is indicated by Eq. (5-9).

Dilution of the Plume. In some instances, estimates of the concentrations existing in the plume itself are required at various distances from the source. These estimates are particularly useful when the plume occasionally intercepts a building or hill or when there is a possibility of explosive flash back in the plume. A Bosanquet (10) expression for the axial plume concentration indicates it to be inversely proportional to the square of the distance downwind:

$$C_A \text{ (ppm)} = \frac{Q10^6}{4\pi p^2 x^2 u} \qquad (5\text{-}23)$$

For average turbulence and assuming uniform distribution of contamination in the plume, the angle of spread of the plume is shown to be approximately 12°. Assuming conical distribution, the angle of spread is about 20°.

The corresponding Sutton (91) expression for peak concentration in the cloud is

$$C_A \text{ (ppm)} = \frac{2Q10^6}{\pi C_y C_z u x^{2-n}} \qquad (5\text{-}24)$$

in which the rate of decrease depends on the parameter n. The results are of the same order of magnitude as those given by the Bosanquet expression. The Sutton relationship for the width of the cloud from a continuous point source is

$$2y_0 = 3.04 C_y x^{1-n/2} \qquad (5\text{-}25)$$

Barad (4) has found that the Sutton expression for gas concentrations at plume level gives concentrations that are not in agreement with those observed during smoke experiments under inversion conditions at Brookhaven. Using the classical work of Roberts (76) as a basis, Barad derives a theoretical equation for the stable case. Instead of considering a point source of infinite concentration as do Bosanquet and Sutton, Barad considers an area source of finite concentration. He also represents diffusion as a two-phase problem—a first or aerodynamic phase in the vicinity of the stack depending on stack parameters such as stack temperature and velocity and a second or meteorological phase beginning when the axis of the plume becomes horizontal.

Since Barad's assumptions concerning the nature of the source are much more realistic than the assumptions of Bosanquet and Sutton, his method has certain theoretical advantages for conditions close to the source and for very stable atmospheres. The Bosanquet and Sutton expressions are preferred for average and unstable conditions at distances not too close to the source. Further verification of all these equations by field tests is required.

Dispersion of Particulate Material. When the free falling speed of particulate matter being discharged cannot be neglected, the problem is somewhat complicated. In such instances—for example, the dispersion of fly ash from a powerhouse—material falls out of the smoke plume and accumulates about the countryside. The area dosages are of interest in these cases.

Bosanquet et al. (11) developed a theoretical method for determining the average rate of dust deposition by accounting for the free falling speeds of the dust fractions and the wind-direction climatology. Theoretically, the position of maximum deposition will get closer to the stack as the free falling speed of particulate matter increases. In verification of this method, Bosanquet found that the predicted results agreed with observation within a factor of 2. Bosanquet also points out (9) that increasing the stack height results in considerable improvement close to the source, but a mile or two away increasing the stack height has little effect.

The average dust-deposition rate F_b over a 45° sector downwind according to Bosanquet et al. (11) is given by the formula

$$F_b = 1.27 \frac{Wbp^2}{H^2} \left[\frac{(f/pu)(H/px)^{2+f/pu} e^{-(H/px)}}{\Gamma(1 + f/pu)} \right] \tag{5-26}$$

where W is the emission rate of the dust from a stack of effective height H (see Calculation of Effective Stack Height on page 5-33), b is the fraction of time that the wind blows toward the 45° sector, and f is the free falling velocity of the dust particle. Values of f can be obtained from Fig. 5-20. The other symbols are as used in previous equations. The gamma (Γ) functions of the values $(1 + f/pu)$ are given in Table 5-4. If f and u are in consistent units, and H and x in feet, then F_b has the

Table 5-4. Gamma (Γ) Functions of $(1 + f/pu)$

f/pu	$\Gamma(1 + f/pu)$	f/pu	$\Gamma(1 + f/pu)$	f/pu	$\Gamma(1 + f/pu)$	f/pu	$\Gamma(1 + f/pu)$
0.00	1.000	0.40	0.887	0.80	0.931	1.20	1.102
0.02	0.989	0.42	0.886	0.82	0.937	1.22	1.114
0.04	0.978	0.44	0.886	0.84	0.943	1.24	1.126
0.06	0.969	0.46	0.886	0.86	0.949	1.26	1.139
0.08	0.959	0.48	0.886	0.88	0.955	1.28	1.153
0.10	0.951	0.50	0.886	0.90	0.962	1.30	1.167
0.12	0.944	0.52	0.887	0.92	0.969	1.32	1.181
0.14	0.936	0.54	0.888	0.94	0.976	1.34	1.195
0.16	0.930	0.56	0.890	0.96	0.984	1.36	1.210
0.18	0.924	0.58	0.891	0.98	0.992	1.38	1.227
0.20	0.918	0.60	0.894	1.00	1.000	1.40	1.242
0.22	0.913	0.62	0.896	1.02	1.009	1.42	1.258
0.24	0.908	0.64	0.899	1.04	1.017	1.44	1.276
0.26	0.904	0.66	0.902	1.06	1.027	1.46	1.294
0.28	0.901	0.68	0.905	1.08	1.036	1.48	1.311
0.30	0.898	0.70	0.909	1.10	1.046	1.50	1.329
0.32	0.895	0.72	0.913	1.12	1.057		
0.34	0.892	0.74	0.917	1.14	1.067		
0.36	0.890	0.76	0.921	1.16	1.079		
0.38	0.889	0.78	0.926	1.18	1.090		

Derived from tables in R. S. Burington, "Handbook of Mathematical Tables and Formulas," 2 ed., Handbook Publishers, Sandusky, Ohio, 1940.

units of W per square foot. For example, if W is tons per month, then F_b is tons per square foot per month. To convert to tons per square mile per month, multiply F_b by 2.79×10^7.

The axial deposition rate F_a downwind from a stack is given by

$$F_a = 0.282 \frac{Wp}{H^2} \left[\frac{(f/pu)(H/px)^{2+f/pu} e^{-(H/px)}}{\Gamma(1 + f/pu)} \right] \tag{5-27}$$

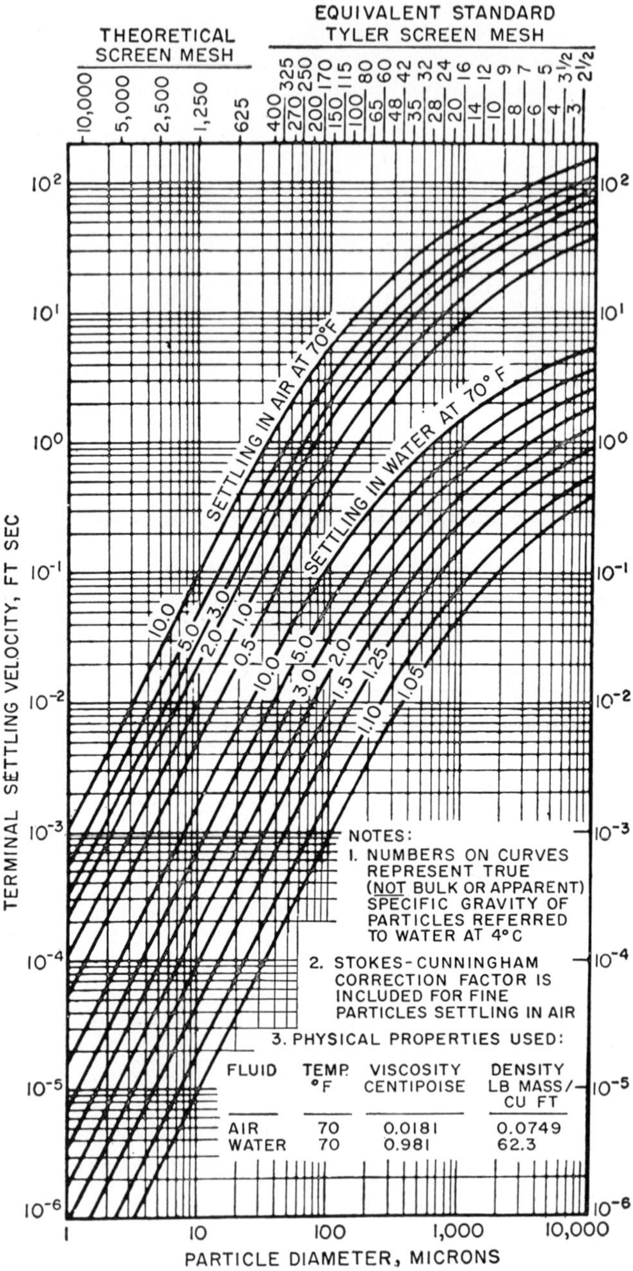

FIG. 5-20. Terminal velocities of spherical particles of different density settling in air and water at 70°F, under the action of gravity. (*By permission, from J. H. Perry, ed., "Chemical Engineers' Handbook," McGraw-Hill Book Company, Inc., New York,* 1950.)

Thus,

$$\frac{F_a}{F_b} = \frac{0.222}{p_b} \tag{5-28}$$

To compute the deposition rate of a stack dust of mixed particle sizes, such as fly ash, the weight–particle-size distribution must be known. Then the total dust-deposition rate is obtained by adding the rates for individual size ranges. For example, a stack emits 150 tons of dust per month. The size distribution and W for use in Eqs. (5-26) or (5-27) are given in the accompanying table.

Particle size range, μ	Mean particle size to compute f, μ	Per cent by weight of dust	W, tons/month, to compute F_a or F_b
0–10	5	40	60
10–20	15	30	45
20–40	30	20	30
40–80	60	10	15
		100	150

The value of F_a or F_b is computed for each mean particle size, and the sum is the total deposition rate.

Sutton's equations also have been modified to account for the finite settling velocity of aerosol particles (6,51). The formulas developed by Baron et al. (6) indicate several theoretical differences between diffusion of a gas and diffusion of aerosols from continuous elevated sources:

1. The maximum concentration downwind from a source of gas is practically independent of atmospheric turbulence (assuming Sutton C_z/C_y constant); but the maximum concentration of an aerosol at ground level and the maximum rate of deposition increase considerably with decreasing turbulence.

2. The maximum ground-level concentration downwind from a source of gas is inversely proportional to the square of the stack height, but the maximum concentration and deposition rate of an aerosol are approximately inversely proportional to the stack height.

In applying diffusion equations for aerosols, the possibility of agglomeration of the particulate matter should not be overlooked.

Calculation of Effective Stack Height. If stack gases are at an elevated temperature or have an appreciable exit velocity from the stack, the effective stack height is greater than the physical stack height because of buoyancy and jetting action. Several theoretical formulas are available for estimating the magnitudes of these effects. The problem is somewhat complicated because the temperature and velocity rise are not constant for given stack conditions but depend on the wind speed and atmospheric-temperature lapse rate.

Equations developed by Bosanquet et al. (11) for the rise caused by the velocity of the stack gases are as follows:

$$h_{vx} = h_{v\,\max}\left(1 - 0.8\,\frac{h_{v\,\max}}{x}\right) \tag{5-29}$$

where $x >$ about $2h_{v\,\max}$

$$h_{v\,\max} = \frac{4.77}{1 + 0.43\,\dfrac{u}{v_s}}\left(\frac{\sqrt{Q_{T1}v_s}}{u}\right) \tag{5-30}$$

Figure 5-21 is a graphical solution of Eq. (5-30) for the maximum velocity rise. To obtain the velocity rise at a distance x downwind from the stack, use Eq. (5-29).

Callaghan and Ruggeri (18) have proposed an expression for velocity rise that yields results comparable to those obtained with the Bosanquet method at low wind speed, but gives considerably higher values at higher wind speeds.

The Bosanquet (11) expression for maximum thermal rise of a plume is as follows:

$$h_{t\ \max} = 6.37g \frac{Q_{T_1}\Delta}{u^3 T_1} Z \qquad (5\text{-}31)$$

where

$$Z = \ln J^2 + \frac{2}{J} - 2 \qquad (5\text{-}32)$$

and

$$J = \frac{u^2}{\sqrt{Q_{T_1} v_s}} \left(0.43 \sqrt{\frac{T_1}{gG}} - 0.28 \frac{v_s}{g} \frac{T_1}{\Delta}\right) + 1 \qquad (5\text{-}33)$$

Equation (5-31) indicates the marked dependence of temperature rise on the wind speed. A twofold increase in wind speed results in approximately an eightfold reduction in temperature rise.

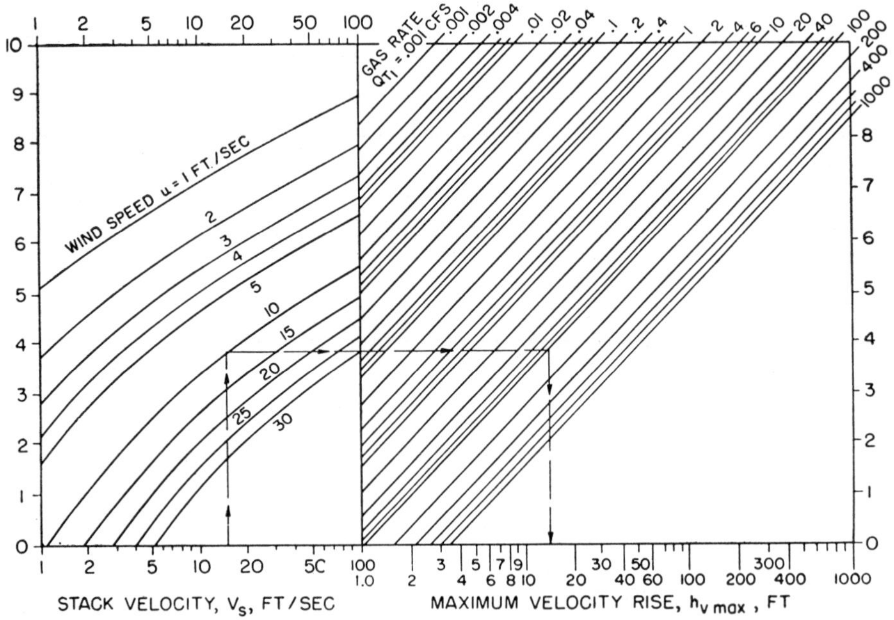

FIG. 5-21. Maximum velocity rise of plume (Bosanquet formula).

The effect of atmospheric-temperature lapse rate is accounted for in Eq. (5-33) by the term G, the gradient of potential atmospheric temperature. The potential temperature of dry air is defined as the temperature the air would assume if brought adiabatically from its actual pressure to a standard pressure of 1,000 mb. Thus, for an adiabatic lapse rate, $G = 0$; for an isothermal lapse rate, $G = 0.003°C/ft$; and for a strong inversion, $G = 0.006°C/ft$. The effect of increasing atmospheric stability is to decrease the maximum rise.

The path of the plume in an adiabatic atmosphere can be found by use of Eqs. (5-34) and (5-35) in conjunction with Fig. 5-22, showing the relation between X and Z:

$$h_{tx} = 6.37g \frac{Q_{T_1}\Delta}{u^3 T_1} Z \qquad (5\text{-}34)$$

$$x = 3.57 \frac{\sqrt{Q_{T_1} v_s}}{u} X \qquad (5\text{-}35)$$

Sutton (92) has also derived an approximate expression for thermal rise based on work by Schmidt (79).

In the calculation of effective stack heights for use in diffusion equations, some question exists concerning the use of maximum velocity and thermal rise. The calculated maximum rise may occur at a greater distance downwind from the source than the point for which the ground-level concentration is being calculated. Consequently, it is reasonable to use a rise somewhat smaller than the maximum in determining effective stack heights. Until further information is developed on this point, the velocity and thermal rise at a reasonable distance downwind (say 200 to 300 ft) or a fixed percentage of the maximum rise (say 50 to 75 per cent) should be used.

When the velocity rise and thermal rise are negligible, maximum ground-level concentrations occur at very low wind speeds [see Eqs. (5-11) and (5-13)]. When the velocity and/or thermal rise cannot be neglected, maximum ground-level concentrations occur at a somewhat higher wind speed because increasing wind speed decreases the effective stack height. The critical wind speed for a given stack can be estimated by calculating maximum ground-level concentrations for a number of different wind speeds, using Eq. (5-11) or (5-13) with appropriate effective stack heights.

Sample Calculation. The following sample problem is illustrative of many that are frequently encountered:

For a 100-ft-high 2-ft-diameter stack discharging 6,000 cfm of air containing 0.5 per cent by volume SO_2 at 200°F, calculate the maximum ground-level concentration of SO_2, the approximate location of the maximum, and the concentration at 5,000 ft downwind under different weather conditions.

Solution:

1. Estimate the velocity rise:

$$v_s = \frac{6{,}000}{60\pi \frac{2^2}{4}} = 32 \text{ fps}$$

assume that atmospheric temperature = 70°F (21°C)

$$Q_{T_1} = \frac{6{,}000}{60} \frac{530}{660} = 80 \text{ cfs}$$

From Fig. 5-21:

Wind Speed u, fps	*Max Vel. Rise* h_v max, ft
2	120
3	80
5	50
7	35
10	22
15	13
20	10

Equation (5-29) can be used to calculate the velocity rise at distance x. For this sample problem the total velocity rise will be used.

2. Estimate the thermal rise:

a. At an arbitrary distance of 1,000 ft in an adiabatic (unstable) atmosphere using Eqs. (5-34) and (5-35) and Fig. 5-22,

$$X = \frac{xu}{3.57\sqrt{Q_{T_1}v_s}} = \frac{1{,}000u}{3.57\sqrt{80 \times 32}} = 5.5u$$

$$h_{tx} = 6.37g\frac{Q_{T_1}\Delta}{u^3T_1}Z$$

$$= 6.37 \times 32.2\frac{80(200 - 70)5/9}{(70 - 32)5/9 + 273}\frac{Z}{u^3} = 4{,}030\frac{Z}{u^3}$$

Wind speed u, fps	X	Z (Fig. 5-22)	u^3	h_t at 1,000 ft
2	11	3.1	8	1,560
3	17	4.0	27	598
5	28	4.9	125	158
7	38	5.4	343	64
10	55	6.1	1,000	25
15	83	6.9	3,380	8
20	110	7.2	8,000	4

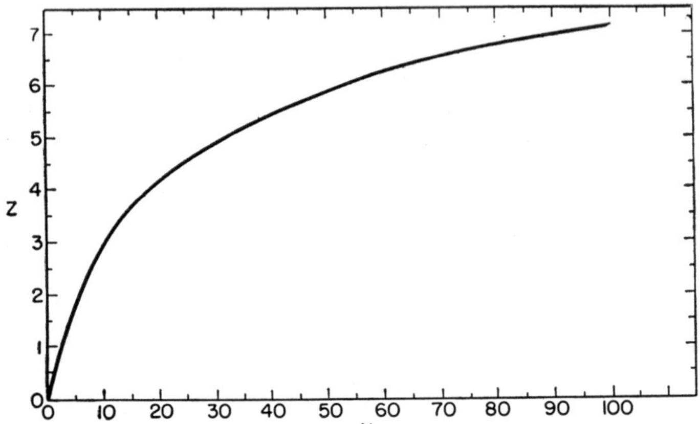

FIG. 5-22. Bosanquet thermal-rise parameters for an adiabatic atmosphere.

b. Estimate the maximum thermal rise with strong inversion when $G = 0.006°C/ft$. using Eqs. (5-31), (5-32), and (5-33):

$$J = \frac{u^2}{\sqrt{Q_{T_1} v_s}} \left(0.43 \sqrt{\frac{T_1}{gG}} - 0.28 \frac{v_s}{g} \frac{T_1}{\Delta} \right) + 1$$

$$J = \frac{u^2}{\sqrt{80 \times 32}} \left(0.43 \sqrt{\frac{294}{32.2 \times 0.006}} - 0.28 \frac{32}{32.2} \frac{294}{72.2} \right) + 1$$

$$J = \frac{u^2}{50.8} (16.8 - 1.1) + 1$$

$$J = 0.31u^2 + 1$$

$$Z = \ln J^2 + \frac{2}{J} - 2$$

$$h_{t\,\max} = 6.37g \frac{Q_{T_1}\Delta}{u^3 T_1} Z = 4{,}030 \frac{Z}{u^3}$$

Wind speed u, fps	J	Z	u^3	$h_{t\,\max}$, ft
2	2.24	0.51	8	257
3	3.79	1.18	27	176
5	8.75	2.56	125	83
7	16.2	3.70	343	43
10	32	4.97	1,000	20
15	71	6.53	3,380	8
20	125	7.62	8,000	4

Comparison of the thermal rises (h_t) for the strong inversion case with the preceding adiabatic case shows that, for all practical purposes, the effect of lapse rate is not particularly important in the resultant thermal rise at higher wind speeds (above 10 fps).

3. Estimate the effective stack heights H, maximum ground-level concentrations from Fig. 5-17, and critical wind speeds:

$$SO_2 \text{ emission rate at } 70°F = (6{,}000)(0.0055^{30}\!/_{660}) = 24 \text{ cfm}$$

assume $$H = h_s + 0.75(h_v + h_t)$$

and $$\frac{p}{q} = 0.5$$

a. For adiabatic atmosphere

Wind speed, fps	u, fpm	H, ft	C_{max}* (Fig. 5-17), ppm	C_{max} (see Art. 5.3.3)		
				30 min,† ppm	1–3 min, ppm	1–10 sec, ppm
2	120	1,360	0.006	0.012	0.12	0.6
3	180	610	0.03	0.04	0.4	2.0
5	300	256	0.16	0.13	1.3	6.5
7	420	174	0.36	0.20	2.0	10.0
10	600	135	0.59	0.24	2.4	12.0
15	900	116	0.80	0.21	2.1	10.5
20	1,200	110	0.89	0.18	1.8	9.0

* For $Q = 10$ cfm, $u = 100$ fpm.
† $[C_{max}\text{ (Fig. 5-17)}]\,\dfrac{24}{10} \times \dfrac{100}{u}$.

Thus, the critical wind speed is about 10 fps (7 mph) and $C_{max} = 0.24$ ppm.

b. For a strong inversion, $G = 0.006°C/ft$.

Wind speed, fps	u, fpm	H, ft	C_{max}* (Fig. 5-17), ppm	C_{max} (see Art. 5.3.3)		
				30 min,† ppm	1–3 min, ppm	1–10 sec, ppm
2	120	383	0.07	0.14	1.4	7.0
3	180	292	0.13	0.17	1.7	8.5
5	300	200	0.27	0.22	2.2	11.0
7	420	166	0.39	0.22	2.2	11.0
10	600	132	0.62	0.25	2.5	12.5
15	900	116	0.80	0.21	2.1	10.5
20	1,200	110	0.89	0.18	1.8	9.0

* For $Q = 10$ cfm, $u = 100$ fpm.
† $[C_{max}\text{ (Fig. 5-17)}]\,\dfrac{24}{10} \times \dfrac{100}{u}$.

Again the critical wind speed is about 10 fps (7 mph) and $C_{max} = 0.25$ ppm. Note that at the lower wind speeds ground-level concentrations are considerably higher with a strong inversion than with an adiabatic atmosphere.

4. Estimate the probable location of the maximum ground-level concentration and the concentration at 5,000 ft downwind. The normal range in distance to the maximum ground-level concentration is $5H$ for turbulent conditions to $20H$ for stable conditions. For average turbulence the distance is about $10H$.

Turbulence	Wind speed, fps	C_{max} 30 min, ppm	H, ft	x_{max}/H	x_{max}	$5{,}000/x_{max}$	(Fig. 5-18) C/C_{max}	C at 5,000 ft 30 min, ppm
Moderate	10	0.24	135	5	675	7.4	0.11	0.03
Average	10	0.245	134	10	1,350	3.7	0.28	0.07
Low	10	0.25	132	20	2,640	1.9	0.55	0.14

The contribution of SO_2 from the particular stack under consideration at a distance of 5,000 ft downwind would be a maximum at a wind speed of about 7 mph and would vary from 0.03 to 0.14 ppm with changes in atmospheric stability.

Line and Multiple Sources. The theoretical equations for a continuous line source of infinite extent are useful in estimating diffusion from a line of stacks and from some area sources. The rate of diffusion is less from a line source than from a point source because dilution is accomplished only laterally along the edges of the cloud. The Bosanquet (9) expression for ground-level concentration due to a line source at finite height is

$$C\ (\text{ppm}) = \frac{Q10^6}{pux} e^{-H/px} \tag{5-36}$$

where the emission rate Q is expressed as cfs per linear foot. The rate of decrease in concentration downwind is inversely proportional to the first power of the distance rather than the square of the distance as in Eq. (5-9). The maximum ground-level concentration occurs when

$$x_{max} = \frac{H}{p} \tag{5-37}$$

and has a value

$$C_{max}\ (\text{ppm}) = \frac{Q10^6}{euH} \tag{5-38}$$

Corresponding Sutton (91) equations for a continuous infinite line source are

$$C\ (\text{ppm}) = \frac{2Q10^6}{\sqrt{\pi}\, C_z u x \frac{2-n}{2}} e^{-H^2/C_z^2 x^{2-n}} \tag{5-39}$$

$$x_{max} = \left(\frac{2H^2}{C_z^2}\right)^{1/(2-n)} \approx \frac{\sqrt{2}\, H}{C_z} \tag{5-40}$$

$$C_{max}\ (\text{ppm}) = \frac{2Q10^6}{\sqrt{2\pi e}\, uH} \tag{5-41}$$

Johnstone, Winsche, and Smith (51) have applied these equations to dispersion of insecticide aerosols from continuous line sources at ground level.

The problem of pollution from multiple sources such as occur in a city or large factory is exceedingly complex, and knowledge concerning this aspect of meteorology is meager. In limited instances, the theoretical formulas for point and line sources can be used to obtain estimates, but other techniques must be resorted to if the number of sources becomes large. Hewson (47) has prepared an interesting review of the knowledge of such weather effects as wind speed and direction, atmospheric stability and its diurnal variation, and topography on pollution in and about cities. He points out that in United States and English cities the daily pollution maximum usually occurs between 6:30 and 9:00 A.M. and suggests that the cause may be the increase in turbulence and the resulting fumigation at this time of day. A secondary pollution maximum occurs in the late afternoon.

That this relation does not always hold is shown in Fig. 5-23. Here the diurnal variations of SO_2 concentration in Leicester, England (38), St. Louis, Mo. (81), Los Angeles, Calif. (17), and Kanawha City, W.Va. (101), are shown. Both Leicester (in winter) and Los Angeles (October) show high concentrations during midday and the afternoon and lowest concentrations in the early morning hours before dawn. St. Louis (winter) and Kanawha City show almost opposite results: high concentration in the early morning and low during the day. While it may be coincidental, it is interesting to note that both Leicester and Los Angeles are located on the western side of continents (maritime climate); St. Louis and Kanawha City are located far from the ocean (continental climate).

Davidson (24) concluded from studies in New York City that, disregarding wind direction and atmospheric stability, dust concentrations near the surface vary

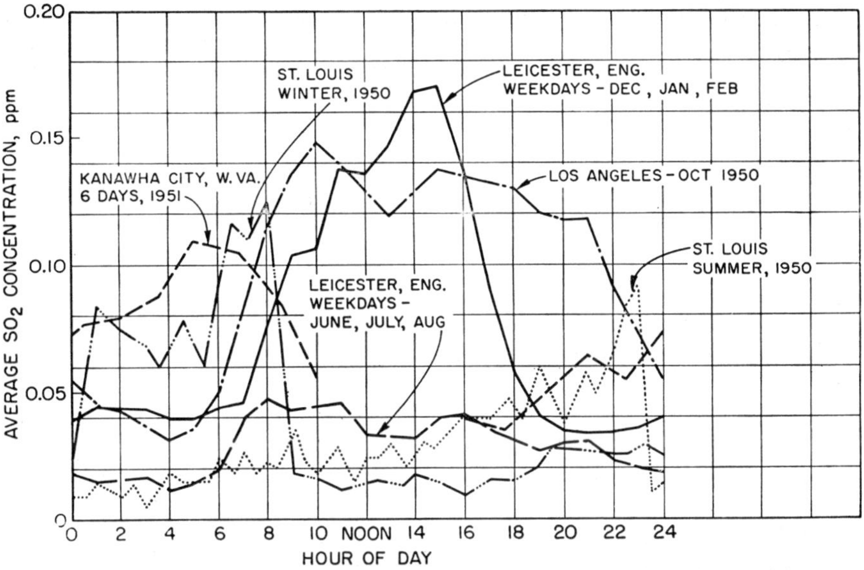

FIG. 5-23. Diurnal variation in sulfur dioxide concentrations for selected cities.

inversely as the square root of the wind speed. Using Davidson's data, Fletcher and Manos (30) made several interesting statistical analyses that yielded significant correlations, considering both wind speed and direction. Other investigators have indicated that surface-smoke concentration varies inversely as the first power of the wind speed. Studies of this sort are important in enabling us to anticipate such disasters as occurred at Donora, Pa., in 1948 and in the Meuse Valley of Belgium in 1930.

Little is known concerning the reduction in ground-level concentrations of pollution downwind from cities or other multiple sources. The survey made in Leicester, England (38), indicated that pollution diminishes as the distance in the interval 4 to 10 miles and as the square of the distance beyond about 10 miles.

5.5 PLANNING AND INTERPRETING AIR POLLUTION SURVEYS

The importance of including weather considerations in the planning of an air pollution survey and of correlating survey results with meteorology cannot be overemphasized. To neglect to do so is to ignore one of the prime variables in atmospheric pollution. By proper consideration of the weather factor, the time required to accomplish the purpose of a survey can be materially reduced and the results obtained will be of considerably more value than they would be otherwise.

Precise rules cannot be given for making air pollution surveys because the requirements and conditions are different for each one. However, certain general principles can be enumerated. In the first place, the design of a survey depends on its purpose. A survey to determine the general level of pollution in a city obviously has different requirements from those of a survey to determine the percentage reduction in discharge from a particular stack required to alleviate a nuisance condition.

5.5.1 Location of Samplers

In general, two different sampling techniques are used in making surveys—stationary sampling and mobile sampling. Different types of information are obtained in each case. With stationary sampling, continuous or intermittent samples are taken at the same locations over an appreciable length of time. The use of a number of dustfall jars at fixed locations throughout a city is an example of this technique. With mobile sampling, samples are taken at different distances upwind, crosswind, and downwind from a stack, plant, industrial area, or city. The locations of the sampling stations depend on wind direction and other weather conditions at the time of sampling.

Fixed-sample locations are of greatest value when the purpose of the survey is one of the following:

1. To determine the general level of pollution in a city or area
2. To provide a basis for the comparison of pollution levels in different sections of a city or in different cities
3. To determine improvement achieved over the years by increased pollution control (or deterioration through lack of it)
4. To determine the correlation between weather and different pollution levels to enable forecasts of air pollution and frequencies of the different levels to be made

In locating long-term fixed-sample stations, special consideration should be given to certain areas:

1. Areas where air pollution would be expected to be most critical, such as residential areas, gardens, parks, etc.
2. Areas of maximum air pollution complaints, to determine cause and justification
3. Areas of theoretical maximum and minimum concentrations of pollution
4. Areas having special topographic features

Care should always be taken to locate sampling stations in representative areas. The location of representative spots sometimes is very difficult and always is a matter of considerable judgment. When sampling stations are placed on the roofs of buildings, they should be located near the center of large, flat sections rather than near the edges where unpredictable eddies are formed. The location of stations in narrow passageways where anomalous wind velocities exist should also be avoided. When the objective is to determine general levels of pollution, all local sources should be noted and samplers located where they will not be unduly affected by them. A detailed description of the significance of deposit-gauge measurements and the importance of location is available (39).

A useful sampler of recent development is the directional dustfall collector used by the Battelle Memorial Institute (67). A different jar is uncovered when the wind is blowing from each major direction so that an analysis of the relative amounts of dirt in the various jars indicates the direction from which the wind was blowing when the major portion of the dustfall occurred. The development of similar equipment to provide directional air analyses for different constituents is to be anticipated. A continuous record of wind velocity at the same location as the directional sampler provides useful supplementary information.

Mobile sampling (selection of sampling location on the basis of the wind direction and atmospheric stability prevailing at the time the sample is taken as opposed to sampling at the same spots day after day) has certain advantages when the primary purpose of the survey is one of the following:

1. To determine the reduction in a specific discharge required to alleviate a nuisance condition
2. To determine the source of a particular air pollution problem
3. To determine the contribution of a specific stack, plant, or area to the general pollution of a region

In the last two instances, a survey utilizing both fixed and variable stations may be the most desirable.

When variable sampling locations are used, the objective is usually the determination of maximum short-period concentrations directly downwind from a particular source. The theory described in the preceding articles on stack meteorology can be used to advantage in locating the maximum ground-level concentration.

Locating a position directly downwind from a stack is often difficult because the region of maximum concentration is quite narrow and moves with turbulent shifting of the wind. The task is simplified when the plume is visible at ground level. Samples should be taken under a variety of atmospheric stability conditions. This is most easily accomplished by sampling during both day and night.

Data taken with each sample should include, as a minimum, the location (often most convenient to locate on a map), distance and direction from pollution source, wind direction and speed, time, time interval for collection of sample, and estimate of atmospheric stability from observation of wind gustiness, cloud types, general weather conditions, and character of smoke plumes. The time required in the collection of a sample is of particular significance in sampling for maximum concentrations because of the extreme temporal variation in ground-level concentration that occurs directly downwind from a stack. Samples should also be taken upwind from major pollution sources to determine background conditions and for comparison with downwind samples.

5.5.2 Types and Locations of Meteorological Instruments

Just as in determining the types and locations of air samplers, the types and locations of meteorological instruments depend on the purpose and area to be covered by the survey. Some meteorological measurements are required in nearly every case. In general, the rougher the terrain, the greater the number of weather-observing stations required to define sufficiently the microclimate of the area. The types of weather instruments used in air pollution work have been discussed in Art. 5.2.4, Meteorological Measurements. The purpose here is to indicate the usual minimum of weather observations required in conjunction with an air pollution survey.

The most important observations are of wind and atmospheric stability. In a survey of appreciable magnitude, a recording anemometer and wind vane at an elevation approximating that of the stack tops is indispensable. Estimates of atmospheric stability can be obtained from the character of the wind-direction recording (Lowry types) or from measurements of vertical wind or temperature gradients, utilizing an existing mast or tower or using a captive balloon. The temperature at times of sampling is often useful information, since it can sometimes be correlated with the residential heating load and, hence, with air pollution from this source. In addition to stationary weather instruments, a pocket compass and portable hand anemometer are helpful in recording surface winds at the time of sampling.

5.6 ANALYSIS AND INTERPRETATION OF DATA

By statistical and comparative means, much can be done to extend the usefulness of survey data. The interpretation of data is most often based on experience. In the case of municipal or similar air pollution surveys, for example, studies of findings of workers in other localities provide bases of comparison. In another instance, where the design of fly-ash collectors for a powerhouse stack is anticipated, representative samples of the existing ash must be obtained and its particle-size distribution carefully determined and statistically evaluated. The selected design is then based on previous experience of the types of equipment best suited economically, as well as technically, to collect the type of ash under consideration.

The method of evaluation depends primarily on the aim of the program. One must consider, for example, the data available, the representativeness of the number of samples taken, the required accuracy of the results, the proper locations of sampling stations in a community survey, or a representative location for stack sampling. It is often stated that the more data or station locations, the better the test, but such philosophy may lead to unnecessary or repetitious work and is not necessarily true. Actually, the amount of data needed depends on the number of variables and the degree of control over their variation.

In air pollution work, data evaluation is divisible into two broad categories: comparative or statistical. These are not mutually exclusive since the spectrum of requirements may find use for both. As an illustration of statistical techniques, a series of air pollutant concentrations may be evaluated in such a manner that the probability of nuisance or toxic values can be determined. In this way, a better estimation of corrective measures can be made, either to eliminate nuisance or meet required standards.

Comparative evaluation is more widely used. It deals with degrees of similarities and differences between values. Of course, in comparative evaluation, statistical techniques may be required to arrange data in forms more readily subject to comparison. Examples of this technique are the comparison of pollution surveys during different periods of the year, comparison of improvements due to process changes or corrective measures made to reduce the emission of contaminants, and the comparison of findings of community surveys with those of other localities to show the relative degree of pollution.

5.6.1 Mathematical and Statistical Techniques

All treatment of data involves some sort of mathematical or statistical technique. It is the purpose here to examine some of the more common mathematical methods used in air pollution analyses and to suggest the more frequent use of still others.

Averages and Tests of Significance. The most common average in use is the arithmetic mean which, for a series of values, is the sum of the values divided by the number of values. This type of data summarization is used widely because of its ease of computation and facility of grasp. Unfortunately, in certain problems of air pollution work, the mean is insufficient in itself since it tells nothing about the range or distribution of data. The average should be accompanied by the range of the data or, in the case of symmetrically distributed data, by the standard deviation (root mean square).

Other types of "averages" occasionally in use are the median and mode. In a series of observations, the value which divides the data in half is the median. Thus 50 per cent of the values are greater than and 50 per cent are less than the median value. The mode of a series of observations is that value which occurs most frequently.

The use of some sort of statistical significance test such as "Student's" t test can prove valuable in comparing means. In such a test, the probability is determined that the difference between two means could be due to chance. According to Velz (97), "any difference in results, no matter how great, can always be ascribed to chance, but a point is reached where the probability that the difference is due to chance becomes so small that the investigator prefers to ascribe the difference to other causes. The decision that the difference is not due to chance and the subsequent search for the underlying *cause* of the difference is completely outside the realm of statistical technique and is entirely a matter of professional judgment of the specialist in the field. The statistical method is only a tool, not a substitute for reasoning."

Dustfall data for Cincinnati, Ohio, collected by the Kettering Laboratory, University of Cincinnati (19), offer an illustration of the t-test technique. Values for two stations, one in a commercial and industrial area, the other in a residential area, are shown in Tables 5-5 and 5-6 for the winter periods of 1946 through 1952.

Examination of the data for the industrial station (Table 5-5) indicates that the winter periods of 1946–1947, 1947–1948, and 1948–1949 appear, in the main, to be higher in dustfall than do the later three winter periods, 1949–1950, 1950–1951, and 1951–1952. While the mean dustfall is 158.7 for the earlier period and 89.7 tons/sq mile/month for the later period—a 43 per cent reduction—the fact that both periods show measurements in range of 100 to 130 tons/sq mile/month may raise a question as to the significance of the difference in the mean values. Student's t test offers a means of testing this significance. The method is briefly as follows.

If $X_1, X_2, X_3, \ldots, X_j$ represent the individual values of a sample of n_x values and $Y_1, Y_2, Y_3, \ldots, Y_j$ represent the values of a sample of n_y values drawn from populations of X and Y, the problem is to determine whether $\bar{X} = \Sigma X/n_x$ is significantly different from $\bar{Y} = \Sigma Y/n_y$. ΣX and ΣY are the sums of all values of X and Y, respectively. Significant difference between $\bar{X}$ and $\bar{Y}$ indicates that the probability is low that the two samples could have been drawn from the same population.

The sum of the squares of the differences between the individual values of X and $\bar{X}$ is represented by $\Sigma(X - \bar{X})^2$. Similarly, $\Sigma(Y - \bar{Y})^2$ is the expression for the Y values. The value t is given by

$$t = \frac{\bar{X} - \bar{Y}}{\sqrt{\dfrac{\Sigma(X - \bar{X})^2 + \Sigma(Y - Y)^2}{n_x + n_y - 2} + \left(\dfrac{1}{n_x} + \dfrac{1}{n_y}\right)}} \tag{5-42}$$

Table 5-5. Application of t (Significance) Test to Winter Dustfall Data for a Station in a Commercial and Industrial Area of Cincinnati, Ohio

Month	Dustfall, tons/sq mile/month					
	1946–1947	1947–1948	1948–1949	1949–1950	1950–1951	1951–1952
November	132.0	124.0	146.5	111.5	94.2	80.5
December	146.7	146.0	257.0	101.0	104.0	71.0
January	125.0	179.0	102.3	76.8	128.0	74.0
February	321.5	187.0	127.0	79.4	70.3	85.5

Mean $= \bar{X} = 158.7$ tons/sq mile/month
$\Sigma(X - \bar{X})^2 = 23{,}866.33$
Number of samples $= 12 = n_x$

Mean $= \bar{Y} = 89.7$ tons/sq mile/month
$\Sigma(Y - \bar{Y})^2 = 3{,}641.88$
Number of samples $= 12 = n_y$

Reduction in 1949–1952 versus 1946–1949 = 43%

$$t = \frac{\bar{X} - \bar{Y}}{\sqrt{\frac{\Sigma(X - \bar{X})^2 + \Sigma(Y - \bar{Y})^2}{n_x + n_y - 2}\left(\frac{1}{n_x} + \frac{1}{n_y}\right)}} = \frac{158.7 - 89.7}{\sqrt{\frac{23{,}866.33 + 3{,}641.88}{12 + 12 - 2}\left(\frac{1}{12} + \frac{1}{12}\right)}} = 4.75$$

Tables of t show that, for 22° of freedom, $t = 2.82$ for $P = 0.01$ (Table 5-7). Thus the difference between $\bar{X}$ and $\bar{Y}$ is highly significant.

Table 5-6. Application of t Test to Winter Dustfall Data for a Station in a Residential Area of Cincinnati, Ohio

Month	Dustfall, tons/sq mile/month					
	1946–1947	1947–1948	1948–1949	1949–1950	1950–1951	1951–1952
November	7.4	4.7	23.8	3.7	2.0	10.7
December	5.7	10.4	....	5.2	1.9	8.9
January	9.6	9.3	6.6	5.3	3.3	9.4
February	13.4	11.1	5.2	4.6	4.5	3.7

Mean = $\bar{X}$ = 9.7 tons/sq mile/month Mean = $\bar{Y}$ = 6.3 tons/sq mile/month
$\Sigma(X - \bar{X})^2 = 291.27$ $\Sigma(Y - \bar{Y})^2 = 110.64$
$n = 11 = n_x$ $n = 12 = n_y$

Reduction in 1949–1952 versus 1946–1949 = 35%

$$t = \frac{9.7 - 6.3}{\sqrt{\frac{291.27 + 110.64}{11 + 12 - 2}\left(\frac{1}{11} + \frac{1}{12}\right)}} = 1.86$$

For 21° of freedom, when $t = 1.86$, P is between 0.05 and 0.1 (Table 5-7). Thus the difference between $\bar{X}$ and $\bar{Y}$ is not significant and may be positive or negative, depending on whether $\bar{X}$ or $\bar{Y}$ is larger.

The value $n_x + n_y - 2$ gives the number of degrees of freedom. Table 5-7 shows values of t for various degrees; the sign of t is neglected in using this table. This table is arranged to give the probability P that t, at various degrees of freedom, will occur if the two samples whose means are $\bar{X}$ and $\bar{Y}$ are drawn from the same normal population. If P is low, say 0.05, then the chance is 19 to 1 against $\bar{X}$ and $\bar{Y}$ coming from the same population. Putting it another way, the chances are 19 to 1 that the value of $\bar{X}$ is significantly different from $\bar{Y}$. If t is such that $P = 0.01$, the chances are 99 to 1 against $\bar{X}$ and $\bar{Y}$ coming from the same population, i.e., the difference in $\bar{X}$ and $\bar{Y}$ is highly significant. The value of $P = 0.05$ is widely adopted by statisticians to designate the threshold of significance. Lower values of P are significant.

A fuller treatment of the t test and its uses may be obtained in many of the standard introductory texts on statistical methods, e.g., Refs. 26, 31, and 48.

It should be noted that statistical significance and practical significance may not be synonymous. For example, dustfall in an area may be found to be 25 per cent lower in one period than in another, and this difference may prove to be statistically significant. But from a practical standpoint, the difference in dustfall effects as noted by reduced complaints, reduced laundry bills, increased days of improved visibility near airports, etc., may not be significant. Thus the practical and realistic application of statistical evaluation must be considered.

Table 5-5 shows an application of the t test in comparing the average winter dustfall at the industrial station of Cincinnati for the periods 1946 to 1949 and 1949 to 1952. For convenience, the values for the period 1946 to 1949 are designated as X and for 1949 to 1952 as Y. The value t computed in Table 5-5 is 4.75. According to tables of t distribution, the probability is less than 0.01 that t will be as large as 4.75 for 22° of freedom or that these samples could have been drawn from the same population. Thus it is concluded that the difference between the mean dustfall for 1946 to 1949 and 1949 to 1952 is highly significant for the station in the commercial and industrial area.

Table 5-7. Table of t[a]

Degrees of freedom	Values of t for			Degrees of freedom	Values of t for		
	P = 0.1	0.05	0.01		P = 0.1	0.05	0.01
1	6.31	12.7	63.7	15	1.75	2.13	2.95
2	2.92	4.30	9.93	16	1.75	2.12	2.92
3	2.35	3.18	5.84	18	1.73	2.10	2.88
4	2.13	2.78	4.60	20	1.72	2.09	2.85
5	2.01	2.57	4.03	22	1.72	2.07	2.82
6	1.94	2.45	3.71	24	1.71	2.06	2.80
7	1.89	2.36	3.50	26	1.71	2.06	2.78
8	1.86	2.31	3.36	28	1.70	2.05	2.76
9	1.83	2.26	3.25	30	1.70	2.04	2.75
10	1.81	2.23	3.17	40	1.68	2.02	2.70
11	1.80	2.20	3.11	60	1.67	2.00	2.65
12	1.78	2.18	3.06	120	1.66	1.98	2.61
13	1.77	2.16	3.01	∞	1.64	1.96	2.58
14	1.76	2.14	2.98				

[a] Taken from more complete tables in H. A. Freeman, "Industrial Statistics," pp. 170–173, John Wiley & Sons, Inc., New York, 1942; and W. J. Dixon and J. F. Massey, Jr., "Introduction to Statistical Analysis," McGraw-Hill Book Company, Inc., New York, 1951.

Table 5-6 contains a similar calculation, comparing winter dustfall at a residential station in Cincinnati for the periods 1946 to 1949 and 1949 to 1952. The mean for 1946 to 1949 was 9.7 tons/sq mile/month and for 1949 to 1952 was 6.3, a 35 per cent reduction. The value of t is 1.86, which for 21° of freedom is not significant, i.e., P is greater than 0.05; statistically, no significance can be attached to the difference in the means of winter dustfall in the residential areas for the two periods.

Since air pollution phenomena represent dynamic conditions, a question can be raised concerning the significance of the "average" in such a situation. The air which transports pollutants is never static, but varies in both horizontal and vertical direction and speed, as well as in stability which determines the extent of vertical mixing of adjacent layers. One has merely to observe the smoke trail from a chimney to see how variable is the smoke density at a point on the ground. To quote from Gosline (37):

"Downwind from a stack, it is seldom, if ever, that there is a contamination at a particular point present all the time. This means that a particular point is alternately exposed to contaminated air and uncontaminated air. The concentration at that point when the contaminant-bearing eddies are present is higher than would be indicated by the time-average concentrations at the same point. Moreover, within the time period that the eddy is present, there are rapid fluctuations in concentration, and within the eddy there are peak concentrations. Consequently, it is important to note that there are three concentrations of concern downwind from the stack: First, is the time average concentration at the point; second, is the average concentration within the eddy at the point when the contaminant-bearing eddy is present at that point; and third, is the absolute peak concentration of a few seconds duration that will be experienced at that point."

The foregoing illustrates that "average" in air pollution phenomena is a concept which must be carefully defined with respect to other dispersion characteristics. For example, in situations involving sensory nuisance problems, the time-average concentration at a given point may have no significance or be completely misleading. Yet the average for the contaminant-bearing eddy or the peak concentration may become important nuisance values.

In view of these various significant "averages," sampling time in air analyses becomes important. The ideal analytical method for a particular atmospheric contaminant is one which continuously gives instantaneous values. By such methods, integrated time-average, eddy-average, or peak concentrations can be observed or computed. Unfortunately, such methods are available for only a few substances and usually involve physical measurements such as light absorption. Usually, air sampling involves collection of contaminants over periods of several minutes, several

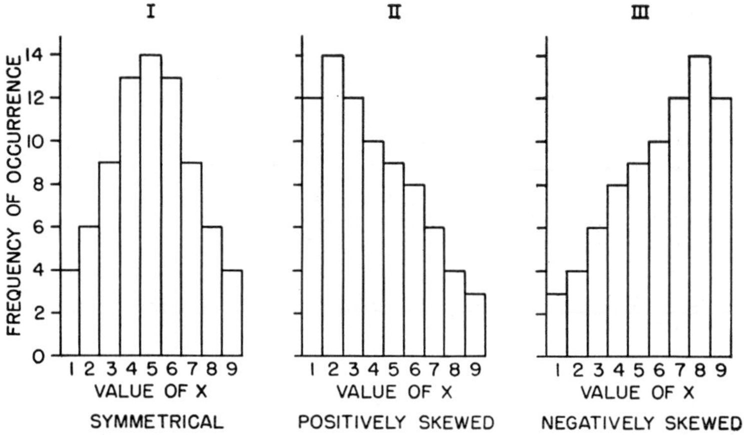

FIG. 5-24. Histograms of fundamental types of frequency distribution.

hours, or even days in certain instances. Thus the concentrations found will already be time-average or, at best, eddy-average values.

Probability Analyses. Probability analyses can frequently be used to advantage in pollution evaluation. This statistical tool is essentially a method of studying the frequency of occurrence of events. Its use, for example, is of fundamental importance in the design of mechanical dust collectors where distribution or frequency of particles of given sizes influences the sizes and capacities of various types of collectors. This type of analysis can be applied in other ways in air pollution work.

The following will but briefly outline the basis and working tools of these types of analyses. Table 5-8, and Fig. 5-24 plotted from it, show three types of distribution. Type I is the symmetrical distribution which, in certain instances, is called the "normal" or Gaussian distribution. Here the frequency of occurrence (ordinate of Fig. 5-24) of values X (abscissa) show an equal occurrence of the values on either side of the most frequently occurring or peak value, the mode. The median is that value dividing the number of cases into two equal parts. The mean (arithmetic average) is obtained by dividing the sum of the values by the total number of values. In a symmetrical distribution, the mean, median, and mode are numerically equal. Symmetrical distributions are likely to be encountered in errors of observation, i.e., by chance; too great and too small values due to observational errors are equally likely to occur.

Table 5-8. Examples of Frequency Distributions

Class X	I Symmetrical		II Positively skewed		III Negatively skewed	
	Number	Cumulative percentage	Number	Cumulative percentage	Number	Cumulative percentage
1	4	5.1	12	15.4	3	3.8
2	6	12.8	14	33.3	4	9.0
3	9	24.4	12	48.7	6	16.7
4	13	41.1	10	61.5	8	26.9
5	14	59.0	9	73.1	9	38.5
6	13	75.7	8	83.3	10	51.3
7	9	87.3	6	91.0	12	66.7
8	6	95.0	4	96.2	14	84.6
9	4	100.0	3	100.0	12	100.0
Total (n)	78		78		78	
Mean = $\bar{X}$	5		4.0		6.0	
Mode = Mo	5		2		8	
Median = Mi	5		3.6		6.3	

A further example of a symmetrical distribution is the spread of material downwind from an elevated point source. Unless influenced by particular terrain or building effects, the ground-level concentration of material downwind will be highest at the center of the dispersion cloud and decrease in a symmetrical manner along a line perpendicular to the mean wind direction.

Type II is known as a positively skewed distribution and is characterized by a longer "tail" toward higher values. Here the mean is greater than the median, which in turn is greater than the mode. Type III is negatively skewed, with the "tail" toward lower values. In this case the mean is less than the median, the latter being less than the mode.

Positively skewed (type II) distributions are characteristic of naturally occurring phenomena such as rainfall intensity, wind force, and stream flow. In air pollution phenomena, a good many data are of this type. Dustfall and atmospheric particulate and gaseous concentrations are typical examples.

Negatively skewed (type III) distributions are less frequently encountered in pollution work. In certain cases, the distribution of fly-ash particle sizes, on a weight basis, may show a greater proportion in the larger size ranges. On a particle-count basis, on the other hand, type II is more representative, i.e., the greater number of individual particles is frequently found in the small-size range.

A useful method of presenting frequency data is on probability paper. This is shown in Fig. 5-25. Here the values X are plotted as ordinates, while the lower abscissa scale is the frequency of occurrence of X on a cumulative basis (data from Table 5-8). Since this type of graph assumes a continuous series of values, the points are plotted so that each value of X covers the interval $X - \frac{1}{2}$ unit to $X + \frac{1}{2}$ unit. Thus, for example, in type I, Table 5-8, the cumulative frequency column indicates that 59 per cent of the values of X are 5 or less. Actually, since a value of 5 units covers the interval 4.50 to 5.49, X of less than 5.50 occurs for 59 per cent of the values. Therefore, the cumulative frequencies in Fig. 5-24 are plotted on the $X + \frac{1}{2}$-unit values.

The symmetrical distribution of type I in Fig. 5-25 appears, essentially, as a straight line over most of the curve. It should be noted that 100 per cent and 0 per cent are at $+\infty$ and $-\infty$, respectively. Since these points cannot be plotted, the curve rapidly becomes asymptotic at $X + 0.5$ and 9.5. A true straight line cannot be obtained for a symmetrical distribution unless there is an infinite number of X values, without finite limits such as 0 and 9.5 in the example cited. The median value 5 occurs at the 50 per cent line. All symmetrical distributions then will be a family of straight lines with the median at 50 per cent on the abscissa.

The skewed distributions, types II and III, appear in Fig. 5-25 as nonlinear curves. The curve for a positively skewed distribution (type II) shows the mean to the right of the 50 per cent line, while the negatively skewed distribution shows the mean to the left of the 50 per cent line.

Gumbel (40,41) has developed a standard skewed distribution which simplifies many statistical problems dealing with a series of extreme values. Gumbel's standard skewed distribution is of the positive type and has proved valuable for plotting naturally occurring extreme phenomena such as floods and droughts. A special

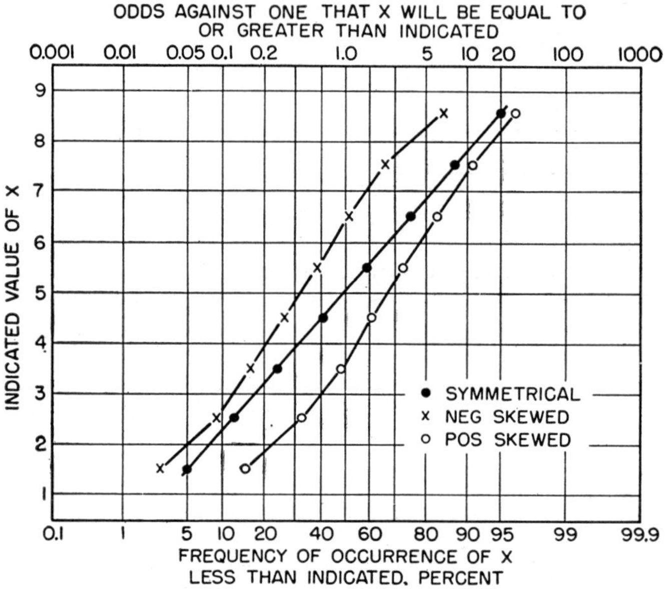

FIG. 5-25. Fundamental types of frequency distribution plotted on probability paper.

probability paper based on this distribution such that skewed data plot as straight lines is discussed by Powell (73) and Velz (98). In air pollution work, a mathematical tool such as this should be useful in working problems of maximum contamination frequencies, smog occurrence, etc.

A plot such as that in Fig. 5-25 can be used to obtain the probability of certain values occurring. The abscissa scale at the top shows the odds against one that X will be equal to or greater than the indicated ordinate value. (Note: If the bottom abscissa scale is plotted as frequency of occurrence of X greater than indicated, then the top scale becomes the odds of X being equal to or less than the indicated ordinate value.)

The foregoing ideas will now be applied to a particular example of an air pollution analysis. Setterstrom and Zimmerman (83) published a list of occurrences of mean and maximum concentrations of SO_2 at the Boyce Thompson Institute for almost the entire period Nov. 1, 1936, to Nov. 1, 1937. A Thomas autometer was used for analysis. The results represent a total sampling time of 2,700 hr. The individual sampling periods were of unequal duration, ranging from 4.3 to 128 hr, with an average sampling period of 19 hr. While this variation in sampling periods will somewhat alter the statistical evaluation, the data present an excellent example of the application of this type of analysis.

Table 5-9 presents a summary of the average SO_2 data in 0.025-ppm class intervals. It will be noted that the cumulative frequency (last column) was computed on the basis of one more than the total number of analyses. This is a statistical device whereby the last point may be plotted on probability graph paper without appreciably affecting the curve for a large finite number of individual cases. Table 5-10 is a similar compilation for the maximum SO_2 concentration in each sampling period.

Table 5-9. Average SO_2 Concentrations at Boyce Thompson Institute

X Avg. conc. during each period, ppm	Number of occurrences, f	Frequency of occurrence, %	Cumulative frequency of occurrence, $\Sigma_0^{n+1} fx$
0.000–0.024	1	0.7	0.7
0.025–0.049	22	15.7	16.3
0.050–0.074	61	43.6	59.6
0.075–0.099	36	25.7	85.1
0.100–0.124	12	8.6	93.6
0.125–0.149	8	5.7	99.3
Total (n)........	140	100.0	

Mean = 0.073.
Mode in interval = 0.050 to 0.074 ppm.
Median in interval = 0.050 to 0.074 ppm.

Table 5-10. Maximum SO_2 Concentrations at Boyce Thompson Institute

X Max conc. during each period, ppm	Number of occurrences, f	Frequency of occurrence, %	Cumulative frequency of occurrence, $\Sigma_0^{n+1} fx$
0.00–0.09	0	0	0
0.10–0.14	47	33.6	33.6
0.15–0.19	35	25.0	57.2
0.20–0.24	19	13.6	71.2
0.25–0.29	13	9.3	80.9
0.30–0.34	12	8.6	89.4
0.35–0.39	6	4.3	93.6
0.40–0.44	2	1.4	95.0
0.45–0.49	1	0.7	95.7
0.50–0.54	2	1.4	97.2
0.55–0.59	1	0.7	97.9
0.65–0.69	1	0.7	98.6
0.75–0.79	1	0.7	99.3
Total (n)........	140	100.0	

Mean = 0.21.
Mode in interval = 0.10 to 0.14 ppm.
Median in interval = 0.15 to 0.19 ppm.

These data are plotted in Fig. 5-26. Both curves indicate a positive skewness since the mean value for each is to the right of the median (50 per cent value). The probability of occurrence of particular values of SO_2 can now be obtained. For example, the chances are more than 1,000 to 1 against the average SO_2 concentration exceeding 0.2 ppm. The probability against the maximum concentration exceeding 1 ppm is also over 1,000 to 1. The minimum value of maximum SO_2 concentration reported by the authors is 0.10 ppm. The odds are about 7 to 1 against the average

SO_2 concentration exceeding this lower maximum value. These interpretations are predicated on essentially static sources of pollution. They must be modified if the sources of pollution change, as, for example, by a rapid increase in use of natural gas or lower sulfur coal in the area either for reasons of economy or because of ordinances.

Analyses such as the foregoing can be applied to other phases of air pollution. These include problems involving the probability of occurrence of nuisance or dangerous concentrations; the relationship of pollution and pollution climatology, such as inversion frequencies, fog frequencies, or persistence of certain wind directions which carry pollution to populated areas; and the design of pollution-abatement equipment in the light of process or fume-discharge variability. Reference is made to Gosline (37) who applied this type of analysis to the relationship of peak ground-level concentrations of gas fumes from short stacks to atmospheric turbulence and wind velocity.

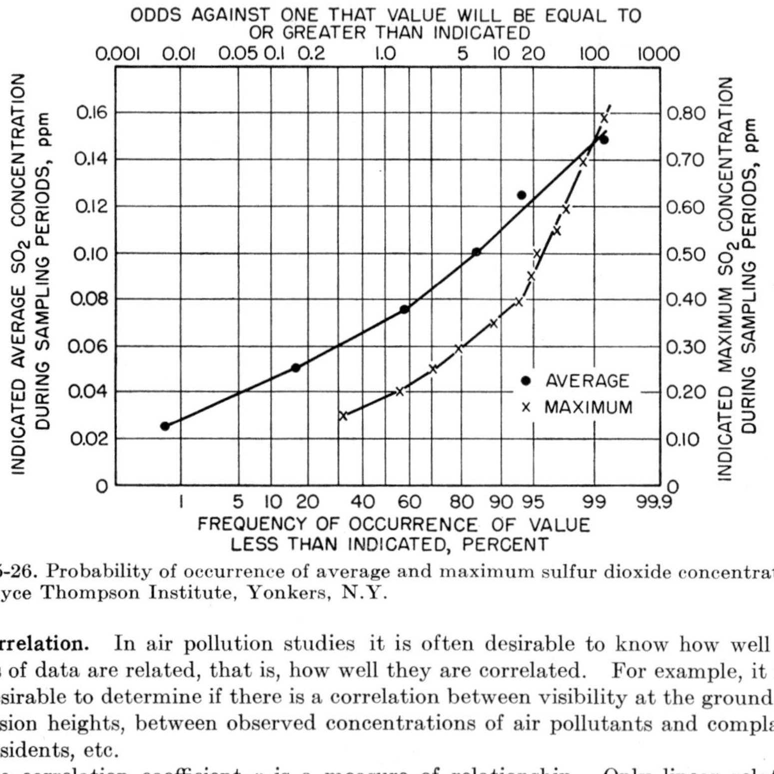

FIG. 5-26. Probability of occurrence of average and maximum sulfur dioxide concentrations at Boyce Thompson Institute, Yonkers, N.Y.

Correlation. In air pollution studies it is often desirable to know how well two series of data are related, that is, how well they are correlated. For example, it may be desirable to determine if there is a correlation between visibility at the ground and inversion heights, between observed concentrations of air pollutants and complaints by residents, etc.

The correlation coefficient r is a measure of relationship. Only linear relationships, i.e., those which can be represented by straight lines, will be discussed here. For other types of relationships, refer to several standard statistical texts.

The correlation coefficient is computed from the following equations:

$$r = \frac{\dfrac{\Sigma XY}{n_{xy}} - \dfrac{\Sigma X}{n_{xy}}\dfrac{\Sigma Y}{n_{xy}}}{\sqrt{\left[\dfrac{\Sigma X^2}{n_{xy}} - \left(\dfrac{\Sigma X}{n_{xy}}\right)^2\right]\left[\dfrac{\Sigma Y^2}{n_{xy}} - \left(\dfrac{\Sigma Y}{n_{xy}}\right)^2\right]}} \quad (5\text{-}43)$$

or

$$r = \frac{\Sigma XY - n_{xy}\bar{X}\bar{Y}}{\sqrt{(\Sigma X^2 - n_{xy}\bar{X}^2)(\Sigma Y^2 - n_{xy}\bar{Y}^2)}} \quad (5\text{-}44)$$

For perfect direct correlation, $r = +1$, i.e., where X increases, Y increases proportionately. For perfect inverse correlation, $r = -1$, i.e., when X increases, Y decreases in an inverse proportion. When $r = 0$, no correlation exists, i.e., X and Y are related only in a random fashion.

Table 5-11 illustrates the calculation to determine the correlation of dustfall and suspended dust based on data for a commercial and industrial location in Cincinnati (19). A value of $r = 0.28$ results. When n_{xy} is small, then r must be corrected for the size of the sample by

$$r_c^2 = 1 - (1 - r^2)\left(\frac{n_{xy} - 1}{n_{xy} - 2}\right) \tag{5-45}$$

For the illustration in Table 5-11, $r_c = 0.23$.

Table 5-11. Correlation Coefficient r for Relation of Dustfall and Suspended Dust for a Commercial and Industrial Station, Cincinnati, Ohio

Month	1948		1949		1950		1951	
	S.D.	D.F.	S.D.	D.F.	S.D.	D.F.	S.D.	D.F.
January	3.74	179.0	3.92	102.3	6.06	76.8	5.72	128.0
February	9.40	187.0	8.40	127.0	6.20	79.4		
March	5.40	175.0	....	.....	4.72	57.0		
April	3.26	159.5	4.10	120.0	2.06	126.5	6.72	106.0
May	1.77	149.0	3.10	122.5	5.20	67.3	3.58	103.0
June	4.84	206.0	3.10	90.5	2.46	81.5	2.50	80.0
July	2.78	127.0	4.07	82.5	2.48	87.0	3.36	69.0
August	2.06	155.0	2.24	106.0	....	.....	1.72	77.0
September	3.46	110.0	2.02	118.0	5.75	60.3	2.34	86.0
October	7.07	152.0	....	.....	....	.....	1.19	89.0
November	5.00	146.5	1.21	111.5	1.68	94.2		
December	5.25	257.0	7.30	101.0				

S.D. = suspended dust, mg/10 cu m = X
D.F. = dustfall, tons/sq mile/month = Y

$$n_{xy} = 39 \text{ pairs of data} = n$$

$$r = \frac{\dfrac{\Sigma XY}{n} - \dfrac{\Sigma X}{n}\dfrac{\Sigma Y}{n}}{\sqrt{\left[\dfrac{\Sigma X^2}{n} - \left(\dfrac{\Sigma X}{n}\right)^2\right]\left[\dfrac{\Sigma Y^2}{n} - \left(\dfrac{\Sigma Y}{n}\right)^2\right]}} \qquad \text{[Eq. (5-43)]}$$

$\Sigma XY = 19{,}109.496$; $\Sigma XY/n = 489.987$; $\Sigma X = 157.21$; $(\Sigma X)/n = 4.031$; $(\Sigma X/n)^2 = 16.249$; $\Sigma Y = 4{,}490.5$; $(\Sigma Y)/n = 115.141$; $(\Sigma Y/n)^2 = 13{,}257.450$; $\Sigma X^2 = 792.454$; $(\Sigma X^2)/n = 20.319$; $\Sigma Y^2 = 598{,}476.60$; $(\Sigma Y^2)/n = 15{,}345.553$

$$r = \frac{489.987 - (4.031)(115.141)}{\sqrt{(20.319 - 16{,}249)(15{,}345.553 - 13{,}257.450)}} = 0.28$$

According to Gavett (32), the degrees of near-linear relationship between variables in terms of the coefficient of correlation are arbitrarily given as:

Value of r_c	*Degree of Relationship*
<0.30	Practically none
0.30–0.50	Moderate
0.50–0.75	Marked
0.75–0.90	High
>0.90	Very high

5.6.2 Atmospheric Analyses

Degree of Atmospheric Pollution. The pollution of a particular area, be it a large community area affected by effluents of large numbers of various municipal or industrial sources or a neighborhood affected by a single source, is judged on the basis of experience. In attacking an air pollution situation, whether due to a large community or caused by a single stack, it is necessary to establish the degree of pollution present. This, of course, may be a particularly difficult task. One need only cite the example of Los Angeles, where years of effort and millions of dollars have been spent in establishing the degree of existing pollution and its causes. As data are accumulated on pollution, the degree of pollution can be established only on the basis of experience. Comparisons can be made with published data on various atmospheric contaminants. Among these are dustfall, suspended particulate material, low visibility or fog frequency, and chemical constituents or physiological effects. It is for these comparative purposes that original data secured in atmospheric test programs ought to be published. Such should include not only average and range values, but frequencies of occurrence, together with associated meteorological and other pertinent information.

It is realized that a variety of factors influence the findings in individual localities. Among these can be cited type of community, type and kinds of industrial establishments, the meteorology or pollution climatology, and terrain.

Adams (1) has summarized the major portion of the United States literature dealing with concentrations of air contaminants found in investigations of air pollution situations (see also Ref. 50). The concentrations given are those which have been found. It should not be inferred that these are necessarily deleterious from a health standpoint. Nevertheless, these concentrations as well as future work can serve as the basis for evaluating degrees of pollution.

In attacking a specific problem, the properties of contaminants which cause difficulty must be considered. Nuisances in a locality may be due to odors, irritating effects on nose and throat membranes, reduction of visibility, corrosive attack on materials of construction, or dustfall (dirtiness). Thus it is important to identify the culprit before effective action can be taken. In many instances this may prove a difficult and expensive task.

Available information on odor, for example, does not cover some materials all too often branded as causative agents of a particular nuisance instance. Odor panels are a means of establishing undesirable levels and concrete goals toward which to work. These panels consist of groups of persons who are not "odor-blind" (analogous to color-blind) and who have been instructed in fundamentals of odor perception. These persons can establish at what concentrations individual or mixtures of odorous materials would become perceptible or reach nuisance values. Once nuisance is established, suitable corrective measures can be sought to reduce pollution to desired levels.

Long- and Short-range Improvement. When pollution-abatement measures have been taken, the resultant improvements ought to be determined. Short-range improvement means results obtained by process changes or installation of abatement facilities in the instances of clearly defined sources such as single plants or specific processes. Both before and after major changes, surveys, under a variety of meteorological conditions, are necessary to determine the degree of improvement.

Such surveys need not necessarily be elaborate. This can be illustrated by a case of dust pollution in the authors' experience. It was found that the dust concentration and dirtfall downwind from a powerhouse stack were causing complaints from nearby residential areas. A survey was made, using impinger and greased-slide deposition techniques, which showed that an 80 to 85 per cent reduction of dust emission would result in atmospheric contamination well below that usually found in residential areas of industrial communities. Such a reduction was feasible by the use of mechanical dust separators. After the equipment was installed and tested for rated efficiency, a field survey was again carried out which proved that the installation actually reduced the dust in the community to the desired level. This is an example of a survey with a short-range objective.

When dealing with large pollution areas such as cities or industrial aggregates, such simple surveys do not suffice. Improvement is slow as sources of pollution are located, studied, and suitable abatement equipment is installed or process improvements are worked out. Continued evaluation of the status of pollution must be carried out to show what improvements are being made, where critical areas are located, and when certain goals have been reached. In certain large industrial areas, complete elimination of pollution might conceivably be a technical possibility, but would probably be economically ruinous. Thus long-range improvement must proceed along lines of the best use of the atmosphere for waste disposal, as in the concepts now being adopted in many instances of stream-pollution abatement.

Two examples of the magnitude of the evaluation of long-range improvement can be cited. First, the pollution-abatement activities of Los Angeles: here several millions of dollars have been spent for installation of industrial pollution-abatement facilities. At the same time, extensive investigations of the pollution climatology, the analysis of the atmosphere for contaminants, and research in the causes of certain pollution effects such as eye irritation, visibility reduction, and crop damage have been carried out. Improvement has been gradual, and only recently have causes of the pollution effects begun to be clarified. The result has been an increasing realization that the problem is highly complex and that domestic sources such as automotive equipment and trash incinerators, as well as industrial plants, play an important part in the atmospheric pollution of the area.

Second, the industrial areas of England have been subject to an extensive, costly, and long-time survey of dust and gaseous contaminants. The data collected form an excellent background against which to measure future improvements of large areas.

Misinterpretation of Data. In evaluating data with respect to any problem, there are always sources of error. If the error is sufficiently large, the possibility of misinterpretation exists. Therefore it is imperative in approaching any problem in air pollution involving collection and analysis of samples that the most nearly correct methods and procedures be used and that differences between various methods be recognized.

In air pollution work with dusts, particles are often collected in a liquid and counted by the following procedure: An aliquot of the liquid is placed in a counting cell, such as a Sedgwick-Rafter cell or a hemocytometer, and a representative number of microscope fields are counted. The mean of the microscope fields counted is then assumed to be the mean for the entire counting cell. The question has been raised as to how well the mean of the fields counted represents the true mean for the entire cell. Moore (64) has discussed this problem in relation to counting plankton in water.

If X represents the counts in each of n_x microscope fields, and $\bar{X}$ is the mean of these counts, the coefficient of variation C_v is the standard deviation s divided by $\bar{X}$. This may be computed by the following equation:

$$C_v = \frac{(\Sigma X^2/\bar{X}^2) - n_x}{n_x - 1} \tag{5-46}$$

The percentage error of the mean is then

$$p = \frac{100C_v}{\sqrt{n_x}} \tag{5-47}$$

The significance of p is as follows: The odds, when $n_x > 30$, are approximately 2 to 1 in favor of the mean of the fields counted being within $\pm p$ per cent of the true mean count for the entire cell and approximately 20 to 1 in favor of the mean of the fields counted being within $\pm 2p$ per cent of the true mean count. When $n_x = 10$, odds are 20 to 1 for $\pm 2.3p$, and when $n_x = 2$, for $\pm 12.7p$.

Table 5-12 contains the particles counted on a filter surface in 40 separate fields of a microscope. What is the precision of such a count? If it is assumed that only the 10 fields in column 1 were counted, then $C_v = 0.27$ and $p = 8.5$ per cent. That is, the mean of the 10 counts, 11.1 particles, is almost certainly within $2.3p$ or 20 per cent of the true mean for the entire filter. If all 40 fields are considered, then p is reduced to 4.1 per cent. The results are approximately twice as precise, although four times as many fields had to be counted.

Table 5-12. Precision of Microscope Particle Counting

(Count, particles/field)

(1)	(2)	(3)	(4)
13	11	11	5
16	11	12	14
15	9	8	9
10	6	13	11
9	9	12	10
8	9	13	8
12	4	12	14
12	8	7	12
8	7	12	10
8	10	12	10

Column 1:
Mean = 11.1
C_v = 0.27 [Eq. (5-46)]
p = 8.5 per cent [Eq. (5-47)]

All columns:
Mean = 10.3
C_v = 0.26 [Eq. (5-46)]
p = 4.1 [Eq. (5-47)]

Three examples of misinterpretation by inadequacy or difference of methods of collection will be cited.

First, the sampling of a moving gas stream for particulate material requires particular care to ensure that the velocity of gas passing into the sampler nozzle is the same as that of the gas stream. This is called isokinetic sampling. If the nozzle velocity is lower than the gas-stream velocity, then the sample of dust collected will show a higher proportion of large particles than the dust in the gas stream. If the nozzle velocity is higher than that of the sampled air stream, the frequency distribution of the particle size of the dust will show too many small particles. Isokinetic sampling is discussed more fully by Magill (59).

Somewhat related to nonisokinetic sampling is the fallacy of obtaining a representative sample for the determination of particle-size distribution by the use of an adhesive material placed in a dust-carrying stream.

An illustration of this is shown in Fig. 5-27. A jet stream (2,500 fpm) of air containing a dust was sampled isokinetically, and the frequency distribution of particle size of the dust was determined microscopically. This distribution is that represented by the line connecting the crosses in Fig. 5-27. The same stream was sampled by allowing the dust to collect on a strip of cellophane tape inserted in the air stream. The frequency-distribution curve obtained is shown by the line connecting the dots in Fig. 5-27. It is obvious that the latter curve exaggerates the percentage of larger particles. Isokinetically determined, the percentage of particles of diameter 20μ or greater is 0.17; the other curve indicates 4.0 per cent, more than a twenty-fold exaggeration.

The second instance of misinterpretation due to collection techniques is that involving wet- and dry-impingement collection of atmospheric dust. Collection by dry impingement and examination with dark-field illumination appear to give a higher particle count than by wet impingement with bright illumination. Thus it is important in evaluating dust surveys that account be taken of the type of collection and analysis.

The third example is a recording analyzer of atmospheric sulfur dioxide based on the measurement of the change in conductivity produced in a solution by the oxidation of absorbed sulfur dioxide to sulfuric acid. Actually, any substance in air which can be dissolved and/or oxidized to produce an electrolyte will register as a sulfur dioxide concentration. While sulfur dioxide is frequently the principal substance present, it is quite likely that in certain instances some other material may cause an apparent SO_2 concentration and lead to data misinterpretation.

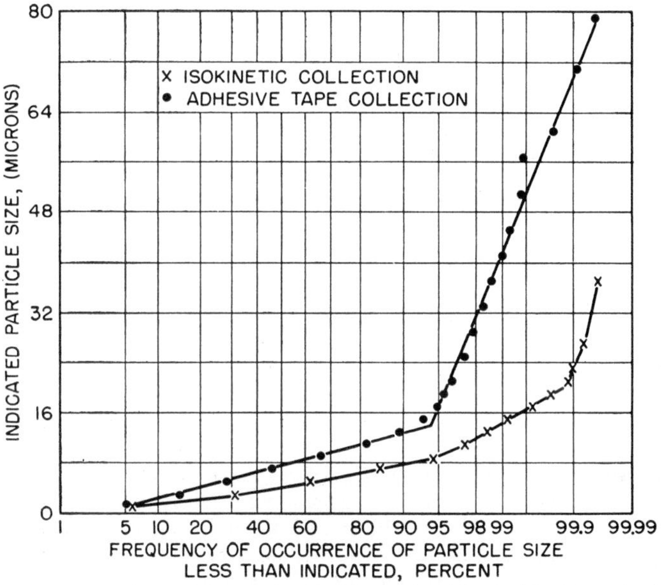

FIG. 5-27. Difference in dust size distribution determined by isokinetic and adhesive-tape collection.

Insufficiency can be a source of incorrect or incomplete data evaluation. The quantity of data required depends, of course, on the investigation. In stack analyses of process effluents, two or three collections under each different operating condition usually suffice. If the process operation is quite variable, more sampling is indicated.

In atmospheric surveys, a more extensive sampling program is required. When dealing with problems arising out of emissions from a single stack, one should be particularly interested in variations of downwind concentrations due to various meteorological conditions and due to process variations. A fairly complete picture can be obtained concerning meteorological conditions by downwind sampling under three types of conditions:

1. Strong turbulence found in the early afternoon of a nearly cloudless day
2. Little turbulence (inversion) found before dawn after a clear night
3. Intermediate turbulence of a cloudy day

Table 5-13. Processes and Other Sources of Pollution

Source of pollution	Substance discharged	Ref.	Process weight, lb/hr	Dust, fume, mist, or gas discharge, lb/hr	Stack loading, grains/cu ft	Dust wt/process wt, %
Backyard incinerator	97.7% of dust < 6μ	56	27.2	0.3	...	1.1
	Aldehydes	56	...	...	21.7 ppm	
Municipal incinerator	92% < 44μ; 24% < 6μ	56	12,000	98	...	0.82
	Aldehydes	56	...	...	3 ppm	
Aluminum furnace	Dust	56	1,857	2.2	0.182	0.12
Rock-wool cupola	Dust and fume	56	3,900	70.1	1.276	1.8
	SO_2	56	...	...	425 ppm	
	Aldehydes	56	...	...	36 ppm	
Asphalt plant	Dust and fume	56	200,000	84	0.802	0.04
Glass furnace	Dust	56	10,250	24.1	...	0.24
Waste-heat wood burner	100% < 6μ	56	372	0.49	0.077	0.13
	Total aldehydes	56	...	...	2.19 ppm	
Electric furnace	72% < 6μ	56	28,823	110	0.537	0.38
Steel, electric	...	62	3,325	...	0.306–1.17	0.29–1.02
Gray-iron cupola	71% < 44μ 18% < 6μ	56	11,013	76.3	1.09	0.69
		62	8,350–18,900	...	0.798–1.604	0.56–1.45
Blast-furnace gas	Dust mostly iron oxide	80	...	150	0.77	
Blast-furnace stove stack	Iron, sulfur, and chlorine compounds	80	...	2.9	0.002	
Blast-furnace sinter-plant stack	Iron and sulfur compounds	80	...	650	0.08	
Blast furnace (iron)	Iron ore and coke dust	52	...	...	3–24	
Blast furnace (sec. lead)	Lead compounds	52	...	...	2–6	
Open-hearth stacks	Iron and lead compounds	80	...	88	0.00004	
Steel, open-hearth (oxygen lanced)	Iron oxide	52	...	...	1–6	
Steel, open-hearth (scrap furnace)	Iron and zinc oxides	52	...	...	0.5–1.5	
Wire mill (nail galvanizing)	Zinc compounds	80	...	4.2	0.09	
Zinc sinter plant	Zinc, lead, cadmium compounds	80	...	120	0.24	
Zinc smelter plant	Zinc, lead, cadmium compounds	80	...	1,770	0.06	
Cadmium plant (after collectors)	Cadmium compounds	80	...	0.5	0.04	
Lead smelting	...	62	5,000	...	0.355	1.80–3.01
Reverberatory lead furnace	Lead and tin compounds	52	...	...	1–3	
Brass smelting	...	62	630–2000	...	0.39–0.985	1.26–3.95
Red brass	...	62	435–1620	...	0.25–0.90	0.79–1.92
Yellow brass	...	62	470–535	...	0.41–0.78	1.27–2.57
Brucker roaster	$BaSO_4$ + coal → BaS	22	250–6,000 lb/day	...	0.26–1.04	
Copperas roasting kiln	H_2SO_4	52	...	...	3.1–4.4	
Acid concentrator	H_2SO_4	52	...	...	2.1	

Table 5-13. Processes and Other Sources of Pollution *(Continued)*

Source of pollution	Substance discharged	Ref.	Process weight, lb/hr	Dust, fume, mist, or gas discharge, lb/hr	Stack loading, grains/cu ft	Dust wt/ process wt, %
Chlorosulfonic acid plant	H_2SO_4	52	...	...	11.7	
Phosphoric acid plant	H_3PO_4	52	...	...	3.0	
Roaster gases	H_2SO_4	28	...	...	2.5–11.3	
Superphosphate den and mixer	Fluorine compounds	52	...	...	4.8	
Black-liquor recovery furnace (kraft)	Na_2SO_4 and Na_2CO_3	52	...	...	1–2.5	
Dry-ice plant	Monoethanolamine	52	...	...	0.38	
Wood-distillation plant	Tar and acetic acid	52	...	...	14.5–16.7	
Sulfuric acid contact plant	H_2SO_4 mist	55	...	...	0.0026–0.12	
	SO_2	55	...	...	3.3–98 mg/cu ft	

The reference numbers refer to the references at the end of this section.

The three sampling conditions at wind velocities of 0 to 5, 5 to 15, and over 15 mph will give a fairly complete coverage of meteorological conditions affecting dispersion. In certain instances, peculiar microclimate or terrain characteristics of the locality may dictate special meteorological conditions for study. Valley and drainage effects are among these.

When dealing with pollution surveys for large industrial areas, even more extensive sampling programs are required; the extent depends upon the scope of the problem. Certainly a minimum study would evaluate pollution under conditions of worst nuisance and immediately upon the end of nuisance conditions. Such a survey would indicate existing conditions as well as the improvement required in order to avoid nuisance, i.e., make use of the atmosphere as a waste-disposal system.

5.6.3 Comparative Data for Evaluation

The comparative scarcity of readily available published quantitative data on the effluents of various municipal and industrial sources of pollution is a reflection of the relatively recent lay and technical interest in air pollution. More accent on atmospheric analyses appears in the literature than on the analyses of the sources of pollution. Indeed, such information on individual sources is often unavailable or is incompletely published, so that comparative data cannot be computed, i.e., on the basis of stack rates of emission or rate of emission per unit of product produced or unit of raw material processed.

Table 5-13 summarizes readily available information on dust emission from various processes. The list is not meant to be complete. Process-weight rate, dust-emission rate, and stack-dust loadings are given when available. In some instances dust weight as a percentage of process weight has been published. In others this quantity has been computed on the basis of the other data available in the reference. Variations among processes are reflected both in the process-weight rate and in the percentage of process weight leaving the stack as dust.

The dust emissions from various types of coal-fed furnaces have been summarized by Major (60) and O'Mara (71). Of course, the dust loading of boiler stacks varies considerably with the type of fuel used, manner of fuel feed, gas velocity, and furnace operation. Ranges of dust loading are given in Table 5-14.

Table 5-14. Range of Dust Loading

Type of Furnace	*Dust Loading, grains/cu ft*
Pulverized coal (71)	1–12 (1.4 avg)
Pulverized coal (12)	0.8–1.95
Spreader stokers (71)	2–12
Spreader stokers (60)	1–3 (0.7 avg)
Spreader stokers (12)	0.16–1.98
Underfeed stokers (71)	0.5–1.5
Underfeed stokers (60)	0.15–1.3
Underfeed stokers (12)	0.1

The numbers in parentheses refer to the references at the end of this section.

Table 5-15. Furnace Losses Based on Process or Coal Firing Rate

Type of Furnace	*Dust Removed by Furnace Gases as Per Cent of Ash in Coal*
Pulverized coal (12)	80–140
Spreader stoker (5)	20–40
Spreader stoker (12)	20–95
Spreader stoker (78)	20–25
Traveling-grate stoker (78)	15
Underfeed stoker (78)	15
Underfeed stoker (5)	5–30
Cyclone burner (53)	16

The numbers in parentheses refer to the references at the end of this section.

The figures in Table 5-14 are general and represent findings under a variety of conditions of operation and design.

Furnace losses have been summarized on the basis of process or coal firing rate. These are given in Table 5-15.

5.7 SITE SELECTION AND AIR POLLUTION CLIMATOLOGY

Air pollution climatology might be defined as putting air pollution on a probability basis for a given region or site. Knowledge of air pollution climatology is of particular value in site selection and in arriving at a satisfactory and economical solution for control problems. For some time, site investigations have included consideration of stream flow both from the viewpoint of water supply and ability of the stream to absorb liquid wastes without interference with subsequent reasonable use. Sites are now being evaluated on the basis of potential air pollution problems and the ability of the atmosphere to absorb wastes without the creation of a nuisance or hazard.

The primary reasons for evaluating the air pollution climatology of a potential or existing site are:

1. To compare alternative sites under consideration from the cost standpoint of potential air pollution problems
2. To anticipate air pollution problems of the site selected
3. To provide basic data for use in plant layout
4. To provide a basis for design of stacks, scrubbers, etc., to obtain the most economical solution of air pollution control problems

An adequate site survey should include information on the following:

1. Air pollution history of the region. The past experience of air pollution in the area under consideration can be of appreciable value, particularly in highly industrialized areas. A survey of vegetation is a useful and rapid method of assessing existing conditions since vegetation may indicate past history more faithfully than a few spot checks on atmospheric contamination.

2. Relationship of site to populated areas, fertile agricultural land, and other critical areas. In this connection, prevailing wind directions and regional topography should not be overlooked.

3. Regional climatology. Regional climatology is best evaluated by statistical analysis of Weather Bureau and other weather data. A method for using mean pressure maps as aids in evaluating regional characteristics has been discussed previously. On the other hand, Fletcher and Manos (30) have demonstrated a significant correlation between low smoke concentration and anticyclonic curvature of isobars (associated with low-pressure areas) and between high smoke concentration and cyclonic curvature (high-pressure areas) in New York City. Wind roses (and, more particularly, wind roses for light winds), frequency of calms and different ranges of wind speeds, and fog frequency and duration are all useful in evaluating regional air pollution characteristics. Comparisons of the frequency and duration of surface-temperature inversions in different regions or analyses of inversion probability on the basis of wind direction are of appreciable value.

4. Microclimate of site. Representative information of this type is probably the most difficult to obtain in a short period of time. Nevertheless, much can be accomplished in estimating local topographic effects by a comparison of short-term observations made at the site with longer-term records of the nearest weather station. Observations of wind velocity and character at the site and determination of air flow around hills and other topographic features by means of smoke or no-lift balloons with different wind and stability conditions are of particular value. Possibilities of

seasonal variations and local effects such as sea and land breezes and drainage winds should also be considered.

A pollution climatology defining the frequencies and locations of various ranges of concentrations in the vicinity of a single source of pollution has been proposed by Lowry (58). The only parameters required are the mean hourly wind speed and direction at emission level, the type of direction distribution prevailing (A, B, C, or D, see Table 5-1), and the occurrence of high morning concentrations caused by "fumigation." These factors are used to define a pollution index a_m/u for each hour of record, the value of which is called the "smoke hour." Multiplication of the index by $(0.23Q10^6)/H^2$ [see Eq. (5-21)] or by the appropriate constants from Eq. (5-11) or (5-13) gives the actual mean concentration for 1 hr at the point of maximum ground-level concentration. By analysis of past wind records, a "smoke rose" or pollution climatology can be developed for the site.

A useful method for estimating the frequency and duration of future wind velocities has been described by Sherlock (86). His primary concern was the probability of eddies at the top and in the wake of a stack at high wind speeds, but the statistical method of analysis is applicable to other pollution problems involving wind speed and direction.

The work of Fletcher and Manos (30) contains some interesting examples of correlation of pollution with meteorological parameters.

5.8 METEOROLOGICAL CONTROL

Meteorological control is the control of contaminant discharge in accordance with meteorological conditions. When conditions are favorable for good atmospheric diffusion, higher rates of discharge are possible without creation of undesirable conditions. Some successful examples of meteorological control are the control of sulfur dioxide emission from the smelter at Trail, B.C. (46), the control of the nuclear reactor at Brookhaven National Laboratory (88), and the proposed control of stack velocities under adverse conditions of wind speed and direction at the Consolidated Edison Astoria Plant (25).

The successful use of meteorological control methods requires flexibility of operations and constant attention to the weather. Since most chemical-manufacturing units must be operated at maximum capacity and with a minimum number of shutdowns to be economical, meteorological control has only limited applications in this industry. Also, many operations are such that pollution is greatest during start-up and shutdown. Gosline (36) contends that meteorological control is usually a last resort for the following reasons:

1. It is not always a positive correction since dispersion forecasts are likely to be wrong about 15 per cent of the time.
2. It cannot be applied to continuous operations.
3. It is usually not economical.
4. It is applicable only to single plants in relatively isolated areas.

The foregoing statements are not without exception, however, and in some instances, meteorological control may be the most economical solution to the problem at hand. Situations where meteorological control may be economical are:

1. Scheduling auxiliary operations, for which continuous production is not required, for periods of favorable weather.
2. Operation of collectors during periods of poor atmospheric diffusion but not when diffusion is good, in order to save power, cost of scrubbing medium, or disposal cost for the material collected.
3. Scheduling collection equipment shutdowns for repairs for periods of good diffusion so that operations may be continued.

5.9 NOMENCLATURE

Symbol	Definition	System of Consistent Units
a_m	Mode of the frequency distribution of 10-sec mean wind directions taken at 3° intervals (after Lowry); per cent of period wind blew in the mean downwind direction	Dimensionless
b	Fraction of time wind blows toward a 45° sector	Dimensionless
C_A	Axial concentration in plume:	
	For gas	ppm by vol
	For particulate matter	mg/cu ft
C	Ground-level concentrations:	
	For gas	ppm by vol
	For particulate matter	mg/cu ft
$C_{\max}$	Maximum average ground-level concentration:	
	For gas	ppm by vol
	For particulate matter	mg/cu ft
C_v	Coefficient of variation	Dimensionless
C_y	Generalized eddy-diffusion coefficient in the horizontal, cross-wind direction (after Sutton)	$ft^{n/2}$
C_z	Generalized eddy-diffusion coefficient in the vertical (after Sutton)	$ft^{n/2}$
e	Natural or Napierian logarithmic base	2.718
F_a	Axial dust-deposition rate	tons/sq ft/month
F_b	Average dust-deposition rate over 45° sector	tons/sq ft/month
f	Free (terminal) falling velocity of dust particle	fps
G	Gradient of potential atmospheric temperature	°C/ft
g	Acceleration due to gravity	32.17 ft/sec^2
g'	Gustiness parameter	Dimensionless
g_x', g_y', g_z'	Gustiness parameters in rectangular coordinates	Dimensionless
H	Effective stack height	ft
h_s	Stack height	ft
h_v	Maximum velocity rise	ft
h_{vx}	Velocity rise at distance X	ft
h_t	Maximum thermal rise	ft
h_{tx}	Thermal rise at distance X	ft
J	Parameter for calculation of thermal rise (after Bosanquet)	Dimensionless
M	Emission rate (for particulate matter)	kg/sec
n	Turbulence parameter (0 to 1) (Sutton)	Dimensionless
n_x	Number of X variates	Dimensionless
n_y	Number of Y variates	Dimensionless
n_{xy}	Number of pairs of X and Y variates	Dimensionless
P	Probability	Dimensionless
p	Percentage error of the mean	Dimensionless
p	Diffusion coefficient (Bosanquet)	Dimensionless
q	Diffusion coefficient (Bosanquet)	Dimensionless
Q	Emission rate at atmospheric temperature (for gas)	cfs

Symbol	*Definition*	*System of Consistent Units*
Q_{T_1}	Total gas-emission rate at temperature T_1	cfs
R	Ratio $x/x_{\max}$	Dimensionless
r	Correlation coefficient	Dimensionless
r_c	Correlation coefficient corrected for values of n_{xy}	Dimensionless
S	Ratio $C/C_{\max}$	Dimensionless
s	Standard deviation	Dimensionless
$S_{y=0}$	Value of S directly downwind from emission source ($y = 0$)	Dimensionless
T	Atmospheric temperature	°C
T_1	Temperature at which stack gas density equals that of the atmosphere	°abs
t	Value to test for significance	Dimensionless
u or $\bar{u}$	Mean wind speed	fps
u'	Eddy velocity in direction of mean wind (x direction)	fps
v'	Eddy velocity in the horizontal crosswind direction (y direction)	fps
v_s	Stack exit velocity	fps
W	Emission rate of dust from stack	tons/month
w'	Eddy velocity in the vertical (z direction)	fps
X	Parameter for calculation of thermal rise (after Bosanquet)	Dimensionless
$X, X_1, X_2, \ldots$	Variates	
$\bar{X}$	Mean value of X variates	
x	Downwind distance from emission source; direction of mean wind	ft
$x_{\max}$	Distance from stack to maximum ground-level concentration	ft
$Y, Y_1, Y_2, \ldots$	Variates	
$\bar{Y}$	Mean value of Y variates	
y	Distance measured perpendicular to x in the horizontal (crosswind direction)	ft
y_0	Distance from center line to edge of plume	ft
$y_{0.1}$	Cross-wind distance from point $C_{\max}$ ($R = 1$) to value of $S = 0.1$	ft
z	Vertical distance	ft
Z	Parameter for calculation of thermal rise (after Bosanquet)	Dimensionless
Γ	Gamma function	
Σ	Indicates a summation	
γ	Temperature lapse rate of atmosphere	°C/ft
γ_d	Dry-adiabatic lapse rate of atmosphere	°C/ft
Δ	Temperature difference between stack temperature and T_1	°C
ν	Kinematic viscosity of air	$\sim$0.162 10^{-3} sq ft/sec at ground level
σ	Standard deviation of the 10-sec mean wind direction in degrees of direction	Dimensionless

5.10 ACKNOWLEDGMENTS

The authors wish to express appreciation to the Manufacturing Chemists' Association for permission to exercise free use of parts of their Air Pollution Abatement Manual. Chapters 8 and 11 of this manual formed the basis of most of the present text.

REFERENCES

1. Adams, E. M.: Physiological Effects, in chap. 5, "Air Pollution Abatement Manual," Manufacturing Chemists' Association, Washington, D.C. 1951.
2. Anderson, L. J.: Captive-balloon Equipment for Low-level Meteorological Soundings, *Bull. Am. Meteorol. Soc.*, **28,** 356–362 (1947).
3. Anderson, P. A., and others: The Captive Radiosonde and Wiresonde Techniques for Detailed Low-level Meteorological Sounding, *NDRC Rpt.* 14-192, NDRC Project *PDRC*-647, *Rpt.* No. 3, October, 1943, 11 pp., PB-32743.
4. Barad, M. L.: Diffusion of Stack Gases in Very Stable Atmospheres, *Meteorol. Monographs,* **1,** No. 4, 9–14 (1951).
5. Barkley, J. F., and R. E. Morgan: Fuel-burning Equipment Dimensions Required by Smoke Abatement Ordinances, *U.S. Bur. Mines, Inform. Circ.* 7557, March, 1950.
6. Baron, T., E. R. Gerhard, and H. F. Johnstone: Dissemination of Aerosol Particles Dispersed from Stacks, *Ind. Eng. Chem.*, **41,** 2403–2408 (1949).
7. Berry, F. A., Jr., E. Bollay, and N. R. Beers: "Handbook of Meteorology," McGraw-Hill Book Company, Inc., New York, 1945.
8. Best, A. C.: Transfer of Heat and Momentum in the Lowest Layers of the Atmosphere, Geophysical Memoir No. 65, Great Britain Meteorological Office, London, 1935.
9. Bosanquet, C. H., and J. L. Pearson: The Spread of Smoke and Gases from Chimneys, *Trans. Faraday Soc.*, **32,** 1249–1264 (1936).
10. Bosanquet, C. H., W. F. Carey, and E. M. Halton: Dust Deposition from Chimney Stacks—Appendix VII (Note G), Institute of Mechanical Engineers, London, 1949.
11. Bosanquet, C. H., W. F. Carey, and E. M. Halton: Dust Deposition from Chimney Stacks, *Proc. Inst. Mech. Engrs. (London)*, **162,** No. 3, 355–365, disc. 365–367 (1950).
12. Brown, L. J., L. L. Falk, C. A. Gosline, and E. N. Helmers: Private communications, 1950, 1951, 1952.
13. Brunt, D.: "Physical and Dynamical Meteorology," 2d ed., Cambridge University Press, New York, 1939.
14. Bryant, L. W.: The Effects of Velocity and Temperature of Discharge on the Shape of Smoke Plumes from a Tunnel or Chimney—Experiments in a Wind Tunnel, *Nat. Phys. Lab., Aerodynamics Div.*, NPL Adm., 66, January, 1949.
15. Burington, R. S.: "Handbook of Mathematical Tables and Formulas," 2d ed., Handbook Publishers, Sandusky, Ohio, 1940.
16. Byers, H. R.: "General Meteorology," McGraw-Hill Book Company, Inc., New York, 1944.
17. California State Assembly: Final Summary Report of Assembly Interim Committee on Air and Water Pollution, 1951.
18. Callaghan, E. E., and R. S. Ruggeri: Investigation of Penetration of an Air Jet Directed Perpendicular to an Air Stream, *Nat. Advisory Comm. Aeronaut., Tech. Note No.* 1615, 1948.
19. Cholak, J.: Private correspondence, 1952.
20. Church, P. E.: Dilution of Waste Stack Gases in the Atmosphere, *Ind. Eng. Chem.*, **41,** 2753–2756 (1949).
21. Church, P. E., and C. A. Gosline, Jr.: Characteristics of Mixing and the Dilution of Waste Stack Gases in the Atmosphere, U.S. Atomic Energy Commission, Document MDDC-73, April 17, 1946.
22. Coal Producers' Committee for Smoke Abatement: A Study of Smoke and Air Pollution in the Kanawha Valley, Cincinnati, Ohio, 1949.
23. Davidson, W. F.: Studies of Stack Discharge under Varying Conditions, *Combustion*, **23,** 49–51 (1951).
24. Davidson, W. F.: A Study of Atmospheric Pollution, *Monthly Weather Rev.*, **70,** No. 10, 225–241 (1942).
25. Davidson, W. F.: Studies of Stack Discharge under Varying Conditions, *Combustion*, **23,** 49–51 (1951).

26. Dixon, W. J., and J. F. Massey, Jr.: "Introduction to Statistical Analysis," McGraw-Hill Book Company, Inc., New York, 1951.
27. Eisenbud, M., and W. B. Harris: Meteorological Techniques in Air Pollution Surveys. *Arch. Ind. Hyg. & Occupational Med.*, **3,** 90–97 (1951).
28. Ekman, F. O., and H. F. Johnstone: Collection of Aerosols in a Venturi Scrubber, *Ind. Eng. Chem.*, **43,** 1358–1363 (1951).
29. Falk, L. L.: Personal communication, 1952.
30. Fletcher, R. D., and N. E. Manos: The Importance of Several Meteorological Elements in the Spreading of Smoke, *Bull. Am. Meteorol. Soc.*, **31,** 365–370 (1950).
31. Freeman, H. A.: "Industrial Statistics," pp. 170–173, John Wiley & Sons, Inc., New York, 1942.
32. Gavett, G. I.: "A First Course in Statistical Methods," p. 244, McGraw-Hill Book Company, Inc., New York, 1937.
33. Geiger, R. (trans. by M. N. Stewart): "The Climate near the Ground," Harvard University Press, Cambridge, Mass., 1950.
34. Gleeson, T. A.: On the Theory of Cross-valley Winds Arising from Differential Heating of the Slopes, *J. Meteorol.*, **8,** No. 6, 398–405 (1951).
35. Gosline, C. A.: Dispersion from Short Stacks, *Chem. Eng. Progr.*, **48,** 165–172 (1952).
36. Gosline, C. A.: Meteorology Applied to Air Pollution Abatement, *Meteorol. Monographs*, **1,** No. 4, 4–5 (1951).
37. Gosline, C. A.: Air Pollution Abatement in Chemical Plants, *Am. Ind. Hyg. Assoc. Quart.*, **11,** 21–29 (1950).
38. Great Britain: Atmospheric Pollution in Leicester—A Scientific Survey, *Dept. Sci. Ind. Research (Brit.), Atm. Pollution Research, Tech. Paper No.* 1, 1945.
39. Great Britain: The Investigation of Atmospheric Pollution. Report on Observations in the Year Ended 31st March, 1937, *Dept. Sci. Ind. Research (Brit.), 23d Rept.*, 1938.
40. Gumbel, E. J.: The Return Period of Flood Flows, *Annals Math. Statistics*, **12,** No. 2, 163–190 (1941).
41. Gumbel, E. J.: Simplified Plotting of Statistical Observations, *Trans. Am. Geophys. Union*, **26,** Part I, 69–82 (1945).
42. Hand, I. F.: Atmospheric Contamination over Boston, Massachusetts, *Bull. Am. Meteorol. Soc.*, **30,** 252–254 (1949).
43. Haurwitz, B.: "Dynamic Meteorology," McGraw-Hill Book Company, Inc., New York, 1941.
44. Hewson, E. W.: Recommendations of the Meteorology Panel, United States Technical Conference on Air Pollution, *Bull. Am. Meteorol. Soc.*, **31,** 264–265 (1950).
45. Hewson, E. W.: Research on Turbulence and Diffusion of Particulate Matter in the Lower Layers of the Atmosphere, Progress Report No. 5, Massachusetts Institute of Technology Round Hill Field Station, South Dartmouth, Mass., Nov. 1, 1949 to Jan. 31, 1950.
46. Hewson, E. W.: The Meteorological Control of Atmospheric Pollution by Heavy Industry, *Quart. J. Roy. Meteorol. Soc.*, **71,** 266–282 (1945).
47. Hewson, E. W.: Atmospheric Pollution, in "Compendium of Meteorology," American Meteorological Society, Boston, Mass., 1951.
48. Hoel, P. G.: "Introduction to Mathematical Statistics," pp. 250–253, John Wiley & Sons, Inc., New York, 1947.
49. Instructions for Airway Meteorological Service, U.S. Dept. of Commerce, Weather Bureau, Circular N, 5th ed., 1941.
50. Jenkins, G. F.: Bibliography, "Air Pollution Abatement Manual," chap. 12, Manufacturing Chemists' Association, Washington, D.C. 1952.
51. Johnstone, H. F., W. E. Winsche, and L. W. Smith: The Dispersion and Deposition of Aerosols, *Chem. Revs.*, **44,** 353–371 (1949).
52. Jones, W. P., and A. W. Anthony, Jr.: Pease-Anthony Venturi Scrubbers, chap. 39, in Louis C. McCabe, ed., "Air Pollution," McGraw-Hill Book Company, Inc., New York, 1952.
53. Kaiser, E. R.: Factors Affecting Dust Emission from Boiler Furnaces, *Proc. Air Pollution and Smoke Prevention Assoc. Amer.*, **44,** 65–74 (1951).
54. Landsberg, H.: Climatology and Its Part in Air Pollution, *Meteorol. Monographs*, **1,** No. 4, 7–8 (1951).
55. Lombardo, J. B.: Analysis of Sulfuric Acid Contact Plant Exit Gas, *Anal. Chem.*, **25,** 154–160 (1953).
56. Los Angeles County Air Pollution Control District, Los Angeles, Calif., First Technical and Administrative Report on Air Pollution Control in Los Angeles County. 1949–1950.
57. Lowell, P. D., W. Hakkarinen, and D. L. Randall: National Bureau of Standards

Mobile Low-level Sounding System, *J. Research Nat. Bur. Standards*, **50**, 7–17 (1953); *Research Paper* 2381.
58. Lowry, P. H.: Microclimate Factors in Smoke Pollution from Tall Stacks, *Meteorol. Monographs*, **1**, No. 4, 24–29 (1951).
59. Magill, P. L.: Sampling Procedures and Measuring Equipment, chap. 6 in "Air Pollution Abatement Manual," Manufacturing Chemists' Association, Washington, D.C., 1952.
60. Major, W. S.: Small Industrial Plants Can Abate Smoke and Dust, *Plant.*, **1**, No. 3, 39 (1950).
61. Malone, T. F., ed.: "Compendium of Meteorology," American Meteorological Society, Boston, Mass., 1951.
62. McCabe, L. C., A. H. Rose, W. J. Hamming, and F. H. Viets: Dust and Fume Standards, *Ind. Eng. Chem.*, **41**, 2388–2390 (1949).
63. McElroy, G. E., et al.: Dilution of Stack Effluents, *U.S. Bur. Mines, Tech. Paper* 657, 1944.
64. Moore, E. W.: The Precision of Microscopic Counts of Plankton in Water, *J. Am. Water Works Assoc.*, **44**, No. 3, 208–216 (1952).
65. Mossman, C. A., J. Lundholm, Jr., and P. E. Brown: A Thermistor Temperature Recorder for Meteorological Survey, ORNL Report 556, Oak Ridge National Laboratories, Oak Ridge, Tenn., May 17, 1950.
66. Munger, H. P.: Techniques for the Study of Air Pollution at Low Altitudes, *Chem. Eng. Progr.*, **47**, 436–439 (1951).
67. Munger, H. P.: Meteorological Methods for Studying Air Pollution, 12th Intern. Congr. Pure and Appl. Chem., New York City, Sept. 10–13, 1951.
68. Munger, H. P.: The Engineering Approach to Air Pollution, *Am. J. Public Health*, **42**, 936–946 (1952).
69. Neuberger, H.: Condensation Nuclei—Their Significance in Atmospheric Pollution, *Mech. Eng.*, **70**, 221–225 (1948).
70. Neuberger, H., and M. Gutnick: Experimental Study of the Effect of Air Pollution on the Persistence of Fog, *Proc. 1st Nat. Air Pollution Symposium*, Stanford Research Institute, Los Angeles, Calif., 1949.
71. O'Mara, R.: Current Trends in Fly Ash Recovery, *Combustion*, **21**, No. 10, 38-43 (1950).
72. Petterssen, S.: "Weather Analysis and Forecasting," McGraw-Hill Book Company, Inc., New York, 1940.
73. Powell, R. W.: A Simple Method of Estimating Flood Frequencies, *Civil Eng.*, **13**, No. 2, 105–107 (1943).
74. Randall, D. L., and M. Schulkin: Survey of Meteorological Instruments Used in Tropospheric Propagation Investigations, *Nat. Bur. Standards Rept.* C.R.P.L. 2-1, July 21, 1947.
75. Richardson, L. F.: Supply of Energy from and to Atmospheric Eddies, *Proc. Roy. Soc. (London), Ser. A.*, **97**, 354–373 (1920).
76. Roberts, O. F. T.: The Theoretical Scattering of Smoke in a Turbulent Atmosphere, *Proc. Roy. Soc. (London), Ser. A.*, **104**, 640–654 (1923).
77. Rouse, H.: Model Techniques in Meteorological Research, in "Compendium of Meteorology," pp. 1249–1254, American Meteorological Society, Boston, Mass., 1951.
78. Rowley, L. N., and L. C. McCabe: Clean Air—The Engineer's Way, *Power*, **94**, 49 (1950).
79. Schmidt, W.: Turbulent Propagation of a Stream of Heated Air, *Z. angew. Math. Mech.*, **21**, 265–278, 351–363 (1941).
80. Schrenk, H. H., H. Heimann, G. D. Clayton, W. M. Gafafer, and H. Wexler: Air Pollution in Donora, Pa., *Public Health Bull.* 306, 1949.
81. Schueneman, J. J.: Atmospheric Concentrations of Sulfur Dioxide in St. Louis, 1950, *Am. Ind. Hyg. Assoc. Quart.*, **12**, 30–36 (1951).
82. Scrase, F. J.: Some Characteristics of Eddy Motion in the Atmosphere, Geophysical Memoir No. 52, Great Britain Meteorological Office, London, 1930.
83. Setterstrom, C., and P. W. Zimmerman: Sulfur Dioxide Content of Air at Boyce Thompson Institute, *Contribs. Boyce Thompson Inst.*, **9**, 171–178 (1938).
84. Sheppard, P. A.: The Aerodynamic Drag of the Earth's Surface and the Value of von Kármán's Constant in the Lower Atmosphere, *Proc. Roy. Soc. (London), Ser. A*, **188**, 208–222 (1947).
85. Sherlock, R. H., and E. A. Stalker: The Control of Gases in the Wake of Smokestacks, *Mech. Eng.*, **62**, 455–458 (1940).
86. Sherlock, R. H.: Analyzing Winds for Frequency and Duration, *Meteorol. Monographs*, **1**, No. 4, 42–49 (1951).

87. Smith, M. E.: Meteorological Factors in Atmospheric Pollution Problems, ***Am. Ind. Hyg. Assoc. Quart.***, **12,** 151–154 (1951).
88. Smith, M. E.: The Forecasting of Micrometeorological Variables, *Meteorol. Monographs*, **1,** No. 4, 50–55 (1951).
89. Sutton, O. G.: "Atmospheric Turbulence," Methuen & Co., Ltd., London, 1949.
90. Sutton, O. G.: The Theoretical Distribution of Airborne Pollution from Factory Chimneys, *Quart. J. Roy. Meteorol. Soc.*, **73,** 426–436 (1947).
91. Sutton, O. G.: The Problem of Diffusion in the Lower Atmosphere, *Quart. J. Roy. Meteorol. Soc.*, **73,** 257–281 (1947).
92. Sutton, O. G.: The Dispersion of Hot Gases in the Atmosphere, *J. Meteorol.*, **7,** 307–312 (1950).
93. Sutton, O. G.: "Micrometeorology," McGraw-Hill Book Company, Inc., New York, 1953.
94. Taylor, G. I.: Statistical Theory of Turbulence I–V, *Proc. Roy. Soc. (London), Ser. A*, **151,** 421–424, 444–454 (1935).
95. Thiessen, A. H.: "Weather Glossary," U.S. Weather Bureau, Washington, D.C., 1946.
96. Thomas, M. D., G. R. Hill, and J. N. Abersold: Dispersion of Gases from Tall Stacks, *Ind. Eng. Chem.*, **41,** 2409–2417 (1949).
97. Velz, C. J.: Graphical Approach to Statistics. Part V(A)—Tests for Statistical Significance, *Water & Sewage Works*, **98,** 262–265 (1951).
98. Velz, C. J.: Graphical Approach to Statistics. Part III—Use of Skewed Probability Paper, *Water & Sewage Works*, **97,** 393–400 (1950).
99. von Hohenleiten, H. L., and E. F. Wolf: Wind Tunnel Tests to Establish Stack Height for Riverside Generating Station, *Trans. Am. Soc. Mech. Eng.*, **64,** 671–683 (1942).
100. West Virginia Department of Health, Bureau of Industrial Hygiene: Atmospheric Pollution in the Great Kanawha River Valley Industrial Area, Feb., 1950–Aug., 1951, Charleston, W. Va., 1952.
101. Willett, H. C.: "Descriptive Meteorology," Academic Press, Inc., New York, 1944.
102. Willett, H. C.: Meteorology as a Factor in Air Pollution, *Ind. Med. and Surg.*, **19,** No. 3, 116–120 (1950).

SECTION 6

VISIBILITY AND AIR POLLUTION

BY CARSTEN STEFFENS

6.1 INTRODUCTION

Both pollutants and the natural components of the atmosphere may affect the passage of light through it and thereby cause various optical effects. The most important of these effects are the decreased visual range through the polluted atmosphere and the obvious appearance of certain pollutants themselves, especially as they emerge from a stack. The air pollution engineer is interested in these phenomena because of a hope that they may yield answers to such questions as:

1. How much pollution is present if the visual range has some observed value?
2. If a specified pollutant is emitted at a specified rate under specified conditions, what will be its effect on the visibility?

This interest seems to arise because the visibility is so natural, quick, and obvious an observation to make, and one can see the qualitative effects so readily. The quantitative relationships are, however, quite complicated, and, in fact, observations of visual range alone do not suffice to answer the first question above.

In addition to visibility, there are more subtle optical effects that may sometimes be useful in characterizing pollutants or in estimating the amount of pollution. These optical effects have rarely been applied. Nevertheless, they continue to interest the air pollution engineer because they are capable of yielding information about the pollutant as it exists in the atmosphere *in situ*, avoiding, in principle, the step of collection with all its attendant doubts.

This section is concerned (1) with the transmission of light through a turbid atmosphere, (2) with some physiological problems associated with observations of visual range, (3) with the interaction of light with suspended particles, (4) with methods of observation and measurement of the attenuation coefficient and the visual range, (5) with the observation and effect of a single source of smoke, and (6) with some examples of the relation between pollution and the other quantities mentioned.

6.2 TRANSMISSION OF LIGHT THROUGH A TURBID ATMOSPHERE

The treatment of this problem in this section starts from Eq. (4-23) and Fig. 4-1 of Sec. 4:

$$\frac{dI_\nu(\mu,\phi)}{\rho\, ds} = -k_\nu I_\nu(\mu,\phi) + \frac{k_\nu^{(s)}}{4\pi}\int_{-1}^{+1}\int_0^{2\pi} P(\mu,\phi;\,\mu',\phi')I_\nu(\mu',\phi')\,d\mu'\,d\phi' \qquad (6\text{-}1)$$

or

$$\frac{dI}{\rho\, ds} = -kI + k\mathfrak{J} \qquad (6\text{-}2)$$

The symbols are defined in Sec. 4, except for $\mathfrak{J}$, which has been substituted for $k_\nu^{(s)}/4\pi k\nu$ times the double integral of the last term. $\mathfrak{J}$ is called the source function.

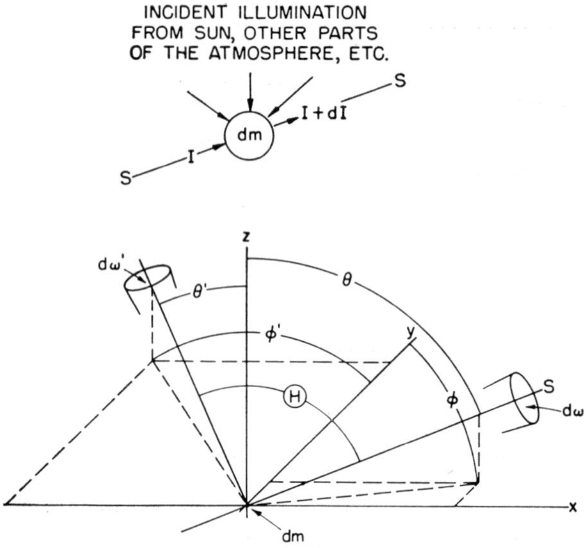

FIG. 6-1. Scattering of light by an element *dm* of a turbid atmosphere.

Equation (6-1) is quite general. It describes the change in the intensity I of the light as it passes through an infinitesimal mass dm along the optically straight line s (a geodesic for the light wave) (see Fig. 6-1). The position of the mass dm and the direction of the line s are arbitrarily chosen; they can be any point and any direction. The intensity changes on passing through dm because of two effects, which contribute the two terms on the right side of Eq. (6-1). Some of the light that enters dm in the direction s is removed from the beam, either by absorption or by scattering into other directions. These losses lead to the term $-kI$, where k is the mass attenuation coefficient. At the same time, the beam gains some flux, as some of the light that strikes dm from other directions is scattered toward the observer. The amount of this added flux is $k\mathfrak{J}$, and since it is contributed by the turbid atmosphere, it is often called the "air light" or "space light." The light that strikes dm may have come directly from such primary sources as the sun or may have been reflected or scattered one or more times before.

The quantities I, $\mathfrak{J}$, P, k, and ρ, which appear in Eq. (6-1), are all functions of position and time. I, $\mathfrak{J}$, P, and k are also functions of direction, frequency, and the state of polarization of the light (see Sec. 4). The functional dependence will not, however, be explicitly shown in this section, except when it seems to be necessary to avoid confusion.

The mass dm can, in the general case, contribute flux by radiation. Since such contributions are unimportant for this section (they are invisible, for telluric atmospheres), they have been omitted from this discussion.

The variables μ and ϕ in Eq. (6-1) describe direction at any given point. ($\mu = \cos\theta$, where θ is the zenith angle. ϕ is the azimuth.) $I_\nu(\mu',\phi')$ (in $\mathfrak{J}$) describes the intensity of the light that strikes dm from all directions, without regard to the history of this light. Accordingly, the integrals of $\mathfrak{J}$ include all directions around dm, but do not include a radial coordinate.

The quantities ρk and $\rho k^{(s)}$ (Sec. 4) are often replaced by β and σ, respectively, called the "attenuation coefficient" and the "scattering coefficient," and this substitution will be used in this section. However, the separate symbols are apt to be clearer in problems involving nonuniform atmospheres.

The identity of the several coefficients for directed light from an object with those for the air light is still questioned (37).

Equation (6-1) is a first-order linear differential equation and can be integrated (Chandrasekhar) (10) to give

$$I(s) = I(0)\exp\left(-\int_0^s \beta\,ds\right) + \int_0^s \mathfrak{J}(s')\exp\left(-\int_{s'}^s \beta\,ds\right)\beta\,ds' \qquad (6\text{-}3)$$

or

$$I(s) = \int_{-\infty}^s \mathfrak{J}(s')\exp\left(-\int_{s'}^s \beta\,ds\right)\beta\,ds' \qquad (6\text{-}4)$$

These equations give the light intensity at any point along the path s. The intensity at any point and in a given direction results from the contributions by all anterior points s', and by the intensity $I(0)$ already present at the origin of s, all reduced by the factor $\exp\left(-\int_{s'}^s \beta\,ds\right)$ for attenuation by the intervening atmosphere. The second form arises if the integration is to minus infinity instead of to the origin of s.

Before these equations can be used, more information is needed about the attenuation coefficient β, the phase function P, and the intensity of illumination along the path $I(\mu',\phi')$. The attenuation coefficient β depends on the optical properties, and hence on the chemical constitution, of the particles in the atmosphere; it also depends on their concentration, size, shape, and orientation with respect to the beam. The optical properties vary with temperature and with the wavelength and state of polarization of the incident light. The phase function P depends on the variables listed above and also on the direction of the incident light. If there is a material object at the origin (or anywhere else along the path), the intensity there will be determined by the optical properties of the object and by the illumination on it.

The set of conditions that has received the most attention is that corresponding to the definition of visibility in meteorology. The definition is (52): "'Visibility' is the mean greatest distance toward the horizon that prominent objects, such as mountains, buildings, towers, etc., can be seen and identified by the normal eye unaided by special optical devices, such as binoculars, telescopes, glare-eliminating goggles, etc., and which distance must prevail over a range of half or more of the horizon." Among the instructions following the definition is the statement, "For accurate determinations during daylight hours it is advisable to confine the choice of marks to black, or nearly black, objects against the horizon sky, rejecting light-colored marks and those appearing against terrestrial backgrounds."

The usual, corresponding, theoretical analysis assumes:

1. Liminal contrast between the test object and the adjacent horizon sky.
2. A perfectly black test object, i.e., $I = 0$ at the object.
3. Lines of sight to object and sky in substantially identical directions; i.e., the zenith angles and azimuths are the same for the two lines.
4. Horizontal lines of sight; i.e., $\theta = \pi/2$ and $\mu = 0$.
5. An atmosphere that is uniform between object and observer and far enough beyond the object to justify integrating to infinity in that direction under this same assumption; i.e., $\beta \neq \beta(s)$.
6. Uniform illumination of the elements of the path (this includes uniform illumination by primary sources like the sun, uniform air light, and uniform reflection by the ground); i.e., $\mathfrak{J} \neq \mathfrak{J}(s)$.
7. The object is of such size that it does not affect the light path by shading or otherwise and is such that visual acuity is not significantly greater for larger objects.

Items 5 and 6 are more restrictive than the definition of visual range. The restrictions are necessary for a simple mathematical treatment; it may ultimately be possible to relax them, but the more realistic cases do not seem to have been discussed so far.

Under restrictions 5 and 6, Eqs. (6-3) and (6-4) reduce to

$$I(\mu,\phi,s) = I(\mu,\phi,0) \exp(-\beta s) + \mathfrak{J}(\mu,\phi)[1 - \exp(-\beta s)] \tag{6-5}$$

and

$$I(\mu,\phi,s) = \mathfrak{J}(\mu,\phi) \tag{6-6}$$

The physical interpretation of these equations is simple. $\mathfrak{J}$ is evidently the intensity from the direction of the horizon sky, and for an object at the (arbitrary) zero of s, $I(\mu,\phi,0)$ is the light intensity at the object itself.

Equation (6-6) can also be derived from (6-5) by letting $s \to \infty$. This derivation brings out the fact that the characteristics of an object of finite brightness at an infinite distance do not affect the result; the result depends only on the air light when the distance approaches infinity.

Item 4 in the list of assumptions has not been used. As a practical matter, however, the atmosphere is usually so markedly stratified that the uniformity required by 5 and 6 can be attained only by using a horizontal path. Such a path will therefore be assumed for the rest of this discussion.

Since an object can be distinguished visually from surroundings of the same color only by the contrast in brightness, Duntley (18,19) has rewritten the equations in terms of contrast. The contrast C between two lines of sight is defined as

$$C = \frac{I_1 - I_2}{I_2} \tag{6-7}$$

if Eq. (6-5) is applied to the two lines of sight, under the conditions that zenith angles, azimuths, and distances are all the same for the two lines, then

$$C = \frac{I_1(\phi,s) - I_2(\phi,s)}{I_2(\phi,s)} = \frac{I_2(\phi,0)}{I_2(\phi,s)} \frac{[I_1(\phi,0) - I_2(\phi,0)]}{I_2(\phi,0)} \exp(-\beta s)$$

$$= \frac{I_2(\phi,0)}{I_2(\phi,s)} C(0) \exp(-\beta s) \tag{6-8}$$

where $C(0)$ is the contrast at the origin. If the objects are at the origin, $C(0)$ is Duntley's "inherent contrast."

This equation or something equivalent to it is the basis for the usual discussion of the ease with which objects can be seen under various circumstances. Although in the derivation of this equation the optical characteristics of the object and its background are not restricted to the case of a black object against the horizon sky, the atmosphere must still be uniform in composition and illumination between the object and the observer and for some distance beyond the object if object 2 is the horizon sky.

$$C = C(0) \exp(-\beta s) \tag{6-9}$$

For this case, "the apparent contrast is exponentially attenuated."

If the object 1 is perfectly black, so that $I_1(0)$ is zero (this is the first use of condition 2), Eqs. (6-5), (6-8), and (6-9) become

$$I(\phi,s) = \mathfrak{I}(\phi)[1 - \exp(-\beta s)]$$

or

$$\frac{I(\phi,s)}{\mathfrak{I}(\phi)} = 1 - \exp(-\beta s) \tag{6-10}$$

$$C = -\frac{I_2(\phi,0)}{I_2(\phi,s)} \exp(-\beta s) \tag{6-11}$$

$$C = -\exp(-\beta s) \tag{6-12}$$

The minus signs in (6-11) and (6-12) indicate that the object 1 is darker than its background.

Two special cases of attenuation should be mentioned. First, fogs, clouds, and water haze (in clean air), plumes of steam, and other systems that show no significant absorption in the visible spectrum affect the visual range by scattering alone. The attenuation coefficient k then reduces to the scattering coefficient $k^{(s)}$. The forms of the equations given above remain unaltered for this case.

Second, if there are no significant external light sources and if secondary scattering can be neglected, then the source function $\mathfrak{I}$ becomes zero. This situation arises in measuring the transmittance of a light beam through the smoke in a stack and in some other applications in air pollution. For this case, $\mathfrak{I} = 0$ and Eqs. (6-3) and (6-5) reduce to

$$I(s) = I(0) \exp\left(-\int_0^s \beta\, ds\right) \tag{6-13}$$

$$I(s) = I(0) \exp(-\beta s) \tag{6-14}$$

The equations for the contrast are less useful for this case.

Several methods have been used for measuring the attenuation coefficient β. All depend on making enough independent observations or assumptions to determine the value of the constants in Eq. (6-5) or its equivalent. For example, the light flux may be measured from the object at zero distance (where it is zero for a black object), at some intermediate distance where the object can be seen but its luminance has been affected by the haze, and at infinity where the light flux is the light from the horizon sky. These three independent observations determine the three constants β, $I(0)$, and $\mathfrak{I}$.

In a transmissometer, the ratio of the light fluxes before and after the beam passes through a known thickness x of the atmosphere is measured. By using some means of distinguishing the light from this source from that from any other, e.g., by using an intermittent beam or by excluding other light, the source function $\mathfrak{I}$ becomes zero, neglecting secondary scattering. Equation (6-14) can be solved for the attenuation coefficient β to give

$$\beta = -\frac{1}{s} \ln \frac{I(s)}{I(0)} = -\frac{1}{s} \ln T \tag{6-15}$$

The ratio $I(s)/I(0) = T$ is called the transmission over the distance s.

Another possible arrangement is the use of two approximately black targets at different distances (see Fig. 6-2). Equation (6-5) applies to each of the objects, and if $\mathfrak{J}$ is eliminated between the two equations,

$$\frac{I_1(s_1) - I_1(0)\exp(-\beta s_1)}{I_2(s_2) - I_2(0)\exp(-\beta s_2)} = \frac{1 - \exp(-\beta s_1)}{1 - \exp(-\beta s_2)} \tag{6-16}$$

For the special case of two black objects [both $I(0)$'s equal to zero], this equation reduces to

$$\frac{I_1(s_1)}{I_2(s_2)} = \frac{1 - \exp(-\beta s_1)}{1 - \exp(-\beta s_2)} \tag{6-17}$$

Only a measurement of the ratio of the two light fluxes is now required, in addition, of course, to knowing how far away the two objects are.

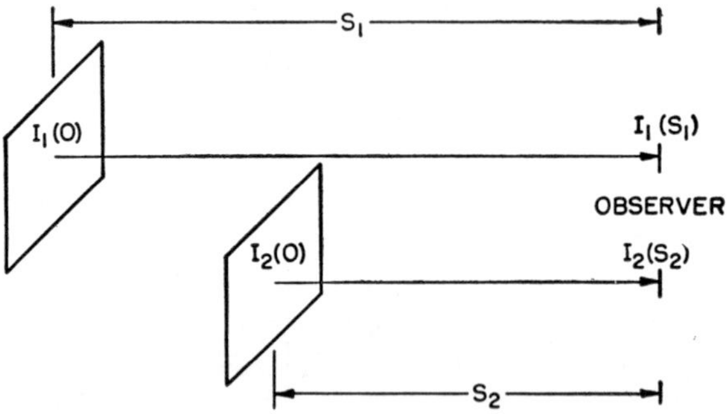

FIG. 6-2. Light intensity from objects at differing distances.

All these methods of measurement can be adapted to determining the attenuation coefficient as a function of wavelength. It is only necessary to make the measurements in two or more narrow bands of wavelengths instead of in white light.

6.3 PHYSIOLOGICAL EFFECTS AND THE VISUAL RANGE

Since the visual range is defined as the distance at which the standard object is just identifiable, the minimum contrast that the eye can distinguish must now be introduced into the analysis. (In order to eliminate the factors of search and recognition, with which this treatment is not concerned, "identified" has been replaced by "distinguished.") This contrast limen (the "psychophysical constant") is denoted by ϵ and the corresponding visual range by v. For a black object at its visual range against the horizon sky, from Eq. (6-12)

$$\epsilon = -\exp(-\beta v) \qquad \text{or} \qquad v = -\frac{1}{\beta}\ln(-\epsilon) \tag{6-18}$$

The attenuation coefficient β can be substituted into Eq. (6-18) to give the visibility if this is desired. The contrast limen has been defined by reference to Eq. (6-7) and is therefore negative for an object that is darker than its background.

The contrast limen ϵ is often taken as a defined constant, equal to -0.02 (35). The corresponding visual range is called the standard visibility. It is related to the attenuation coefficient, from Eq. (6-18), by

$$v_2 = -\frac{1}{\beta}\ln 0.02 = \frac{3.912}{\beta} \tag{6-19}$$

In meteorological work, ϵ is often taken as -0.055 or -0.05, because the resulting visual range, which for $\epsilon = -0.05$ is

$$v_5 = -\frac{1}{\beta} \ln 0.05 = \frac{2.996}{\beta} \tag{6-20}$$

often corresponds more closely to that reported by field observers.

Each of these visual ranges (v_2 and v_5) has been tabulated against the attenuation coefficient β in Tables 6-1 and 6-2. The same data have also been plotted in Fig. 6-3 for convenient use when the precision of the tables is not needed.

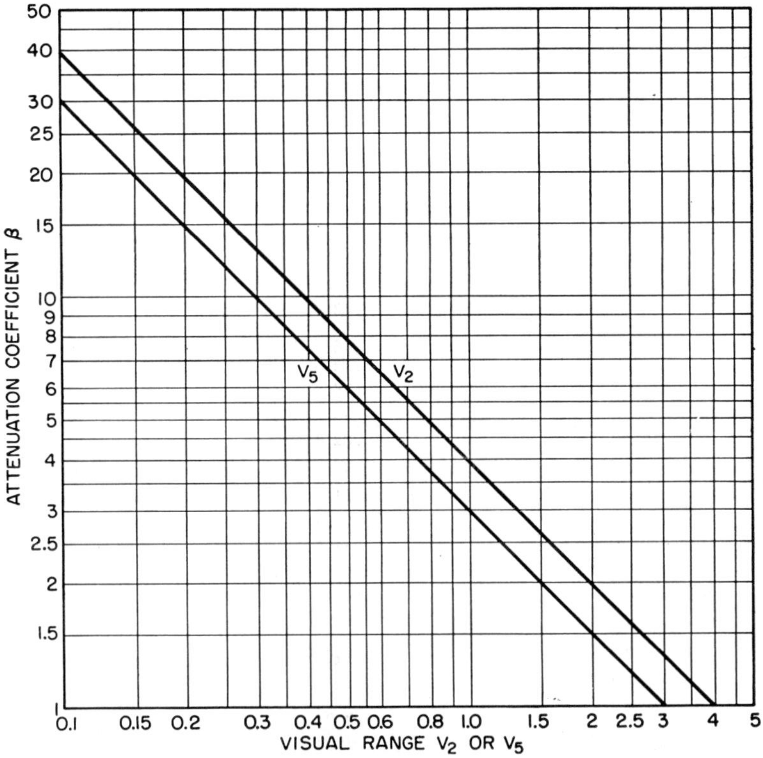

FIG. 6-3. Attenuation coefficient as a function of visual range. Any convenient units of length, e.g., miles, can be used for the unit of visual range. The attentuation coefficient will then be in the reciprocal of the corresponding units, e.g., per mile^{-1}.

The conditions that are incorporated into the foregoing theoretical treatment do not correspond to those in the polluted atmosphere above a city. In particular, the atmosphere is not homogeneous. Local sources, such as chimneys, are usually obvious, and even when these sources have merged to give a continuous cloud, the cloud is not uniform. In Los Angeles, Rubin* has observed that the regions of approximate uniformity are only about a quarter of a mile across. The equations should be applied with the assumptions in mind. See below, however, under Coleman's telephotometer, page 6-27.

Figure 6-4 (on page 6-10) is a graph of Eq. (6-10). It shows that the air light obscuring distant objects arises mostly from the haze in the immediate vicinity of the observer. The air light from more distant parts of the haze is itself attenuated by the

* Private communication.

haze between its point of origin and the observer. If the visual range through a uniform fog were 100 yd, half of the obscuration would be produced by the 18 yd of fog nearest the observer, and the fog beyond 60 yd (to infinity) would produce only 10 per cent of the effect. This strong weighting of the material near the observer is probably one reason why the theory works as well as it does. The observer tends to require uniform conditions throughout the region that he can see fairly clearly; the conditions beyond have little influence (32,41).

Table 6-1. Attenuation Coefficient β and Visual Range v_2 for $\epsilon = 0.02$[a]

Visual range	0.000	0.001	0.002	0.003	0.004	0.005	0.006	0.007	0.008	0.009
0.10	39.12	38.73	38.35	37.98	37.62	37.26	36.91	36.56	36.22	35.89
0.11	35.56	35.24	34.93	34.62	34.32	34.02	33.72	33.44	33.15	32.87
0.12	32.60	32.33	32.07	31.80	31.55	31.30	31.05	30.80	30.56	30.33
0.13	30.09	29.86	29.64	29.41	29.19	28.98	28.76	28.55	28.35	28.14
0.14	27.94	27.74	27.55	27.36	27.17	26.98	26.79	26.61	26.43	26.26
0.15	26.08	25.91	25.74	25.57	25.40	25.24	25.08	24.92	24.76	24.60
0.16	24.45	24.30	24.15	24.00	23.85	23.71	23.57	23.43	23.29	23.15
0.17	23.01	22.88	22.74	22.61	22.48	22.35	22.23	22.10	21.98	21.85
0.18	21.73	21.61	21.49	21.38	21.26	21.15	21.03	20.92	20.81	20.70
0.19	20.59	20.48	20.37	20.27	20.16	20.06	19.96	19.86	19.76	19.66
0.20	19.56	19.46	19.37	19.27	19.18	19.08	18.99	18.90	18.81	18.72
0.21	18.63	18.54	18.45	18.37	18.28	18.20	18.11	18.03	17.94	17.86
0.22	17.78	17.70	17.62	17.54	17.46	17.39	17.31	17.23	17.16	17.08
0.23	17.01	16.94	16.86	16.79	16.72	16.65	16.58	16.51	16.44	16.37
0.24	16.30	16.23	16.17	16.10	16.03	15.97	15.90	15.84	15.77	15.71
0.25	15.65	15.59	15.52	15.46	15.40	15.34	15.28	15.22	15.16	15.10
0.26	15.05	14.99	14.93	14.87	14.82	14.76	14.71	14.65	14.60	14.54
0.27	14.49	14.44	14.38	14.33	14.28	14.23	14.17	14.12	14.07	14.02
0.28	13.97	13.92	13.87	13.82	13.77	13.73	13.68	13.63	13.58	13.54
0.29	13.49	13.44	13.40	13.35	13.31	13.26	13.22	13.17	13.13	13.08
0.30	13.04	13.00	12.95	12.91	12.87	12.83	12.78	12.74	12.70	12.66
0.31	12.62	12.58	12.54	12.50	12.46	12.42	12.38	12.34	12.30	12.26
0.32	12.22	12.19	12.15	12.11	12.07	12.04	12.00	11.96	11.93	11.89
0.33	11.85	11.82	11.78	11.75	11.71	11.68	11.64	11.61	11.57	11.54
0.34	11.51	11.47	11.44	11.41	11.37	11.34	11.31	11.27	11.24	11.21
0.35	11.18	11.15	11.11	11.08	11.05	11.02	10.99	10.96	10.93	10.90
0.36	10.87	10.84	10.81	10.78	10.75	10.72	10.69	10.66	10.63	10.60
0.37	10.57	10.54	10.52	10.49	10.46	10.43	10.40	10.38	10.35	10.32
0.38	10.29	10.27	10.24	10.21	10.19	10.16	10.13	10.11	10.08	10.06
0.39	10.03	10.01	9.980	9.954	9.929	9.904	9.879	9.854	9.829	9.805

Visual range	0.00	0.01	0.02	0.03	0.04	0.05	0.06	0.07	0.08	0.09
0.40	9.780	9.541	9.314	9.098	8.891	8.693	8.504	8.323	8.150	7.984
0.50	7.824	7.671	7.523	7.381	7.244	7.113	6.986	6.863	6.745	6.631
0.60	6.520	6.413	6.310	6.210	6.112	6.018	5.927	5.839	5.753	5.670
0.70	5.589	5.510	5.433	5.359	5.286	5.216	5.147	5.081	5.015	4.952
0.80	4.890	4.830	4.771	4.713	4.657	4.602	4.549	4.497	4.445	4.396
0.90	4.347	4.299	4.252	4.206	4.162	4.118	4.075	4.033	3.992	3.952
1.00	3.912	3.873	3.835	3.798	3.762	3.726	3.691	3.656	3.622	3.589

[a] The visual range can be in any convenient units of length, e.g., miles, and the attenuation coefficient will then be in the reciprocal of the same units, e.g., miles^{-1}.
This table is adapted from the Smithsonian Meteorological Tables.

The attenuation coefficient can be measured over short distances, as mentioned above. From such measurements, the standard visibility can be calculated for a hypothetical atmosphere that has the same characteristics everywhere as it has in the region that was measured. There is some advantage to expressing the results in terms of standard visibility, because this concept is more easily grasped by the layman than that of the attenuation coefficient.

Table 6-2. Attenuation Coefficient β and Visual Range v_5 for $\epsilon = 0.05$[a]

Visual range	0.000	0.001	0.002	0.003	0.004	0.005	0.006	0.007	0.008	0.009
0.10	29.96	29.66	29.37	29.09	28.81	28.53	28.26	28.00	27.74	27.49
0.11	27.24	26.99	26.75	26.51	26.28	26.05	25.83	25.61	25.39	25.18
0.12	24.97	24.76	24.56	24.36	24.16	23.97	23.78	23.59	23.41	23.22
0.13	23.05	22.87	22.70	22.53	22.36	22.19	22.03	21.87	21.71	21.55
0.14	21.40	21.25	21.10	20.95	20.81	20.66	20.52	20.38	20.24	20.11
0.15	19.97	19.84	19.71	19.58	19.45	19.33	19.21	19.08	18.96	18.84
0.16	18.73	18.61	18.49	18.38	18.27	18.16	18.05	17.94	17.83	17.73
0.17	17.62	17.52	17.42	17.32	17.22	17.12	17.02	16.93	16.83	16.74
0.18	16.64	16.55	16.46	16.37	16.28	16.19	16.11	16.02	15.94	15.85
0.19	15.77	15.69	15.60	15.52	15.44	15.36	15.29	15.21	15.13	15.06
0.20	14.98	14.91	14.83	14.76	14.69	14.61	14.54	14.47	14.40	14.33
0.21	14.27	14.20	14.13	14.07	14.00	13.93	13.87	13.81	13.74	13.68
0.22	13.62	13.56	13.50	13.43	13.38	13.32	13.26	13.20	13.14	13.08
0.23	13.03	12.97	12.91	12.86	12.80	12.75	12.69	12.64	12.59	12.54
0.24	12.48	12.43	12.38	12.33	12.28	12.23	12.18	12.13	12.08	12.03
0.25	11.98	11.94	11.89	11.84	11.80	11.75	11.70	11.66	11.61	11.57
0.26	11.52	11.48	11.44	11.39	11.35	11.31	11.26	11.22	11.18	11.14
0.27	11.10	11.06	11.01	10.97	10.93	10.89	10.86	10.82	10.78	10.74
0.28	10.70	10.66	10.62	10.59	10.55	10.51	10.48	10.44	10.40	10.37
0.29	10.33	10.30	10.26	10.23	10.19	10.16	10.12	10.09	10.05	10.02
0.30	9.987	9.953	9.921	9.888	9.855	9.823	9.791	9.759	9.727	9.696
0.31	9.665	9.633	9.603	9.572	9.541	9.511	9.481	9.451	9.421	9.392
0.32	9.363	9.333	9.304	9.276	9.247	9.218	9.190	9.162	9.134	9.106
0.33	9.079	9.051	9.024	8.997	8.970	8.943	8.917	8.890	8.864	8.838
0.34	8.812	8.786	8.760	8.735	8.709	8.684	8.659	8.634	8.609	8.585
0.35	8.560	8.536	8.511	8.487	8.463	8.439	8.416	8.392	8.369	8.345
0.36	8.322	8.299	8.276	8.253	8.231	8.208	8.186	8.163	8.141	8.119
0.37	8.097	8.075	8.054	8.032	8.011	7.989	7.968	7.947	7.926	7.905
0.38	7.884	7.864	7.843	7.822	7.802	7.782	7.762	7.742	7.722	7.702
0.39	7.682	7.662	7.643	7.623	7.604	7.585	7.566	7.547	7.528	7.509

Visual range	0.00	0.01	0.02	0.03	0.04	0.05	0.06	0.07	0.08	0.09
0.40	7.490	7.307	7.133	6.967	6.809	6.658	6.513	6.374	6.242	6.114
0.50	5.992	5.875	5.762	5.653	5.548	5.447	5.350	5.256	5.166	5.078
0.60	4.993	4.911	4.832	4.756	4.681	4.609	4.539	4.472	4.406	4.342
0.70	4.280	4.220	4.161	4.104	4.049	3.995	3.942	3.891	3.841	3.792
0.80	3.745	3.699	3.654	3.610	3.567	3.525	3.484	3.444	3.405	3.366
0.90	3.329	3.292	3.257	3.222	3.187	3.154	3.121	3.089	3.057	3.026

[a] The visual range can be in any convenient units of length, e.g., miles, and the attenuation coefficient will then be in the reciprocal of the same units, e.g., miles^{-1}.
This table is adapted from the Smithsonian Meteorological Tables.

Subjective estimates of the visual range have been used to obtain values of the attenuation coefficient and corresponding estimates of the characteristics of the material suspended in the air. Since the visual range is related to these physical quantities through the contrast limen, the effect of altered conditions of observation can be reduced to altered numerical values for this limen.

The contrast limen depends on the angular size of the object, its shape, the level of luminance to which the observer's eyes are adapted, and the conditions and technique of observing. These factors were investigated by the Camouflage Section (16.3) of the NDRC. The results have been reported by Blackwell (7) and Duntley (18,19). Since Duntley has reduced the results to a convenient nomographic form, his charts are given here (Figs. 6-5 to 6-13, pages 6-11 to 6-19).

Figures 6-5 to 6-13 relate the liminal target distance, the meteorological range (the standard visibility), the contrast of target and background, the area of the target, and the level of illumination to one another for circular targets. The results are not very sensitive to the shape of the target.

Liminal values are commonly measured by requiring the observer to respond to a measured stimulus and taking the value of the stimulus at which half of these forced responses are correct and half are wrong. The observer has no confidence that he is

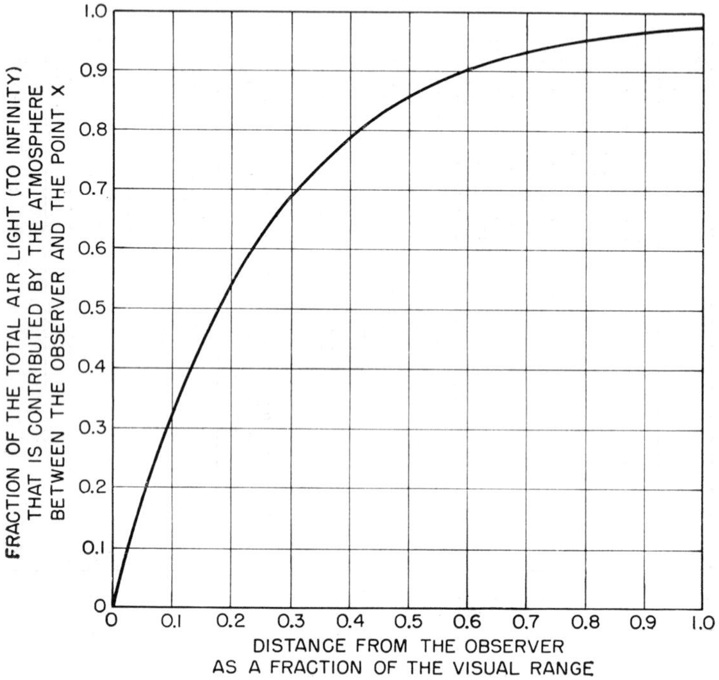

Fig. 6-4. Fraction of air light contributed by the atmosphere between the observer and point x.

right any of the time at the limen, but in most cases, if the stimulus is doubled, he becomes confident of his own correctness. Figure 6-14 (on page 6-20) shows the relation between relative contrast and the probability of a correct response for this particular type of stimulus. If the relative contrast is twice its threshold value, the probability of the observer's being correct when he thinks he sees the object is 98 per cent, and this point is close to the value that most observers will pick as the limit of visibility if left to themselves.

Figure 6-15 shows the relation between threshold contrast, visual angle, and brightness level. For daylight observations on targets larger than about 0.5° diameter, the liminal contrast is independent of the size of the object, within the limits of the chart. For targets smaller than this, however, the liminal contrast changes rapidly. This fact is the basis for the lower limit of target size for visual observations.

The upper limit of permissible size for a target for estimating visual range is usually set at about 5° for a physical reason that applies to instrumental as well as visual observations. A target with the sun behind it necessarily shades part of the optical path, thereby increasing the visual range. For accurate observation, this effect must be negligibly small, and it is negligible, in view of other uncertainties, if the target is smaller than 5°.

The curves of Fig. 6-15 seem to be approaching a vertical asymptote of −2.6 at the left side of the figure. This value corresponds to $\epsilon = 0.0025$ for experimental conditions like those of the meteorological definition of visual range, except that the factors of certainty and recognition differ. These measurements can be compared with values of the contrast limen calculated from Eq. (6-18) from simultaneous obser-

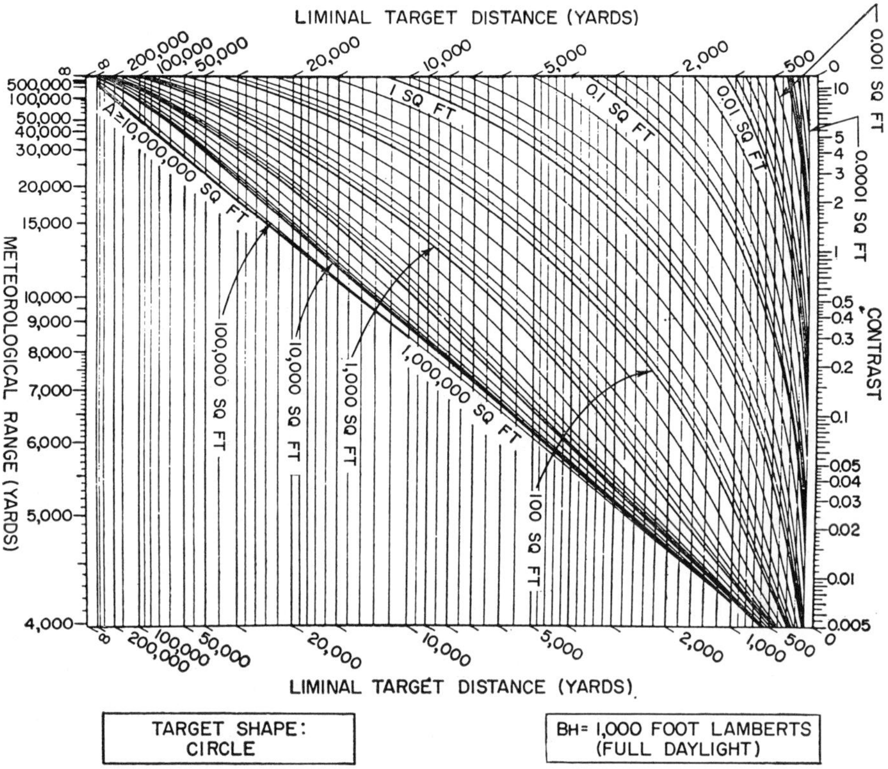

FIG. 6-5. Liminal target distance as affected by size of target and level of illumination. (See Ref. 19.)

vations of the visual range and the attenuation coefficient. Houghton (27) and Douglas and Young (17) obtained in this way values of the contrast limen five to ten times as large as Blackwell's.

Middleton and Mungall (38) reported measurements of the contrast of the marks that meteorological observers actually chose as being at the visual range in the normal process of observing visibility. These measurements also were reported as corresponding values of ϵ (which should probably not be called the contrast limen in this context, since it does not conform to the definition given above). The mode of the markedly skewed distribution was 0.024; more than 10 per cent of the observations gave values greater than 0.070, and more than 5 per cent gave values greater than 0.100.

These several sets of operations define different quantities; so it is not surprising

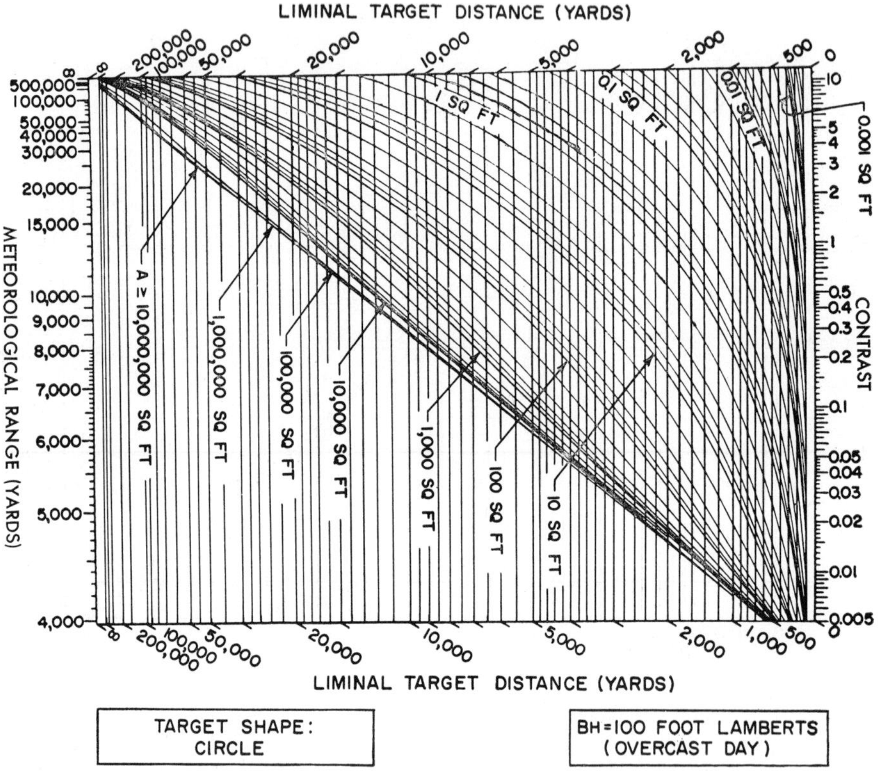

FIG. 6-6. Liminal target distance as affected by size of target and level of illumination. (See Ref. 19.)

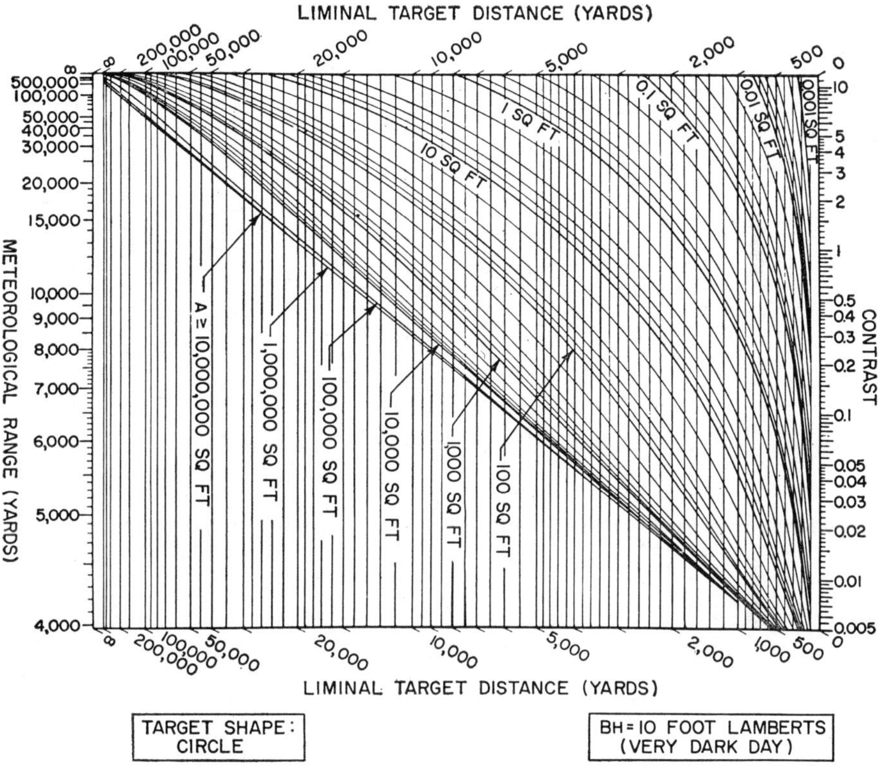

FIG. 6-7. Liminal target distance as affected by size of target and level of illumination. (See Ref. 19.)

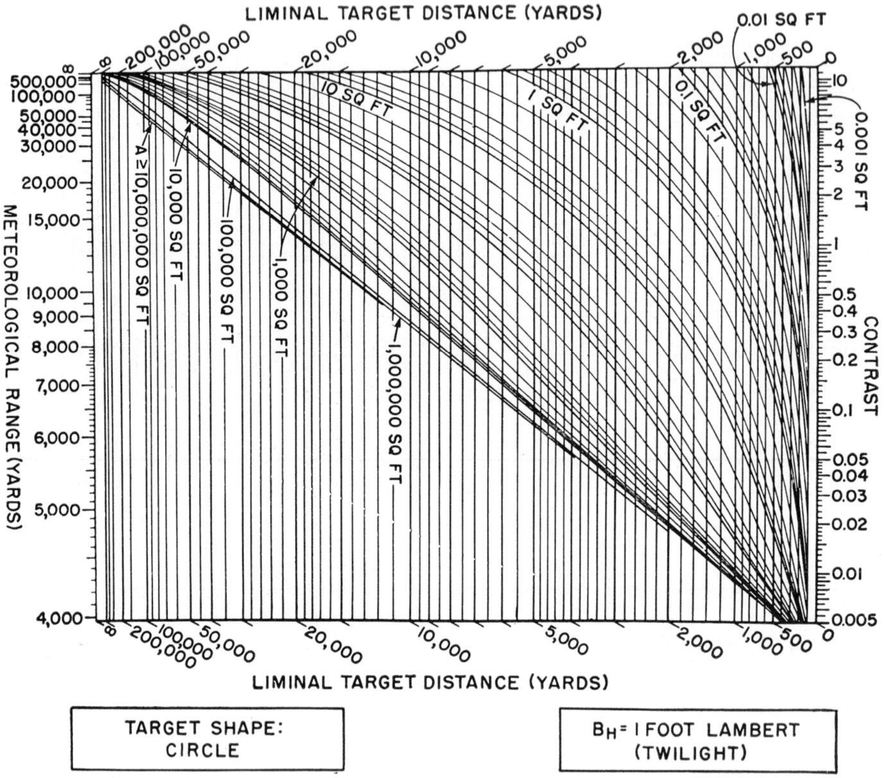

FIG. 6-8. Liminal target distance as affected by size of target and level of illumination. (See Ref. 19.)

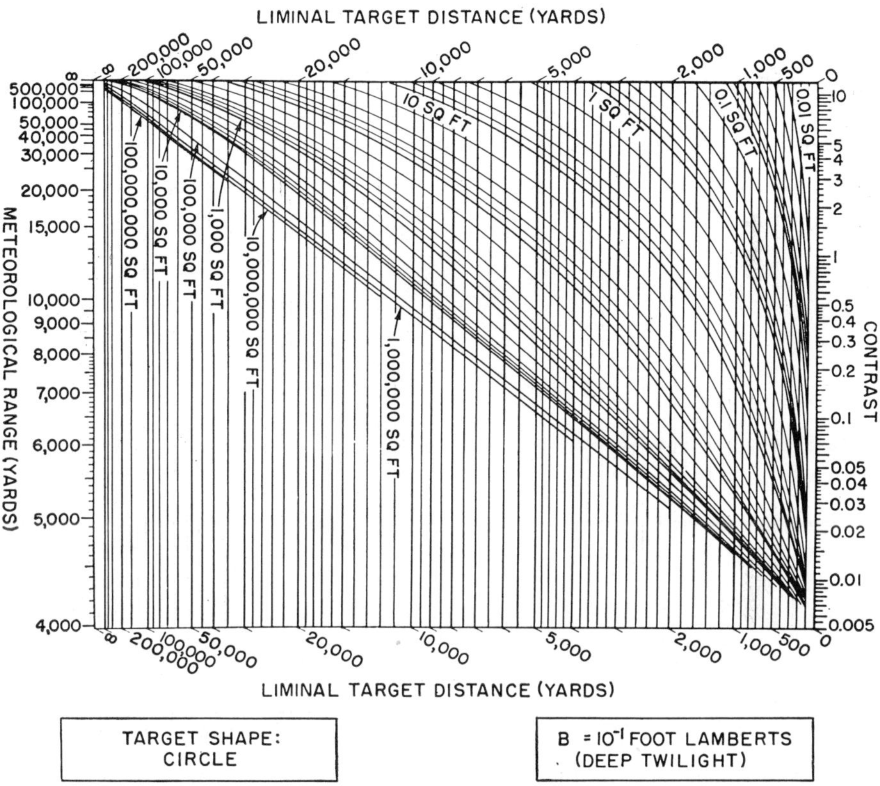

FIG. 6-9. Liminal target distance as affected by size of target and level of illumination (See Ref. 19.)

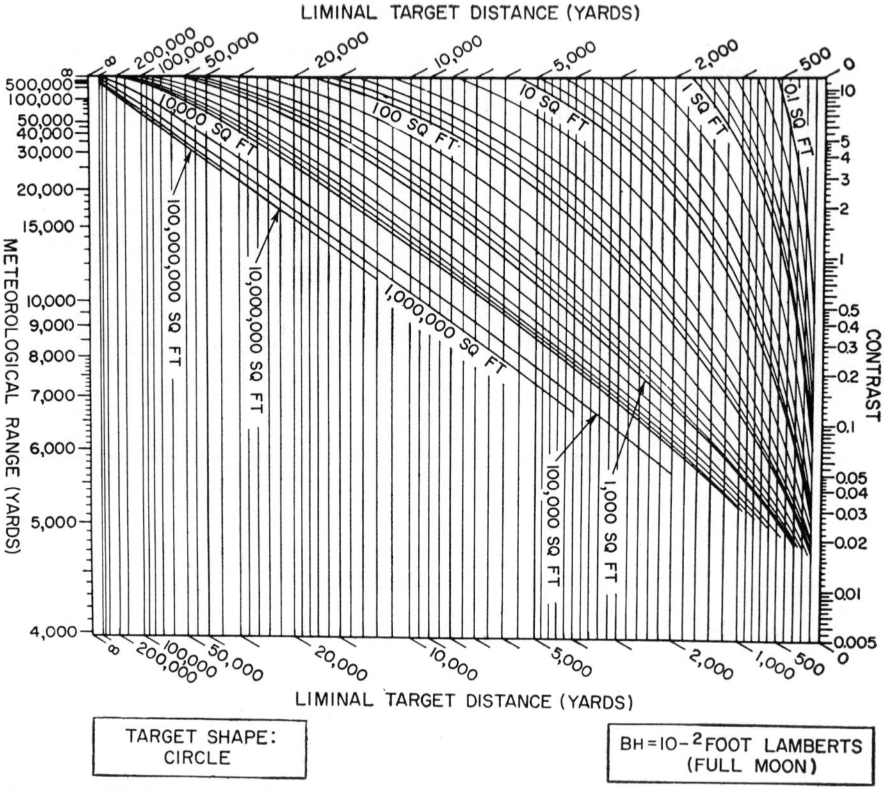

FIG. 6-10. Liminal target distance as affected by size of target and level of illumination. (See Ref. 19.)

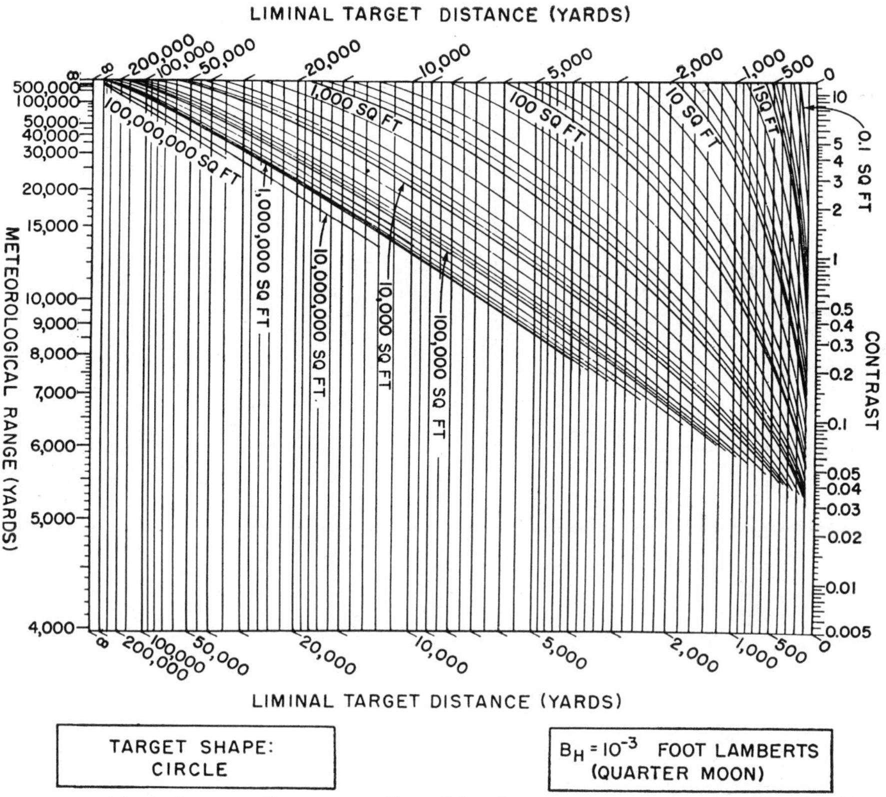

FIG. 6-11. Liminal target distance as affected by size of target and level of illumination. (See Ref. 19.)

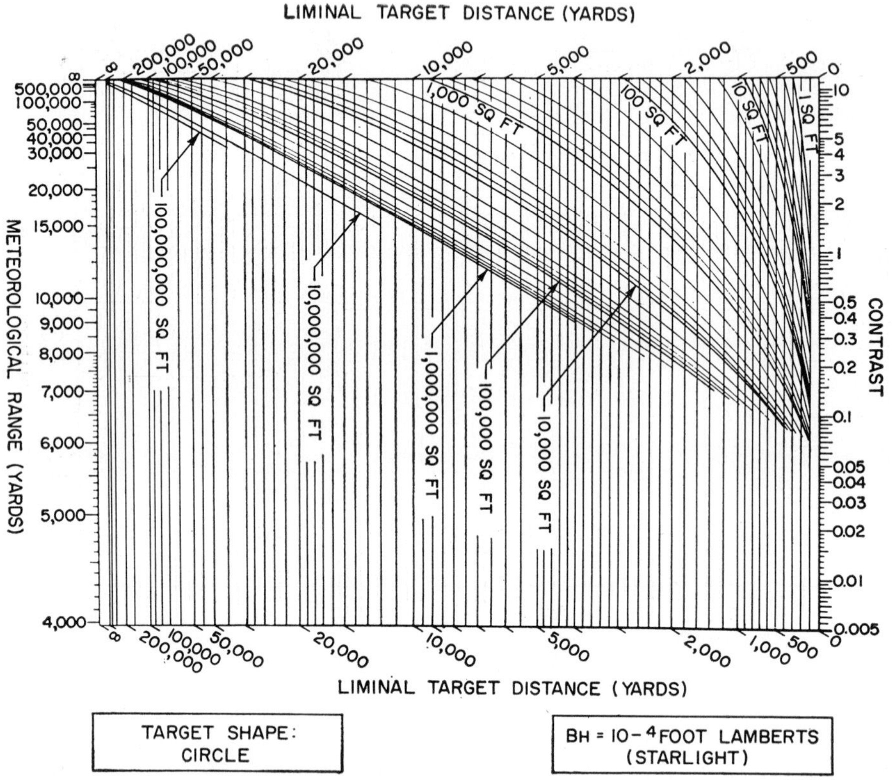

FIG. 6-12. Liminal target distance as affected by size of target and level of illumination. (See Ref. 19.)

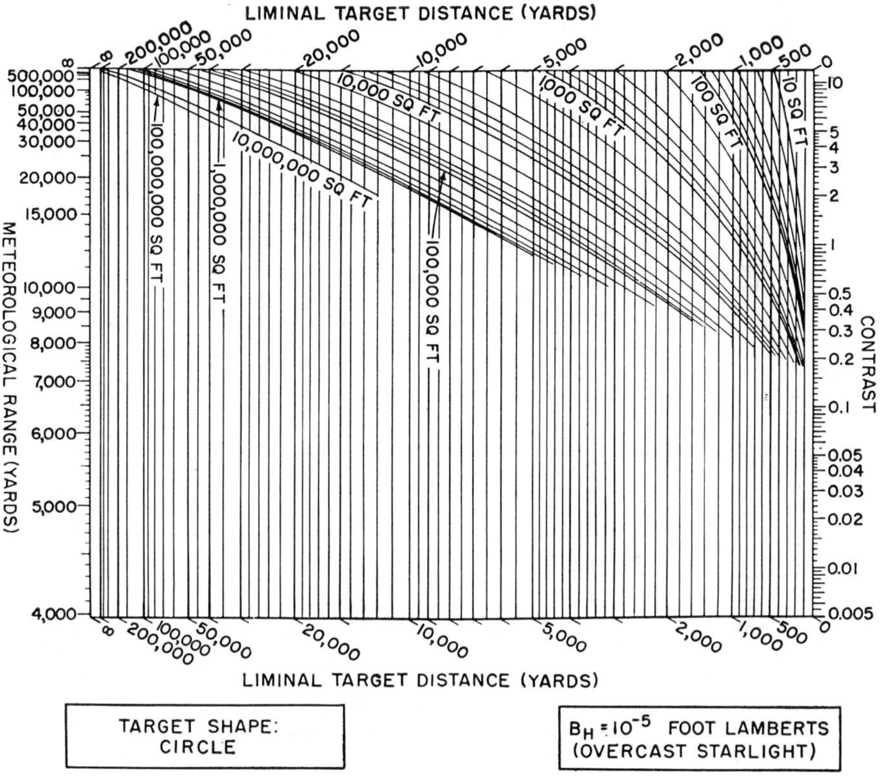

FIG. 6-13. Liminal target distance as affected by size of target and level of illumination. (See Ref. 19.)

that the numerical measures also differ. (The 0.02 of Wright and Löhle is a defined constant, used to convert measurements of attenuation into a defined visual range that has qualitative characteristics similar to those of the meteorological visual range.)

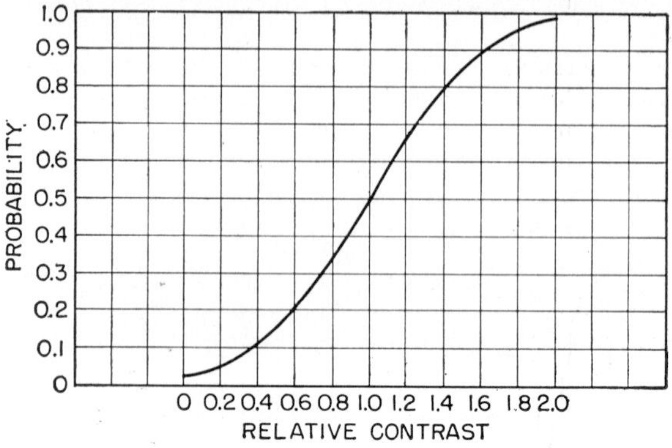

FIG. 6-14. Average probability curve. (See Ref. 7.)

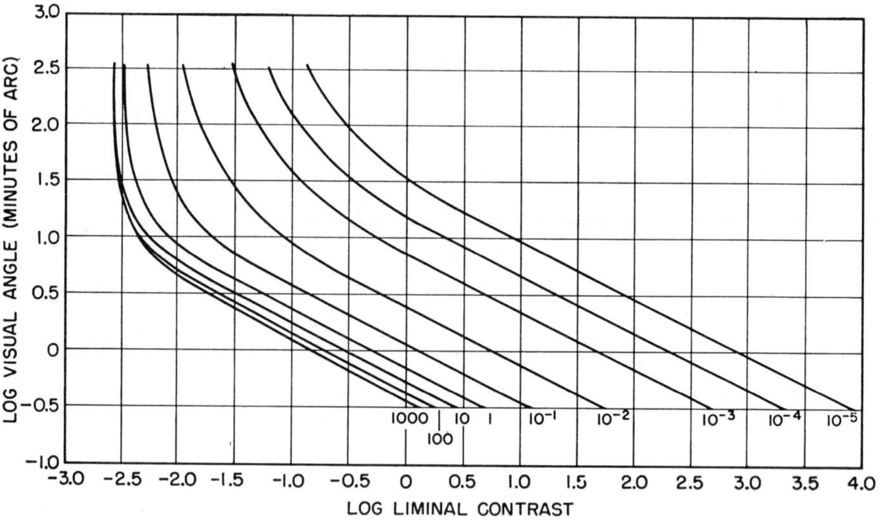

FIG. 6-15. The relation between threshold contrast and stimulus area for a given adaptation brightness. (See Ref. 7.)

If subjective observations of the visual range are to be used for determining the attenuation coefficient, the value of ϵ should be so chosen as to conform to the particular conditions of observation. Blackwell's data are particularly useful in suggesting how the selection should change with the conditions. The high variance of the values of ϵ reported by Middleton and Mungall emphasizes the need for caution in interpreting such observations.

6.4 THE INTERACTION OF LIGHT WITH SUSPENDED PARTICLES

Phenomenological theories like the one outlined above give no information about the interaction of the light with individual particles. The effect of altering these interactions is regarded merely as altering the numerical values of the constants (β

and 5). A more searching analysis, however, relates these numerical values to the physical and chemical characteristics of the particles. (See section on Physics of the Atmosphere in this handbook.) Such relations have interested some air pollution engineers as a way of determining what is in the air by measurements on the atmosphere itself.

The interaction of a plane electromagnetic wave with a sphere is the classical problem that was solved by Mie (39) and Debye (16). The most accessible treatment is by Stratton (51). A given sphere behaves as though it had different cross-sectional areas for different wavelengths of light. The factor by which the geometrical cross-sectional area of a sphere must be multiplied to give the effective cross-sectional area (i.e., effective in intercepting light) is called the scattering-area coefficient K_s.

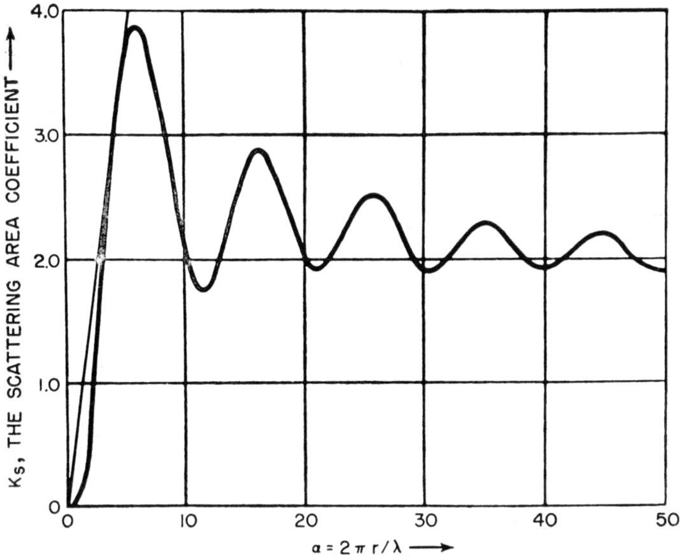

FIG. 6-16. The scattering-area coefficient K_s as a function of the radius α. This is the curve as smoothed by Houghton and Chalker, after the suggestion of Van de Hulst, for spherical drops of a dielectric of refractive index 1.33 (water). (See Ref. 40.)

The analysis and its results apply also to assemblages of randomly placed spheres, provided that the scattered light is incoherent (i.e., provided that the phases of the light scattered from the several spheres have no relation to one another).

In order to decrease the number of variables to be tabulated, it is customary to use a variable α, defined as $2\pi r/\lambda$, where λ is the wavelength of the light in question and r is the radius of the sphere. α is the number of times one wavelength of light will fit onto the circumference of the particle; it is the circumference in terms of the wavelength of the light used. Table 6-3 (48) and Fig. 6-16 (28) give the scattering-area coefficient K_s as a function of α for spheres having an index of refraction equal to that of water (1.33). Figure 6-17 (29) gives the corresponding plot for spherical particles of iron (index of refraction $1.27 - 1.37i$).

For an assemblage of such spheres, all of the same size, suspended in a total volume of 1 m cube,

$$\beta\, ds = \pi r^2 \bar{n} K_s\, ds = \frac{3WK_s\, ds}{4rD} = \frac{3\pi W K_s\, ds}{2D\lambda\alpha} \tag{6-21}$$

where $\bar{n}$ is the number of spheres, W is the total weight of the spheres, each per meter cube of the suspension, and D is the density of the condensed water or iron.

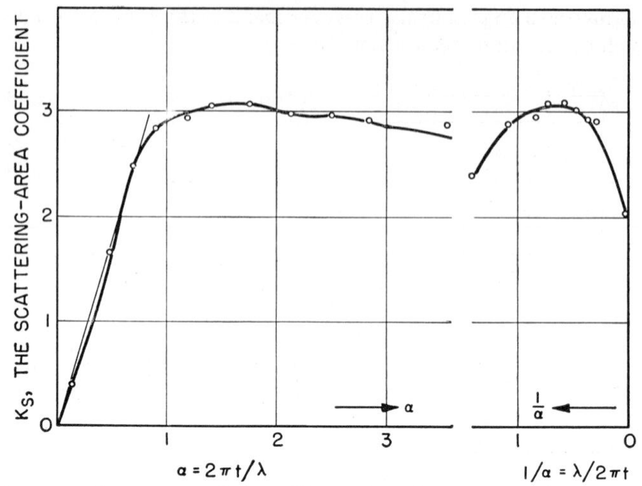

FIG. 6-17. The scattering-area coefficient K_s as a function of the radius α, for iron spheres. (See Ref. 29.)

The question is often raised as to the amount of material that is causing a particular lessened visibility. This question cannot be answered exactly unless both the material and its size distribution are known for the particular suspension being considered. The approximations that are usually made should be applied only with great caution in problems of air pollution, for the various components, both natural and man-made, interact with one another. Thus, the size of the droplets condensed on hygroscopic nuclei depends on the relative humidity of the air, and the effect on light of a soot particle imbedded in a water droplet is not simply the sum of the effects of the soot particle and the water droplet separately. Nevertheless, approxi-

Table 6-3. Scattering-area Coefficients for Water Drops in Air[a]

α	K_s	α	K_s	α	K_s	α	K_s
0.5	0.00676	8.0	3.282	12.125	1.776	17.0	2.632
0.6	0.0138	9.0	2.738	12.250	1.892	17.125	2.704
1.0	0.0938	9.5	2.394	12.333	1.936	17.25	2.822
1.2	0.171	10.0	2.152	12.5	1.938	17.375	2.820
1.5	0.322	10.25	2.052	12.6	1.822	17.5	2.738
1.8	0.522	10.5	1.886	12.75	1.850	18.0	2.598
2.0	0.710	10.625	1.918	13.0	2.012	18.5	2.432
2.4	1.126	10.75	1.930	13.5	2.192	19.0	2.218
2.5	1.212	10.875	1.832	14.0	2.474	19.25	2.090
3.0	1.754	11.0	1.740	14.5	2.528	19.5	1.998
3.6	2.376	11.125	1.728	15.0	2.744	19.75	1.976
4.0	2.826	11.333	1.734	15.5	2.740	20.0	2.092
4.8	3.490	11.4	1.768	16.0	2.870	20.25	2.180
5.0	3.592	11.5	1.858	16.25	2.872	20.5	2.078
6.0	3.888	11.75	1.758	16.5	2.850	21.0	1.834
.0	3.722	12.0	1.670	16.75	2.810	22.0	1.922
						24.0	2.433

[a] The scattering-area coefficient K_s as a function of the radius α in terms of the wavelength of the light for spherical drops of a dielectric of refractive index 1.33 (water).

Data from H. G. Houghton and W. R. Chalker, The Scattering Cross Section of Water Drops in Air for Visible Light, *J. Opt. Soc. Amer.*, **39**, No. 11, 955–957 (1949); "Tables of Scattering Functions for Special Particles," National Bureau of Standards, Applied Mathematics Series No. **4**, 1949; and "Smithsonian Meteorological Tables," 6th ed., rev., Smithsonian Institution, Washington, D.C. (1951).

mations are sometimes helpful, and Tables 6-4 and 6-5 and Figs. 6-18 and 6-19 simplify answering such questions in some cases.

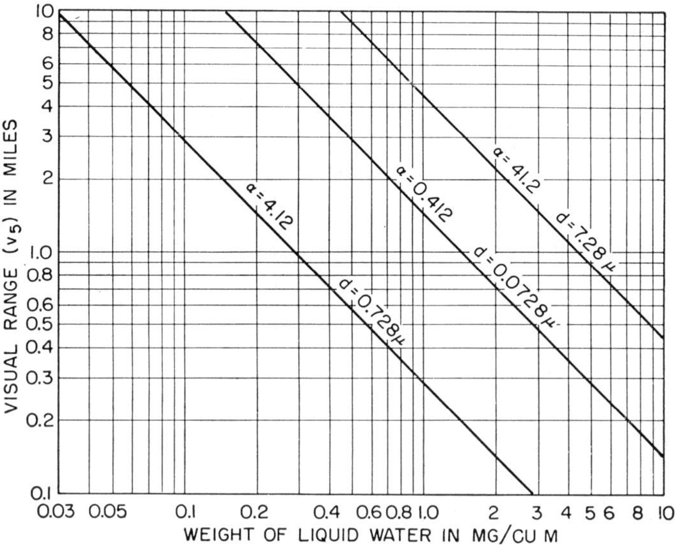

FIG. 6-18. Visual range as a function of weight of liquid water in the atmosphere.

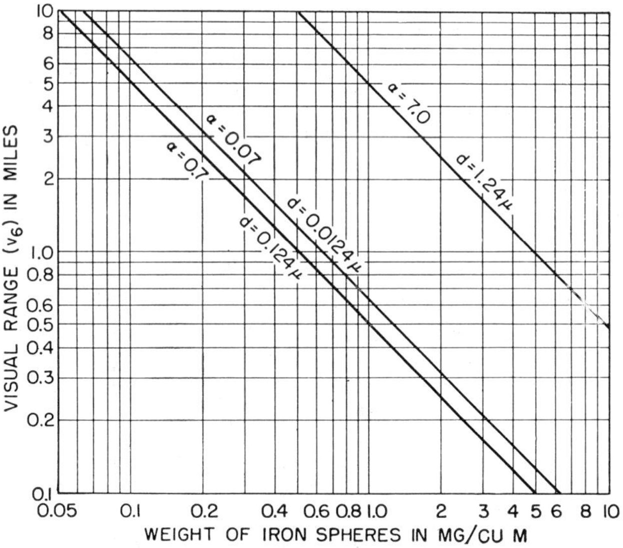

FIG. 6-19. Visual range as a function of weight of iron spheres in the atmosphere.

Equations (6-20) and (6-21) can be combined and solved for W to give

$$W = \frac{2\alpha\lambda \ln \epsilon}{3\pi K_s} \frac{1}{v} = 0.2194 \frac{\alpha}{K_s} \frac{1}{v_5} \tag{6-22}$$

where W is the milligrams of pollutant in the spheres that are suspended in a cubic meter of air and the visual range v_5 is in miles. These tables and graphs are con-

Table 6-4. Weight of Water, in Droplets of Certain Sizes, Corresponding to Particular Visual Ranges

Visual range v_5, miles	Weight of liquid water W, mg/cu m		
	α = 0.412 d = 0.0728μ	α = 4.12 d = 0.728μ	α = 41.2 d = 7.28μ
0.1	14.79	2.86	45.2
0.2	7.40	1.429	22.6
0.3	4.93	0.953	15.05
0.4	3.70	0.715	11.29
0.5	2.96	0.572	9.03
0.6	2.47	0.476	7.53
0.7	2.11	0.408	6.45
0.8	1.849	0.357	5.64
0.9	1.643	0.318	5.02
1.0	1.479	0.286	4.52
2.0	0.740	0.1429	2.26
3.0	0.493	0.0953	1.505
4.0	0.370	0.0715	1.129
5.0	0.296	0.0572	0.903
6.0	0.247	0.0476	0.753
7.0	0.211	0.0408	0.645
8.0	0.1849	0.0357	0.564
9.0	0.1643	0.0318	0.502
10.0	0.1479	0.0286	0.452

Table 6-5. Weight of Iron, in Spheres of Certain Sizes, Corresponding to Particular Visual Ranges

Visual range v_5, miles	Weight of iron W, mg/cu m		
	α = 0.07 d = 0.0124μ	α = 0.7 d = 0.124μ	α = 7.0 d = 1.24μ
0.1	6.3	5.0	49.
0.2	3.2	2.5	25.
0.3	2.1	1.68	16.4
0.4	1.58	1.26	12.3
0.5	1.26	1.01	.9
0.6	1.05	0.84	8.2
0.7	0.90	0.72	7.0
0.8	0.79	0.63	6.2
0.9	0.70	0.56	5.5
1.0	0.63	0.50	4.9
2.0	0.32	0.25	2.5
3.0	0.21	0.168	1.64
4.0	0.158	0.126	1.23
5.0	0.126	0.101	0.99
6.0	0.105	0.084	0.82
7.0	0.090	0.072	0.70
8.0	0.079	0.063	0.62
9.0	0.070	0.056	0.55
10.0	0.063	0.050	0.49

structed for a wavelength of 5,550 A, which is close to that for maximum sensitivity of the eye. The contrast limen is taken as 0.05 because this value corresponds approximately to the results of many field observations. Water is chosen as a typical substance that might be present in the air as droplets. Although water by itself is not usually considered an air pollutant, water condensed on hygroscopic nuclei is a characteristic component of urban atmospheres, and the index of refraction of such droplets is presumably not greatly different from that of water. However, even droplets of such a liquid as kerosene would not have a refractive index so different from that of water as to vitiate the very rough estimates that are under consideration.

A minimum value of W for a given visual range corresponds to maximum effectiveness of the dispersed material. W will be a minimum, from Eq. (6-22), when α/K_s is a minimum or K_s/α is a maximum. K_s/α is a maximum, from Fig. 6-16, at $K_s = 3.16$ and $\alpha = 4.12$. For a given quantity of material, subdivision into particles of this size gives the maximum attenuation and the minimum visual range that can be attained under the conditions set for this problem. Table 6-4 is calculated for three values of α, viz., this most efficient size, ten times as large, and one-tenth as large. Natural suspended material shows a wide range of particle sizes, but calculations based on these formulas give only the effects to be expected if all the particles are of one size.

If one wishes to take account of the effect of various particle sizes, it is necessary to rewrite Eq. (6-21) as

$$\beta\, ds = \pi(\Sigma r_j^2 \bar{n}_j K_{sj})\, ds \qquad (6\text{-}23)$$

It is now no longer possible to introduce W in the simple manner of Eq. (6-21), for

$$W = \tfrac{4}{3}\pi \Sigma \bar{n}_j r_j^3 D_j = \tfrac{4}{3}\pi D \Sigma \bar{n}_j r_j^3 \qquad (6\text{-}24)$$

if we neglect the small variation of density with particle size. If ψ is defined as the ratio $\Sigma \bar{n}_j r_j^3 / \Sigma \bar{n}_j r_j^2 K_{sj}$, we can rewrite Eq. (6-23) as

$$\beta\, ds = \frac{3}{4}\frac{W}{\psi D}\, ds \qquad (6\text{-}25)$$

$$W = \frac{4}{3}\psi D \beta = -\frac{4}{3}\frac{D}{v} \ln - \epsilon \qquad (6\text{-}26)$$

This equation reduces to a slightly simpler form, which is, however, much easier to apply, if K_s is not a function of r. This assumption is permissible in the range of particle sizes of major interest in clouds and fogs, where K_s is approximately 2.0. This method has accordingly been used by the meteorologists (27,31), but in this simple form the equation is inapplicable to the problems of air pollution.

Even without extending the theory to heterogeneous distributions of suspended material, one might anticipate that the visual range would be inversely proportional to the concentration of the suspended material, so long as other conditions were not much changed. This anticipation is borne out by the calculation of the regression of visibility on smoke at Leicester (22). When the concentration of smoke doubled, the visual range decreased by the factor 0.57 ± 0.07. This factor does not differ significantly from 0.50, and it did not change significantly from summer to winter.

The attenuation coefficient for opaque conducting particles is a complex number because of the complex index of refraction of conducting materials. Table 6-5 and Fig. 6-17 (29) give the scattering-angle coefficient K_s as a function of α for a fairly extreme case, that of metallic iron. Carbon would be more interesting to the air pollution engineer, but does not seem to have been worked out completely. The index of refraction of carbon has been measured by Senftleben and Benedict (45), with the results shown in Table 6-6. The first few coefficients needed in the calculation are given by Ruedy (43,44).

Table 6-5 is calculated in the same fashion as Table 6-4. The only striking difference is that, for conducting particles, the visual range resulting from a given quantity of the substance suspended per square meter is substantially independent of the size of the particles, if they are smaller than about 0.1μ.

Half a gram to a gram per square meter of either water or iron in spheres of the most effective particle size suffices to keep one from seeing through the cloud. Since particles much larger than the optimum tend to settle, and some of those much smaller tend to agglomerate by collisions, such suspensions as concern the air pollution engineer are often found to be near the most effective particle size. Thus it was reported at Leicester that 0.5 g of their pollutant (presumed to be smoke) per square meter corresponds to the visual range.

Table 6-6. Index of Refraction of Carbon

Wavelength, μ	*Refractive Index*[a]
0.436	1.90–0.68i
0.492	1.94–0.66i
0.546	1.96–0.66i
0.578	1.97–0.65i
0.623	2.00–0.66i

[a] The estimates of the standard deviations for these figures are 0.016 for the real part and 0.0088 for the imaginary.

Data from R. Ruedy, Absorption of Light and Heat Radiation by Small Particles. I. Absorption of Light by Carbon Particles, *Can. J. Research*, **19A**, 117–125 (1941); II. Scattering of Light by Small Carbon Particles, *ibid.*, **20A**, 25–32 (1942).

6.5 SIZE DISTRIBUTION FROM MEASUREMENTS OF ATTENUATION

The techniques for measuring the optical properties of atmospheres have been elaborated principally by the astrophysicists, and their equipment is out of reach for most air pollution engineers, because of both expense and complexity. The method for finding some characteristics of the size distribution of the suspended material can, however, be readily adapted, since it merely involves measuring the attenuation coefficient β at two or more wavelengths.

Following Henyey and Greenstein (25), suppose that the aerosol of Fig. 6-1 contained $n(r)$ particles of radius r per cubic centimeter and that the aerosol is so dilute that the particles do not overlap significantly. For purposes of this discussion, let $\mathfrak{J} = 0$, i.e., assume that the suspension does not itself contribute flux. Then Eq. (6-2) becomes

$$\frac{dI}{I\,dx} = -\int \pi r E_1(r) n(r)\,dr = -\int \beta_r\,dr = -\beta \qquad (6\text{-}27)$$

where $E_1(r)$ is the "extinction," defined as the ratio of the energy lost by the incident wave to the energy of the pencil that is geometrically obstructed. $E_1(r)$ is a function of the wavelength of the light and of the refractive index of the particle as well as of its radius.

A distribution function that has been useful in the astrophysical analogue of this problem as well as elsewhere, and which we shall adopt, is

$$n(r) = n_0 r^{-q} \qquad (6\text{-}28)$$

where q is to be determined from the observational data. Then

$$\beta = \int_0^\infty \pi r^2 E_1(r) n_0 r^{-q}\,dr \qquad (6\text{-}29)$$

Let $\alpha = 2\pi r/\lambda$, then

$$\beta = \pi n_0 \left(\frac{\lambda}{2\pi}\right)^{3-q} \int_0^\infty \alpha^{2-q} E(\alpha)\,d\alpha \qquad (6\text{-}30)$$

where $E(\alpha)$ is now not a function of wavelength for dielectrics and hence

$$\beta = \pi n_0 K_s \left(\frac{\lambda}{2\pi}\right)^{3-q} \qquad (6\text{-}31)$$

where K_s can be calculated from the Mie theory. For light of two different wavelengths,

$$\frac{\beta_2}{\beta_1} = \left(\frac{\lambda_2}{\lambda_1}\right)^{3-q} \tag{6-32}$$

or

$$q = 3 - \frac{\log (\beta_2/\beta_1)}{\log (\lambda_2/\lambda_1)} \tag{6-33}$$

6.6 INSTRUMENTS FOR MEASURING VISUAL RANGE OR ATTENUATION COEFFICIENT

No instruments are in general use at present for measuring either the visual range or the attenuation coefficient for the purposes of the air pollution engineer or of the meteorologist. Many have been proposed, but the variability of even an unpolluted atmosphere with direction makes it difficult to replace an observer who can use his judgment with a machine that cannot. Moreover, those instruments which are intended merely to aid the observer are seldom used because the difficulty is in the averaging, not in the observing in a particular direction.

There are at least three cases, however, such that instruments are useful. One arises when visibility along a particular line is needed and either when experienced observers are not always available at a particular location or when an objective measurement is needed. An example is a runway for aircraft, when conditions on it cannot be satisfactorily appraised from the observer's station. A second example is a manufacturing plant that emits visible pollution, whose manager wishes to assure himself and others that he is not allowing a plume of pollution to drift onto a particular sensitive area, such as a residential district, an airfield, or a ship channel. An instrument so placed that the plume would necessarily register on it would meet such a situation.

A second case is the measurement of attenuation over a short path to provide a record of conditions in the immediate vicinity of the path or to give information about the suspended material. The air pollution engineer uses such measurements to supplement other means of estimating the suspended material in the air and to investigate its nature when that is in doubt.

The third case is the estimation of the intensity and amount of pollution from a single source. Whether or not a stack is smoking and how badly it smokes if it is smoking are perhaps the most common of all observational problems in air pollution.

Three types of instruments have been used for measuring attenuation coefficients: The first two measure the transmitted light or the scattered light from a beam passing through the atmosphere; the third measures the apparent brightness of an object of known brightness at a known distance. In addition to these instrumental methods, the attenuation coefficient can be calculated from observations of the visual range.

Many instruments have been built to measure the transmission of a beam of light through the atmosphere. Two recently developed ones are that of Bradbury and Fryer (9), as modified by the Stanford Research Institute for laboratory measurements, and that of Douglas and Young (17).

The arrangement of the SRI modification of Bradbury and Fryer's instrument is shown in Fig. 6-20. Mechanically chopped light passes repeatedly through a tube containing the air being tested and falls, finally, on a photocell. The output from the photocell is amplified and recorded. Air is mixed with whatever reagents are to be tested and drawn through the tube. Light is thrown directly from the source onto the photocell for calibration.

The Douglas and Young transmissometer (17) is available commercially at a price of about $3,000 (1952) from the Crouse-Hinds Company of Syracuse, N.Y. The equipment has five major components: an incandescent lamp and projector, a power supply for this source, a receiver, a power supply for the receiver, and a remote-station indicator and recorder. The transmissometer is designed to measure the transmission over a path of 500 ft, and the indicating and recording equipment can be as much as 10 miles away from the observation point. The instrument can be used both by day and by night and would seem to be the best available solution to the first class of applications mentioned above.

All instruments of this and the following types suffer from one theoretical drawback. Scattering by particles that are larger than the wavelength of light is far from uniform in direction; it is concentrated within a few degrees of the direction of the original beam (see also Sec. 4). Some of this scattered light is necessarily included with the unscattered beam in the measurements. The amount is unknown experimentally, and it is not clear how to take account of it theoretically even if the amount were determined.

A second type of instrument measures the scattered light. This type has been used less except for intentionally contaminated air. It is sensitive, but the results cannot be interpreted in terms of visual range unless the absorption coefficient is also known for the suspended material.

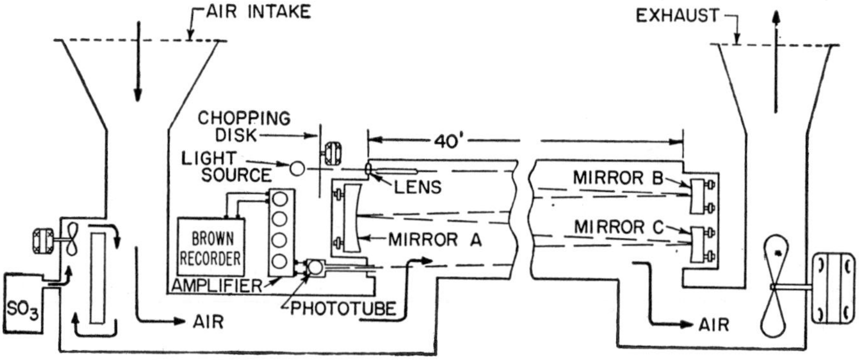

FIG. 6-20. Schematic arrangement of transmissometer. Transmissometer at the laboratories of the Stanford Research Institute used to measure the effect of various substances in decreasing visibility.

It seems to be commonly assumed (6,53) that in unpolluted air the attenuation is caused practically exclusively by scattering. Since this assumption is clearly not valid for dense black smoke of relatively large particle size, a combination of measurements of attenuation and scattering on the same sample of air might be used to estimate the degree of such contamination, but such measurements do not seem to have been reported.

A forward-scattering tyndallometer diagrammed in Fig. 6-21 has been used by Sinclair and LaMer (47) to measure 10^{-9} g/liter (0.001 mg/cu m) of stearic acid smoke. S is an automobile headlight bulb; L_1 and L_2 are aspheric condenser lenses of about 2¼-in. focal length; Sc is a screen with a 3⁄16-in. diameter hole; D is an opaque disk about ½ in. in diameter; C is either a visual or photoelectric measuring device; and W is a window that is smaller than D.

Figure 6-22 diagrams a related instrument (11) that operates photographically to record particles down to 0.3μ.

A similar instrument has been used to count suspended particles down to 0.6μ at 1 to 1,000 per minute (20,23,24).

Beutell and Brewer (6) report the design and test of several models of nephelometric instruments. Some of these are, or could readily be adapted to be, useful either day or night and for all visual ranges from clear air to dense fog. The sample path is a few centimeters long.

Berek, Männchen, and Schäfer (4) describe a commercially made portable tyndallometer (Leitz) for measuring the concentrations of dusts that might be encountered in industrial hygiene work (Fig. 6-23). The calibration covers the range 10 to 220 mg/cu m. The price (1953) with accessories is $1,740. A simplified instrument (the tyndallscope) is also available.

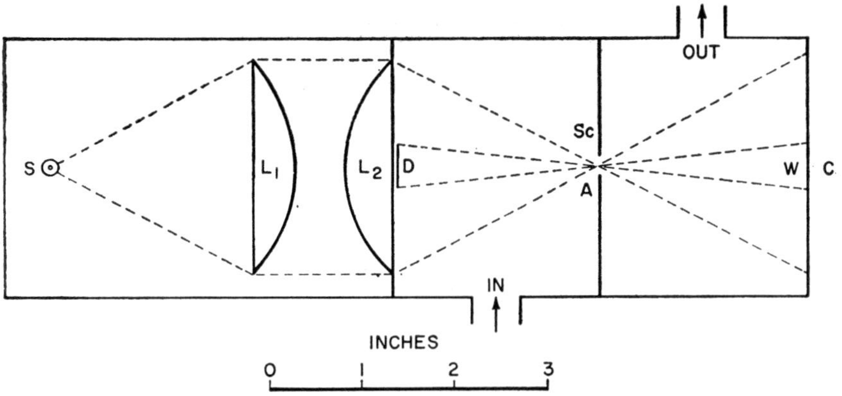

FIG. 6-21. Forward-scattering tyndallometer.

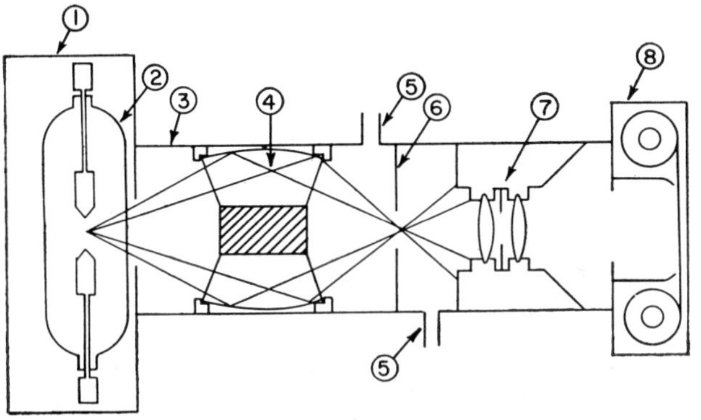

1. FLASH TUBE AND TRIGGER CIRCUIT BOX
2. GE FT-230 FLASH TUBE
3. TUBULAR BODY
4. LUCITE ELLIPTICAL REFLECTING CONDENSER
5. AIR FLOW INLET AND OUTLET
6. FIELD DEFINING BAFFLE
7. 25mm PHOTOGRAPHIC OBJECTIVE
8. 35mm CAMERA BOX

NOTES:
IMAGE OF FLASH DISCHARGE IS FOCUSED ON DEFINING BAFFLE BY ANNULAR LUCITE CONDENSER IN CONICAL BEAM WHICH DOES NOT ENTER OBJECTIVE LENS

DEFINING BAFFLE AND FILM ARE AT CONJUGATE FOCI OF 25mm OBJECTIVE, IMAGING SCATTERED LIGHT FROM PARTICLES WITHIN DEPTH OF FOCUS OF LENS

FIG. 6-22. Diagram of portable tyndallometer used as "dust camera."

The Rubicon Company of Philadelphia, Pa., markets a portable instrument for visual examination of the Tyndall beam. It is not intended to be quantitative, but it is fairly easy to learn to distinguish between acceptable and objectionable concentrations if the same pollutant is repeatedly examined.

Such instruments as these seem especially adapted to observing conditions at a particular location.

The third type of instrument, which measures the apparent brightness of objects of known brightness, has been extensively used, probably because it is a natural instrumental extension of the usual visual method of estimating the visual range. Much early work with instruments of this type was vitiated, however, by inadequate regard for some of the limitations laid by the theory or by inadequate technique. The most recent extensive investigation, which is free from these objections, is that by Coleman and his collaborators (12,13,14).

Coleman measured the attenuation of contrast by the atmosphere. He used a set of seven targets, together with the horizon sky. Each target had a black circular area, which was actually the opening into a black box, and a white (painted) area beside it. The targets were so graduated in size that each black area subtended 1.86 min at the observation point. All were close to the horizon. The brightness of

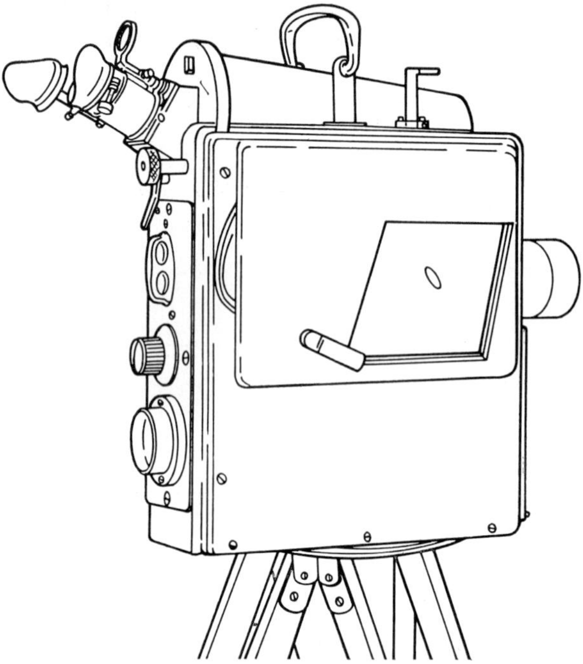

FIG. 6-23. Leitz tyndallometer. (*Courtesy of E. Leitz, Inc.*)

each of the targets and of the horizon sky was measured through a telescope in which the stray light was reduced practically to zero.

Three types of equipment were used for the actual measurements, viz., a photocell, photographic photometry, and visual photometry. Diagrams of the equipment are shown in Figs. 6-24 to 6-26. The results with the different types of equipment agreed with one another.

Five filters isolated spectral regions for measurement. Four were interference filters with maximum transmissions at 4,170, 4,980, 5,670, and 6,540 A. The width of the transmitted band at half the maximum was 160 to 180 A. The fifth filter was so matched to the photomultiplier tube as to approximate the spectral sensitivity of the eye.

The data were plotted as shown in Fig. 6-27. A dashed line, parallel to the line through the observations on the black targets, through the intercept

$$4.6052 \ (= \log_e 100)$$

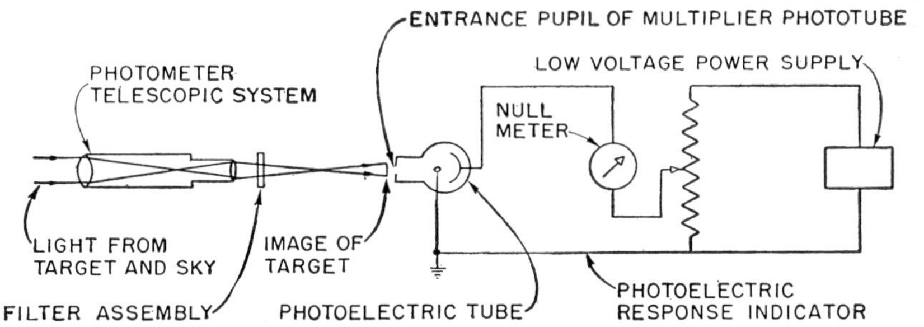

FIG. 6-24. Schematic diagram of the photoelectric telephotometer. (See Ref. 12.)

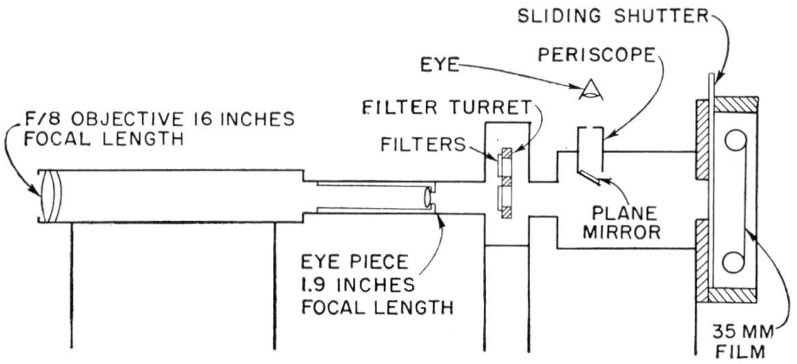

FIG. 6-25. Coleman's photographic telephotometer. (See Ref. 13.)

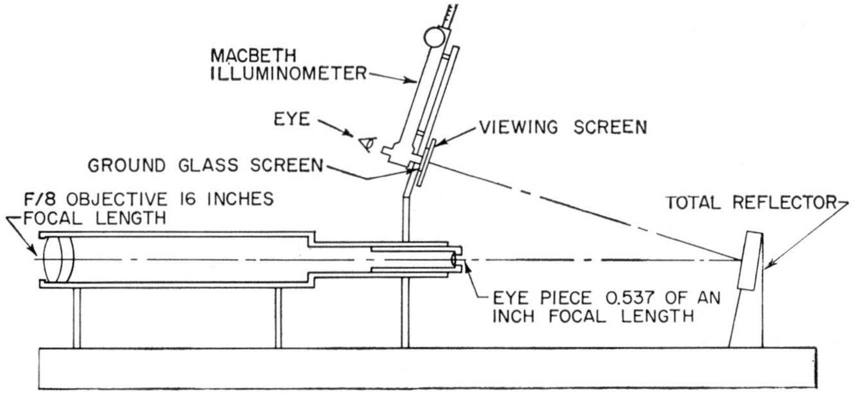

FIG. 6-26. Coleman's visual telephotometer. (See Ref. 14.)

gives the results that would be expected for perfectly black targets. This diagram is a plot of Eq. (6-9) in the form

$$\ln C = \ln C(0) - \beta s \tag{6-34}$$

The slope gives the value of the attenuation coefficient directly. No correction for incomplete blackness of target is needed as long as all the targets have the same blackness.

Figure 6-28 shows the lines obtained by using different filters in the telephotometer.

Coleman's data show that the laws of attenuation derived above hold for many more cases of interest than the quite restrictive assumptions made in the derivation

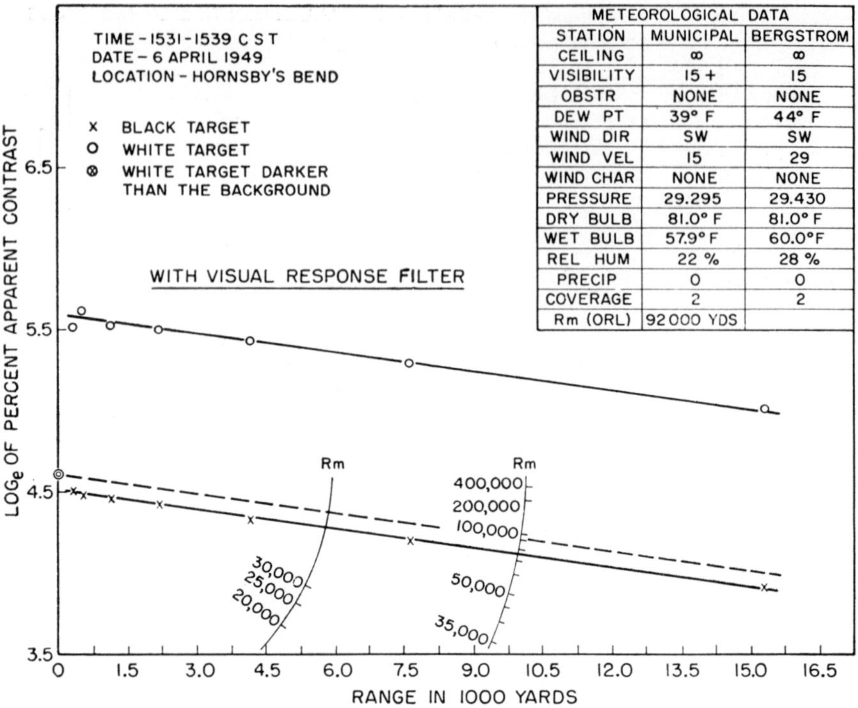

FIG. 6-27. Log$_e$ of per cent apparent contrast vs. range. (See Ref. 12.)

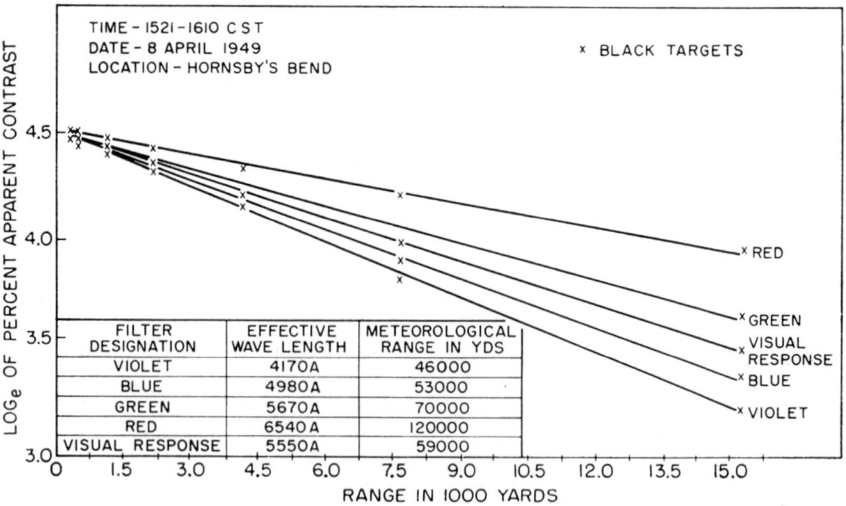

FIG. 6-28. Log$_e$ of per cent apparent contrast vs. range using different filters. (See Ref. 12.)

would indicate. The attenuation coefficient and the meteorological range are surprisingly insensitive to many conditions that might be expected to affect measurements in the field.

A second method of this class that has been used in air pollution studies involves photographing a black object, at a known distance and near the horizon sky, and measuring the density of the images of the object and the horizon sky with such an instrument as the Weston photographic analyzer (11,49,50).

The theory as given above (for all methods of this class) requires a black target. Since a real target will not be black, it is of some interest to ask how much error is introduced thereby if it is not eliminated by Coleman's or a similar method. It is not very difficult to set down equations for a target of specified initial brightness—the difficulties arise in using them experimentally. The brightness of any target except a black one varies with its illumination and must therefore be measured or calculated for each observation. As a practical matter, therefore, the target must be so dark that the assumption that it is black does not introduce significant error. As an additional precaution, most targets must be in the shade, although this precaution is not necessary for such unusual targets as an abandoned adit to a coal mine. Table 6-7 gives the albedo of various objects, as a guide for selection.

Table 6-7. The Albedo of Various Objects

	Per Cent
Forest, green	3–10
Forest, snow-covered ground	10–25
Ground, bare	10–20
Black mold, dry	14
Black mold, wet	8
Sand, dry	18
Sand, wet	9
Grass, dry	15–25
Grass, high dry	31–33
Grass, high fresh	26
Grass, wet	22–37

Data from miscellaneous sources.

Assuming that the object is black is equivalent to assuming that $I(0) = 0$. The fractional error in the visual range that arises from neglecting $I(0)/\mathfrak{I}$ ranges from 0.02 at $I(s)/\mathfrak{I} = 0.9$ to 0.1 at $I(s)/\mathfrak{I} = 0.4$, if $I(0)/\mathfrak{I}$ is really 0.05. The objects that are used must evidently be quite dark if this term is to be neglected. The ratio $I(0)/\mathfrak{I}$ has been calculated (21) to be half the albedo of the object, if the object is lit by a uniformly illuminating sky. The visual range, under a clear sky, of a gray object of albedo 0.25, in the shade, against the horizon sky is over 98 per cent of that of a black object (54). $I(0)/\mathfrak{I}$ seems likely, therefore, to be about one-quarter to two-thirds of the albedo for objects that are not in direct sunlight.

Color, in targets that are dark enough to be acceptable, is not of much importance in visual observations. Close to the visual range, the target is overlain by so much haze that its own color is almost entirely lost. For instrumental observations, where the target is not close to the visual range, the effect of color may have to be included in the observation calibration, but it is probably always of minor importance compared to other errors.

6.7 THE MEASUREMENT OF THE INTENSITY AND EFFECT OF A SINGLE SOURCE OF SMOKE

The commonest way of describing the black smoke from a source is to compare it to the shades of gray of the Ringelmann chart. The following description of the chart and its use is quoted from *Information Circular* 6888 of the U.S. Bureau of Mines (33).

"The Ringelmann system is virtually a scheme whereby graduated shades of gray, varying by five equal steps between white and black, may be accurately reproduced by means of a rectangular grill of black lines of definite width and spacing on a white background. The rule given by Professor Ringelmann by which the cards may be reproduced is as follows:

Card 0—All white.

Card 1—Black lines 1 mm thick, 10 mm apart, leaving white spaces 9 mm square.

Card 2—Lines 2.3 mm thick, spaces 7.7 mm square.

Card 3—Lines 3.7 mm thick, spaces 6.3 mm square.

Card 4—Lines 5.5 mm thick, spaces 4.5 mm square.

Card 5—All black.

"Copies of the chart may be obtained free upon request of the Director, Bureau of Mines, Washington, D.C.

"To learn to use the chart, it is hung on a level with the eye, about 50 feet from the observer, as nearly as possible in line with the chimney. The observer glances from the smoke as it issues from the chimney and notes the number of the chart most nearly corresponding with the shade of the smoke, and records this number with the time of observation. Observers with proper experience find it unnecessary to continue to refer to the chart. A clear chimney is recorded as No. 0, and 100 per cent black smoke as No. 5. Observations are repeated at one-fourth or one-half minute intervals. The readings are then reduced to the total equivalent of No. 1 smoke as a standard. No. 1 smoke being 20 per cent dense, the percentage density of the smoke for the entire period of observations is obtained by the formula:

$$\frac{\text{Equivalent units of No. 1 smoke} \times 0.20}{\text{Number of observations}} = \text{percentage smoke density.''}$$

See also Figs. 6-29 and 6-30.

For use by people other than trained official inspectors, an observation card like that shown in Fig. 1-7, Sec. 1, is helpful.

These measurements are direct and require no equipment in the field. The results are, of course, in arbitrary units and become somewhat doubtful if the smoke is not black. Nevertheless, it is an important method and seems likely to remain one because its use is specified in most legal codes. "Many objections have been raised to the shade method of judging the amount of smoke coming from a stack, such as the effects of the diameter of the stack, the cloud and wind conditions, and the human element in judging shade, all of which are considered to affect the result. These comments and objections, however accurate, become inconsequential when the entire purpose of the readings is considered. The purpose is not merely to fine some violator because smoke of a certain shade has been emitted, concerning which it may be argued just what percentage of blackness occurred or how many black particles were emitted. The purpose is to lessen smoke emission and to improve cooperation in making a cleaner neighborhood. The smoke-chart scheme has proven effective and is simple; moreover, no more practical method has yet been devised" (3). For such objections, backed by observations, see Ref. 36.

Such remarks may constitute adequate justification for using the Ringelmann chart for legal control; they do not offer any means of using such observations to give the quantity of pollution. Neither theory nor experiment has advanced far enough as yet for us to do this. In such observations, the smoke is regarded as an object of variable inherent density. This case is quite different from that discussed in the theoretical sections above and does not seem to have been treated. Experimentally, the variables that must be taken into account are now quite well understood (2), but there do not seem to have been studies on smokes in which all were controlled or observed.

FIG. 6-29. Ringelmann's scale No. 4 for grading the density of smoke. Complete chart consists of five blocks similar to the one above but of differing degrees of blackness. It has lines and space of the sizes illustrated, but each panel is about 5¾ by 8½ in.

The measurement of transmission through the smoke in a stack, by instruments described elsewhere in this handbook, is also capable in principle of yielding information on the pollution contributed by such a source and its effect on the visual range. Again, the intermediate investigations that are needed for such a use of the observations are lacking.

LOCATION. . . . .

HOUR. 9:00 – 10:00 A.M. . . . . DATE. . . . .

9	0	1/4	1/2	3/4		0	1/4	1/2	3/4
0	—	—	—	—	30	1	1	1	1
1	—	—	—	—	31	1	1	1	1
2	—	—	—	—	32	—	—	—	—
3	1	1	1	1	33	—	—	—	—
4	1	1	1	1	34	—	—	—	—
5	2	2	2	2	35	1	1	1	1
6	2	3	3	3	36	1	1	1	1
7	3	3	3	3	37	1	1	1	1
8	2	2	1	1	38	1	1	—	—
9	1	1	—	—	39	—	—	—	—
10	—	—	—	—	40	—	—	—	—
11	—	—	—	—	41	—	—	—	—
12	—	—	—	—	42	—	—	—	—
13	—	—	—	—	43	—	—	—	—
14	—	—	—	—	44	1	1	2	2
15	—	—	—	—	45	2	2	3	3
16	—	—	—	—	46	3	3	3	3
17	—	—	—	—	47	3	3	4	3
18	—	—	—	—	48	2	2	2	2
19	2	2	2	2	49	2	2	2	2
20	2	2	2	2	50	2	1	1	1
21	2	2	2	2	51	1	1	1	1
22	3	3	3	3	52	1	1	1	—
23	3	4	4	4	53	—	—	—	—
24	4	5	5	5	54	—	—	—	—
25	5	5	5	5	55	—	—	—	—
26	4	4	3	3	56	—	—	—	—
27	3	3	3	3	57	—	—	—	—
28	2	2	1	1	58	—	—	—	—
29	1	1	1	1	59	—	—	—	—

POINT OF OBSERVATION

. . . .

DISTANCE TO STACK. . . . . . .

DIRECTION OF STACK. . . . . . .

DIRECTION OF WIND. . . . .

VELOCITY OF WIND. . . . .

		EQUIV. NO. 1 UNITS
7	UNITS NO. 5	35
7	UNITS NO. 4	28
27	UNITS NO. 3	81
34	UNITS NO. 2	68
52	UNITS NO. 1	52
113	UNITS NO. 0	0
240	UNITS	264

$\frac{264}{240} \times 20\% =$

. . . 22% . . . SMOKE DENSITY

OBSERVER. . . . .

CHECKED BY. . . . .

FIG. 6-30. Ringelmann chart reading.

There are also instruments, described elsewhere in this handbook, that deposit pollutants from the atmosphere onto a collecting surface. The color or transmission of such deposits is measured, and some effort has been made to correlate such measures with the visibility. Aside from the usual sampling problems that are associated with the requirement that the equipment be efficient for small particles, one must consider the effects caused by the proximity of the particles to one another in the collected sample.

6.8 OBSERVED VISUAL RANGE AND POLLUTION

Before proceeding with this article, it may perhaps be well to trace the argument connecting visual range and air pollution. The visual range can be defined, somewhat arbitrarily, as was shown above, in terms of the attenuation coefficient and a selected value for the psychophysical constant. Moreover, the visual range so defined seems to correspond reasonably well with the meteorologically defined visual range in the cases where they can be compared, although the latter has a greater variance, which is not matched by the former. This greater variance is presumed to be caused by the many factors that influence the estimation of the meteorological visual range and that are excluded from the definition of the standard visibility. The next step in the reasoning, the relation between the attenuation coefficient and the amount and kind of material suspended in the atmosphere, has likewise been discussed above and is fairly well understood for simple cases. In contrast to these, the final step, the relation between the amount and kind of suspended material and the "air pollution," can at present be discussed only in a vague and qualitative manner. The reason is, of course, that we do not have any generally accepted, universally applicable, quantitative definition of air pollution. All that can be done at present, then, is to present a few examples to show the manner in which various investigators have sought to close this gap at particular places and times, and with reference to particular complexes of pollutants. "The contribution of visibility studies (to the problem of atmospheric pollution) has to date been strictly limited" (26).

The conditions in a polluted urban atmosphere are not uniform, either in space or time. The observer takes the best average he can of a visual range that differs with direction, and he does not usually, of course, have marks in all directions. The observed visual ranges change hour by hour, and sometimes in a few minutes. These changes are superimposed on the already large fluctuations of visibility at unpolluted sites. The resulting fluctuations are so great that conclusions can usually be drawn only from series of observations that extend over many years.

In spite of these difficulties, observations of visual range or related quantities are recorded because of practical interest in the subject and because they are often the best that can be made under existing conditions. Various techniques have been used in interpreting and presenting such data, and some of these will be given in the following examples.

Observations at Prague, in Czechoslovakia, from 1800 on, all taken at the same position and by the same rules, show an average of 82 days of fog per year from 1800 to 1880. Since that time, the average number of foggy days per year has nearly doubled, and it is since 1880 that the city has become markedly industrialized.

Table 6-8 shows the summary of the U.S. Weather Bureau hourly observations of smoke in Pittsburgh from 1946 to 1953. These observations were all taken from the same location—the top of a tall building in the downtown part of the city. In spite of the fluctuations, the trend is clear (42). It seems that such data could be discussed statistically to yield more quantitative statements.

Allix (1) and Besson (5) have presented similar, but less extensive, evidence that the atmospheric pollution has increased at Lyons and Paris during the past few decades as more and more coal has been used.

Bonacina (8) and Coste (15) discuss the changing visual range in London with the seasons. At one observation point, for example, it varies from about 1 mile in midwinter to about 7 miles in midsummer.

If the meteorological conditions are uniform over an area large compared to that affected by serious pollution, it is sometimes possible to compare the records at two different locations (46). Figures 6-31 and 6-32 show the result of such a comparison of the data for Los Angeles Airport with that for San Diego. It illustrates one way of presenting such data.

If the meteorological conditions associated with poor visibility are known for a particular location, current observations can be compared with those taken under analogous conditions in preceding years. The reliability of the comparison is thereby considerably improved. Unfortunately, visibility is not one of the usual synoptic elements; consequently the necessary meteorological discussion must usually be developed for the particular location in question. This method has been applied by the Los Angeles County Air Pollution Control District, as shown in Fig. 6-33 (34). Weather Bureau data were used. Average monthly visibility for selected surface wind speeds is plotted as a function of time. This approach has been rejected by some meteorologists because of the difficulty of finding, or even of defining precisely, the "analogous" conditions. Visual range correlates more or less with several of the synoptic elements, and meteorology is not yet an exact science in the sense that the visual range can be calculated if the usual synoptic elements are given.

Table 6-8. Hours of Smoke in Pittsburgh, Pa.
(1946–1953)

	Hours of moderate smoke								Hours of heavy smoke							
	1946	1947	1948	1949	1950	1951	1952	1953	1946	1947	1948	1949	1950	1951	1952	1953
Jan.	52	29	22	19	61	11	15	18	29	8	9	9	7	0	0	1
Feb.	28	49	40	52	17	11	20	7	14	15	23	16	6	2	3	0
Mar.	60	67	23	32	32	15	19	11	44	28	8	7	3	3	0	2
Apr.	30	43	6	49	6	15	6	18	17	24	1	17	4	7	2	0
May.	57	23	28	65	49	20	13	46	2	6	2	43	3	6	0	0
June.	49	18	29	17	22	23	7	53	10	10	13	12	3	4	0	0
July.	59	18	6	35	54	32	20	32	7	4	0	13	0	13	0	0
Aug.	72	50	48	54	33	39	47	32	34	49	13	9	6	5	1	0
Sept.	47	28	34	23	47	40	48	20	55	13	8	8	3	7	2	0
Oct.	75	42	68	24	30	20	33	44	44	39	29	10	8	0	9	8
Nov.	103	4	44	37	27	14	29	20	37	1	15	6	2	2	4	5
Dec.	75	29	27	40	15	16	6	2	5	39	11	12	11	2	0	0
Mod.	707	400	375	437	393	256	263	303	298	236	132	162	56	51	21	16
Heavy.	298	236	132	162	56	51	21	16								
Total mod. and heavy.	1,005	636	507	599	449	307	284	319								

Reduction of heavy smoke in 1953 over 1946 is 94.4 per cent.
Reduction of total smoke in 1953 over 1946 is 69.4 per cent.
Data from Pittsburgh, Pa., Department of Public Health, Annual Report for 1953 (compiled by S. B. Ely), Bureau of Smoke Prevention, Pittsburgh, Pa., 1953.

The need for statistical criticism is shown by the Leicester study. The data on visibility at Leicester (22) correlated less well with smoke than might have been anticipated. Visual range was estimated in the usual way from a station at one edge of the city. Four marks from 1¼ to 6¼ miles, which covered 90 per cent of the recorded ranges, were in or across the city. Smoke was measured by photometering the stain on filter paper that resulted from drawing about 50 cu ft of air per 24 hr through a circle 1 to 2 in. in diameter. The intake of the filter probably accepted only particles smaller than 10μ in diameter. Several of these filters were located at points in and around the central part of the city. The correlation coefficient for 199 sets of observations in June, July, and August was -0.37, and for 148 sets in December, January, and February it was -0.46. Both figures are significant, but they indicate that a fifth to a sixth of the variance of visibility across Leicester was associated with variations in the concentration of smoke near ground level in the central part of the city The relatively low correlation would prevent single obser-

FIG. 6-31. Difference between visibilities at Los Angeles and San Diego. Per cent of time that Los Angeles visibility is 2 miles or less, minus the per cent of time that San Diego visibility is 2 miles or less. Figures and squares show the percentage of hours, and contour lines join points of equal percentage. (See Ref. 46.)

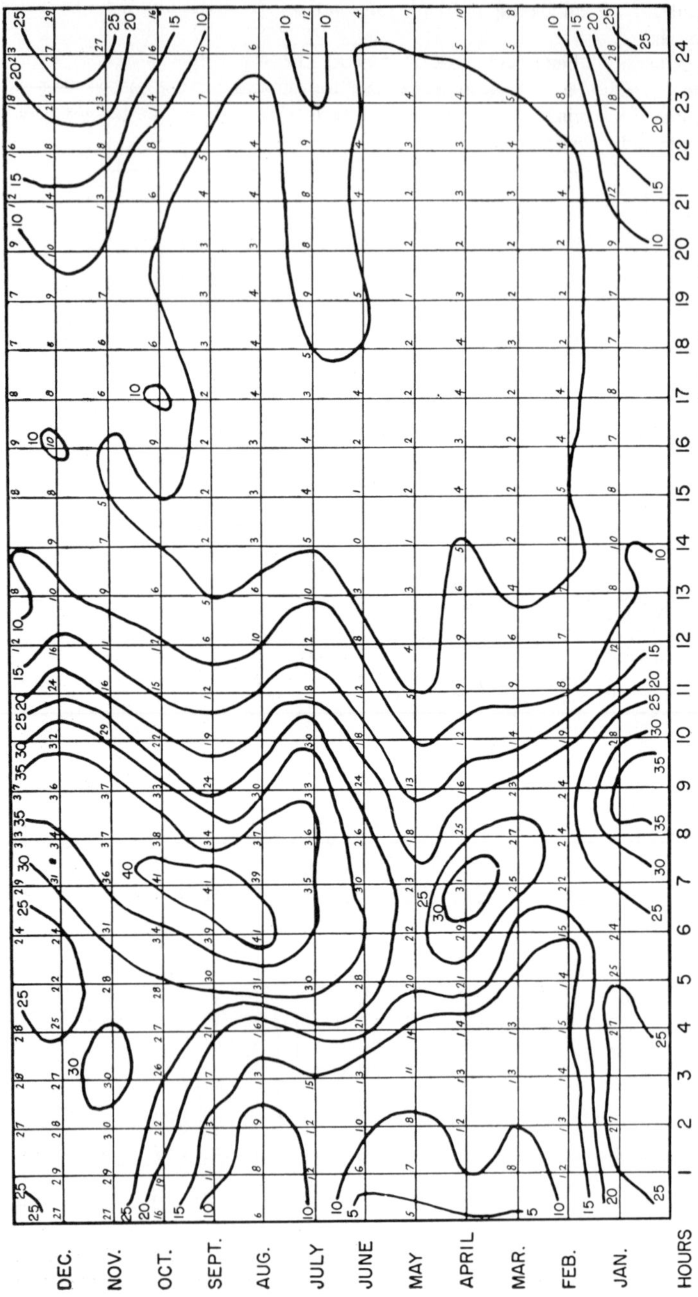

Fig. 6-32. Visibility in Los Angeles. Per cent of time that the Los Angeles visibility is 2 miles or less. Figures and squares show the percentage of hours, and contour lines join points of equal percentage. (See Ref. 46.)

vations of visibility from being used for estimating the concentration of smoke, at least at Leicester.

There seems to have been little use of optical measurements in characterizing the air pollutants. Johnston (30) comments on "sulfuric acid mist which is one of the principal causes of low visibility." It would be interesting to see whether this statement is true at such a location as Leicester.

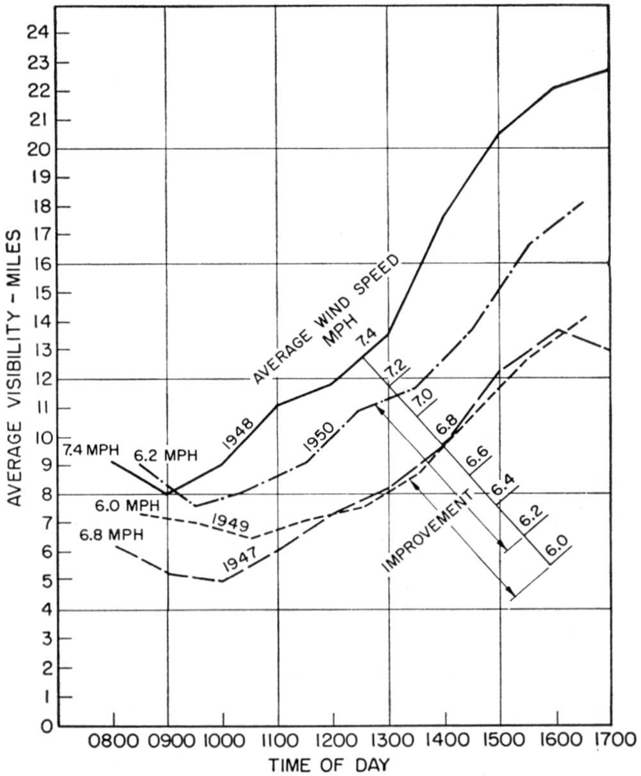

FIG. 6-33. Average day of visibility for the month of April. (See Ref. 34.)

The Los Angeles County Air Pollution Control District introduced the "relative pollution index" (RPI) to show the relative importance of various pollutants in decreasing the visual range. The RPI is defined as follows:

$$\text{RPI} = \frac{NP}{\text{cc}} = \frac{6}{D}\frac{W}{V}$$

where NP/cc is the number of particles per cubic centimeter, D is the density of the material in the particles in grams per cubic centimeter, and W is the weight in micrograms of material caught in sampling V cu m of air. All the material is assumed to be in particles of 0.682μ diameter. This figure is an average effective diameter for particles of the various compositions (and therefore various indices of refraction) that were found. It had already been shown that "95 per cent of the particles in the smog are below 1 micron in diameter" and that "as the visibility decreases in Los Angeles County, the particles in the 0.5- to 0.8-micron range increase more rapidly than in other size ranges." Table 6-9 shows the results of such calculations. "It is apparent in this evaluation that the weights of materials found in the atmosphere are not

Table 6-9. Relative Pollution Index

Pollutant	*w/v*, micrograms/cu m	*D*, g/cu cm	RPI
Ether-soluble aerosols	120	0.8	904
Sulfuric acid mist	110	1.4	461
Carbon	132 (53)[a]	2.1	378 (151)[a]
Silicon[b]	28	2.4	69
Lead[b]	42	9.1	28
Aluminum[b]	8	2.0	24
Calcium[b]	7	2.8	14
Iron[b]	10	5.1	12

[a] It is known by analysis and calculations that approximately 60 per cent of the carbon is found in the ether-soluble aerosol; therefore the RPI for the carbon should be reduced as indicated.
[b] As determined by flame spectrophotometric analysis.
Data from G. P. Larson, First and Second Technical and Administrative Reports on Air Pollution Control in Los Angeles County, Air Pollution Control District, Los Angeles County, Calif., 1949–1950 and 1950–1951.

necessarily significant unless studied on this or on a similar basis to determine their importance."

REFERENCES

1. Allix, A.: Obscurcissement progressif de l'atmosphère lyonnaise: La visibilité des Alpes, *Compt. rend.*, **195,** 1301–1303 (1932).
2. Axford, D. W. E., K. F. Sawyer, and T. M. Sugden: The Physical Investigation of Certain Hygroscopic Aerosols, *Proc. Roy. Soc. (London), Ser. A* **195,** 13–33 (1948).
3. Barkley, J. F.: Air-pollution Prevention in the United States, *Mech. Eng.*, **73,** No. 4, 284–288 (1951).
4. Berek, M., K. Männchen, and W. Schäfer: Measurement of Atmospheric Dust Content, *Z. Instrumentenk.*, **56,** 49–56 (1936).
5. Besson, L.: Sur la visibilité et la teneur de l'air en poussières à Paris, *Compt rend.*, **186,** 882–885 (1928).
6. Beuttell, R. G., and A. W. Brewer: Instruments for the Measurement of the Visual Range, *J. Sci. Instr.*, **26,** No. 11, 357–359 (1949).
7. Blackwell, H. R.: Contrast Thresholds of the Human Eye, *J. Opt. Soc. Amer.*, **36,** 624–643 (1946).
8. Bonacina, L. C. W.: London Visibility (letter to the editor), *Weather*, **1,** 83 (1946).
9. Bradbury, N. E., and E. M. Fryer: A Photoelectric Study of Atmospheric Condensation Nuclei and Haze, *Bull. Am. Meteorol. Soc.*, **21,** No. 10, 391–396 (1940).
10. Chandrasekhar, S.: "Radiative Transfer," Oxford University Press, New York, 1950.
11. Chaney, A. L.: A Recording Visibility Meter, in L. C. McCabe, ed., "Air Pollution," 679–682, McGraw-Hill Book Company, Inc., New York, 1952.
12. Coleman, H. S., F. J. Morris, and H. E. Rosenberger: A Photoelectric Method of Measuring the Atmospheric Attenuation of Brightness Contrast along a Horizontal Path for the Visible Region of the Spectrum, *J. Opt. Soc. Amer.*, **39,** No. 7, 515–521 (1949).
13. Coleman, H. S., and H. E. Rosenberger: A Comparison of Photographic and Photoelectric Measurements of Atmospheric Attenuation of Brightness Contrast, *J. Opt. Soc. Amer.*, **39,** No. 12, 990–993 (1949).
14. Coleman, H. S., and H. E. Rosenberger: A Comparison of Visual and Photoelectric Measurements of the Attenuation of Brightness Contrast by the Atmosphere, *J. Opt. Soc. Amer.*, **40,** No. 6, 371–372 (1950).
15. Coste, J. H.: London Smoke, *Weather*, **1,** No. 2, 53 (1946).
16. Debye, P.: Der Lichtdruck auf Kugeln von beliebigem Material, *Ann. Physik* (Series 4), **30,** 57–61 (1909).
17. Douglas, C. A., and L. L. Young: Development of a Transmissometer for Determining Visual Range, *Civil Aeronaut. Adm., Tech. Div. Rept.* 47, 1945.
18. Duntley, S. Q.: The Reduction of Apparent Contrast by the Atmosphere, *J. Opt. Soc. Amer.*, **38,** 179–191 (1948).
19. Duntley, S. Q.: The Visibility of Distant Objects, *J. Opt. Soc. Amer.*, **38,** 237–249 (1948).
20. Ferry, R. M., L. E. Farr, Jr., and M. G. Hartmann: The Preparation and Measurement of the Concentration of Dilute Bacterial Aerosols, *Chem. Rev.*, **44,** No. 2, 389–417 (1949).
21. Foitzik, L.: Sichtweite bei Tag and Tragweite bei Nacht, *Meteorol. Z.*, **49,** 134–139 (1932).

22. Great Britain: Atmospheric Pollution in Leicester—A Scientific Survey, *Dept. Sci. Ind. Research (Brit.), Atm. Pollution Research, Tech. Paper No.* 1, 1945.
23. Gucker, F. T., Jr.: Sensitive Photoelectric Photometer, *Electronics*, **20**, 106–110 (1947).
24. Gucker, F. T., Jr., H. B. Pickard, and C. T. O'Konski: A Photoelectric Instrument for Comparing the Concentrations of Very Dilute Aerosols and Measuring Low Light Intensities, *J. Am. Chem. Soc.*, **69**, 429–438 (1947)
25. Henyey, L. G., and J. L. Greenstein: Theory of Colors of Reflection Nebulae, *Astrophys. J.*, **88**, 580–604 (1938).
26. Hewson, E. W.: Atmospheric Pollution, in "Compendium of Meteorology," p. 1140, American Meteorological Society, Boston, Mass., 1951.
27. Houghton, H. G.: On the Relation between Visibility and the Constitution of Clouds and Fog, *J. Aeronaut. Sci.*, **6**, No. 10, 408–411 (1939).
28. Houghton, H. G., and W. R. Chalker: The Scattering Cross Section of Water Drops in Air for Visible Light, *J. Opt. Soc. Amer.*, **39**, No. 11, 955–957 (1949).
29. Hulst, H. C. van de: "Optics of Spherical Particles," N. V. Drukkerij, J. F. Duwaer & Zonen, Amsterdam, Netherlands, 1946.
30. Johnstone, H. F.: Technical Aspects of the Los Angeles Smog Problem, *J. Ind. Hyg. Toxicol.*, **30c**, 358–369 (1948).
31. aufm Kampe, H. Joachim, and H. K. Weichmann: Trabert's Formula and the Determination of the Water Content in Clouds, *J. Meteorol.*, **9**, No. 3, 167–171 (1952).
32. Koschmieder, H. H., and H. Ruhle: Danziger Sichtmessunger I. Forschungsarb. Staatliches Observatorium Danzig No. 2, 1930.
33. Kudlich, R.: Ringelmann Smoke Chart, *U.S. Bureau of Mines, Inform. Circ.* 6888, revised 1941.
34. Larson, G. P.: First and Second Technical and Administrative Reports on Air Pollution Control in Los Angeles County. Air Pollution Control District, Los Angeles County, Calif., 1949–1950 and 1950–1951.
35. Lohle, F. F.: Sichtschatzung und Luftlichtmessung, *Z. angew. Meteorol.*, **53**, 71–82 (1936).
36. Marks, L. S.: Inadequacy of the Ringelmann Chart, *Mech. Eng.*, **59**, 681–685 (1937).
37. Middleton, W. E. K.: Visibility in Meteorology, in "Compendium of Meteorology," pp. 92–93, American Meteorological Society, Boston, Mass., 1951.
38. Middleton, W. E. K., and A. G. Mungall: On the Psychophysical Basis of Meteorological Estimates of "Visibility," *Trans. Am. Geophys. Union*, **33**, No. 4, 507–512 (1952).
39. Mie, G.: Beitrage zur Optik truber Medien, speziell kolloidaler Mettallosungen, *Ann. Physik* (4th *ser.*), **25**, 377–445 (1908).
40. National Bureau of Standards: "Tables of Scattering Functions for Spherical Particles," National Bureau of Standards, Applied Mathematics, Series No. 4, 1949.
41. Neuberger, H.: "Introduction to Physical Meteorology," School of Mineral Industries, Pennsylvania State College, 1951.
42. Pittsburgh, Pa., Department of Public Health: Annual Report for 1953 (compiled by S. B. Ely), Bureau of Smoke Prevention, Pittsburg, Pa., 1953.
43. Ruedy, R.: Absorption of Light and Heat Radiation by Small Particles. I. Absorption of Light by Carbon Particles, *Can. J. Research*, **19A**, 117–125 (1941).
44. Ruedy, R.: Absorption of Light and Heat Radiation by Small Spherical Particles. II. Scattering of Light by Small Carbon Spheres, *Can. J. Research*, **20A**, 25–32 (1942).
45. Senftleben, H., and E. Benedict: Optical Constants and Radiation Laws of Carbon, *Ann. Physik*, **54**, 65–78 (1918).
46. Showalter, A. K.: Reported in Stanford Research Institute, The Smog Problem in Los Angeles County, First Interim Report, p. 32, Western Oil and Gas Association, Los Angeles, Calif., 1948.
47. Sinclair, D., and V. K. La Mer: Light Scattering as a Measure of Particle Size in Aerosols: The Production of Monodisperse Aerosols, *Chem. Rev.*, **44**, 245–267 (1949).
48. Smithsonian Institution: "Smithsonian Meteorological Tables" (prepared by Robert J. List), 6th ed., rev., Smithsonian Institution, Washington, D.C., 1951.
49. Steffens, C.: Measurement of Visibility by Photographic Photometry, *Ind. Eng. Chem.*, **41**, 2396–2399 (1949).
50. Steffens, C., and S. Rubin: Visibility and Air Pollution, *Proc. 1st Nat. Air Pollution Symposium*, pp. 103–108, Stanford Research Institute, Los Angeles, Calif., 1949.
51. Stratton, J. A.: "Electromagnetic Theory," McGraw-Hill Book Company, Inc., New York, 1941.
52. U.S. Weather Bureau: Instructions for Airway Meteorological Service, effective June, 1939, *U. S. Weather Bureau, Aerological Div., Circ.*, 1939.
53. Waldram, J. M.: Measurement of the Photometric Properties of the Upper Atmosphere, *Trans. Illum. Eng. Soc. (London)*, **10**, 125–130 and 147–187 (1945).
54. Wright, H. L.: Atmospheric Opacity: A Study of Visibility Observations in the British Isles, *Quart. J. Roy. Meteorol. Soc.*, **65**, 411–442 (1939).

SECTION 7

THE EPIDEMIOLOGY OF AIR POLLUTION

BY JOHN J. PHAIR

7.1 INTRODUCTION

It is generally accepted today that one of the most important and vexing challenges remaining to be met in the field of public health is the proper control of air pollution in cities and industrial areas. Safe and satisfactory disposal of airborne wastes, particulate or gaseous, is a field equaling and usually exceeding in complexity that dealing with disposal of liquid or solid matter. Materials from a tremendous number and a wide variety of sources, both in home and industry, are discharged continually in enormous amounts into the atmospheres of communities as smokes, fumes, vapors, mists, and dusts. Many of these compounds in high concentrations or following long exposures have been known to produce physiological changes in man, animals, and plants.

In human beings, reactions such as local irritation of the respiratory tract or, if the pollutants are absorbed into the blood, tissue changes in the brain, liver, kidneys, and other tissues have been described. Some materials will sensitize an exposed individual so that a subsequent contact will give rise to allergic reactions, such as asthma or localized edema of the skin and mucous membranes. In animals, essentially the same picture is found as in man, but with one very important additional factor. Materials ingested during the consumption of contaminated vegetation may constitute a far greater hazard than those brought into contact with the mucous membranes of the upper respiratory tract. The vegetation itself is likewise susceptible to injury. The effect may be local, with the reaction confined to the area in contact, or general, in that the plant as a whole is affected adversely. In many instances, plant damage is a far more sensitive indication of the degree of air pollution than any test, chemical or otherwise, particularly when certain compounds are involved. Some workers have even suggested that plant tolerance might serve as the basic standard in justifying controls for given areas or kinds of pollution.

Pollution of the ambient air, at least to some degree, has been universal since the beginning of time and is certain to continue in the future. Although much has been accomplished by various investigators in detecting and measuring air burdens, only scanty and fragmentary evidence is available regarding the possible injurious effect on living organisms following long exposure by inhalation or deposition of small or even trace amounts of the various contaminants emitted by modern industrial processes. Even with severe pollution, when the amount of material discharged can be measured in tons, the atmosphere of an industrial area will contain only trace quantities far below the values customarily accepted as harmful for man. It must be recognized, however, that pollutants will be present in complex mixtures which may have an additive or synergistic effect. However, the subject as a health problem should not be discounted or avoided because it is complex, but rather the possible physiological effects of the various situations and conditions on man, animals, and plants should be intensively studied and properly evaluated. With field and laboratory methods which would permit a careful and critical appraisal of the factors involved in any given situation, it should be possible to justify at least reasonable control, if not total prevention, of harmful atmospheric pollution.

Man must constantly contend against a wide variety of forces in his environment to protect and maintain his health and well-being. For example, heat and cold play a large role in determining the land areas used for living and growing required food. Man counters the elements by constructing adequate shelters and providing proper clothing. At the same time, he must defend himself not only against other animals, but also against man. He is forced to resist and tolerate the microparasites which gain entrance to his body in his food, drink, and environmental contacts. And, finally, he must neutralize and eliminate toxic or noxious materials absorbed by contact, ingestion, or inhalation.

Tremendous strides have been made in solving the problems of climate, and each year populations spread into areas that were once unsuitable because of heat, cold, or other environmental factors. New methods of housing and unusual materials for clothing are being brought forward from all sides to meet man's needs. Additional sources of food supplies have been and are being developed either by selecting or adapting plants and animals better suited for this purpose.

The efforts to eradicate, control, or minimize disease hazards arising from living in a universe teeming with microorganisms, viruses, and other pathogens have met with varying degrees of success. The control of food and drinking water has brought a striking reduction in intestinal illnesses. This is the one area to which the professional public-health workers point with particular pride. On the other hand, control of the airborne bacteria and viruses that are the causative agents of the diseases of the upper respiratory tract has been less successful. With these latter agents, the extreme host-parasite adaptations customarily found create almost insurmountable obstacles in the development of practical control procedures.

As defenses have been raised and methods found to meet or minimize various health hazards, others have grown in importance. With the increase in industrialization, the aggregation of human populations, and the recognition of the special problems of urbanization, standards of living have been raised. At the same time, the transmission of some parasites is facilitated and exposure to various toxic or noxious materials increased. While making changes in his environment to withstand these forces, man too has changed, so that resistant stocks have been developed by a process of selection. The healthy adult today can meet most of the common health hazards, including severe exposures to relatively high concentrations of toxic materials, without clinical reaction. As a matter of fact, he may even develop a readily demonstrable tolerance to his environment, an attribute common to nearly all living organisms. It is worthwhile to note at this point that man usually is more vulnerable when very young or old or when his defenses are weakened by malnutrition or intercurrent causes.

Each of these areas is under constant study by investigators interested in the control and prevention of human diseases. However, this section must deal solely with the evidence which has been collected and the measures which have been undertaken by both health departments and private investigators to evaluate the effect on man of contaminants discharged into the atmosphere from sources either in the home or industrial plant.

7.2 THE PROBLEM

The pollution of the atmosphere by various irritating or toxic materials is neither a new nor an avoidable annoyance or hazard for man, either as an individual or as a member of a community. "Pure air" has been crudely defined as atmosphere containing only its usual chemical components in their customary proportions and free from all other airborne substances, a state, if this be perfection and desirable, approached only in mid-ocean and on mountaintops. This description was given long before the atomic age and the recognition of cosmic dust. Also, it is obviously of little value in outlining the problems of modern civilization, since the demands of our industrial economy make some pollution inevitable. As a matter of fact, no generally acceptable concept of "physiological air" has even been proposed. Man has adapted to his atmosphere, but it is very possible that varying the constituents and the concentrations would give more desirable conditions. This is done in the oxygen tent, in meeting the problems of aviation, and in other similar situations.

What is needed by the community, its industries and health administrations, therefore, is a practical working definition of "safe and tolerable air." At the same time, it must be realized that any compromise which is worked out will leave much to be desired by both the community and industry. Furthermore, it must be noted that any solution proposed cannot hold for all areas, seasons, and future growth.

The so-called "extraneous compounds," particulate or gaseous, organic or inorganic, which may annoy or affect man, animals, or plants, reach the air from innumerable sources, some natural and others artificial. Dusts, fumes, and vapors are given off by volcanoes, geysers, hot springs, and swamps. Smoke, occasionally in tremendous quantities, is produced by forest and grass fires. Sand and soil are frequently gathered up by the winds over deserts or drought areas to be carried considerable distances to the discomfort and distress of great numbers of peoples. Pollens and similar allergens derived from vegetation are a seasonal, if not constant, nuisance. All these "natural" contaminants have a role in determining health and well-being and frequently cannot be controlled by any practical community-wide procedure.

With the rise of our modern civilization, other materials, particulate and gaseous, derived at first from the simple combustion of the fuels employed to provide heat and power and later including those released during certain industrial processes, particularly in the metallurgical and chemical fields, were added to the group of pollutants derived from natural sources. In the past 50 years, the rapid expansion of our industrial life and the extraordinary accomplishments of modern technology have brought benefits seen everywhere in the normal activities of the present-day community. However, this increased rate of growth in the industrial pattern and the adaptation of new substances to meet needs have brought the problems of air contamination to a point where they cannot be ignored by either the community or those agencies charged with the problem of promoting the health and well-being of its individual members.

It is essential to remember that the only acceptable objectives, from the point of view of the community, are the definition of air pollution controls required in a given situation and the justification for their use by appropriate health agencies. In no aspect do these problems differ greatly from those encountered in water pollution, which has been intensively attacked during the past half century. Liquid and solid wastes, human and industrial, must be disposed of in such a fashion that man, animals, and plants are not adversely affected. Even with the tremendous efforts devoted to this particular task, much remains undone and unsolved. However, the concepts, patterns, and methods developed in the search for control of water contamination should not be forgotten since many are equally applicable to the disposal of airborne wastes.

With solid or liquid materials, either human or industrial in origin, three methods of disposal are available:

1. Deposition on or burial in the soil
2. Dilution in streams, lakes, or oceans
3. Treatment to produce a material which can be removed or will not constitute a hazard

This broad classification of waste-disposal procedures represents ascending levels of control. It also indicates a progressive increase in the technical difficulties and in the cost. In the establishment of standards of water safety, therefore, the goal is not to prevent all contamination at all times, but rather to select the most practical method of meeting the requirements of an area, a population, and an industry. It is accepted that a certain amount of pollution is unavoidable and may even be necessary. In this case, the problem is usually to hold the concentration of the contaminant at a level which can be tolerated by man, animals, and plants. The choice of the method, of necessity, always requires exact definition of the extent of the hazard which will be created and the requirements of a given locale. In other words, a balance must be struck after consideration of all variables, and the selection must represent a "calculated risk."

Translating this approach to the control of airborne wastes, particulate or gaseous, two disposal methods are available:

1. Dilution in the atmosphere
2. Trapping and treatment to produce a material which can be disposed of without hazard

Again, there is a significant difference in the technical problems and the costs of the procedures. Consequently, in the quest for standards of air safety, the goal, of necessity, cannot be to prevent all contamination, but rather to select the most practical method of meeting the requirements of a given situation. It must be accepted that some degree of air contamination is unavoidable in an industrial civilization and in highly aggregated populations. The solution is to determine within limits the proper "tolerance level." This requires balancing three major variables:

1. Ability of the atmosphere to dilute the airborne wastes or residues, which includes consideration of the terrain and meteorological factors
2. Availability of treatment procedures, including consideration of cost and efficiency
3. Ability of man to tolerate the presence of the airborne wastes

The capacity of the atmosphere to dilute (Sec. 5) falls into the realm of the engineer and meteorologist, but it is important in determining the amount of air pollution to be tolerated in any given situation. The ability of the ambient air to receive and disperse to safe limits particulate and gaseous wastes will differ according to terrain, local meteorological conditions, latitude, extent of industrialization, density of population, and season of the year. With the improvement in technical methods, it should be possible to predict with considerable accuracy the amount and kind of contamination which can be absorbed under various circumstances.

The treatment of airborne wastes is an engineering subject and is discussed in Sec. 13. However, it too must be mentioned as one of the major variables to be considered in any review of air pollution and its effect on human health. It should be obvious to everyone that plant location and design are governed and limited by the extent to which wastes can be safely dispersed. Labor, transport, and power are always thought of in industrial planning. Good engineering practice today, in view of community needs, must include studies of waste disposal and not permit complacent reliance on simple dilution of wastes of any kind.

Man is a resistant, resilient variable when placed in any environment, favorable or unfavorable. He can readily withstand and neutralize many noxious contacts. This ability to adapt will vary greatly with race, age, sex, state of nutrition, and the presence of certain acute or chronic conditions. Under stress, there will be adaptation through natural selection of resistant stocks. This has been noted with regard to tuberculosis, malaria, and other diseases. To measure his tolerance to atmospheric pollutants of various kinds in differing concentrations, properly designed morbidity studies under defined exposures must be carried out. These must include not only clearly recognizable illnesses but also the less serious effects which can only be described as detrimental to the "mental and social well-being" of the exposed population.

The Public Health Service report describing the Donora episode and the reports contributed by the Kettering Laboratory, the Pennsylvania State Board of Health, and others illustrate only too well how few answers are available regarding the possible effects on human health of transient or persistent air pollution in industrial communities. These several investigations also, as well as the prior and subsequent reports of many essentially similar situations, point up the many obstacles encountered in seeking to demonstrate a relationship between air contaminants and the incidence of certain human reactions.

In considering where and how to demonstrate a causal relationship between air pollution and the reactions observed in man, animals, or plants, the question of what will be regarded as "adequate proof" assumes a position of paramount importance. It must be recognized that if pollution problems are to be solved effectively and fairly, both from the point of view of the community and its industries, there must be an acceptable factual basis for any conclusions or actions. On the other hand, since, all too frequently, claims and suits are based upon inadequate evidence or the interpretation of results is biased because of selfish motives, general agreement as to what is useful evidence is not readily reached. Usually, the motivation of everyone is so colored that the situation becomes confused, and it is very difficult or impossible to reach an objective definition. This is true even when the question of air pollution is brought up for discussion on a general basis without reference to a specific condition or location.

When the mechanisms of bacterial infection are studied, the kind of evidence which justifies the conclusion that a disease entity is caused by a given parasite can be best summarized by a series of conditions, commonly called in most bacteriological texts "Koch's postulates." They are as follows:

1. The organism should be found in all cases of the disease in question, and its distribution in the body should be in accordance with the lesions observed.

2. The organism should be cultivated outside the body of the host, in pure culture, for several generations.

3. The organism so isolated should reproduce the disease in other susceptible animals.

Additional postulates have been added or suggested later, such as the demonstration of specific antibodies or unusual resistance to a challenge infection.

It is important to note, however, that these conditions cannot be satisfied with every specific disease or pathogenic agent. The organism may be so small and its form or structure so doubtful that it cannot be demonstrated with any certainty in the tissues or lesions. The agent may not grow in vitro because of a variety of reasons. In other instances, the third postulate may prove the obstacle in that there is no adequate experimental animal immediately available or that the lesions in the common laboratory animals do not resemble those which characterize the disease in man.

In the investigation of mechanisms of bacterial infection, the causative role is accepted customarily, therefore, in the absence of complete proof, recognizing that any omission in the chain of evidence involves a risk of error. Today subclinical infections are recognized as part of the total picture, and an occasional patient will show symptoms with little resemblance to the classical reaction. It is known that the lesions in the experimental animal do not need to be and frequently will not be identical with those described in man. This attitude is in accord with the understanding that the pathologic condition produced by any infective process depends upon the host tissues as well as the activities of the parasite, and in this particular, as in other respects, one animal species differs from another. For some syndromes (for example, syphilis or diphtheria) a single specific parasite is responsible. However, the more significant symptoms of a disease may frequently be caused by several species of bacteria. Secondary pneumonias and bacillary dysenteries are good examples. In these cases, attention must be focused on the organisms rather than on the clinical reactions.

When the problems of air pollution are considered, there are no "Koch's postulates" to be satisfied that have been accepted by the various investigators, the involved communities, or industry. Adams, discussing possible physiological effects in a chapter of the "Air Pollution Abatement Manual" of the Manufacturing Chemists' Association, stated that, "regardless of the nature of the problem, the causal relationship is to be proved by establishing the following four points:

1. A toxic effect has occurred;
2. A pollutant capable of causing this effect is (has been) present;
3. There is (has been) an excessive exposure to the pollutant;
4. Other causes of the toxic effect do not exist."

By "toxic effect" is meant real injuries which are self-evident or readily demonstrable by an appropriate medical examination. The determination of the presence of the toxic material is to be based upon the results of the examination of stacks and other sources and sampling of the atmosphere. On the other hand, a simple demonstration of an increased incidence of nonspecific illnesses, which may take many forms, such as bronchitis, sinusitis, and other diseases of the respiratory tract, is not entirely acceptable because of the variation in severity and incidence which is known to occur all the time, regardless of the amount of air pollution. The implications to be drawn from the presence of the pollutant will depend upon associations of both time and location, of the occurrence of the toxicant and the toxic effect, and the elimination of other causative agents. Furthermore, the mere presence of an atmospheric pollutant, even with the coincidental occurrence of a toxic effect, is not considered necessarily adequate evidence of a causal relationship. There must be excessive exposure, i.e., an exposure of sufficient intensity, concentration, and duration that clear-cut toxic effects can occur within the exposure period.

Adams's attempt to define or set down conditions which must be met to establish the validity of evidence, pro or con, is a step toward clarification of the problem. However, in this form, as with Koch's original postulates, the conditions are too rigid, and it is extremely unlikely that they can be satisfied in the usual urban setting. They have not been accepted by all investigators, involved communities, or industry in their common quest for a mutually satisfactory description of "safe air." These conditions, although not entirely satisfactory, have been described as illustrative of the kind of evidence which is required to justify control. Unless the studies of the future follow some such logical pattern, nothing much will be gained.

7.3 PRESENT KNOWLEDGE

In the search for adequate proof of correlation of human disease with air pollution, three avenues have been most accessible. Unfortunately, although they are not wholly adequate or pertinent when the usual community air pollution problems are considered, all the presently available evidence, either directly or indirectly bearing upon this question, has been obtained from these sources. The first is the broad field of industrial toxicology. The second is the group of careful studies of the "epidemic" incidents when, for a variety of causes, large numbers of people are subjected to an extraordinary exposure. The third is the comparison of morbidity and mortality records collected routinely by health departments and other official agencies.

7.3.1 Industrial Toxicology

As listed, the first of the principal sources of evidence regarding the effect of air pollution on man is found in the many reports on clinical observations of industrial workers. Occupational diseases have been subjected to extensive and painstaking investigations for many years, and serious attempts have been made to reduce the hazards of employment. In these efforts, the reaction of living organisms to varying degrees of exposure to a great many potentially toxic compounds and materials has been studied under plant conditions. Acceptable now without question is the fact that various elements and compounds, organic and inorganic, when present in the atmosphere of the shop or workroom, will cause clinical reactions in certain exposed employees.

Such illnesses may result from absorption through the skin or respiratory tract of gases, vapors, mists, fumes, and dusts. The number and kind of effects will be dependent upon the form and concentration of the toxic material and the nature, degree, and length of the period of exposure. The reaction, regardless of the system involved, must be described as a chemical injury. It may be classified as acute or chronic, specific or nonspecific. Many toxicants will produce apparently the same response and will occasionally simulate an infectious disease. For some substances, exposure will be followed by an irritation of the upper respiratory tract; others will cause lesions of the liver and kidney; and some will produce changes in the central nervous system.

With the desire to lower the incidence of industrial disease and to promote the health of workers, concentrations of toxic materials which may be considered relatively safe have been defined within reasonably narrow limits. These tolerance levels, labeled "maximum allowable concentrations" or, as usually abbreviated, M.A.C. values, have been established and are published at periodic intervals by the American Standards Association and the American Association of Governmental Industrial Hygienists. They represent the consensus of the published and accepted results of laboratory and field studies assembled by competent workers. In their opinion, these are desirable working goals by which physicians and hygienists interested in environmental sanitation can measure the effectiveness of various control procedures.

As distributed by these two groups, the table of values has no official standing, nor can it be accepted as a legal standard. Correctly employed, however, these definitions of probably safe limits permit the relatively clear differentiation of safe and dangerous exposures in the plant. They do not represent safe limits for all persons under every condition, nor do they signify necessarily that quantities in excess of the published figures will be injurious under any or all circumstances. These values simply represent concentrations of specified substances to which workmen may be exposed for the customary working periods with a relatively low probability that undesirable reactions will follow. Even in the enclosed and controlled environment, they cannot be applied equally to the young and old, to the disabled and diseased, as well as to normal and healthy workers.

Unfortunately, these levels have been incorporated frequently as "standards" into state and local factory hygiene codes. This has been followed by the establishment of rules, regulations, and, in some instances, laws which industry must accept and conform to without adequate proof of the validity of the M.A.C. values. Too often the aim has become the attainment of a certain atmospheric concentration (with a total disregard of the individual) rather than the prevention of disease and promotion of health of the employee. With this misuse, the "acceptable" concentrations, in many instances crudely defined, are considered to be the end and not the means of attaining the goal of a healthy working environment.

Some authorities have resorted to these definitions in considering the possible hazard of a number of substances which occur with some regularity in the atmosphere of modern towns and cities. When the M.A.C. limits are employed in predicting possible community hazards, it must be recognized that, even under the very unusual conditions found in certain areas such as Donora, the toxic and irritant materials which give the most concern in industrial processes are not found in sufficient amounts to justify any great hope of discovering in the community the human reactions seen inside the plant. These values may be employed for this purpose only if one can accept cautiously the generalization that the working environment has a far higher concentration of possibly toxic materials for long periods—hence greater hazards to the health of the worker. It must be noted too that the heavier and the acute exposures are also followed by more recognizable clinical reactions. Furthermore, the amounts and kinds of industrial pollutants in the atmospheres of the plants can be measured with far greater accuracy for correlation with the responses of the workers as determined by the clinical studies. For these reasons, as well as others discussed earlier, it must be recognized that the industrial values as defined cannot be translated readily or directly to the conditions found in the usual urban setting.

Even though the concentrations of possibly toxic materials ordinarily discovered in the atmosphere of industrial cities are not sufficient, apparently, to justify any unusual apprehension about their effect upon the exposed population, it is not wise to be too dogmatic about this point. Unfortunately, the various studies upon which the M.A.C. values are based are frequently deficient in many important aspects. Many of the probably toxic materials have not been tested in even an exploratory fashion either in the field or laboratory. Another possible variable which must be considered is the possible additive effect of a combination of potentially injurious materials. There is evidence to indicate that solid particulate matter may act as a vehicle for a toxic vapor, bringing the injurious compound into more intimate contact with the pulmonary tissues than ordinarily would be the case.

The principal obstacle encountered in the utilization of the results of laboratory studies in industrial toxicology is that most of the efforts to quantify or anticipate human reactions to an industrial plant exposure have been based on animal reactions. Unfortunately, many of the compounds do not produce the same lesions in test animals as in man. The differences in the anatomical structures and the metabolic patterns of various species apparently have a tremendous role in determining the result of exposure to possible toxic materials. It must be recognized and accepted that there is definite and easily demonstrable species susceptibility to various substances, such as that observed when dealing with pathogenic parasites. This compels the use of much caution in the interpretation of laboratory findings and restricts the employment of such results in the prediction of the probable reaction in man, even if comparable conditions occur.

However, one should not be completely pessimistic or negativistic about the use of toxicological studies to provide valuable information regarding the possible effect of air contamination on human health under community conditions. Nearly all that can be said today with reasonable certainty about some compounds has stemmed from clinical studies of workers and the experimental investigations of industrial hazards. It is certain also that much information for the solution of community problems, as well as those of industry, can still be obtained in the future through well-designed experiments in both the laboratory and the plant.

7.3.2 Unusual or Accidental Exposures

The second approach to the definition of the hazards of air pollution is the careful study and description of the incidents where numbers of people are subjected to extraordinary exposures to toxic or irritating airborne substances, giving rise to readily demonstrable reactions. Some of these result from various situations in plants which arise from (1) inadequate provision for the control of materials, particularly in new industrial operations; (2) temporary failure of plant equipment; (3) mistakes of operating personnel; and (4) unusual weather conditions.

Other essentially similar situations are found in urban areas when (1) there is a rapid expansion of industrial production, with its concomitant and occasionally relatively greater increase in the amount and dissemination of atmospheric pollutants; (2) the introduction of new processes, particularly in the chemical field; (3) the introduction of new industries; or, conversely, (4) the extension of residential developments into areas which had been confined previously to industrial plants. Comparable situations and problems have been studied from time to time in essentially rural neighborhoods usually close to mining operations, smelting, and similar manufacturing processes. Usually, in these latter instances, the detrimental effect of the materials on the soil, crops, and domestic animals has overshadowed the potentially harmful exposures of man.

The group of "disaster" incidents which can be utilized to furnish useful information contains those found when there is a catastrophic localized or widespread exposure of a population. Three of the most outstanding are the extraordinary pollution of the Meuse Valley of Belgium in 1930, of Donora, Pa., during October of 1948, and of London in December, 1952. Unusual and accidental exposures involving only a single material have occurred also from time to time. The accidental release of chlorine in Brooklyn; mercury poisoning of a large group following a fire in a mercury mine; and, most recently, the illness and deaths following excessive exposure to hydrogen sulfide due to equipment failure in Poza Rica, Mexico, are prime examples. In seeking an understanding of the possible dangers of ordinary air contamination through studies of gross pollution of the atmosphere, it must be recognized that conditions of type and concentration of contaminant are greatly different. In situations caused by accidents or carelessness in the operation of industrial processes, only one or two materials have been involved. At the same time, extraordinary concentrations were reached—far beyond the values found ordinarily in the atmosphere of an industrial community, even though those values may be very high.

On the other hand, the factors which have given rise to the conditions described in Belgium, Donora, and London must be considered extremely uncommon. In all instances the investigations called attention to the fact that the weather conditions were unique in severity and duration. Usually, the combination of pollutants involved is dissipated promptly and satisfactorily. In the recent Donora episode, a thermal inversion, associated with extraordinarily slow air movement, permitted the accumulation of the ordinary chimney and stack effluents in a bowllike valley for a period of about four days. An almost identical situation occurred in the Meuse Valley. These disasters were carefully investigated, and much useful evidence was collected. In neither instance, however, was it possible to describe any one agent, or a combination of specific agents, which could be blamed for the illness and deaths reported in these areas.

Under the circumstances mentioned, the atmosphere did become irritating to a large proportion of those exposed. Some individuals with severe cardiac or respiratory difficulties could not withstand the impact, and there were an unusual number of deaths. The significance of this occurrence has been discussed elsewhere and need not be dealt with in great detail here, but it should be emphasized that it is not possible to apply these findings readily to the usual urban problem. This combination of heavy industrialization, extreme air pollution, mountainous terrain, and rare meteorological conditions occurs infrequently. It is unlikely that the factors seen in these disasters will be reproduced in many areas of this world.

To emphasize the uniqueness of the observed combination of circumstances, the most significant constituents commonly blamed in industrial plant exposures were not proved to be present in concentrations above those which cause only minor or temporary reactions in normal human beings. According to the findings of the various investigations, it is possible that it was the combination of pollutants which brought about the injurious effect. Some workers have said that possibly the particulate materials served as a means of bringing irritants deep into the lungs when ordinarily the reaction would have been localized in the upper respiratory tract.

If it is accepted that these episodes represent an extreme, the amount of utilizable information contributed by such studies to the solution of the day-by-day problems is necessarily small. An epidemic is customarily defined as a relatively unusual and unpredictable episode in the expected occurrence of a disease. The smog outbreaks and industrial accidents, therefore, must be considered in the same relationship to the usual picture of air pollution as is the epidemic occurrence of the common communicable diseases to the expected or endemic incidence. The acute problems occurring where there is an accidental release of toxic materials or an extraordinary community exposure to airborne wastes justify comprehensive investigation of the meteorological, physiological, and engineering aspects of these unusual instances of air pollution. However, because they are so rare and the conditions so difficult to duplicate, the results of these studies offer little information of immediate value for dealing with ordinary urban air contamination. In other words, it is impossible to predict, on the findings of such investigations, the effect of air pollution on man in usual urban developments. It must be admitted, however, that endemic or chronic exposures and possible human reactions have not been studied carefully in relation to the kind or degree of air contamination found in heavily industrialized areas.

7.3.3 Excess Morbidity and Mortality Rates

The third major source of information regarding the possible hazard of pollution of the atmosphere of industrial cities has been the series of studies which attempt to correlate excess crude morbidity and mortality rates, based on the reports of physicians, with the kind and amount of contaminants known to be present in the atmosphere under a variety of conditions. In the Belgium "epidemic," 63 persons died within a few days, 20 succumbed in the Donora incident, and about 4,000 in the 1952 London episode. In these incidents, many people became severely ill. It has been reported that the number of deaths represented a tenfold increase over the expected mortality for those areas. A number of investigators using the morbidity-mortality approach have sought to provide support for claims that ordinary air contamination has contributed significantly to an increased incidence of disease, acute, chronic, and fatal, in urban centers. They have studied primarily the reactions in the upper respiratory tract, such as pneumonia and pulmonary cancer. Unfortunately, these attempts to estimate the effects of low-grade pollution of the atmosphere of communities with any one material, or a combination of materials, have not been fruitful.

Earlier observers elsewhere had drawn attention to the fact that periods of severe fog were customarily accompanied by an excess number of reported deaths from respiratory disease. There are several such occurrences on record: London, in 1880, 1892, 1948, and 1952; Glasgow, in 1907 and 1925; and Dublin, in 1941. All these reports have certain features in common: the weather was cold; most of the deaths were among individuals over sixty; and a large number were chronic sufferers from asthma, bronchitis, or cardiac disease. In 1913, a similar close correlation between mortality rates from respiratory disease and sootfall was found in Pittsburgh when recorded deaths were studied according to wards. Mills has repeatedly shown that sootfall in various geographical divisions of many American cities will show an apparent correlation with what seems to be excessive death rates in the group of respiratory diseases. Unfortunately, in all these studies, no satisfactory method was found to correct for various other important variables, such as age, sex, occupation, socioeconomic factors, and general environmental conditions.

Two principal difficulties, other than those mentioned above, are encountered in this technique. The first of these is the lack of a clear-cut and established disease entity to form the numerator of the proposed rates. Without knowledge of either incidence or prevalence, because of the inability to list the number of accepted clinical reactions, it is impossible to set up an epidemiological investigation which will explain the physical, biological, and social components. Attempts have been made in these reports, as quoted, to compare the number and severity of acute and chronic respiratory diseases, such as sinusitis, bronchitis, bronchial asthma, tuberculosis, and pneumonia, in the general population under varying conditions. It has long been known that certain chronic respiratory diseases which fall within these categories occur under conditions of prolonged exposure to significant concentrations of specific industrial emissions. However, when this approach is used for a general population, the lesser degrees of pollution cannot be directly related to recognizable and specific pathological lesions. Likewise, individual variation in susceptibility becomes far more important and far more difficult to evaluate.

Even with all the problems encountered in determining the numerator, as indicated above, it is yet more difficult to estimate the denominator. Socioeconomic factors, environmental conditions, and host habits vary so greatly that any study of crude rates, either mortality or morbidity, is almost always doomed to fail in securing acceptance. What is required is careful investigation of the entire subject by more precise and specific methods in order to obtain definitive and unequivocal answers.

7.4 FUTURE STUDIES

The justification for laboratory and field appraisals of the many problems brought about by the air pollution of urban areas is apparent. Such appraisals can be supported not only on the grounds of health, but also by the generalization to which all subscribe: "It is not proper to dump garbage on your neighbor's lawn." In many areas, the failure to consider the limitations of the atmosphere in coping with extraordinary burdens of particulate and gaseous waste is forcing the imposition of controls, even though harmful reactions in man cannot be readily demonstrated. In other words, industries have had to install mechanisms and devices designed to prevent or minimize the discharge of airborne materials into the atmosphere without proof of a harmful effect on men, animals, or plants. This is a reasonable requirement when this question is looked at from the point of view of being a good neighbor; but, all too frequently, not all the wastes can be trapped, or in many instances, the process becomes uneconomic. Then the nonspecific control does not suffice, since the community must now decide whether or not the waste must be tolerated, i.e., shall it take "a calculated risk."

At this point, there must be differentiation between the concepts of general and specific control. To illustrate, when planning the control of malaria, a community usually discovers that use of general measures, an attack on all fronts, is impossible or very expensive and that species control of the important vector is the most practical and economical approach. However, under stress, such as war, or if the particular mosquito involved has not been identified, general measures must be instituted, to be modified as conditions change. In applying this concept to the solution of air pollution problems, it may be necessary, occasionally, to require general control, but the search should go on for specific hazards in order to predict within practical limits the parameters of a given situation.

Workers in the field of infectious disease have long recognized certain fundamental approaches. The first is the exhaustive study of the sick individual, either in the clinic or at the bedside. The second is the investigation of the problems under the controlled conditions made possible in the modern laboratory. The third is to record the reactions as they occur under natural conditions in whole populations under significantly different degrees of exposure. Clinical studies in the field of air pollution find their principal application at the level of the individual under plant exposures. Experimental laboratory studies offer the only practical procedure for studying tissue changes resulting from controlled contacts with possibly noxious materials.

The investigations of the past which have used the epidemiological method have been limited usually to measuring the simple incidence and prevalence of certain diseases which might have causal relationship with the atmospheric condition. This is an elementary approach, and what is required is the extension and refinement of the data of incidence and prevalence through comparing different divisions of a population exposed to various environmental conditions. This is not an easy task.

The customary starting point of epidemiological studies designed to demonstrate a causal relationship between a suspected agent and human disease usually has been an established, readily identifiable clinical reaction. The first steps in the investigation involve a search for the agent or, if it is known, the demonstration of its character and mode of transmission to man. The second phase is to determine the host, parasite, and environmental factors which influence the incidence and prevalence of the specific illness.

The inability to begin with a recognized disease entity forces the investigator charged with the solution of this problem to rely for his indices on those illnesses which might have any possible relationship, however tenuous, to the degree and kind of atmospheric pollution found in the particular area. This means immediately, regardless of the methods or procedures employed, that any inferences which may be drawn will be received and treated with doubt and suspicion, depending upon the selfish interests or motives of the individuals concerned. Since it is unlikely that the presence of contaminants can be shown as the primary cause but must be evaluated only in terms of their being possible contributing factors, the results will never be judged entirely objectively. The trained physician and epidemiologist learns early that all disease is the resultant or summation of a complex of host, agent, and environmental factors. The community and industry want exact quantification and appraisal of single variables. It is obvious that the desired answers can be obtained only when these costly time-consuming large-scale studies are undertaken with adequately trained personnel, both in the laboratory and the field.

In his foreword to "Air Pollution in Donora, Pennsylvania," a report dealing with the clinical reactions which followed what was an unusual and extraordinarily heavy concentration of air contaminants in that highly industrialized area, Dr. Leonard Scheele, Surgeon General of the U.S. Public Health Service, points out that this epidemiological study must be considered as the beginning of a major effort for improving generally the nation's health by attacking this special and very complex field. In noting that there had been only one similar incident recorded earlier in the medical literature (the fog in the Meuse Valley, Belgium, which was followed by 63 deaths), he called attention to the fact that this investigation of the Donora episode was the first comprehensive effort to consider the effect of air contaminants on apparently healthy individuals as well as establishing, as in the European episode, probable causal relationships with certain of the reported deaths.

Dr. Scheele emphasizes the lack of basic research in the laboratory and field, especially in the clinical and epidemiological aspects of this question. He deals particularly with the pressing need for valid evidence regarding the effect of various air pollutants at the relatively low concentrations found in the usual urban situation. The constitution of the atmosphere not only must be related to the well-being of healthy, strong adults, but must also take into account the physiological effects on individuals of the community who may be extraordinarily susceptible, such as the old, the very young, and those with certain chronic or degenerative diseases. The possible additive effect of the materials found in the atmosphere on concurrent illnesses must be appraised. Furthermore, it must be determined if any role is played in disease of systems other than the respiratory tract.

This approach for quantifying air pollution hazards in community studies, unless the clinicians or the laboratory investigators can demonstrate an entirely new and more delicate test, is visualized as a longitudinal morbidity survey related to the degree and kind of air pollution. Such activities are not entered into lightly. It will be necessary to relate the incidence, severity, and outcome of a group of diseases having a variety of causal agents to the presence or absence of a combination of pollutants which may vary, absolutely and relatively, from time to time. Not many detailed longitudinal morbidity studies have been made, and none of them has attempted to relate the incidence of disease to air pollution. However, it is still an inescapable fact that this is the only way to an objectively reached solution.

Careful planning is essential, because this will determine the feasibility of the investigation from an economic as well as other points of view. It will be necessary to secure the most advantageous population samples for the survey. Areas with unusual topographical features must be sought to ensure the greatest possible variation in the amount of exposure studied in the initial trials. Critical skills for the laboratory phases and the epidemiological investigations must be assembled. Forms must be designed and mechanisms arranged for the tabulation and analysis of the data. Competent field teams must be recruited, and the cooperation of the study population must be secured.

Finally comes the choice of the indices which might be employed. The difficulties which will be encountered in the utilization of the morbidity and mortality rates for the common respiratory and cardiac diseases have been reviewed above. Since it is impossible to determine the resistance or susceptibility of a relatively normal man, the use of these rates, even with adjustment for all the possible variables, will always be open to question. It is apparent that a more objective measurement is necessary if an answer to this problem is to be found. A number of possible valid differences can be readily listed, such as the average duration of upper respiratory disease or asthmatic attacks, the mortality rate of individuals with cardiac or pulmonary lesions, and similar factors. Whether they will give the necessary evidence must await the results of a careful study. At the moment, there is no satisfactory answer.

The need for further study of the many problems of atmospheric pollution as related to the health of man is clear. This need has been described by many others during recent years. It is hoped that this restatement of the problem will support efforts to secure consideration of the problem from an ecological point of view.

BIBLIOGRAPHY

Recent Bibliographies and Reports

Campbell, I. R., M. R. Christian, and E. Widner: "Classified Bibliography of Publications Concerning Fluorine and Its Compounds in Relation to Man, Animals and Their Environment, Including Effects on Plants," Kettering Laboratory Library, Department of Preventive Medicine, University of Cincinnati, Cincinnati, Ohio, 1950.

Davenport, S. J.: "Bibliography of Bureau of Mines Publications Dealing with Health and Safety in the Mineral and Allied Industries," U.S. Bureau of Mines, 1946.

Davenport, S. J., and G. G. Morgis: Air Pollution: A Bibliography, *U.S. Bur. Mines, Bull.* 537, 1954.

Jenkins, G. F.: Air Pollution Bibliography in "Air Pollution Abatement Manual," chap. 12, Manual Sheet P-13, pp. 1–57, Manufacturing Chemists' Association, Inc., Washington, D. C., 1952.

Kramer, H. P., and M. Rigby: Cumulative Annotated Bibliography on Atmospheric Pollution by Smoke and Gases, *Am. Meteorol. Abstr.*, **1,** 46–71 (1950).

Roth, H. P., and E. A. Swenson: "Air Pollution: An Annotated Bibliography," School of Medicine of Southern California, Los Angeles, Calif., 1947.

Comprehensive Articles

Adams, E. M.: Physiological Effects in "Air Pollution Abatement Manual," chap. 5, Manual Sheet P-6, pp. 1–28, Manufacturing Chemists' Association, Inc., Washington, D. C., 1951.

Batta, G., J. Firket, and E. Leclerc: "Les problèmes de pollution de l'atmosphère," Masson et Cie., Paris, 1933.

Great Britain: Interim Report of the Committee on Air Pollution, H. Beaver, chairman, Her Majesty's Stationery Office, London, 1953.

Bloomfield, J. J.: Health Implications of Air Pollution, *Proc. Air Pollution Smoke Prevention Assoc. Amer.*, 1950. Also *Proc. Am. Soc. Civil Engrs.*, vol. 77 (Separate No. 73), 1951.

Drinker, P.: Atmospheric Pollution, *Ind. Eng. Chem.*, **31,** 1316–1320 (1939).

Heimann, H., H. M. Brooks, Jr., and D. G. Schmidt: "Biological Aspects of Air Pollution: An Annotated Bibliography," Federal Security Agency, Public Health Service, Division of Industrial Hygiene, April, 1950.

Hemeon, W. C. L.: Scientific Boundaries in Air Pollution Studies, *Am. Ind. Hyg. Assoc. Quart.*, **14,** No. 1, 35–40 (1953).

Johnstone, H. F.: Properties and Behavior of Air Contaminants, *Ind. Med. and Surg.*, **19,** 107–115 (1950).

Kehoe, R. A.: Air Pollution and Community Health, *Proc. 1st Nat. Air Pollution Symposium*, pp. 115–120, Stanford Research Institute, Los Angeles, Calif., 1949.

Lanza, A. J.: Health Aspects of Air Pollution, *Combustion*, **21,** No. 1, 56–57 (1949).

Larson, G. P.: Medical Research and Control in Air Pollution, *Am. J. Public Health*, **42,** 549–556 (1952).

Logan, W. P. D.: Fog and Mortality, *Lancet*, **256,** 78 (1949).

Marsh, A.: "Smoke: The Problem of Coal and the Atmosphere," pp. 72–73, Faber & Faber, Ltd., London, 1947.

McCabe, L. C., P. P. Mader, H. E. McMahon, M. V. Hamming, and A. L. Chaney: Industrial Dusts and Fumes in the Los Angeles Area, *Ind. Eng. Chem.*, **41,** 2486–2493 (1949).

McCord, C. P.: The Physiological Aspects of Atmospheric Pollution, *Ind. Med. and Surg.*, **19,** 97–101 (1950).

McDonald, J. C., P. Drinker, and J. E. Gordon: The Epidemiology and Social Significance of Atmospheric Smoke Pollution, *Am. J. Med. Sci.*, **221,** 325–342 (1951).

New York Academy of Medicine: Report of the Committee on Public Health Relations: Effect of Air Pollution on Health, *Bull. N. Y. Acad. Med.*, 2d ser., **7,** 751–775 (1931).

Russell, W. T.: The Influence of Fog on Mortality from Respiratory Diseases, *Lancet*, **207,** 335–339, (1924).

Stanford Research Institute: The Smog Problem in Los Angeles County, Western Oil and Gas Association, Los Angeles, Calif., 1954.

Section 8

THE EFFECTS OF AIR POLLUTANTS ON FARM ANIMALS

BY P. H. PHILLIPS

8.1 INTRODUCTION

In considering the effect of air pollutants on farm animals, it is necessary to keep one fact always in mind. The mechanism by which an animal can become poisoned is entirely different from that by which human beings exposed to deleterious atmospheres are poisoned. In the case of farm animals, we are really concerned with a two-step process: the accumulation of the airborne contaminant in vegetation and forage and the subsequent poisoning of the animals when they eat the contaminated vegetation. In the case of human beings working in contaminated atmospheres in industrial plants, the concern is for the deleterious substances that are directly inhaled. The reason for the difference in point of view is apparent if one considers that 3 ppm of fluoride in the air is the maximum permissible tolerance for men working in confined spaces for a period of 8 hr, whereas the concentrations of fluoride in the air in areas where cattle poisoning may develop will be in the order of a few parts per billion, or fractions thereof. In the case of cattle, the hazard obviously is not the result of inhaling the polluted air, but rather the ingestion of forage which has become contaminated with fluorine from the air.

Air pollutants that present a hazard to livestock, therefore, are those which contaminate the forages in a toxic form or those which are taken up by vegetation and react in the plant to form toxic materials. There are several common examples of the former, although none is now known where a toxic material is formed by reaction of a plant with air pollutants. Sulfur dioxide, for example, which is a common air contaminant and can be quite damaging to vegetation itself, can be present in the air in sufficient quantity to cause damage to 25 per cent of alfalfa leaves, and yet this alfalfa will cause no damage to cattle that consume it (11).

Arsenic and lead are poisons which pose a problem so far as livestock is concerned. Both may originate from industrial sources or from dusting and spraying. The problem of industrial control of air pollution by these substances has been solved to a considerable degree in the past twenty years. However, both elements still present a considerable problem when they originate from dusting or spraying operations (75). Undoubtedly a number of other organic dusts and sprays, such as DDT, may present a toxicity problem so far as animals are concerned. There are very few cases reported in the scientific literature on the subject, and none of the sprays can be considered as a common contaminant in an air pollution problem, except on a very localized and restricted basis.

The three pollutants responsible for most livestock damage are arsenic, fluorine, and lead (Table 8-1). In all cases, the hazard does not exist because the materials

Table 8-1. Pathological Effects of Arsenic, Fluorine, and Lead

	Gross pathology		Histopathology	
	System	Effect	Organ or tissue	Structural effects
Arsenic	Respiratory Gastrointestinal	Inflammation Inflammation	Liver: acute	Injury to reticuloendothelial cells; hepatic necrosis; yellow atrophy
			chronic	Glycogen disappears; fatty degeneration; hepatic necrosis; bile duct congestion; periportal hyperplasia
	Epidermis	Scleroderma Alopecia	Skin	Thickened stratum corneum Scleroderma and alopecia
	General	Depressant Coma Horse-face expression Cachexia	Kidney	Hair-follicle atrophy
Fluorine	Dentition	Enamel hypoplasia Dentine hypoplasia Mottling Dentomalacia Dentalgia	Teeth	Defective calcification Irregular calcification
	Skeleton	Exostosis Ankylosis Chondrodynia Osteomalacia	Bone	Osteoblastic activity, periosteal calcification Osteoclastic removal of bone of marrow cavity
	Soft tissues	Emaciation Inanition Cachexia	Kidney Thyroid	
	Gastrointestinal	No diarrhea		
Lead	Blood formation	Anemia	Blood cells	Reticulocytosis Basophilic leukocytosis (polychromasia)
			Bone marrow	Increased hematopoiesis
	Colic	Gastric disturbances Spastic atony		

are present in the air in amounts large enough to cause trouble on inhalation via the lungs, but because they are deposited on or taken in by forage and vegetation which is subsequently consumed by farm animals.

Fluorine currently presents the greatest pollution hazard so far as cattle are concerned. Although one fluorine-containing compound (cryolite) is used as an insecticidal spray on apple and pear trees, the major air pollution problem has been with various industrial processes that emit fluorine-containing gases and fumes to the atmosphere.

8.2 ARSENIC

Arsenic occurs as an impurity in many ores and in coal. It has been reported to cause poisoning of livestock near various industrial processes and smelters. Grain dried by direct contact with flue gases has been reported to have an arsenic content of 0.2 ppm when the coke used to produce the flue gases contained 80 ppm of arsenic (33). In common with most other industrial air pollutants, arsenic may spread over a considerable area from a stack source. Measurable amounts have been found at distances over 6 miles from the source (45).

Arsenic is used in some insecticides, in the form of arsenic trioxide and lead arsenate. The use of dusts or sprays of such materials on plants can lead to poisoning of cattle, even acute poisoning, although the affected area from a single source will usually be much smaller than that encountered when damaging amounts of arsenic are emitted from an industrial process of any magnitude.

8.2.1 Symptoms of Acute Arsenic Poisoning

In acute cases the symptoms are those of severe colic, i.e., salivation, thirst, vomiting, great uneasiness, feeble and irregular pulse and respiration. The animal stamps, lies down, and gets up. The odor of garlic may be detected on the breath. There is diarrhea, and the feces have a garlic odor and are sometimes bloody. Coldness of the ears may be noticed, and sometimes there is abnormal temperature, trembling, stupor, and convulsions. The animal may become exhausted and collapse. Death may come in a few hours or in several days. Horses may have a raised red line at the base of the incisor teeth, erosion of the outer side of the gums, ulcers of the nose, puffiness above the eyes, dilation of the pupils, difficult breathing, and partial paralysis of the hind legs (26,75).

8.2.2 Symptoms of Chronic Arsenic Poisoning

Arsenic appears to have a depressing effect upon the central nervous system, but in the early stages there may be some evidence of hyperirritability. The animal will eventually become dull and exhibit a lack of appetite, with a resulting loss in weight. There may be a chronic cough. Continual diarrhea may occur. There may be a chronic eczema, thickening of the skin, anemia, and abortion or sterility. Chronic poisoning can result in eventual paralysis and death (26,61).

8.2.3 Pathological Effects of Arsenic Poisoning

In animals, arsenic may cause death of portions of the liver (liver necrosis), inflammation of the stomach and intestines and death of the cells lining the wall of the intestines (enteritis with necrosis), anemia and destruction of red blood cells, and damage to the kidney (35,49,58,63,72).

8.2.4 Tolerance

All soluble arsenic-containing compounds, as well as the element itself, are extremely poisonous. Sheep have been poisoned by as little as 0.25 to 0.50 g of arsenic daily, although cattle and horses may tolerate 1.3 to 1.9 g daily. Cattle have been reported to sicken on alfalfa containing 40 to 45 ppm arsenic, although they survived (61). In the laboratory, 10 mg of arsenic per kilogram of body weight per day has produced chronic arsenic poisoning in guinea pigs.

Laboratory work with rabbits indicates that arsenic-poisoned muscle contains less phosphoric acid ester than normal muscle (47).

8.3 FLUORINE

Fluorine is never found in its elemental state in nature, but occurs commonly as an impurity in vegetation and various minerals. Fluorine occurs in trace to substantial amounts in most ores, coals, clays, and soils. It is present in the form of apatite in phosphate rock. Fluoride minerals such as cryolite and fluorspar, as well as sodium fluoride, are used as fluxes in various metallurgical processes.

Fluorine is volatilized by calcining processes into a gaseous form, such as hydrofluoric acid (HF), silicon tetrafluoride (SiF_4), or fluosilicic acid (H_2SiF_6). Tiny particles are mechanically carried out through the stacks in particulate form, such as the various sodium aluminum fluorides, apatite, calcium fluoride, iron fluoride, and sodium fluoride.

Various processes such as the manufacture of aluminum, clay bricks, phosphate chemicals and fertilizers, and steel have emitted fluorides to the atmosphere (1,2,4,5, 6,22,42,65,73,74,76). Since fluorine occurs to some degree in coal and other fuels, fluorides may also be emitted from ordinary combustion operations, and measurable amounts of fluorine may be found in the air over any large coal-burning city during the winter.

So far as fluorosis of cattle and other farm animals is concerned, agricultural sprays and dusts present a minor problem. The only use of fluoride-containing chemicals for this purpose is in the spraying or dusting of apples and pears with cryolite, as already mentioned.

Cattle and sheep are the most susceptible to fluorine toxicosis of all farm animals. Swine occupy a second rank, although fluorosis of swine in the United States has not been a major problem (18,34). Horses appear to be quite resistant to fluorine poisoning, and authenticated cases of fluorosis of horses in the United States are rare. Poultry are probably the most resistant to fluorine of all farm animals and present no problem so far as fluorosis is concerned (25).

Most of the fluorosis problems in sheep have been reported from North Africa, Australia, and England (27,52,53,69,77). Since fluorosis of sheep in the United States is relatively rare, and since the tolerances and symptoms for cattle and sheep are similar, with much more data for cattle being available, the discussion of tolerances and symptoms that follows will be largely related to cattle.

8.3.1 Symptoms of Acute Fluorine Poisoning

As is the case with any toxic material, sufficient fluoride may be fed to cattle over a short period of time to cause illness and death. In air pollution cases, however, this would be unlikely. Lack of appetite, rapid loss in body weight, a rapid decline in health and vigor, lameness, periodic diarrhea, muscular weakness, and death characterize the acute form of fluorine poisoning (16,62). A measurable increase of bone fluorine may occur in acute fluorine toxicosis. Urine analyses would show greatly increased concentrations of urinary fluorine.

8.3.2 Symptoms of Chronic Fluorine Poisoning

Fluorine is a cumulative poison under conditions of continuous exposure to subacute doses. The toxicity of fluorides depends upon their solubility and subsequent absorption in the gastrointestinal tract. Because of its cumulative characteristics the degree of toxicity is a function of both ingestion level and length of exposure. It has been demonstrated that levels which are marginally toxic for short periods of exposure may become toxic if the period of exposure is sufficiently lengthened. Under chronic exposure to fluorides, soft tissues react in such a manner as to support the concept that fluorine is a protoplasmic poison, a fact which finds its explanation in the blocking of certain essential enzyme systems.

Teeth in the process of formation are easily affected by fluorine. Hence, tooth symptoms are a sensitive and unique criterion of chronic fluorosis. Fluorine ingestion causes a disturbed calcification of the growing tooth, which results in incomplete formation of the enamel, the dentine, or the tooth itself. The normal translucency of the enamel is replaced in part or in toto by a white chalky enamel, referred to as mottled enamel.

Teeth are also softened by too much fluorine, and as a result (in addition to mottling) there may be excessive wear and staining. When the teeth, especially the molars, are soft enough to wear rapidly, it may be assumed that exposure to fluorine has been sufficient to cause interference (8,57).

Reliance upon mottling of the teeth as the sole criterion of fluorine toxicosis has some limitations. The mottling, regardless of degree, occurs only during the period of formation of the tooth. Hence it reflects the disturbance caused by the ingested fluorine only during this period and may not reflect at all the state of the animal at the time of inspection. Mottling is valuable in diagnosing fluorosis in the bovine that ingested excessive amounts of fluorine when less than five years of age, but is of little or no value if the animal is over five years of age when first exposed to excessive fluorine. In the adult bovine, dental fluorosis records only the chronological evidence of previous exposure to fluorine, which in cattle is estimated to take place 9 to 15 months prior to eruption of the affected teeth. Another limitation is that low calcium rations have been reported to produce dental effects grossly indistinguishable from those of dental fluorosis (19). Dental changes caused by fluorine are permanent.

Although in mild or marginal fluorine toxicosis skeletal effects cannot be observed with precision, certain symptoms may be observed in more advanced cases. A bony "overgrowth" (exostosis) in advanced cases may be observed on the leg bones, jawbone, and ribs. Lameness may occur as a result of this overgrowth on the leg bones or of calcification of the ligaments in the legs. This may be followed by stiffness, and calcification of the joints. Effects upon skeletal structures are usually considered permanent, although in some cases bony overgrowths have been observed to diminish when excessive fluorine is removed from the animal's diet (1).

Animals showing clinical fluorosis will invariably exhibit lack of appetite, emaciation, and general ill health due to malnutrition, all of which will adversely affect milk production and retard growth. The systemic reaction of a fluorine-poisoned animal is primarily one of self imposed starvation, and the clinical symptoms seen, aside from the dental and skeletal effects, are identical with those of starvation. The systemic effects of excessive fluorine ingestion are not permanent, and they are readily amenable to dietary therapy with low fluorine rations (2,4,43). Mottled teeth remain unaffected and bony overgrowth little affected by such therapy.

The symptoms and pathologic changes associated with fluorosis given here are those of cattle and sheep. The symptomatology developed in other species such as swine, rabbits, horses, and poultry closely follows the pattern presented here for ruminants, i.e., mottling, staining and wearing of the teeth (exception, poultry), bony overgrowths on the skeleton, stiffness and lethargy, emaciation and general ill health from starvation. The monogastric species, with their higher tolerance levels to fluorine and lesser dependence upon roughage for their total dietary needs, are thus less likely to be poisoned by fluorides.

8.3.3 Diagnosis of Chronic Fluorine Poisoning

Many of the symptoms of fluorosis may also be seen in other toxicoses, diseases, or dietary deficiencies. The diagnostician is fortunate, therefore, in having at his disposal chemical methods for fluorine analyses which can be used to confirm or support his clinical and field diagnoses.

Analysis of the urine provides a ready means of detecting current ingestion of fluorides. Skeletal fluoride concentrations provide reliable data with which to detect fluorosis as well as to determine its severity (4,7,10,13,32,34,55,65). Table 8-2 presents the fluorine analyses of urine and various parts of the skeletons of cattle for normal animals and for those suffering from marginal, mild, and classical (clinical) fluorosis. Table 8-3 gives fluorine-excretion levels in urine as related to the dietary levels observed in the forage on which the cattle fed.

Table 8-2. Concentrations of Fluorine (in PPM) in or Excreted by Cattle Fed Different Levels of Fluorine

(Bone and tooth values on dry fat-free basis)

Material analyzed	Normal	Degree of toxicity		
		Marginal	Mild	Clinical fluorosis
Metacarpal bone	400–1,000	Up to 3,000	3,000–4,000+	5,000–10,000+
Spongy bone (rib)	400+	1,000+	3,500	9,000+
Teeth:				
Enamel	250–500	3,000	3,000	7,000
Dentine	450–1,000	4,000	5,000	8,000+
Heart muscle[a]	0.25	0.60	0.75	0.90
Urine (at 1.04 sp gr)	<10	18–20	25–30	30–40+

[a] Low concentrations of fluorine in soft tissue should not be used as a diagnostic aid.
Data from Refs. 4, 7, 10, 13, 32, 36, 55, and 65 at the end of this section.

Table 8-3. Data on the Urinary Excretion of Ingested Fluorine by Cattle Fed Contaminated Forage or Sodium Fluoride

Investigator	Source of fluorine	F in forage, ppm dry wt basis		F in urine, ppm at 1.04 sp gr	
		Toxic	Nontoxic	Fluorosis	Normal
Blakemore et al.	Brick kilns	90	...	16–68	<10
Blakemore et al.	Alumina reduction	35+	9	15–48	<10
Miller and Phillips	Sodium fluoride	40+	20	13–39	6
Hobbs et al.	Sodium fluoride	30	25	.....	<10
Krueger	Chemical industry	100+	100	21–89	4

Data from Refs. 4, 5, 29, 36, and 43 at the end of this section.

Such analytical data on bones, teeth, and urine provide sound and impersonal criteria of fluorosis as well as support for clinical diagnoses of the disease. It must be emphasized that symptoms used as evidence for diagnosis of fluorosis should be susceptible of proof by chemical analysis of bones and urine. Similar analytical data are available for sheep (52), swine (34), rabbits (8), and laboratory animals.

Once the fluorine is deposited in the bone, elimination may occur from the so-called spongy bone by feeding rations low in fluorine, but repeated experiments with rats have shown that only about 60 per cent of the bone fluoride is thus removed in periods extending to nearly a third of the life span of the species. A cow with abnormal amounts of fluorine in the skeleton was also shown to have retained large quantities for many months (1) although considerable amounts were excreted in the urine for a long time after she was placed on a diet containing normal amounts of fluorine.

8.3.4 Tolerance

The toxicity of the fluorine in various fluorides will vary somewhat with the solubility of the fluoride in the gastrointestinal tract. At low or trace levels, soluble and insoluble fluorides show similar physiological effects. At ingestion levels of fluorine where fluorosis occurs, however, there will be a difference in the toxicity of soluble and relatively insoluble fluorides.

In the bovine, 10 to 15 gal of saliva containing sodium bicarbonate are secreted daily. This secretion is sufficient to convert any volatile fluoride (whether the acid HF, fluosilicic acid, or silicon tetrafluoride) to the sodium salt, and thus the toxicity of volatile fluorides would be expected to be identical to that of sodium fluoride, and the effects upon the animal would be identical with those obtained by the experimental feeding of sodium fluoride.

At levels where fluorosis occurs, the relative toxicity of soluble and insoluble fluorides is approximately as follows:

F as sodium fluoride: F as calcium fluoride or fluorspar = 2:1
F as sodium fluoride: F as apatite (rock phosphate) = 2:1
F as sodium fluoride: F as cryolite = 2:1

The fluorine tolerances of species differ. In order to make comparisons between species it is necessary to use the same source of fluorine and the same criterion to measure the effect. Such a comparison is possible from studies at the Wisconsin Experiment Station where the source of fluorine was raw rock phosphate and where growth retardation was used as an indication of toxicity. Under these conditions the relative tolerance to apatite fluoride among species was as follows:

Species	*Mg of F per Kg Body Wt to Cause Retardation of Growth*
Cattle	2–3
Rabbits	11
Swine	8–10
Rats (white)	8–10
Guinea pigs	18
Chickens	35–70

Cattle will withstand ingestion of fluorine in insoluble forms such as fluoapatite at levels approximating 75 ppm in the total feed (17,57).

The amount of soluble fluorine (as in sodium fluoride) necessary to mottle teeth of cattle faintly is near 20 ppm (dry-weight basis). Two to three times this amount of soluble fluoride may cause retardation of growth and reduction of milk production (28,57,65). In studies of the long-time effect (2 to 5 years) of soluble fluorides on cattle, the work of Hobbs and coworkers (28) at the Tennessee Experiment Station, Schmidt and Newell at the Stanford Research Institute (66), and Miller and Phillips at the Wisconsin Experiment Station (43,55) indicates that the tolerance level of fluorine in sodium fluoride is approximately 40 ppm (dry-weight basis). At this level of ingestion there is some evidence of dental fluorosis, but this is unaccompanied by retarded growth rates or the depression of milk production. Added dietary fluorine does not increase the fluorine content of the milk in chronic fluorosis.

Drinking-water supplies may contain unduly high levels of fluorides, as in the case of the work reported by Neeley et al. (46) at Lubbock, Tex. Water-borne fluoride is reported to be five to ten times more toxic than the equivalent amount of fluorine present in dry feeds. Furthermore, mineral supplements may also contribute appreciable amounts of fluorine to the diet. Since the total fluorine ingested is the critical test for tolerance, it is apparent therefore that farm animals may be exposed to toxic amounts of fluorine other than airborne fluorine. Hence a check of all possible sources must be made when investigating fluorosis.

Table 8-4. Airborne Fluorine Toxicity Data for Cattle and Sheep

Investigator and species	Source of fluorine	Forage F concentration, ppm dry-wt basis		Degree of mottling	F content of leg bones, ppm	Condition of animals
		Toxic	Nontoxic			
Blakemore et al. (cattle)	Brick	25–90	14–16	Mild	....	Emaciated
	Ironstone (calcining)	490+	.....	....	10,500–15,500[a]	Emaciated
	Aluminum	35+	9	Well defined	9,600[a]	Emaciated
Harris et al. (cattle)	Steel	30+	30	Mild+	6,983[b]	Emaciated
Meyn et al. (cattle)	Industrial plant	60+	.....	Well defined	3,460[b]	Emaciated
Blakemore et al. (sheep)	Aluminum	44	.....	Mild	8,170[a]	Emaciated

[a] Ash basis.
[b] Dry fat-free basis.
Data from Refs. 4, 5, 42, and 76 at the end of this section.

Table 8-5. Experimental Fluorine Toxicity Data for Livestock

Investigator and species	Source of fluorine	Ingestion levels, ppm dry-wt basis		Mottling of teeth (marginal level)	Fluorine content of leg bone, ppm	
		Toxic	Nontoxic (marginal level)		Normal	Fluorosis
Peirce (sheep)	Rock phosphate	160	60	Mild	640[a]	8,700[a]
Kick et al. (swine)	Rock phosphate	330+	330	Mild	600[a]	6,200[a]
	NaF	290+	290	Mild	600[a]	6,700[a]
Briggs and Phillips (rabbits)	Rock phosphate	220+	220	Mild	275[b]	4,360[b]
	NaF	210+	210	Mild	275[b]	4,834[b]
Phillips et al. (cattle)	Rock phosphate	80+	80	Mild	584[b]	5,354+[b]
Miller and Phillips (cattle)	NaF	50	30	Mild		

[a] Bone ash.
[b] Fat-free dry bone.
Data from Refs. 8, 34, 43, 52, and 57 at the end of this section.

Tables 8-4 and 8-5 present information on toxic and nontoxic amounts of fluorine for various species, both as observed in the field and in controlled feeding experiments. Accompanying information is also given on teeth condition and fluorine content of the leg bones.

8.3.5 Alleviators

Hobbs and his coworkers (28) of the Tennessee Experiment Station have studied the use of supplemental aluminum salts under a variety of conditions to determine if fluorosis could be prevented or reduced thereby. There was a slight beneficial effect when these salts were used as alleviators. Peirce (52), Lawrenz and Mitchell (37), Majumdar et al. (41), and others (12,57) have demonstrated that high dietary calcium depresses assimilation of fluoride and thereby reduces the skeletal deposition. The use of supplemental calcium carbonate is therefore indicated in diets high in fluorine. Miller and Phillips (43) made use of this principle and measured the reduction in urinary fluorine excretion of cows fed sodium fluoride. There was a measurable reduction in the fluorine content of the urine as well as in biopsy rib samples.

Fargo et al. (18) have shown that pigs on pasture nearly doubled their tolerance to apatite fluoride, thus suggesting a marked alleviating effect of pasture grasses and legumes. A mitigating effect upon acute fluorine toxicity was obtained by feeding orange juice to young rats (56), and an increased survival time was observed in fluorine-poisoned guinea pigs fed high levels of orange juice (54). Pandit and Rao (51) and Wadhwani (78) found that the administration of ascorbic acid lessened the severity of fluorosis in monkeys and increased survival time in monkeys fed acutely toxic amounts of sodium fluoride.

8.4 LEAD

Lead-containing ores are smelted to recover metallic lead, along with other metals. Lead is also an impurity in some coals, which may contain as much as 54 ppm (15). In addition to possible industrial sources of lead contamination such as smelters, coke ovens, and other coal-combustion processes, lead is also used in dusts and sprays containing lead arsenate. Other lead compounds of importance in toxicology are lead oxide, white lead, and lead acetate.

8.4.1 Symptoms of Acute Lead Poisoning

In cases of acute lead poisoning the onset is sudden and the course relatively short. Characteristic symptoms are nervous, excitable jerking of muscles, frothing at the mouth (salivation) followed by delirium, stupor, and collapse. Prostration, staggering, and inability to rise are prominent symptoms. The pulse is always fast and weak, and the extremities are cool. In addition, the animal may stand with the head pressed against a wall. Attacks resembling epilepsy or muscle twitchings are occasionally observed. Other nervous symptoms in cattle are grinding of the teeth, rapid chewing of the cud, and a "kink" in the neck from contraction of the lateral neck muscles. Some animals may fall suddenly, stiffen the legs, and have convulsions. There is complete loss of appetite, paralysis of the digestive tract, and diarrhea. The symptoms of acute lead poisoning should not be confused with the symptoms of tetany in calves.

Symptoms of lead poisoning in horses are complete loss of appetite, nervous depression, lethargy, stupor, and death, with or without manifestations of gastritis and diarrhea.

8.4.2 Symptoms of Chronic Lead Poisoning

Chronic lead poisoning (saturnism) has been observed frequently in horses that have been grazing on forage near smelters, lead mines, and in orchards that have been sprayed. Paralysis of the muscles of the larynx and difficulty in breathing are the principal symptoms. Paralysis of the larynx is the most constant symptom of chronic lead poisoning in the horse, although there may be other symptoms. Convulsions may occur from the paralysis of the throat, and the difficulty in breathing may be unusually severe and persistent, during and after exercise. Another complication is mechanical pneumonia, resulting from paralysis of the pharynx or larynx and causing abscesses in the lungs (39).

8.4.3 Tolerance

Lead is a cumulative poison. As a result, the continuous ingestion of very small daily doses will ultimately be as effective as one toxic dose. Depending on the amount of lead that may be deposited from dusts or sprays, poisoning and death can take many months if slightly contaminated hay is fed or can occur within 24 hr in animals feeding in or near orchards which have just been heavily sprayed (39). In sheep, lead passes through the placental membranes into the fetal lamb and it is secreted in the milk of the ewe (38).

In the vicinity of a coke oven, the herbage was found to contain 25 to 46 ppm of lead; this was sufficient to cause poisoning of cattle and sheep (15). The feeding of 1 g of lead arsenate daily to a cow caused symptoms of lead poisoning after 26 days and death in 40 days (50).

Inadequate dietary calcium increases the retention of lead in the animal body. Low-calcium diets may result in lead storage of as much as five times more than that found in animals that receive adequate amounts of calcium. Supplemental amounts of calcium above the amount considered adequate will not offer additional protection against poisoning (3,9,21,24,70).

REFERENCES

1. Agate, J. N., G. H. Bell, G. F. Boddie, R. G. Bowler, M. Bucknell, E. A. Cheeseman, T. H. J. Douglas, H. A. Druett, J. Garrad, D. Hunter, K. M. A. Perry, J. D. Richardson, and J. B. de V. Weir: Industrial Fluorosis—A Study of the Hazard to Man and Animals near Fort William, Scotland, *Med. Research Council, Mem. No.* 22, 1949.
2. Aston, B. C., and others: Mineral Content of Pastures Research: Fourth Annual Report, *J. Agr. New Zealand,* **42,** 226–230 (1931).
3. Aub, J. C.: The Biochemical Behavior of Lead in the Body, *J. Am. Med. Assoc.,* **104,** 87–90 (1935).
4. Blakemore, F., T. J. Bosworth, and H. H. Green: Industrial Fluorosis of Farm Animals in England, Attributable to the Manufacture of Bricks, the Calcining of Ironstone and to Enameling Processes, *J. Comp. Pathol.,* **58,** 267–301 (1948).
5. Blakemore, F.: Industrial Fluorosis of Animals in England, *Proc. Nutrition Soc.,* **1,** 211–215 (1944).
6. Boddie, G. F.: Effects of Fluorine Compounds on Animals in the Fort William Area. Industrial Fluorosis, *Med. Research Council, Memo.* 22, pp. 32–46, 1949.
7. Bredemann, G.: "Biochemie und Physiologie des Fluors," G. E. Stechert & Company, New York, 1951.
8. Briggs, G. M., and P. H. Phillips: Development of Fluorine Toxicosis in the Rabbit, *Proc. Soc. Expt. Biol. Med.,* **80,** 30–33 (1952).
9. Calvery, H. O.: Chronic Effects of Ingested Lead and Arsenic, *J. Am. Med. Assoc.,* **111,** 1722–1729 (1938).
10. Cohrs, P.: Zur pathologischen Anatomie und Pathogenese der chronischen Fluorvergiftung des Rindes, *Deut. tierärztl. Wochschr.,* **49,** 352–357 (1941).
11. Cunningham, O. C., L. H. Addington, and L. T. Elliot: Nutritive Value for Dairy Cows of Alfalfa Hay Injured by Sulfur Dioxide, *J. Agr. Res.,* **55,** 381–391 (1937).
12. Danckwortt, P. W.: The Detection of Fluorine Poisoning by the Determination of Fluorine Content of Teeth and Bones, *Z. physiol. Chem.,* **268,** 187–193 (1941).
13. Danckwortt, P. W.: Der chemische Nachweis von Fluorvergiftungen, *Deut. tierärztl. Wochschr.,* **49,** 365–366 (1941).
14. DeEds, F.: Chronic Fluorine Intoxication, *Medicine,* **12,** 1–60 (1933).

15. Dunn, J. T., and H. C. L. Bloxam: The Presence of Lead in the Herbage and Soil of Lands Adjoining Coke Ovens and the Illness and Poisoning of Stock Fed Thereon, *J. Soc. Chem. Ind.*, **51**, 100–2T (1932).
16. DuToit, P. J., A. I. Malan, J. W. Groenewald, and G. W. DeKock: The Effect of Fluorine on Pregnant Heifers, *Union S. Africa, Dept. Agr.*, 18*th Rept. Director Vet. Services and Animal Ind.*, pp. 805–817, Aug., 1932.
17. Elmslie, W. P.: Effect of Rock Phosphate on the Dairy Cow, *Proc. Am. Soc. Animal Production*, **29**, 44–48 (1936).
18. Fargo, J. M., G. Bohstedt, P. H. Phillips, and E. B. Hart: The Effect of Fluorine in Rock Phosphate on Growth and Reproduction in Swine, *Proc. Am. Soc. Animal Production*, **31**, 122–125 (1938).
19. Franklin, M. C.: The Influence of Diet and Dental Development in the Sheep, *Australia, Commonwealth Sci. and Ind. Research Organization Bull.* 252, 1950.
20. Gaud, M., A. Charnot, and M. Langlais: Human Darmous, *Bull. inst. d'hyg. Maroc*, Nos. 1 and 2, 1934.
21. Gray, I.: Recent Progress in the Treatment of Plumbism, *J. Am. Med. Assoc.*, **104**, 200–205 (1935).
22. Green, H. H.: An Outbreak in Industrial Fluorosis in Cattle, *Proc. Roy. Soc. Med.*, **39**, 795–799 (1946).
23. Greenwood, D. A.: Fluoride Intoxication, *Physiol. Rev.*, **20**, 582–616 (1940).
24. Hadjioloff, C.: Calcium Therapy in Lead Poisoning, *Arch. Gewerbepathol. Gewerbehyg.*, **10**, 360–369 (1940).
25. Haman, K., P. H. Phillips, and J. G. Halpin: The Distribution and Storage of Fluorine in the Tissues of the Laying Hen, *Poultry Sci.*, **15**, 154–157 (1936).
26. Harkins, W. D., and R. E. Swain: The Chronic Arsenical Poisoning of Herbivorous Animals, *J. Am. Soc. Chem.*, **30**, 928–946 (1908).
27. Hatfield, J. D., C. L. Shrewsbury, and L. P. Doyle: The Effect of Fluorine in Rock Phosphate in the Nutrition of Fattening Lambs, *J. Animal Sci.*, **1**, 131–136 (1942).
28. Hobbs, C. S., S. L. Hansard, E. R. Barrick, D. Sikes, J. L. West, W. H. McIntire, R. P. Moorman, and J. M. Griffith: The Effect of Feeding Various Levels of Fluorine on Cattle, abstract from *J. Animal Sci.*, **9**, 659–660 (1950).
29. Hobbs, C. S., C. C. Chamberlain, J. L. West, L. J. Hardin, W. H. McIntire, and R. P. Moorman: Fluorine Level in Urine as a Diagnostic Measure of Fluorine Toxicosis in Cattle, *J. Animal Sci.*, **10**, 1084–1089 (1951).
30. Huffman, C. F., and O. E. Reed: Results of a Long Time Mineral Feeding Experiment with Dairy Cattle, *Mich. State Coll. Agr., Circ. Bull.* 129, pp. 1–11, 1930.
31. Hupka, E.: Chemical Observation on Intoxications by Vapors of Hydrofluoric Acid, *Berlin u. Münch. tierärztl. Wochschr.*, p. 366, 1941.
32. Hupka, E.: Klinische Beobachtungen über Fluorvergiftungen bei Weidetieren, *Deut. tierärztl. Wochschr.*, **49**, 349–352 (1941).
33. Jones, C. R., and E. C. Dawson: The Arsenic Content of Grain Dried Directly with Flue Gas, *Analyst*, **70**, 256–257 (1945).
34. Kick, C. H., R. M. Bethke, B. H. Edgington, O. H. M. Wilder, P. R. Record, W. Wilder, T. J. Hill, and S. W. Chase: Fluorine in Animal Nutrition, *Ohio Agr. Expt. Sta. Bull.* 558, pp. 3–77, 1935.
35. Kiese, M.: Chronic Hydrogen Arsenide Poisoning, *Arch. exptl. Pathol. Pharmakol.*, **186**, 337–376 (1937).
36. Krueger, E.: Nachweis Fluorhaltiger Industrieexhalationen durch Analyse von Rinderharn, *Deut. tierärztl. Wochschr.*, **56**, 325 (1949).
37. Lawrenz, M., and H. H. Mitchell: The Effect of Dietary Calcium and Phosphorus on Assimilation of Dietary Fluorine, *J. Nutrition*, **22**, 91–101 (1941).
38. McDougall, E. I.: The Copper, Iron and Lead Contents of a Series of Livers from Normal Fetal and New Born Lambs, *J. Agr. Sci.*, **37**, 337–341 (1947).
39. MacKintosh, P. G.: Clinical Manifestations and Surgical Treatment of Lead Poisoning in the Horse, *J. Am. Vet. Med. Assoc.*, **74**, 193–195 (1929).
40. Majumdar, B. N., S. N. Ray, and K. C. Sen: Fluorine Intoxication of Cattle in India, *Indian J. Vet. Sci.*, **13**, 95–107 (1943).
41. Majumdar, B. N., and S. N. Ray: Fluorine Intoxication of Cattle in India. II. Effect of Fluorosis on Mineral Metabolism, *Indian J. Vet. Sci.*, **16**, 107–112 (1946).
42. Meyn, A., and K. Viehl: Chronic Fluorine Poisoning in Cattle, *Arch. wiss. u. prakt. Tierhielk.*, **76**, 329–339 (1941).
43. Miller, R., and P. H. Phillips: University of Wisconsin, unpublished work, 1952.
44. Mitchell, H. H., and M. Edman: The Fluorine Problem in Livestock Feeding, *Nutr. Abstr. & Revs.*, **21**, 787–804 (1952).
45. Muhlsteph, W.: Chemical Detection of the Dissemination of Arsenic by Flue Gases, *Tharandt. forstl. Jahrb.*, **87**, 239–277 (1936).

46. Neeley, K. L., and F. G. Harbaugh: The Effects of Fluoride Ingestion on a Herd of Dairy Cattle in the Lubbock, Texas, Area, *J. Am. Vet. Med. Assoc.*, **124,** 344–350 (1954).
47. Nonnenbruch, W., Z. Stary, A. Bareyther, and H. Thelen: The Muscle Metabolism of Rabbits Poisoned with Arsenic, *Arch. exptl. Pathol. Pharmakol.*, **180,** 437–439 (1936).
48. Ockerse, T.: Endemic Fluorosis in the Pretoria District, *South African Med. J.*, **15,** 261–266 (1941).
49. Ostrovskaya, I. S.: Pathomorphology of Poisoning with Arsine, *Arkh. Patol.*, **11,** 78–80 (1949).
50. Paige, J. B.: Cattle Poisoning from Arsenate of Lead, *Mass. Agric. Expt. Sta.*, 21*st Ann. Rept.*, part II, pp. 183–199, 1909.
51. Pandit, C. G., and D. Narayana Rao: Endemic Fluorosis in South India. Experimental Production of Chronic Fluorine Toxicosis in Monkeys, *Indian J. Med. Research*, **28,** 559–574 (1940).
52. Peirce, A. W.: Observations on the Toxicity of Fluorine for Sheep, *Australia, Commonwealth Council Sci. and Ind. Research, Bull.* 121, pp. 1–35, 1938.
53. Peirce, A. W.: Chronic Fluorine Intoxication in Domestic Animals, *Nutr. Absts. & Revs.*, **9,** No. 2 (1939).
54. Phillips, P. H.: The Manifestations of Scurvy-like Symptoms Induced by the Ingestion of Sodium Fluoride, *J. Biol. Chem.*, **100,** 79–80 (1933).
55. Phillips, P. H.: The Development of Chronic Fluorine Toxicosis and Its Effect on Cattle, *Proc. 2d Nat. Air Pollution Symposium*, pp. 117–121, Stanford Research Institute, Los Angeles, Calif., 1952.
56. Phillips, P. H., and C. Y. Chang: The Influence of Chronic Fluorosis upon Vitamin C in Certain Organs of the Rat, *J. Biol. Chem.*, **105,** 405–410 (1934).
57. Phillips, P. H., E. B. Hart, and G. Bohstedt: Chronic Toxicosis in Dairy Cows Due to the Ingestion of Fluorine, *Wisconsin Expt. Sta. Research Bull.* 123, pp. 1–30, 1934.
58. Prell, H.: Poisoning of Animals at a Distance within Industrial Fume, *Arch. Gewerbepathol. Gewerbehyg.*, **7,** 656–670 (1937).
59. Rand, W. E., and H. J. Schmidt: The Effect upon Cattle of Arizona Waters of High Fluorine Content, *Am. J. Vet. Research*, **13,** 50–61 (1952).
60. Ranganathan, S.: Calcium Intake and Fluorine Poisoning in Rats, *Indian J. Med. Research*, **29,** 693–697 (1941).
61. Reeves, G.: The Arsenical Poisoning of Livestock, *J. Econ. Entomol.*, **18,** 83–89 (1925).
62. Roholm, Kaj.: "Fluorine Intoxication," H. K. Lewis and Co., London, 1937.
63. Rössing, P.: Liver Injury in Arsenic Poisoning, *Arch. Gewerbepathol. Gewerbehyg.*, **11,** 131–142 (1941).
64. Schaaf, E., and J. Maurer: Arsenic Content of the Guinea Pig, *Z. physiol. Chem.*, **280,** 65–75 (1944).
65. Schmidt, H. J., and W. E. Rand: A Critical Study of the Literature on Fluorine Toxicology with Respect to Cattle Damage, *Am. J. Vet. Research*, **13,** 38–49 (1952).
66. Schmidt, H. J., G. W. Newell, and W. E. Rand: The Controlled Feeding of Fluorine as Sodium Fluoride, to Dairy Cattle, *Am. J. Vet. Research*, **15,** 232–239 (1954).
67. Schour, I., and M. C. Smith: The Histologic Changes in the Enamel and Dentin of the Rat Incisor in Acute and Chronic Experimental Fluorosis, *Univ. Ariz. Agr. Expt. Sta., Tech. Bull.* 52, 1934.
68. Schucht, F., H. H. Baetge, and M. Duker: Survey of Smoke-injured Area of the Oker Foundry, *Landwirtsch. Jahrb.*, **76,** 51–98 (1932).
69. Seddon, H. R.: Chronic Endemic Dental Fluorosis in Sheep, *Australian Vet. J.*, **21,** 2–8 (1945).
70. Shields, J. B., and H. H. Mitchell: Effect of Calcium and Phosphorus on the Metabolism of Lead, *J. Nutrition*, **21,** 541–552 (1941).
71. Sikes, D., and M. E. Bridges: Experimental Production of Hyperkeratosis of Cattle with Chlorinated Naphthalene, *Science*, **116,** 506–507 (1952).
72. Sinitsina, T. A.: Toxic Enteritis (Changes in the Intestinal Mucosa Caused by Arsenic Poisoning), *Arch. sci. biol.* (*U.S.S.R.*), **56,** 34–42 (1939). In England, 1942.
73. Slagsvold, L.: Fluorine Poisoning, *Norsk. Vet.-Tidsskr.*, **46,** 2–16, 61–68 (1934).
74. Stas, M. E.: Fluorine Poisoning Due to Flue Gases and Dust from Super Phosphate works. Determinations of the Fluorine Content of Deposits on Plants, in Water, Hay and Bones, *Chem. Weekblad*, **38,** 585–593 (1941).
75. U.S. Department of Agriculture, Bureau of Animal Industry: Special Report on Diseases of Cattle, Rev., 1942.
76. Utah State Agricultural College Agricultural Experiment Station: Recommended Practices to Reduce Fluorosis in Livestock and Poultry, *Circ.* 130. pp. 1–16, 1952.
77. Velu, H.: Le darmous fluorose spontanée des zone phosphatées, *Arch. inst. Pasteur d'Algérie*, **10,** 41–118 (1932).
78. Wadhwani, T. K.: Mitigation of Fluorosis, *Indian Med. Gaz.*, **87,** No. 1, 5–7 (1952).

SECTION 9

EFFECT OF AIR POLLUTION ON PLANTS

BY MOYER D. THOMAS AND RUSSEL H. HENDRICKS

9.1 INTRODUCTION

Injury to vegetation caused by air pollution from cities and factories has been well recognized because of many botanical and chemical investigations during the past hundred years. These studies have demonstrated conclusively that the injuries are caused by a small number of contaminants, present in the air in relatively low concentration, of which the principal ones are (1) sulfur dioxide; (2) hydrogen halides and halogens, particularly hydrogen fluoride, though hydrogen chloride and chlorine may be occasional toxic agents; (3) smog gases containing certain organic compounds, particularly the olefins, together with oxidizing agents such as ozone and nitrogen oxides to make them highly reactive; (4) many organic compounds such as aldehydes, ketones, organic acids, and chlorinated compounds, which can cause leaf injury in a few hours at about 1 ppm (singly they seldom reach concentrations sufficient to cause plant damage, but together they may do so); and (5) 2-4D and other similar herbicidal sprays. Water-soluble arsenicals are powerful herbicides when applied as a spray to foliage, but insoluble arsenicals are practically nontoxic to most vegetation. As contaminants of the atmosphere the arsenicals have not been reported to cause injury to plants, though they have impaired the value of the vegetation for animal feed, if deposited on the plants in sufficient amount. In enclosed places like greenhouses, ethylene, hydrogen cyanide, and mercury vapor have caused injury to vegetation. There are a number of other substances which are relatively unimportant because they are unlikely to be found in the atmosphere in sufficient concentration to be toxic to vegetation, except perhaps accidentally. This latter group includes hydrogen sulfide, nitrogen oxides above the natural level, ammonia, and carbon monoxide.

It has usually been assumed that coal smoke does not have much effect on vegetation beyond that caused by the accompanying sulfur and/or fluoride. It was suggested (38,43,85) that soot and tars, either in the air or deposited on the surfaces of leaves, might absorb enough light to reduce photosynthesis, but it was not certain that there was any effect. However, Ruston's (87) work on the smoke in Leeds, England, suggested, and recent work by Bleasdale (7) in Manchester seems to confirm, the possibility that coal smoke or its oxidation products in the atmosphere contain organic constituents that may interfere measurably with plant growth without necessarily causing leaf lesions. This has also been observed with smog (44,52). Road dust and cement dust present a simpler problem. In the absence of published data on dusty plants, it can probably be assumed that these materials, by reducing the amount of light reaching the leaf and also by clogging stomata, may reduce carbon dioxide uptake to some extent, thus interfering with photosynthesis without being otherwise harmful to the plants.

Some idea of the extent of mechanical interference with photosynthesis may be obtained from the work of Barr (4), who applied a waxy coating consisting of an emulsion of paraffin and vegetable oils to corn plants. Leaf samples showed that sugar accumulation in the sprayed leaves was reduced 38 per cent after 1 day and 20 per cent after 21 days. The waxy coating was shrunken and cracked at the latter time. Of interest also in this connection is the work done at Cornell (15,18), where apple trees were dusted and sprayed with various forms of sulfur dust and lime sulfur sprays. The subsequent effect of the treatment on photosynthesis was then measured. In general these materials had little or no effect on assimilation, though some of the lime sulfur sprays inhibited photosynthesis for a time, probably more because of chemical than physical action.

There is no published evidence that the vegetation near any factory processing radioactive elements has been injured by the radioactivity in any element absorbed by or deposited on the vegetation.

In the early days of the chemical industry, it was not uncommon to eliminate into the atmosphere large quantities of waste material in concentrated form near ground level. For example, in the Leblanc soda process, all the hydrogen chloride formed by treating salt with sulfuric acid was wasted into the air, with resulting severe injury to vegetation in the neighborhood of the plant. Further, in the smelting industry the method of "heap roasting" of sulfide ores was practiced. This consisted of piling up alternate layers of ore and wood and igniting the mass. The sulfur dioxide and fumes from this combustion spread over the surrounding country, causing severe injury to vegetation and consequent erosion, of which the Copperhill district in southeastern Tennessee is an outstanding example.

These practices have long ago been corrected, either by removing the offending materials, both solids and gases, at the source, or by eliminating the smoke—after precipitating or filtering out the solids—at elevated temperatures from tall stacks so that potential and actual ground concentrations of the gases and solids have been radically reduced. Smoke damage to vegetation is now, therefore, comparatively light and generally results from occasional unfavorable meteorological conditions.

Even so, the growth of the chemical industry and the concentration of population in many areas have been so great that air pollution problems of importance to agriculture are still with us. In fact, new problems have arisen during the past decade, which were either nonexistent or subacute and unrecognized earlier, including the smog in Los Angeles and elsewhere, and a number of fluoride problems in different parts of this country, though the latter have been recognized in Europe for a long time (11,12,86).

Damage to crops in Los Angeles County has been estimated at $500,000 yearly (72), while claims against various industries for fluoride damage have totaled millions of dollars.

9.2 SULFUR DIOXIDE

Sulfur dioxide is probably the most widespread, and it has been the most intensively studied, air pollutant. It is produced during the combustion of nearly all fuels, and especially in the roasting of sulfide ores or the burning of high-sulfur coal. Its analytical determination in the atmosphere is simple, and automatic methods are available for the purpose. This makes it comparatively easy to evaluate ground concentrations injurious to vegetation.

Extensive observations have been made on the effects of sulfur dioxide on vegetation, dating back to the studies in Europe described in books by Haselhoff (36), Haselhoff and Lindau (38), Haselhoff, Bredemann, and Haselhoff (37), Schroeder and Reuss (88), Sorauer (93), Stoklasa (100), Wieler (120), and Wislicenus (121). In this country, the Report of the Selby Smelter Commission (42) was one of the first outstanding contributions. Other contributions include the reports of the work of the Canadian Research Council on the Trail smelter (25,49,77) and numerous papers from The Boyce Thompson Institute for Plant Research (90,91,115,125), and also from the American Smelting and Refining Company (39,103 to 114). The latter include a review of the literature on gas damage to plants (104).

9.2.1 Description of Lesions

Sulfur dioxide produces two types of injury on the leaves of plants, acute and chronic. The former is characterized by the killing of sharply defined marginal or interveinal areas of the leaf. Immediately following the fumigation these areas take on a dull, water-soaked appearance. They subsequently dry up and usually bleach to an ivory color, though some species finally assume a brown or reddish-brown color if anthocyanins are present in appreciable amounts. The leaf tissue surrounding the acutely injured areas is usually green and apparently entirely normal. Chlorotic or chronic markings are caused by the absorption of an amount of gas somewhat less than that necessary to cause acute injury, or they may be caused by the slow, long-continued absorption of sublethal amounts of gas which accumulate until the buffer capacity of the leaf is exceeded or a salt effect is produced. The leaf does not collapse because of this type of injury, but histological examination (105) reveals that some of the mesophyll and palisade cells may be shattered or the chloroplasts in some otherwise intact cells may be ruptured, allowing the grana to fill the cell. Photosynthesis measurements (105) of cotton leaves indicate that the chronically injured areas are about one-half as active as normal areas. Silver leaf, which consists initially of a collapse of the subepidermal cells on the lower surface of the leaf causing a silver (or bronze) metallic sheen, and which is characteristic of smog injury, is rarely encountered in sulfur dioxide fumigations, except with cotton, which frequently shows extensive silvering. Typical acute and chronic injuries due to sulfur dioxide on a number of types of vegetation are illustrated in Figs. 9-1 and 9-2. In Fig. 9-3 are shown sulfur dioxide-like markings due to other causes.

The conifers (48,49,77) present a special problem because the needles remain on the trees for several years, thus greatly extending the exposure time beyond that of the annuals. Maximum sensitivity to sulfur dioxide occurs in the late spring or early summer, paralleling the physiological activity of the plant. Acute lesions consist of reddish discolorations of the needles with subsequent shrinkage of the tissue. The discolorations may involve the whole leaf or limited areas of any portion of the leaf. Abscission may follow after a variable interval. In winter the first symptom is a general change from the normal dark green to a lighter green. Then definite areas turn yellow-brown and finally red-brown, giving a banded appearance. Because

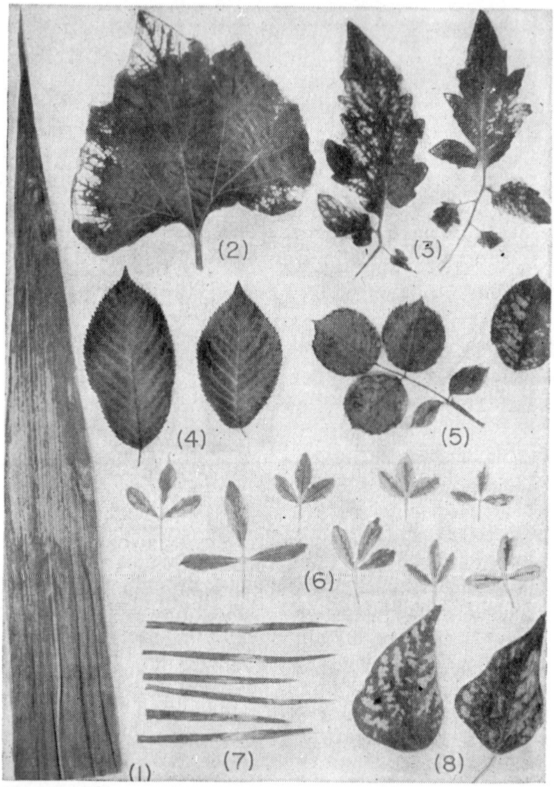

FIG. 9-1. Sulfur dioxide damage: (1) corn, (2) sunflower, (3) tomato, (4) sweet cherry, (5) rose, (6) alfalfa, (7) wheat, (8) beans.

abscission occurs readily, conifers in a smoke area are often deficient in needles. Western larch is the most sensitive conifer in the Northwest to sulfur dioxide, followed by Douglas fir and yellow pine. Red cedar, white pine, hemlock, Engelmann spruce, grand fir, and lodgepole pine are considerably more resistant. In a smoke area, all varieties build up the sulfur content of their leaves from year to year, because of absorption of the gas.

Chronic markings do not appear immediately after a fumigation. Rather, their first appearance is delayed for 2 to 4 days, and they continue to develop for several days. Chronic markings due to long-continued exposure to sublethal concentrations of the gas may require several weeks to develop. A large amount of sulfate is found in the latter leaves (108), whereas leaves that have been injured either acutely or chronically by a single fumigation show only a small increase in the sulfate content.

"Invisible injury" due to sulfur dioxide, which was supposed to be characterized by a reduced vigor of the plants and a reduced level of photosynthetic activity, was postulated by Stoklasa (100). All attempts to demonstrate invisible injury have yielded negative results (77,107,114). There may be a temporary partial inhibition of photosynthesis while the gas is present, but if the leaves are not permanently injured, the normal level, or even a slightly higher than normal level, of photosynthesis is rapidly regained after the fumigation is stopped, so that no appreciable reduction of assimilation is found. Moreover, a large amount of yield data by many investigators

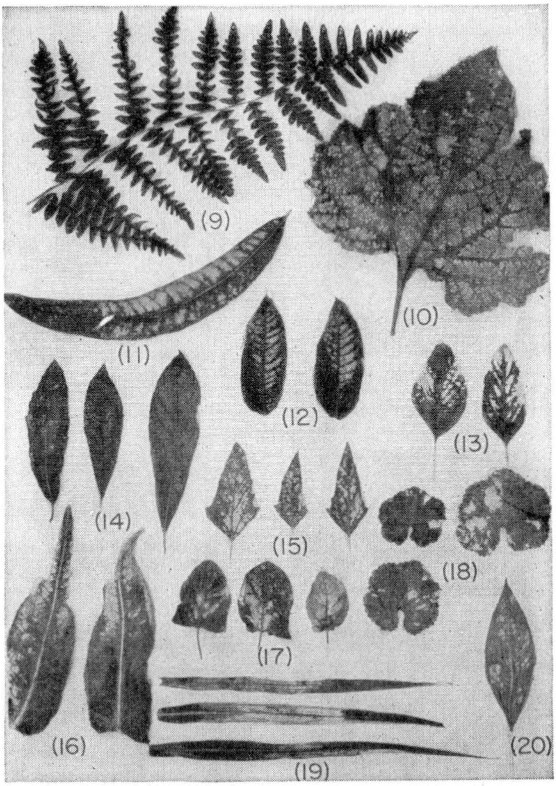

FIG. 9-2. Sulfur dioxide damage: (9) fern, (10) cocklebur, (11) fireweed, (12) ailanthus, (13) redroot (pigweed), (14) alder, (15) lamb's-quarters, (16) prickly lettuce, (17) wild morning glory, (18) mallow, (19) barley, (20) guara parviflora.

indicates that fumigations with sulfur dioxide, which do not injure the leaves visibly, do not reduce the yield (39,42,49,77,91,102,114).

Sulfur is an essential element in plants, because the amino acids cystine and methionine are essential constituents of some plant proteins. Sulfur is also a constituent of many plant hormones and enzymes such as thiamin biotin, enzyme A, lipoic acid, cytochrome C, and others. The organic sulfur level (108) of the leaves of most broad-leaved plants ranges from about 0.15 to 0.3 per cent on a dry basis. Conifer needles contain about 0.1 per cent and some plants, such as the crucifers, contain as much as 0.6 per cent (110). Little or no sulfate is present if the leaf contains only its normal quota or less of organic sulfur, and a large excess of sulfate does not raise the organic sulfur level appreciably above normal.

Sulfur deficiency in plants, manifested by a general yellowing of the leaves and retarded growth rate, can be corrected (107) by supplying the sulfur as sulfur dioxide to the leaves, but because the element thus introduced is not readily translocated from the leaves it is less effective to supply the sulfur in this way than as sulfate through the roots. Tracer experiments (99,103,106) using S^{35} have shown that there is no difference in the ultimate disposition of the sulfur, whether it is supplied as sulfate to the roots or as sulfur dioxide to the leaves, provided, of course, that no visible lesions are caused by the gas.

A number of suggestions have been made as to the mechanism of sulfur dioxide injury. Haselhoff and Lindau (38) assumed that the gas reacts with such compounds

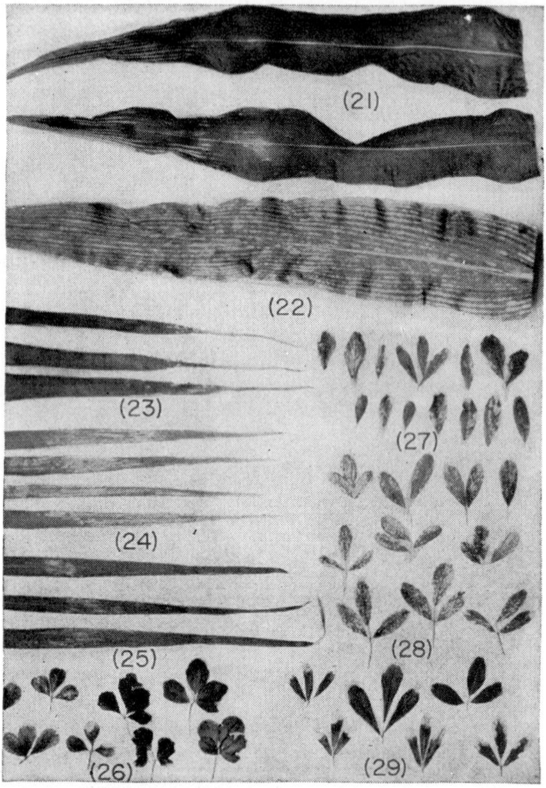

FIG. 9-3. Markings resembling sulfur dioxide damage: (21) corn (frost injury), (22) corn (probably nutritional deficiency), (23) barley (terminal bleach), (24) barley (salt injury), (25) barley (rust injury), (26) alfalfa (frost injury), (27) alfalfa (occurs along with white spot), (28) alfalfa (white spot), (29) alfalfa (tipburn).

as aldehydes and sugars in the leaf, forming addition products which slowly release sulfurous or sulfuric acid, causing injury to the cells. They were uncertain whether the addition products themselves could cause injury. This mechanism cannot be more than partly correct because injury due to a given quantity of absorbed gas is greater in the morning, when the sugar concentrations are low, than in the afternoon, when they are high. Presumably the carbohydrates have some protecting effect. Noack (79) suggested that the sulfur dioxide inactivates the iron in the chloroplasts, causing interference with its catalytic properties in assimilation. Then secondary changes affect the photooxidative processes of the leaf so that bleaching and death of the cells result. Dorries (27) suggested that the absorbed acidity could decompose chlorophyll, liberating magnesium and forming pheophytin. This reaction will be discussed later. However, it is evident that the toxicity of sulfurous acid is related primarily to its reducing properties rather than to its acidity because it is thirty-fold more toxic than sulfuric acid (107).

9.2.2 Proof of Sulfur Dioxide Injury

Unfortunately there is no objective method of identifying with certainty sulfur dioxide injury on isolated leaves. Observations in the field, while the markings are fresh, particularly if the leaves are seen in the early stages of collapse before bleaching, offer convincing evidence of gas injury, provided there is evidence from odor, air analysis, or meteorological conditions that sulfur dioxide was present. A characteristic odor is always noted when there is enough sulfur dioxide present in the field to injure vegetation, but this disappears soon after the fumigation is finished. It is generally possible to find characteristic markings on a number of different species of sensitive plants, including sensitive weeds, if one of them is injured. Sulfur dioxide-like markings on a single species without a wider distribution needs careful scrutiny, because there are many lesions that resemble more or less those due to the gas, for example, white spot, sunscald, tipburn, and frost. In this case it may be advisable to search for similar lesions in a comparable smoke-free area. Massey (71) has discussed the similarity between disease symptoms and chemically induced plant injury.

Chemical analysis of the leaves for total and sulfate sulfur offers a possible method of diagnosis, particularly if the leaves had been analyzed before the fumigation. An increase in the total sulfur, together with an equal increase in the sulfate fraction of 0.1 to 0.2 per cent sulfur or more, would then suggest that the injury was probably due to sulfur dioxide. A large amount of sulfate sulfur would suggest a long-continued low-concentration fumigation, which might induce sulfate toxicity. However, it must be noted that a similar sulfate build-up in the leaves can be due to soluble sulfate in the soil (110). Gypsum is very effective in supplying luxury sulfate to the plants. Without a knowledge of the sulfur level in the leaves before fumigation, the analysis after fumigation would have little diagnostic value for acute injury, because there is a wide range of sulfur content in "normal" uninjured leaves. There is one case in which the sulfate analyses would have positive significance for at least a limited time after the lesions appeared, namely, if no sulfate were present, the lesions could not be due to sulfur dioxide.

Dorries (27) claimed that leaves injured by sulfur dioxide could be identified by the presence of pheophytin, which would be formed in the leaves by the acidification of chlorophyll. He made an acetone extraction of the leaves and examined the solution with a spectrograph. Leaves injured by frost, insects, and diseases did not give the pheophytin reaction. Attempts to repeat this work (105) have not been successful. Alfalfa leaves severely injured by sulfur dioxide gave absorption spectra in the Beckman spectrophotometer that were practically identical with the spectra of uninjured leaves. Possibly separation of the pigments chromatographically would lead to a more successful method for pheophytin.

9.2.3 Relative Susceptibility

Different species of plants and even different varieties of a species may vary considerably in their susceptibility to sulfur dioxide. Moreover, the same plant grown in different environments or at different stages of growth may show differences in susceptibility due to differences in cuticle and cellular densities and stomatal numbers. Nevertheless, it is advisable to tabulate a few of the more important cultivated and native plants in order of their resistance to the gas. O'Gara (82) worked for several years to determine resistance factors for over 300 species and varieties. His method consisted of fumigating the plants for 1 hr with a measured concentration of sulfur dioxide just sufficient to cause traces of injury. The relative humidity was measured, and the concentration corresponding to 100 per cent relative humidity was calculated from an equation based on data given later. Finally, this calculated value

Table 9-1. Relative Sensitivity of Cultivated and Native Plants to Injury by Sulfur Dioxide

(Determined by O'Gara)

Sensitive		Intermediate		Resistant	
		Cultivated Plants			
Alfalfa	1.0[a]	Cauliflower	1.6[a]	Gladiolus (1.1–4.0)[b]	2.6[a]
Barley	1.0	Parsley	1.6	Horse-radish	2.6
Endive	1.0	Sugar beet	1.6	Sweet cherry	2.6
Cotton	1.0	Sweet William	1.6	Canna	2.6
Four o'clock	1.1	Aster	1.6	Rose	2.8–4.3
Cosmos	1.1	Tomato (1.3–1.7)[b]	1.7	Potato (Irish)	3.0
Rhubarb	1.1	Eggplant	1.7	Castor bean	3.2
Sweet pea	1.1	Parsnip	1.7	Maple	3.3
Radish	1.2	Apple	1.8	Boxelder	3.3
Verbena	1.2	Catalpa	1.9	Wisteria	3.3
Lettuce	1.2	Cabbage	2.0	Mock orange	3.5
Sweet potato	1.2	Hollyhock	2.1	Honeysuckle	3.5
Spinach	1.2	Peas	2.1	Hibiscus	3.7
Bean	1.1–1.5	Gooseberry	2.1	Virginia creeper	3.8
Broccoli	1.3	Zinnia (1.2)[b]	2.1	Onion	3.8
Brussels sprouts	1.3	Marigold	2.1	Lilac	4.0
Pumpkin	1.3	Hydrangea	2.2	Corn	4.0
Table beet	1.3	Leek	2.2	Cucumber	4.2
Oats	1.3	Begonia	2.2	Gourd	5.2
Bachelor's-button	1.4	Rye (1.0)[b]	2.3	Chrysanthemum	5.3–7.3
Clover	1.4	Grape	2.2–3.0	Snowball	5.8
Squash (1.1–1.4)[b]	1.4	Linden	2.3	Celery	6.4
Carrot	1.5	Peach	2.3	Citrus	6.5–6.9
Swiss chard	1.5	Apricot	2.3	Cantaloupe (muskmelon)	7.7
Turnip	1.5	Kale	2.3	Arborvitae	7.8
Wheat	1.5	Nasturtium	2.3	Currant blossom	12.0
		Elm	2.4	Live oak	14.0
		Birch	2.4	Privet	15.0
		Iris	2.4	Corn silks and tassels	21.0
		Plum	2.5	Apple blossoms	25.0
		Poplar	2.5	Apple buds	87.0
		Native Plants			
Gaura (1.0)[b]	......	Dandelion	1.6	Purslane	2.6
Tobacco tree (*N. glauca*)	1.0	Orchard grass	1.6	Sumac	2.8
June grass (*B. tectorum*)	1.0	Rough pigweed (redroot)	1.7	Shepherd's purse	3.0
Prickly lettuce	1.0	Black mustard	1.7	Milkweed	4.6
Mallow	1.1	Smartweed	1.8	Salt grass	4.6+
Ragweed	1.1–1.2	Lamb's-quarters	1.8	Pine[c]	7–15.0
Curly dock	1.2	Sweet clover	1.9		
Buckwheat	1.2–1.3	Nightshade	2.0		
Bouncing bet	1.3	Hedge mustard	2.1		
Plantain	1.3	Cocklebur	2.3		
Sunflower	1.3–1.4	Tumbling mustard	2.4		
Rye grass	1.4				

[a] Factors of relative resistance compared with alfalfa as unity.
[b] More probable factors based on later experience.
[c] Data for pine represent October fumigations in Palo Alto, Calif., of Monterey, white, Allepo, and Coulter varieties. O'Gara factors calculated from the data of Katz and McCallum (Effect of Sulfur Dioxide on Vegetation, *Nat. Research Council Can., Pub.* 815, 1939) are as follows: larch in May, 1.5; Douglas fir in May, 2.3; yellow pine (year-old) seedlings in May, 1.6, in August, 2.4 to 4.7.

was divided by 1.25 ppm, which was the concentration required for incipient marking on alfalfa in 1 hr. The factors so determined represent the relative resistance of the plants compared with alfalfa as unity. A number of fumigation treatments were applied in each case, and the minimum factors were selected. The greater the number of fumigations, the more likely the correct factor will be found.

Table 9-1 summarizes over 100 of O'Gara's factors, most of which have been published earlier (109). An arbitrary susceptibility rating has been made, based on these factors. In some cases, conditions for maximum suceptibility may not have existed during the treatments and the factors will be high. However, experience during nearly 30 years has demonstrated the essential correctness of most of the factors. In a few cases a value considered to be somewhat more probable is inserted along with O'Gara's factor. For example, young rye in the field is considerably more sensitive than wheat and is comparable to barley. The O'Gara factor for wheat refers to spring wheat. Very young plants, in the two- to three-leaf stage, are practically as sensitive as barley. Later the plants become more resistant. On the other hand, winter wheat in the early spring is very resistant. Later the plants become more sensitive.

An extreme case of varietal variability has been noted in a field fumigation of gladiolus. Here the leaf destruction on the different varieties ranged from 2 to 60 per cent, and plants from bulblets had as much as 80 per cent blade injury. If the O'Gara factors are calculated for this fumigation, they range from about 1.1 to 4.0, with an average of 2.4—in close agreement with O'Gara's value of 2.6. By contrast, O'Gara found that 12 varieties of apple had the same factor. Most of this work was done near Salt Lake City, Utah. In more humid or drier climates the leaves would have different morphology, and somewhat different factors could be expected. For these reasons the O'Gara factors should be considered as suggestive only and subject to change for different environments, varieties, and stages of growth.

9.2.4 Effect of Environment

The primary factor which controls gas absorption by the leaves is the degree of opening of the stomata. Loftfield (56) worked out and illustrated the diurnal changes of the stomatal apertures of a number of plants. These changes are known to correspond to the usual diurnal changes in sensitivity of the plants to injury by sulfur dioxide. Katz and Ledingham (77) confirmed these observations and conclusions for alfalfa. When the stomata are wide open, absorption is maximum, and vice versa. Consequently those conditions which induce the stomata to open (light, adequate moisture supply for the roots of the plant, high relative humidity, moderate temperatures) predispose the plant to injury by sulfur dioxide. Most plants regularly close their stomata at night; consequently, they are much more resistant at night than in the daytime. However, plants like the potato, which do not close their stomata at night, are about as sensitive in the dark as in the light. Similarly, individual alfalfa leaves, observed to have their stomata open in the dark, are readily injured by the gas (109). Since one of the essential functions of the stomata is to regulate evaporation (56), it follows that moisture stress in the leaves causes stomatal closure and leaves that show the slightest signs of wilting are highly resistant to the gas (125). Conversely, with adequate moisture supply in the daytime and particularly with high relative humidity to reduce evaporation, the stomata will open and susceptibility to the gas will approach the maximum.

The effect of relative humidity is summarized by O'Gara (82) in Table 9-2, which gives the relative sensitivity factors and their reciprocals (resistance factors) of alfalfa and other plants at lowered humidities compared with a sensitivity or resistance of unity at 100 per cent relative humidity.

Table 9-2. Effect of Relative Humidity on Alfalfa and Other Plants

Relative humidity, %	Relative sensitivity	Relative resistance
100	1.00	1.00
80	0.89	1.12
60	0.77	1.30
50	0.69	1.45
40	0.54	1.85
30	0.31	3.2
20	0.18	5.5
10	0.13	7.7
0	0.10	10.0

If a standardized fumigation of, for example, 1 hr duration at 4 to 5 ppm of SO_2 is applied to a series of similar alfalfa plots throughout a cloudless day, keeping the relative humidity constant, only slight injury will be produced in the early morning. Injury will increase rapidly to a maximum at about 11 A.M.; then it will fall off appreciably until midafternoon and decrease rapidly to a low value or zero in the late afternoon. Stomatal movement is an important factor in this diurnal pattern, but the afternoon effects are probably caused partly by accumulated carbohydrates in the leaf, which reduce to some extent the toxicity of the gas, possibly because of the sulfite-aldehyde reaction.

9.2.5 Time-concentration Equations

Considering actual fumigation conditions, it is evident that injury to the plants will be more severe, the greater the concentration of the sulfur dioxide in the air and the more protracted the exposure. These relationships were expressed by O'Gara (81) as follows:

$$(C - C_0)t = K \tag{9-1}$$

where C is the concentration necessary to produce a definite percentage of markings on the leaves in time t, C_0 is a limiting concentration that can be endured indefinitely, and K is a constant.

Under conditions of maximum susceptibility the time-concentration relationships for producing incipient injury on alfalfa become

$$(C - 0.33)t = 0.92 \tag{9-2}$$

suggesting that this alfalfa could be injured by about 1.25 ppm in 1 hr and that 0.33 ppm could be tolerated indefinitely without causing injury. Subsequent work has shown the essential correctness of this equation, but the limiting concentration for very long fumigation treatments appears experimentally to be about 0.1 to 0.15 ppm, and at times incipient injury on alfalfa has been observed in the field with somewhat smaller exposure than that indicated by the equation.

Thomas, Hendricks, and Hill (109,112,113) generalized these equations for alfalfa so that the time-concentration relationships could be calculated for any degree of injury. First it was found that leaf destruction due to sulfur dioxide is proportional to the quantity of the gas that is absorbed by the leaves in a specified exposure time and is expressed by a linear equation of the form $x = a + by$, where y is the percentage of leaf destruction and x is the percentage of sulfur dioxide absorbed in the leaves. An appreciable amount of gas, represented by a, can be absorbed without causing any injury; thereafter the injury-absorption line has a slope, b. Then two or more such equations, each representing a different exposure period, were solved simultaneously, giving linear time-absorption equations which were calculated for 0 (traces), 50, and 100 per cent leaf destruction. Rate of absorption can vary over a wide range under different environmental conditions, but if a definite rate of absorption is specified, for example, a maximum rate associated with maximum susceptibility, the equations are readily reduced to the O'Gara equation and satisfactory agreement with O'Gara's constant is obtained:

$$(C - 0.24)t = 0.94 \qquad \text{traces of leaf destruction} \tag{9-3}$$

$$(C - 1.4)t = 2.1 \qquad 50\% \text{ leaf destruction} \tag{9-4}$$

$$(C - 2.6)t = 3.2 \qquad 100\% \text{ leaf destruction} \tag{9-5}$$

The fact that the equations contain a term indicating a limiting concentration which could be tolerated for protracted periods suggests that the sulfur dioxide, after absorption, is being changed into a less toxic form, namely, sulfate. The latter is the toxic agent in long fumigations in which lethal concentrations of sulfur dioxide are never accumulated in the leaves. It was thus shown that sulfite is about thirty times as toxic as sulfate. It was also shown that, if the sulfur dioxide could be added rapidly enough so as to build up a toxic concentration without any oxidation, a concentration of about 1,350 ppm, calculated as sulfur in the dry tissue, would constitute a toxic dose in the leaves of alfalfa. If the reactive gas-absorbing cells represent 50 per cent of the leaf substance, the toxic dosage for these tissues would be about 2,700 ppm. Further, it appeared that a similar toxicity applies to reactive cells of many other species of plants, including water plants.

9.2.6 Yield–Leaf-destruction Equations

From a practical point of view it is important to have a criterion by which economic crop damage can be reliably estimated, in order to settle the claims of farmers whose crops have been injured by sulfur dioxide. Such a criterion may be had if the relationship between the extent of leaf destruction and crop yield is known, since the former can readily be measured in the field immediately following a fumigation.

A few crops have been adequately investigated from this point of view. Many more have also been studied, but the results are not yet available for publication. In general, and at least as a first approximation, the leaf-destruction–yield functions are straight lines, starting with 100 per cent yield y at 0 per cent leaf destruction x and ending with a definite decrease in yield at 100 per cent leaf destruction. The equations take the form

$$y = a - bx$$

where a approximates 100 per cent and b is the yield as a fraction of the control yield when the leaf tissue is totally destroyed by sulfur dioxide. Some investigators express the leaf destruction as a percentage of the total functional area at the time of fumigation. Brisley and Jones (14) calculate the leaf destruction as a percentage of the total leaf area produced during the season. The two methods of expressing the injury are not greatly different in mature plants, but in young developing plants, the former method emphasizes and the latter minimizes the leaf injury. Each method has advantages for different crops.

Because of the variability of small field plots, an appreciable error is always involved in determining individual plot yields. It is, therefore, necessary in working out the yield–leaf-destruction equations to deal with a sufficient number of field plots with injury ranging from zero to nearly 100 per cent, in order to obtain statistically significant equations. In the absence of sulfur deficiency, in which fumigations that do not cause serious injury actually increase the yield, the equations indicate that even a small amount of leaf destruction would cause a calculable decrease in yield. This is not rigidly proved experimentally, but merely represents a reasonable interpretation of the data and equations. As a practical matter, however, the reduction in yield is difficult to measure experimentally if leaf destruction is less than about 5 per cent. This explains the conclusion of Katz (77) that injury up to about 5 per cent has no effect on the yield of alfalfa.

The following is a summary of the published data on the yield–leaf-destruction relationships for different crops.

Alfalfa. Hill and Thomas (39) fumigated alfalfa plots (6 by 6 ft) with sulfur dioxide one, two, or three times during the growth of the crop, allowing 10 to 14 days to elapse between successive fumigations and between the final fumigation and the harvest of the crop. They determined the extent of leaf destruction x by counting the injured and uninjured leaves on representative samples of the vegetation and estimating the average area of injury on the marked leaves. Leaf destruction was always measured within a few days after the fumigation. In double and triple fumigations on the same crop, the leaf-destruction values are the average of the individual treatments, each measured separately. The crops were harvested at the normal intervals, and the dry-weight yield of each plot was expressed as a percentage y of the yields of several similar adjacent check plots. The results were summarized as follows:

1. Single fumigation at early, medium, or late stage, representing either 25, 50, or 80 per cent of the growth period of crop:

$$\begin{aligned} y &= 99.5 - 0.30x \\ n &= 96 \\ r &= 0.64 \pm 0.06 \\ Sy &= 7.4\% \end{aligned} \tag{9-6}$$

2. Double fumigation at early and medium, early and late, or medium and late stages in the growth of the crop:

$$\begin{aligned} y &= 95.5 - 0.49x \\ n &= 34 \\ r &= 0.79 \pm 0.07 \\ Sy &= 8.2\% \end{aligned} \tag{9-7}$$

3. Triple fumigation at early, medium, and late stages in the growth of the crop:

$$\begin{aligned} y &= 96.6 - 0.75x \\ n &= 12 \\ r &= 0.98 \pm 0.014 \\ Sy &= 4.1\% \end{aligned} \tag{9-8}$$

where n = number of plots fumigated
r = correlation coefficient
Sy = standard deviation of individual yields from regression line

Experiments in which various percentages of the leaf area of the plants were clipped off gave yield–leaf-destruction relationships closely approximating the preceding equations. This suggests that reduction in yield of alfalfa due to sulfur dioxide fumigation is caused by a reduction in the leaf area of the plant, without other effects such as might be caused by a systemic poison. There was no significant difference in the reduction in yield due to a given percentage of leaf destruction, whether the plants were one-fourth or three-fourths grown at the time of fumigation or clipping, even though three times as much leaf tissue was removed in the later as in the earlier treatment.

The Canadian National Research Council (77) gives data from which the following yield–leaf-destruction relationship for alfalfa can be calculated:

$$\begin{aligned} y &= 99 - 0.37x \\ n &= 103 \\ r &= 0.48 \\ sY &= 8.8\% \end{aligned} \tag{9-9}$$

Notations have the same significance as in Eq. (9-6).

In the Hill and Thomas alfalfa experiments, most of the fumigations were of 1 to 2 hr duration. In the Canadian experiments most of the fumigations were more protracted. Of the latter plots, 35 received fumigations of 10 to 53 hr duration and 21 had fumigations of 60 to 600 hr. Considering the differences in fumigation treatments, the two sets of data are in satisfactory agreement.

Barley. The Selby Smelter Commission (42) in 1914 conducted an extensive program of sulfur dioxide fumigations on barley, including single fumigations at various stages of growth and multiple fumigations ranging throughout the whole growth period of the plant. The single fumigations were usually of 1-hr duration at concentrations of 5 to 20 ppm. The multiple fumigations had durations of 2 to 240 min and concentrations of 0.5 to 8 ppm applied 2 to 198 times. Leaf destruction was estimated immediately following the treatments in the case of the single or more limited multiple fumigations. With the more protracted series of fumigations, leaf destruction was estimated during the development of the head, before incipient maturity. At harvest, measured areas of the fumigated plots and closely comparable check plots were cut. The total yield was obtained; then all the heads were counted, discarding those which were moldy. Only the good heads were weighed and threshed. Yield of grain was calculated as the average weight of grain per good head. The yield of the fumigated plot was divided by that of the check to give the percentage yield due to the treatment. The Selby Commission did not calculate the yield–leaf-destruction relationships, but the data give the following statistics:

Fumigations in early stage only (less than 25 cm height) before April 20:

$$\begin{aligned} y &= 98 - 0.06x \\ n &= 18 \\ r &= 0.13 \\ Sy &= 12.2\% \end{aligned} \tag{9-10}$$

Fumigations after heading out began, April 28 and later (some plots were fumigated earlier also, as well as throughout the ripening period):

$$\begin{aligned} y &= 98 - 0.40x \\ n &= 60 \\ r &= 0.74 \\ Sy &= 10.2\% \end{aligned} \tag{9-11}$$

The statistics were not altered appreciably if the yields were figured on an area basis as weight per good head times total number of heads per unit area.

These data indicate that the fumigation of barley in the early vegetative state (before the boot stage) does not reduce the yield more than slightly, even though a large part of the leaf area on the plant at that time is destroyed. Later, during the period of head formation and development, there is reduction of yield proportional to leaf destruction. These relationships are confirmed by the American Smelting and Refining Company's published data on oats (112), which indicate as well that fumigations during the final ripening have little effect on yield. The latter investigators also found similar relationships for irrigated barley and wheat.

Wheat. Brisley and Jones (14) conducted an extensive study of Sonora wheat in Arizona from 1941 to 1944. They measured the leaf area destroyed by fumigation with sulfur dioxide and expressed the result as a percentage of the total leaf area developed during the life of the plant. A leaf-destruction value greater than about 67 per cent was not possible because there was never a greater percentage of functional tissue than this on the plant at one time. The equation for the single fumigations in 1941 and 1944 was

$$\begin{aligned} y &= 100 - 0.59z \\ n &= 142 \\ r &= 0.93 \\ Sy &= 2.7\% \end{aligned} \tag{9-12}$$

where z designated leaf destruction calculated as percentage of the total leaf area produced by the plant, to distinguish it from x, the leaf destruction calculated as a percentage of the leaf area at the time of fumigation. The 1944 data alone gave the equation:

$$\begin{aligned} y &= 100 - 0.66z \\ n &= 130 \\ r &= 0.95 \\ Sy &= 2.2\% \end{aligned} \tag{9-13}$$

The 130 fumigations in 1944 were applied at the rate of 10 each week for 14 weeks (omitting the third week), from one month after planting when the plants had only two leaves with a total area of 3 sq cm on February 29 until May 29 when the plants were beginning to mature. The fumigations were applied so as to give a wide range of leaf destruction on the 10 plots each week. In the thirteenth week the check plants had 1,796 sq cm of functioning tissue (the maximum) and in the fourteenth week, 1,234 sq cm. Total leaf area produced was 2,660 sq cm, of which 1,796 sq cm is 67 per cent. Equation (9-13) applies equally well to the first eight and last six weeks calculated separately, indicating that a given leaf area destroyed reduced the yield by the same amount regardless of the stage of growth when the fumigation occurred.

However, when the leaf destruction is calculated* as a percentage of the leaf area at the time of fumigation, a somewhat different picture emerges. Dividing the data into three groups, the first seven weeks, the last six weeks, and the last four weeks, omitting the eighth week as a transitional period, the equations become for the first to seventh week—functional leaf area = 0.1 to 11 per cent of total leaf area:

$$\begin{aligned} y &= 99 - 0.01x \\ n &= 60 \\ r &= 0.045 \\ Sy &= 5.0\% \end{aligned} \tag{9-14}$$

for the ninth to fourteenth week—functional leaf area = 37 to 67 per cent of total leaf area:

$$\begin{aligned} y &= 100 - 0.35x \\ n &= 60 \\ r &= 0.95 \\ Sy &= 2.7\% \end{aligned} \tag{9-15}$$

and for the eleventh to fourteenth week—functional leaf area = 50 to 67 per cent of total leaf area:

$$\begin{aligned} y &= 100 - 0.37x \\ n &= 40 \\ r &= 0.97 \\ Sy &= 2.4\% \end{aligned} \tag{9-16}$$

Recent American Smelting and Refining Company data for winter wheat between the boot and milk stages, growing under conditions of average fertility and moisture supply, gave

* Brisley kindly supplied additional data from which these calculations could be made.

$$\begin{aligned} y &= 103 - 0.37x \\ n &= 71 \\ r &= 0.75 \\ Sy &= 10.2\% \end{aligned} \qquad (9\text{-}17)$$

Equations (9-14) to (9-16) support very satisfactorily the Selby and American Smelting and Refining Company data, indicating little or no effect on yield of grain due to severe fumigations in the tillering stages of growth and a coefficient of x of about 0.4 in the heading-out stages. For about three weeks, while rapid elongation of the stems was taking place, intermediate coefficients apply. For example, during the seventh week when the functioning area represented 11 per cent of the total area, total leaf destruction reduced the yield an estimated 8 per cent, and during the eighth and ninth weeks when the functioning areas were 22 and 37 per cent, respectively, the corresponding reductions in yield due to 100 per cent leaf destruction were estimated at 16 and 28 per cent.

In addition to the fumigation experiments already described, Brisley and Jones carried out a series of double and triple fumigations in 1941–1942–1943:

$$\begin{aligned} y &= 98 - 0.26z \\ n &= 90 \\ r &= 0.84 \\ Sy &= 4.1\% \end{aligned} \qquad (9\text{-}18)$$

45 plots were treated twice, first at male anthesis and second at approaching maturity; 46 plots were treated three times, first at vegetative state, second at male anthesis, and third in early dough stage before leaves showed signs of maturity.

It is not entirely clear why Eq. (9-18) is so far out of line with the other wheat data. Brisley and Jones state that 1944 was a very poor crop year compared with 1941, 1942, and 1943. Average foliage areas of the check plants in the four consecutive years were 4,700, 3,950, 3,640, and 2,660 sq cm. Possibly the leaf area was a more critically controlling factor in seed production in 1944 for this reason. Alternatively, they suggest that the shock of a double fumigation would be less than that of two single fumigations that destroy the same total amount of leaf tissue. Careful scrutiny of their data reveals considerable support for the first suggestion and little or no support for the second. If the more severe fumigations cause greater shock than additively equivalent lighter fumigations, Eq. (9-13) should not be linear, but rather should curve downward. Figure 6 of the Brisley and Jones article, in which the data comprising Eq. (9-18) are plotted, shows definite segregation of the values for 1941, 1942, and 1943, with 1941 data having the least slope and 1943 the greatest. Equations are as follows:

$$1941 \quad y = 101 - 0.25z \qquad (9\text{-}19)$$

$$1942 \quad y = 98 - 0.29z \qquad (9\text{-}20)$$

$$1943 \quad y = 98 - 0.49z \qquad (9\text{-}21)$$

The limited data in Brisley and Jones's Table IV suggest the same conclusion. Here multiple fumigations in 1944 fall close to Eq. (9-13), while in 1942 they are close to Eq. (9-18). The 1943 values are intermediate. With wheat, seasonal effects are thus of great importance in determining the reduction in yield due to a given amount of leaf destruction. Other plants would probably be influenced similarly but to different extents, depending on the ability of the plant to develop new leaves after the fumigation.

It has been observed that fumigation of winter wheat in an extremely fertile field, where the yield exceeded 60 bu/acre, did not reduce the yield appreciably at any stage of growth, even though extensive leaf destruction was produced. Evidently leaf destruction causes less reduction of yield in wheat that has adequate or excessive nutrition and moisture than with average or deficient nutrition and moisture.

9.2.7 Sulfuric Acid

Sulfuric acid aerosol appears to be toxic to vegetation only under special circumstances. The Selby Smelter Commission (42) treated barley plants under large glass jars with dense clouds of aerosol (up to 600 ppm) without causing injury, so long as the mixture was continuously circulated. On standing, the small acid droplets coalesced and settled on the leaves, producing necrotic spots. Repetition of this work in field plots (111) showed that sugar beets and alfalfa were not injured by 30 to 65 ppm of sulfuric acid aerosol so long as the particles were not larger than about 1μ in diameter. Larger particles settled on the leaves as a definite "bloom" that could be readily seen from certain angles. These droplets did not wet the leaf but remained as discrete spheres, resting on the leaf hairs or even touching the surface without flattening out. It was only when free water was sprayed or dropped on the leaves that characteristic acid spotting was produced. Evidently the small droplets do not diffuse through the stomata into the interior of the leaf. This is not because they are too large to pass through the pores. Possibly electrostatic forces prevent their entrance. Injury due to sulfuric acid aerosol has never been observed in the field.

9.3 HYDROGEN FLUORIDE

Susceptibility of different species of plants to injury by hydrogen fluoride varies over a wide range. Only a few have marked susceptibility. This is not due entirely to the rate and quantity of absorption of the fluoride, as is the case with sulfur dioxide, but to a number of other factors also, such as (1) the rate and extent of translocation of the fluoride to areas of concentration at the tips and margins of the leaves; (2) the concentration of inorganic ions in the leaf which can inactivate part of the fluoride as insoluble compounds like calcium fluoride or apatite; (3) other poorly understood effects, such as interference with enzyme systems (70,118) and conversion to organic fluorine compounds.

No consideration will be given in this discussion to fluorosis in cattle and sheep, caused by eating, for a sufficient length of time, forage containing more than 30 to 50 ppm fluorine (2). Such forage is usually normal in growth behavior and without any leaf lesions.

Fluorine compounds are evolved from a number of industries, of which those manufacturing certain metals, superphosphates, and ceramics are the most important. Impurities in the raw materials are the principal source of the fluorides, except in the electrolytic aluminum process, in which the cryolite bath is the source of some silicon tetrafluoride and cryolite fumes which may hydrolyze in the air to produce hydrogen fluoride. The treatment of rock phosphate with acid liberates hydrogen fluoride from the fluoride impurities. In the high-temperature reactions of the ceramic and metal industries, either sulfur compounds in the combustion gases or silica may liberate silicon tetrafluoride and hydrogen fluoride.

It is not suggested that hydrogen fluoride is the only gaseous fluorine compound which causes injury to vegetation in the field. As already noted, silicon tetrafluoride is an important effluent from some industrial processes. Some fluosilicic acid may also be evolved or formed from the tetrafluoride. It is usually assumed that the latter hydrolyzes more or less completely to hydrogen fluoride. Recent fumigation experiments with silicon tetrafluoride carried out at the Boyce Thompson Institute (123) suggest that its action on plants is similar to that of hydrogen fluoride. Other fluorine-containing gases such as fluorine oxide or volatile organic fluorine compounds have not yet been studied. Elemental fluorine, if liberated, appears to be so reactive in the atmosphere that it is probably changed to fluoride before reaching the vegetation. It is unlikely that solid fluorine compounds (except perhaps soluble compounds like sodium fluoride or even, in some cases, the slightly soluble cryolite) deposited on the surface of the leaves would cause injury to the vegetation, though such deposits would make forage more liable to cause fluorosis in cattle and sheep that consumed it. It is well known that cryolite has been used extensively as an insecticide, particularly in apple orchards, without causing foliar damage. However, peaches and corn are injured by these sprays (24).

9.3.1 Description of Lesions

The nature of the lesions caused by hydrogen fluoride depends to a considerable extent on the type of fumigation. If the plants are fumigated with about 1 ppm for 1 hr or more, interveinal and marginal acute markings like those caused by sulfur dioxide will be produced, except in very resistant vegetation, such as squash leaves, which may show only a marginal chlorosis after a very long exposure. On the other hand, if sensitive plants are fumigated with 1 to 50 ppb for several hours to several days, the lesions will be more like those encountered in the field. They vary considerably in different species.

With very low concentrations, the tips of leaf blades of gladiolus and iris show first a water-soaked condition. This develops into an ivory-colored or brown necrosis which gradually extends down the blade with a fairly uniform front, sometimes with a band of red or brown tissue separating the necrotic and healthy areas. Isolated areas of necrotic tissue or an irregular front do not often occur except with higher concentration of hydrogen fluoride. Iris is somewhat less sensitive to the gas than gladiolus. These lesions and those on fruit trees are referred to in the literature as "leaf scorch." Photographs of fluorine injury are shown in Fig. 9-4.

In rye, oats, barley, and wheat, injury occurs at first mainly at the tip of the blade and in flax at the tips of the leaf serrations. These plants are less sensitive than the gladiolus, but the pattern of injury is similar. The succulent leaves of young plants are the most sensitive (105).

In pines and other conifers the young needles are very sensitive to hydrogen fluoride, whereas the old needles are very resistant (Table 9–3). The young needles develop yellow tips, which collapse and turn reddish brown. They will also be shortened in length if the injury occurs before they are fully grown. In Washington (1), ponderosa pine needles have been shortened to one-fourth their normal length, and they have been shed in 1 to 2 years instead of the usual 5 to 6 years.

In corn, the response is different from that in the preceding species. Low concentrations of gas produce a chronic mottled type of injury, with little or no acute collapse of the tissue. The injury is usually spread over an extensive area of the leaf as a definite mottling, seen best by transmitted light. It is not confined to the tip or margin. In addition to this mottling, small collapsed areas have been observed (105) which subsequently cleared up, leaving tiny chlorotic spots. These collapsed spots occurred 2 or 3 days after the fumigation, in the early morning. They disappeared in an hour or two. Analyses of corn leaves indicate relatively little translocation of fluoride to the margins, compared with gladiolus. In one experiment the tips had about 50 per cent more fluorine than the remainder of the leaf, after 4 days, and two to three times as much after a month (105).

In apricot, prune, and peach the margins of leaves first become yellow, then necrotic. The width of the marginal necrotic area varies around the margin and depends on the severity of the fumigation. A sharp line of demarcation later develops between the healthy and necrotic tissue, sometimes with a streak of brown or red near the line. In time, the necrotic area falls off, leaving an apparently uninjured leaf, except for the reduced area and the serrated edge. These lesions develop slowly. Sometimes they do not begin to appear until 2 to 4 days following the fumigation. New lesions have been seen in course of development 8 to 10 days after the fumigation treatment. Evidently the absorbed fluoride is translocated from the actively absorbing areas of the leaf to the tips and margins, where it builds up lethal concentrations, a process which requires time. This transport is slower and less complete than in gladiolus. Extensive shedding of the injured leaves of fruit trees has been reported. If half the leaves on a tree are injured, abscission may be as high as 25 to 50 per cent (124). This abscission has been observed in apple, apricot, plum, prune, peach, and fig orchards (26,75), sometimes with serious loss of fruit in the first year and failure to

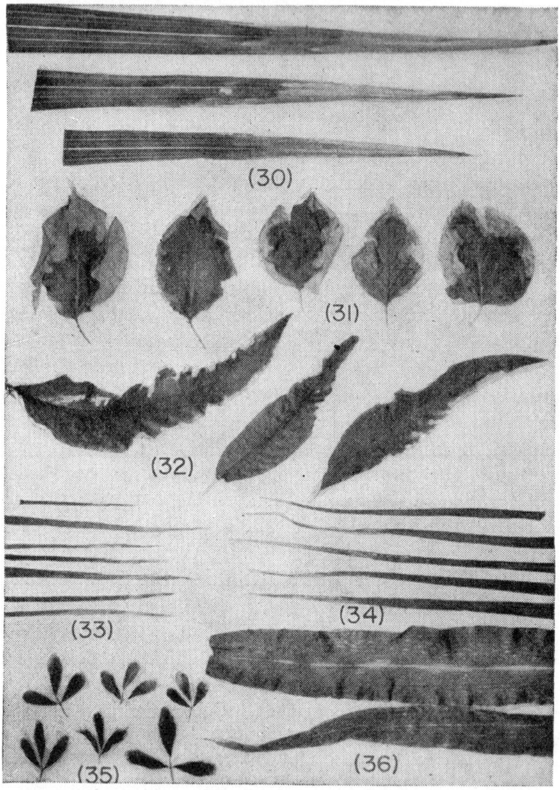

FIG. 9-4. Hydrogen fluoride damage: (30) gladiolus, (31) apricot, (32) peach, (33) rye, (34) barley, (35) alfalfa, (36) corn.

set fruit in subsequent years. Injury to the apical part of the fruit has been ascribed to fluorine in early plum by Kotte (53) and in pear by Radeloff (84). Fruit injury has not been reported in this country.

In the dahlia (105), injury occurs first on the tips and margins of the young leaves that are emerging from their buds or are expanding rapidly. These lesions are dark-colored or almost black. Older leaves are more difficult to injure even though they are actively functional.

Plants vary over a wide range in their susceptibility to injury by hydrogen fluoride. Even within a species, the varietal variation is great. Some varieties of gladiolus, such as Picardy, Aladdin, Surfside, and Shirley Temple (46), may be injured severely, while other varieties like Algonquin, Stoplight, and Commander Koehl, growing alongside in the same field, are injured only slightly. In an apricot orchard the Chinese variety may be severely injured while the Moorpark is only slightly injured. The varieties that are injured the most may absorb the least amount of fluoride. For example (46), Algonquin with 7 per cent leaf injury had 611 ppm fluorine, while Shirley Temple with 54 per cent leaf injury had only 138 ppm fluorine.

Injury to the gladiolus flowers has been observed in fumigation experiments, but a higher concentration of gas is required to injure the petals than the leaves. Petunia petals, on the other hand, are much more sensitive than petunia leaves (105).

Larger differences in susceptibility between species than between varieties would be expected. In the presence of severely injured gladiolus, nearly all the other vegetation usually appears normal. Squash (105) can withstand over a hundredfold the exposure of gladiolus, and it seems remarkable that fumigated cotton leaves can contain as much as 5,000 ppm fluorine without any apparent injury, particularly when it is remembered that cotton is as sensitive to sulfur dioxide as alfalfa. Uptake of fluoride by prune foliage is treated by Miller, Johnson, and Allmendinger (76).

9.3.2 Translocation

Fluoride absorbed from the air into the leaves tends to migrate to the tips and margins. There is very little translocation to other parts of the plant (24). The roots, stems, flowers, and seed generally remain very low in fluorine even when the leaves are high. If soluble fluoride is absorbed through the roots (55,83), both the roots and leaves are high. A method is thus afforded for distinguishing between fluoride absorbed from the air and fluoride from the soil. In general, fluoride in the soil, while appreciable in amount, is very insoluble (62). Absorption through the hair roots into the fleshy roots and tops is slight, particularly if the pH of the soil is not below 6.5, but the tiny roots themselves retain a considerable amount. Soluble fluoride in the soil cannot be excluded in this way, and the fleshy roots retain a considerable part of the fluoride as it is being translocated toward the tops.

Migration of fluoride to the tips and margins of the leaves offers a partial explanation of the toxicity. In gladiolus (20,105) the injured tip of the blade may contain 5 to 10 times as much fluoride as the next 3 in. and 25 to 100 times as much as the basal 3 in. The concentration of the element in the tip causes the injury. Similarly, the marginal injury of the leaves of fruit trees is associated with a two- to ten-fold concentration of fluoride in the injured marginal areas as compared with the uninjured areas.

Romell (86) suggested on the basis of filter-paper models of leaves that the rate of absorption of gases near the edges should be greater than over the body of the leaf, because of a "distortion of the diffusion field in the air around edges and protruding extremities." This effect would be greatest in highly diluted mixtures of a very soluble gas with air, such as the hydrogen fluoride mixtures in the field. Data are lacking to evaluate the Romell effect as a contributing factor in marginal injury, but there can be no doubt that translocation is a very important factor. Compton and Remmert (20) injected fluoride solutions or applied them in lanolin paste or as spray near the base of the gladiolus leaf blade and soon found the fluoride largely in the blade tips where injury was produced. It has also been observed (105) that, a few weeks after a fumigation treatment of gladiolus, practically all the fluoride is found in the tip, the remainder of the blade retaining only a few ppm.

9.3.3 Loss by Volatilization

A series of analyses of vegetation, subsequent to a fumigation, indicates a gradual but extensive lowering of the fluoride content of the leaves with time. While this effect may be due in part to translocation and dilution by growth, careful examination suggests that part of the fluorine is lost by volatilization. Air analyses appear to confirm this type of loss (105). Zimmerman and Hitchcock (128) made earlier and similar observations regarding fluoride taken up from the soil. Evidently the plants do not retain all the fluoride they absorb. The losses by volatilization would offset an appreciable uptake either from the air or the soil. Possibly some organic fluorine compounds are evolved. There are no losses from dry leaf tissue.

9.3.4 Fluoride in the Air

The practical fluoride problem in the field depends on the compounds involved, whether gaseous or particulate, and their concentration in the atmosphere. Particulate fluorides deposited on the leaves would probably not cause foliar injury unless they are water soluble. Gaseous compounds can readily enter the stomata and cause injury.

Only a few analytical results of airborne fluoride measurements have been published. McIntire, Hardin, and Hester (63) reported the fluoride content of rain water and Spanish moss at different locations in Tennessee as an indication of the comparative values of the fluoride content of the atmosphere. Miller and associates (75) studied the ponderosa pine blight near Spokane and defined an area in which they claimed that the trees were injured by fluoride from an aluminum plant. They noted that pine needles of the current year (1949) contained 129 ppm fluoride, while the 1946 needles contained 462 ppm; outside this area the needles contained only 2 to 4 ppm. They reported air analyses ranging from zero to 64 ppb. Adams et al. (1) in a later report made a systematic study of the area by sampling the air continuously at 12 stations, taking three samples each 24 hr. The average concentration at the different stations ranged from 5 to 18 ppb and the maximum from 11 to 147 ppb. Additional sampling with a mobile laboratory was carried out downwind from the plant. Of 1,521 samples, 1,452 ranged from 0 to 5 ppb; 38, from 5 to 10 ppb; 19, from 10 to 20 ppb; 8, from 20 to 50 ppb; and 3 showed 50+ ppb. The maximum was 351 ppb and the next highest 73 ppb. The frequency of occurrence of the higher concentrations decreased as the wind velocity increased.

Largent (54) quoted fluoride analyses of the air near a superphosphate plant ranging from 0 to 29 ppb, and Kehoe (50) found 3 to 6 ppb as a result of systematic analyses in Cincinnati from 1947 to 1949. The Fort William, Scotland, report (2) found total fluoride in the air ranging from 25 to 270 ppb and gaseous fluoride from 9 to 140 ppb in the area 100 yd to 1 mile downwind from the aluminum factory. Fluoride in the vegetation ranged from 5 to 1,000 ppm. The question of foliar injury was not considered in this report nor in those by Largent and Kehoe.

Coal smoke appears to be a general source of fluoride in the air which should not be overlooked. Churchill, Rowley, and Martin (19) reported 85 to 295 ppm in coals from a number of sources. They found 9 to 269 ppm in tree leaves and 53 ppm in grass in the suburban and agricultural areas near Pittsburgh.

9.3.5 Fluoride from the Soil

All agricultural soils contain appreciable amounts of fluoride. Concentrations of 100 to 500 ppm are frequently encountered, and concentrations up to 25,000 ppm have been noted. Uptake of fluoride from these soils is generally slight even from the highest concentrations, because of the slight solubility of the fluorine compounds. McIntire's (62) work in Tennessee indicated that acid soils could render the fluoride slightly soluble and increase the uptake by plants a little but that liming the soil corrected this condition. Addition of phosphate also reduced uptake of fluoride by the plants. Daines et al. (24) observed appreciably more uptake by tomato plants from loam soil at pH 4.5 to 5.5 than at 6.5.

Soluble fluoride (24,83) in the soil can be taken up readily by vegetation. Lesions are produced in the leaves that are quite similar to those caused by gaseous fluorides. The lesions in both cases are caused by about the same amounts of fluorides. The soluble soil fluorides may raise the concentrations in roots to 1,000 to 6,000 ppm, while several hundred parts per million accumulate in the leaves. Airborne fluorides do not reach the roots in appreciable amount.

Work with nutrient solutions (105) indicated that, at 10 ppm in the solution, uptake by gladiolus and other crops was slow, requiring about 10 days to produce slight tip injury. At 40 ppm, 4 to 5 days were required. At 100 ppm, injury occurred the same day and was very severe. The New Jersey Experiment Station (55) found that peach, buckwheat, and tomato absorbed fluoride at different rates. Slight injury was produced in peach and buckwheat in 10 to 13 days at 25 ppm, but tomato required 48 days at 25 ppm and 27 days at 50 ppm to develop slight injury. Buckwheat leaves absorbed twice as much fluoride (533 ppm) as the peach (261 ppm) or tomato (277 ppm) from the 25-ppm solution. The tomatoes, therefore, absorbed at one-fourth the rate of peach and one-eighth the rate of buckwheat. Plants grown in nutrient solution (13) showed maximum fluoride injury when the concentrations of nitrogen, phosphorus, and calcium in the solution were optimum for plant growth, whether the fluoride was supplied from the solution or from the air.

9.3.6 Fumigation Experiments

Hydrogen fluoride is avidly absorbed by all surfaces, and it is difficult to maintain a definite concentration of the gas in a fumigation cabinet. It is necessary to use a large volume of air, sufficient to replace the air in the chamber about two to four times each minute. The earlier experiments by Zimmerman (124) employed concentrations ranging from about 0.05 to 3 ppm for periods up to 8 hr. Typical results indicated slight injury to prune in 4 hr at 0.07 ppm and to buckwheat, sweet potato, and peach in 6.7 hr at 0.085 ppm. These plants and gladiolus, crabgrass, corn, bean, and young pine needles were consistently injured by 0.1 ppm, while tomato required 2.2 hr at 0.67 ppm for slight injury. Severe injury was produced on all these plants in 2.2 hr at 1.48 ppm. Daines et al. (24) applied comparable treatments (time not stated) to produce slight injury as follows: sweet potato, white pine, peach, and gladiolus, 0.01 to 0.10 ppm; tomato, catbriar, smartweed, crabgrass, sorrel, tobacco, begonia, and geranium, 0.1 to 0.4 ppm; spinach, pepper, and corn, 0.40 to 0.50 ppm; bean, aster, poinsettia, ragweed, plantain, zinnia, marigold, and petunia, about 1.0 ppm. Similar experiments have been described by Griffin and Bayles (29). Peach, wild black cherry, buckwheat, tomato, and corn were most sensitive and were slightly injured in a few hours by 0.05 to 0.1 ppm.

Compton and Remmert (20) were the first to explore the low concentration range (0.1 to 14 ppb), which represents the range usually encountered in the field. They employed large portable Vinylite-covered greenhouses 8 by 12 by 8 ft high. Air was exhausted from the house near the ground by a large fan. Fresh air entered the top of the chamber through a short length of 14-in. pipe. Replacement of air occurred at least 1.5 times per minute. A mixture of dry air and hydrogen fluoride containing about 0.1 to 1.0 per cent of the latter was prepared in a large dry bottle. Mineral oil was then pumped into the bottle to dispense the gas at a known constant rate. Modified forms of this equipment have been employed by a number of other investigators in subsequent fumigation work.

Compton and Remmert found that the most sensitive varieties of gladiolus (Picardy and Shirley Temple) were injured more than 4 in. down from the tip by 10 ppb in 22 hr and that the tip 3-in. section contained 279 ppm fluorine. Appreciable injury developed with 1 to 2 ppb in a few days, and 1 in. of tip injury was noted with 0.1

ppb in 5 weeks. The latter blades contained 148 ppm in the tip 3 in. In control experiments, there was no injury and only a few parts per million fluorine in the leaves. In all cases a continuous sampling of the air in the chamber throughout the fumigation period was made, using an alkaline-absorbing solution, and the absorbed fluorine was determined by distillation. The results were said to be about 80 per cent of the values calculated from the amount of hydrogen fluoride added to the air stream. While there may be some uncertainty about the determination of 0.1 ppb, the observations of Compton and Remmert on injury to gladiolus by about 1 to 14 ppb have been confirmed by subsequent work (105,123). The lower limit of concentration, to cause injury in protracted fumigation, has not been determined exactly.

The toxicity of low concentrations of hydrogen fluoride to a few species and the apparent lack of toxicity to most other species raise the question of whether "invisible injury" is produced by this gas, as was claimed by Stoklasa for sulfur dioxide. Exploration of this field is in its initial stages, but photosynthesis data (105) so far obtained with gladiolus, fruit trees, barley, alfalfa, and cotton indicate that invisible injury does not occur until the concentration of the gas exceeds a threshold value for each plant. With higher concentrations there may be a temporary reduction of photosynthesis, without corresponding acute injury as was observed with sulfur dioxide fumigations. However, recovery from the fluoride treatment is much slower than from the sulfur dioxide; therefore the fluoride treatment may cause a small but significant reduction in assimilation. This difference in rate of recovery is probably explained by the fact that, whereas sulfite in the leaves is rapidly oxidized to the relatively nontoxic sulfate, a photosynthesis-inhibiting concentration of fluoride in the tissues can be removed only by the slower processes of translocation, volatilization, and possibly reaction with organic compounds in the leaf.

Many plots of different varieties of gladiolus have been fumigated for a number of days with sublethal amounts of hydrogen fluoride (1 to 10 ppb). So long as no visible injury was produced, there was no decrease in the rate of carbon dioxide assimilation. In another experiment, a plot of the sensitive variety, Snow Princess, was fumigated 6 to 7 hr each day, 5 days a week, for over 2 months. After the first 7 days, injury began to appear at the tips of the blades and gradually extended down the blades until over 40 per cent of the leaf area was destroyed. Careful measurements of the area injured were made at intervals. The rate of photosynthesis and percentage of intact tissue followed each other faithfully throughout this period. For example, 40 per cent leaf destruction reduced photosynthesis about 40 per cent. It is to be noted that most of the absorbed fluorine was found in the injured areas. The functioning areas had much lower concentrations. Higher concentrations (20 to 50 ppb) caused a sharp reduction in photosynthesis followed by a rapid recovery at first, then a gradual increase, until the level was reached corresponding to that expected from the amount of leaf destruction caused by the fumigation. Similar responses were shown by fruit trees at 20 to 50 ppb, by young barley at 40 ppb, and by alfalfa at 250 ppb in 4- to 8-hr fumigations. These effects were equivalent to a total interruption of photosynthesis for a few hours to 2 to 3 days. The photosynthesis of cotton was not significantly affected by concentrations of hydrogen fluoride up to 500 ppb. Higher concentrations (500 to 1,100 ppb) interfered temporarily to some extent, but when the fumigation was stopped before causing injury, the photosynthesis returned to its original level in a few hours.

Susceptibility data based largely on fumigation experiments, but confirmed in a few cases by field observations, are summarized in Table 9-3. The Boyce Thompson data are based on continuous fumigations of 7 to 9 days duration or their equivalent. Plants injured with 5 ppb or less were regarded as sensitive and grouped in classes 1 and 2. Plants injured by 5 to 10 ppb are included in class 3 and by more than 10 ppb in classes 4, 5, and 6. The American Smelting and Refining Company experi-

Table 9-3. Relative Sensitivity of Cultivated and Native Plants to Injury by Hydrogen Fluoride

Sensitive		Intermediate		Resistant	
Cultivated Plants					
Class 1		Class 3		Class 5	
Gladiolus	ABC	Corn (Golden Cross bantam)	B	Columbine and Canterbury bell	A
Pine (young needles)	B	Pepper (Calif. Wonder)	C	Dogwood and lilac	C
Apricot (Chinese)	BR	Raspberry (Washington)	C	Lobelia and petunia	A
Azalea	C	Aster and sweet William	A	Rose	B
Blueberry (Jersey)	C	Dahlia (immature leaves)	A	Apple (red Jonathan)	C
Prune (Italian PRH 1)	C	Dahlia	C	Live oak and pine	B
Tulip	B	Petunia petals	A	Tomato	AB
Class 2		Clover	B	Class 6	
Milo maize	B	Barley, flax, oats, and rye (young plants)	A	Alfalfa (Ranger)	C
Corn (Dixie 17, Funk 134 and 512)	B	Apple (Delicious), cherry (Bing)	C	Corn (Golden Cross bantam)	C
Corn (Spancross)	B	Apple	B	Cotton	A
Corn (bantam)	A	Birch, hawthorne, silver maple, mountain ash, mulberry, sycamore, and yellow willow	C	Tobacco	BC
Sweet potato (Triumph)	B	Peach	AB	Bean and celery	B
Apricot (Moorpark)	ABC	Iris and begonia	B	Cucumber and squash	B
Prune (late Italian)	AB	Class 4		Squash (Hubbard and zucchini)	AC
Peach (some varieties under best conditions)	B	Azalea	BC	Cabbage, cauliflower, eggplant, onion, pepper, and soybean	A
Peach (Elberta, Lovell stock)	C	Begonia (some varieties)	B	Parsnips (hollow crown)	C
Strawberry (Marshall)	C	Rose, geranium, coleus, lilac, spirea	B	Tomato (Marglobe)	C
Grapes (some European varieties)	RW	Verbena	A	Chrysanthemum	AB
Iris	AB	Alfalfa	AB	Marigold, privet, and snapdragon	A
		Clover	B	Sweet pea (Spencer hybrid)	C
		Red clover	C	Rhododendron (Pink Pearl)	C
		Bean (Tendergreen bush)	C	Rose (Talisman)	C
		Carrot (Chantenay)	C	Zinnia	B
		Carrot	A	Citrus	A
		Grape (Concord)	CR	Laurel and locust	C
		Lettuce	C	Pine (old needles)	B
		Sweet potato (Nigger Killer)	B		
		Parsnip, peas, potato, rhubarb, spinach, sugar beet	A		
		Wheat	A		
		Barberry, bridal wreath, honeysuckle, and mock orange	A		
		Apple (Wealthy and Winesap)	C		
		Arborvitae	C		
Native Plants					
Class 1		Class 3		Class 5	
Pine (ponderosa, young needles)	BC	Pokeweed, crabgrass	B	Pigweed, dock	B
Larch	C	Smartweed, Johnson grass	B	Carpet weed	B
Hypericum	B	Oxalis	B	Pine (lodgepole)	C
Class 2		Chickweed, barnyard grass	B	Pine (ponderosa, old needles)	C
Crabgrass	B	Class 4		Class 6	
Cattail	B	Pokeweed, pigweed, lamb's-quarters, dock, plantain	B	Dandelion, plantain, purslane, galinsoga, bidens, nightshade, and ragweed	B
Johnson grass, oxalis, smartweed	B			Douglas fir, grand fir, hemlock, white pine, Engelmann spruce	C
Buckwheat	B				
Pine (ponderosa, 3–4-month-old needles)	C				

Concentrations required to cause slight injury in 7 to 9 days: classes 1 and 2 = 5 ppb or less; class 3 = 5 to 10 ppb; classes 4, 5, and 6 = more than 10 ppb; higher concentrations required proportionally less time.

A = American Smelting and Refining Company; B = Boyce Thompson Institute; C = State College Washington; R = B. L. Richards; W = F. W. Went.

ments were based on concentrations ranging from 1 to 100 ppb and exposure times of 8 hr to 14 days. The concentrations corresponding to 7 to 9 days exposure were estimated.

A wide range of sensitivity is apparent in Table 9-3. Only a small number of plants have thus far been found which are regarded as sensitive. These include practically all the plants on which injury in the field has been observed. Fig and walnut injury along with injury to some broad-leaved plants in a severe industrial exposure have been reported (26), but without a knowledge of the exposure conditions, the sensitivity of these species cannot be stated. The small number of sensitive plants in Table 9-3, and the fact that field injury to vegetation is largely confined to these plants, would indicate that the plant-damage phase of the fluoride problem is not of widespread importance, but rather is confined to areas where these sensitive crops are grown commercially. Of course the indirect injury to sheep and cattle due to too much fluorine in forage poses a different problem. It is of interest to note that a number of fluoride-emitting industries have been able to reduce their emissions to practically innocuous levels. Others are making strong efforts in that direction.

The more resistant plants have more than academic interest. Many of them are very sensitive to sulfur dioxide or smog. They may, therefore, serve as diagnostic aids in determining whether a particular case of crop damage is due to one of these chemical agents or to some other factors. This will be discussed later.

Efforts have been made to reduce fluoride damage to vegetation by spraying the plants with a lime spray (3,75). This treatment reduced the injury to gladiolus from 9 to 18 per cent of the leaf area to 1 to 4 per cent. The absorbed fluoride was found largely on the exterior surface of the leaf; 74 to 94 per cent could be washed off.

9.4 MISCELLANEOUS FUMIGANTS

9.4.1 Chlorine

Apart from the fluorides, the halogens and their volatile compounds are relatively unimportant air pollutants which might cause injury to vegetation in the field, even though chlorine is more toxic to vegetation than sulfur dioxide by a factor of 2 or 3 (115). Lesions are generally marginal and interveinal. They are somewhat similar to those caused by sulfur dioxide, but a silver-leaf effect of the upper epidermis has been observed in sugar beets. Injury to beans and radish (124) occurred in 0.5 hr at 1.3 ppm, to roses in 0.5 hr at 1.5 ppm, to buckwheat in 1 hr at 0.46 ppm, and to peach in 3 hr at 0.56 ppm. Damage to vegetation caused by chlorine is rare, and all reported cases are due to accidents or excessive use of the gas for sterilizing, such as an accident in a chemical plant or gas from a heavily chlorinated swimming pool near New York and from a sewage-disposal plant in California (101).

9.4.2 Hydrogen Chloride

Hydrogen chloride is considerably less toxic to vegetation than sulfur dioxide. A century ago large amounts of hydrogen chloride were evolved into the atmosphere in the Leblanc soda process when salt was treated with sulfuric acid. This caused severe injury to vegetation near the factory. However, equipment for the recovery of the gas was developed between 1836 and 1863, and in 1874 it became illegal in Great Britain to emit more than 0.45 mg hydrogen chloride per cubic meter of stack gas (approximately 0.35 ppm). Thereafter hydrogen chloride has played a minor role as an air pollutant. The threshold concentration is about 10 ppm for a few hours fumigation. The older literature indicated that about 50 to 100 ppm was required to cause injury, but these determinations were done in a static atmosphere.

The gas is rapidly absorbed through the stomata, and considerable build-up of chloride occurs in the leaves. The chloride, like the fluoride, tends to travel to the margins of the leaves where it accumulates, causing first a chlorotic margin which may become necrotic. At higher concentrations, lesions like acute sulfur dioxide markings are produced. The threshold marking concentration appears to be about 10 ppm for a few hours exposure (105). The low toxicity of the chloride is indicated by the fact that, whereas a few hundred ppm of fluoride in the leaf may cause injury, 10,000 ppm or more of chloride has often been found in apparently normal leaves.

9.4.3 Nitric Oxides

Injury due to nitric acid vapors (37) has been observed near factories handling large amounts of this acid. Symptoms include brown margins and brown to brownish-black spots on the leaves. The blades of grain plants and the needle tips of conifers may assume a bright yellow color. Concentrations of 25 ppm have been indicated as needed to cause these effects. However, very limited work has been done in this field.

The gases are normally present in the atmosphere to the extent of about 0.015 ppm, and a maximum concentration of 0.4 to 0.5 ppm has been found in the Los Angeles smog. Nitrogen oxides are important in the photochemical reactions which cause the smog.

9.4.4 Ammonia

Ammonia is a gas of intermediate toxicity (115,124). Tomato, sunflower, buckwheat, and coleus were injured as follows: 40 ppm for 1 hr, definite injury; 16.6 ppm for 4 hr, slight injury; 8.3 ppm for 5 hr, slight injury or none at all. "Nearly all parts of the leaf assumed a cooked green appearance, becoming brown upon drying." Slight injury was marginal. The variegated leaves of coleus "lost their brilliant color, appearing green thereafter." It is of considerable interest that hydrogen chloride and ammonia have about the same toxicity. Possibly these gases exert their effects as acidity or alkalinity, uncomplicated by other toxic properties.

9.4.5 Hydrogen Sulfide

It is improbable that hydrogen sulfide could ever reach a concentration in the open atmosphere sufficient to cause plant damage, except as a result of an accident which would probably be lethal to animals. However, lime sulfur sprays on apple trees (15,18,60,61) sometimes cause leaf injury similar to that caused by hydrogen sulfide, and the injury due to dusting elemental sulfur (116) on citrus trees when the temperature is high may also be due in part to hydrogen sulfide and in part to sulfur dioxide.

McCallan et al. (60) observed slight injury to a few species of plants fumigated with 20 to 40 ppm for 5 hr. Even at 400 ppm for 5 hr, some species escaped injury. Injury was more severe, the higher the temperature. The lesions consisted of scorching of the youngest, rapidly elongating leaves, while the mature, actively functioning leaves were generally uninjured. In this way hydrogen sulfide injury differs sharply from sulfur dioxide injury.

The most sensitive plants were cosmos, radish, clover, tomato, poppy, salvia, cucumber, soybean, and aster; intermediate sensitivity was shown by cornflower, buckwheat, nasturtium, sunflower, gladiolus, castor bean, and pepper; while resistant species were coleus, peach, strawberry, cherry, apple, purslane, and carnation.

9.4.6 Hydrogen Cyanide

Hydrogen cyanide is used to fumigate greenhouses and individual trees in orchards for scale and other resistant pests. Sometimes this treatment injures the vegetation.

Bartholomew et al. (5) applied 1,100 ppm for 40 min to Valencia orange trees. Green fruits absorbed 5.4 times as much gas as the mature fruits and were more readily injured. Cyanide could be recovered from the leaves for 60 hr, from green fruit for 35 to 40 hr, and from mature fruit for 20 to 25 hr after a fumigation. Absorption and injury were greater in the day than at night, and well-irrigated plants were more sensitive than drier plants.

Hydrogen cyanide (41) has been found in artificial gas to the extent of 200 to 300 ppm, and it has caused root injury to plants when it leaked into greenhouses from the underground supply pipes. Initial toxicity effects were noted with 4 to 10 mg of cyanide per 500 g of soil. Studies with natural gas (30) showed that cyanide and other materials toxic to plant roots were virtually absent.

9.4.7 Mercury

It has been known for over 150 years that in enclosed places plant injury could be caused by mercury vapor. This subject was investigated by workers at the Boyce Thompson Institute (126) who found that mercury salts mixed with the soil in a commercial greenhouse to kill earthworms cause injury to roses. The lesions consist of brown spotting of the older active leaves. "The petals from partially opened buds were brown, the corollas of younger buds had turned brown and abscissed without opening, stamens were killed, and peduncles were turned dark brown or nearly black in places."

The mercury vapor is liberated by contact between the mercury salt and the organic matter in the soil. This reaction also occurs when the organic vapors given off by a plant come in contact with the mercury salt. The reaction is favored by raising the temperature, which increases the vapor pressure of the mercury, causing an increase in the toxicity of the element. From the vapor-pressure curve of mercury it may be estimated that the concentration may have ranged from 0.64 ppm at 10°C to 8 ppm at 40°C. Little or no injury is noted at 10°C or below. Uptake of mercury by plants is substantial. After 8 days in a closed chamber having soil moistened with 1 per cent mercuric chloride, the leaves of roses took up 317 ppm mercury, causing severe injury, while tobacco leaves absorbed 4,757 ppm with only slight injury.

9.4.8 Ethylene

Ethylene in high dilution (22,23,51) causes injury to leaves of sensitive plants. The effects consist of epinasty, curling, inhibition of nutation, chlorosis and premature shedding of leaves, inhibition of shoot elongation, and retardation of growth. As little as 0.1 ppm ethylene in the air causes epinasty in sweet peas and tomatoes and 0.05 ppm in buckwheat and sunflowers. Other plants are less sensitive. Of 202 species tested, 50 per cent showed no epinasty at 10 ppm, but nutation was inhibited. The young leaves recovered rapidly from epinasty in uncontaminated air, but the older leaves did not fully recover. Shoot-bud formation was often stimulated by the treatment.

Injury by ethylene (40,129) has been observed in greenhouses with leaking gas pipes. Epinasty was approximately proportional to the ethylene content of the gas. Natural gas, which is free of ethylene, does not injure the elm (30). In some cases, leaking gas pipes under the ground have caused injury in nearby gardens. Injury to the roots (41) often occurred along with top injury, but the root injury appeared to be due to hydrogen cyanide. Greenhouse plants that have been economically injured by ethylene include lilacs, narcissus, tulips, and roses.

Acetylene and carbon monoxide (122,127) affect vegetation like ethylene. The toxicity of the higher olefins decreases rapidly with the number of carbon atoms in the molecule. For example, the concentrations required to cause similar injury to tomatoes are ethylene 0.1 ppm, acetylene 50 ppm, propylene 50 ppm, and butylene 50,000 ppm, or carbon monoxide 500 ppm. None of these gases is likely to be an important air pollutant except in confined spaces.

9.4.9 Herbicides

2-4D. Economic injury to sensitive vegetation may be caused in the field through uncontrolled or careless use of weed killers like 2-4 dichlorophenoxyacetic acid (2-4D) in its various spray forms, particularly the volatile esters. Ditch banks or weedy grain fields have sometimes been sprayed when there was wind, and the spray has been carried for several miles in sufficient concentration to injure such crops as cotton and tomatoes. For example, in 1947, in two Texas counties (28) 10,000 acres of cotton suffered loss of 20 per cent, or $200,000, because of airplane spraying of rice fields at least 15 to 20 miles from the nearest cotton. Drift of spray this distance was probably the cause of the injury, though it is possible that some of the 2-4D may have come from leaking spray equipment in planes that passed closer to the cotton. More limited injury (45) may be caused by slow evaporation of the ester from a sprayed area and its dispersion on nearby vegetation over a period of several days.

Broad-leaved plants are generally sensitive to 2-4D, though their sensitivity varies over a fairly wide range, whereas the grains and grasses are very resistant. Cotton (98) is one of the most sensitive plants to 2-4D. The effects are first noted near the terminal growing points as a rolling and ruffling of the outer margins of the leaves. Later the new leaves are narrow, closely veined, and deeply lobed. The flowers are elongated and narrowed. The bracts are also deeply lobed, elongated, and tend to form a sheath around the developing boll. In the field, the first symptoms may not appear until 2 to 5 weeks after exposure. Following is a description (28) of more severe injury to cotton due to the dusting in May or early June of the Texas rice fields referred to earlier. The leaves developed with scalloped margins, narrow lamina, and long tentaclelike teeth which persisted until mid-August. The stems were swollen, and the cortex was broken open. Spherical gall-like stem swellings occurred near the ground. Squares and bolls were malformed; some developed on one side only, with one to two instead of five to six locules. New growth formed with lateral branching, and the main stem often died. The new lateral branches from the lower nodes had normal leaves. Germination of seed was reduced. Emergence was retarded, and the tips of the emerging hypocotyls showed swellings.

Many other plants exhibit somewhat similar symptoms, e.g., roses, phlox, zinnia, petunia, tomato, cabbage, pepper, grapes, and tobacco. Malformed leaves; swollen stems, petioles, and root primordia; and lateral branching are common symptoms of 2-4D injury in these plants. Some tomato plants produce seedless fruit.

9.5 SMOG

In the complex environment of large industrial and urban areas, many pollutants of an organic nature in addition to soot and tars are emitted into the atmosphere along with the more common inorganic pollutants. Following usage in Los Angeles, Calif., this mixture is now referred to as "smog" (smoke and fog). However, fog is not an essential constituent of smog, since more often than not, low humidity conditions prevail and fog is absent. In Los Angeles, where gas and oil, rather than coal, are the fuels employed, the smog is light-colored, consisting of solid and liquid aerosols in a fine state of subdivision, together with numerous organic and inorganic gases. Coal smoke would make the smog dark, because of soot and tars.

Three properties of severe smog are clearly recognized: (1) reduced visibility, (2) lachrymation, and (3) characteristic injury to the leaves of certain vegetation. Only the third will be discussed here.

The degree of plant injury caused by smog depends not only on the quantity of pollutants emitted into the air but also to a large degree on meteorological factors. In the Los Angeles district, a greatly expanded industrial activity (58) during the past decade has combined with persistent periods of low wind velocity and atmospheric stability lasting days or weeks to produce sufficient concentrations of pollutants to cause economic crop damage. In recent years the damage in Los Angeles County alone has been estimated at over $500,000 per year (72).

It has been postulated (31,32,57,58) that the phytotoxic materials are not all emitted directly into the atmosphere but that some of the more important constituents of the mixture are formed by comparatively slow photochemical reactions in the atmosphere between nontoxic substances such as unsaturated hydrocarbons and small concentrations of ozone or nitrogen oxides. A time lag would, therefore, be involved in the build-up of harmful concentrations of these toxic substances, which occur only in daylight and appear nearly simultaneously over extensive areas of the valley, as if their precursors* were already widely distributed.

The details of the photochemical reactions involved in smog formation and the production of phytotoxic compounds have not been fully worked out. Suggested mechanisms have been advanced by Blacet (6), Cadle and Johnston (16), Los Angeles County Air Pollution Control District (58), and Haagen-Smit, Bradley, and Fox (34). They are discussed in Sec. 4.

A good correlation has been established between the smog intensity in the Los Angeles area (reduced visibility, lachrymation, and plant damage) and the oxidant content of the atmosphere (67,96) as measured by iodine liberation from neutral, buffered potassium iodide. The Stanford Research Institute has developed automatic recording analyzers (69) to apply this reagent, which is nonspecific for the different oxidizing compounds, but which is thought to record principally ozone. Reduced phenolphthalein (86) is also an excellent colorimetric reagent for oxidants. On smoggy days (69,97) the oxidant level rises in the morning after sunrise to a maximum at about noon, where it remains or falls off slowly until 4 P.M., then falls rapidly to about zero at 7 P.M.

If the air stream (97) is passed through a 50-liter flask irradiated with ultraviolet lamps before entering the oxidant recorder, the oxidant pattern is changed radically. At night the oxidant level of artificially irradiated air may rise as high as or higher than the daytime maximum of naturally irradiated air, usually at about 9 P.M. Then the level may fall to a minimum at about 4 A.M. and again rise to a maximum at about 9 A.M., where it may remain essentially constant or fall off appreciably because of daytime turbulence of the air, being at approximately the same level as the naturally irradiated air. It therefore appears that the "oxidants" in the atmosphere are formed by photochemical reactions from so-called "precursors" which have a widespread distribution. There is no evidence that the precursors cause plant damage (34).

Haagen-Smit et al. (34) consider that "this oxidizing effect of smog is due to the combined action of nitrogen oxides, peroxides, and ozone counteracted by the reducing action of sulfur dioxide, which is present in concentrations of 0.1 to 0.2 ppm." The peroxides referred to are probably organic.

Plant injury similar to that due to smog has been observed by Went (119) in or near a number of cities, for example, London, Paris, São Paulo, Philadelphia, Baltimore, and San Francisco. However, the observed injury was not extensive. Middleton, Kendrick, and Darley (72) also report smog injury in the San Francisco area. Possi-

* A precursor is a substance which will form an oxidant, presumably ozone, upon irradiation with ultraviolet light.

bly the damage to vegetation due to coal smoke in Leeds, Manchester, and other industrial cities, as described by Ruston (87), Bleasdale (7), and others, is caused in part by the same (or similar) compounds as those responsible for the Los Angeles smog.

Haagen-Smit et al. (35) noted a depressing effect on the growth of plants due to sublethal amounts of smog which produced no visible lesions. This observation was confirmed by Hull and Went (44), who found that the growth of alfalfa, sugar beet, endive, oats, spinach, and tomato was materially retarded by sublethal fumigations with smog. They also reported interference with the "avena" test for growth hormones. This interference constitutes an extremely sensitive biological test for smog. Closure of the stomata and reduction of both water uptake and transpiration in tomato were observed by Koritz and Went (52) as immediate responses to "synthetic smog" (1-n-hexene plus ozone). Possibly photosynthesis was reduced, as might be expected if the stomata closed. Loss of fresh weight was somewhat greater than loss of dry weight, which in typical experiments amounted to about 25 per cent. Considerably more injury occurred at 30°C than at 17°C (44). There was still less injury at 5°C. These temperature effects were noted after prefumigation and postfumigation temperature treatments when all the plants were fumigated at the same temperature and also when similarly treated plants were fumigated at different temperatures. The cotyledons were more easily injured than the first leaves and young plants more easily than old plants. By contrast, sulfur dioxide injury is independent of the temperature above 5°C, and the leaves are more sensitive than the cotyledons. Both smog and sulfur dioxide fumigations are less effective at low relative humidity and in the dark than at high humidity and in the light. Low soil moisture increases resistance in both cases. Fertilizer nitrogen (72) added to the soil at the rate of 45 lb/acre increased injury to spinach 40 per cent and to Romaine lettuce 80 per cent above the unfertilized check plants. Fertilized barley and oats were damaged only slightly more than the controls.

9.5.1 Symptoms

Two types of smog injury to vegetation have been recognized in Los Angeles, one due to gases (smog gas) and the other due to deposition on the leaves of fog droplets (smog fog).

The leaf lesions due to smog gas are fairly characteristic, but all plants do not show the same qualitative response. In spinach, the undersurface of the leaf assumes at first an oily or water-soaked appearance, followed in a day or two by a light-colored sheen which is described as silver leaf or bronzing. The lower epidermis at first glance appears to be detached, but microscopic examination shows that the spongy cells and the lower epidermal cells are collapsed, leaving large air spaces below the epidermis. Soon the upper epidermis opposite the silver-leaf areas becomes light-colored and the collapse extends through the leaf. This pattern of injury is also shown by beets, Romaine lettuce, celery, and many other plants.

Bobrov (8,9) has discussed in detail, with many excellent photomicrographs and drawings, the effects of smog on a number of different types of vegetation, including corn and oats, with their long narrow leaves; spinach, a fleshy leaf with palisade cells; and endive, without palisade cells. She used hand sectioning of the living tissue, followed by staining with dilute thionine or other dyes to study the responses of leaves to smog. As the first response (58): "The lower epidermis is raised in tiny blisters, which on close inspection may be seen with the naked eye. In the region of some, but not all, of the stomata, the guard cells, together with several surrounding epidermal cells, are engorged with water and push up above the substomatal chamber, thus forming a blister. . . . The epidermal cells swell in length and width, become turgid, and the walls lose their fluted appearance and become regular. The guard cells increase in width, but not in length, opening the stomata wide. The attached epidermal cells then collapse. . . ." When the water-soaked stage appears, the cells lining both lower and upper substomatal chambers are the first to show plasmolysis. The stomata probably close, as suggested by Koritz and Went (52). The chloroplast membranes may be disrupted. The cell walls shrink but maintain connections with the neighboring cells. Water enters the cavities.

If the exposure to smog is relatively light, and the chloroplasts remain intact, the injured cells may recover and evidence of injury may disappear. This is "transient" injury. If only a limited number of cells are killed so that no macroscopic collapse occurs, there will be lightening of the color and the injury is "chronic." At higher concentrations of smog, injury may extend to adjacent cells, killing all in a limited or extensive area and causing "acute" injury. The silver-leaf stage is due to the dehydration, death, and shrinkage of many mesophyll cells, forming enlarged air-filled intercellular spaces, which give the lower epidermis a detached appearance.

Nielsen, Benedict, and Holloman (78) have observed that the smog-injured areas of leaves fluoresce bright bluish white under near ultraviolet light from a mercury lamp. This fluorescence is not apparent for several hours after the fumigation treatment until the injured areas begin to dry out. Then it increases to a maximum and fades out over a period of about one week. It is suggested that the process may be related to the so-called "browning reaction" in foods which fluoresce bright bluish white in ultraviolet light just before they assume the characteristic dark brown color on exposure to oxidation by the air. Possibly products of smog oxidation in the leaf behave similarly. Ozone might be expected to cause similar oxidation, but fluorescence in this case is questionable. This test appears to be specific for smog injury as distinct from gas injury due to inorganic fumigants such as sulfur dioxide, hydrogen fluoride, chlorine, etc., which do not cause fluorescence.

Figure 9-5 shows photographs of smog-damaged plants.

In heavy smog, endive with its more delicate leaves often shows immediate and complete collapse, except for part of the midrib. According to Noble (80) in lighter smog, when the exposure is near the threshold, injury occurs on the young newly developed leaves that are just starting to function actively. The injury may occur only as a band across the most active part of the leaf. If the fumigation is repeated at intervals, similar lesions will appear in different positions on the same or successively younger leaves, resulting in a series of bands on different leaves of the plant. This banding of the plants has also been observed on petunia, wild oats, and many others. In broad-leaved plants like spinach the injured portion is not a band but rather an irregular-shaped area of the intravenous tissue at various distances down the leaf from the tip. The blades of wild oats have been seen showing banding from a single fumigation at different positions on several successive blades, suggesting that the most active area moved from the tip of the blade to the base as the blade matured.

A possible explanation of banding may be the suberization of the older cells that has been described by Scott (89) and others. Very young leaves are resistant to gas injury until they are fully developed structurally and functionally. At this time the cells at the tip of the leaf have maximum sensitivity. Later the process of suberization develops in these cells, which protects the absorbing surfaces from the gas and increases resistance of the tip while the cells lower down develop maximum sensitivity. Observations by Bobrov (8,9) support this explanation.

Alfalfa with stomata on both sides of the leaf may show a chronic injury on either the upper or lower surface or on both sides at once.

Table beet shows a pattern somewhat like spinach, but the upper surface often develops a weathered appearance with reddening, while the lower surface may develop a layer of brown cork in the region of the lower epidermis (58). Chrysanthemum likewise develops a weathered appearance, but without silver leaf. Red and bronze shades may appear in severe smog, and the surface is speckled with rust-colored spots. The weed malva also exhibits a spotted and weathered appearance on the upper surface of the leaf. Cauliflower has been observed with slight silver leaf and brown

FIG. 9-5. Smog damage: (*a*) Endive, healthy and damaged. (*b*) Romaine lettuce, healthy and damaged. (*c*) Spinach, healthy and damaged.

spots on the underside. Even spinach exhibits at times a weathered appearance of the upper surface.

It is possible that some of the weathered and spotted effects are due to smog fog, rather than to gas injury. If during a period of heavy smog there is also a dense fog, droplets of moisture may be deposited on the leaf, wetting the surface, either uniformly or with discrete drops. This liquid may be slightly acid. A pH of 3.0 has been observed by pressing indicator paper against the surface (111). With the dispersal of the fog and drying of the leaf, a shot-hole type of injury may develop on table beets along with a weathered appearance. The shot-hole injury may be due to coalesced droplets on the surface, while the weathered appearance may be due to liquid that remained more or less uniformly spread over the surface until it dried up. It must be emphasized, however, that pitting and shot-hole injury to the leaf is not necessarily due to smog fog. Injury of this type sometimes occurs because of sulfur dioxide and presumably might also be caused by smog gas.

Finally, ozone alone can cause characteristic injury to many plants. The lesions consist of a bleached appearance of the surface of the leaf without collapse and differ considerably from typical smog injury. In the Earhart Laboratory (35) these lesions have been produced on spinach, endive, and alfalfa by a 5-hr fumigation with 0.2 ppm ozone.

9.5.2 Susceptibility to Smog

Tables 9-4 and 9-5 classify a number of cultivated and native plants as sensitive, intermediate, or resistant, according to published literature and according to the opinions of six observers who have studied the vegetation near Los Angeles intensively for several years. These observations were all made in areas remote from sources of inorganic pollutants such as sulfur dioxide or hydrogen fluoride.

In most cases, there is good agreement in the classification. When observers disagree as between sensitive and intermediate classification, or as between intermediate and resistant, it is possible that two additional classifications, sensitive intermediate and intermediate resistant, could have been set up. This is not always true. For example, some observers place alfalfa at the top of the sensitive list; others consider it of intermediate sensitivity. The few cases of wider disagreement among the observers are probably due to special or unusual observations or even varietal differences, or the plants may not have been encountered when they were most sensitive. Middleton, Kendrick, and Darley (72) use only two classifications, susceptible and resistant. Table 9-6 reproduces their data with the addition of notations to indicate the corresponding classifications in Tables 9-4 and 9-5. In nearly all cases the sensitive plants, according to Middleton et al., are listed as sensitive to intermediate by the other observers, and the resistant plants are listed as the intermediate to resistant. It is of particular interest to note the differences in susceptibility between different varieties of the same species.

Considerable variability in the relative responses to smog of different species has been observed from time to time. For example, at one time the only crop plant injured in a given area will be spinach; at another time only beets will be injured; still again the most prominent injury may be found on radish. Recently, injury to cabbage, so severe that it could be seen for half a mile, was found by W. W. Jones (47) in the Santa Ana Valley. Noble (80) has seen similar injury. Plants in about the five-leaf stage were most severely injured. Broccoli was also noticeably injured, and the sensitive weeds were severely bleached, but there was little or no injury on oats, barley, and alfalfa. All observers agree that cabbage and broccoli are very resistant to smog. Evidently there are certain conditions or stages of growth in which resistant plants may become sensitive. Malva is much more sensitive in the spring than in the summer (80). Change in susceptibility at different stages of growth is, therefore, an important cause of this variability. Susceptibility to sulfur dioxide injury varies similarly. Alternatively, the composition of the smog may be variable, and this may account in part for the variable effects noted on the plants.

Table 9-4. Sensitivity of Various Cultivated Plants to Injury by Smog in Southern California

Plant	No. of observers	No. of observers who classified species as		
		Sensitive	Intermediate	Resistant
Endive	6	6		
Spinach (Viroflay and Nobel)[a]	6	6		
Romaine lettuce	6	6		
Petunia (some varieties)	3	3		
Barley (young)	3	3		
Alfalfa	6	4	2	
Oats	6	4	2	
Swiss chard	6	4	2	
Table beet	5	4	1	
Celery	5	3	2	
Chinese cabbage	4	2	2	
Sugar beet	5	2	3	
Barley	4	1	3	
Snapdragon	3	1	2	
Parsley	5	1	3	1
Eggplant	4	1	2	1
Radish (white varieties)[a]	5	..	5	
Mustard	4	..	4	
Aster	2	..	2	
Sweet pea	1	..	1	
Chrysanthemum	1	..	1	
Onion	5	..	4	1
Head lettuce	5	..	4	1
Turnip	5	..	4	1
Potato	5	..	3	2
Pole bean	3	..	2	1
Parsnip	3	..	2	1
Rhubarb	5	..	2	3
Squash	5	..	2	3
Cucumber	4	..	1	3
Corn	3	..	1	2
Wheat	3	..	1	2
Pansy	3	..	1	2
Cabbage[a]	5	..	..	5
Cantaloupe (muskmelon)	5	..	..	5
Carrot	4	..	..	4
Broccoli	4	..	..	4
Cauliflower	4	..	..	4
Gladiolus	4	..	..	4
Bean (most varieties)	3	..	..	3
Kale	3	..	..	3
Shasta daisy	2	..	..	2
Pumpkin	2	..	..	2
Brussels sprouts	2	..	..	2
Collards	2	..	..	2

[a] The Nobel variety of spinach is definitely less sensitive than Viroflay. New Zealand spinach has intermediate sensitivity (Hull) or is resistant (Noble). The white radish is distinctly more sensitive than the common red variety.

Cabbage in about the five-leaf stage appears to be sensitive. According to Went, yellow pines are sensitive, white pines are intermediate, and two-neeedle pines are resistant. Mangle and coriander (Preston) and cineraria, lobelia, primula, and calendula (Holloman) are listed as sensitive; pea, Dutch iris, eucalyptus, and hypericum (Preston), ranunculus and stock (Holloman) are listed as intermediate.

Observers: A. J. Haagen-Smit, E. F. Durley, M. Zaitlin, H. Hull, and W. Noble, Investigation on Injury to Plants from Air Pollution in the Los Angeles Area, *Plant Physiol.*, **27,** 18–34 (1952). Arthur Holloman, Jr. H. M. Hull. W. W. Jones. J. T. Middleton, J. B. Kendrick, Jr., and H. H. Schwalm, Smog in the South Coastal Area, *Calif. Agr.*, **4** (11), 7–10 (1950); and Injury to Herbaceous Plants by Smog or Air Pollution, *Plant Disease Reptr.*, **34,** 245–252 (1950). W. M. Noble. D. A. Preston. F. W. Went.

Table 9-5. Sensitivity of Various Native Plants to Smog in Southern California

Plant	No. of observers	No. of observers who classified species as		
		Sensitive	Intermediate	Resistant
Annual bluegrass (*Poa annua*)	6	6		
Chickweed (*Stellaria media*)	5	5		
London rocket (*Sisymbrium irio*)	5	5		
Cheeseweed (*Malva parviflora*)	5	5		
Curly dock (*Rumex crispus*)	4	4		
Wild oat (*Avena fatua*)	4	4		
Dwarf nettle (*Urtica urens*)	4	4		
Sowthistle (*Sonchus oleraceus*)	3	3		
Jimson weed (*Datura meteloidses*)	2	2		
Knotweed (*Polygonum aviculare*)	2	2		
Barley grass (*Hordeum sp.*)	2	2		
Wild buckwheat (*Eriogonum fasciculatum*)	1	1		
Sweet basil (*Ocimun basicum*)	1	1		
Bush sunflower (*Encelia californica*)	1	1		
Quickweed (*Galinsoga parviflora*)	5	4	1?	
Nettle-leaf goosefoot (*C. murale*)	5	4	1	
Bristly oxtongue (*Picris echioides*)	3	2	1	
Rough pigweed (*Amaranthus retroflexus*)	4	2	2	
Canary grass (*Phalaris canariensis*)	2	1	1	
Bur clover (*Medicago hispida*)	2	1	1	
Annual yellow sweet clover (*M. indica*)	3	2	1	
White sweet clover (*Melilotus alba*)	3	1	2	
Wild barley (*Hordeum murinum*)	3	1	2	
Orchard grass (*Dactylis glomerata*)	4	1	2	1
White stem filaree (*Erodium moschatum*)	5	1	2	2
Lamb's-quarters (*Chenopodium album*)	6	2	1	3
Mexican tea (*Chenopodium abrosioides*)	3	..	3	
Brome grass (*Bromus sp.*)	3	..	3	
Darnel (*Lolium temulentum*)	2	..	2	
Nightshade (*Solanum sp.*)	2	..	2	
Field mustard (*Brassica campestris*)	2	..	2	
Tobacco tree (*Nicotiana glauca*)	1	..	1	
Red goosefoot (young)(*Chenopodium rubrum*)	1	..	1	
Crabgrass (*Syntherisma sanguinalis*)	1	..	1	
Rye grass (*Elymus sp.*)	1	..	1	
Rabbit foot grass (*Polypogon monspeliensis*)	1	..	1	
Ground cherry (*Physalis sp.*)	1	..	1	
Wild radish (*Raphanus sativa*)	4	..	3	1
Dandelion (*Taraxacum vulgare*)	3	..	2	1
Jointed wild radish	2	..	1	1
Black mustard (*Brassica nigra*)	2	..	1	1
Sunflower (*Helianthus annuus*)	4	..	1	3
Bermuda grass (*Cynodon dactylon*)	5	..	..	5
Tumbling mustard (*Sisymbrium altissimum*)	3	..	..	3
Alkaline clover (*Cressa truxillensis*)	3	..	..	3
Purslane (*Portulaca oleracea*)	3	..	..	3
Castor bean (*Ricinus communis*)	2	..	..	2
Johnson grass (*Sorghum halepense*)	1	..	..	1
Sumac (*Rhus ovata*)	1	..	..	1
Buckhorn plantain (*Plantago sp.*)	1	..	..	1
Cocklebur (*Xanthium canadense*)	1	..	..	1
Cattail (*Typha latifolia*)	1	..	..	1

Lady's thumb has intermediate sensitivity (Jones), while love grass, turkey mullein, Australian brass buttons, giant reed, and California brome (Preston) and redscale (Jones) are resistant.

Observers: Arthur Holloman, Jr. H. M. Hull. W. W. Jones. J. T. Middleton, J. B. Kendrick, Jr., Smog in the South Coastal Area, *Calif. Agr.*, **4** (11), 7–10 (1950); and Injury to Herbaceous Plants by Smog or Air Pollution, *Plant Disease Reptr.*, **34,** 245–252 (1950). W. M. Noble. D. A. Preston. F. W. Went.

Table 9-6. Sensitivity of Plants to Injury by Smog in Southern California

(Listing by Middleton, Kendrick, and Darley)
(Classifications in Tables 9-4 and 9-5 are indicated)

Susceptible		Resistant	
Tree Crops			
		Grapefruit	
		Lemon	
		Orange	
Field Crops			
Alfalfa	S[a]	Sweet clover	SI[a]
Oats	S	Barley	I
Sudan grass		Mustard—black	IR
Sugar beet	SI	—white	I
		Wheat	R
		Blackeyed bean	
		Vetch	
Vegetable Crops			
Bean—common		Bean—common	R[b]
Golden Cluster		Bountiful	
Pink		Kentucky Wonder	
Pinto		Bean—lima	
Small white		Concentrated Fordhook	
Bean—lima		Westan	
Fordhook 242		Chinese cabbage	SI
Beet	S	Eggplant	I
Endive	S	Lettuce, head	I
Lettuce, Romaine	S	Mustard	I
Spinach	S	Pea	I
Swiss chard, Lucullus	S[b]	Radish	I
Celery	SI	Rhubarb	IR
Onion	I	Tomato	IR
Parsley	I	Leek	
Parsnip	I	Pepper	
Turnip	I	Potato	
		Swiss chard (large ribbed)	
		Broccoli	R
		Cabbage	R
		Cauliflower	R
		Corn	R
		Muskmelon	R
Ornamental Crops			
Petunia	S	Calendula	S
Chrysanthemum (some varieties)	I[b]	Lobelia	S
Grass—annual rye		China aster	I
Perennial rye	I[b]	Chrysanthemum (most varieties)	I[b]
Snapdragon	I	Sweet pea	I
Larkspur		Grass—Bermuda	R
		Kentucky blue	
		Pansy	R
		Stock	I
		Dahlia	
		Gaillardia	
		Viola	
		White clover	

[a] Classification in Tables 9-4 and 9-5—S, sensitive; SI, sensitive-intermediate; I, intermediate; IR, intermediate-resistant; R, resistant.

[b] Variety not known.

9.5.3 Smog Composition

The composition of the Los Angeles smog has been intensively studied for 5 or 6 years by the Stanford Research Institute (16,68,69,94,95,96,97) and the Los Angeles County Air Pollution Control District (57,58,59,64,65,66,92,117). It has been shown that the smog is a complex mixture of inorganic and organic pollutants of which over 50 have been identified by chemical methods and by the mass spectrograph. No single constituent is likely to exceed 0.5 ppm, except perhaps close to a source of emission, but it might be possible for a group of similar substances to reach a concentration of 1 ppm or more.

Early work by the Stanford Research Institute (67) indicated that the lachrymatory effects and presumably also the phytotoxic effects could be due to a number of constituents acting jointly, such as sulfur dioxide, aldehydes, ketones, acids, and chlorinated compounds. Difficulty was encountered when the attempt was made to locate the sources of these materials in the effluents of factories, incinerators, automobile exhausts, etc. None of the pollutants was emitted in sufficient amount to cause the widespread effects complained of, and even in combination adequate sources could not be found to explain crop injury, especially in remote places.

Another approach to the problem, already mentioned, is due to Haagen-Smit and his coworkers (31,34,58), who suggested that a number of olefins, which were themselves without lachrymatory properties or phytotoxicity, could be oxidized through a series of steps by the ozone and nitrogen oxides in the air, to give powerfully lachrymatory and phytotoxic compounds. When the oxidation was carried to completion with the formation of aldehydes, carboxylic acids, or polymerized compounds, the harmful properties were greatly reduced or lost. Haagen-Smit and coworkers observed that 1-n-hexene and a number of other similar olefins having four to nine carbon atoms would react with nitrogen oxides in sunlight or with ozone without irradiation to give substances with typical smog properties even in concentrations estimated to be less than 0.1 ppm.

The extensive controlled fumigation experiments of Haagen-Smit, Went, et al. (35, 44,52) were carried out in the Earhart Laboratory of California Institute of Technology as a cooperative project between the Los Angeles County Air Pollution Control District, the University of California, and the California Institute of Technology. Spinach, endive, beets, oats, and alfalfa were the test plants. The latter, grown under artificial light in a controlled environment, were very lush and appeared to be more susceptible to injury by smog than plants grown outdoors. Many organic compounds were tested. The most reactive, viz., acrolein, chloracetone, trichloroacetaldehyde, monochloroacetic acid, α-chloropropionic acid, and chloroacetophenone, caused injury to some of the test plants at about 0.1 to 0.8 ppm in 3 to 9 hr, but except in some cases with alfalfa the symptoms were not typical of smog injury. Other alcohols, aldehydes, and acids, including chlorinated compounds, did not show any effects at 1 ppm in 5 hr. Formic acid required 2 ppm or more for 5 hr to cause severe atypical markings on the vegetation. Its threshold concentration was about 1 ppm and that of formaldehyde about 2 ppm.

Mader (66) has found by chromatographic analysis that smog-damaged leaves of spinach and some other crops may have up to three times the formic acid of undamaged leaves. Fumigation with artificial smog caused a similar elevation of the formic acid level. Other acids showed only small changes. In view of the fact that Mader has observed leaf injury with less than 1 ppm formic acid and has found 0.4 ppm in heavy smog, it is quite possible that this acid is an important contributor to smog damage, though by itself it causes lesions that are unlike typical smog injury.

Twenty unsaturated hydrocarbons (35) ranging from ethene to 1-n-decene and

1-n-tetradecene, including benzene, tetralin, and turpentine, were treated in concentrations of 2 to 7 ppm with about 0.2 ppm ozone. An excess of hydrocarbon assured that no ozone would remain to injure the vegetation directly. Atypical lesions were obtained with ethene, propene, and tetralin. Cis- and trans-butene, 2-n-pentene, the decenes, and turpentine failed to cause any lesions. All the others gave typical lesions of varying severity on most of the plants, including slight injury with 1-n-butene and isobutene. Benzene required 0.6 ppm ozone to produce typical smog lesions on spinach, endive, and alfalfa, but the lesions on beets were not typical. The most reactive substance appeared to be 1-n-hexene, which gave slight but typical lesions with an initial concentration of 0.05 ppm ozone and 4.0 ppm olefin. Analysis of the gas mixture showed 0.02 ppm aldehyde, 0.01 ppm organic acid, and 0.01 ppm organic peroxide. The corresponding values for butadiene were 0.15 ppm ozone, 2.6 ppm hydrocarbon, 0.11 ppm aldehyde, 0.02 ppm acid, and 0.01 ppm peroxide. Gasoline vapor, particularly the fraction boiling between 59 and 69°C, also reacted readily with ozone to give phytotoxic compounds.

In these experiments, the organic peroxide concentration as measured for the different compounds bore no obvious relationship to type or severity of the plant injury. However, when 1-n-hexene was treated with three different ozone concentrations, it gave three peroxide levels. These caused typical lesions, the severities of which were in line with the peroxide concentration. Presumably the peroxides of the different olefins vary in toxicity, if they are principal toxic materials, as the authors believe, though there is no general agreement among the various investigators on this matter. The authors suggest that this may be due to differences in stability of the primary ozonides and their peroxide degradation products. Those of ethene and propene are very unstable and quickly form aldehydes and acids which do not cause typical smog injury. Those of the decenes are too stable to be very reactive. Further, the length of chain of the degradation products affects the activity: for example, C_4 reaction products from 1-n-pentene appear to be more active than C_2 and C_3 products from 2-n-pentene.

Nitrogen dioxide can produce phytotoxic oxidation products with the olefins under the influence of sunlight or ultraviolet light. At 0.4 ppm nitrogen dioxide, typical smog lesions were obtained with gasoline and its fraction boiling at 59 to 69°C; these also occurred with 1-n-hexene. The high-boiling gasoline fractions did not cause any injury. However, it was noted that oats and alfalfa were more severely damaged in each case than spinach and at times than endive or beets. This is the reverse effect of some of the experiments with ozone. Evidently the reaction products in the different mixtures vary appreciably either qualitatively or quantitatively or both.

These oxidations were carried out experimentally in rather concentrated mixtures which were subsequently diluted. Difficulty has been encountered in controlling the oxidation, which in some cases was too slow, in others too rapid or too productive of polymerization products. Cann, Noble, and Larson (17) have made mixtures of hydrocarbons and ozone at high dilution, allowing the gas to remain in sunlight for about 17 min in order to react, thus simulating more closely the conditions of smog formation. Automobile exhaust gases gave typical smog injury to vegetation under these conditions but not when reacted with ozone in concentrated mixtures. It is possible that other organic compounds in addition to the intermediate olefins will be found to be reactive to vegetation when the oxidation occurs in high dilution as in the open atmosphere, rather than in concentrated mixtures. Longer reaction times than 17 min may be necessary to simulate adequately the conditions outside.

It is interesting to note that photochemical oxidations, catalyzed by nitrogen oxides and leading to ozone formation, occur at high dilution with saturated as well as unsaturated organic compounds. In the latter case, the ozone may be absorbed at the double bond by the excess unsaturated compound (33,69). Haagen-Smit, Bradley, and Fox (34) irradiated with sunlight mixtures of nitrogen oxides with hydrocarbons, alcohols, aldehydes, or acids. Pyrex glass flasks of 2 to 12 liters capacity were used. In most cases the concentration of nitrogen oxides was 0.4 ppm and of the organic compound 1 ppm or 0.1 ppm. The rate of rubber cracking after irradiating for 5 min to 4 hr was measured and expressed as the average rate of ozone formation in parts per million per hour. These experiments simulated the concentrations encountered in the open atmosphere. Using 3-methylheptane over a wide range of concentrations, 70 ppm gave no rubber cracking and 0.1 ppm gave a maximum rate equivalent to 1.3 ppm ozone per hour. The reaction appears to be quenched by large amounts of hydrocarbon unless sufficient nitrogen dioxide is present to support the reaction. In dynamic systems small amounts of nitrogen dioxide produce ozone with relatively large concentrations of 3-methylheptane. With a series of saturated fatty acids at 1 ppm, formic acid gave no rubber cracking and acetic only a slight reaction. Maximum cracking rate equivalent to about 1.0 ppm ozone per hour occurred with valeric to caprylic acids. Various other acids, aldehydes, alcohols, and hydrocarbons, including gasoline fractions, mesitylene, and branched chain compounds, gave similar rates which were not appreciably different at 0.1 and 1.0 ppm of the organic compound. Diketones, particularly diacetyl, which readily form free radicals, gave rubber cracking at these low concentrations with nitrogen oxides, but they also gave rubber cracking without nitrogen oxides at concentrations above 40 ppm.

In the foregoing experiments the rubber-cracking material (10) was assumed to be ozone, and in some cases the presence of ozone was established by freezing out the gas at −180°C and testing it chemically and spectrometrically. It should be mentioned that doubt has been cast on the specificity of rubber cracking as a test for ozone, since Crabtree and Biggs (21) have found that products formed by the photolysis of organic peroxides, presumably free radicals, crack rubber in a manner indistinguishable from the action of ozone. The matter needs further study. However, free radicals could hardly account for rubber cracking equivalent to 1 ppm ozone per hour when only 0.1 ppm of the organic compound was present originally, unless there was a chain reaction of great length.

The phytotoxicities of the products of these oxidations were not studied, but it seems probable that some of them might cause smog effects. If so, the experiments support the suggestion that the high dilution reactions in the open atmosphere may involve many organic compounds besides olefins, in the production of phytotoxic substances, and that the reactions go better at great dilutions.

9.6 DIAGNOSIS OF GAS DAMAGE

Determination of the agent causing gas damage may present considerable difficulty in an area where a number of different effluents are discharged into the atmosphere. The uncomplicated case of sulfur dioxide injury has already been discussed. Similarly, fluorine damage is often found in areas free from other toxic agents. In this case chemical analysis of the leaves would furnish definite evidence as to whether or not the injury was caused by fluorine compounds (11,12). Smog injury could be established by observing the fluorescence under ultraviolet light (78). In all cases a screening must be applied to exclude such nongas damage as nutritional and physiological injury, bacterial and virus diseases, insect attack, or wind and weather damage.

Observations of symptoms on all the vegetation, including native plants within the smoke zone and similar observations in comparable areas outside the zone, constitute the obvious first approach. Data on relative susceptibility of the plants, such as that given in Tables 9-1 to 9-6, with due allowance for the current weather conditions and soil moisture, may then serve to suggest the toxic agent or agents. In Table 9-7 are listed a number of plants sensitive to sulfur dioxide, hydrogen fluoride, and smog, respectively, and so far as possible their relative sensitivity to the other two agents. Several differences in susceptibility that might have diagnostic value are apparent. For example, gladiolus is very sensitive to hydrogen fluoride but resistant to sulfur dioxide and smog; alfalfa and barley are sensitive to sulfur dioxide and smog but have intermediate sensitivity to hydrogen fluoride; cotton is sensitive to sulfur dioxide but resistant to hydrogen fluoride; corn is sensitive to hydrogen fluoride but resistant to sulfur dioxide. Evidently these tables of relative susceptibility can serve a useful diagnostic purpose, if used with skill, judgment, and a knowledge of their limitations, provided a good representation of known plant species is present in the area. The weeds are often particularly useful in this connection. Finally, there should be a diminution in the intensity of damage with distance from the suspected source of pollution.

Table 9-7. Relative Susceptibility of a Few Plants to Injury by Sulfur Dioxide, Hydrogen Fluoride, and Smog

Agent	Sensitive	Intermediate	Resistant
SO_2	Alfalfa Barley Endive Cotton Gladiolus Sweet pea Rhubarb Radish Spinach Lettuce, head Sweet potato Broccoli Squash Table beet Oats Buckwheat Clover Carrot Wheat Larch	Dandelion Cabbage Apricot Peach Prune Gladiolus	Gladiolus Corn Grapes
HF	Gladiolus Apricot Prune Larch Sweet potato (some varieties) Corn Grapes (some European varieties) Peach Buckwheat	Alfalfa Barley Buckwheat Carrot Clover Lettuce, head Oats Rhubarb Spinach Sweet potato (some varieties) Wheat	Cabbage Cotton Dandelion Squash Sweet pea
Smog	Endive Spinach Romaine lettuce Barley (young) Alfalfa Table beet Oats Buckwheat	Dandelion Lettuce, head Radish Squash Sweet pea	Broccoli Cabbage Carrot Corn Gladiolus Rhubarb Squash Wheat

No consideration is given in this discussion to possible synergistic effects of mixtures of gases. This is a subject on which there has been much speculation but little or no experimental work. A discussion would therefore not be profitable at this time since there is no published evidence of such effects. Reference may be made again to the antagonistic action of sulfur dioxide and the oxidizing substances in smog.

9.7 ACKNOWLEDGMENTS

Part of the discussion of sulfur dioxide and hydrogen fluoride, including data in Tables 9-1 and 9-3, is based on unpublished material from the Department of Agricultural Research, American Smelting and Refining Company. We are indebted to the company for permission to include this material.

We are also indebted to the following persons who supplied unpublished information concerning the susceptibility of vegetation or otherwise assisted us in the preparation of this manuscript: D. F. Adams, R. A. Bobrov, H. R. Brisley, O. C. Compton, A. J. Haagen-Smit, G. R. Hill, A. Holloman, H. H. Hull, W. W. Jones, G. P. Larson, F. E. Littman, P. P. Mader, P. L. Magill, J. T. Middleton, Jr., W. N. Noble, D. A. Preston, B. L. Richards, F. W. Went, and P. W. Zimmerman.

REFERENCES

1. Adams, D. F., D. J. Mayhew, R. N. Gnagy, E. P. Richey, R. K. Koppe, and I. W. Allen: Atmospheric Pollution in the Ponderosa Pine Blight Area, Spokane County, Washington, *Ind. Eng. Chem.*, **44**, 1356–1365 (1952).
2. Agate, J. N., G. H. Bell, G. F. Boddie, R. G. Bowler, M. Buckell, E. A. Cheeseman, T. H. J. Douglas, H. A. Druett, J. Garrad, D. Hunter, K. M. A. Perry, J. D. Richardson, and J. B. de V. Weir: Industrial Fluorosis—A Study of the Hazard to Man and Animals near Fort William, Scotland, *Med. Research Council Mem. No.* 22, 1949.
3. Allmendinger, D. F., V. L. Miller, and F. Johnson: The Control of Fluorine Scorch of Gladiolus with Foliar Dusts and Sprays, *Proc. Am. Soc. Hort. Sci.*, 56, 427–432 (1950).
4. Barr, C. G.: Photosynthesis in Maize as Influenced by a Transpiration-reducing Spray, *Plant Physiol.*, **20**, 86–97 (1945).
5. Bartholomew, E. T., W. B. Sinclair, and D. L. Lindgren: Measurements on Hydrocyanic Acid Absorbed by Citrus Tissues during Fumigation, *Hilgardia*, **14**, 373–409 (1942).
6. Blacet, F. E.: Photochemistry of the Lower Atmosphere, *Ind. Eng. Chem.*, **44**, 1339–1342 (1952).
7. Bleasdale, J. K. A.: Atmospheric Pollution and Plant Growth, *Nature*, **169**, 376–377 (1952).
8. Bobrov, R. A.: Effect of Smog on Anatomy of Oat Leaves, *Phytopathology*, **42**, 558–563 (1952).
9. Bobrov, R. A.: The Anatomical Effects of Air Pollution on Plants, *Proc. 2d Nat. Air Pollution Symposium*, pp. 129–134, Stanford Research Institute, Los Angeles, Calif., 1952.
10. Bradley, C. E., and A. J. Haagen-Smit: The Application of Rubber in the Quantitative Determination of Ozone, *Rubber Chem. and Technol.*, **24**, 750–755 (1951).
11. Bredemann, G., and H. Radeloff: Zur Diagnose von Fluorrauchschäden, *Phytopathol. Z.*, **5**, 195–206 (1932).
12. Bredemann, G., and H. Radeloff: Über Fluorrauchschäden (Absorption of Fluorine by the Bark and Shoots and Its Action). *Angew. Botan.*, **19**, 172–181 (1937).
13. Brennan, E. G., I. A. Leone, and R. H. Daines: Fluorine Toxicity in Tomato as Modified by Alterations in the Nitrogen, Calcium, and Phosphorus Nutrition of the Plant, *Plant Physiol.*, **25**, 736–747 (1950).
14. Brisley, H. R., and W. W. Jones: Sulfur Dioxide Fumigation of Wheat with Special Reference to Its Effect on Yield, *Plant Physiol.*, **25**, 666–681 (1950).
15. Brody, H. W., and N. F. Childers: The Effect of Dilute Liquid Lime Sulfur Sprays on the Photosynthesis of Apple Leaves, *Proc. Am. Soc. Hort. Sci.*, **36**, 205–209 (1939).
16. Cadle, R. D., and H. S. Johnston: Chemical Reactions in the Los Angeles Smog, *Proc. 2d Nat. Air Pollution Symposium*, pp. 28–34, Stanford Research Institute, Los Angeles, Calif., 1952.
17. Cann, G. R., W. M. Noble, and G. P. Larson: Detection of Smog Forming Hydrocarbons in Automobile Exhaust Gases Using Plants as Indicators, *Air Repair*, **4**, 83–86 (1954).
18. Christopher, E. P.: Comparison of Lime Sulfur and Flotation Sulfur Sprays on Apple Trees, *Proc. Am. Soc. Hort. Sci.*, **40**, 63–67 (1942).

19. Churchill, H. V., R. J. Rowley, and L. N. Martin: Fluorine Content of Certain Vegetation in a Western Pennsylvania Area, *Anal. Chem.*, **20,** 69–71 (1948).
20. Compton, O. C., and L. R. Remmert: Private communication, 1950.
21. Crabtree, J., and B. S. Biggs: Cracking of Stressed Rubber by Free Radicals, *J. Polymer Sci.*, **11,** 280–281 (1953).
22. Crocker, W., A. E. Hitchcock, and P. W. Zimmerman: Similarities in the Effects of Ethylene and the Plant Auxins, *Contribs. Boyce Thompson Inst.*, **7,** 231–248 (1935).
23. Crocker, W., P. W. Zimmerman, and A. E. Hitchcock: Ethylene Induced Epinasty of Leaves and Relation of Gravity to It, *Contribs. Boyce Thompson Inst.*, **4,** 177–218 (1932).
24. Daines, R. H., I. Leone, and E. Brennan: The Effect of Fluorine on Plants as Determined by Soil Nutrition and Fumigation Studies, chap. 9 in L. C. McCabe, ed., "Air Pollution," McGraw-Hill Book Company, Inc., New York, 1952.
25. Dean, R. S., and R. E. Swain: Report Submitted to the Trail Smelter Arbitral Tribunal, *U.S. Bur. Mines Bull.* 453, 1944.
26. de Ong, E. R.: Injury to Apricot by Fluorine Deposit, *Phytopathology*, **36,** 469–471 (1946).
27. Dorries, W.: Über die Brauchbarkeit der spectroskopischen Phäophytinprobe in der Rauchschaden-Diagnostik, *Z. Pflanzenkrankh. u Gallenkunde*, **42,** 257–273 (1932).
28. Dunlap, A. A.: 2,4-D Injury to Cotton from Airplane Dusting of Rice, *Phytopathology*, **38,** 638–644 (1948).
29. Griffin, S. W., and B. B. Bayles: Some Effects of Fluorine Fumes on Vegetation, chap. 10 in L. C. McCabe, ed., "Air Pollution," McGraw-Hill Book Company, Inc., New York, 1952.
30. Gustafson, F. G.: Is the American Elm (*Ulmus americana*) Injured by Natural Gas? *Plant Physiol.*, **25,** 433–440 (1950).
31. Haagen-Smit, A. J.: The Air Pollution Problem in Los Angeles, *Eng. and Sci.*, **14** (3), 7–13 (1950).
32. Haagen-Smit, A. J.: Chemistry and Physiology of Log Angeles Smog, *Ind. Eng. Chem.*, **44,** 1342–1346 (1952).
33. Haagen-Smit, A. J., C. E. Bradley, and M. M. Fox: Formation of Ozone in Los Angeles Smog, *Proc. 2d Nat. Air Pollution Symposium*, pp. 54–56, Stanford Research Institute, Los Angeles, Calif., 1952.
34. Haagen-Smit, A. J., C. E. Bradley, and M. M. Fox: Ozone Formation in the Photochemical Oxidation of Organic Substances, *Ind. Eng. Chem.*, **45,** 2086–2089 (1953).
35. Haagen-Smit, A. J., E. F. Durley, M. Zaitlin, H. Hull, and W. Noble: Investigation on Injury to Plants from Air Pollution in the Los Angeles Area, *Plant Physiol.*, **27,** 18–34 (1952).
36. Haselhoff, E.: "Grundzüge der Rauchschadenkunde," Verlagsbuchhandlung Gebrüder Borntraeger, Berlin, 1932.
37. Haselhoff, E., G. Bredemann, and W. Haselhoff: "Entstehung, Erkennung und Beurteilung von Rauchschäden," Verlagsbuchhandlung Gebrüder Borntraeger, Berlin, 1932.
38. Haselhoff, E., and G. Lindau: "Die Beschädigung der Vegetation durch Rauch," Verlagsbuchhandlung Gebrüder Borntraeger, Leipzig, 1903.
39. Hill, Geo. R., Jr., and M. D. Thomas: Influence of Leaf Destruction by Sulfur Dioxide and by Clipping on Yield of Alfalfa, *Plant Physiol.*, **8,** 223–245 (1933).
40. Hitchcock, A. E., W. Crocker, and P. W. Zimmerman: Effect of Illuminating Gas on Lily, Narcissus, Tulip and Hyacinth, *Contribs. Boyce Thompson Inst.*, **4,** 155–176 (1932).
41. Hitchcock, A. E., W. Crocker, and P. W. Zimmerman: Toxic Action in Soil of Illuminating Gas Containing Hydrocyanic Acid, *Contribs. Boyce Thompson Inst.*, **6,** 1–30 (1934).
42. Holmes, J. A., E. C. Franklin, and R. A. Gould: Report of the Selby Smelter Commission, *U.S. Bur. Mines Bull.* 98, 1915.
43. Höricht, W.: Forest Devastation by Smoke, *Kranke Pflanze*, **15,** 90 (1938).
44. Hull, H. M., and F. W. Went: Life Processes of Plants as Affected by Air Pollution, *Proc. 2d Nat. Air Pollution Symposium*, pp. 122–128, Stanford Research Institute, Los Angeles, Calif., 1952.
45. Johnson, E. M.: Injury to Plants by Minute Amounts of 2,4-dichlorophenoxyacetic acid, *Phytopathology*, **37,** 367–369 (1947).
46. Johnson, F., D. F. Allmendinger, V. L. Miller, and C. J. Gould: Leaf Scorch of Gladiolus Caused by Atmospheric Fluoric Effluents, *Phytopathology*, **40,** 239–246 (1950).
47. Jones, W. W.: Private communication, 1953.

48. Katz, Morris: The Effect of Sulfur Dioxide on Conifers, Chap. 8 in L. C. McCabe, ed., "Air Pollution," McGraw-Hill Book Company, Inc., New York, 1952.
49. Katz, Morris: Sulfur Dioxide in the Atmosphere and Its Relation to Plant Life, *Ind. Eng. Chem.*, **41,** 2450–2465 (1949).
50. Kehoe, Robert A.: Air Pollution and Community Health, *Proc. 1st Nat. Air Pollution Symposium*, pp. 115–120, Stanford Research Institute, Los Angeles, Calif., 1949.
51. Knight, L. J., and W. Crocker: Toxicity of Smoke, *Botan. Gaz.*, **55,** 337–371 (1913).
52. Koritz, H. G., and F. W. Went: The Physiological Action of Smog on Plants. I. Initial Growth and Transpiration Studies, *Plant Physiol.*, **28,** 50–62 (1953).
53. Kotte, W.: Rauchschäden an Steinobstfrüchten, *Nachrbl. deut. Pflanzenschutzdienst*, **9,** 91–92 (1929).
54. Largent, Edward J.: Effects of Fluorides on Man and Animals, *Proc. 1st Nat. Air Pollution Symposium*, pp. 129–134, Stanford Research Institute, Los Angeles, Calif., 1949.
55. Leone, I. A., E. G. Brennan, R. H. Daines, and W. R. Robbins: Some Effects of Fluorine on Peach, Tomato and Buckwheat When Absorbed through the Roots, *Soil Sci.*, **66,** 259–266 (1948).
56. Loftfield, J. V. G.: Behavior of Stomata, *Carnegie Inst. Wash. Publ. No.* 314, 1921.
57. Los Angeles County Air Pollution Control District, Los Angeles, Calif.: First Technical and Administrative Report on Air Pollution Control in Los Angeles County, 1949–1950.
58. Los Angeles County Air Pollution Control District, Los Angeles, Calif.: Second Technical and Administrative Report on Air Pollution in Los Angeles County, 1950–1951.
59. McCabe, L. C., P. P. Mader, H. E. McMahon, W. J. Hamming, and A. L. Chaney: Industrial Dusts and Fumes in the Los Angeles Area, *Ind. Eng. Chem.*, **41,** 2486–2493 (1949).
60. McCallan, S. E. A., A. Hartzell, and F. Wilcoxon: Hydrogen Sulfide Injury to Plants, *Contribs. Boyce Thompson Inst.*, **8,** 189–197 (1936).
61. McCallan, S. E. A., and F. Wilcoxon: Fungicidal Action of Sulfur. II. Production of Hydrogen Sulfide by Sulfured Leaves and Spores and Its Toxicity to Spores, *Contribs. Boyce Thompson Inst.*, **3,** 13–38 (1931).
62. MacIntire, W. H., and Associates: Effects of Fluorine in Tennessee Soils and Crops, *Ind. Eng. Chem.*, **41,** 2466–2475 (1949).
63. MacIntire, W. H., L. J. Hardin, and W. Hester: Measurement of Atmospheric Fluorine, *Ind. Eng. Chem.*, **44,** 1365–1370 (1952).
64. Mader, P. P., M. W. Heddon, R. T. Lofberg, and R. H. Koehler: Determination of Small Amounts of Hydrocarbons in the Atmosphere, *Anal. Chem.*, **24,** 1899–1902 (1952).
65. Mader, P. P., R. D. MacPhee, R. T. Lofberg, and G. P. Larson: Composition of Organic Portion of Atmospheric Aerosols in the Los Angeles Area, *Ind. Eng. Chem.*, **44,** 1352–1355 (1952).
66. Mader, P. P.: Private communication, 1954.
67. Magill, P. L.: The Los Angeles Smog Problem, *Ind. Eng. Chem.*, **41,** 2476–2486 (1949).
68. Magill, P. L., and R. W. Benoliel: Air Pollution in Los Angeles—Contribution of Combustion Products, *Ind. Eng. Chem.*, **44,** 1347–1351 (1952).
69. Magill, P. L., and F. E. Littmann: Air Pollution in Los Angeles, *Am. Soc. Mech. Eng. Paper* 53-A-163, delivered in New York, December, 1953.
70. Massart, L., and R. Dufait: Potassium Cyanide and Sodium Fluoride Inhibition of Fermentation with Special Consideration of Metals as Activators of Enzymes, *Z. physiol. Chem.*, **272,** 157–170 (1942).
71. Massey, L. M.: Similarities between Disease Symptoms and Chemically Induced Injury to Plants, chap. 3 in L. C. McCabe, ed., "Air Pollution," McGraw-Hill Book Company, Inc., New York, 1952.
72. Middleton, J. T., J. B. Kendrick, Jr., and E. F. Darley: Air Pollution Injury to Crops, *Calif. Agr.*, **7,** (11) 11–12 (1953).
73. Middleton, J. T., J. B. Kendrick, Jr., and H. W. Schwalm: Smog in the South Coastal Area, *Calif. Agr.*, **4** (11), 7–10 (1950).
74. Middleton, J. T., J. B. Kendrick, Jr., and H. W. Schwalm: Injury to Herbaceous Plants by Smog or Air Pollution, *Plant Disease Reptr.*, **34,** 245–252 (1950).
75. Miller, V. L., and Associates: The Effect of Atmospheric Fluoride on Washington Agriculture, chap. 11 in L. C. McCabe, ed., "Air Pollution," McGraw-Hill Book Company, Inc., New York, 1952.
76. Miller, V. L., F. Johnson, and D. F. Allmendinger: Fluorine Analysis of Italian Prune Foliage Affected by Marginal Scorch, *Phytopathology*, **38,** 30–37 (1948).

77. National Research Council of Canada: Effect of Sulfur Dioxide on Vegetation, *Publ.* 815, 1939.
78. Nielson, J. P., H. M. Benedict, and A. J. Holloman: Fluorescence as a Means of Identifying Smog Markings on Plants, *Science*, **120**, 182–183 (1954).
79. Noack, Kurt: Damage to Vegetation from Gases in Smoke, *Z. angew. Chem.*, **42**, 123–126 (1929).
80. Noble, W. N.: Private communication, 1953.
81. O'Gara, P. J.: Abstract of paper, Sulfur Dioxide and Fume Problems and Their Solutions, *Ind. Eng. Chem.*, **14**, 744 (1922).
82. O'Gara, P. J.: Unpublished data in files of American Smelting and Refining Company.
83. Prince, A. L., F. E. Bear, E. G. Brennan, I. A. Leone, and R. H. Daines: Fluorine, Its Toxicity to Plants and Its Control in the Soil, *Soil Sci.*, **67**, 269–277 (1949).
84. Radeloff, H.: Untersuchung und Begutachtung von Rauchschäden, *Jahresber. Hamburgischen Inst. Angew. Botan.*, **55**, 119–120 (1938); **56**, 126–127 (1939).
85. Rhine, J. B.: Clogging of Stomata of Conifers in Relation to Smoke Injury and Distribution, *Botan. Gaz.*, **78**, 226–232 (1924).
86. Romell, Lars-Gunnar: Localized Injury to Plant Organs from Hydrogen Fluoride and Other Acid Gases, *Svensk Botan. Tidskr.*, **35**, 271–286 (1941).
87. Rushton, A. G.: The Plant as an Index of Smoke Pollution, *Ann. Appl. Biol.*, **7**, 390–402 (1921).
88. Schroeder, J. V., and C. Reuss: "Beschädigung der Vegetation durch Rauch und Oberharzer Huttenrauchschäden," Paul Parey, Berlin, 1883.
89. Scott, F. M.: Internal Suberization of Tissues, *Botan. Gaz.*, **111**, 378–394 (1950).
90. Setterstrom, Carl, and P. W. Zimmerman: Factors Influencing Susceptibility of Plants to Sulphur Dioxide Injury. I. *Contribs. Boyce Thompson Inst.*, **10**, 155–181 (1939).
91. Setterstrom, Carl, P. W. Zimmerman, and W. Crocker: Effect of Low Concentrations of Sulphur Dioxide on Yield of Alfalfa and Cruciferae, *Contribs. Boyce Thompson Inst.*, **9**, 179–198 (1938).
92. Shepherd, M., S. W. Rock, R. Howard, and J. W. Stormes: Isolation, Identification, and Estimation of Gaseous Pollutants of Air—Examination of Los Angeles County, Calif., Smog, *Anal. Chem.*, **23**, 1431–1440 (1951).
93. Sorauer, P.: "Mikroskopische Analyse rauchbeschädigter Pflanzen," Paul Parey, Berlin, 1911.
94. Stanford Research Institute: The Smog Problem in Los Angeles County, First Interim Report, Western Oil and Gas Association, Los Angeles, Calif., 1948.
95. Stanford Research Institute: The Smog Problem in Los Angeles County, Second Interim Report, Western Oil and Gas Association, Los Angeles, Calif., 1949.
96. Stanford Research Institute: The Smog Problem in Los Angeles County, Third Interim Report, Western Oil and Gas Association, Los Angeles, Calif., 1950.
97. Stanford Research Institute: The Smog Problem in Los Angeles County, Western Oil and Gas Association, Los Angeles, Calif., 1954.
98. Staten, Glen: Contamination of Cotton Fields by 2,4-D or Hormone-type Weed Sprays, *J. Am. Soc. Agron.*, **38**, 536–544 (1946).
99. Steward, F. C., J. F. Thompson, F. K. Millar, M. D. Thomas, and R. H. Hendricks: The Amino Acids of Alfalfa as Revealed by Paper Chromatography with Special Reference to Compounds Labelled with Sulfur, *Plant Physiol.*, **26**, 123–135 (1951).
100. Stoklasa, J.: "Die Beschädigung der Vegetation durch Rauchgase und Fabriksexhalation," Urban & Schwartzenberg, Berlin, 1923.
101. Stout, G. L.: Chlorine Injury to Lettuce and Other Vegetation, *Calif. Dept. Agr. Monthly Bull.*, **21**, 340–344 (1932).
102. Swain, Robert E., and A. B. Johnson: Effect of Sulphur Dioxide on Wheat Development, *Ind. Eng. Chem.*, **28**, 42–47 (1936).
103. Thomas, M. D.: Proc. Auburn Conference on Use of Radioactive Isotopes in Agricultural Research, 1947, pp. 103–117, Alabama Polytechnic Institute, 1948.
104. Thomas, M. D.: Gas Damage to Plants, *Ann. Rev. Plant Physiol.*, **2**, 293–322 (1951).
105. Thomas, M. D., and R. H. Hendricks: Unpublished data.
106. Thomas, M. D., R. H. Hendricks, L. C. Bryner, and G. R. Hill: A Study of the Sulfur Metabolism of Wheat, Barley and Corn Using Radioactive Sulfur, *Plant Physiol.*, **19**, 227–244 (1944).
107. Thomas, M. D., R. H. Hendricks, T. R. Collier, and G. R. Hill: The Utilization of Sulfate and Sulfur Dioxide for the Sulfur Nutrition of Alfalfa, *Plant Physiol.*, **18**, 345–371 (1943).
108. Thomas, M. D., R. H. Hendricks, and G. R. Hill: Some Chemical Reactions of Sulfur Dioxide after Absorption by Alfalfa and Sugar Beets, *Plant Physiol.*, **19**, 212–226 (1944).

109. Thomas, M. D., R. H. Hendricks, and G. R. Hill: Sulfur Metabolism of Plants. Effects of Sulfur Dioxide on Vegetation, *Ind. Eng. Chem.*, **42,** 2231–2235 (1950).
110. Thomas, M. D., R. H. Hendricks, and G. R. Hill, Sulfur Content of Vegetation, *Soil Sci.*, **70,** 9–18 (1950).
111. Thomas, M. D., R. H. Hendricks, and G. R. Hill: Some Impurities in the Air and Their Effects on Plants, chap. 2 in L. C. McCabe, ed., "Air Pollution," McGraw-Hill Book Company, Inc., New York, 1952.
112. Thomas, M. D., R. H. Hendricks, and G. R. Hill: The Action of Sulfur Dioxide on Vegetation, *Proc. 1st Nat. Air Pollution Symposium*, pp. 142–147, Stanford Research Institute, Los Angeles, Calif., 1949.
113. Thomas, M. D., and G. R. Hill: Absorption of Sulfur Dioxide by Alfalfa and Its Relation to Leaf Injury, *Plant Physiol.*, **10,** 291–307 (1935).
114. Thomas, M. D., and G. R. Hill: Relation of Sulfur Dioxide in the Atmosphere to Photosynthesis and Respiration of Alfalfa, *Plant Physiol.*, **12,** 309–383 (1937).
115. Thornton, N. C., and C. Setterstrom: Toxicity of Ammonia, Chlorine, Hydrogen Cyanide, Hydrogen Sulfide, and Sulfur Dioxide Gases on Green Plants, III, *Contribs. Boyce Thompson Inst.*, **11,** 343–356 (1940).
116. Turrell, F. M.: Physiological Effects of Elemental Sulfur Dust on Citrus Fruits, *Plant Physiol.*, **25,** 13–62 (1950).
117. Viets, F. H., G. I. Fischer, and A. P. Fudurich: Atmospheric Pollution from Hydrocarbons in Automobile Exhaust Gases, Los Angeles County Air Pollution Control District Publication 43, Apr. 23, 1953.
118. Warburg, O., and W. Christian: Chemical Mechanism of Fluoride Inhibition of Yeast, *Naturwissenschaften*, **29,** 590 (1941).
119. Went, F. W.: Private communication, 1950.
120. Wieler, A.: "Untersuchungen über die Einwirkung schwefliger Säuren auf die Pflanzen," Verlagsbuchhandlung Gebrüder Borntraeger, Berlin, 1905.
121. Wislicenus, H.: "Sammlung von Abhandlungen über Abgase und Rauchschäden," Paul Parey, Berlin, 1914.
122. Zimmerman, P. W.: Anaesthetic Properties of Carbon Monoxide and Other Gases in Relation to Plants, Insects, and Centipedes, *Contribs. Boyce Thompson Inst.*, **7,** 147–155 (1935).
123. Zimmerman, P. W.: Private communication, 1953.
124. Zimmerman, P. W.: Impurities in the Air and Their Influence on Plant Life, *Proc. 1st Nat. Air Pollution Symposium*, pp. 135–141, Stanford Research Institute, Los Angeles, Calif., 1949.
125. Zimmerman, P. W., and W. Crocker: Toxicity of Air Containing Sulfur Dioxide Gas, *Contribs. Boyce Thompson Inst.*, **6,** 455–470 (1934).
126. Zimmerman, P. W., and W. Crocker: Plant Injury Caused by Vapors of Mercury and Compounds of Mercury, *Contribs. Boyce Thompson Inst.*, **6,** 167–187 (1934).
127. Zimmerman, P. W., W. Crocker, and A. E. Hitchcock: Initiation and Stimulation of Roots from Exposure of Plants to Carbon Monoxide Gas, *Contribs. Boyce Thompson Inst.*, **5,** 1–17 (1933).
128. Zimmerman, P. W., and A. E. Hitchcock: Fluorine Compounds Given Off by Plants (abstract), *Am. J. Botany*, **33,** 233 (1946).
129. Zimmerman, P. W., A. E. Hitchcock, and W. Crocker: Effect of Ethylene and Illuminating Gas on Roses, *Contribs. Boyce Thompson Inst.*, **3,** 459–481 (1931).

Section 10

SAMPLING PROCEDURES

BY R. D. CADLE, P. L. MAGILL, A. A. NICHOL, H. C. EHRMANTRAUT, AND G. W. NEWELL

10.1 AIR-SAMPLING PROCEDURES

10.1.1 Introduction

The solution of most air pollution problems requires sampling and analysis of the contaminated atmospheres. From the standpoint of air pollution, the purpose of collecting samples is generally to determine types and properties of impurities, degree of contamination, and presence or absence of toxic or harmful substances. Samples also provide quantitative information for the development of remedial measures, evidence for legal purposes, a basis for police or control action, standards of equipment performance, and a basis for future reference.

Sampling processes used in air pollution studies can be classified in at least three ways:

1. Sampling for particulate impurities vs. sampling for gaseous impurities
2. Stack, duct, or flue sampling vs. open-air sampling
3. Continuous vs. intermittent sampling

Each of these classes involves unique difficulties which must be overcome. Many of the sampling problems encountered during investigations of air pollution are strictly comparable to those encountered during many other types of investigations. These problems arise in attempting to choose representative samples of the air mass being studied. However, sampling contaminated atmospheres usually includes the removal of the contaminants from the sample. This removal process introduces many chances for error. Another source of error not usually encountered in sampling operations arises when attempting to sample air containing suspended particulate material. It is difficult to sample such suspensions without altering the concentration of particulate material in the sample.

The purpose of this section is to discuss various procedures used for sampling contaminated atmospheres. Specific objectives are itemized below:

1. To describe various types of sampling equipment
2. To discuss briefly the principles on which the sampling techniques are based
3. To describe general applications of the sampling methods
4. To discuss the advantages and disadvantages of the various methods.

Much of the information included in this chapter is summarized in Table 10-1.

10.1.2 General Considerations

The importance of the correct sample cannot be overemphasized. Samples should be collected with knowledge of how results are to be used and of the accuracy required. These requirements are subject to many variations. All too frequently, samples are collected that, regardless of the analytical techniques applied, will not provide the information that is necessary.

The purpose of this article is to describe some of the general principles to be followed in collecting samples for the purpose of studying air pollution problems.

Statistical Approach to Obtaining Representative Samples. Statistical techniques are being used increasingly as tools in planning research. They help to determine the amount of data needed to reach valid conclusions. They are useful in designing experiments. They enable the research worker to obtain the most information from the results. They help to increase the precision of scientific work.

Applied statistics is a highly specialized field. All too frequently, the engineer and the chemist have been too preoccupied with other problems of their work to acquire an up-to-date knowledge of statistics. Usually a statistician should be consulted to help plan experiments and analyze the results. The help of a statistician becomes particularly important when problems involve numerical values arrived at by subjective judgments, as is frequently the case in air pollution.

The statistician obtains the answers to his problems in terms of probability. He does not usually admit the possibility of a 100 per cent correct answer. Rather he would say that there is a 99 to 1, a 1,000 to 1, or perhaps a 10 to 1 chance that the answer is correct. Given a particular set of data, the statistician's problem is to determine the probability of its accuracy within certain limits.

Although it is not recommended that the investigator of air pollution problems attempt to become a statistician, a general knowledge of what statistics can accomplish, and also its limitations, is very useful. Recent books on the subject by Gore (53) and by Youden (153) are recommended reading.

Size of Sample. Aside from the statistical dictates of how large and how frequent samplings are necessary, additional facts must be considered to ensure results representative of the average. Obviously, the sample must be large enough for the analyst to analyze. Frequently, the samples are not procured by the analyst. Under these conditions close working arrangements between the sampler and the analyst are necessary. The container, the manner of sampling, and the size of the sample should all be discussed with the analyst before the sample is collected.

In some instances only a few milligrams are necessary for the analyst, but in outdoor air sampling it may be next to impossible to collect even this small quantity. Under these circumstances either new methods must be devised by which larger samples can be obtained or new techniques must be developed which permit the analysis of the quantity available.

Alteration of Sample during and after Collection. Samples collected for air pollution study are especially susceptible to alteration. At low concentrations many substances are capable of existing simultaneously as true identities in a gas stream without appreciable reaction, but upon collection by any means, they may react to form substances not existing as such in the gas stream or the open atmosphere.

Particulate matter is subject to agglomeration or fracture upon collection so that the size distribution of the collected sample may bear little resemblance to that in its suspended state. Fine dust may agglomerate and defy efforts to redisperse it in order to obtain a true average particle size.

These possibilities of alteration should constantly be borne in mind when interpreting the results.

Continuous vs. Intermittent Sampling. An increasing proportion of the sampling and measuring equipment developed during recent years is designed to give continuous records of the pollutants in stacks, flues, and the open atmosphere. The use of automatic and continuously recording equipment usually decreases the number of man-hours required for sampling and analysis. Even more important is the fact that such equipment, if sufficiently sensitive to momentary changes in the concentrations of pollutants, records peak concentrations which otherwise would escape attention. Such peak concentrations may be very important with respect to certain unpleasant effects of polluted atmospheres, such as the damage they do to vegetation. Continuously recording equipment can also be used to establish important correlations. For example, by use of a recording ozone analyzer and a recording visibility meter, Littman and coworkers (89,98) showed that a strong correlation existed between decrease in visibility and the ozone content of the Pasadena atmosphere.

Many of the continuously operating and recording instruments combine the sampling and analyzing operations in the same piece of equipment. For such devices, it is quite artificial to attempt to differentiate between sampling and analysis.

Sampling Volatile Constituents. Primarily, the problem of sampling volatile constituents of contaminated atmospheres is one of efficient collection, concentration,

Table 10-1. Comparative Summary of Sampling Devices

Type of contaminant	Principle of method	Instruments	Method of quantitation	Skill in operation	Skill in quantitation	Ref.[a]	Remarks
Gaseous	Absorption	Bubbler trains	Chemical analysis	Some		48, 49, 111, 141, 154	Efficient for collection of reactive gases. Efficiency low for fine particulate matter and gases of low or slow solubility. Simple commercially available equipment
		Spray contactor	Chemical analysis	Considerable		35, 94, 131	Efficiency low for gases of low or slow solubility. More efficient than bubblers for particulate matter. Permits large ratio of volume gas scrubbed to liquid used, increasing sensitivity of test. Spray losses should be considered. Equipment not commercially available
		Continuous operation	Continuous chemical	Some	Little (automatic)	90, 129, 137, 138, 144	Specific for certain types of gases. Provides more useful information than batch samplers. Some types of equipment commercially available. Original equipment expensive, but saves manpower
	Adsorption	Tubes or cartridges filled with charcoal, silica, or alumina gel	Weight, chemical analysis, change in color of adsorbed reagents	Considerable	Considerable for trace quantities Less for gross amounts	23, 33, 41, 108, 132, 135, 139, 140	Highly efficient for collection of wide variety of gases. Chemical analysis may require desorption under pretested conditions. Equipment must be constructed to meet test conditions. Commercially available for specific purposes
	Condensation	Freeze-out	Chemical analysis	Considerable		24, 71, 102, 121	Highly efficient for vaporized liquids. Requires low-temperature refrigeration. Apparatus may be assembled from commercially available components
	Mechanical retention	Bottle collection	Chemical analysis	Some		31, 95, 110	Quick and convenient method. No power source needed at sampling position. Gas volumes limited to size of available containers. Equipment readily available
		Continuous operation	Physical analysis, such as absorption spectroscopy	Considerable	Considerable, although automatic	38, 91, 92, 97, 113, 145	Some types can be made specific for certain types of gases. Provides more useful information than batch samples. Some types of equipment are commercially available
Particulate	Sedimentation	Dust jars, screens, glass plates, boxes	Count, weight, chemical analysis	Considerable	Considerable	47, 109, 120, 122	Simple equipment readily available. Useful for qualitative examination and collection for chemical identification and size analysis. Long settling times needed for submicron particles. Accepted method for pollen collection
	Impingement	Cascade impactors	Count, chemical, micrurgic, weight	Some	Considerable	87, 115, 116, 147	Provides rapid means of fractioning aerosols. Not well adapted to high concentrations if particles are to be counted. Glass or machined parts may be used

Table 10-1. Comparative Summary of Sampling Devices (*Continued*)

Particulate	Impingement	Single-jet impactors	Count, chemical, micrurgic, weight	Some	Considerable		Provides efficient collection for particles down to 1μ. Collects smaller particles less efficiently. Useful for distinguishing between solid and liquid aerosols. Glass or machined parts may be used
		Greenburg-Smith impinger	Count, chemical analysis	Some	Considerable	57, 73	Can be used for soluble gases and particulate material. Commercially available in compact portable form
		Fine fiber impinger	Count, chemical, micrurgic	Considerable	Considerable	15, 20, 39	Simple equipment, light weight and compact. Highly useful for distinguishing between liquid and solid aerosols
	Filters	Sugar or salycylic acid, or naphthalene, packed containers	Count, chemical analysis	Some	Some	63, 64, 65	Provides means of separating filter body from collected material. Not suitable for wet or hot atmospheres. Sugar and salycylic acid removed by dissolving, naphthalene by evaporation
		Paper thimbles or disks or continuous operation	Weight, chemical analysis	Some	Considerable for trace quantities Less for gross amounts and weight	12, 21, 45, 54, 56, 59, 72, 93, 103, 104, 119, 123, 124, 128, 130, 136, 142, 143	Can measure quantitatively by discoloration. Inexpensive, readily available. Can be impregnated with indicators to show acid mists
		Cotton, glass, asbestos, or wool mats					Adaptable to wide range of aerosol separation from gases, for weight analysis, and under some conditions chemical analyses. Difficulty frequently encountered in separation of aerosol from filter body for detailed analysis. May be built for large or small sampling rates. Inexpensive, readily available
	Centrifugal	Cyclones	Weight, chemical analysis Count, screen analysis	Some	Considerable	78, 79, 133	May be built in wide range of sizes and of temperature and corrosion-resistant materials. Collected material readily removed. Inexpensive, readily available
	Precipitation	Thermal precipitator	Count, microscopic measurement	Considerable	Considerable	14, 27, 28, 43, 75, 81, 117	Highly efficient for collection of submicron particles. Temperature effects may modify liquid aerosols. Slow sampling rate. Useful for microscopic counting. Commercially available, medium expensive
		Electrostatic precipitator	Weight, chemical analysis	Some	Considerable	6, 8, 19, 28, 37, 40, 52, 68, 74, 118, 150	Highly efficient for collection of submicron particles for total weight determination. Low pressure drop. Hazardous in explosive atmospheres. Microscopic counting difficult. Commercially available, medium expensive

[a] References are given at the end of this section.

and removal for analysis. The analysis almost always deals with trace quantities of impurities, and for this reason large samples of air containing the trace quantities are usually necessary. No universal method of sampling such constituents has yet been developed.

Sampling Particulate Constituents. Sampling particulate material suspended in gases involves several requirements which are not present when gaseous impurities are sampled. One of these is the necessity for sampling isokinetically when the particles are over a few microns in diameter. Strictly speaking, isokinetic sampling occurs only when the suspension enters the orifice of the sampling device without disturbance or acceleration of any kind. There must be no eddy formation near the orifice and no change of direction as the suspension enters the orifice. The air carrying the particles must be flowing uniformly. Anisokinetic sampling produces errors which result from failure of the larger particles to follow changes in the direction and velocity of air flow. Consider, for example, the sampling of stack gases. If the probe is pointing upstream and the velocity of gases in the probe is slower than in the stack, much of the gas approaching the probe orifice must be diverted around it. Larger particles, because of their momentum, do not follow the flow of gas, but enter the orifice instead. As a result, calculated concentrations of the larger particles are erroneously high. If the probe is pointed downstream, too few of the larger particles are sampled.

The amount of error resulting from isokinetic conditions depends upon the size of the particles collected (78,79). The error is not only in the total particulate loading, but also in the size distribution, since the anisokinetic conditions produce classification of the particles according to size.

In practice, strictly isokinetic sampling does not seem possible, since the presence of a probe always has some effect on the flow of gases. However, by keeping the probe pointed upstream and sampling at essentially the same velocity as the gases being sampled, the error from anisokinetic sampling conditions becomes insignificant.

Watson (146) recently developed a tentative semiempirical theory of isokinetic sampling. He developed the equation

$$\frac{C}{C_0} = \frac{U_0}{U}\left\{1 + f(p)\left[\left(\frac{U}{U_0}\right)^{1/2} - 1\right]\right\}^2 \qquad (10\text{-}1)$$

where C = concentration measured
C_0 = true concentration
U_0 = stream velocity
U = mean air velocity at sampling orifice
p = $d^2\rho U_0/18\eta D$
D = diameter of orifice
ρ = specific gravity of particles
d = diameter of particles
η = viscosity of fluid carrying the aerosol

Experiments were conducted in a wind tunnel to determine $f(p)$. The powders used were two varieties of spores, lycopodium powder ($d = 32\mu$) and *Lycoperdon giganteum* ($d = 4\mu$). The experimentally determined values of $f(p)$ are shown in Fig. 10-1. Figure 10-2 was derived from Fig. 10-1 in order to give some idea of the magnitude of C/C_0 for a range of particle sizes using spheres of unit density.

Another difficulty (often encountered when sampling particulate material suspended in air) is preventing the particles from collecting on the internal surfaces of probes or other ducts leading to the collection device. The surface on which collection is desired (such as the surface of a filter) should be placed as close as is reasonable to the air inlet. In fact, some filter holders are designed so that the air enters the filter directly, with-

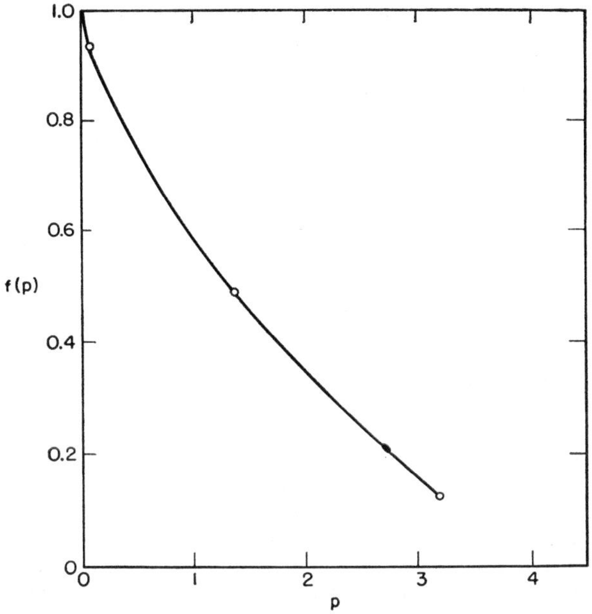

FIG. 10-1. Relationship between $f(p)$ and p. *(Courtesy of the American Industrial Hygiene Association.)*

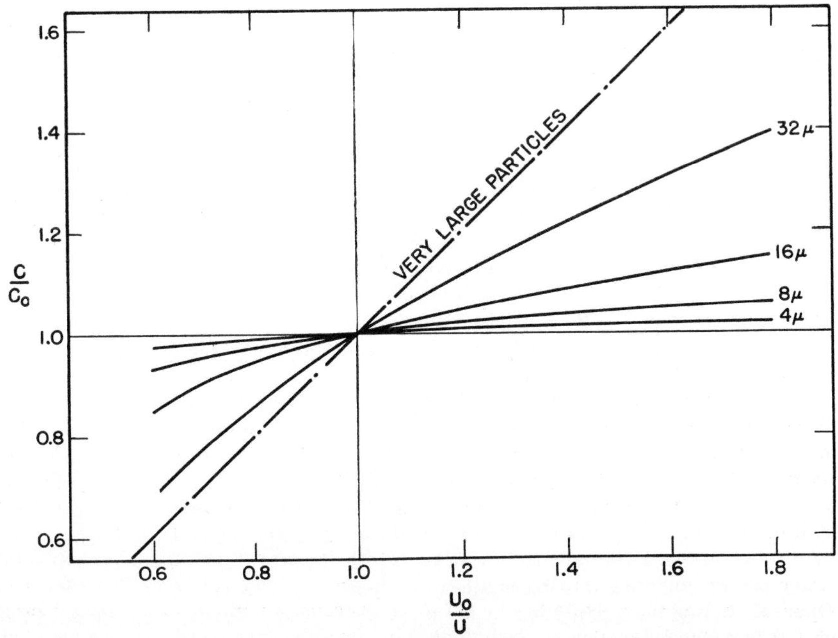

FIG. 10-2. Values of C/C_0 calculated from Eq. (10-1) for spheres of unit density and a number of diameters as indicated. *(Courtesy of the American Industrial Hygiene Association.)*

out having to pass through a probe. There should be no sharp bends and as few bends as possible in the tubing leading from the probe to the collection device.

Sampling devices which remove particulate material from the air are never 100 per cent efficient for all the sizes of particles which might be encountered. Thus a collection technique should be chosen which is efficient over the particle size range of interest.

Sampling Stack Gases. Stack-gas sampling involves a number of problems which are not usually encountered in open-air sampling. The gases being in a stack or flue, some sort of opening must be provided or prepared. The gases are often at a sufficiently high temperature that certain types of sampling equipment, such as paper filters, cannot be used. On the other hand, the gases must not be cooled below their dew point, since condensing moisture may carry with it some of the gases to be analyzed. Flow rates and concentrations are generally not identical across the stack. Thus samples should be obtained at a number of positions within the stack.

Smith and his coworkers (126) have discussed methods they have used in overcoming the problems encountered in sampling open-hearth stacks. Temperatures were as high as 1100°F, and oxidizing conditions introduced the problem of corrosion. Sampling at locations at least 10 stack diameters from the nearest obstruction was recommended, although even at the 10-stack-diameter level the flow was still not entirely uniform (because of entry of gases and steps in the stack lines). When a location even lower down the stack was chosen, the flow on one side was actually negative. Smith recommended pitot tubes fabricated from standard 18-8 stainless steel for making velocity traverses and recommended the use of this same material for constructing nozzles, probes, and other items for sampling. However, he pointed out that corrosion is severe on this type of steel when used in open-hearth stacks unless the material has been stabilized with titanium or columbium, as in AISI types 321 or 347. Greenburg-Smith impingers made of Monel metal were used to collect gaseous compounds of silicon.

Fitton and Sayles (46) attempted to develop methods for simplifying the collection of a representative flue-dust sample. They set up a model duct which was 10 in. square. It was provided with a 90° bend near the air inlet and a weighing feeder to inject a controlled amount of dust. Seventeen restricting devices to produce uniform dust distribution in the duct were tested. Various types of baffles were unsuccessful, but square nozzles blanking off one-half to two-thirds of the duct area were helpful. By using such a nozzle combined with baffles, and by sampling from the center of the cross section, the authors were able to obtain measured dust loadings within 10 per cent of those calculated from the input rate.

The British Standards Institution has described a code for the sampling and analysis of flue gases (17). Methods for sampling and analyzing flue gases from a number of operations are given. The British Standards method calls for a minimum of 24 sampling points; according to Fitton and Sayles it is laborious and unsuited to routine operation.

Pitot tubes are commonly used to determine the gas velocities in stacks and flues in order to attain isokinetic sampling. When the gases contain high concentrations of dusts, as in the primary air supply to a pulverized-fuel burner, a method of preventing the dust from blocking the pitot tube must be provided. Hughes and Sayles (70) have described a pitot-tube purging apparatus for this purpose. A diagrammatic sketch of their equipment is shown in Fig. 10-3. The apparatus is designed so that air is forced outward through the pitot tube into the main gas stream except when velocity measurements are being made.

Open-air Sampling. Sampling in the open air (either indoors or outdoors) differs from flue and stack sampling in several respects. Temperatures are usually relatively low. Concentrations of contaminants are usually lower in outside air than in ducts.

It is often necessary to sample much larger volumes of outside air than of stack gases in order to collect sufficient material for an analysis. Isokinetic sampling is difficult because of the rapidly changing wind direction and velocity.

The impure outdoor air to be sampled may cover areas as large as cities or counties. Thus it is practically impossible to obtain a single sample which is sufficiently large

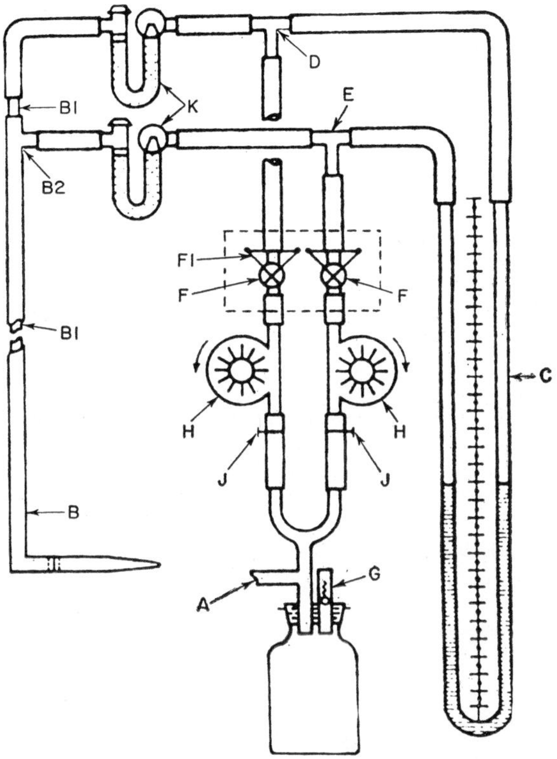

A-PURGING AIR INLET
B-PITOT TUBE
BI-PITOT TUBE, TOTAL HEAD LINE
B2-PITOT TUBE, STATIC HEAD LINE
C-"U" GAUGE
D & E-"T" CONNECTIONS
F-CHANGE-OVER COCKS
FI-OPERATING LINKAGE
G-RELIEF VALVE
H-ANEMOMETERS
J-CONSTRICTING CLIPS
K-FILTERS

FIG. 10-3. Diagrammatic sketch of pitot-tube purging apparatus. (*Reproduced by permission of the controller, H. M. Stationery Office.*)

to be representative. One of the most effective ways of sampling and analyzing contaminants over such wide areas is to use continuously recording instruments. A number of instruments analyzing for the same contaminant are operated simultaneously throughout the area. Simultaneous and continuous measurements of wind velocity and direction at each station may yield important information concerning the source of the contaminants. The work of Littman and coworkers (89,98) referred to above provides an interesting example of the advantage of such techniques over spot sampling.

The selection of a sampling location outdoors, the time at which the sample is taken, and the duration of the sampling period are affected markedly by meteorological conditions and topography.

The two most important meteorological factors are the wind and the stability of the atmosphere. The wind acts on effluents from an emission source in several ways. The effect of the direction is obvious in that effluents will travel with the wind. Moreover, variations in wind direction are important in preventing a plume from continually sweeping over the same area.

A stable atmosphere is one in which little turbulent mixing occurs. The stability is usually brought about by a mass of cold air resting under a warmer mass of air.

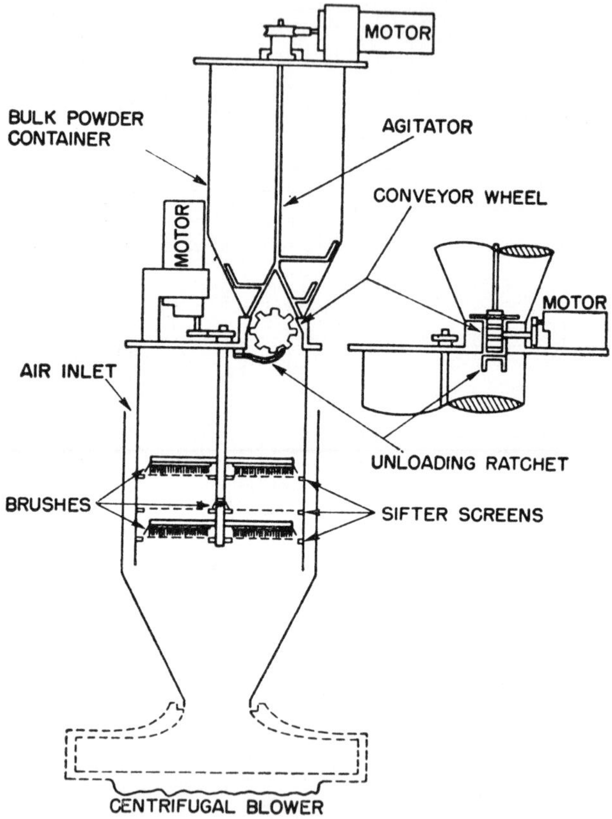

Fig. 10-4. Blower disperser.

The degree of stability controls how far away from the stack a plume will reach the ground and, in combination with the wind speed, controls the rate of diffusion of the effluent.

In recent years mathematical methods have been developed for calculating the theoretical ground concentration of effluents, either gaseous or particulate, evolved from single point sources of emission (7,32,44,60,107). These methods are highly useful in translating the results obtained under one set of atmospheric conditions to those likely to be obtained under some other set of conditions. They are sometimes difficult to apply because of inability to obtain all the meteorological data necessary for use in the formulas.

The local topography is important. Mountain barriers or river valleys may be very effective in restricting the volume of air into which waste effluent is discharged. Under stable atmospheric conditions with little wind movement along a valley, a source which may not be noticeable under many conditions could become a nuisance.

Methods of Following Air Masses. Sampling for impurities in the atmosphere is usually carried out with the philosophy that one will determine first which impurities are present and later where they come from. Equally important is the problem of determining the fate of effluents from a given source. When there is only one source of a given contaminant, the problem becomes one of analyzing the air for that contaminant at various distances from the source. However, there are often two or more sources of a given contaminant. In such cases it is often desirable to follow the air masses leaving the source in question. If, for example, the source is a stack, a convenient method is to introduce some material into the stack gases which can later be collected from the atmosphere and identified.

Several papers have appeared during the last few years which deal with methods of following air masses. Perkins and coworkers (112) described a fluorescent atmospheric tracer technique. They used two fluorescent pigments chiefly, a zinc cadmium sulfide made by the New Jersey Zinc Co. and a zinc silicate made by du Pont. These substances had mass median diameters of about 2.0 and 1.5μ, respectively, and appeared to be reasonably stable to chemical or photochemical action while airborne. These materials were dispersed into the air with the apparatus shown in Fig. 10-4. The air was sampled either with molecular (membrane) filters or with a drum impactor. The collected particles were counted microscopically, using a 0.25 numerical-aperture objective and a magnification of 100X. Illumination was by means of ultraviolet lamps. Braham and coworkers (13) described a technique for tagging and tracing air parcels which involved dispersal of a fluorescent pigment, sampling of the dispersed aerosols by means of airborne impactors, and microscopic identification of collected particles with the use of ultraviolet illumination. Holden and coworkers (62) described statistical methods for counting particles. The methods permit control of the probable error for each sample. A preliminary count of particle density on each filter provides data for estimating the number of fields to be counted to yield the desired statistical control. Counting procedures and methods for training personnel to count fluorescent particles were also described.

10.1.3 Sampling Equipment

The selection of the proper sampling method is always a problem that confronts the experimenter. So many sampling methods have been published that, without knowledge of the advantages and limitations of each, the selection of the best method for the problem at hand becomes most confusing.

As in many other fields of endeavor, the particular tool may not be so important as the experience and knowledge of the operator. Individual preferences and availability of equipment are also important. There are many factors to be considered. In the articles that follow, an attempt has been made to describe the principal types of sampling equipment and some of their limitations.

Probes. Probes are used whenever it is not desirable to place the collecting equipment directly in the stream of gas being sampled. The most common use is in sampling ducts or stacks. A special situation sometimes arises where it is necessary to sample at higher than ground elevations, and for this purpose probes can be tied to towers or be elevated by balloons.

Ducts and Stacks. The construction and placement of probes for sampling gases free from particulate matter are usually simpler than for gases that contain finely suspended solids or liquids. For example, gas tends to be more thoroughly mixed than particles in a flowing system. Also, isokinetic sampling is required only when particles are present.

The materials of probe construction must not react with the materials being sampled. Stainless steel (18-8 alloy) is adequate for most purposes. Unusual temperature and corrosive conditions may dictate the use of such materials as glass and quartz.

The length of the probe must be sufficient to reach all parts of the duct or stack from which the sample is removed. There must be sufficient projection outside the equipment to permit manipulation and connection to sample-collecting equipment. This frequently requires a sampling probe of a length greater than the distance across the duct or stack. This will almost inevitably be true if flow velocities are not similar at all positions being sampled.

The diameter of the probe should be no larger than necessary to conduct the sample at the selected rate to the sampling equipment. It usually is not necessary to exceed 2 in. inside diameter. Sizes smaller than ¼ in. diameter may be subject to frequent plugging if particulate matter is present. To increase rigidity during manipulation, probes may be reinforced by metallic members of proper design, provided the reinforcing members do not interfere with maintaining isokinetic sampling.

The over-all construction of the probe must permit easy cleaning. When particulate matter is being sampled, or if the probe cannot be maintained above the dew point of the least volatile material in the system, it may be impossible to prevent some material from collecting in the probe. When this occurs, the probe must be cleaned after each sampling and the cleanings considered part of the sample.

The probe should be pointed upstream. The sampling head should be interchangeable so that the probe can be used over a large range of gas velocities. Isokinetic conditions at the entrance to the probe are achieved by adjusting the size of the entrance opening and the sampling velocity which is measured as volume per unit time, so that the linear sampling velocity at the entrance is the same as the gas velocity in the duct or stack.

If the gas velocities in the flowing system have been measured and found to be nonuniform across the duct or stack and do not vary appreciably with respect to time, then the sampling rate can be readily adjusted with respect to flows at each location. These velocities may already have been measured to estimate the total volumetric flow rates in the system.

It may be found that the gas velocities are not the same from one moment to the next, and under these conditions frequent changes in sampling rates may be necessary. Various sampling probes are shown in Fig. 10-5 (2).

A sampling nozzle which will permit maintaining isokinetic conditions where flow rates in the system are not known, or where they change frequently, was designed by Lapple (80) and is illustrated in Fig. 10-6. In practice this may be difficult to use because the manometer must be protected from large temperature changes.

Blackie (11) and Bailey (4) developed devices to obtain an average sample when the gas flow is pulsating.

Sampling at Locations above Ground Level. It may be desirable to:

1. Measure the vertical distribution of contaminants
2. Obtain samples from stacks after emission where the effluent stream does not reach ground level at a convenient position

In some instances it is feasible to construct a vertical tower to carry a sampling probe to the necessary elevations. This is justified only in extreme cases and has been used in some of the atomic-energy projects.

A method commonly considered and sometimes used by industry is sampling from an airplane. In using this technique a number of special problems arise, some of which are as follows:

1. The horizontal velocity of the airplane makes it difficult to obtain an adequate sample at any one location.

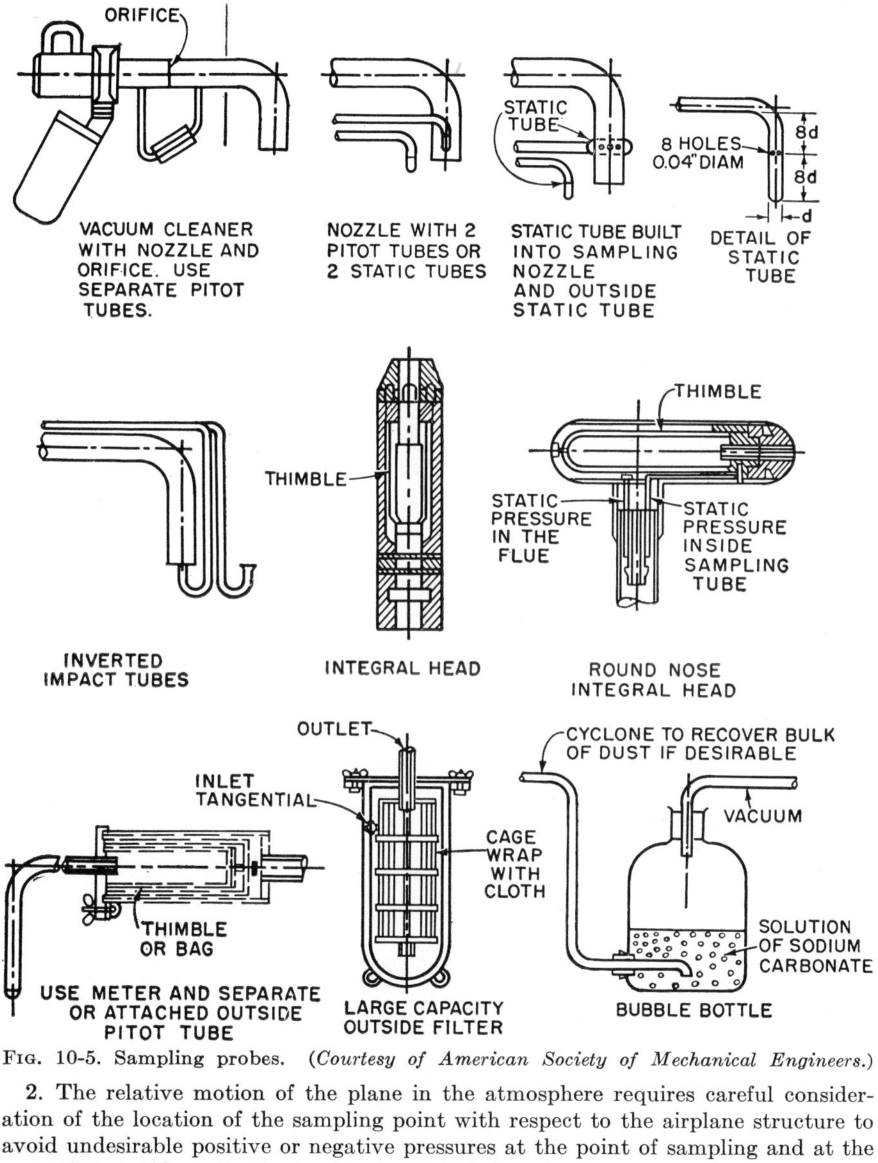

FIG. 10-5. Sampling probes. (*Courtesy of American Society of Mechanical Engineers.*)

2. The relative motion of the plane in the atmosphere requires careful consideration of the location of the sampling point with respect to the airplane structure to avoid undesirable positive or negative pressures at the point of sampling and at the same time avoid contamination by the plane's exhaust.

3. If a light plane is used, the weight of the apparatus may have to be modified to match the carrying capacity of the plane.

4. Local CAA regulations may not permit flying at the elevations where it is desirable to collect the sample.

5. Power-driven equipment must match the plane's power supply, or auxiliary power equipment must be provided.

Crozier and Seely (36) used the velocity of the air stream caused by the plane's motion to operate an impactor for collecting samples.

H. P. Munger (106) apparently has overcome many of the difficulties attendant upon sampling above ground level by utilizing balloons to carry instruments and

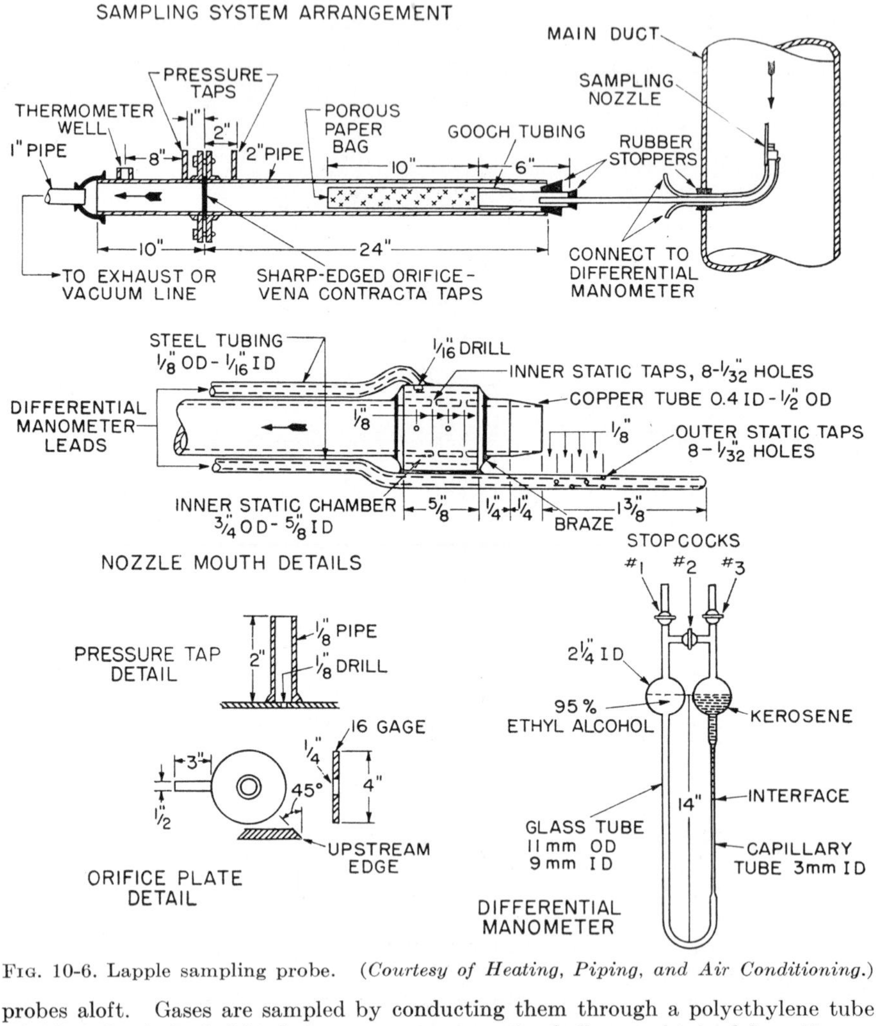

Fig. 10-6. Lapple sampling probe. (*Courtesy of Heating, Piping, and Air Conditioning.*)

probes aloft. Gases are sampled by conducting them through a polyethylene tube raised to the desired altitude by one or more captive balloons, obtained from Dewey and Almy, Cambridge, Mass., or the Seyfang Corp., Long Island, N.Y. This tubing may be maintained at the desired altitude for long periods of time while the gas is being sampled continuously. Rubber tubing is not satisfactory for this because of its permeability and reactivity with some materials, notably ozone and nitrogen dioxide.

For aerosol sampling the collecting head of the electrostatic precipitator is carried aloft. Suction is provided through polyethylene tubing which also carries the electric conductors that provide the high-voltage power for the electrostatic head. Other measurements such as those pertaining to relative humidity, temperature, or smoke-stream trajectories can be made with similar equipment.

An ingenious sampling device for ozone, hydrogen peroxide, and nitrogen dioxide for use with balloons was employed by Pring (114).

Prior to any use of anchored balloons, CAA regulations should be consulted.

Metering Devices. The accuracy of an analysis can be no greater than the volume of gas which is measured. According to ASME Power Test Codes (2): "All devices installed in the sampling lines, whether merely as a guide in regulating the sampling rate or for measuring the quantity of gas sampled, shall be calibrated with a high-grade gas meter or gasometer. It is not sufficiently accurate to use average values for the coefficient of discharge when the orifices are small. Thin-plate orifices are usual, although thick orifices with well-rounded inlets may be employed if desired. The size of the orifices depends on the rate of sampling. In general, the orifice should be small enough to give a deflection which can be read accurately on the type of manometer used and yet not so small that the added pressure drop will adversely affect the operation of the exhausting device. If desired, the orifices can be made and used with the same inclined manometers as with the Pitot tubes. For instructions regarding the design, installation, and use of these orifices, refer to 'I. and A.,' Part 5, Chapter 4. Flow Measurements by Means of Standardized Nozzles and Orifice Plates, ASME Power Test Codes."

A more sensitive device than the standard pitot tube is the pitot-venturi element, obtainable from the Taylor Instrument Company, Rochester, N.Y., which provides ten times the differential produced by a conventional pitot tube for velocities exceeding 1,000 fpm. According to the manufacturer, it combines the advantages of both the venturi and pitot tubes. The small size of the unit adapts it for use in congested areas in any pipe of 3 in. or larger. Construction is of brass, type 304 or 316, 18-8 steel, or monel. The unit is calibrated when supplied, and the manometer range is adjusted in accordance with flow conditions encountered.

Sensitive manometers for use with small flow rates may be built, as described by Lapple (80), for which deflection of 1 in. is equivalent to 0.01 in. of water. A Wahlen gauge, sometimes known as an Illinois micromanometer (148), may also be used.

Wet test meters are often very convenient for estimating the total amount of air which has been sampled. Furthermore, they are often used for calibrating other meters. Air drawn through the device rotates a drum which is sealed with liquid. The drum operates a pointer mounted above a dial which is calibrated in terms of volume. Water is usually used as the sealing liquid. The surface of the liquid inside the meter should be at the same level as when the meter was calibrated. The surface of the liquid is adjusted by means of a sight glass. The instrument is provided with crossed spirit levels, and it should always be operated in a level position. Such meters should be operated at close to atmospheric pressures. Goldman and Jacobs (51) suggest that for work in the field of industrial hygiene it is desirable to have one meter whose dial indicates 3 liters or 0.1 cu ft per revolution and perhaps two additional meters, one reading 1 liter per revolution and the other 1 cu ft per revolution.

Dry gas meters, such as those used for metering illuminating gas into homes, are also useful. They have an accuracy of about ± 1 per cent when properly calibrated. Like the wet test meters, they are more useful for indicating a volume of gas sampled than the rate of sampling.

Orifice flowmeters are commonly used for determining the rate of sampling. There are numerous modifications, but probably the simplest is that shown in Fig. 10-7. The pressure drop is a function of gas velocity. Thus the manometer can be calibrated in terms of rate of flow. Such flowmeters are popular because they are simple to construct and calibrate. By using interchangeable orifice tubes, a wide range of flow rates can be measured accurately. Orifice meters have the disadvantages that they are fragile when constructed of glass and that there is usually a risk of blowing or sucking the manometer fluid into the main line. The latter difficulty can be largely overcome by the use of traps or check valves in each arm of the manometer.

Rotameters are also very popular for determining sampling rates. They consist essentially of a float in a vertical tapered tube. They are sturdier than the orifice meters and can readily be permanently mounted. Rotameters using glass or metal spheres as floats are easily constructed in the usual glass-blowing shop. The tapered tubes are prepared by shrinking glass tubing onto a tapered steel mandrel. Rotameters can be purchased in sets covering a wide range of gas velocity.

Suction Devices. Almost all techniques for sampling impurities in contaminated atmospheres require some device for drawing air through a sampling instrument.

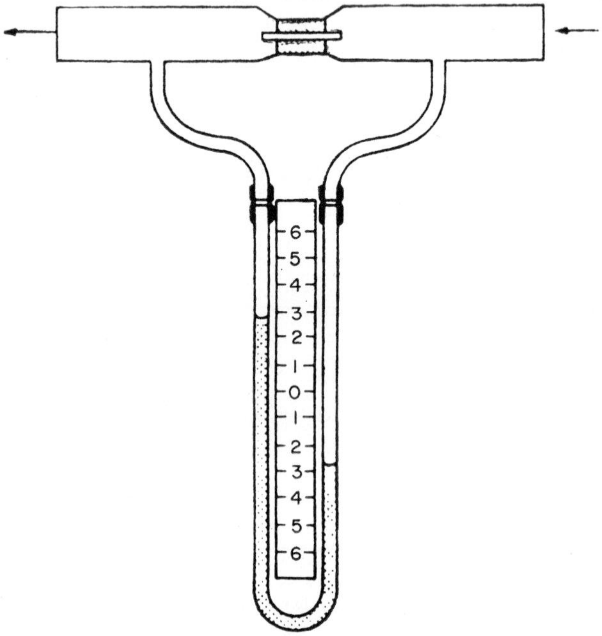

FIG. 10-7. Orifice flowmeter.

Many of these are quite simple, such as the rubber aspirator bulb on the Mine Safety Appliance carbon monoxide detector or the piston syringe on the Owens dust sampler. However, most of the sampling operations require that the air be drawn through the sampling device for a prolonged period of time. One of the more convenient pumps for carrying out such operations is a portable blower and vacuum pump which is sold by most scientific supply houses. Motor and pump are sealed into a single unit. The free air capacity is about 1 cfm and provides vacuum up to 27 in. of mercury. It is satisfactory for operating impactors, which require pressure drops across the jets up to ½ atm. The Gast Manufacturing Corp., Benton Harbor, Mich., sells a "portable sampler" which includes a flowmeter and filter-paper holder, in addition to a small air pump. The maximum capacity of the pump is about 0.6 cfm, and the maximum vacuum is about 15 in. of mercury. Both a-c and d-c motors are available for this unit. The Jordan Pump Co., Kansas City, Mo., manufactures a small pump which provides a vacuum of 22 in. of mercury.

Leach and coworkers (83) have developed a semiportable air-sampling system which consists of a vacuum-pressure pump, valve, rotameter, and manometer. The unit was constructed for aerosol sampling with the cascade impactor or with filter paper.

Water and gas ejectors are sometimes used. Carbon dioxide, from a cylinder of liquid carbon dioxide, can be used as an ejector fluid when the common sources of power are not available. A simple gas ejector for air sampling, similar to one designed by Wilson (151), is shown in Fig. 10-8.

Some operations require higher sampling speeds than are furnished by the pumps described above. Vacuum cleaners with the filtering unit removed are convenient for high-speed sampling. A typical air velocity for a vacuum cleaner is 30 cfm. The Staples Co., of Brooklyn, N.Y., manufactures a "high-volume air sampler" designed for use with various filters.

Hand pumps are convenient for many purposes. They are usually lighter than motor-driven pumps and of course do not require electricity. Such pumps are usually calibrated to deliver (or draw) a specified volume of air with each stroke of the handle or turn of the crank.

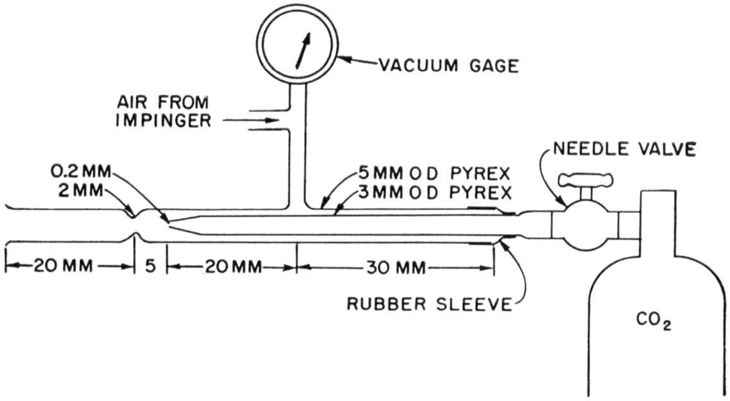

FIG. 10-8. Glass laboratory ejector. (*Courtesy of Manufacturing Chemists' Association.*)

Gas Samples. CONTINUOUS SAMPLERS. The growing interest in continuous samplers and analyzers was mentioned above. Since these devices involve sampling as well as analysis, a few of them will be briefly described.

Two methods are commonly used for continuously analyzing for sulfur dioxide. One involves the use of the well-known Thomas Autometer (137,138). The sulfur dioxide is absorbed in slightly acidulated distilled water containing hydrogen peroxide. The apparatus continuously measures the electrical conductivity of the resulting sulfuric acid solution. This analyzer is commercially available from the Leeds and Northrup Co., Philadelphia, Pa. The other method involves the use of the Titrilog (144). Air is bubbled through a bromide solution in the inner compartment of an electrolytic cell which has a "sensor" electrode and a "generator" electrode for production of bromine. Sulfur dioxide and many other oxidizable substances in the air are oxidized by the bromine. The momentary decrease in bromine concentration changes the potential at the sensor electrode and unbalances the system so that an electric current passes through the cell and more bromine is generated. The generating current is continuously recorded and is a measure of the amount of sulfur dioxide absorbed. The instrument is manufactured commercially by the Consolidated Engineering Co., Pasadena, Calif.

There is a marked correlation between ozone concentration and incidence of smog in Los Angeles County. Because of this correlation and also because of the interest of meteorologists in ozone concentrations, a number of continuous ozone analyzers have been described in the literature. Littman and Benoliel (90) have described such an analyzer. Ozone reacts with potassium iodide in neutral buffered solutions according to the equation

$$O_3 + 2KI + H_2O \rightarrow O_2 + I_2 + 2KOH$$

The amount of iodine liberated is measured photoelectrically and continuously recorded. Stair et al. (129) developed a continuous ozone analyzer which is based on the optical absorption characteristics in the Hartley and Huggins ultraviolet bands. The sensitivity is a few tenths of one part of ozone per 100 million parts of air.

The infrared gas analyzer is potentially a very useful tool for continuously determining the concentration of certain contaminants in the atmosphere. This device has been described by Luft (92) and others. It is commercially available from the Applied Physics Corp., Pasadena, Calif. The equipment consists essentially of a

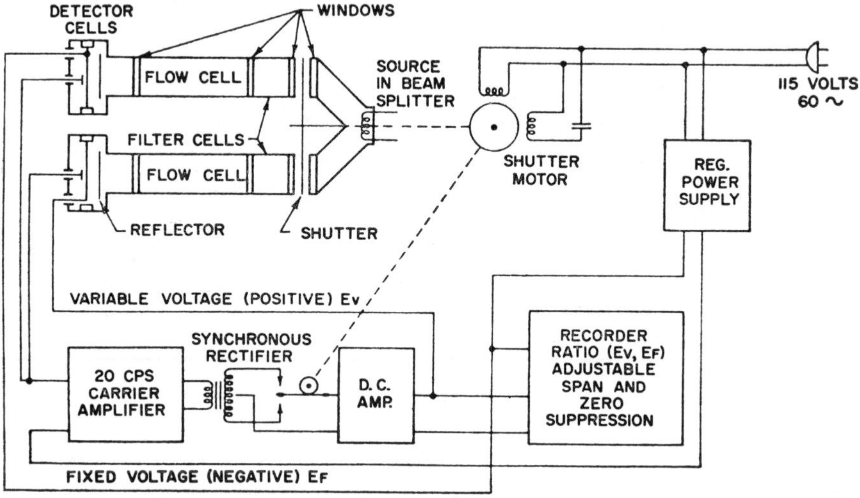

Fig. 10-9. Infrared analyzer.

source of infrared radiation, an identical twin-cell assembly, and two gas-filled microphone detector cells (Fig. 10-9). The infrared radiation is chopped to 20 cycles/sec by a rotating shutter. Contaminated air is drawn through one of the twin cells. The absorption of the intermittent radiation by the contaminants causes a rhythmical movement of the microphone membrane, which is translated to electrical energy and amplified. The two detector cells are connected in series opposition, and the ratio of their output is recorded by a modified Brown potentiometer. This arrangement is more sensitive than the customarily used thermopiles or bolometers and is capable of a certain degree of specificity. If, for example, the detector cells are filled with carbon dioxide, any carbon dioxide in the sample will cut out exactly the wavelength to which the detector is sensitive. Other gases, although also absorbing infrared radiation, will interfere only to the extent that their absorption spectra overlap that of carbon dioxide. Further desensitization to interfering compounds can be achieved by placing filter cells containing the interfering substances ahead of the sample cell.

Watkins and Gemmill (145) described an inexpensive infrared analyzer for the detection of carbon dioxide at concentrations of 0.01 to 0.07 per cent. The apparatus can be used for the analysis of other gases absorbing in the near infrared. Martin (97) has discussed various applications of the infrared analyzer.

The Stanford Research Institute has used the infrared analyzer for monitoring organic substances in the atmosphere (91). A useful sensitivity (about one scale division of the recorder per part per million of butane) was achieved by operating the instrument at the limit of sensitivity and incorporating a number of refinements. The major difficulty was the presence of moisture in the atmosphere. This was largely overcome by removing moisture from the air by passing it through a trap packed with stainless-steel helices and cooled with a mixture of dry ice and isopropanol. The entire instrument had to be maintained at a constant temperature to prevent a drift in the zero concentration readings.

The Liston-Becker Co. has designed an infrared analyzer for the continuous analysis of two components in a complex gas stream. Two detector cells are used in series, each charged with one of the two sensitizing gases. A single amplifier is used, and the results are indicated on a multipoint recorder.

DePiccolellis (38) investigated the continuous analysis of blast-furnace top gas, using a thermal-conductivity-type gas analyzer manufactured by Cambridge Instrument Co., New York. Good results were obtained, and the method was adopted by the Republic Steel Corp.

Changes in the surface potential of specially prepared plates have been used to indicate the presence of impurities in air streams (113). Concentrations as low as 0.1 ppm of polar vapors can be detected by this method. The technique is recommended for use in alarm systems and in certain manufacturing processes requiring a controlled atmosphere.

Chaikin, Glassbrook, and Parks (29) described an instrument for the continuous analysis of atmospheric fluorides. The method is based on measurement of the quenching by hydrogen fluoride of the fluorescence of filter-paper tape impregnated with magnesium oxinate. The impregnated tape is pulled through a sampling device where air is continuously drawn through the paper. The paper is then illuminated with ultraviolet light, and the emitted light is focused on a photocell. The intensity of the emitted light is continuously recorded.

A continuous hydrogen sulfide sampler is manufactured by the Research Appliance Company, Pittsburgh, Pa. It utilizes filter-paper tape impregnated with lead acetate. The air stream passes first through a filter to remove suspended solids and then through the impregnated tape, producing a dark spot on the tape when hydrogen sulfide is present in the air. The optical density of the spot is proportional to the amount of hydrogen sulfide in the air sample.

INTERMITTENT SAMPLING. The most common method for sampling gaseous contaminants in air is to scrub the air with some liquid. This method is often used in connection with both continuous and intermittent or batch sampling. Most of the batch scrubbers involve either drawing the air through sintered glass disks submerged in the liquid or impinging the air against the bottom of the vessel containing the liquid.

For collecting readily soluble gases, standard ASTM sulfur-determination absorbers serve as an efficient and convenient means for collecting specific gaseous components of an atmosphere when used with appropriate absorbing reagents (Fig. 10-10). In use, a measured volume (usually 50 to 100 ml) of reagent is added above the disk, and the gas sample is drawn into the U tube, up through the disk and the absorbing solution, and out through the entrainment trap. In passing through the porous disk, the gas is broken into fine bubbles and thus comes into intimate contact with the absorbing solution. The rate of sampling may be varied from 0.01 to 0.3 cfm. Groups of absorbers may be operated in parallel, to sample for several components simultaneously, or in series to ensure complete absorption. Typical of the compounds which may be collected by this means are hydrogen cyanide, ozone, nitrogen dioxide, hydrogen sulfide, sulfur dioxide, and ammonia.

Analysis of the absorber solution may be done colorimetrically, gravimetrically, polarographically, amperometrically, or by other appropriate means. By choosing the proper reagent, volume of sample, and analytical method, concentrations of gaseous components below 0.1 ppm may be measured.

For some analyses where the material being collected is present in trace amounts and the sensitivity of the analytical method is near the threshold of that which can be collected in the sampling train, it is desirable to decrease the ratio of the liquid used for scrubbing to the gas volume as compared with the above-mentioned bubbling train. This may be done by recirculating a small quantity of scrubbing liquid and using it in the form of a spray to contact the gas. Such a contactor may be made of glass. It is illustrated by a modification of the equipment used by Crabtree and Kemp (35) for the measurement of ozone in the atmosphere. This contactor consists of two bulbs and an air jet so arranged that, when vacuum is applied to the upper bulb, liquid in the lower bulb is drawn upward through a connecting tube while air is drawn through the jet into the lower bulb. A side arm on the jet enters the bottom of the upper bulb, and the aspirating action of the jet draws the liquid down and

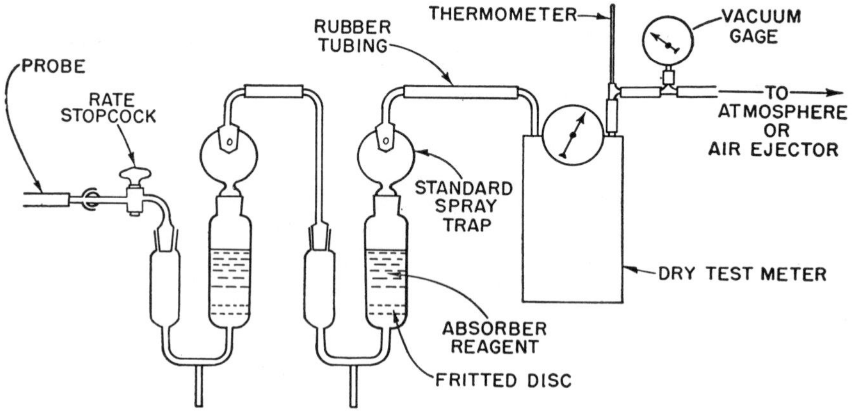

FIG. 10-10. Absorption train and gas-metering system for gas-sample collection. (*Courtesy of Manufacturing Chemists' Association.*)

discharges it in a fine spray into the lower bulb where it is again lifted by vacuum and the cycle repeated. The operation is shown in the diagrammatic sketch (Fig. 10-11). The illustration shows the contactor equipped with electrodes used for measuring atmospheric ozone or other soluble oxidizing agents (131).

In Fig. 10-12 another modification of a spray contactor is shown which may be refrigerated and is useful for scrubbing an atmosphere using hydrocarbons or other solvents for infrared or ultraviolet spectrographic analysis. Still another modification is the venturi scrubber (94).

A few recent published applications of absorption techniques might be mentioned. Urone and Druschel (141) collected chlorinated hydrocarbon vapors by drawing air through a "midget" impinger containing isooctane. Gage (48) developed an absorber of the impinger type with the jet modified so that the air entrains the absorbing liquid as it leaves the jet. An intimate mixture of air and liquid impinges on the base. Ammonia, o,o-diethyl-o-p-nitrophenyl thiophosphate mist, and potassium permanganate dust were sampled at rates up to 9 liters/min, with sampling efficiencies in excess of 90 per cent. Pearce and Schrenk (111) absorbed sulfur dioxide in an aqueous solution of iodine, potassium iodide, and starch in the flask of a conventional midget impinger. Young et al. (154) absorbed ethylene from an air stream in 2 ml of water saturated with butyl alcohol, prior to determining the concentration of ethylene manometrically. Gisclard and coworkers (49) used a device consisting of a frame which held side-arm test tubes and a 100-ml syringe to supply suction.

The collection of samples by adsorption on solid adsorbents such as activated charcoal, silica gel, or activated alumina offers many possibilities for concentrating contaminants from diluted atmospheres. The use of these methods should always

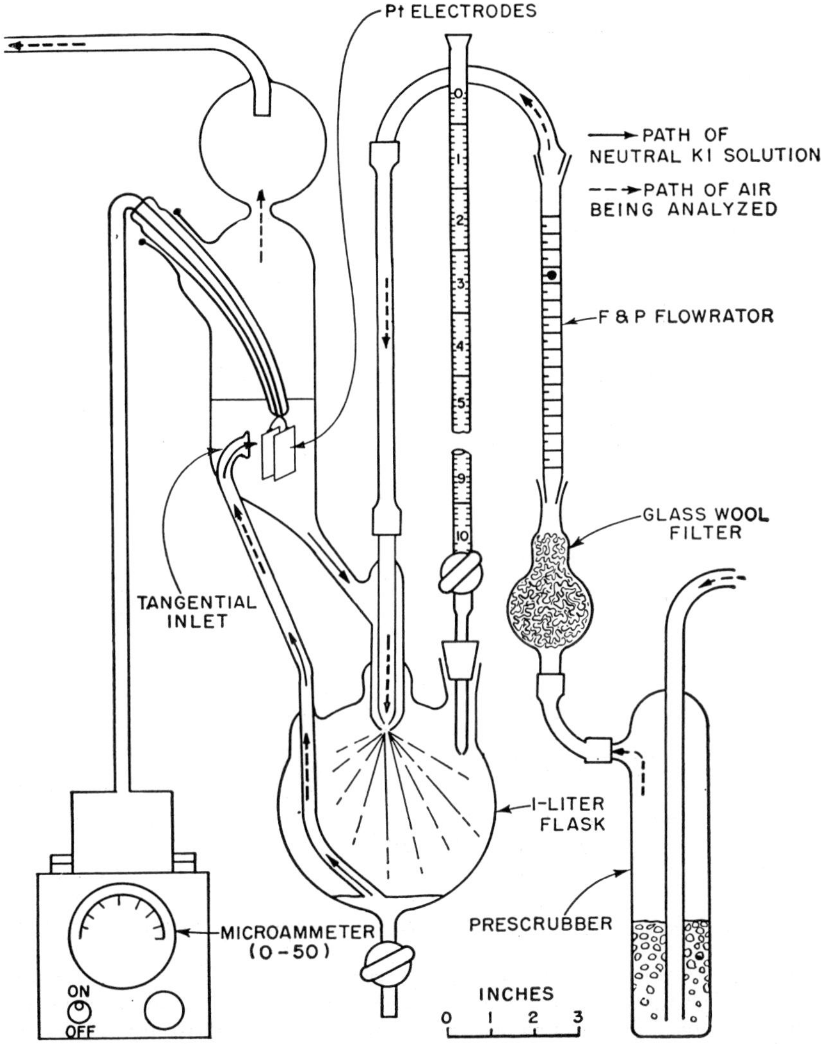

FIG. 10-11. Spray contactor. (*Courtesy of Manufacturing Chemists' Association.*)

be preceded by enough exploratory work to be certain that the material can be removed quantitatively and unaltered from the solid adsorbent. This is frequently quite difficult to accomplish. With proper precautions and selection of adsorbent, highly useful results have been obtained. Edgar and Paneth (41) collected ozone from an atmosphere containing 2 to 3 pphm by adsorption on specially prepared silica gel at the boiling point of liquid air. The ozone was removed by warming the silica gel to −120°C. Ozone has been identified in Los Angeles smog by collecting it on silica gel, desorbing it from the gel, and measuring the amount of desorbed ozone spectroscopically (23,132).

Stitt and Tomimatsu (135) studied the adsorption and recovery of traces of ethylene in air. Silica gel at 0°C almost completely removed ethylene from air at concentrations of 0.02 ppm or higher. The gel was heated to release the adsorbed ethylene.

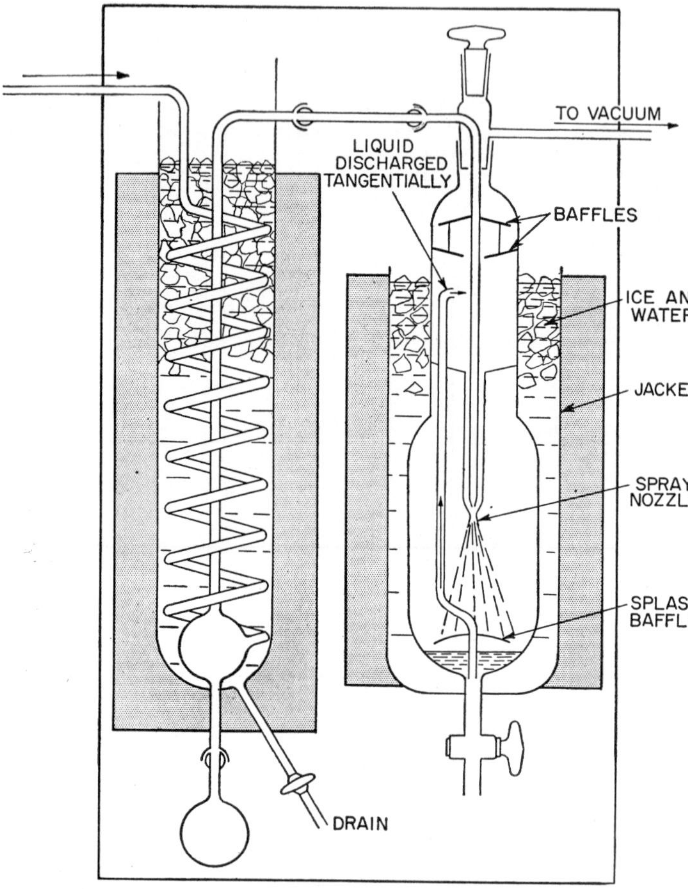

Fig. 10-12. Refrigerated spray contactor. (*Courtesy of Manufacturing Chemists' Association.*)

A number of workers have investigated the separation of gas and vapor mixtures by fractional desorption from an adsorbent. Claesson (33) gives a list of references. Two general methods have been employed. In the first, the gases are removed from the adsorbent by moving a heater along the column. This method has been applied by Turner (140), and an instrument employing this principle is stated to be available for the analysis of gas mixtures containing light hydrocarbons (139). In the second method, an inert carrier gas is passed through the column of adsorbent and plays a part similar to that played by the solvent in liquid chromatography. The various components of the mixture are desorbed one by one, and an "elution" curve is plotted. This method has been improved by Claesson (33), who employs a carrier gas containing a constant concentration of a vapor more strongly adsorbed than any of the components of the mixture.

A popular method for sampling and analyzing impurities in the atmosphere is to adsorb them on silica gel impregnated with a suitable indicator. The Mine Safety Appliance Co. carbon monoxide analyzer and sulfur dioxide analyzer are based on this principle. Patterson and Mellon (108) have investigated the determination of sulfur dioxide by color-changing gels. The sulfur dioxide is adsorbed on silica gel impregnated with ammonium vanadate. Sulfur dioxide in the air changes the color of the impregnated gel from yellow to green or blue. The limit of sensitivity is about 10 ppm.

Condensation on cold surfaces can often be used to concentrate volatile materials from contaminated atmospheres. The equipment for accomplishing the condensation has been made in several forms, depending upon the nature of the material being collected, the quantity, and the precision required.

Ice, solid carbon dioxide, liquid nitrogen, and liquid oxygen are commonly used as refrigerants. The temperature must be sufficiently low that the vapor pressure of the volatile material is very much less than the partial pressure of the vapor in the contaminated air. The equipment must be so constructed that water and/or carbon dioxide from the atmosphere does not cause stoppage of the gas flow before sampling is completed.

It has been pointed out by several investigators (24,71,121) that the quantitative collection of volatile materials by freezing methods is sometimes vitiated by formation of fine crystals in the freezing apparatus that blow on through and subsequently volatilize in the warmer parts of the apparatus. Cadle and coworkers (24) minimized this effect by packing the freeze-out traps with metallic packing. Shepherd and coworkers (121) employed a glass filter in the cold zone. The freeze-out technique has been useful in the collection and identification of odors (102).

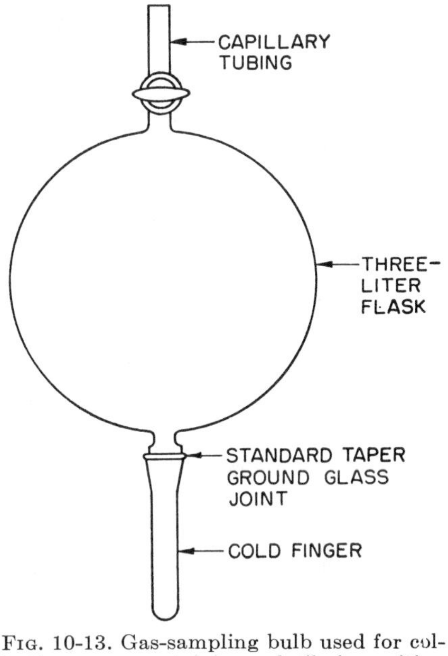

FIG. 10-13. Gas-sampling bulb used for collecting gases containing volatile impurities.

The sampling methods described above have involved almost immediate removal of the contaminants from the sampled air. However, under some circumstances it is advisable to collect the air at the sampling point and transport it to the laboratory for analysis. The usual procedure is to use some type of evacuated flask which is opened to obtain the sample. If the gases are hot and contain large amounts of water vapor, it may be necessary to close the flask before atmospheric pressure is reached. This will prevent condensation of water vapor when the flask cools. An interesting gas-sampling bulb for collecting gases from automobile "blow-by" prior to analyzing them with the mass spectrometer is shown in Fig. 10-13 (110). The bulb is evacuated, and the sample is obtained as usual through the capillary tubing. The tip of the cold finger is then immersed in a liquid nitrogen bath. The liquid nitrogen bath is gradually raised over a 6-hr period until the cold finger is completely immersed. When condensation has ceased, the remaining gases are pumped off, removing about 95 per cent of the original volume. The condensed materials are allowed to expand into a flask of smaller volume and are then introduced into the mass spectrometer. A system for sampling automobile exhaust gases prior to analyzing the gases with a mass spectrometer is shown in Fig. 10-14 (95). The exhaust gases are brought to a trailer in a double tube to guard against condensation of water. The sample is taken from the inner tube in a 250-ml sample bottle which has been evacuated. The pressure is allowed to rise to about 100 mm of mercury. The blower shown in Fig. 10-14 purges the line of unwanted exhaust.

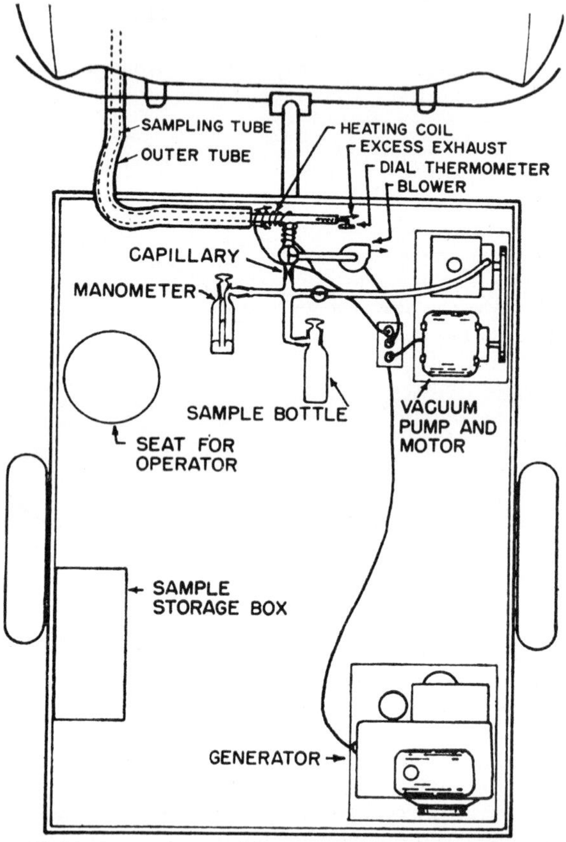

Fig. 10-14. Apparatus used for automobile-exhaust sampling.

Some gases are difficult to absorb in aqueous solutions. It is convenient to collect air containing such gases in an evacuated bottle. A 5-gal bottle enclosed in a can or a box is suitable. The reagent to be used is first placed in the bottle, and all parts of the bottle are contacted with the reagent to clean the surface of any adsorbed materials. A portion of the solution is then withdrawn for a blank analysis. Pressure in the bottle is then reduced to 200 mm of mercury or less, and the pressure is recorded. The sampling line to the bottle is flushed with the sample gas, and while the gas is flowing, it is connected to the bottle. The gas pressure in the bottle is then allowed to reach atmospheric pressure. The bottle is disconnected and shaken with the reagent intermittently until the gas is absorbed. For nitric oxide this requires approximately 2 hr. The gas volume is then calculated from the volume of the bottle and the pressure change (31).

Particulate Matter. FILTRATION. In the filtration of aerosols through porous media, several mechanisms come into play. The relative importance of these depends primarily upon particle size. Table 10-2 shows the mechanisms involved and the size range over which each is most effective (123). Very small particles will be removed effectively by the diffusion mechanism and large particles by inertial or screening effects, depending upon the ratio of particle size to pore size. Electrostatic attraction may play an important part in the filtration of very small aerosol particles.

Table 10-2. Mechanisms Involved in Collection of Aerosols by Filtration
(After Silverman)

Mechanism	*Particle Diameter, μ*
Diffusion	0.3
Interception	1 (depends on fiber size)
Inertial (impingement)	0.5
Electrostatic attraction	0.01–1.0
Sieving	0.1 (depends on pore size)

"Molecular" or membrane filters are very useful for the collection of many types of aerosols. The filters are composed of a mixture of cellulose nitrate and acetate. They filter by a screening action rather than by impaction; so the particles are collected on the surface. The collected particles are visible under the microscope, using incident illumination. Fine particles penetrate into most other filtering media and cannot be observed in this manner. Membrane filters are soluble in many organic liquids, such as acetone and the Cellosolves. The filters can be dissolved in an appropriate solvent and the particle size distribution of the collected particles determined by sedimentation (21). The filter material can be dissolved away as a step in the preparation of electron micrographs of particles collected on the filters (45,72).

Membrane filters become transparent when mounted in a medium having approximately the same refractive index as the filters (about 1.50). Transparent mounts of the filters on microscope slides can be prepared by placing a drop of cedarwood oil on the slide and laying a small portion of the filter over the drop. A cover glass placed over the filter completes the mount. The filters become transparent when mounted in polymethylmethacrylate. A technique for collecting, mounting, and sectioning airborne particulate material which takes advantage of this property of membrane filters has been described by Cadle and his coworkers (25). General discussions of the use of membrane filters for sampling dusts and aerosols have been written by Goetz (50) and by First and Silverman (45).

Membrane filters are commercially available as three types: HA (hydrosol assay), AA (aerosol assay), and AP (aerosol protective). Estimated pore sizes of the filters are shown in Table 10-3. Filters which have been dyed black are also commercially available. These are useful for studying the particles by incident illumination, as when making fluorescent particle tracer studies (62,112). Membrane filters are sold under the trade names Millipore and Isopore filters by the Lovell Chemical Co., Watertown, Mass., and the A. G. Chemical Co., Pasadena, Calif., respectively. Auxiliary equipment such as filter holders is sold by the Leslie R. Burt Co., Arcadia, Calif.

Table 10-3. Pore Size of Membrane Filters

Filter Type	*Approximate Pore Size, μ*
HA	0.1–0.3
AA	0.5–0.7
AP	1.2–1.6

Filter samplers designed to use membrane filters are manufactured by the Ralph M. Parsons Co., Pasadena, Calif. The air-flow rate can be varied between 5 and 12 liters/min. The vacuum pump is battery-powered.

Filters consisting of soluble crystals such as sugar or salicylic acid have been used for many years. When sampling is completed, the filter is dissolved in a suitable solvent. The resulting suspension can be studied microscopically or by sedimentation techniques. Holt (63,64,65) has suggested the use of naphthalene filters for determining the mass concentration of dust in industrial atmospheres. A measured volume of air is drawn through a naphthalene pad. The pad is transferred to a weighed platinum or aluminum dish, and the naphthalene is evaporated by heating. The dust is then weighed.

An air sampler employing glass wool was described by Sehl and Havens (119). The sampler is recommended for collecting toxic particulate material such as lead oxide. Blasewitz and coworkers (12) have reported considerable experimental data concerning the permeability and efficiency of glass-fiber filters.

Filters made of certain plastics and resins, or of various fibers coated with resins, develop sufficient electrical charge to attract particles of dust. Filters designed to operate in this manner have been described by Thomas (136) and by Van Orman and Endres (142). Other materials frequently used as filter bodies are porous paper; porous ceramics; cotton, wool, and glass cloths; and packed asbestos mats. For high efficiencies some of the recently developed papers containing asbestos permit collection of particles down to 0.1μ in diameter (128). Where an ash-free medium is needed for air analysis, certain grades of chemical filter paper can be used. Efficiencies for recovery of submicron particles are low for these papers but improve with increasing air velocity.

The collection of particulate matter in large amounts can be accomplished by using a high-volume sampler developed at the Harvard School of Public Health (124). This filter collects a large volume in a short time because of its rapid sampling rate, 50 to 70 cfm. Enough material can be obtained for a weight analysis on normal city air in an hour or less. Special techniques are used in weighing.

For duct and stack analyses, porous cups small enough to fit inside sampling probes are useful (104).

For analyses of sulfuric acid mists, a series of four Gooch crucibles packed with asbestos mats is recommended (130). This method has the disadvantages of a slow sampling rate and difficulties due to cracking of the asbestos mat, which permits escape of the aerosol particles. Mader, Hamming, and Bellin (93) recommend the use of specially prepared Whatman No. 4 filter paper, clamped between two ground-glass joints. Sampling rates as high as 1 cfm were used, and recoveries of acid as high as 91 to 97 per cent were reported. The sulfuric acid may subsequently be removed from the filter paper by digestion with water.

Numerous filtering operations have been placed on a continuous basis. One of the commonest methods for estimating the concentration of soot in city air involves drawing the air through filter papers. The darkening of the paper indicates the concentration of contaminant. Hall (54) has described an automatic, semicontinuous "auto-sampler." One sample is taken each hour over a 24-hr period. The darkening is measured by the light reflected from the deposit and is recorded automatically. A recorder which operates on somewhat the same principle was developed by von Brand (143). This device employs a continuously moving tape for collecting fine particles. The tape is compared with a standard color chart. Hemeon, Haines, and Ide (59) have discussed in considerable detail the determination of smoke concentrations by filter-paper samplers.

Continuous filtering devices are used for sampling and measuring radioactive atmospheric pollution (56,103).

Impaction. A number of the simplest and most useful methods for sampling airborne particles are based on the fact that the particles in a moving air stream have a greater momentum than the volume of air displaced by the particles. When a rapidly flowing aerosol encounters an obstacle, the direction of flow changes near the obstacle. The particles are driven toward the surface by the resulting centrifugal force. If this force is sufficiently great, many or all of the particles reach the surface and cling to it. This is the principle of operation of the various "impactors" and "impingers."

Cascade impactors are devices for simultaneously collecting particles and classifying them into several size ranges. They consist of sets of jets and slides arranged in series (99,127). A drawing of an impactor constructed of plastic is shown in Fig. 10-15. It is built from three pieces of machined Lucite. The steel pieces forming the jets are cemented in place and the Lucite pieces cemented together. The jet sizes are 8, 3, and 1.5 mils. This impactor is especially useful when the particulate material is to be examined with an electron microscope. The particles must be collected on small collodion-coated screens mounted on the microscope slides. Since the impactor is transparent, the screens can easily be placed directly beneath the jets.

The jets through which the aerosol is drawn become progressively smaller, and the speed of air through the jets becomes progressively greater. As a result, the particles are classified according to size, the largest ones being collected on the slide opposite the first (largest) jet and the smallest particles on the slide opposite the last jet. Motor or hand-driven vacuum pumps are used to draw the aerosol through the jets.

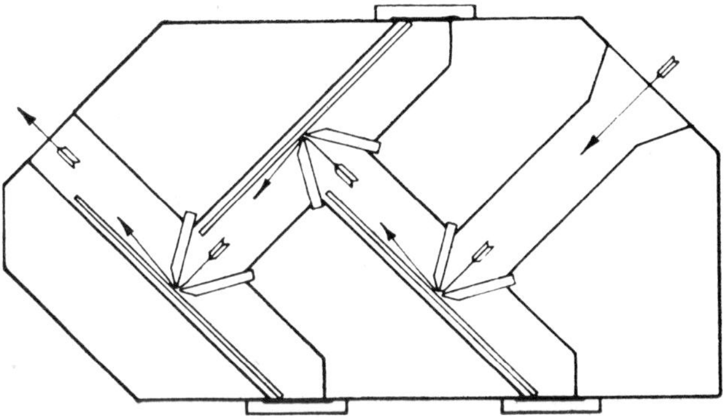

FIG. 10-15. Drawing of plastic jet impactor.

When the pressure drop across the last jet is more than ½ atm, the velocity of air through this jet is equal to the speed of sound.

Impactors are capable of very high efficiencies. An efficiency of 100 per cent for particles as small as 0.6μ has been obtained.

Ranz and Wong (115) have made a theoretical and experimental study of the collection of aerosol particles by impaction. They proposed the use of the dimensionless factor ψ to characterize the inertial method of collection, where

$\psi = C_p U_0 D_p^2 / 18 \mu D_c$

C_p = empirical correction factor for resistance of a gas to movement of small particles, dimensionless $C_p = 1 + (2L/D_p)[1.23 + 0.4 \exp(-0.44 D_p/L)]$ for $0.1 < (2L/D_p) < 134$

D_c = width or diameter of aerosol jet, cm

D_p = diameter of aerosol particle or diameter of equivalent spherical particle in case of irregular particles, cm

L = mean free path of gas molecule, cm

U_0 = initial or undisturbed velocity of aerosol stream, cm/sec

μ = gas viscosity, poise

A theoretical relationship between ψ and the efficiency of impaction was developed. This relationship is shown in graphical form in Fig. 10-16. Ranz and Wong also determined impaction efficiency experimentally. The results are shown in Fig. 10-17. The theoretical and experimental efficiency agreed very well. Figure 10-17 can be used to predict the behavior of existing equipment for a given aerosol and can also form the basis for the development of new impactor designs.

Before use, an impactor should be calibrated to determine the size of particles which are sufficiently small that they are collected with only 50 per cent efficiency

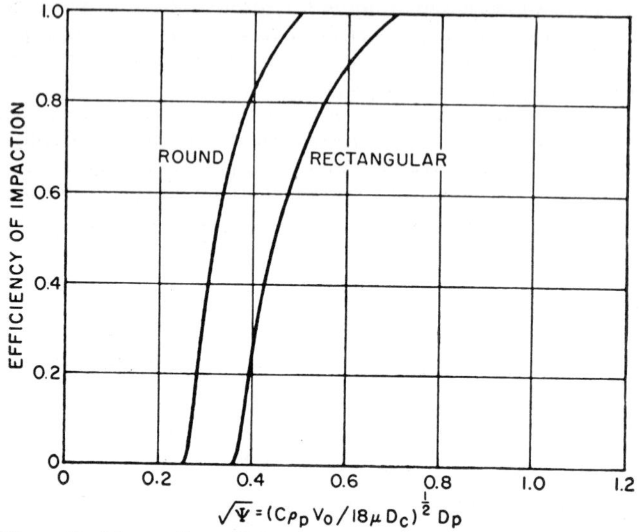

FIG. 10-16. Theoretical impaction efficiencies of aerosol jets. (*Reproduced by permission from Industrial and Engineering Chemistry.*)

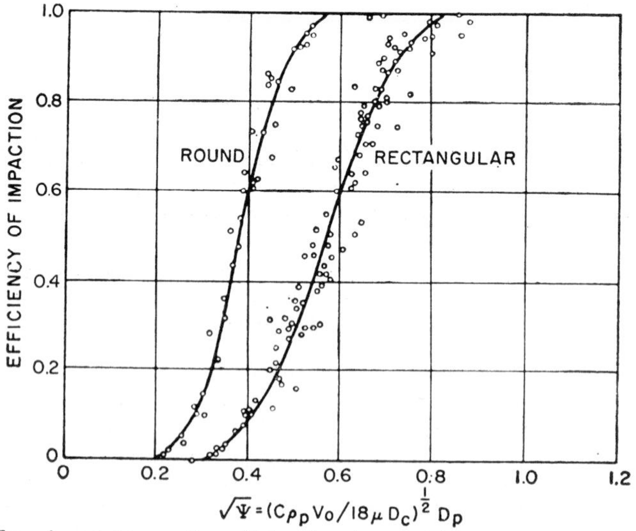

FIG. 10-17. Experimental impaction efficiencies of aerosol jets. (*Reproduced by permission from Industrial and Engineering Chemistry.*)

at each stage. After the dust is collected, the weight of material collected at each stage is determined and the results are plotted as cumulative weight-distribution curves. Ranz and Wong, in another publication (116), have recommended the operation of such impactors with the stages in parallel instead of in series.

Levine and Kleinknecht (87) used a cascade impactor in connection with their studies of the sizes of droplets in clouds. Sampling was effected from an airplane. Droplets entering the impactor impinged on slides coated with magnesium oxide. The relation between the size of the droplet impressions and the droplet size was evaluated so that droplet-size distributions could be determined. A five-stage cascade impactor has been designed to collect particles having a wide size range (0.2 to 100μ) (147). A single-stage impactor, such as that shown in Fig. 10-18, is often useful because of its simplicity of operation.

A continuous sampling device using a single jet can be made by installing a jet facing a rotating drum driven by clockwork, the whole apparatus being placed in a vacuum chamber. A 2-qt pressure cooker makes a convenient vacuum chamber. The jet may be caused to impinge upon white paper, leaving black or gray marks, the shaded intensity of which will correspond to the concentration of the colored

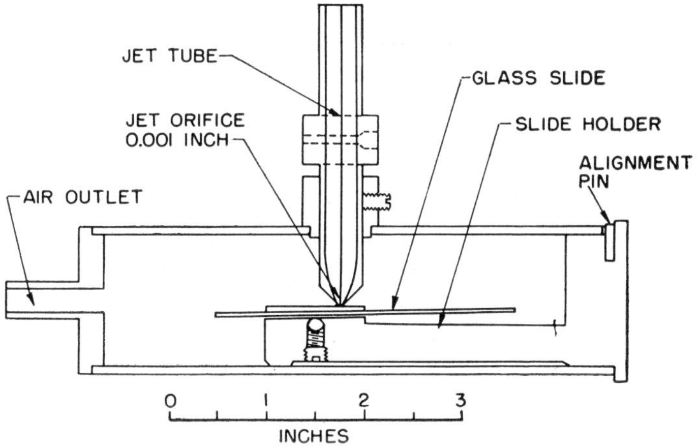

FIG. 10-18. Single-stage impactor. (*Courtesy of Manufacturing Chemists' Association.*)

material in the atmosphere. It may be used with impregnated paper using, for example, bromcresol green for indicating the comparative concentrations of sulfuric acid or other particulate acid material in the atmosphere.

A commercial instrument containing a single jet, pump, microscope, and light source is manufactured by Bausch and Lomb Optical Company, Rochester, N.Y. Twelve samples may be taken without exposing the collecting plate to contamination. This device is rather inefficient, but it has been successfully used for comparative purposes. The Greenburg-Smith impinger employs a high-velocity air jet that impinges on a glass plate immersed in water or alcohol. The action of the nozzle causes the particles to be collected in the surrounding fluid. Various factors affecting the efficiency of such instruments were investigated quantitatively by Katz and coworkers (73) and later by Hatch, Warren, and Drinker (57).

Optimum values for practical operation in industrial plants have been incorporated in the modern instrument which is shown in Fig. 10-19. The sampling rate is 1 cfm. This impinger is highly useful for insoluble dusts larger than 2μ in diameter. The dust normally remains dispersed (if the proper liquid is used) and can be transferred to a microscope slide.

The midget Greenburg-Smith impinger is similar to the instrument just described, but the flow rate is 2.8 liters of air per minute through an orifice 1 mm in diameter. A manually operated pump is usually used.

It should be mentioned that May (99) used the term "impactor" to indicate that velocities through the cascade impactor are slower than through the "impingers."

However, most cascade impactors are now operated using high air velocities, and this differentiation has lost much of its significance.

Fine fibers as collecting devices have been used by Dessens (39), Bricard (15), and Brun (20). Dessens obtained his fibers by causing the smallest spider he could find to traverse repeatedly a small aperture, leaving her web behind to form a grid. Brun employed what appears to be a more controllable method by creating fine fibers from an aniline solution of Plexiglas. A drop of the aniline solution was placed on a piece of paper of a porosity determined by trial. A rubber eraser or the tip of the finger was touched to the solution, and, if the right stage of evaporation had been obtained, withdrawal of the contacting object produced many fine fibers less than a micron in diameter. These were then transferred to a small loop, and the process was repeated until a parallel grid of fibers had been produced. With this as a collecting device in an air stream, the authors state that submicron particles may be sampled with high efficiency.

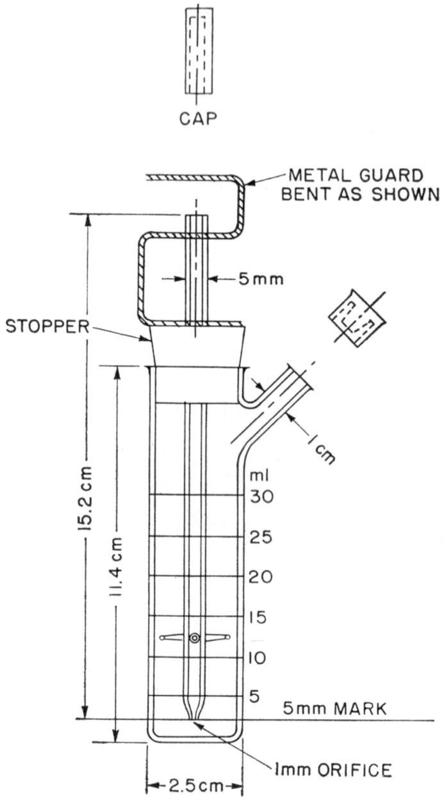

FIG. 10-19. Greenburg-Smith impinger. (*Courtesy of Manufacturing Chemists' Association.*)

THERMAL PRECIPITATION. A thermal precipitator is a device which takes advantage of the repulsion of dust particles from a hot surface. The hot surface sets up a temperature gradient in the air. Particles suspended in the air undergo a more violent bombardment by air molecules on the side toward the hot surface than on the opposite side and are repelled from the hot surface. Cawood (27) developed the following theoretical relation for thermal repulsion:

$$t = \frac{12 \frac{dT}{dx} X^2 + 2T_g X}{P\lambda d \left(\frac{dT}{dx}\right)\left(1 + \frac{2A\lambda}{d}\right)} \qquad (10\text{-}2)$$

where t = time (sec) required for a particle to move a distance X (cm) in a thermal gradient

μ = viscosity of gas, poises

dT/dx = temperature gradient, °C/cm

d = diameter of the particle, cm

T_g = temperature of gas, °K

P = gas pressure, dynes/sq cm

λ = mean free path of gas molecules, cm

A = a constant, dimensionless

The second term in the last factor in the denominator is the Cunningham correction to Stokes' law. It corrects for "slippage" between molecules and can be neglected for particles greater than about 3μ in diameter.

Epstein (43) derived an equation for thermal repulsion which takes into account the heat conduction by the particle, which is assumed to be a sphere:

$$F_t = -9\pi a \frac{k_a}{2k_a + k_i} \frac{\mu^2}{\rho T_g} \frac{dT}{dx} \tag{10-3}$$

where F_t = thermal force, dynes
a = radius of particle
k_a = thermal conductivity of gas, cal/(sec)(cm)(°K)
k_i = thermal conductivity of sphere, cal/(sec)(cm)(°K)
ρ = gas density, g/cu cm
$\frac{dT}{dx}$ = temperature gradient, °C/cm

Saxton and Ranz (117) undertook an experimental investigation of the thermal forces on an aerosol in a temperature gradient. Their experimental results agreed much more closely with Epstein's equation than with Cawood's.

Thermal precipitators have very high collection efficiencies, especially for particles less than 1μ in size. They have the disadvantage that they generally sample at a very slow rate. For instance, the Casella instrument, sold in this country by the Mine Safety Appliance Co., samples at about 4 ml/min. Figure 10-20 is a diagram of this instrument.

Joseph D. Ficklen III, Pasadena, Calif., manufactures several types of thermal precipitators. The warmer body in the case of the continuous and oscillating precipitators is a nichrome resistance wire. The "gravimetric precipitator" uses an electrically heated plate, which permits more rapid sampling. The manufacturer recommends a sampling rate of 600 ml/min with the gravimetric precipitator.

A thermal precipitator designed specifically for the gravimetric estimation of solid particles in flue gases was described by Bredl and Grieve (14). The flue gases are drawn through an annular gap formed by a truncated conical core. This core is surrounded by a cylindrical sleeve bisected along its length. The core contains a heating element. The split sleeve makes a sliding fit in the water-cooled body of the thermal precipitator. The temperature of the water must be above the flue-gas dew point. The two halves of the split sleeve are lined with aluminum foil on which the particles precipitate.

The aluminum liners are weighed before and after the sampling; the weight of material collected is obtained by difference. Sampling rates as high as about 0.1 cfm were used by Bredl and Grieve. The maximum ratio of sampling rate to temperature difference for all the deposit from flue gases to be collected on the 10-cm length of aluminum foil was 0.043 cu ft/(hr)(°C).

When aerosol passes between a hot wire and a cold surface, the particles are precipitated according to size in the direction of air travel. If one wishes to determine the particle-size distribution microscopically, the entire deposit must be examined. Laskin (81) among others avoided this difficulty by providing the plates with an oscillating mechanism so that the particles are spread over a larger area with more uniform distribution. Cember, Hatch, and Watson (28) accomplished the same thing by rotating the collection plate. According to Cember and his coworkers, the instrument is easier to make and has smaller dimensions than the oscillating variety.

Kethley and coworkers (75) have designed a thermal precipitator for collecting bacteria from the atmosphere. The hot surface is a metal disk which is a commercial heating element. Particles are collected on cover slips which are placed on a water-cooled chamber. Preliminary tests indicated that the precipitator is effective at flow rates as high as 400 ml/min and that particles are uniformly precipitated over the surface of the cover slip on which they are collected.

Electrostatic Precipitation. Electrostatic precipitators are often used for sampling the particulate material in contaminated atmospheres. They have the advantage over many types of equipment for sampling particulate material that the air-flow rates are large and weighable amounts of material can be collected rapidly

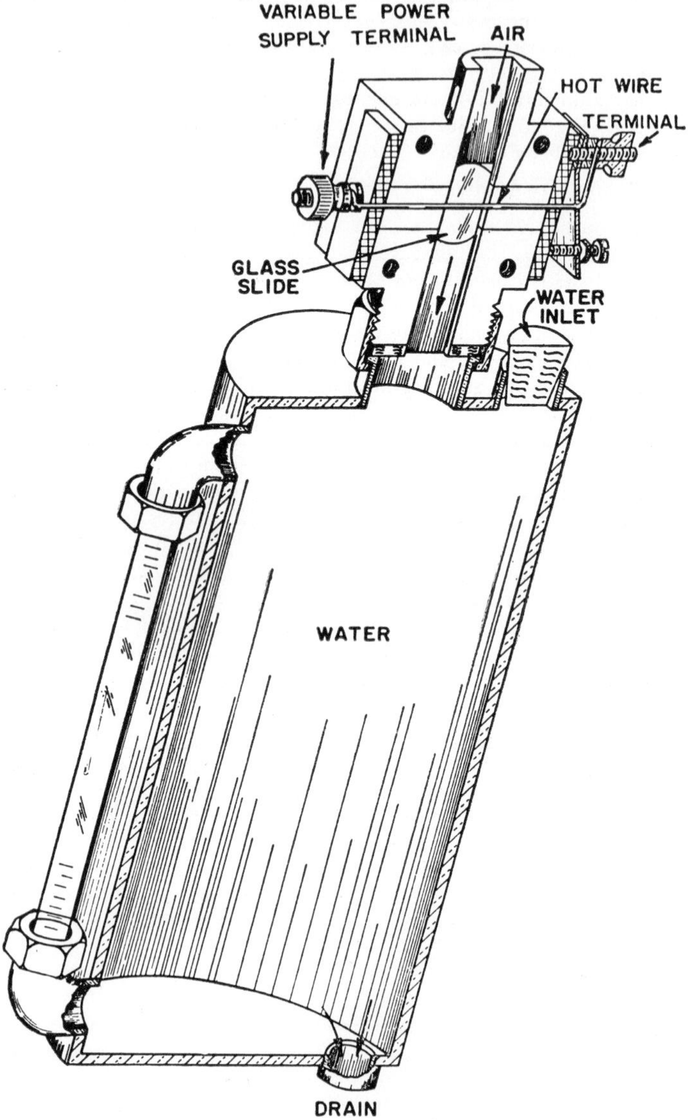

Fig. 10-20. Thermal precipitator.

The weight of the collecting electrode can be made small to permit accurate weighing of the collected materials.

Electrostatic precipitators have the disadvantage that sparking may occur. They should not be used in explosive atmospheres. The particles are collected in such a manner that they are not readily examined by microscopic means for particle-size determination because of their agglomeration and uneven distribution on the collecting medium.

Probably the first published observations of electrical precipitators were those of Beccaria (8), who noted certain phenomena of electric discharges in smoke as early as 1771. Since that time many shapes, forms, and sizes of electrical precipitators have been built for laboratory and commercial sampling. Reference to only a few is necessary (19,40,74,77,118) to illustrate the types used. For most field and laboratory work, a commercial model such as that manufactured by the Mine Safety Appliance Co., Pittsburgh, Pa., is adequate. A typical all-glass model which may be constructed in the laboratory is described by Wilner (150).

Goodale, Carder, and Evans (52) recently developed an electrostatic precipitator for isokinetically sampling dust particles from an airplane. The sampler is designed to achieve high collection efficiency over a wide range of altitudes and air velocities. It is mounted on gimbals and is vaned to maintain its axis parallel to air-flow lines. The particles are collected on glass slides, electron-microscope screens, and selected metallic surfaces.

Another recently described electrostatic precipitator is portable and operates from either 110 volts alternating current or 6 volts direct current (68). The device collects samples on either glass slides or metal plates. It has a lightweight sampling head with built-in flowmeter. The complete unit weighs less than 17 lb.

A useful review of electrostatic methods of sampling particulate material has been written by Barnes (6). The history of electrostatic precipitation is briefly presented, and a number of electrostatic precipitators, both alternating-current and direct-current, are described. The theory of electrostatic precipitation has recently been reviewed by Davies (37). He concludes that, given a reasonable length of gas path, it should be possible to collect completely particles over 0.5μ in radius. Very minute particles may also be readily removed. However, Davies suggests that the collection efficiency of particles in the region of 0.1μ radius may be low because of the difficulty of charging and the low limiting charge.

Settling Techniques. One of the simplest and most effective ways of collecting particulate material is to allow it to settle out on microscope slides. The method is effective for particles as small as about 1.0μ and possibly for even smaller particles. The settling process is usually carried out in a small chamber containing the aerosol.

Many sizes and shapes of settling chambers have been used. A convenient chamber used at the Stanford Research Institute consists of a wooden box about 1.5 ft square with a door occupying all of one side. The interior is varnished so that particles collected on the walls can easily be wiped off. Microscope slides are taped to the floor, ceiling, and walls of the box. About four slides are usually taped on each surface, making a total of 24 slides. If the air to be sampled occupies a very large volume, the box can be filled by passing it manually back and forth through the air or the air can be gently fanned into the box. The particles are generally allowed to settle overnight, that is, for about 16 hr.

Often nearly as many particles are collected on the ceiling and walls of the box as on the floor. There are at least two causes for this. Submicron particles may diffuse to the walls faster than they settle. Convection currents may bring the particles close to the walls where they are intercepted by the surfaces or are impinged on the surfaces by inertial effects. Convection currents are usually set up by temperature differences within the box; these can be minimized by thermally insulating the box. Examination of the slides is, of course, greatly simplified if most of the particles are collected on a few slides on the bottom of the box.

Particles collected on microscope slides by settling techniques are often in a more or less agglomerated form. The agglomerates were not necessarily present in the contaminated atmosphere when originally introduced into the settling chamber but may have occurred during the settling process. The extent of the agglomeration which occurred during settling can be estimated by using equations developed in Sec. 3.

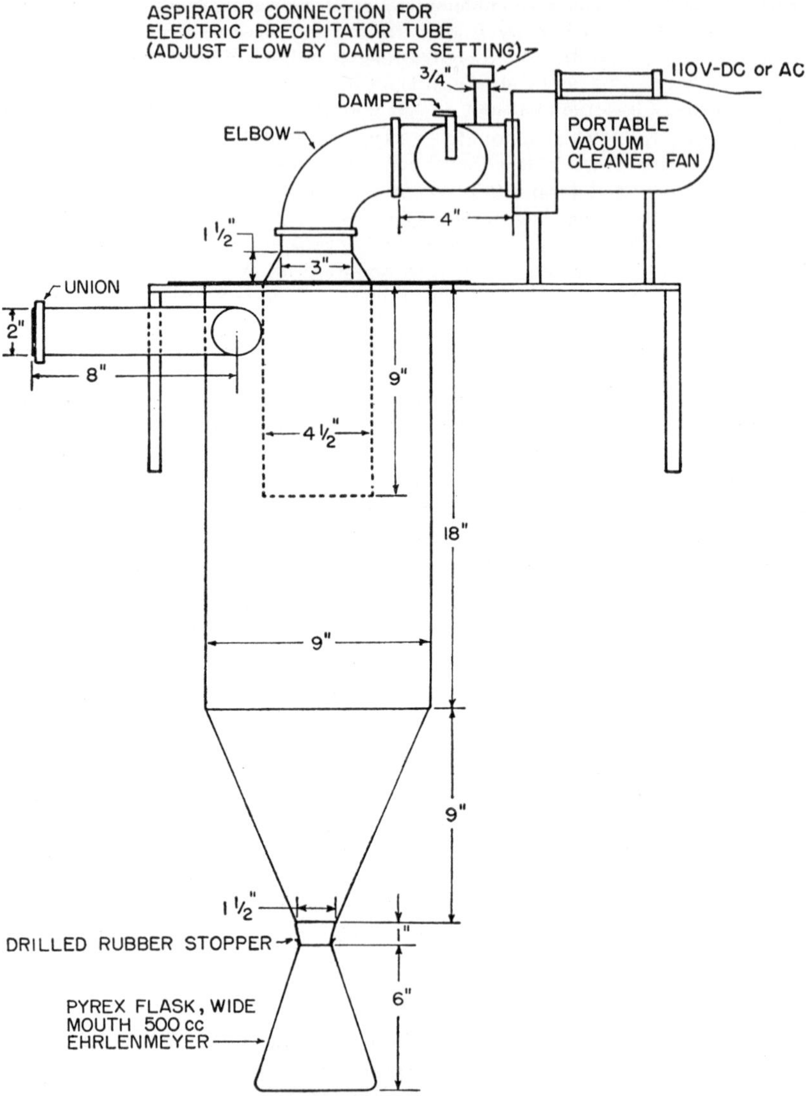

FIG. 10-21. Cyclone sampler. (*Courtesy of the Combustion Publishing Co.*)

Clean microscope slides are often used to collect solid particles by the sedimentation method. If the collected material is to be examined using an electron microscope, small collodion-covered screens are mounted on the slides. Specially prepared slides are sometimes useful for collecting droplets. Hydrophobic coatings such as petroleum jelly help prevent spreading of aqueous droplets and can be used to preserve the droplets.

Droplets greater than 5μ in diameter sometimes are collected on slides which are coated with carbon by holding them over a benzene flame or coated with magnesium oxide by holding them over a burning ribbon of magnesium. The droplets penetrate the coating, forming patterns which are related to the original size of the droplets.

The collection of samples by sedimentation is widely used for the collection of pollen grains from the atmosphere, employing the so-called 24-hr gravity slide as developed by the Committee on National Pollen Survey of the American Academy of Allergy. The equipment consists of a 1- by 3-in. petroleum-jelly-coated glass slide placed horizontally in a metal shelter. The shelter is made from two 9-in. stainless-steel disks set horizontally 3 in. apart and held by three strut separators. The glass slide is placed in a holder and set 1 in. above the lower disk. The shelter is carried on a 30-in. rod above a supporting base usually located on top of a tall building. Slides are stained and pollen grains identified and counted under a microscope. Concentrations are reported as number of grains per square centimeter of slide collected in 24 hr. The method is simple and inexpensive but is far from satisfactory for quantitative estimation of pollen concentrations. Other methods of collecting pollen are discussed by Fletcher and Velz (47).

The dust-jar collection method is frequently used. It is simple and inexpensive and is claimed to give comparative measures of the amount of dustfall in a particular area over a period of time. This method consists of placing a jar or a bucket at the location of interest and weighing the amount of solid materials collected over a known period of time. Sometimes water is placed in the bottom of the jar to retain the samples (120,122).

Instead of jars, sometimes 6-in.-square glass plates, lightly greased and placed at strategic locations, are used to estimate dustfall (109). Petroleum jelly is satisfactory for most purposes to hold particles on the plates. Again the results are primarily useful for comparative purposes.

Cyclones. Cyclones consist essentially of chambers having circular cross sections. Particle-laden gases moving at a moderately high velocity are led tangentially into the chambers. The resulting whirling motion produces centrifugal forces which bring the particles to the walls from which they usually drop to a collection chamber. Cyclones have the advantages that they are simple to construct, can be built of temperature-resistant materials, and collect samples rapidly. They have the disadvantage that they generally collect particles less than 5μ in diameter very inefficiently. Cyclones are usually used for sampling the particulate material in stacks, ducts, and similar systems.

Stern (133) built and calibrated the cyclone sampler shown in Fig. 10-21. Similar devices may be purchased from the Thermix Corporation, Greenwich, Conn. Stern claimed nearly 100 per cent efficiency (on a weight basis) for materials such as fly ash. However, he did not estimate the lower size limit for efficient collection of the particles.

Typical cyclone designs have been discussed by Lapple (78) and Lapple and Shepherd (79).

10.2 VEGETATION AND SOIL-SAMPLING PROCEDURES

10.2.1 Introduction

Purpose and Requirements of Sample. The purpose of any sampling program is to obtain fractions of material which are truly representative of the entire category being sampled. By the projection of the information obtained from these fractions, an approximation may be made of properties of the entire category. Since the results of the final approximation will depend upon the reliability of the data obtained from the samples, the importance of truly representative samples cannot be overestimated.

Samples should be collected with knowledge of how the results are to be used and of the accuracy required. These requirements are subject to many variations. All too frequently, samples are collected that will not provide the information that is needed, regardless of how accurately analytical techniques are applied. The range of variability of samples is always important and sometimes becomes quite large when dealing with vegetation. For instance, growing plants show great selectivity in ability to absorb and concentrate gaseous impurities from the atmosphere with regard to (1) various parts of the same plant, (2) different varieties within a species, and (3) different species. In apparently identical plants, uptake of pollutant may vary with such factors as intensity or duration of light, temperature, relative humidity, soil moisture or composition, age, and so forth. It is, therefore, always necessary to take sufficient samples over an adequate pattern so that reliable statistical measures of variability may be calculated. The reader is referred to the article on Statistical Approach to Obtaining Representative Samples, page 10-2, for a discussion of the use of statistics as a tool in sampling problems.

Equipment for Sampling. The equipment required for sampling vegetation is relatively simple and consists essentially of maps, shears, containers, and notebook and pencils. Exploratory work will have located the boundaries of the area to be sampled, and maps covering this area should be obtained or made. If available, U.S. Geological Survey quadrangle sheets are excellent because of their large working scale. The number and size of sample containers, usually paper bags, depend upon the number of sites and type of vegetation, respectively, but in the latter respect, the bags should be sufficiently large to hold 75- to 100-g samples. In addition to the bags, a large box or a square of canvas should be provided for mixing comminuted plant tissue, prior to taking the field sample. The box, which is preferable for plant-sample mixing, should be approximately cubical in shape, and about 18 in. on each edge, although the size and shape are not critical. Sheep shears appear to be most satisfactory for cutting and comminuting the samples, although for some materials a paper cutter may serve better for the comminution. Good-quality shears are important in obtaining efficient operation from a field crew.

For collecting soil samples one needs an ordinary 3-in. fence-post auger about 4 ft long, No. 4 paper bags, a 1-sq-ft sieve of 16-gauge screen, a garden trowel, and a piece of heavy canvas about 2 ft square.

Preservation of Sample. Fresh plant samples are subject to enzyme action, such as fermentation or autolysis, which may significantly alter their composition. Mold growth on samples can also introduce errors. If the samples are to be analyzed within a few days, they may be kept simply in glass jars with screw tops. If the analyses will be delayed longer, the samples should be kept in a deep freeze. If this has been done, analysis should be run shortly after thawing the samples, as they are then particularly susceptible to decomposition.

10.2.2 Preliminary Sampling

Every air pollution problem has its unique combination of factors, and hence no two sampling programs are quite identical. The general principles of proper design and statistical control remain the same, but each new program will be concerned with different species of crops and plants, topography, meteorology, and types of contaminants.

Because conditions are never the same, an exploratory sampling is recommended. This can be achieved most effectively by plotting a few radial transects from the suspected source of pollutant and taking samples at specified intervals along these transects. The directions and distances to which the transects should be run depend on several aspects of the particular problem but can be approximated satisfactorily from the meteorology of the area, stack physics, and the nature of the pollutant, if known.

An example may illustrate the general approach. Assume that an industrial operation is located in a valley devoted primarily to farming and that some of the crops in the area are allegedly either damaged directly or are causing damage when fed to livestock. A preliminary analysis of the meteorological summaries shows mainly a two-directional wind pattern with average velocities ranging from 6 to 10 mph. In this situation, the radial transect pattern adopted might contain four radii originating from the source of fume, with the two longest following the two main wind plumes and two shorter ones following intermediate paths. Sampling sites should be placed at approximately equal intervals along these transects; the transects should be extended far enough so that, in the judgment of the sampler, the level of concentration of the contaminant at the terminal site approaches the normal background level for the general area. This judgment must be based on calculations of the nature and volume of contaminant, stack heights, etc. The distance between sampling sites depends largely on the length decided upon for the transects. Ten or twelve sites along a main transect and six to eight along a secondary one might be selected.

In the exploratory sampling, at least three sites should be chosen where the variability of the material and of the sites can be determined. To continue with the above-postulated example, three sites along one of the primary transects would be selected, at one of which the concentration of the pollutant would be expected to be high, at another intermediate, and at the third low, representing increasing distances from the source. At each of these sites, a sufficient number of samples should be taken, within 5 to 10 ft of each other, to permit a statistical evaluation of the range of variability of the material.

The information obtained during the preliminary sampling will permit the planning of a sound program for the over-all sampling effort.

10.2.3 Detailed Sampling

General Considerations. Before going into the field, an idealized grid pattern of sites at fixed intervals around the suspected source should be plotted on the map. The actual dimensions of the grid will depend on the data from the preliminary survey. The grid may well have ½-mile intervals close to the suspected source, with increases to perhaps 2-mile intervals on the periphery. It is essential to have enough points to plot properly the concentrations of pollutants found in the area.

After the map has been marked with the idealized grid pattern, it should be taken into the field to select the sites actually available. Obviously, the sites will not conform to the grid in many places. Therefore, the available site nearest the grid site should be located and the change indicated on the map. If the nearest available site is more than halfway to the next grid point, it may be skipped.

When the field sites have been established, the collection of samples can be initiated. The location of the exact place at a site where a given sample will be taken can be predetermined by the use of a table of random numbers which serves to decide how far and in what direction the sampler walks into the field to find the center point of the area from which the sample will be collected.

There is considerable uniformity in field-grown crops such as vegetables and grains and in single-species hay, such as alfalfa. Therefore, the need of rigid randomization is not so great as with crops of mixed composition such as pastures. However, certain cautions should be observed even in sampling uniform crops. The sample should be taken far enough away from roadsides so that there can be no contamination from road dust or fume effects from passing automobiles, trucks, trains, etc. The sites should have comparable exposures. Thus, samples should not be collected under trees at one site and in an open exposure at another.

Samples from all sites should be made as representative as possible by taking the same fraction of the plants and a sufficient number of each. (For a detailed example, refer to the next article, on the sampling of green hays.) Finally, in any sampling concerned with isoconcentrations, it is essential that in every sample only the same variety of plant be collected. Most domesticated species, i.e., corn, alfalfa, wheat, gladioli, apricots, etc., have a large number of varieties in any one species. Some have several hundred. Many look so similar that even horticultural specialists may have difficulty identifying the variety in the field. This identification is very important, however, since it has been found that two varieties of the same species that are indistinguishable in the field may exhibit a tenfold difference in ability to concentrate pollutant under identical conditions.

Detailed notes of the condition of the sample and/or environment, when there appears to be any departure from the normal, greatly aid in the final interpretation of the data.

Forages and Feeds of Livestock. Green Hays. Crops such as alfalfa represent relatively straightforward sampling problems. Directions for a typical sampling run on a field follow:

"Walk into a field a distance of 45 paces, on a diagonal from one corner, carrying the shears and collection box. At that point, drop the collection box; this becomes the center point of the sampling site. With the shears, collect the top 6 in. of the plants passed while walking away from the box a distance of about 10 ft. Turn 120° in either direction and continue collecting along another 10 ft. At this point, again turn 120° in the same direction, until facing the box, and collect back to it. You will thus have completed collection over an equilateral triangle. Continuation of this pattern on the opposite side of the box will produce an hour-glass shaped pattern, with the box at the constriction. While executing the pattern, when a comfortable handful of material has been gathered, return with it and comminute into the box in ¼- to ½-in. lengths. This will result in several hundred plants being represented in the collecting box. The number of individual pieces will run from 75,000 to over 100,000.

"These must then be mixed thoroughly. This is done most effectively by placing both hands in the box and executing a twirling motion with both wrists, while keeping the fingers open, stopping occasionally to draw in the material lodged in the corners of the box. Finally, a grab sample of the order of 40 to 50 g can be taken from the material and placed in the sample container. A good starting approximation for ensuring that the sample will be representative is to gather and comminute about ten times as much material as is actually required for the sample."

Good comminuting and thorough mixing will result in the 40- to 50-g sample being representative of the area sampled. The large collection from which the small aliquot for analysis is taken is necessary in order to reduce the variation inherent in plants even of the same variety. All parts of the same plant are not equally responsive to the pollutant. The parts of the plant exposed directly to the atmosphere consist of the stem, young leaves, mature leaves, senile leaves, and flowers and fruits in season. In the case of alfalfa, it has been found that taking the top 6 or 7 in. gives material of excellent comparative value and, at the same time, representative of the entire portion of the plant that would be harvested for feed. A sample made up of stem and leaf bits tends to stratify upon mixing, with the denser stem fragments settling to the bottom of the box. This must be guarded against since mature leaves may carry eight to ten times the amount of pollutant found in the stems.

Pastures. Pastures are almost invariably of mixed composition and include both palatable and unpalatable grasses and weeds. There is also great variation in the dominant species of different pastures located in the same general area. However, good values can be obtained by following the usual sampling procedure, but increasing the number of samples.

One method that can be recommended is to select six to eight pastures in an area which seem to be highly representative in terms of species, stage of growth, etc. They should also be located at varying distances and directions from the source of the suspected pollutant, their selection being guided by the meteorological data available and the information gained from the exploratory survey. In each pasture, choose six sites, strictly at random, and collect the vegetation as nearly as possible as it is taken by the livestock. This requires skipping the unpalatable species which the animals avoid and cutting the palatable plants to the same height that they are eaten by the stock. In a lush growing pasture this will mean taking the top 5 or 7 in. of the palatable plants, while in a tightly grazed pasture the vegetation will have to be taken as close to the ground as it can be taken by the animals. In very short pasture, one is apt to get some soil on the samples, but this is still representative of what the animal ingests.

Sampling should continue at intervals of every 2, 3, or 4 weeks, depending upon local grazing practice, during the season that the pastures are utilized by livestock. At the end of the season, dependable values will be found on the amount of pollutant, and because of the appearance of patterns with respect to distance and direction, it becomes possible to estimate closely the conditions in the unsampled pastures.

In sampling green forages that in themselves become injured by fumes but are noninjurious to livestock if ingested, the so-called quadrant method of sampling can be employed to advantage. This employs a light steel wire or wooden frame (usually a meter square) that is randomly dropped throughout the area to be sampled. The vegetation that it encloses represents the sample at that site. The sample is then assessed for the percentage of the plant tissue that has been damaged by fumes, either by leaf count or estimation of the fraction of the total affected.

Since disease, drought, insects, senility, etc., can cause symptoms similar in appearance to fume damage, it is necessary that the operator be able to recognize and take into account these other factors. As the season advances, necrotic conditions attributable to factors other than fumes increase. Therefore, surveys made earlier in the season can be done more quickly and more accurately.

DRY FEEDS AND ENSILAGE. Usually a different problem in sampling is presented by this category. Grains, hay, and ensilage have been transported from the site on which they were grown and supposedly contaminated. The purpose of the sampling here is to obtain a true measure of the amount of toxic material that is present in the feeds. A physical problem is likely to arise, in that it is impractical to do all the sampling at one time. Several trips to the same site (or silo) may have to be made during the feeding of the contents so that sample collections will be taken from enough different levels to result in valid data. The same may be true for large amounts of stacked or baled hay, where the samples become accessible only as the hay is consumed. Here possible changes in concentration with time must also be considered.

Crops for Human Consumption. The samples included in this grouping are usually obtained to measure the influence the pollutant has directly on the species itself. However, cases arise where information is wanted on the amount of material present that could be harmful if consumed. Sampling under the first condition usually means a visual measurement or assessment of the selected sample, whereas in the second the sample is collected for quantitative chemical analyses.

VEGETABLES. Many vegetables are so large that by the time an adequate sample is collected the bulk becomes unwieldy. Reduction in the field is possible without impairing the value of the sample. This can be done by making wedge-shaped sections of each item, in which case the ratio of outer surface to any inner surface remains the same as in the whole. Thus, cabbage, lettuce, potato tubers, root crops, and so forth can be reduced to a practical and workable size. Ears of corn can be reliably sampled by taking only a few kernels from each ear. Most vegetables are planted as row crops, and good samples can be obtained by taking every tenth plant in every other row of ten rows.

One caution applies to all sampling, but it should be mentioned here since the hazard is greatest with vegetables, and that is that care should be exercised in preventing contamination of the samples by soil. A general rule that can be followed is to prepare the sample in the manner in which it is more or less standardly prepared for consumption.

FRUITS. It may be desired to sample the fruit or the leaves or both. Trees are large units with many genetic and individual variations, and whether fruit or leaves are sampled, it is necessary to include in any one sample material from at least 10 to 14 trees at each site. This can be done by using a table of random numbers and selecting at random the trees to be sampled. This is a recommended way since many variations in soils, exposures, etc., are sampled.

If leaves are being sampled for comparative data on accumulation of the pollutant, leaves of as nearly the same age as possible should be sampled each time. Good comparative values can be obtained by limiting the collection to those leaves which have first appeared on last season's growth of wood. Care in taking each succeeding collection from this same age wood and not including leaves from the current season's growth of wood is important. When the sampling is adequate in quantity and distribution, data on the rate and total amount of accumulation through the season will be obtained.

With vine fruits, such as grapes, which produce the canes before the leaves, the best uniformity is obtained by limiting the collection to the first 10 to 15 leaves to appear on the cane. As the season progresses, the earlier leaves may drop because of maturity or senility, and any change in the collection caused by changes in the plant should be noted.

In small fruits, such as cherries and raspberries, the entire fruit can be used in the sample, but larger fruits, such as apples, peaches, etc., can be reduced to practical sample size in the field by cutting out wedge-shaped sections as suggested for vegetables.

FIELD CROPS. Cereals and grains often need to be sampled. Where the species is small and fine-stemmed, such as wheat or oats, the entire plant may easily be taken. In coarse species, however, such as sugar cane and sorghum, better results are usually obtained if only certain leaves are collected from each plant, such as the fourth to eighth. If only comparative values are wanted, the fourth to eighth leaves from the top will suffice; if total accumulation of pollutant in the leaves is wanted, then a grouping of the older leaves should be decided upon. If the pollutant concentration of an entire plant is wanted, stem fragments must also be included. These should be analyzed separately from the corresponding leaf fragments if a ratio between the two is desired. Never fewer than 35 plants, preferably more, should be represented in any one sample collection.

If it is desired to sample the grains of these field crops, they should be separated from the bracts and chaff so as to appear as they would when this is done in actual practice.

10.2.4 Soil Sampling

Sampling Pattern. Soils are very complex and variable; vertical profiles may change in type, texture, and composition in the matter of a few feet. Therefore, adequate statistical design and testing of the data are a necessity to guard against erroneous conclusions.

Where it is suspected that air pollutants are being absorbed by the soil or are damaging it, transect sampling often will obtain the data needed with a minimum expenditure of time and money. For example, if there is mainly a two-directional wind over the suspected source, a radial transect pattern of four radii, with the two longest following the two main wind-plume patterns, can be set up. In soil sampling, gaseous matter is rarely involved, and transects long enough to reach the boundaries of the particulate fall are all that is necessary, the distance limitations being calculated reasonably from the meteorology, stack physics, and nature of pollutant.

It can be expected that the concentration change will be greatest near the source, and therefore the sample sites should be closer together near the suspected point of discharge. Sampling sites located at intervals of 2,000 ft to ½ mile for the first six or seven stations, increasing to ¾ or 1 mile for the next eight to ten stations in the prevailing wind directions, will usually produce enough points so that valid curves can be drawn for the contaminant in question.

Sampling Procedure. If profile samples are wanted (and they give the most complete data), it is ideal to dig a hole at the sample site and take an undisturbed slice from the side of the hole. This requires a great deal of labor, however, and in most cases is unnecessary.

Good valid samples can be collected with an ordinary 3-in. fence-post auger. The materials needed for this type of sampling are the 3-in. auger (about 4 ft long), No. 4 Kraft paper bags, a 16-gauge screen 12 by 12 in. and about 5 in. deep, a garden trowel, and a 2-by-2-ft square of heavy canvas.

Since the surface layer will probably be most important, it is customary to take the top 2 in. of soil with the trowel, place it in the screen over the canvas, and sift out the organic matter, stones, and lumps. Then, using the auger, additional samples may be taken to the depth desired. Two or three samples are taken at 6-in. intervals and perhaps two more at a 1-ft interval. Each of these is likewise screened and thoroughly mixed in the canvas and the size aliquot taken that is needed for analysis. Restraint should be exercised in filling the auger head so that it is not so full that there will be spillage back into the hole. Also, caution should be taken when removing the auger not to scrape the sides of the hole. Precision comes soon with a little practice, and when on occasion the edge of the auger head shaves the sides, it is a simple matter to remove a sufficient amount from the top of the sample and discard it without affecting the validity of the sample.

The variability in the soil complex makes it desirable to take paired samples either in transect or area-wide sampling. In transect sampling, these are taken about 50 ft apart, perpendicular to the transect. In area-wide sampling, the area is divided into a statistically sufficient number of plots and the paired samples are randomly located within each plot. In some cases these plots may be arbitrary grid patterns. In other cases, gross differences in topography or evident differences in soil characteristics may require a conformity of the plots to these differences. In any case the pattern and numbers must meet the statistical needs of the problem.

Avoid recently plowed fields or disturbed ground. If this is impossible, the time and nature of the disturbance should be made a matter of record in order that false comparisons are not made. Make a classification of each sample's texture. Make pH readings if possible. All these help the statistician and the sampler or investigator in interpretation of the data.

10.3 ANIMAL-SAMPLING PROCEDURES

10.3.1 Introduction

A study of the damage done to animals by air pollutants requires the sampling and analysis of various body components. In animals, as opposed to air and plants, only specific portions of the "whole" need be analyzed. Knowledge, therefore, of what tissues may be affected by a particular pollutant is necessary for proper diagnosis of the condition. From an economic standpoint, damage to farm animals is of paramount importance and interest; so the sampling procedures discussed here will apply primarily to these animals. The common air pollutants which affect farm animals, as was pointed out in Sec. 8, accumulate in the body principally as a result of consuming forage which has been contaminated with toxic materials from the air.

From the standpoint of the effect of air pollutants on animals, the objectives in collecting samples are to determine the nature of the toxic material and the concentration of this material in the body. These samples also supply factual records for legal procedures, quantitative information on the degree of injury, and reference data for comparison after remedial steps have been undertaken.

Sampling of animals is usually accomplished in one or more of the following ways:

1. Collection of body fluids and excreta, such as blood, urine, and feces
2. Biopsy, wherein a section of a tissue or bone is removed without seriously impairing the normal functions of the animal
3. Slaughter of a suspected animal and sampling of the tissue and bones

In Sec. 8, The Effects of Air Pollutants on Farm Animals, it was pointed out that the three most frequently occurring pollutants which cause damage to farm animals are arsenic, fluorine, and lead. Although other materials, such as insecticide sprays and dusts, are occasionally responsible for toxic effects, they do not occur generally or frequently enough to be considered as major factors in air pollution problems.

10.3.2 General Principles

The proper collection and preservation of animal samples is of greatest importance if reliable data are to be obtained. It is necessary to know whether chemical, physical, or histopathological analyses are to be used. Each of these approaches may involve variations in size of sample, multiple collections (if possible), different preservatives, etc.

The following paragraphs describe some of the principles which must be considered in obtaining samples from animals in the study of air pollutants.

Statistical Considerations. In the planning of biological experiments, the application of statistical procedures has lagged behind other fields of research. In controlled laboratory experiments, proper statistical design allows an investigator to obtain a maximum of information from his results. Animal air pollution problems, on the other hand, allow an investigator little latitude in planning his sampling work since the situation is predetermined. With proper planning, even with this limitation, the resultant data can be analyzed by various procedures to determine the probability of accuracy within certain specific limits. Unless one is familiar with statistics, however, it is best to discuss the problem with a statistician before any field work is undertaken. The books on statistics as applied to biological studies by Emmens (42) and by Burn et al. (22) are suggested for an understanding of the subject.

Sample Size, Packaging, and Shipping. Very often the collection of samples is made by a veterinarian or field representative and not by the persons carrying out the analytical determinations. It is thus important that a common understanding exist between the two groups involved. Agreement should be reached on the method of sampling, type of container, size of sample, and manner of sample preservation.

When collecting urine or feces from a herd of cattle, individual samples may be required. Under certain conditions the collections may be pooled and representative aliquots taken for analysis.

Each sample should be properly identified. It should list owner, description of animal (breed, sex, age, number or name of animal), and a clinical history of the animal's condition.

Biological specimens will spoil and become unsuitable for examination or analysis unless proper measures are taken for preservation. Common problems of improperly preserved specimens are hemolysis of blood samples, tissue decomposition as a result of bacterial growth, autolysis or digestion of a tissue by its own enzymes, and drying or fragmentation of a specimen. Refrigeration is the best method of specimen preservation. Either natural ice or dry ice may be used, depending upon the temperature desired and the distance the sample must be transported. Certain chemicals have specific applications as preservatives. Samples to be studied histopathologically are placed in a 10 per cent formalin solution immediately upon removal from the animal. Special stain preservatives may be necessary when a specific tissue or type of cell must be examined microscopically. Solutions of 0.5 per cent phenol or 1:1,000 merthiolate may be utilized as bactericidal agents for blood serum preservation, while formaldehyde is a standard preservative for fecal specimens.

For more detailed information on this subject, the reader is referred to a recent book by Coffin (34).

Single or Intermittent Sampling. In certain situations, the collection of a single sample from any one animal may be all that is possible. This condition is obvious when only necropsy samples can be obtained. However, when biopsies or blood, urine, or fecal collections can be made, repeated sampling at selected intervals can often develop a truer picture of the suspected condition. If multiple sampling is feasible, prior statistical planning will assure the procurement of the maximum information with the least amount of effort.

10.3.3 Sampling of Soft Tissue

Soft-tissue samples are usually collected at autopsy. Such material is best taken by a veterinarian or animal husbandryman who is familiar with animal anatomy. Proper selection of the sample will ensure a minimum of damage to the carcass if some economic recovery is desired. In some cases the organ to be analyzed may be so large that only a portion of it need be sampled, i.e., the liver or kidney. The sample, however, should be large enough to be representative of the particular tissue or organ. Other organs, such as the adrenals or thyroid, should be removed in their entirety. Again, the amount needed for analysis should be known before any sampling is begun. In general, the amount needed depends upon the range of concentration expected in the tissue and the sensitivity of the analytical method.

When tissue biopsy samples are taken, they are usually so small that only enough material is removed to allow for gross or microscopic examination.

10.3.4 Sampling of Bone

Bone samples are often collected in cases of air pollution where fluorine toxicosis is suspected. Biopsy samples of the twelfth or thirteenth rib can be obtained for chemical analysis without severe damage to the animal. The knowledge of an experienced and competent technician is required in the removal of a bone section; the removal is best handled by a practicing veterinarian.

If it is necessary to slaughter an animal, a metacarpal, metatarsal, rib, or the mandible is usually saved for analysis in cases of fluorine poisoning. If the mandible is removed, the teeth may be left intact since they can serve as a subjective indicator of fluorosis. The teeth should be coated with a plastic spray shortly after removal of the mandible to prevent cracking and splitting which may occur if the teeth are allowed to dry. Bones to be examined microscopically should be preserved as indicated in Art. 10.3.8, Sampling for Histopathological Examination.

10.3.5 Sampling of Blood

Blood from large farm animals is most easily obtained from the jugular vein. Clip the hair from the side of the neck and swab with alcohol. Apply pressure with the thumb until the vein becomes prominent. Perform the venipuncture by inserting a sterile dry 16- or 18-gauge needle attached to a sterile dry syringe into the vein and collect 5 to 60 ml of blood into a dry sterile container. If blood counts are desired, a 5-ml sample should be collected in a properly prepared tube. A mixture of dry ammonium oxalate 6 mg and of potassium oxalate 4 mg for each 5 ml of blood will prevent shrinkage or swelling of the cellular constituents. For preparation of such blood tubes, see Wintrobe (152). If blood counts are to be made, the sample should be delivered to the laboratory within 24 hr. In addition to total counts of red and white cells, blood counts usually include determination of the number of different kinds of leukocytes (white cells). For this purpose a drop of blood is smeared evenly on a clean microscope slide and allowed to dry. The slide can then be returned to the laboratory to be stained and examined under a microscope. For best results the smear should be made of unoxalated blood and at the time the blood sample is obtained. Kolmer, Spaulding, and Robinson (76) describe the technique required in making blood smears.

When chemical analyses are to be performed on blood serum, the collected anticoagulant-free blood is allowed to coagulate and is then sent to the analyst. The stoppered tube should be sealed with paraffin or Scotch tape to prevent evaporation.

10.3.6 Sampling of Urine

A chemically clean container must be used. Although glass bottles are often used, at the Stanford Research Institute screw-top polyethylene containers have been used for several years. They have the advantages of less inherent contamination and elimination of breakage. With farm animals, the sample may be collected while the animal is urinating or it may be collected by catheterization. The latter specimens are preferable since there is less contamination from urethral or vaginal discharge. Sampling may be hastened by aspiration of the urine with a large hypodermic syringe fitted to the catheter. The samples should be taken by a specialist trained in animal work. Beesen, Pence, and Holm (9) and Hobbs, Hansard, and Barrick (61) have devised collecting equipment for separating urine and feces. Their methods allow the collection of a complete 24-hr voiding but require that the animal be confined to a stanchion. In the collection of single samples, the flow of urine may often be stimulated with gentle massage of the skin on either side of and just below the vulva. Single samples are most conveniently caught in a wide-mouthed jar; the first 50 to 100 ml of voided urine is discarded and then 250 to 500 ml saved for analyses.

Urine decomposes rapidly and should be treated with a preservative at the time of collection. A thin layer of toluene, sufficient to cover the surface, is a very satisfactory preservative for routine use. Formaldehyde, at the rate of two drops per 50 ml of urine, is equally satisfactory (58).

Single specimens collected at varying times of the day may yield divergent results. Thus, single specimens should be labeled with the time the urine is voided.

10.3.7 Sampling of Feces

Fresh specimens should be collected in screw-top glass, tin, or polyethylene containers. If analysis of the sample is to be delayed for several days, refrigeration plus a few drops of formalin or an alcoholic solution of thymol should be used as a preservative.

With dairy cattle, a common procedure is to have an attendant follow the animal and collect the fecal matter as it is voided (100). Such a method, however, is tedious and expensive. McCall, Clark, and Patton (101) have used fecal collection bags attached to the animal in pasture studies. Such a bag is satisfactory for steers but cannot be used for female animals because of the difficulty in separating the feces from the urine. Metabolism cages for pigs (10,55), sheep (16,96), and cattle (61) have been devised for the separation and qualitative collection of urine and feces. Special equipment of this nature is probably not practical in the usual farm situation.

10.3.8 Sampling for Histopathological Examination

The objectives of microscopic tissue examination are to allow detailed observations of the component cellular structure as it existed in the animal body. Since autolytic changes take place rapidly in cells after the animal has been slaughtered, it is most important that the tissue specimens be taken as promptly as possible after the cessation of circulation. These specimens should be placed immediately in a fixative solution.

Tissue blocks should be cut with a sharp knife and should be thin enough that the fixing fluid readily penetrates throughout in a short time. A general rule is to cut sections not more than ¼ in. thick. The volume of fixing fluid used should be fifteen to twenty times that of the tissue fixed. A container with a wide mouth should always be used so that the specimen will not be bent or folded.

For routine work, the most commonly used fixing solutions are 10 per cent formalin and Zenker's solution. Other fixatives may be desired by the pathologist for special study of the tissue. Thus, a close working relationship must be established between the one who performs the autopsy and the microtomist in the laboratory.

Bones and teeth also need adequate fixation, and this requires the preparation of sections thin enough (2 to 3 mm) for rapid penetration of the fixative. A hacksaw with relatively fine teeth or a jeweler's saw will serve as a satisfactory cutting tool for small bones. Large pieces are best cut on a band saw. An aqueous solution of 10 per cent formalin is the best fixative for bone specimens.

For more detailed descriptions of preparing tissue and bone specimens for histological study the reader is referred to texts by Gatenby and Beams (84), Lillie (88), and Stitt, Clough, and Branham (134).

10.3.9 Sampling in Arsenic Poisoning

Chronic arsenic poisoning in farm animals is difficult to diagnose since it may be confused with lead poisoning. Consequently, chemical analysis of the body tissue and fluids is the only method for positive diagnosis. Milks (105) has listed the kidney, heart, and liver as the tissues which store arsenic to the greatest extent. After oral ingestion arsenic is mainly excreted in the feces, with some found in the urine. A photometric method for the determination of arsenic in biological materials has been developed by Hubbard (69). It is sensitive for the detection of 0 to 100μg of arsenic in 50-ml samples of urine or 5-g samples of blood or tissue. Other methods for the chemical analysis of arsenic in biological samples are listed for reference (5,30, 125,155). Organs which may be examined for microscopic study include the kidney, liver, skin, and bone marrow.

10.3.10 Sampling in Fluorine Poisoning

Analysis of the urine is the most frequently used method of determining the animals' current ingestion of fluorine. Rib biopsies may be used to determine the amount of fluorine deposited in the bones. Physical examination of the teeth and long bones, a form of "sampling," has been reviewed by Phillips in Sec. 8. At autopsy, bone specimens, such as the metacarpal, metatarsal, rib, and jawbone, as well as the teeth, should be saved for chemical and histological study. Fecal samples are collected when total excretion values of fluorine are desired. The soft tissues are not appreciably affected by fluorine, except possibly the kidney. Fluorine is not concentrated in the blood or milk to a sufficient extent for these fluids to be used as reliable indicators of fluorosis.

The analytical method of Willard and Winter (149) as modified by the Aluminum Company of America (1) is frequently used for the analysis of bones and feces. The method as outlined by the 1950 Association of Official Agricultural Chemists (3) is widely used for fluorine analysis of urine and of soft tissue. A slight modification of this procedure is followed at the Stanford Research Institute: 500 cu cm instead of 250 cu cm is collected in the distillation step and then a 200-cu cm aliquot taken for direct titration.

10.3.11 Sampling in Lead Poisoning

In cases of suspected current ingestion of lead, the urine and feces should be sampled, since they are reliable indicators of the intake. In chronic poisoning, lead is deposited in the bones, teeth, liver, kidney, spleen, and bone marrow (105). Blood samples can be obtained to determine the occurrence of red blood cell stippling (a basophilic staining reaction with methylene blue) (18) and anemia. Histopathological changes are often found in chronic lead poisoning in the form of tubular damage to the kidney (26).

Various methods are available for the determination of lead in biological materials. Letonoff and Reinhold (85,86) have developed a colorimetric procedure which is applicable in the analysis of blood or serum, urine, feces, bone or tissue. Recently, spectrophotometric (66) and polarographic (67) methods have become available for the specific determination of lead in biological specimens.

10.3.12 Miscellaneous Sampling

Although this section is directed toward the technics of collecting animal samples, other sources which may contribute toxic materials, and which also should be considered in air pollution problems concerning farm animals, should be mentioned briefly. Drinking water should be sampled routinely since it may contain fluorine or lead in appreciable amounts. Grain, mixed feed, salts, and silage should also be analyzed for the particular pollutant under study.

The collection of photographic data is a "sampling" procedure of documentary importance. Black and white or colored pictures can often describe a pathologic condition more vividly than can words or analytical data. At the Stanford Research Institute a photographic technique developed by Lawton (82) for the production of detailed Kodachrome transparencies of cows' teeth, plant leaves, etc., has been used successfully for several years. A description of the equipment and its assembly is as follows (Fig. 10-22):

A 35-mm camera with internal flash synchronization and with a removable back for ground-glass focusing is the basic unit. A jig made of lightweight metal is attached to the base of the camera through the tripod screw. This jig holds a ¼-in. rod below and in front of the camera on the lens axis and also supports the flash unit on the lens axis behind the camera. The rod angle is adjusted to hold an identification number, in a clip at the end of the rod, in the field of view but itself below the field. The rod's length is adjusted to indicate the object plane; so when the identification number is held against the subject, an accurate focus is automatically accomplished. Since the field size is determined when the unit is assembled, no manual focusing is necessary.

Supplementary lenses are added to increase detail. A general recommendation is two Portra +3 lenses, a neutral density filter, a 1A Skylight filter, and a hood. These lenses may be varied according to the field coverage desired and the light compensation necessary.

The flash reflector is covered inside with black photographic Scotch tape to reduce highlight reflections and reduce intensity. Best color reproduction has been obtained with Daylight Type Kodachrome film used with No. 5B or No. 25B flash bulbs. The

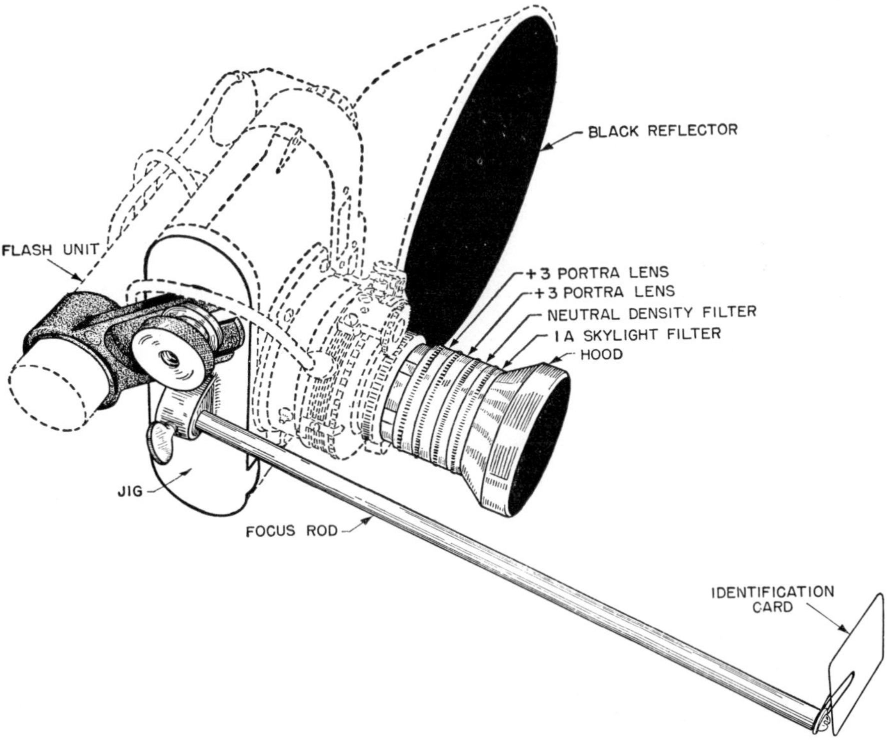

FIG. 10-22. Equipment used to make close-up 35-mm transparencies.

camera focus is adjusted for its closest setting, with the shutter speed set at 1/25 sec. Flash synchronization is minimized by using B-C units which fire the bulbs surely and rapidly.

REFERENCES

1. Aluminum Company of America: Determination of Fluoride, Aluminum Research Laboratories, Tech. Paper 914, 1947.
2. American Society of Mechanical Engineers: Test Code for Dust-separating Apparatus, Power Test Codes, PTC-21, 1941.
3. Association of Official Agricultural Chemists: Official and Tentative Methods of Analysis, 1950.
4. Bailey, Neil P.: Pulsating Air Velocity Measurement, *Trans. Am. Soc. Mech. Engrs.*, **61,** 301–308 (1939).
5. Barner, R. D., and B. L. Smits: Clinical Application of the Reinsch Test: A Rapid Preliminary Method for the Identification of Arsenic, *Vet. Med.*, **45,** 111–113 (1950).
6. Barnes, E. C.: Atmospheric Sampling by Electrostatic Precipitation, in L. C. McCabe,

ed., "Air Pollution," pp. 547–555, McGraw-Hill Book Company, Inc., New York, 1952.
7. Baron, T., E. R. Gerhard, and H. F. Johnstone: Dissemination of Aerosol Particles Dispersed from Stacks, *Ind. Eng. Chem.*, **41**, 2403–2408 (1949).
8. Beccaria, Giovanni Battista: "Elettricismo artificiale," Nella Stamperia Reale, Torino, 1772.
9. Beeson, W. M., J. W. Pence, and G. C. Holm: Urinary Calculi in Sheep, *Am. J. Vet. Research*, **4**, No. 11, 120–126 (1943).
10. Bell, J. M.: An Adjustable Cylindrical Cage for Use in Metabolism Studies with Young Pigs, *J. Nutrition*, **35**, 365–369 (1948).
11. Blackie, A.: Apparatus for Taking an Average Sample of a Variable Flow of Gas, *J. Soc. Chem. Ind. (London)*, **58**, 293–296 (1939).
12. Blasewitz, A. G., R. V. Carlisle, and others: "Filtration of Radioactive Aerosols by Glass Fibers," U.S. Atomic Energy Commission, Hanford Works, HW-20847, Apr. 16, 1951.
13. Braham, R. R., B. K. Seely, and W. D. Crozier: A Technique for Tagging and Tracing Air Parcels, *Trans. Am. Geophys. Union*, **33**, No. 6, 825–833 (1952).
14. Bredl, J., and T. W. Grieve: A Thermal Precipitator for the Gravimetric Estimation of Solid Particles in Flue Gases, *J. Sci. Instruments*, **28**, 21–23 (1951).
15. Bricard, J.: Propagation of Visible and Infra-red Radiations through Fog, *Roy. Meteorological Soc., Centenary Proc.*, pp. 36–42, 1950.
16. Briggs, H. M., and W. D. Gallup: Metabolism Stalls for Wethers and Steers, *J. Animal Sci.*, **8**, 479–482 (1949).
17. British Standards Institution: Code for the Sampling and Analysis of Flue Gases, British Standard No. 1756, 1952.
18. Brookfield, R. W.: Blood Changes Occurring during the Course of Treatment of Malignant Disease by Lead, *J. Pathol. Bacteriol.*, **31**, 277–301 (1928).
19. Brown, J. K., A. D. Hosey, and H. H. Jones: A Lightweight Power Supply for an Electrostatic Precipitator, *Arch. Ind. Hyg. Occupational Med.*, **3**, 198–203 (1951).
20. Brun, E., L. Demon, and M. Vasseur: Captation mecanique de corpuscules en suspension dans l'air, *Recherches aéronautiques (Paris)*, **1**, 15–19 (1948).
21. Burke, W. C., Jr.: Size Determination of Silica Particles Collected on Membrane Filters, *Am. Ind. Hyg. Assoc. Quart.*, **14**, 299–302 (1953).
22. Burn, J. H., D. H. Finney, and L. G. Goodwin: "Biological Standardization," Oxford University Press, New York, 1950.
23. Cadle, R. D., and H. S. Johnston: Chemical Reactions in Los Angeles Smog, *Proc. 2d Nat. Air Pollution Symposium*, pp. 28–34, Stanford Research Institute, Los Angeles, Calif., 1952.
24. Cadle, R. D., M. Rolston, and P. L. Magill: Cold-surface Collection of Volatile Atmospheric Contaminants, *Anal. Chem.*, **23**, 475–477 (1951).
25. Cadle, R. D., A. G. Wilder, and C. F. Schadt: A Technique for Collecting, Mounting, and Sectioning Airborne Particulate Material, *Science*, **118**, 490–491 (1953).
26. Calvery, H. O.: Chronic Effects of Ingested Lead and Arsenic, *J. Am. Med. Assoc.*, **111**, 1722–1729 (1938).
27. Cawood, W.: The Movement of Dust or Smoke Particles in a Temperature Gradient, *Trans. Faraday Soc.*, **32**, 1068–1073 (1936).
28. Cember, H., T. Hatch, and J. A. Watson: Dust Sampling with a Rotating Thermal Precipitator, *Am. Ind. Hyg. Assoc. Quart.*, **14**, No. 3, 91–94 (1953).
29. Chaikin, S. W., C. E. Glassbrook, and T. D. Parks: An Instrument for the Continuous Analysis of Atmospheric Fluoride, presented at the 123d meeting, American Chemical Society, Los Angeles, Calif., Mar. 15 to 19, 1953.
30. Chaney, A. L., and H. J. Magnuson: Colorimetric Microdetermination of Arsenic, *Ind. Eng. Chem., Anal. Ed.*, **12**, 691–673 (1940).
31. Cholak, J.: Kettering Laboratory, University of Cincinnati, Cincinnati, Ohio. Private communication with Stanford Research Institute, Los Angeles, Calif.
32. Church, P. E.: Dilution of Waste Stack Gases in the Atmosphere, *Ind. Eng. Chem.*, **41**, 2753–2756. (1949).
33. Claesson, S.: Studies on Adsorption and Adsorption Analysis with Special Reference to Homologous Series, *Arkiv. Kemi, Mineral. Geol.*, vol. 23A, No. 1, 1946.
34. Coffin, D. L.: "Manual of Veterinary Clinical Pathology," Comstock Publishing Associates, Inc., Ithaca, N.Y., 1953.
35. Crabtree, J., and A. R. Kemp: Accelerated Ozone Weathering Test for Rubber, *Ind. Eng. Chem., Anal. Ed.*, **18**, 769–774 (1946).
36. Crozier, W. D., and B. K. Seely: Some Techniques for Sampling and Identifying Particulate Matter in the Air, *Proc. 1st Nat. Air Pollution Symposium*, pp. 45–49, Stanford Research Institute, Los Angeles, Calif., 1949.

37. Davies, C. N.: The Separation of Airborne Dust and Particles, *Proc. Instn. Mech. Engrs. (London)*, **1B**, No. 5, 185–198; communs. 199–213 (1952).
38. dePiccolellis, J.: Continuous Gas Analyzer Serves High-pressure Stack, *Steel*, **128**, No. 19, 136, 138 (1951).
39. Dessens, H.: The Use of Spiders' Threads in the Study of Condensation Nuclei, *Quart. J. Roy. Meteorol. Soc.*, **75**, 23–26 (1949).
40. Drinker, P., and R. M. Thomson: Determination of Suspensoids by Alternating-current Precipitators, *J. Ind. Hyg.*, **7**, 261–272 (1925).
41. Edgar, J. L., and F. A. Paneth: The Determination of Ozone and Nitrogen Dioxide in the Atmosphere, *J. Chem. Soc.*, pp. 519–527 (1941).
42. Emmens, C. W.: "Principles of Biological Assay," Chapman & Hall, Ltd., London, 1948.
43. Epstein, P. S.: Zur Theorie des Radiometers, *Z. Physik*, **54**, 537–563 (1929).
44. Etkes, P. W., and C. F. Brooks: Smoke as an Indicator of Gustiness and Convection, *Monthly Weather Rev.*, **46**, 459–460 (1918).
45. First, M. W., and L. Silverman: Air Sampling with Membrane Filters, *Arch. Ind. Hyg. Occupational Med.*, **7**, 1–11 (1953).
46. Fitton, A., and C. P. Sayles: The Collection of a Representative Flue-dust Sample, *Engineering*, **173**, 229–230 (1952).
47. Fletcher, A. H., and C. J. Velz: Pollens: Sampling and Control, *Ind. Med. and Surg.*, **19**, 129–140 (1950).
48. Gage, J. C.: High-speed Absorber for the Determination of Toxic Substances in Air, *J. Sci. Instruments*, **29**, 409 (1952).
49. Gisclard, J. B., J. H. Rook, W. V. Andresen, and W. R. Bradley: A Simple Device for Air Analysis, *Am. Ind. Hyg. Assoc. Quart.*, **14**, 23–25 (1953).
50. Goetz, A.: Application of Molecular Filter Membranes to the Analysis of Aerosols, *Am. J. Pub. Health*, **43**, 150–159 (1953).
51. Goldman, F. H., and M. B. Jacobs: "Chemical Methods in Industrial Hygiene," Interscience Publishers, Inc., New York, 1953.
52. Goodale, T. C., B. M. Carder, and E. C. Evans: Dust Particles in High Velocity Air Streams—Representative Sampling, *Am. Ind. Hyg. Assoc. Quart.*, **13**, 226–231 (1952).
53. Gore, W. L.: "Statistical Methods for Chemical Experimentation," Interscience Publishers, Inc., New York, 1952.
54. Hall, S. R.: Evaluation of Particulate Concentrations with Collecting Apparatus, *Anal. Chem.*, **24**, 996–1000 (1952).
55. Hansard, S. L., M. P. Plumlee, C. S. Hobbs, and C. L. Comar: The Design and Operation of Metabolism Units for Nutritional Studies with Swine, *J. Animal Sci.*, **10**, 88–96 (1951).
56. Harris, W. B., and H. D. LeVine: Sampling and Measurement of Radioactive Atmospheric Pollution, *Air Repair*, **3**, No. 1, 17–21 (1953).
57. Hatch, T., H. Warren, and P. Drinker: Modified Form of the Greenburg-Smith Impinger for Field Use, with a Study of Its Operating Characteristics, *J. Ind. Hyg. Toxicol.*, **14**, 301–311 (1932).
58. Hawk, P. B., B. L. Oser, and W. H. Summerson: "Practical Physiological Chemistry," 13 ed., The Blakiston Division, McGraw-Hill Book Company, Inc., New York, 1954.
59. Hemeon, W. C. L., G. F. Haines, Jr., and H. M. Ide: Determination of Haze and Smoke Concentrations by Filter Paper Samplers, *Air Repair*, **3**, No. 1, 22–28 (1953).
60. Hill, George R.: The Place of Tall Stacks in Air Pollution Control, in L. C. McCabe. ed., "Air Pollution," McGraw-Hill Book Company, Inc., New York, 1952.
61. Hobbs, C. W., S. L. Hansard, and E. R. Barrick: Simplified Methods and Equipment Used in the Separation of Urine and Feces Eliminated by Heifers and Steers, *J. Animal Sci.*, **9**, 565–570 (1950).
62. Holden, F. R., F. W. Dresch, and R. D. Cadle: Statistical Control of Fluorescent Particle Tracer Studies, *Arch. Ind. Hyg. Occupational Med.*, **9**, 291–296 (1954).
63. Holt, P. F.: Study of Dusts in Industrial Atmospheres. IV. Determination of Mass Concentration, *Metallurgia*, **43**, 309–310 (1951).
64. Holt, P. F.: Dusts in Industrial Atmospheres. V. Determination of Mass Concentration by the Volatile Filter Method, *Metallurgia*, **44**, 52–54 (1951).
65. Holt, P. F.: The Determination of the Mass Concentration of Air-borne Dusts; Electrical Sampling Pump for Use with Volatile Filters, *Inst. Mining Met., Bull.* 539, pp. 15–20, October, 1951; Discussion, *Trans. Inst. Mining Met.*, **61**, Part 3, 127–128 (1951–1952).
66. Horiuchi, K., M. Miki, and H. Murata: Industrial Lead Poisoning. I. Spectrophotometric Microdetermination of Lead in Biological Materials, *Igaku to Seibutsugaku*. **25**, 152–154 (1952).
67. Horiuchi, K., and N. Ida: Industrial Lead Poisoning. III. A Modified Polarographic

Determination of Lead in Biological Materials, *Igaku to Seibutsugaku*, **26,** 98–100 (1953).

68. Hosey, A. D., and H. H. Jones: Portable Electrostatic Precipitator Operating from 110 Volts Alternating Current or 6 Volts Direct Current, *Arch. Ind. Hyg. Occupational Med.*, **7,** 49–57 (1953).
69. Hubbard, D. M.: Determination of Arsenic in Biological Material, *Ind. Eng. Chem., Anal. Ed.*, **13,** 915–918 (1941).
70. Hughes, T. H., and C. P. Sayles: The Use of a Pitot Tube in a Dust-laden Gas Stream, *Engineer*, **191,** 503–505 (1951).
71. Jacobs, M. B.: "The Analytical Chemistry of Industrial Poisons, Hazards, and Solvents," 2d ed., Interscience Publishers, Inc., New York, 1949.
72. Kalmus, E. H.: Preparation of Aerosols for Electron Microscopy, *J. Appl. Phys.*, **25,** 87–89 (1954).
73. Katz, S. H., and others: Comparative Tests of Instruments for Determining Atmospheric Dusts, *U.S. Public Health Bull.* 144, 1925.
74. Keenan, R. G., and L. T. Fairhall: Absolute Efficiency of the Impinger and of the Electrostatic Precipitator in the Sampling of Air Containing Metallic Lead Fumes, *J. Ind. Hyg. Toxicol.*, **26,** 241–249 (1944).
75. Kethley, T. W., M. T. Gordon, and C. Orr, Jr.: A Thermal Precipitator for Aerobacteriology, *Science*, **116,** 368–369 (1952).
76. Kolmer, J. A., E. H. Spaulding, and H. W. Robinson: "Approved Laboratory Technic," Appleton-Century-Crofts, Inc., New York, 1951.
77. Lamb, A. B., G. L. Wendt, and R. E. Wilson: Portable Electrical Filter for Smokes and Bacteria, *Trans. Am. Electrochem. Soc.*, **35,** 357–369 (1919).
78. Lapple, C. E.: Mist and Dust Collection in Industry and Buildings, *Heating, Piping, Air Conditioning*, **16,** 410–414 (1944); **17,** 611–615 (1945); **18,** 108–113 (1946).
79. Lapple, C. E., and C. B. Shepherd: Calculation of Particle Trajectories, *Ind. Eng. Chem.*, **32,** 605–617 (1940).
80. Lapple, C. E.: Mist and Dust Collection, *Heating, Piping, Air Conditioning*, **16**, 578–581 (1944).
81. Laskin, S.: Oscillating Thermal Precipitator, University of Rochester Atomic Energy Project, July 24, 1950 (UR-126).
82. Lawton, W.: Personal communication. Stanford Research Institute, Los Angeles, Calif.
83. Leach, L. J., R. H. Wilson, K. E. Lauterbach, C. J. Spiegl, and S. Laskin: Semiportable Air-sampling System, *Arch. Ind. Hyg. Occupational Med.*, **8,** 382–383. (1953).
84. Lee, A. B.: "The Microtomist's Vade-Mecum," 11th ed., Gatenby and Beams, eds., The Blakiston Division, McGraw-Hill Book Company, Inc., New York, 1950.
85. Letonoff, T. V., and J. G. Reinhold: Colorimetric Determination of Lead Chromate by Diphenylcarbazide: Application of a New Method to the Analysis of Lead in Blood, Tissues and Excreta, *Ind. Eng. Chem., Anal. Ed.*, **12,** 280–284 (1940).
86. Letonoff, T. V., and J. G. Reinhold: Precautions in the Determination of Lead in Biological Materials by Diphenylcarbazide, *Anal. Chem.*, **25,** 838–839 (1953).
87. Levine, J., and K. S. Kleinknecht: Adaptation of a Cascade Impactor to Light Measurement of Droplet Size in Clouds, U.S. N.A.C.A. Research Memo. E51G05, Sept. 18, 1951.
88. Lillie, R. D.: "Histopathologic Technic and Practical Histochemistry," The Blakiston Division, McGraw-Hill Book Company, Inc., New York, 1954.
89. Littman, F. E., and P. L. Magill: Some Unique Aspects of Air Pollution in Los Angeles, *Air Repair*, **3,** No. 1, 29–34 (1953).
90. Littman, F. E., and R. W. Benoliel: Continuous Oxidant Recorder, *Anal. Chem.*, **25,** 1480–1483. (1953).
91. Littman, F. E., and J. Q. Denton: Development of an Infrared Spectrometric Method for the Monitoring of Organic Substances in the Atmosphere, presented at the 123d meeting, American Chemical Society, Los Angeles, Calif., Mar. 15 to 19, 1953.
92. Luft, K. F.: A New Recording Method of Gas Analysis by Means of Infrared Absorption without Spectral Splitting, *Z. tech. Phys.*, **24,** 97–104 (1943).
93. Mader, P. P., W. J. Hamming, and A. Bellin: Determination of Small Amounts of Sulfuric Acid in the Atmosphere, *Anal. Chem.*, **22,** 1181–1183 (1950).
94. Magill, P. L., M. Rolston, J. A. MacLeod, and R. D. Cadle: Sampling Certain Atmospheric Contaminants by a Small Scale Venturi Scrubber, *Anal. Chem.*, **22,** 1174–1177 (1950).
95. Magill, P. L., D. H. Hutchison, and J. M. Stormes: Hydrocarbon Constituents of Automobile Exhaust Gases, *Proc. 2d Nat. Air Pollution Symposium*, pp. 71–83, Stanford Research Institute, Los Angeles, Calif., 1952.

96. Marston, H. R.: The Technique Employed for Determining the Utilization of Foodstuffs, *J. Agr. Sci.*, **25,** 103–112 (1935).
97. Martin, A. E.: The Infra-red Gas Analyzer and What It Can Do, *Research*, **6,** 172–176 (1953).
98. Marynowski, C. W., and F. E. Littman: A Continuous Recording Visibility Meter for Air Pollution Studies, *Air Repair*, **3,** No. 1, 45–50 (1953).
99. May, K. R.: The Cascade Impactor: An Instrument for Sampling Coarse Aerosols, *J. Sci. Instruments*, **22,** No. 10, 187–195 (1945).
100. Maynard, L. A.: "Animal Nutrition," 3d ed., McGraw-Hill Book Company, Inc., New York, 1951.
101. McCall, R., R. T. Clark, and A. R. Patton: The Digestibility and Nutritive Value of Several Native and Introduced Grasses, *Mont. Agr. Expt. Sta., Tech. Bull* 418, 1943.
102. McCord, C. P., and W. N. Witheridge: "Odors: Physiology and Control," McGraw-Hill Book Company, Inc., New York, 1949.
103. McKenzie, A. A.: Area Monitoring by AEC, *Electronics*, **25,** No. 6, 131–133 (1952).
104. Methods for Determination of Velocity, Volume, Dust and Mist Content of Gases, 4th ed., Western Precipitation Corp., Los Angeles, Calif., Bull. WP-50, 1948.
105. Milks, H. J.: "Practical Veterinary Pharmacology, Materia Medica and Therapeutics," Alexander Eger, Inc., Chicago, 1949.
106. Munger, H. P.: Air Pollution at Low Altitudes: Techniques for Study, *Chem. Eng. Prog.*, **47,** 436–439 (1951).
107. O'Gara, P. J., and E. P. Fleming: American Smelting and Refining Co., unpublished data.
108. Patterson, G. D., Jr., and M. G. Mellon: Determination of Sulfur Dioxide by Color-changing Gels, *Anal. Chem.*, **24,** 1586–1590 (1952).
109. Paxton, R. R.: Measuring Rate of Dustfall, *Rock Products*, **54,** No. 2, 114, 116, 118 (1951).
110. Payne, J. Q., and H. W. Sigworth: The Composition and Nature of Blowby and Exhaust Gases from Passenger Car Engines, *Proc.* 2d *Nat. Air Pollution Symposium*, pp. 62–70, Stanford Research Institute, Los Angeles, Calif., 1952.
111. Pearce, S. J., and H. H. Schrenk: Determination of Sulfur Dioxide in Air by Means of the Midget Impinger, *U.S. Bur. Mines*, Rept. invest. 4282, 1949.
112. Perkins, W. A., P. A. Leighton, S. W. Grinnell, and F. X. Webster: A Fluorescent Atmospheric Tracer Technique for Mesometeorological Research, *Proc. 2d Nat. Air Pollution Symposium*, pp. 42–46, Stanford Research Institute, Los Angeles, Calif., 1952.
113. Phillips, G.: An Electronic Method of Detecting Impurities in the Air, *J. Sci. Instruments*, **28,** 342–347 (1951).
114. Pring, J. N.: The Occurrence of Ozone in the Upper Atmosphere, *Proc. Roy. Soc. (London)*, **A90,** 204–219 (1914).
115. Ranz, W. E., and J. B. Wong: Impaction of Dust and Smoke Particles on Surface and Body Collectors, *Ind. Eng. Chem.*, **44,** 1371–1381 (1952).
116. Ranz, W. E., and J. B. Wong: Jet Impactors for Determining the Particle-size Distributions of Aerosols, *Arch. Ind. Hyg. Occupational Med.*, **5,** 464–477 (1952).
117. Saxton, R. L., and W. E. Ranz: Thermal Force on an Aerosol Particle in a Temperature Gradient, *Univ. Illinois Eng. Expt. Sta.* Contract AT(30-3)-28, *Tech. Rept.* 6, Dec. 31, 1951; *J. Appl. Phys.*, **23,** 917–923 (1952).
118. Schadt, C., P. L. Magill, R. D. Cadle, and L. Ney: An Electrostatic Precipitator for the Continuous Sampling of Sulfuric Acid Aerosols and Other Air-borne Particulate Electrolytes, *Arch. Ind. Hyg. Occupational Med.*, **1,** 556–564 (1950).
119. Sehl, F. W., and B. J. Havens, Jr.: A Modified Air Sampler Employing Fiberglas, *Arch. Ind. Hyg. Occupational Med.*, **3,** 98–100 (1951).
120. Shaw, C. F., and J. S. Owens: "The Smoke Problem of Great Cities," Constable & Co., Ltd., London, 1925.
121. Shepherd, M., S. M. Rock, R. Howard, and J Stormes: Isolation, Identification, and Estimation of Gaseous Pollutants of Air, *Anal. Chem.*, **23,** 1431–1440 (1951).
122. Siegel, J., and B. Feiner: Sootfall Studies for New York City. Part III of Air Pollution Survey Report, *Heating, Piping, Air Conditioning*, **17,** No. 9, 495–501 (1945).
123. Silverman, L.: Filtration through Porous Materials, *Am. Ind. Hyg. Assoc. Quart.*, **11,** 11–20 (1950).
124. Silverman, L.: Sampling of Industrial Stacks and Effluents for Atmospheric-pollution Control, *Proc.* 1*st Nat. Air Pollution Symposium*, pp. 55–60, Stanford Research Institute, Los Angeles, Calif., 1949.
125. Smales, A. A., and B. D. Pate: Determination of Arsenic in Biological Materials, *Analyst*, **77,** 196–202 (1952).

126. Smith, J. H., G. L. Rounds, and H. J. Matoi: Some Problems Encountered in Sampling Open Hearth Stacks, *Air Repair*, **3**, No. 1, 35–40 (1953).
127. Sonkin, L. S.: A Modified Cascade Impactor: A Device for Sampling and Sizing Aerosols of Particles Below One Micron in Diameter, *J. Ind. Hyg. Toxicol.*, **28**, No. 6, 269–272 (1946).
128. Stafford, E., and W. J. Smith: Dry Fibrous Air Filter Media: Performance Characteristics, *Ind. Eng. Chem.*, **43**, 1346–1350 (1951).
129. Stair, R., T. C. Bagg, and R. G. Johnston: Continuous Measurement of Atmospheric Ozone by an Automatic Photoelectric Method, *J. Research Nat. Bur. Standards*, **52**, 133–139 (1954).
130. Standard Method for the Determination of the Efficiency of Acid Mist Precipitators, Western Precipitation Corp., Los Angeles, Calif., *Tech. Bull.* 3-C, July, 1931.
131. Stanford Research Institute: The Smog Problem in Los Angeles County, Third Interim Report, Western Oil and Gas Association, Los Angeles, Calif., 1950.
132. Stanford Research Institute: The Smog Problem in Los Angeles County, Western Oil and Gas Association, Los Angeles, Calif., 1954.
133. Stern, A. C.: The Measurement and Properties of Cinders and Fly-ash, *Combustion*, **4**, No. 12; **5**, No. 1, 35–47 (1933).
134. Stitt, E. R., P. W. Clough, S. E. Branham, and others: "Practical Bacteriology, Hematology, and Parasitology," 10th ed., The Blakiston Division, McGraw-Hill Book Company, Inc., New York, 1948.
135. Stitt, F., and Y. Tomimatsu: Removal and Recovery of Traces of Ethylene in Air by Silica Gel, *Anal. Chem.*, **25**, 181–183 (1953).
136. Thomas, D. J.: Fibrous Filters for Fine Particle Filtration, *J. Inst. Heating Ventilating Engrs.* (*London*), **20**, 35–55; disc. 55–70 (1952).
137. Thomas, M. D., J. O. Ivie, and T. C. Fitt: Automatic Apparatus for Determination of Small Concentrations of Sulfur Dioxide in Air, *Ind. Eng. Chem., Anal. Ed.*, **18**, 383–387 (1946).
138. Thomas, M. D., J. O. Ivie, J. N. Abersold, and R. H. Hendricks: Automatic Apparatus for Determination of Small Concentrations of Sulfur Dioxide in Air, *Ind. Eng. Chem., Anal. Ed.*, **15**, 287–290 (1943).
139. Turner-Burrell Adsorption Fractionator, Burrell Supply Co., Pittsburgh, Pa., *Bull.* 205, 1946.
140. Turner, N. C.: Development in Analysis of Hydrocarbon Gases by Adsorption Fractionation, *Oil Gas J.*, **41**, No. 51, 48, 51–52, 69 (1943).
141. Urone, P. F., and M. L. Druschel: Infrared Determination of Chlorinated Hydrocarbon Vapors in Air, *Anal. Chem.*, **24**, 626–630 (1952).
142. Van Orman, W. T., and H. A. Endres: Self-charging Electrostatic Air Filters, *Heating, Piping, Air Conditioning*, **24**, 157–163 (1952).
143. von Brand, E. K.: Continuous Record for Air Pollution Control, *Air Repair*, **2**, No. 3, 113–115 (1953).
144. Washburn, H. W., and R. R. Austin: The Continuous Measurement of SO_2 and H_2S Concentrations by Automatic Titration, in L. C. McCabe, ed., "Air Pollution," McGraw-Hill Book Company, Inc., New York, 1952.
145. Watkins, J. M., and C. L. Gemmill: Infrared Gas Analyzer for Low Concentrations of Carbon Dioxide, *Anal. Chem.*, **24**, 591 (1952).
146. Watson, H. H.: Errors Due to Anisokinetic Sampling of Aerosols, *Am. Ind. Hyg. Assoc. Quart.*, **15**, 21–25 (1954).
147. Wilcox, J. D.: Design of a New Five-stage Cacade Impactor, *Arch. Ind. Hyg. Occupational Med.*, **7**, 376–382 (1953).
148. Willard, A. C., A. P. Kratz, and V. S. Day: Investigation of Warm-air Furnaces and Heating Systems, *Univ. Illinois, Eng. Expt. Sta. Bull.* 120, March, 1921.
149. Willard, H. H., and O. B. Winter: Volumetric Method for Determination of Fluorine, *Ind. Eng. Chem., Anal. Ed.*, **5**, 7–10 (1933).
150. Wilner, T.: Electric Gas Filter for Analytical Purposes, *Am. Ind. Hyg. Assoc. Quart.*, **12**, 115–116 (1951).
151. Wilson, L. D.: A Gas Ejector for Air Sampling, *Am. Ind. Hyg. Assoc. Quart.*, **12**, 58 (1951).
152. Wintrobe, M. M.: "Clinical Hematology," 3d ed., Lea & Febiger, Philadelphia, 1951.
153. Youden, W. J.: "Statistical Methods for Chemists," John Wiley & Sons, Inc., New York, 1951.
154. Young, R. E., H. K. Pratt, and J. B. Biale: Manometric Determination of Low Concentrations of Ethylene, with Particular Reference to Plant Material, *Anal. Chem.*, **24**, 551–555 (1952).
155. Zaikovskii, F. V.: Determination of Arsenic in Biological Objects, Aptechnoe Delo No. 4, 37–42 (1952).

Section 11

ANALYTICAL METHODS

BY JACOB CHOLAK

11.1 INTRODUCTION

The atmosphere contains such a great variety of materials in such variable concentrations from time to time that its exact composition will always be somewhat indeterminate. Comprehensive analyses, as we know them today, involve only those impurities which are known or suspected to have some harmful effect on living things or inanimate property. These represent only a small fraction of the total number of foreign substances which are present in the atmosphere. Analytical techniques suitable for the special forms and low levels of concentration of many impurities in the atmosphere are still to be developed, but much useful immediate information concerning many pollutants can be obtained by the use of analytical methods developed in other fields. Certain of the problems are similar to those encountered within industry, and some of the "standard methods" which are useful in this field and described in various publications (38,99,101,108,166,245,273) can be adapted, with little change, to the analysis of the external atmosphere. The success of the application of such methods is apparent from a number of special articles and reports on community air pollution which have appeared during the last few years (63,140,156, 176,195,219,265,292,293,340).

11.2 GENERAL CONSIDERATIONS

The determination of the concentration of an atmospheric pollutant is an essential requirement in defining an air pollution problem, whether the aim of the analysis is to establish base-level concentrations for legislative purposes, to measure the effectiveness of control equipment, to determine the effect of air pollution on the health of people or animals, or to determine the levels of concentration damaging to vegetation and property. The analytical requirements are not likely to be identical for each of these aims, so that a variety of procedures may be needed to study even a single pollutant at various levels of concentration. In cursory investigations, a quick answer is often possible by the use of simple methods and a minimum of special equipment, but in comprehensive examinations or when it is necessary to identify an unrecognized pollutant, the solution is more difficult and may require the application of specialized techniques, often instrumental in nature. An important consideration, therefore, is the provision of a well-equipped laboratory to handle the varied analytical problems encountered. The nature and cost of the special equipment depends on the problem, and the equipment of a laboratory can be a very costly affair if any or all of the useful analytical tools such as spectrographic, spectrophotometric (visible, ultraviolet, infrared), X-ray diffraction, electron microscopic, or mass spectrographic equipment are found to be necessary.

An analysis and the conclusions which can be drawn from it will have significance only as certain requirements are satisfied. These requirements include the control of the sampling-analytical setup and imply that the over-all efficiency of the sampling equipment is known, that a sufficient amount of impurity has been collected to permit the choice of a method in the desired range of concentration, and that the analytical method chosen possesses satisfactory sensitivity and reliability. Since the physical state of a pollutant is often an important consideration, the selection of a sampling technique which will provide the analyst with the unaltered pollutant is also desirable but not always possible. In fact, the sampling-analysis relationship is so close that it is not possible to discuss one without the other. For this reason the material in the following articles will be found to overlap that dealt with in Sec. 10.

The problem of contamination is such an important consideration that misleading results will be obtained if the analyst is not on constant guard against its occurrence. Satisfactory procedures for the determination of traces of many toxicants—lead and beryllium for example—have been developed only after years of effort in this type of analysis, so that the uninitiated investigator may have difficulty in choosing a satisfactory method and may waste much time in trying out the various procedures. Therefore, it is a wise policy for those contemplating this type of work for the first time to familiarize themselves with the special techniques through a short period of training in a laboratory where there is a clear understanding of the conditions and facilities required for carrying out satisfactory microanalytical work.

The analyst must be extremely critical of a single high concentration, and he must exercise great care when reporting such a finding. Such a finding can lead to nothing but confusion, since it is likely to be accepted as a valid determination by the uninformed. Repeated check analysis or the statistical evaluation of the order and range of concentrations in a sample series generally will screen out unusual findings for more careful scrutiny.

Analytical results should be so reported that there is no confusion or misunderstanding as to what was determined. A radical, functional, or elemental group is frequently determined, calculated, and reported as a specific contaminant, which, unless qualified, is regarded as the precise substance subjected to analysis. Comparison of such results with those obtained by more specific methods is not uncommon and tends to contribute to confusion, especially when the results obtained in different communities are set side by side.

Questions of specificity of detection and of interference follow naturally from the foregoing. These factors impose serious limitations on the selection of analytical methods and on the interpretation of the findings, but can hardly be discussed in a general manner at this time. Specific interferences and questions of specificity and limitations will, therefore, be reserved for those articles of this section in which specific methods are described.

The great variety of methods which have been applied or have possible application to the analysis of atmospheric samples may be divided into three groups:

1. Chemical methods
2. Instrumental methods
3. Biological methods

In the consideration of these three groups, attempts will be made to illustrate applications by examples from actual field use. Representative references to the literature are also included, so that the reader can obtain more details than the scope of this section warrants. No claim is made for the completeness of the citations, since these have been selected mainly as examples of representative applications.

11.3 CHEMICAL METHODS

11.3.1 Gravimetric Methods

The simplest and most rapid of the analytical methods employs the principle of weighing. All quantitative analyses require the determination or calculation of weight, but the direct weighing of an impurity to determine its concentration in the atmosphere has been applied principally to particulate matter.

Settled Dust. Settled-dust (sootfall, dirtfall, dustfall, or particle-fall) tests averaged over several years and considered in relation to variations in weather as well as the sites of collection are used to compare the cleanliness of different communities and of different sections of the same community. This type of test, which determines the quantity of material settling from the atmosphere due to the action of gravity, is generally performed by setting out cylindrical jars at selected points for a 30-day period of time. The usual collecting agent is water, to which are added certain materials to inhibit the growth of molds in the warmer months or to prevent freezing during the cold seasons. When brought to the laboratory, these samples are first filtered through a 60-mesh screen to remove bugs, leaves, twigs, and other forms of biological life. The suspended matter is then separated by filtration, dried, and weighed. This procedure, most commonly employed when glycerol is the additive to prevent freezing during the winter season, gives only the weight of the insoluble matter and does not record the soluble material discarded with the filtrate.

A more representative measurement of the total particle fall can be obtained by evaporating the entire sample to dryness in a tared beaker. This is possible when ethyl, isopropyl, or propyl alcohol is used as the antifreeze agent. The final weight obtained after the residue has been heated to constant weight at 105°F is a measure of the total particle fall, including the alterations due to the chemical reactions which have occurred while collecting the sample and carrying out the analysis. The total residue obtained is usually expressed as tons of dirtfall per square mile per month. A simple formula for converting the grams of particulate matter collected in a cylindrical jar to tons per square mile is as follows:

$$\text{Tons per square mile} = \frac{5{,}660 \text{ (grams of particulate matter)}}{\text{(diameter of jar in inches)}^2}$$

The dried, weighed residues and certain filtrates obtained in the tests are frequently subjected to a variety of treatments, each terminating in a weighing step in order to derive additional information concerning the nature of the dirtfall. Such tests include the procedures for measuring combustibles, ash, solubles, tarry matter, and a number of inorganic materials (223).

Suspended Matter. Suspended matter (aerosols), solid or liquid, collected by dynamic sampling equipment is also determined gravimetrically. The amount of this material is usually expressed as milligrams of particulate matter per cubic meter of air. If collected by an impinger in water, the mixture may be evaporated to dryness in a tared beaker for weighing, as in the case of dustfall determinations. A good microbalance is needed if the suspended matter is removed from less than 1 cu m of air. Since impingers do not collect particles below 1μ in diameter efficiently, the weights are likely to be low and complicated by chemical alterations and losses of volatiles resulting from the use of water and the heating steps.

Samples collected by filtration through papers are rarely weighed because of difficulties due to the adsorption of moisture. An exception is the large cellulose (84 sq in.) fluted filter used in high-volume sampling (276). In this case the adsorption of moisture is minimized by drying the filters in metal salve cans which are sealed immediately after removal from the drying oven operated at 110°C (179). An ordinary analytical balance may be used since the quantity of dust collected with this filter frequently amounts to several grams (179,276).

Molecular membrane filters (also known by the trade names of Millipore and Isopore filters) may be used for gravimetric dust determinations (111). These cellulose ester filters are close to 100 per cent efficient for particles less than 0.1μ in size and are, therefore, promising media for collecting atmospheric dusts. However, the filters need to be well supported when used, since they have high air-flow resistances and crack easily when flexed. Two-inch disks, when properly supported, can be operated with sampling rates of 1 cfm and can be used to indicate weight gains of 1 mg by means of a sensitive balance. Particulate matter collected on Millipore filters can be counted directly on the filter by either the light-field or the dark-field technique (111). The chemical and physical methods described in the following articles may also be used to examine the dusts which have been collected on these filters (also see Sec. 3).

The most accurate information concerning the absolute quantity of suspended matter in the atmosphere is obtained with the electrostatic precipitator. A microbalance may be necessary to weigh the particulate matter in the collecting tube itself or after its removal to a tared 10-ml beaker. As little as 0.1 mg per 10 cu m of air can be determined by the technique (63) when the Mine Safety Appliance Company's electrostatic precipitator is used. The maximum rate of sampling with this equipment is 3 cfm, with approximately 100 per cent efficiency for collecting particles in the submicron range. This equipment is suitable only for short periods of operation (several hours), and the quantity of material collected may therefore be too small for comprehensive analysis. A high-volume electrostatic precipitator for use in air pollution work, weighing only 50 lb, which can be operated continuously for periods up to 3 weeks without impairment of its efficiency, has recently been described by Clayton (73).

The quantity of organic material in suspended matter may also be estimated gravimetrically. The organic materials may be removed by extracting 0.5 g or more of the dust with various solvents in a Soxhlet apparatus. The solvent used most commonly is benzene (63), but selective solvents may be employed to distinguish between proteins and other organic material. The solvent is then evaporated, and the residue is weighed on a suitable balance. (Since the usual procedure is to use a temperature slightly above the boiling point of the solvent used to extract the sample, some volatile materials may be lost during the drying step.) The organic residue can then be examined by appropriate chemical or physical-chemical methods, infrared spectrophotometry being very useful for this purpose.

A simple weight determination may also be employed to determine the quantities of airborne dust collected on glass plates which have been coated with a thin covering of petroleum jelly. In this case, the plate may be immersed in alcohol or benzene to dissolve the petroleum jelly, and the dust is removed for weighing (or for chemical analysis) by filtering it through a tared, fine-pored, sintered-glass filter crucible. The glass slides may be set out to collect dust on the vertical or the horizontal surface, and thus they can be used to furnish an approximate differentiation between the quantities of the coarser particles which settle out of the air because of gravity and the suspended particles which can be removed only by filtration or impaction. The difference between the quantities collected on the two surfaces has been used to yield information on the average wind speed at the sampling site (190). The weights expressed per unit of area of collecting surface are empirical relative measurements of the cleanliness of the various sites. The slides must be exposed for a sufficient period of time to collect weighable quantities of material. Dust loadings should also be compared for equal periods of time. Dust particles collected on vertical or horizontal slides may be counted and may also be examined for the distribution of the sizes of the particles by any of the standard dust-counting and size-measuring techniques (166). For these purposes, it is necessary to employ unfiltered aliquots of the alcohol suspensions.

A modification of the plate collector may be used as an approximate directional dust collector, roughly similar in principle to Munger's mechanical directional sootfall collector (232). For this purpose, a cylinder of suitable size (2 to 6 in. in diameter) is divided into four (or more) equal vertical areas by means of strips of adhesive tape and the areas between the strips are thinly coated with a mixture of petroleum jelly in glycerol (166). After exposing the cylinder for any desired period of time, the petroleum-jelly layer between each pair of adhesive strips, corresponding to a quadrant of the compass, may be removed separately and analyzed gravimetrically as described above. The jar may be set in a large petri dish, and the dust may be removed from each quadrant in turn by wetting the surface with alcohol and careful policing.

As was the case with sootfall samples, the information obtained from slides is most valuable when numerous tests for each site are averaged.

Other Materials. The gravimetric procedure may also be used to measure the quantities of a number of gaseous impurities which can be collected by adsorption on activated silica, carbon, or alumina. This method is not suitable for low concentrations and is nonspecific since the increase in weight gives the total mass of adsorbable material which will include water vapor unless special precautions are employed to prevent its adsorption. Other gravimetric methods for analyzing grossly contaminated air or particulate matter removed from large volumes of air employ a precipitant which can be weighed, for example, barium sulfate for the determination of sulfur dioxide and silver chloride for measuring chlorine or chloride. English investigators employ this principle when they use the lead peroxide instrument to measure sulfur dioxide (223,240,278). The lead peroxide and the lead sulfate formed on exposure to sulfur dioxide are dissolved, and the sulfate is precipitated as barium sulfate which is weighed. It has been determined that the yield of lead sulfate is proportional to the concentration of sulfur dioxide for all concentrations up to 1 part in 1,000 parts of air under uniform conditions, but that the rate of reaction is dependent on a number of factors such as wind force and direction, rainfall, humidity, temperature, and the physical condition of the lead peroxide surface which is exposed (240).

11.3.2 Volumetric Methods

Volumetric methods of analysis may be quite specific in special situations, but generally this technique is more useful for obtaining information for a type of pollutant. Volumetric procedures involving alkalimetry, acidimetry, and iodometry have been applied to the determination of a number of atmospheric impurities.

Sulfur Dioxide. A standard method for measuring atmospheric sulfur dioxide consists of titrating with standard alkali the sulfuric acid produced when sulfur dioxide is trapped in a solution of hydrogen peroxide (240,279). Large samples are required for the estimation of trace quantities of sulfur dioxide, and the results are likely to be high because of the absorption of other acids.

Another method for determining sulfur dioxide consists of acidifying the alkaline collecting medium and titrating the liberated sulfurous acid with a standard iodine solution (166,284). This method is satisfactory for low concentrations and for special situations when autoxidation to sulfuric acid is eliminated (see colorimetric methods). A number of other titrimetric procedures employing iodine have also been employed for the determination of sulfur dioxide. Some of these are satisfactory for the direct measurement of sulfur dioxide when interferences are absent. In one procedure, a standard iodine solution (0.04 *N*) used to collect the sulfur dioxide is titrated with standard 0.04 *N* sodium thiosulfate. The difference between titrations of equal volumes of control iodine and the iodine used to collect the sulfur dioxide is a measure of the quantity of sulfur dioxide (166). Another method employs an alkaline solution of iodine (0.05 *N* in 5 per cent sodium hydroxide) with the thiosulfate titration noted above (265) and is said to eliminate errors due to the oxidation of iodine by the oxygen of the air (127). The alkaline iodide method has been used successfully in the author's laboratory for the detection of sulfur dioxide in stack gases. In this application, the same solution can be used to determine the partition between sulfur trioxide or sulfuric acid and the sulfur dioxide in an atmosphere. This is accomplished by titrating an aliquot of the standard alkaline iodine absorbing solution for sulfur dioxide and then obtaining the total sulfate in a suitable aliquot of the remainder. The difference between the two analyses can be used to measure the quantities of sulfuric acid, sulfur trioxide (or sulfate salts), which were present. Since many materials are also oxidized by iodine, any method which employs iodine as a reagent will lack specificity and may give erroneous sulfur dioxide concentrations for the atmosphere of certain communities.

Sulfuric Acid. Sulfuric acid (sulfur trioxide) may be estimated titrimetrically after collecting the mist on specially prepared papers or on fritted glass disks (129,212). As little as 0.009 ppm sulfuric acid has been measured in the atmosphere of London by this method (129).

Aldehydes. Aldehydes as a class may be identified by the Goldman-Yagoda method, employing the aldehyde-bisulfite decomposition at pH 6 to 7 and the subsequent titration of the liberated sulfite with standard iodine solution (128). The results obtained by this type reaction are commonly calculated and reported as acrolein or formaldehyde. Ketones interfere, but the interference by acetone may be eliminated by treatment with sodium bicarbonate (128). Specific lower aldehydes can be identified by procedures described in several of the following articles. A concentration of aldehyde of the order of 0.005 ppm or less can be measured by the sulfite-iodine-thiosulfate titration method, and the method has been widely used for the determination of total aldehyde (reported as formaldehyde or acrolein) in the air of a number of communities (50,64,194,292).

Chlorine. The quantities of chlorine in the atmosphere can be established with a procedure employing standard solutions of sodium iodide and sodium thiosulfate (109, 166). In this method the iodine liberated from the sodium iodide used to collect the chlorine is titrated with standard thiosulfate solution. Like most iodometric methods, the sensitivity is high, but the result may be dubious if other oxidizing materials are present in the atmosphere.

Ozone. The method described above with a number of modifications to improve sensitivity and specificity is the one generally recommended for the determination of ozone (25,315). The addition of 5 ml of a solution containing 5 g of aluminum chloride and 1 g of ammonium chloride in 1 liter of water to each 100 ml of 2 *N* KI solution makes it unnecessary to acidify before titrating with thiosulfate and permits the detection of a much smaller quantity of ozone than is possible otherwise (315). Interference by hydrogen peroxide and by the oxides of nitrogen is said to be eliminated by bubbling the test air through three scrubbers containing chromic acid and potassium permanganate before allowing it to come into contact with the iodide solution (315).

As is indicated in the preceding article, chlorine in the atmosphere will be recorded along with ozone when the latter is determined from the amount of iodine released from neutral or buffered potassium iodide. Peroxides will cause a similar interference, while sulfur dioxide or any reducing substance present in the atmosphere will use up liberated iodine and will tend to give low ozone values. Since the background level of ozone in the atmosphere is of the order of 2 to 5 pphm parts of air, the presence of large quantities of sulfur dioxide may result in a zero ozone finding (66). The interference by sulfur dioxide may be avoided if the air sample is collected in alkaline potassium iodide to which a little hydrogen peroxide has been added (286). It is claimed that alkaline iodide solutions are not affected by hydrogen peroxide, so that if a little of this material is added to the scrubber it will preferentially oxidize the sulfur dioxide to sulfate and eliminate the interference. The hydrogen peroxide must then be removed by boiling the solution before acidifying it for titration by thiosulfate. The titration technique requires such a large volume of air to give a detectable iodine titer change that it is not recommended for the determination of background ozone levels in the atmosphere.

Ammonia. A volumetric method also may be used to identify ammonia or its compounds when they are collected in weak acid media. A direct titration with standard alkali is sometimes adequate, but usually ammonia is aerated from the alkalinized medium used to collect it. The evolved ammonia is then recollected in standard acid (usually sulfuric), and the amount of ammonia is determined from the decrease in the quantity of free acid in the collecting medium. Obviously, this method gives total ammonia liberated from gaseous ammonia and solid ammonium compounds present in the sample. As little as 0.01 ppm total ammonia can be determined by sampling 30 cu ft of air.

11.3.3 Colorimetric Methods

Colorimetric methods as a class are among the most satisfactory of analytical procedures. So many elements, radicals, and organic compounds form colored complexes suitable for analytical purposes that texts have been written on each type of application. Colorimetric methods for the determination of many trace metals have been described by Sandell (261), while Snell and Snell give colorimetric tests for many inorganic, organic, and biological materials (288). The publications on microchemical analysis, particularly those by Feigl (105,106), also describe sensitive colorimetric tests for many materials. Descriptions of principles and of equipment suitable for colorimetry are not within the scope of this section, but all procedures involving the use of colorimeters; visual photometers, filter or otherwise; and photometric devices employing photocells or phototubes, whether indicating or recording, are considered together. The theoretical considerations and the equipment requirements may be obtained from the standard texts dealing with colorimetry (121,225).

Sulfur Dioxide. Atmospheric sulfur dioxide has been determined colorimetrically by means of acid-bleached fuchsin solution (12,136,292,294,297,322). Various materials, particularly protein materials, thiols, or thiosulfates, which interfere may be removed by treatment of the solution with mercuric chloride or the chlorides of platinum or palladium II and filtering off the resultant precipitate before developing the color (136,297,322). Autoxidation of sulfur dioxide to sulfuric acid in the solution used to collect the gas must be avoided by adding glycerol or benzyl alcohol (12,292,294,322). The method is suitable for traces of sulfur dioxide, microgram quantities being easily determined with it.

A very sensitive colorimetric method based on the violet color produced by reacting benzidine sulfate with *N* (1-naphthol-ethylenediamine hydrochloride) has also been used for the detection of small quantities of sulfur dioxide (181). As little as 0.011 mg sulfate ion could be detected; chlorides and pnosphates interfere (181).

A colorimetric procedure for the determination of sulfur dioxide, employing color-changing gels impregnated with ammonium vanadate, iodate, or periodate, or other common heteropoly compound constituents, has also been suggested (243). The vanadate-silica gels were found to be the most promising, and ability to detect 10 ppm sulfur dioxide with a 150-ml sample is claimed. Data on colors resulting with larger samples and lower concentrations were not determined, but the method was considered promising provided that interference effects are determined. A new indicating-gel sulfur dioxide detector has also been announced by the Mine Safety Appliances Company, Pittsburgh, Pa. The instrument is designed for the 0 to 50 ppm concentration range of sulfur dioxide with small volumes (150 ml) of the working atmosphere. There is the possibility of increasing the sensitivity with larger air samples. The gel is blue in color and changes to white in the presence of sulfur dioxide. The length of the white area is an index of the sulfur dioxide concentration.

Some of the methods employing indicating gels are not sufficiently sensitive for air pollution work. These colorimetric methods generally lack specificity, and therefore they are not so reliable as the alkaline iodide or the total sulfate methods.

Formaldehyde. Colorimetric methods using Schiff's reagent or chromotropic acid have been proposed as specific tests for formaldehyde (26,32,93,98). The chromotropic acid method appears to be the more specific since only formaldehyde forms a purple color with it (32). However, acetaldehyde, acrolein, and higher aldehydes, when present in appreciable quantities, may interfere with the color development. A sensitivity of 1μg per 1 ml of solution and an accuracy of ± 5 per cent have been claimed for the method (32). In the case of Schiff's reagent, the magenta color due to formaldehyde is said to be the only color remaining if the color is allowed to set for 6 hr before measuring it (26). In ordinary work the limit of sensitivity is 0.02 mg, but it is claimed that as little as 0.02 μg can be detected in colorimetric capillaries (26).

Oxides of Nitrogen. Two colorimetric procedures are commonly used to determine the oxides of nitrogen other than nitrous oxide. One method depends on the diazotization of sulfanilic acid followed by coupling with alpha-naphthylamine (138). This is an excellent method for nitrogen dioxide since, in aqueous solution, this gas yields equal quantities of nitrite and nitrate ions. Nitric oxide as such, nitric acid, or the nitrate salts are not measured by this method. Nitric oxide, however, is oxidized to the dioxide in the atmosphere, so that in practice the procedure measures this oxide along with the dioxide of the atmosphere.

A determination of total nitrate, which will account for some of the chemical changes occurring in the atmosphere, may also be used to furnish information concerning the oxides of nitrogen. Samples collected for this purpose are first oxidized to produce nitrate ions which are determined by the colorimetric phenoldisulfonic acid method (54,115,332). Results are calculated as parts per million nitrogen dioxide but include free nitric acid and any salts of nitrous and nitric acids which were present in the atmosphere. The method is satisfactory for low concentrations of nitrate ion, but great care must be taken to avoid accidental contamination by nitrates.

Other colorimetric methods suitable for nitrate ion detection employ 2:4-xylenol (19) and the production of chloranil by the oxidation of pentachlorophenol (22,54), but these techniques offer no improvement of sensitivity or specificity over the phenoldisulfonic acid procedure. A combination of the phenoldisulfonic acid and the sulfanilic acid–alpha-naphthylamine method should be useful to differentiate between nitrogen dioxide and nitric oxide, nitric acid, or nitrate salts in samples collected properly for this type of examination.

Ammonia. The colorimetric method employing Nessler's reagent is sufficiently reliable and sensitive to detect and estimate 0.5μg of ammonia (277). Generally, ammonia is distilled from the alkalinized solution used to collect the sample in order to eliminate interference from other nitrogenous compounds. Ammonium salts and distillable nitrogenous materials, collected while sampling, will also be indicated by this method.

Hydrogen Sulfide. The most specific method for the detection of hydrogen sulfide in the atmosphere makes use of the methylene blue color developed when p-aminodimethyl aniline and ferric chloride are mixed with the sulfide which has been collected in 1 per cent alkaline zinc acetate solution (112,262). A sensitivity of 3μg of hydrogen sulfide has been claimed for this procedure (112). Experience has indicated that the odor of hydrogen sulfide can be detected at such a low level of concentration that positive results with air samples are obtained only when the odor persists during a considerable portion of the sampling period.

Chlorine. The yellow color produced by the interaction between chlorine and *o*-tolidine has been made the basis for the colorimetric determination of free chlorine in the atmosphere (137,133). A comparison with standards prepared from potassium dichromate forms the basis of the evaluation in one test (137), while in another method the color is measured spectrophotometrically in order to improve the sensitivity of detection in the range below 0.10 ppm (333). Other oxidizing materials (nitrogen dioxide) interfere with the test.

Ozone. The amount of iodine liberated by ozone from neutral buffered solutions of potassium iodide or following the acidification of solutions of alkaline potassium iodide may be determined photometrically and related to the "oxidant" or ozone concentration in the atmosphere (204,286,287). This procedure is much more sensitive than the titration method described in an earlier article. Ozone concentrations as low as 1 to 2 pphm are detectable with 60-liter samples of air and the measurement of the liberated iodine at 352 mμ in a Beckman spectrophotometer (286). The methods for eliminating interferences described in the titration method apply to the colorimetric method.

Another colorimetric method based on the oxidation of phenolphthalin to phenolphthalein has also been proposed as a sensitive method for the determination of ozone or total oxidants (148). The color developed is measured in a Klett-Summerson colorimeter, using a green filter (No. 54). Hydrogen peroxide and nitrogen dioxide interfere. No statements concerning the effects of sulfur dioxide were made. The method does not appear to offer any improvement in sensitivity or specificity over the various iodide methods.

Fluoride. The most sensitive analytical methods for fluorides employ color reactions or the bleaching of certain colors by the fluoride. Among the colorimetric methods proposed are those using thoron–thorium nitrate (158), peroxidized titanium (86, 230), alizarin-thorium (87,306), alizarin-zirconium (11,189), aluminum hematoxylin (250), and the lake produced by aluminum chloride and eriochromecyanine (316). The differences in sensitivities between the methods are so slight that no preference can be indicated. All have been used to detect fluorides at the microgram level of concentration. A very satisfactory method used in the author's laboratory may be called a titrimetric-colorimetric method; it is mentioned here because it depends on the matching of alizarin–thorium nitrate colors obtained by back-titrating with a standard fluoride solution (87).

The colorimetric estimation of fluoride is more reliable when it is used with material obtained as a result of distillation to remove interfering ions and when sample preparation prior to distillation includes a fusion step (7,343). Samples containing aluminum fluoride or large quantities of silica should be evaporated to dryness, and the residue should be fused with sodium hydroxide before distilling out the fluorine (7). The normal fluoride content of the atmosphere is extremely low (2 to 8 ppb) (50,64, 134,195,265,292); so large volumes of air must be sampled in order to obtain a quantity of fluoride significantly larger than the amount normally associated with the reagents which are employed.

Carbon Monoxide. The quantities of carbon monoxide in the atmosphere are measured most conveniently by a colorimetric technique based on the use of the NBS indicating gel (271). The NBS indicating gel is an improved version of the gel developed in 1941 by the Royal Aircraft Establishment, Farnborough, England, in which the length of stain was measured when air containing carbon monoxide was passed through a silica gel which was impregnated with ammonium molybdate, sulfuric acid, and palladium chloride. In the NBS modification, the palladium chloride was replaced by a sulfuric acid digest of palladium metal or palladium oxide, and it was then found that the gel could be used colorimetrically for matching the clear green to bluish-green colors which developed with different concentrations of carbon monoxide. The gel will detect less than 1 part of carbon monoxide in 500 million parts of air. These indicating gels have been incorporated into an aspirator bulb-type field tester which can be obtained from several supply houses.

The field-type carbon monoxide detector employing an indicating gel has been designed for use with 50- to 250-ml samples of air (one to five squeezes of the aspirator bulb). The lowest concentration detectable with the maximum volume is 0.001 per cent (10 ppm). However, by employing 10 squeezes (500 ml) this limit may be reduced to 5 ppm. The aspirator-bulb method, therefore, is inconvenient when measuring very low concentrations (less than 5 ppm) of carbon monoxide in the atmosphere. However, very low concentrations of carbon monoxide can be detected if the indicating gel is arranged in an apparatus so that air at a rate of 80 to 100 ml/min can be passed through it by means of a small pump. As little as 0.5 to 1.0 ppm carbon monoxide in the atmosphere can be detected if the test is continued until the first appearance of color. The efficiency of reaction depends on the temperature, and the test should preferably be carried out at a temperature of not less than 70°F. The equipment must be calibrated with known concentrations of carbon monoxide which may be obtained from the National Bureau of Standards.

Another colorimetric method for carbon monoxide makes use of the blackening which results when air containing carbon monoxide is passed over a paper strip coated with selenium sulfide. This reaction has been incorporated into a portable instrument said to be useful with an average error of 10 per cent (23) in the range from several parts per million to 3 per cent carbon monoxide.

Carbon Dioxide. Although determination of carbon dioxide concentrations in the atmosphere is seldom required, a sensitive colorimetric method has been used to obtain the quantities of this pollutant (50,64). The method depends on the bleaching of the phenolphthalein color of a standard alkali solution after shaking fixed volumes of the solution and air (289). The sensitivity of this method is adequate for most purposes, while its simplicity makes it very convenient for use in the field. The absorption of other acidic impurities does not seriously invalidate the analysis because the concentration of carbon dioxide in the air is generally several hundred times greater than that of any other common gaseous impurity.

Particulate Matter. Colorimetric trace analysis is frequently employed to determine the composition of particulate matter. Descriptions of modern colorimetric methods for the analysis of many elements when present in traces have been published (166,261), and new applications are appearing constantly in the current literature. Highly sensitive and specific methods have been developed for the determination of aluminum (53,241), antimony (117,217), arsenic (33,159,180), beryllium (4), bismuth (160,196), boron (14,100,208,260), cadmium (57,68), chromium (321), cobalt (178, 222), lead (49,60), magnesium (124,207), manganese (76,119), mercury (58,227), molybdenum (95,236), nickel (5), silicon and silica (3,28), sulfur (246,289), tellurium (92), thallium (1), thorium (307,314), titanium (118), uranium (268), zinc (55,326), and zirconium (139). The references are far from complete, and those selected are merely representative of the many applications. Many of the references cited refer to the analysis of biological material, where the requirements for sensitivity and specificity are as exacting as those required in air pollution investigations. These methods should be applicable to the analysis of dusts removed from the atmosphere.

11.3.4 Nephelometry or Turbidimetry

In this article are included those procedures in which precipitates are examined either by the methods of nephelometry (348) or by transmitted light. Methods for the determination of two of the more common atmospheric pollutants, sulfur dioxide and chlorides, are the best examples of this technique. In the case of sulfur dioxide (or total sulfur), sulfurous material is first oxidized to the sulfate which is then precipitated as barium sulfate. The quantity of barium sulfate is then determined by measuring the transmission of the solution at any desired wavelength (540 mμ being very convenient) (318). Quantities of sulfur dioxide as low as 0.01 ppm can be detected when 30 cu ft of air is sampled and when the measurement is made with a 25-mm cell. The turbidimetric procedure for sulfur dioxide has been used in a number of atmospheric-pollution investigations (50,64,195,265,292). The method is useful for the detection of total sulfur in the atmosphere and is ideal for use when it is desired to obtain information concerning the chemical reactions or conversions of sulfur dioxide in the atmosphere. If sulfur dioxide per se is wanted at any time, the air may be sampled with a sampling train consisting of an electrostatic precipitator followed by a scrubber to collect the sulfur dioxide. Material collected by each portion of the train is analyzed independently in order to differentiate between materials containing sulfur in the gaseous and particulate states.

The turbidimetric method for measuring chloride for the reasons cited above, except in special situations, also gives more representative data concerning the chloride content of the air than can be obtained by any other procedure. Total chloride of the order of 0.01 ppm has been determined (50,64) by measuring the transmission power of solutions from which the chloride has been precipitated as the silver salt (206).

The scrubber and electrostatic precipitator sampling train described above may also be used to distinguish between particulate and gaseous chlorine compounds. In the latter case, differentiation between chlorine and hydrochloric acid will require some knowledge of the particular situation.

11.3.5 Chromatography

The many applications of chromatography suggest the possible satisfactory use of this technique for air pollution investigations. The general attractiveness of the method is indicated by the fact that, in 1951 alone, Strain and Murphy list 385 publications in which this method of analysis was described (303). Columnar ion exchange and paper methods have been developed for the separation and identification of a great many materials. Paper chromatography is particularly suited for the separation and the identification of the components of complex mixtures where the components may be isolated without chemical change (301). Traces of inorganic ions may be identified by paper chromatography, and the results are considered comparable to orthodox chemical analysis but with the advantage of economy of time and material (188,248). The method has been applied to many diverse fields which are reviewed in a number of publications (27,45,197,301,302,303). Of special interest is a paper dealing with the benzpyrene content of town air in which chromatography was used to separate the benzpyrene from unwanted material when coal smoke was extracted with various solvents (334).

11.3.6 Other Methods

In this category one must place those subjective methods which include corrosion studies by exposure of test panels, the testing for damage to painted surfaces such as blackening caused by hydrogen sulfide, and similar tests (345). A recent development has been the introduction of a subjective test for the determination of on-the-spot ozone concentrations. In this method advantage is taken of ozone's capacity for cracking stretched, specially compounded rubber. First suggested by Reynolds (255), the method has been described and used by a number of investigators (31, 80,81). Claims for this method suggested that ozone was the only atmospheric component that developed the type of cracking seen with stretched rubber (80). However, more recent work has shown this to be in error, since a number of free radicals cause the same phenomenon (215).

11.4 INSTRUMENTAL METHODS

Problems of specificity and sensitivity, at times, demand the application of methods based on practically every known physical principle and the instrumental methods developed therefrom. Instruments designed for laboratory and field use are considered and discussed together except for the number of field recorders which are treated separately as a class.

Principles of theory and descriptions of equipment and procedures are obviously beyond the scope of this book; these can be obtained from the appropriate literature citations in the following articles.

11.4.1 Spectrographic Methods

Spectrographic methods include all techniques whether based on emission (including fluorescence) or absorption phenomena in the visible, ultraviolet, or infrared regions. All are potentially useful for the analysis of material removed from the atmosphere.

Emission Spectrography. Emission spectrography is widely used to establish the elemental composition of mixtures and is indispensable for the exploratory examination of particulate matter. Of the known elements, about 70 are susceptible to spectrographic analysis, with large differences in the various sensitivities of detection. The alkalies and the alkaline earths exhibit very high sensitivities of detection (0.00001 per cent) followed by about 40 or 50 elements with moderate sensitivities (0.001 to 0.0001 per cent). The rare earths, the platinum groups, and the uranium series show poorer sensitivities in the range of 0.05 to 0.001 per cent (269), while a group of gaseous elements and metalloids require special handling and are rarely determined spectrochemically. The techniques of spectrochemical analysis are well established, and these, as well as the instrumental requirements, may be obtained from a number of excellent texts dealing with the subject (35,153,233).

The spectrochemical approach to be used depends on the aim of the investigation and the amount of material available. If more than 10 mg of particulate matter has been collected, then two or three exposures of 5 to 10 mg in the d-c arc in selected spectral regions with an instrument of large dispersion will be sufficient to give an exploratory analysis of the material. A semiquantitative analysis can be made by comparing the intensities of the characteristic spectrum lines with similar lines from graded standards photographed on the same plate. Accurate quantification is more difficult and requires the control of a number of factors in addition to the use of chemical separation and concentration (52). A complete exploratory examination is not possible if less than 5 mg of material is available.

Keenan and Byers (179) describe a semiquantitative method for 21 elements in dust collected by a large-volume sampler. Their samples average 0.7 g removed from 2,832 cu m of air during a 24-hr sampling period. The entire sample was first digested with acid, and the solution and residual were made up to a volume of 10 ml. Graphite rods impregnated with this mixture were burned in a d-c arc with exposures in three different regions to cover the desired spectral range. Quantification was by densitometric comparison of the characteristic spectrum lines with similar lines for graded standard solutions of the elements investigated. This procedure was used to examine the particulate matter collected from the atmospheres of Windsor, Ontario, Detroit, Mich., and Charleston, W.Va. (72,172,179).

Other comprehensive spectrochemical analyses of aerosols and settled dusts are described by Cholak (50) and Cholak, Schafer, and Hoffer (63,64). In these investigations a variety of spectrochemical methods combining chemical separations and concentrations with a special technique of excitation were used to improve the sensitivities of detection. An acid digest of the dust was made up to a definite volume, and aliquots placed on graphite rods were exposed in the d-c arc. Aluminum, copper, iron, lead, manganese, silver, tin, titanium, and vanadium were determined from a single exposure (52), while portions of the dust sample were treated separately to determine antimony (59), beryllium (61,62,65), and cadmium (57). In the case of beryllium, as little as 0.0002 μg was detected when using the cathode-layer method of excitation (62,65).

Concentrations of 22 elements varying from more than 10 per cent to less than 0.001 per cent have also been detected spectrographically in the particulate matter of the Los Angeles atmosphere (292).

A number of spectrographic methods which were developed for the limited examination of dusts and fumes in industrial atmospheres can be applied with little or no change to the analysis of the particulate matter of the external atmosphere. Included in this classification are procedures for beryllium (18,238,285), cadmium (237), lead (13,49,187,237), palladium (114), platinum (114), thallium (219), and zinc (237). Of special interest are two methods for the direct measurement of lead in air (13,187) and a very sensitive method for molybdenum in soils (219). A sensitivity of detection of 0.05μg of molybdenum in the arc was claimed for the latter method (219).

Flame Photometry. An account of emission spectrography would not be complete without the mention of the use of flame photometry or spectrography. This method of analysis has been widely accepted and is gaining favor in this country for the analysis of the alkalies and the alkaline earths. A fine mist of the test solution is burned in an air-acetylene, air–coal-gas, or oxygen–coal-gas flame, and the flame is analyzed spectrographically or spectrophotometrically. In the former method an internal standard is added to the solution and a small spectrograph is employed to photograph the spectrum lines, which are measured with a densitometer for the determination of the quantities of the elements producing the lines (56). A filter photometer for the determination of sodium and potassium has also been developed (21), while attachments for the Beckman spectrophotometer convert this instrument into a flame photometer of superior performance.

Because of the lower energy of excitation, the applicability of flame photometry or spectrography is limited to approximately 30 elements over a wide range of sensitivities (56). However, the favorable high sensitivities of detection of the alkalies and the alkaline earths make the flame method attractive for investigating these elements. Calcium, magnesium, potassium, and sodium in the atmosphere of a number of communities have been detected by this method (50,63,64,195). Flame spectrographic methods were said to have been used to determine aluminum, iron, lead, and silicon in smog of Los Angeles (195), but this may be in error since the sensitivity of detection of lead by this method is not sufficient for the small quantities involved, while aluminum and silicon are not susceptible to detection by flame photometric methods (56).

Although the direct application of emission spectrography to air pollution studies is limited to the cases cited in the foregoing discussion, it should be pointed out that this method has been applied to analytical problems in practically every field of science, and it is likely that a number of the applications can be transferred to air pollution analysis without too much revision of the procedures. The many spectrochemical applications have been reviewed to the year 1945 by Meggers and Scribner (224) and for the period 1945 to 1950 by Scribner and Meggers (270).

Spectrographic equipment is expensive, and the costs are further increased by the necessity of providing a satisfactory working space protected by an efficient dust-removal system. The cost for equipping such a laboratory may approach $25,000, of which $5,000 to $8,000 would represent the cost of a spectrograph suitable for emission work with complex spectra, $1,500 to $8,000 would be spent for a densitometer, and the balance for such necessities as excitation sources, dark-room facilities, and a ventilation system provided with efficient dust-removal equipment.

Flame spectra may be obtained with less expensive equipment, such as a small quartz or grating spectrograph which may be obtained for $500 to $1,500. Attachments for the Beckman D.U. spectrophotometer, costing approximately $700, will convert this instrument to a very sensitive flame photometer.

11.4.2 Spectrophotometry

Spectrophotometry in the visual, ultraviolet, and infrared regions is widely used in analytical chemistry and therefore is potentially useful for air pollution analysis. Some of the applications in the visible region have already been indicated in the article on colorimetric methods. A bibliography of photoelectric-spectrophotometric methods for determining inorganic ions chiefly in the visible region has been published by Stillman (299), and another list of references for the photometric analysis of many inorganic and organic constituents has been given by Moss (225). Photometric methods in the visible region are applicable to the analysis of both particulate and gaseous material and are used in elemental and molecular analysis. Applicable instruments vary from the simple colorimeter to the recording photoelectric spectrometer. Wavelength regions for measurement may be isolated with filters or more accurately with prisms or gratings. Descriptions of various types of photometers and their uses are given in a number of publications (35,121,153,225).

Ultraviolet. Absorption by a colorless molecule in the ultraviolet region is frequently used to characterize organic material, but the method is not entirely satisfactory for trace analysis. Difficulties in the practical application are the insufficient sensitivity and the interference by numerous coexisting materials in the atmosphere. Satisfactory results, however, have been obtained when the method was applied to determine impurities in industrial-plant atmospheres and in effluent stacks when the amount of the impurity was measured directly in the atmosphere or following collection by a standard sampling procedure using gas-washing bottles.

Acrylonitrile was determined by collecting it in water in a train or two gas-washing bottles and measuring the amount of absorption at 210 mμ (34). Other materials absorbing in the same region interfere. Absorption in the ultraviolet region has also been used to measure benzene and hexachlorobenzene following collection in liquid absorption media (10,89); many other examples can be obtained from a number of publications on ultraviolet spectrophotometry (35,153,225).

Ultraviolet spectrophotometry may be used to measure the quantities of some gaseous impurities directly in the atmosphere. One instrument, which was designed to give a continuous spectrum with a hydrogen tube and which has been used to measure very small quantities of benzene in toluene (75), has possibilities for this purpose. Another instrument measures the absorption of the mercury radiation at 2,537 A and is very sensitive to vapors of mercury (0.0001 ppm) and of tetraethyl lead (0.13 ppm) (151). Since many compounds absorb in the 2,537 A region, special measures must be taken when mixtures of such compounds are present. For example: aniline and xylene and the monochlorobenzenes absorb strongly in the region, while materials like acetone, gasoline, and phosgene show absorption only when present to the order of 5 ppm or more (151). A similar instrument provided with an automatic recorder has been used in field studies to study the travel of gas clouds (182), while another is commercially available for mercury-vapor analysis (120). The latter instrument, when provided with suitable means for removing interferences, can be applied to the investigation of other air pollutants. Absorption in the ultraviolet region has also been used for the direct measurement of ozone in the atmosphere (96, 304) and to study the dispersions of nitric oxide from short stacks (126,133). Equipment for use in the ultraviolet region can be obtained for costs ranging from $400 to $1,500.

Infrared. The infrared region has been successfully applied to the analysis of solid, liquid, and gaseous materials (20). Most of the applications have been for control purposes or for standardization, but the advantages of infrared analysis are so remarkable that the technique is certain to attract widening attention. The possibilities may be gauged from the following brief summary of advantages and disadvantages taken verbatim from a paper by Gore and Petersen (132).

"ADVANTAGES

1. All molecules possess infrared spectra, whether solids, liquids, gases, or in solution.

2. A single infrared spectrogram may give clues to the identity of constituent radicles, substituents, and the positions in the molecule.
3. The spectrum is a finger print of the molecule, useful in both qualitative and quantitative analysis.
4. This spectrogram gives an indication of the over-all purity of the preparation.
5. The techniques are simple.
6. Relatively small samples of pure materials are sufficient (several milligrams). Samples may be recovered unchanged.

Disadvantages

1. Although all molecules possess infrared spectra, with present instruments, as the molecular weight increases the clarity and uniqueness of the spectrum decreases.
2. The present knowledge of infrared spectra is insufficient to account for and predict all absorption bands. This makes it necessary, as in the case of mixed melting point determinations, to obtain known compounds when identifying an unknown one.
3. Traces of impurities, unless they are exceedingly strong absorbers, are not easily detected.
4. Dilute solutions are handled with difficulty, unless strong absorbers are present."

Instruments in general use today employ prisms to isolate the desired energy region and generally are set up so that samples collected in the field must be returned to the laboratory for analysis. Considerable chemical handling is necessary for air pollution work, and with few exceptions gaseous samples per se are unsatisfactory. Concentration of impurities by freezing out or absorption or a combination of the two is generally employed to provide sufficient materials for infrared analysis. Washburn and Austin describe a technique to determine ethylene at a concentration of 3 parts in 1 billion parts of air. They also mention the difficulties in applying infrared methods to air pollution investigations (336).

The infrared absorption method was employed to study the organic contaminants in the atmosphere of a warehouse where apples were stored. Absorptions in regions typical of a mixture of esters, aldehydes, and perhaps alcohols were obtained. A very large sample was required and was obtained by adsorption on carbon followed by desorption and collection in freezing traps immersed in an ice-salt mixture and dry ice. The combined material was extracted with 1 ml of carbon tetrachloride, which was subjected to the infrared analysis (319). Absorption of energy in the infrared region has also been used to identify 15 chlorinated hydrocarbons removed from the air by a cooled scrubber technique. A sensitivity of 0.04 mg per milliliter of sampling solvent was claimed (323). In another case a 100-cm cell was used to measure 0.1 ppm benzene and small quantities of hydrocarbons from air streams and polluted areas when 60 liters of air was collected by the Shepherd sampler (213). Techniques for concentrating organic material removed by carbon adsorption for infrared analysis which are of practical value for the investigation of air pollution are also described in a recent paper (320).

The direct field use of infrared analysis has appeared attractive, and a number of instruments have been developed to detect certain contaminants in industrial atmospheres (77). At the present time, with the exception of an analyzer for carbon dioxide (337), none of the instruments appears to have the sensitivity required for the analysis of the trace quantities of impurities encountered in the external atmosphere (103,226,344). The method was applied to air samples containing ethylene glycol ethers and quantities of carbon monoxide, carbon dioxide, hydrogen cyanide, and water at levels of 0.005, 0.002, 0.03, and 0.01 per cent, respectively (103,235). Water vapor and ozone in the atmosphere were also measured directly in the atmosphere with infrared photometers (113,304).

An instrument suitable for a specific application may be fabricated in the laboratory at a moderate expense, but elaborate research equipment will cost \$5,000 to \$15,000. The latter is the average cost for a compensating recording infrared spectrophotometer. The need for locating this equipment in a space where humidity is controlled and the need for such ancillary equipment as various types and sizes of cells add to the expense of operating infrared equipment. A 1-m cell will cost about \$400, while another cell which can be adjusted to give light paths varying from 1.25 to 10 m has recently been announced by the Perkin-Elmer Corporation, Norwalk, Conn., for \$1,715. This latter cell should be particularly useful for determining trace components in the atmosphere.

Fluorescence. Fluorescence is useful for the analysis of trace amounts of a variety of materials, but in some cases the removal of interfering ions is required to ensure specificity of detection. Beryllium can be estimated in extremely low concentrations by reacting it with morin following separation from the many ions which interfere (261). A satisfactory procedure for eliminating all interferences except aluminum employs cathodic electrolysis and the production of beryllium acetyl acetonate, which can be extracted with benzene (317). Quinizarn (a-4-dihydroxyanthraquinone) and 1 amino-4-hydroxyanthraquinone have also been used for the quantitative estimation of beryllium by fluorescence (165), but these methods are inferior in sensitivity to the morin method.

Fluorescence has been applied to the determination of traces of hydrogen fluoride. In this case a decrease in fluorescence is measured when hydrogen fluoride reacts with the strongly fluorescing magnesium oxinate (107).

Fluorescence is also potentially useful for the analysis of uranium in the particulate matter of the atmosphere. Fractions of a microgram of uranium may be detected by fusing a solid sample with sodium fluoride and examining the melt under ultraviolet light (329). Interferences are eliminated by isolating the uranium with egg albumin (329). Applications of fluorescence to the analysis of materials in many fields are listed in a number of books and special articles (251,253,341).

11.4.3 X-ray Diffraction

X-ray diffraction techniques may be considered as supplementary analytical tools when used in conjunction with an elemental analysis, such as emission spectrography or chemical methods, and are extremely useful for identifying crystalline materials in particulate matter and other materials. Particularly valuable is the fact that the samples are available for subsequent analysis by other methods. The method most widely used is generally referred to as the Debye-Scherer powder diffraction method in which the crystalline material is identified from its characteristic diffraction pattern, which is photographed on a film bent around 360° with the specimen at the center (71). Identification is made by comparison with reference data for several thousand compounds published by the American Society for Testing Materials (9). Generally, 5 to 10 mg of a powdered material suffice for the identification of the crystalline material. In addition to identifying the material, the character of the pattern also gives some information as to the size of the particles (71). Quantification makes use of the intensities of the diffracted lines, which are measured after photography on a film (15,70,71,259) or by automatic devices in which the film is replaced by a Geiger-Müller tube or other device, the output being recorded automatically on a strip chart (44,177,183,184,254,266). Many of the applications have been to the determination of quartz in industrial atmospheres and in the dust of city air (50,63,172,177,184, 254,266,292), but X-ray diffraction methods have also been used to identify calcium carbonate, iron oxide, and aluminum metal in the atmosphere (172).

A recent paper describes the application of a recording type of X-ray diffraction instrument to the determination of various toxicants in the industrial environment and for the identification of quartz, calcite, gypsum, Fe_2O_3, and Fe_3O_4 in the suspended matter of the atmosphere in Windsor and Ottawa, Canada (198).

Special applications of interest have been the identification of inclusions as small as 0.1μg in various metals and studies of the rare earths when present in 1 part in 10,000 parts of other material by the use of samples of 0.5 mg. It is of interest to record that, in the Los Angeles area, X-ray diffraction patterns showed a number of lines at positions which indicated the presence of long-chain alkyl-type organic liquids (292). General information on equipment, techniques, and applications is given in a number of recent publications and reviews (74,175,283).

X-ray diffraction equipment varies in cost from $5,000 to $15,000. The latter is the cost for the automatic recording instruments which are the most convenient and accurate for quantitative work. Accessory equipment for film processing, grinding of materials, etc., will add considerably to the cost of the equipment.

11.4.4 Mass Spectrometry

Mass spectrometry has attracted considerable attention in recent years and is particularly popular in the petroleum industry because of its relative speed and accuracy for refinery control and the handling of special analytical problems involving complex mixtures of gases and vapors (258,335). The method is quite sensitive and is of value since all materials and unexpected components are indicated. For qualitative purposes a large library of mass spectra must be available for comparison, and these are being accumulated gradually. Rock (258) summarizes the data for 279 compounds and the American Petroleum Institute (API Project 44) is accumulating more data (8). Methods for identifying mass peaks and details of sampling and analysis are described in a number of papers (41,258,295,335). The application of the method to the investigation of industrial-hygiene problems has been described for a number of toxic gases present in concentrations below 100 ppm and with the use of only 50-ml samples of air (152). As was true for the infrared method, trace quantities of atmospheric pollutants must be isolated and concentrated. Turk (319) used a carbon-absorption method for concentrating apple vapors resulting from the storage of apples. Shepherd and coworkers (272) removed air pollutants with a filter at liquid oxygen temperatures, separated the isolated frozen concentrate by isothermal distillation or sublimation at low temperatures and pressures, and estimated and identified the distillate with a mass spectrometer. A sensitivity of 10^{-4} ppm for some of the pollutants was obtained, and the total gas phase of the pollutants in Los Angeles smog was estimated to be of the order of 0.5 ppm. About 60 chemical compounds or families of compounds were identified or tentatively identified. It was shown that Los Angeles smog was primarily a mixture of hydrocarbons and of hydrocarbons combined with chlorine, nitrogen, and oxygen.

Suitable equipment for quantitative analytical purposes is said to cost about $30,000 (167). Many analytical applications of interest are given in the review by Dibeler and Hipple (94).

11.4.5 Polarographic Methods

Polarographic methods are particularly satisfactory for the analysis of dilute solutions and are employed for the analysis of materials in many fields. These methods are applicable to the determination of any reducible or oxidizable substance and can, therefore, be applied to the determination of a great many inorganic and organic materials. The sensitivities of detection under controlled conditions equal those of most chemical and physical methods, and the instrument, therefore, is enjoying some popularity in the industrial-hygiene field. Although it is possible to employ the method for the simultaneous determination of a number of materials, chemical manipulations, isolations, and concentrations are required for maximum sensitivity and specificity of detection. The theory, instrumentation, and methodology of polarography have been given in a number of reviews and books, among which may be mentioned those by Heyrovsky (157), Kolthoff and Lingane (185,186), Lingane (193, 201,202), Parks and Lykken (242), Muller (231), and Wawsonek (339).

Among the materials which are of interest because of their pollution potentials and which have been analyzed polarographically are antimony (200), arsenic (16,200), benzene (192), bismuth (200), cadmium (57,104,199,200,274), chromium (199,324), cobalt (203), copper (161,200), formaldehyde (138,30,205), lead (51,104,191,199, 200), manganese (199), nickel (203), the oxides of nitrogen (54), thallium (346), tin (200), titanium (135), vanadium (239), and zinc (55,191,200). The polarograph may be used to distinguish between a number of aldehydes which have been collected in a semicarbazide solution (205). A sensitivity of 0.1μg of aldehyde per milliliter of solution has been claimed for the method, which has proved satisfactory for such compounds as formaldehyde, acetaldehyde, butyraldehyde, crotonaldehyde, acetone, methylisobutyl ketone, and cyclohexanone (204). A review paper lists 146 references to the polarography of organic materials, many of which are potential air pollutants (339).

A manually operated polarograph can be assembled in the laboratory from standard components for a few hundred dollars. This type of equipment is satisfactory for occasional use, but the automatic recording types are recommended when many samples are handled daily. The approximate cost of the latter equipment is $1,500.

11.4.6 Microscopic Methods

Microscopy has played an important role in advancing our knowledge concerning the physical state of living and inanimate materials encountered in many fields of science, and its usefulness has been widened further by the introduction of the electron microscope. Theoretical considerations, equipment requirements, and general applications are presented in a number of publications (6,47,167).

The principal applications of microscopy in the field of air pollution have been to particle counting, the measurement of particle sizes, and the detection of quartz in atmospheric dust (37,39,40,97,166,176,245). Visual microscopy has also been employed to investigate particles of all types, including the smog of Los Angeles. In this application, physical and chemical techniques were employed to identify tarry materials, oily and aqueous droplets, crystals (mainly ammonium sulfate), nitrate, sulfate, halide, and ammonium carbonate (42,43). Limitations of some of the chemical tests were given, and the observation was made that much of the crystalline material in the air of Los Angeles was 5μ or larger and was chiefly ammonium sulfate (42). Various chemical techniques suitable for the detection of airborne particles by visual microscopy are reviewed by Crozier and Seely (84). Improved methods for characterizing airborne particulates by a surface-coating technique which extends the practical resolution limits of the visual microscope from 0.8μ to the theoretical limit of 0.2μ have also been described (300). Extension of the resolution limit to 0.2μ can also be obtained with dark-field illumination.

The electron microscope, which extends the visibility to 50 A (0.005μ), is an attractive tool for studying particles of submicron sizes collected from the atmosphere. Since the images are caused by scattering of the electron beam and show little index of refraction or color absorption, it is not possible to determine the optical properties of crystals (221). Nevertheless, useful and important contributions have been made, as in the identification of free silica by the combined chemical and electron-microscopic examination of certain dusts in the submicron size range (125). Another application has been to show that particles of submicron size may be responsible for Shaver's disease (168). The electron microscope was used to investigate the Los Angeles smog (43) and is being employed in conjunction with x-ray diffraction and microchemical techniques to examine small particles from the Windsor-Detroit atmosphere (176). A method for counting particles and determining the particle-size distribution of particles collected on Millipore filters by electron microscopy has been described as an absolute method for measuring solid airborne particulates (116). The techniques and basic problems involved in the proper use of the electron microscope for examining powdered materials are described by Watson (338) and by Schuster and Fullam (267). The electron microscope may also be used to obtain electron diffraction patterns of crystalline material, and this application opens a promising field for investigating trace quantities of impurities.

Electron microscopes suitable for detecting and counting extremely small particles (0.001μ and smaller in size) cost approximately $20,000. Table models ($6,000) are useful for specific applications but are not suitable for general research work.

11.4.7 Interferometry

Interferometry, because of insufficient specificity and sensitivity, is of little value for general air pollution work. The method is, however, useful for locating sourcee of pollution when only one pollutant is involved and for the analysis of a single gas in an otherwise stable emission. The method has found some use in the industrial-hygiene field and in toxicological investigations (36,75,154,244,291). The units of refractivity change for many compounds in air have been given in a number of publications (82,244,291).

11.4.8 Thermal Conductivity or Combustion Methods

Thermal conductivity or combustion methods are also of limited value in general air pollution investigations because of insufficient sensitivity and specificity of detection. In special situations (stack effluents or near immediate sources) these principles may be combined with other principles and with recorders to measure satisfactorily some low levels of industrial air pollution.

In most cases, the combustion instruments used as explosimeters or combustible-gas detectors depend on the disturbance of the balance of a Wheatstone-bridge arrangement when the resistance of a heated filament in one arm is altered by the burning of the combustible material on its surface. Other detectors "burn" the gas chemically and measure the temperature difference by means of thermocouples connected to a potentiometric device. The various instruments are described by Jacobs (166) in the bulletins of the manufacturers (90,228) and in the literature describing special applications such as the determination of carbon monoxide (24,77,173), oxygen, and carbon dioxide (150). The carbon monoxide detectors have been of value in general air pollution work, and they have been used successfully for the measurement of carbon monoxide in vehicular tunnels (77,110,174,216).

Instruments combining combustion and electrical-conductivity principles have also been developed for air analysis. Obviously such instruments are not specific and give results which should be considered only as an index of pollution. Such instruments have found application for the detection of hydrogen sulfide, mercaptans, and other sulfur and chlorine compounds in industrial atmospheres (311).

11.4.9 Miscellaneous Instrumental Methods

A number of specialized instrumental methods based on certain physical principles which do not fit into any of the foregoing articles are included in this article.

Radioactivity. Very sensitive and specific analytical methods employing radioactivity have been used for the detection of traces of a number of materials or impurities (29,131,141,162,347). Methods for measuring radioactivity following "dilution" or "activation" techniques have been employed. In the dilution method, a labeled (radioactive) isotope material is added to the inactive isotope and the specific activity of the quantitatively separated diluted material is measured with the proper counting equipment. This method requires the isolation of a weighable or measurable quantity of material and is very susceptible to contamination and purity errors. This technique has been widely used in biological investigations (169,252,256,347) to determine the vaporization rates of mercury (130) and in the air pollution field to study the efficiency of smoke filters (145).

The "activation" method (29) is probably the most sensitive analytical tool available to the analyst and depends on the artificial production of a radioactive isotope and the subsequent counting of its decay disintegrations. The method is free from interference because of the unique character of the decay periods of the radioisotopes, and its broad field of applicability may be appreciated when it is considered that approximately 700 isotopes have been identified for the known elements. Special counting equipment and the need for chemical separations are important considerations for this work. Since activation sources are quite expensive, a wider use of the technique will depend on the development of cheap sources. The possibilities of the technique are evident from the following reported work. One part of antimony in 10^9 parts of other material has been determined in the analysis of pure materials for antimony (163). Rock and mineral samples as small as 0.5 g were analyzed for as little as 0.0003 per cent uranium (280). The arsenic content of sea water was measured with 10-ml samples when the level was only 2.6μg/liter (282), and arsenic at a concentration level of 10^{-10} g has been determined in other materials (281).

Sonic Absorption. Sonic absorption has been applied to the measurement of particle size. The basis for this application is that particles in a sound field of any particular frequency will show a difference in the amplitude of vibration which can be correlated with particle size. The instrument used for this purpose consists of a sonic generator which can be regulated to produce any desired frequency, a thin observation cell in which the intense sound can be set up, an illuminating system (dark field), and means to observe the cell or to photograph it. In the case of photography, the tracks of the particles are employed in the measurement. A limitation is the small number of particles which can be photographed at one time (142).

Proton Scattering. Proton scattering is another unique instrumental method for analyzing minute samples of particulate matter. This technique was used in the Los Angeles area to investigate the nature of smog particles collected by means of the impactor (292). Peaks for C^{12}, O^{16}, S^{32}, and Pb^{207} were identified, although the latter identification was only tentative. This type of measurement shows considerable promise for identification and analysis of substances, and the Stanford Research Institute is conducting a research program on proton scattering.

11.4.10 Continuous Recorders

Continuous recorders are considered separately, although some applications have already been indicated in the foregoing articles. Automatic samplers and analyzers are particularly useful for the continuous monitoring of air pollution and for those situations where a great many analyses are needed at one or more spots for comparative purposes. Since the recorders presently available are nonspecific in performance, the records generally give only a measure of the pollution load, which should be amplified by a more specific analysis of the suspected impurity collected by an intermittent sampling technique. Well-designed surveys, therefore, will make use of information obtained by continuous and intermittent sampling procedures. Continuous recording instruments have been employed to measure a number of gaseous pollutants and the particulate matter in the atmosphere.

Gaseous Pollutants. Since sulfur dioxide is one of the more important impurities of the atmosphere, considerable effort has gone into the development of automatic continuous recorders for estimating this material. The Thomas "autometer," one of the earliest of the automatic and continuous analyzers, measures the electrolytic conductivity of sulfuric acid produced by the oxidation of sulfur dioxide as it is absorbed in slightly acidulated water containing hydrogen peroxide (311,312). "Accumulating" and "instantaneous" types have been used. The accumulating type records the conductivity continuously during the absorption period of 30 min, while the instantaneous type has been designed to record average concentrations for periods of time of 2 min and to give average values for periods of time of 30 min when an accumulating cell is added to the equipment (308). The latter instrument is available from the Leeds and Northrup Company, Philadelphia, Pa. Since any gas or aerosol producing an electrolyte in solution will also be recorded as sulfur dioxide, it is obvious that the autometer, except in special situations, measures a pollution load rather than sulfur dioxide specifically. Some of the problems and difficulties encountered in the field use of the autometer have been made the subject of special articles (122,247).

Another analyzer, the Titrilog, sensitive to 0.02 ppm sulfur dioxide, measures the current flow depending on the amount of sulfur dioxide bubbled through a bromide solution in an electrolytic cell (308). As in the case of the autometer, the Titrilog is not specific for sulfur dioxide, being sensitive to many oxidizable gases such as hydrogen sulfide, mercaptans, and unsaturated hydrocarbons.

Still another automatic sulfur dioxide recorder has been described in which the fading color of a starch-iodine solution is measured photometrically (171). This instrument is also subject to interference by hydrogen sulfide, etc., but, it is claimed, to a lesser extent than the Titrilog because neutral iodine is less reactive than bromine (308).

As is obvious from the foregoing discussion, other pollutants can also be detected in limited investigations by the autometer or the Titrilog. The Titrilog may be used for hydrogen sulfide or mercaptans, while the autometer can be used for any material forming an electrolyte in water directly or after combustion. Thus the autometer has been used to detect as little as 0.01 ppm hydrogen sulfide following combustion to sulfur dioxide on a platinum wire heated electrically to 550°C (308,311). The autometer has been used to measure hydrogen fluoride in the atmosphere, but the interference by other naturally occurring pollutants seriously limits the value of the record (313).

The selective quenching of the fluorescence of magnesium oxinate by hydrogen fluoride (107) is the principle employed in a continuous analyzer for hydrogen fluoride which is being developed at the Stanford Research Institute (46). A sensitivity of 0.1 part of hydrogen fluoride per billion parts of air has been obtained with the instrument. The instrument also has value for detecting free hydrogen fluoride or fluoride ions in special situations. Its value for the determination of total fluoride has not been established.

Recorders or continuous analyzers for hydrogen sulfide employing reflectance or transmissivity measurements of a lead acetate impregnated tape have also been developed (308). A hydrogen sulfide recorder for the range of 0.1 to 500 ppm is available from the Rubicon Company of Philadelphia. Such instruments are particularly useful for the control of plant atmospheres or of stack emissions, since they may be designed to sound a warning or even turn off a process if the concentration in the air exceeds a predetermined level.

Other applications of continuous recording devices for the analysis of gaseous contaminants are the use of "oxidant" recorders to determine ozone and nitric oxides (also free chlorine), either electrometrically or photometrically, by measuring the amount of iodine liberated from potassium iodide solutions (308).

An oxidant recorder which measures the quantity of iodine liberated from a buffered neutral potassium iodide solution was built by Littman and Benoliel (204), and an improved model based on the same principle was built in the author's laboratory (67). The latter instrument permits the recording of four events every minute through the use of a four-point Brown Elektronik model chart recorder. One of the channels records the transmittancy of the buffered potassium iodide solution before it passes to the scrubbers, thereby continuously indicating the zero level of the instrument. Two of the other channels measure the transmittancies of solutions after they have passed through two scrubbers. One of the scrubbers removes oxidant from the air, while the other scrubber removes oxidant from air after the air has been irradiated by ultraviolet light. Thus the one instrument measures the oxidant per se and any precursors which may be present. The other channel of the recorder is intended to be used to record the concentrations of sulfur dioxide on the same chart with the oxidant or ozone concentration. An oxidant recorder based on the oxidation-reduction of phenolphthalin (148) is being built by Beckman Instruments, Inc., of South Pasadena, Calif.

The recorders based on the use of neutral buffered potassium iodide solution will record oxidants at a level of 1 pphm of air when sampling at the rate of 5 liter/min. The interferences have been discussed earlier in the articles dealing with the titration and colorimetric methods.

A highly sensitive automatic instrument for the continuous detection and measurement of microquantities of gaseous impurities has recently made its appearance. The instrument, called the Microsensor, is said to be capable of detecting many gases in quantities as low as $0.1\mu g$/liter (328). The instrument is essentially a color comparator and is used to detect a color change produced on chemically treated paper tape. The instrument operates in cycles the lengths of which depend on the type of chemical action involved but do not exceed 10 sec. In its present form the instrument is designed to measure single contaminants, the specificity of detection depending on the microchemical reaction employed.

Particulate Matter. Automatic analyzers may be used for the direct analysis of aerosols. The smaller particles (less than 5μ) are considered to be of greater importance than the coarse material because they are believed to be the principal contributors to soiling and the reduction of visibility. They are also of greater hygienic importance than the larger particles, and therefore improved methods for their collection and measurement are being sought. Although many types of autosamplers (see above) have been developed, instruments for the continuous measurement of particulate matter in the air are very limited in number and type. One instrument measures the absorption or scattering of light in the atmosphere by recording the visibility directly on a recorder (48). The principle is the same as that used by Steffens (296) except that the camera is replaced by a photocell and amplifier to operate the recorder. The recorded visibility can be correlated with concentration and size of the particulate matter by the proper calibration of the instrument. The principle of light scattering has also been used for the photoelectronic counting of particles. Light scattered by the moving stream of smoke or dust is collected by a lens and falls on the surface of a photosensitive cell. The electric pulses from this cell are then amplified and pass to the grid of a thyratron trigger circuit which can be adjusted to cut out all pulses below a predetermined value (143,144). Various types of instruments for measuring the light absorptivity or scattering of light by smoke or other aerosols and adaptable to recording devices have been described in a number of reviews (78,143,325). The sensitivity of the direct photoelectric counting is limited by many factors, and the applicability of these instruments to counting dust in the field needs further study.

Particulate matter has also been counted electrostatically (146). In this instrument, air is drawn through an 0.8-mm orifice into an evacuated space where the particles impinge on a metal wire maintained at a suitable potential. The individual particles produce tiny electrical impulses on the wire, which are amplified by a factor of 100,000. The output pulses then pass through a discriminating potentiometer to a thyratron tube to actuate a mechanical counter (or recorder). The original instrument was sensitive to particles 2.5μ or larger in diameter. Sensitivities as to size may be improved, but the practical lower limit is considered to be 1μ in size (143).

Devices for continuously measuring sulfuric acid or other airborne particulate electrolytes have been developed by the Stanford Research Institute and the American Smelting and Refining Company (264,313). In the Stanford Research Institute instrument, the aerosol is deposited by electrostatic precipitation on a rotating stainless-steel disk. The disk dips into a flowing stream of water which washes off the electrolyte, dissolves it, and measures it continuously as the rinse water flows through a conductivity cell. The American Smelting and Refining Company instrument also measures the electrolyte by means of a conductivity cell but employs a modified impactor to trap the aerosol. Obviously these instruments measure all the dissolved electrolytes, but they should be quite useful in special locations downwind from known sources of sulfuric acid mist or other electrolyte.

Another example of an automatic sampler and analyzer is the instrument designed to monitor natural atmospheric radioactivity. In this instrument, the aerosol is deposited electrostatically on a moving metal tape which passes under a counter, the count being recorded continuously on a chart (342).

A discussion of continuous recorders would hardly be complete without some mention of the several modifications of the Owens automatic atmospheric analyzer which have been introduced during the past few years (149,155,313,330). In all of these, dust is deposited on a filter-paper disk or tape, which is then returned to the laboratory for examination by transmitted or reflected light. In the Owens instrument circular stains ⅛ in. in diameter are produced near the periphery of a 7-in. circle of white filter paper. The stains are then compared with standard shades and colors (223). The modifications have been introduced to increase the sampling rate (up to 40 cfm) (149) and to provide larger areas for measuring. Although the author knows of no instrument where the measurement is recorded automatically, it should not be too difficult to design equipment to accomplish this. In most of the instruments, the paper is moved only after definite periods of time, but in one instrument the paper moves continuously at any desired rate (330). Records by these instruments are of value for comparative purposes in a single community since they measure, primarily, the blackness due to the carbon in the particulate matter. The instruments must, therefore, be calibrated for each community, and comparisons of results for different communities are difficult if not impossible. Interpretation as to mass likewise depends on careful calibration and standardization, with the same limitations cited for the counts.

11.5 BIOLOGICAL METHODS

11.5.1 Effect on Vegetation and Animals

The identification of a pollutant and the pinpointing of the source responsible for its emission are frequently possible through analysis of vegetation and of biological fluids from animals in an affected area. In the case of exposed or fatally poisoned animals, the body tissues and organs become useful material for establishing the existence of a pollution load, although the interpretations based on such analytical results are not always clear-cut because of the inadequate nature of much of the information concerning the effects of pollutants on living things and, what is more basic, because of the lack of adequate information concerning the distribution of certain materials in normal plants or animals. Of the very great number of impurities present in the atmosphere, only a few have been investigated even sketchily, and much remains to be done in this field.

Exposure of vegetation to some pollutants will produce damage (85,147,170,234, 309,310,349) which can be recognized by skilled plant physiologists and can be used to establish the existence and identity of an air pollutant in an area. In certain other situations, only a chemical analysis will be helpful, as when investigating the effect of particulate matter emitted in a particular area. Historic examples of damage to livestock caused by the emissions of arsenic and lead compounds into the atmosphere have been described (305), and since our knowledge of the toxicity of both of these metals is reasonably satisfactory, quantification of suitable biological samples by any of the methods given in the previous articles can be used to establish the facts concerning these materials.

A considerable fund of information on the effect of fluoride on plants and animals is also available (69,164,194,229,257) so that chemical analysis of suitable biological material will be helpful in determining whether an area is contaminated with fluoride-bearing material. An interesting application of the use of biological material for determining air pollution is illustrated by the use of Spanish moss to measure pollution by fluoride. In one instance it was demonstrated that the initial fluoride content of unsheltered samples of this plant had been enhanced fourfold by exposure to atmospheric conditions over the period of 3 months (209). Rain waters have also been used to measure the periodic washdown of fluoride and to establish the origin and extent of dispersion of fluoride-containing contaminants (2,209).

The effects caused by the lack or excess of certain other elements in plants and animals are described in a number of monographs (79,123,298). Analytical methods for these elements may also be chosen from the various references in this article.

11.5.2 Sensory Responses

A number of organoleptic methods have found use in determining the extent and severity of atmospheric pollution. Odor, lachrymation, and respiratory irritant effects are factors of importance in the sensory judging of the severity of air pollution. In all cases, quantification is severely limited by the many individual and physiological factors which affect perception. Nevertheless these methods are not to be discounted in obtaining a rapid appraisal of a situation.

In the appraisal of odor, which has been investigated most thoroughly, the trend is to eliminate the factor of human variability in both acuity and specificity of sensory perception, by developing impersonal devices based on physical or chemical phenomena. The use of mass spectrography or the absorption of specific wavelengths of the electromagnetic spectrum by the odorant are promising in special situations. In other cases, selective chemical methods appear to be satisfactory. For example, determination of the volatile products of the spoilage of fish has been developed in the form of an iodometric measurement of the volatile reducing substances which could be trapped in a standard solution of potassium permanganate in sodium hydroxide (102). A chemical method would also be satisfactory for any material which can be identified by its characteristic odor and for which a specific chemical method is available.

In spite of its limitation, a subjective test of odor, based on threshold values, provides useful preliminary information concerning limiting values when the method is employed by well-trained personnel. A serious defect in this approach is the scantiness and the contradictory quality of the published data on the threshold values for human perception (82,83,220). In the case of several common impurities, also, threshold values for olfactory detection are recorded at concentrations greatly in excess of those which occur in the general atmosphere. A few examples are chlorine 3.5 ppm, aldehydes 0.2 to 3 ppm, ozone 0.5 ppm, phenol 1 ppm, phosgene 5.6 ppm, and sulfur dioxide 3 to 4 ppm (166,331). Extremely low quantities of odorant may sometimes be concentrated by the use of "accumulators." In this method, petri dishes containing an absorbent (water, glycerol, activated carbon, etc.) are permitted to stand in selected spots for 24 hr or more. The dishes are then examined in an odorless room. Frequently, information as to the nature of an odorous material, as well as its distribution and relative intensity at various points, can be obtained (83). The physiology and control of odors have been reviewed in an excellent monograph by McCord and Witheredge (220).

11.5.3 Lachrymation and Irritation

Lachrymation and irritation of the mucosa of the respiratory tract are other sensory responses which have been used to estimate the concentrations of atmospheric pollutants. Except for certain war gases, however, data concerning irritant effects are either lacking or involve concentrations not likely to occur in the external atmosphere (166,210,211,249,263,275,327,331). Some of the evidence indicates that relatively high concentrations of a single material may be required to cause any irritant effect. For example, 8 to 12 ppm of sulfur dioxide is required to cause coughing, while 20 ppm is needed to cause irritation of the eyes (166). Hydrogen sulfide will cause some conjunctivitis at a level of 10 ppm (91), and phosgene must be present at a level of 3.1 ppm to affect the throat or at 4 ppm to cause irritation of the eyes (166). One part of acrolein per million parts of air will irritate the eyes, while only one-tenth of this concentration of ozone is said to have a powerful irritating action on the mucous membrane (166).

A limited amount of experimental work has been carried out in attempts to establish the relation of certain materials to irritation of the eyes (88,214,292). Acrolein, formaldehyde, hydrogen persulfide, elemental sulfur, oxides of nitrogen, sulfur dioxide, and sulfur trioxide have been tested, and many of the suspected irritants have been eliminated as causes of the lachrymation induced by smog in Los Angeles (214). Of special interest were the low concentrations of acrolein (0.5 to 1.5 ppm), hydrogen persulfide (0.2 ppm), and elemental sulfur (0.1 ppm) which caused lachrymation. Additive or combined effects were observed to be minute in most cases, but a surprising enhancement of the irritant effects was observed to be caused by fine oil droplets. The concentrations in the Los Angeles atmosphere of the material tested were generally below the threshold concentrations noted above (214).

11.6 SUMMARY OF ANALYTICAL PROCEDURES

Method	Pollutant or class	Sensitivity range	Limitations or advantages	Examples and remarks
A. Chemical				
1. Gravimetry (11.3.1)	Gases	Low (mg)	Generally nonspecific; frequently requires large samples	Applied to material absorbed on silica gel, activated carbon, etc.; weighing of precipitates obtained following the use of special reagents. Best example: determination of SO_2 from total SO_4 precipitated as $BaSO_4$.
	Dirtfall Sootfall	Low (mg)	Nonspecific. An index of pollution usually expressed as tons of dirtfall per sq mile/month. Long periods of collection (30 days). Best as average value from repeated tests	Collected in cylindrical jars on slides or on surface of cylinders. Can be used for chemical analysis
	Airborne particulate matter	0.1 mg	May require microbalance. Nonspecific index expressed as mg/cu m of air. Relatively large samples	Collected on filter paper, by impinger, or by electrostatic precipitator. Material can be used for number counting, particle-size determination, chemical examination
2. Volumetry (11.3.2)				
a. Acidimetry Alkalimetry	Acid and basic materials	High to low (0.01 ppm lower limit)	Large samples, nonspecific determination	Chiefly for industrial environments. Examples: SO_2 oxidized to H_2SO_4 and ammonia by direct titration, sulfuric acid
b. Oxidation and reduction (iodometry)	Gaseous materials generally	Very high, frequently less than 0.01 ppm	Large samples, nonspecific	Specially valuable for use in known atmospheres. Many methods satisfactory for screening purposes. Examples: SO_2, chlorine aldehydes, oxidants, ozone. Probably most sensitive and best method for oxidants
3. Colorimetry (11.3.3)	Particulate matter, gases	Many tests in the submicrogram range	Many tests suitably specific. Frequently requires large samples, isolation and concentration of test material	Includes some of the better tests. Particularly satisfactory in the case of metal analyses, formaldehyde, NO_2, NH_3, Cl_2, H_2S, F, CO, CO_2. Inorganic and organic materials. Can be used for SO_2
4. Nephelometry or turbidimetry (11.3.4)	Particulate matter, gases	SO_4, 0.005 ppm Cl_2, 0.005 ppm	Specific. Moderate-size sample. Simple procedure	Preferred by author for determination of index of total SO_2 and Cl_2 from turbidity of total sulfate and chloride

11.6 SUMMARY OF ANALYTICAL PROCEDURES (*Continued*)

Method	Pollutant or class	Sensitivity range	Limitations or advantages	Examples and remarks
5. Chromatography (11.3.5)	All classes of materials	Extremely high. Many in the submicrogram range	Tests are often specific. Useful for metals, inorganic and organic compounds	Has been used to identify traces of materials in biological investigations. Not fully investigated for air pollution studies but is a promising tool. Used to determine benzpyrene in the urban atmosphere
B. Instrumental				
1. Spectrography (11.4.1)	Elements, metals	Varies; very high for alkalies and alkaline earths	Most specific method for metals and elements. 72+ elements are detectable. Quantification requires individual study and frequently isolation and concentration of test material. Simultaneous determination of several elements possible. Equipment is expensive	Chiefly used for exploratory work. Much used for dusts, particulates, most specific for Be, most sensitive for Be, Na, K, Ca, Pb, Cd, Sb, among others
2. Spectrophotometry (11.4.2)				
a. Visible region	All classes of materials	Variable, depends on material	Specificity frequently satisfactory. Requires isolation, concentration	Best used in colorimetric evaluations
b. Ultraviolet region	Organic material	Low sensitivity generally	Usually requires large samples, isolation and concentration. Large collection of pure material for reference required. Equipment cost moderate	Best use is for industrial atmosphere. Benzene, aromatics, etc.
c. Infrared region	Organic and inorganic materials	Rarely better than 0.01% (100 ppm)	Equipment quite costly. Requires large samples, isolation and concentration. Specific, provided adequate reference data are available. May require establishment of own reference data	Very valuable for determining nature of organic material from large samples. Gives nature of molecules and hence is useful for screening purposes
d. Fluorescence	Organic and inorganic materials	Frequently high, of the order of a microgram or less	Rarely specific. Requires isolation and concentration	Used for Be, Sb, HF, spot tests, etc.
3. X-ray diffraction (11.4.3)	Crystalline organic and inorganic materials	Rarely less than 1%	One of the best methods to identify crystalline material. Requires large samples and large reference library. Equipment is very expensive	Used to determine nature of crystalline compounds. Widely used for quartz, etc., in air samples

11.6 SUMMARY OF ANALYTICAL PROCEDURES (*Continued*)

Method	Pollutant or class	Sensitivity range	Limitations or advantages	Examples and remarks
4. Mass spectrometry (11.4.4)	Organic and inorganic material	Sensitivity varies; as low as 0.0001 ppm possible	Specific. Requires moderately large samples, isolation and concentration. Materials must be gasified. Equipment is expensive	Used to identify compounds, elements, and structure. Has had some application to air pollution investigation but chiefly used in petroleum industries
5. Polarography (11.4.5)	Organic and inorganic material	Many tests in the microgram range	Good specificity. Good accuracy of analysis. Frequently permits simultaneous determinations of a number of materials but more often requires isolation and concentration. Equipment cost is moderate	Recommended for every laboratory as a screening and exploratory tool which can be obtained at moderate cost
6. Microscopic (11.4.6)				
a. Visual microscopy	Organic and inorganic material	Extremely sensitive, frequently in the microgram range	Requires isolation and concentration. Complex micro manipulations at times. Nonquantitative except for number counts and particle-size determination. Moderately expensive	Should be available for examination of particulate matter
b. Electron microscopy	Organic and inorganic material	High sensitivity	Best for number counting of particles in the submicron range. Preparation of material difficult and specialized. Electron diffraction can be used like X-ray diffraction but at much lower concentration. Equipment costly	Very useful when available. Gives absolute analysis of particulate matter. Special training required
7. Interferometry (11.4.7)	Gases, liquids	Low; in the per cent range	Not specific for use in complex atmospheres. Equipment moderately expensive	Chiefly for industrial application. Useful for known contaminants. Good for cage and effluent stack-gas analysis
8. Thermal conductivity; combustion (11.4.8)	Organic and inorganic material	Generally low sensitivity	Nonspecific. Requires decomposition by combustion	Of little value for air pollution work. Chiefly useful for in-plant and stack analysis, sewer repair, etc.
9. Radioactivity (11.4.9)	Organic and inorganic material	Most sensitive	Specific, requires considerable handling and preparation. Requires activation source	Promising approach in laboratories provided with personnel trained in this type of analysis

11.6 SUMMARY OF ANALYTICAL PROCEDURES (*Continued*)

Method	Pollutant or class	Sensitivity range	Limitations or advantages	Examples and remarks
10. Continuous recorders (11.4.10)				
a.	Gases	Frequently highly sensitive	Generally nonspecific but excellent as monitors to determine pollution loads, peak intensities, and duration. Some of the equipment is quite inexpensive. Many commercially available	Necessary for adequate investigation of air pollution. Should be used in conjunction with intermittent sampling by methods of suitable specificity. Examples: total SO_2, oxidant, H_2S, HF
b.	Particulate matter	High sensitivity	Generally same as 10*a*. Can be standardized against weight of particles. Standards apply to each community or situation only. Generally measures soiling property of dust or reduction of visibility. Commercially available at moderate cost	Same as 10*a*. Material can be used for chemical analysis. Machines available for determination of particle sizes, visibility, total particulate matter, etc.
C. Biological				
1. Effect on vegetation and animals (11.5.1)	Gases, particulate matter	Many tests extremely sensitive. Be, 0.0003μg; Pb, 0.01μg	Inadequate information concerning effect of some pollutants on living things. May require large samples. Isolation and concentration frequently necessary. Many tests highly specific	Knowledge concerning Pb, As, F, satisfactory
2. Sensory responses (11.5.2)				
a. Irritation *b.* Odor	Gases, particulate matter	Low order of sensitivity	Affected by physiological and individual factors. Nonquantitative	For preliminary appraisal by well-trained personnel

REFERENCES

1. Ackerman, H. H.: The Determination of Thallium in Urine, *J. Ind. Hyg. Toxicol.*, **30,** 300–302 (1948).
2. Adams, D. F., D. J. Mayhew, R. M. Gnagy, E. P. Richey, R. K. Koppe, and I. W. Allen: Atmospheric Pollution in the Ponderosa Pine Blight Area, *Ind. Eng. Chem.*, **44,** 1356–1365 (1952).
3. Adams, M. F.: Determination of Small Amounts of Silica, *Ind. Eng. Chem., Anal. Ed.*, **17,** 542–543 (1945).
4. Aldridge, W. N., and H. F. Liddell: A New Method for the Microdetermination of Beryllium with Particular Reference to Its Determination in Biological Materials, *Analyst*, **73,** 607–613 (1948).
5. Alexander, O. R., E. M. Godar, and N. J. Linde: Spectrophotometric Determination of Traces of Nickel, *Ind. Eng. Chem., Anal. Ed.*, **18,** 206–208 (1946).
6. Allen, R. M.: "The Microscope," D. Van Nostrand Company, Inc., New York, 1940.
7. Aluminum Research Laboratories: Determination of Fluorine, Standard Methods of Analysis, No. 914 (12–44), Aluminum Company of America.
8. American Petroleum Institute Research Project 44: Collection, Calculation and Compilation of Data on the Physical, Thermodynamic and Spectral Properties of Hydrocarbons and Related Compounds, Including Catalog of Mass Spectral Data, Carnegie Institute of Technology, Pittsburgh, Pa.
9. American Society for Testing Materials: X-ray Diffraction Data, Philadelphia, Pa., 1950.
10. Andrews, H. L., and D. C. Peterson: Study of the Efficiency of Methods for Obtaining Vapor Samples in Air, *J. Ind. Hyg. Toxicol.*, **29,** 403–407 (1947).
11. Anselm, C. D., and R. J. Robinson: The Spectrophotometric Determination of Fluoride in Sea Water, *J. Marine Research* (*Sears Foundation*), **10,** 203–214 (1951).
12. Atkin, S.: Determination of Sulfur Dioxide in Presence of Sulfur Trioxide, *Anal. Chem.*, **22,** 947–948 (1950).
13. Aughey, H.: A Rapid Mobile Analyzer for Minute Amounts of Lead in Air, *J. Opt. Soc. Amer.*, **39,** 292–293 (1949).
14. Austin, C. M., and J. S. McHargue: The Determination of Total Boron in Plant Material with Chromotrope-B, *J. Assoc. Offic. Agr. Chemists*, **31,** 427–431 (1948).
15. Ballard, J. W., and H. H. Schrenk: Routine Quantitative Analysis by X-ray Diffraction, *U.S. Bur. Mines, Rept. Invest.* 3888, 1946.
16. Bambach, K.: Polarographic Determination of Arsenic in Biological Material, *Ind. Eng. Chem., Anal. Ed.*, **14,** 265–267 (1942).
17. Barnes, E. C., and H. W. Speicher: The Determination of Formaldehyde in Air, *J. Ind. Hyg. Toxicol.*, **24,** 10–17 (1942).
18. Barnes, E. C., W. E. Piros, T. C. Bryson, and G. W. Wiener: Spectrochemical Determination of Beryllium in Microquantities, *Anal. Chem.*, **21,** 1281–1283 (1949).
19. Barnes, H.: A Modified 2,4-Xylenol Method for Nitrate Estimations, *Analyst*, **75,** 388–391 (1950).
20. Barnes, R. B., R. C. Gore, U. Liddel, and V. Z. Williams: "Infrared Spectroscopy, Industrial Applications and Bibliography," Reinhold Publishing Corporation, New York, 1944.
21. Barnes, R. B., D. Richardson, J. W. Berry, and R. L. Hood: Flame Photometry. A Rapid Analytical Procedure, *Ind. Eng. Chem., Anal. Ed.*, **17,** 605–611 (1945).
22. Barreto, A.: A New Method for Qualitative and Quantitative Analysis of Nitric Acid and Nitrates with the Aid of Chloranil, *Rev. quím. ind.* (*Rio de Janeiro*), **10,** No. 115, 12 (376) (1941).
23. Beckman, A. O., J. D. McCullough, and R. A. Crane: Microdetermination of Carbon Monoxide in Air. A Portable Instrument, *Anal. Chem.*, **20,** 674–677 (1948).
24. Berger, L. B., and H. H. Schrenk: Methods for the Detection and Determination of Carbon Monoxide, *U.S. Bur. Mines, Tech. Paper* 582, 1938.
25. Birdsall, C. M., A. C. Jenkins, and E. Spadinger: Iodometric Determination of Ozone, *Anal. Chem.*, **24,** 662–664 (1952).
26. Blaedel, W. J., and F. E. Blacet: Colorimetric Determination of Formaldehyde in the Presence of Other Aldehydes, *Ind. Eng. Chem., Anal. Ed.*, **13,** 449–450 (1941).
27. Block, R. J., et al.: "Paper Chromatography," Academic Press, Inc., New York, 1952.
28. Bodnár, J., and T. Török: Photometric Determination of Silicic Acid in Biological Materials, *Hoppe-Seyler's Z. physiol. Chem.*, **261,** 258 (1939).
29. Boyd, G. E.: Method of Activation Analysis, *Anal. Chem.*, **21,** 335–347 (1949).

30. Boyd, M. J., and K. Bambach: Polarographic Determination of Formaldehyde in Biological Material, *Ind. Eng. Chem., Anal. Ed.*, **15,** 314–315 (1943).
31. Bradley, C. E., and A. J. Haagen-Smit: The Application of Rubber in the Quantitative Determination of Ozone, *Rubber Chem. and Technol.*, **24,** 750–755 (1951).
32. Bricker, C. E., and H. R. Johnson: Spectrophotometric Method for Determining Formaldehyde, *Ind. Eng. Chem., Anal. Ed.*, **17,** 400–402 (1945).
33. Bricker, C. E., and P. B. Sweetser: Photometric Determination of Microquantities of Arsenic, *Anal. Chem.*, **24,** 409–411 (1952).
34. Brieger, H., F. Rieders, and W. A. Hodes: Acrylonitrile: Spectrophotometric Determination, Acute Toxicity and Mechanism of Action, *Arch. Ind. Hyg. and Occupational Med.*, **6,** 128–140 (1952).
35. Brode, W. R.: "Chemical Spectroscopy," 2d ed., John Wiley & Sons, Inc., New York, 1943.
36. Brodskii, A. I.: The Use of the Interferometer in Industrial Analysis, *Khim. Referat. Zhur.*, **1940,** No. 2, 56–57 [*Chem. Abstracts*, **36,** 1257 (1942)].
37. Brown, C. E., and W. P. Yant: A Microprojector for Determining Particle-size Distribution and Number Concentration of Atmospheric Dust, *U.S. Bur. Mines, Rept. Invest.* 3289, 1935.
38. Brown, C. E., and H. H. Schrenk: Standard Methods of Measuring Extent of Atmospheric Pollution, *U.S. Bur. Mines, Inform. Circ.* 7210, 1942.
39. Brown, C. E., and H. H. Schrenk: Relation between and Precision of Dust Counts (Light- and Dark-field) from Simultaneous Impinger, Midget-impinger, Electric-precipitator, and Filter Paper Samples, *U.S. Bur. Mines, Rept. Invest.* 4568, 1949.
40. Brown, C. E., M. Fisher, and F. F. Boyer: The Size of Smallest Particles Determined in Impinger Dust Counting Methods, *U.S. Bur. Mines, Rept.* 4802, 1951.
41. Burk, R. E., and O. Grummitt: "Recent Advances in Analytical Chemistry," Interscience Publishers, Inc., New York, 1949.
42. Cadle, R. D.: Determination of Composition of Air-borne Particulate Material, *Anal. Chem.*, **23,** 196–198 (1951).
43. Cadle, R. D., S. Rubin, C. I. Glassbrook, and P. L. Magill: Identification of Particles in Los Angeles Smog by Optical and Electron Microscopy, *Arch. Ind. Hyg. and Occupational Med.*, **2,** 698–715 (1950).
44. Carl, H. F.: Quantitative Mineral Analysis with a Recording X-ray Diffraction Spectrometer, *Am. Mineralogist*, **32,** 508–517 (1947).
45. Cassidy, H. G.: "Technique of Organic Chemistry," vol. V, Adsorption and Chromatography, Interscience Publishers, Inc., New York, 1951.
46. Chaikin, S. W., C. I. Glassbrook, and T. D. Parks: An Instrument for the Continuous Analysis of Atmospheric Fluoride, abstracts of papers presented at the 123d meeting of the American Chemical Society, p. 21B, March, 1953.
47. Chamot, E. M., and C. W. Mason: "Handbook of Chemical Microscopy," 2d ed., John Wiley & Sons, Inc., New York, 1938.
48. Chaney, A. L.: Direct Photography of Aerosol Suspensions, in L. C. McCabe, ed., "Air Pollution," pp. 603–606, McGraw-Hill Book Company, Inc., New York, 1952.
49. Cholak, J., et al. (Subcommittee on Chemical Procedures): Methods for Determining Lead in Air and in Biological Materials, American Public Health Association, New York, 1944. (Revised edition in press.)
50. Cholak, J.: The Nature of Atmospheric Pollution in a Number of Industrial Communities, *Proc. 2d Nat. Air Pollution Symposium*, pp. 6–15, Stanford Research Institute, Los Angeles, Calif., 1952.
51. Cholak, J., and K. Bambach: Determination of Lead in Biological Material. A Polarographic Method, *Ind. Eng. Chem., Anal. Ed.*, **13,** 583–587 (1941).
52. Cholak, J., and R. V. Story: Spectrochemical Determination of Trace Metals in Biological Material, *J. Opt. Soc. Amer.*, **31,** 730–738 (1941).
53. Cholak, J., D. M. Hubbard, and R. V. Story: Determination of Aluminum in Biological Material, *Ind. Eng. Chem., Anal. Ed.*, **15,** 57–60 (1943).
54. Cholak, J., and R. R. McNary: Determination of the Oxides of Nitrogen in Air, *J. Ind. Hyg. Toxicol.*, **25,** 354–360 (1943).
55. Cholak, J., D. M. Hubbard, and R. E. Burkey: Determination of Zinc in Biological Material. Photometric and Polarographic Methods Following Extraction with di-B-naphthylthiocarbazone, *Ind. Eng. Chem., Anal. Ed.*, **15,** 754–759 (1943).
56. Cholak, J., and D. M. Hubbard: Spectrochemical Analysis with the Air-acetylene Flame, *Ind. Eng. Chem., Anal. Ed.*, **16,** 728–735 (1944).
57. Cholak, J., and D. M. Hubbard: Determination of Cadmium in Biological Material. Spectrographic, Polarographic and Colorimetric Methods, *Ind. Eng. Chem., Anal. Ed.*, **16,** 333–336 (1944).
58. Cholak, J., and D. M. Hubbard: Microdetermination of Mercury in Biological Material, *Ind. Eng. Chem., Anal. Ed.*, **18,** 149–151 (1946).

59. Cholak, J., and D. M. Hubbard: The Microdetermination of Antimony in Biological Material. A Spectrographic Method, *J. Ind. Hyg. Toxicol.*, **28,** 121–124 (1946).
60. Cholak, J., D. M. Hubbard, and R. E. Burkey: Microdetermination of Lead in Biological Material. With Dithizone Extraction at High pH, *Anal. Chem.*, **20,** 671–672 (1948).
61. Cholak, J., and D. M. Hubbard: Spectrographic Determination of Beryllium in Biological Material and Air, *Anal. Chem.*, **20,** 73–76 (1948).
62. Cholak, J., and D. M. Hubbard: Spectrochemical Determination of Beryllium. Increased Sensitivity of Detection in the Cathode Layer, *Anal. Chem.*, **20,** 970–972 (1948).
63. Cholak, J., L. J. Schafer, and R. F. Hoffer: Collection and Analysis of Solids in Urban Atmospheres, *Arch. Ind. Hyg. and Occupational Med.*, **2,** 443–453 (1950).
64. Cholak, J., L. J. Schafer, and R. F. Hoffer: Results of a Five-year Investigation of Air Pollution in Cincinnati, *Arch. Ind. Hyg. and Occupational Med.*, **6,** 314–325 (1952).
65. Cholak, J., and D. M. Hubbard: An Improved Method for Beryllium Analysis in Biological and Related Materials, *Am. Ind. Hyg. Assoc. Quart.*, **13,** 125–128 (1952).
66. Cholak, J., L. J. Schafer, and D. Yeager: Unpublished data, 1953.
67. Cholak, J., and W. J. Younker: An Improved Oxidant Recorder, unpublished data, 1953.
68. Church, F. W.: A Mixed Color Method for the Determination of Cadmium in Air and Biological Samples by the Use of Dithizone, *J. Ind. Hyg. Toxicol.*, **29,** 34–40 (1947).
69. Churchill, H. V., R. J. Rowley, and L. N. Martin: Fluorine Content of Certain Vegetation in a Western Pennsylvania Area, *Anal. Chem.* **20,** 69–71 (1948).
70. Clark, G. L., and D. H. Reynolds: Quantitative Analysis of Mine Dusts. An X-ray Diffraction Method, *Ind. Eng. Chem.*, *Anal. Ed.*, **8,** 36–40 (1936).
71. Clark, G. L.: "Applied X-rays," 4th ed., McGraw-Hill Book Company, Inc., New York, 1955.
72. Clayton, G. D.: Epidemiologic Approach, *Proc. 2d Nat. Air Pollution Symposium*, pp. 110–116, Stanford Research Institute, Los Angeles, Calif., 1952.
73. Clayton, G. D.: Determination of Atmospheric Contaminants, American Gas Association, New York, 1953.
74. Clifton, D. F., and C. S. Smith: Microsampling and Microanalysis of Metals, *Rev. Sci. Instr.*, **20,** 583–586 (1949).
75. Cole, P. A., and D. W. Armstrong: The Analysis of Toluene or Benzene by an Ultraviolet Absorption of Vapor Mixtures and the Determination of Concentration of Toluene in Air by Use of the Rayleigh-Jeans Interference Refractometer, *J. Opt. Soc. Amer.*, **31,** 740–742 (1941).
76. Committee on Standard Methods, American Conference of Governmental Industrial Hygienists (Recommended Method): Determination of Manganese in Air. Periodate Oxidation Method, July 29, 1949.
77. Cook, W. A.: A Review of Automatic Indicating and Recording Instruments for Determination of Industrial Atmospheric Contaminants, *Am. Ind. Hyg. Assoc. Quart.*, **8,** 42–48 (1947).
78. Coolidge, J. E., and G. J. Schulz: Photoelectric Measurement of Dust, *Instruments*, **24,** 534 (1951).
79. Corrie, F. E.: Some Elements of Plants and Animals, *Fertiliser J.*, London, 1948.
80. Crabtree, J., and A. R. Kemp: Accelerated Ozone Weathering Test for Rubber, *Ind. Eng. Chem.*, *Anal. Ed.*, **18,** 769–774 (1946).
81. Crabtree, J., and R. H. Erickson: Atmospheric Ozone—A Simple Approximate Method of Measurement, *India Rubber World*, **125,** 719–720 (1952).
82. Cralley, L. V.: "Air Pollution Abatement Manual," chap. 7, Analytical Methods, Manufacturing Chemists' Association, Inc., Washington, D.C., 1951.
83. Crocker, E. C., and L. B. Sjöström: Odor Detection and Thresholds, *Chem. Eng. News*, **27,** 1922–1925, 1949.
84. Crozier, W. D., and B. K. Seely: Some Techniques for Sampling and Identifying Particulate Matter in the Air, *Proc. 1st Nat. Air Pollution Symposium*, Stanford Research Institute, Los Angeles, Calif., 1949.
85. Czech, M., and W. Notdurft: Investigation on Injuries of Agricultural and Horticultural Crop Plants by Gas of Chlorine, Nitrogen Dioxide, and Sulfur Dioxide, *Landwirtsch Forsch.*, **4,** 1–36 (1952).
86. Dahle, D.: A New Procedure for Determination of Fluorine by the Peroxidized Titanium Method, *J. Assoc. Offic. Agr. Chemists*, **20,** 505–516 (1937).
87. Dahle, D., R. U. Bonnar, and H. J. Wichmann: Titration of Small Quantities of Fluorides with Thorium Nitrate. I. Effect of Changes in the Amount of Indicator and Acidity, *J. Assoc. Offic. Agr. Chemists*, **21,** 459–467 (1938).

88. Dautrebande, L., R. Capps, and E. Weaver: Studies on Aerosols. IX. Enhancement of Irritating Effects of Various Substances on the Eye, Nose, and Throat by Particulate Matter and Liquid Aerosols in Connection with Pollution of the Atmosphere, *Arch. intern. pharmacodynamie*, **82**, 505–528 (1950).
89. Davidow, B., and G. Woodard: An Ultraviolet Spectrophotometric Method for the Determination of Benzene Hexachloride, *J. Assoc. Offic. Agr. Chemists*, **32**, 751–758 (1949).
90. Davis Emergency Equipment Co.: Davis Vapotester, Newark, N.J.
91. Davis, R. F., and M. A. Elliott: The Removal of Aldehydes from Diesel Exhaust Gas, *Trans. Am. Soc. Mech. Engrs.*, **70**, 745–750 (1948).
92. DeMeio, R. H.: Microdetermination of Tellurium V, *Anal. Chem.*, **20**, 488–489 (1948).
93. Deniges, G.: The Detection of Traces of Formaldehyde in the Presence of Acetaldehyde by Fuchsin Bisulfite, *Compt. rend.*, **150**, 529–531 (1910).
94. Dibeler, V. H., and J. A. Hipple: Mass Spectrometry, *Anal. Chem.*, **24**, 27–31 (1952).
95. Dick, A. T., and J. B. Bingley: The Determination of Molybdenum in Plant and in Animal Tissues, *Australian J. Exptl. Biol. Med. Sci.*, **25**, 193–202 (1947).
96. Dobson, G. M. B.: A Photoelectric Spectrophotometer for Measuring the Amount of Atmospheric Ozone, *Proc. Phys. Soc.*, **43**, 324–339 (1931).
97. Drinker, P., and T. Hatch: "Industrial Dust," 2d ed., Hygienic Significance, Measurement, and Control, McGraw-Hill Book Company, Inc., New York, 1954.
98. Eegriwe, E.: Reactions and Reagents for the Detection of Organic Compounds, *Z. anal. Chem.*, **110**, 22–25 (1937).
99. Elkins, H. B.: "The Chemistry of Industrial Toxicology," John Wiley & Sons, Inc., New York, 1950.
100. Ellis, G. H., E. G. Zook, and O. Baudisch: Colorimetric Determination of Boron Using 1,1′-dianthrimide, *Anal. Chem.*, **21**, 1345–1348 (1949).
101. Fairhall, L. T.: "Industrial Toxicology," The Williams & Wilkins Company, Baltimore, 1949.
102. Farber, L.: Chemical Evaluation of Odor Intensity, *Food Technol.*, **3**, 300–304 (1949).
103. Fastie, W. G., and A. H. Pfund: Selective Infrared Gas Analyzers, *J. Opt. Soc. Amer.*, **37**, 762–768 (1947).
104. Feicht, F. L., H. H. Schrenk, and C. E. Brown: Determination with the Dropping-mercury-electrode Procedure of Lead, Cadmium and Zinc in Samples Collected in Industrial-hygiene Studies, *U.S. Bur. Mines, Rept. Invest.* 3639, 1942.
105. Feigl, F.: "Qualitative Analysis by Spot Tests," New York-Nordemann Publishing Co., New York, 1939.
106. Feigl, F.: "Chemistry of Specific, Selective, and Sensitive Reactions" (trans. by R. E. Oesper), Academic Press, Inc., New York, 1949.
107. Feigl, F., and G. B. Heisig: Analytical Aspects of the Chemical Behavior of 8-Hydroxyquinoline, *Anal. Chim. Acta*, **3**, 561–566 (1949).
108. Ficklen, J. B.: "Manual of Industrial Health Hazards," Service to Industry, West Hartford, Conn., 1940.
109. Fieldner, A. C., G. G. Oberfell, M. C. Teague, and J. N. Lawrence: Methods of Testing Gas Masks and Absorbents, *J. Ind. Eng. Chem.*, **11**, 519–540 (1919).
110. Fieldner, A. C., S. H. Katz, and E. G. Meiter: Continuous CO Recorder in the Liberty Tunnels, *Eng. News-Record*, **95**, 423–424 (1925).
111. First, W. M., and L. Silverman: Air Sampling with Membrane Filters, *Arch. Ind. Hyg. and Occupational Med.*, **7**, 1–11 (1953).
112. Fogo, J. K., and M. Popowsky: Spectrophotometric Determination of Hydrogen Sulfide–Methylene Blue Method, *Anal. Chem.*, **21**, 732–734 (1949).
113. Foster, N. B., and L. W. Foskett: A Spectrophotometer for the Determination of the Water Vapor in a Vertical Column of the Atmosphere, *J. Opt. Soc. Amer.*, **35**, 601–610 (1945).
114. Fothergill, S. J. R., D. F. Withers, and F. S. Clements: Determination of Traces of Platinum and Palladium in the Atmosphere of a Platinum Refinery by a Combined Chemical and Spectrographic Method, *Brit. J. Ind. Med.*, **2**, 99–101 (1945).
115. Francis, A. G., and A. T. Parsons: Determination of Oxides of Nitrogen (Except Nitrous Oxide) in Small Concentration in the Products of Combustion of Coal Gas and in Air, *Analyst*, **50**, 262–272 (1925).
116. Frazer, D. A.: Absolute Method of Sampling and Measurement of Solid Air-borne Particulates, *Arch. Ind. Hyg. and Occupational Med.*, **8**, 412–419 (1953).
117. Fredrick, W. G.: Estimation of Small Amounts of Antimony with Rhodamine B, *Ind. Eng. Chem., Anal. Ed.*, **13**, 922–924 (1941).
118. Gardner, K.: Photometric Determination of Small Amounts of Titanium with 8-Quinolinol, *Analyst*, **76**, 485–488 (1951).

119. Gates, E. M., and G. H. Ellis: A Microcolorimetric Method for the Determination of Manganese in Biological Materials with bis-(4-Dimethylaminophenyl) Methane, *J. Biol. Chem.*, **168**, 537–544 (1947).
120. General Electric Company: Instantaneous-type Mercury-vapor Detector, Catalogue No. 8257667 Gl.
121. Gibb, T. R. P., Jr.: "Optical Methods of Chemical Analysis," McGraw-Hill Book Company, Inc., New York, 1942.
122. Giever, P. M.: Problems Encountered in Field Use of Thomas Autometer, *Arch. Ind. Hyg. and Occupational Med.*, **6**, 445–449 (1952).
123. Gilbert, F. A.: "Mineral Nutrition of Plants and Animals," University of Oklahoma Press, Norman, Okla., 1948.
124. Gillam, W. S.: Photometric Method for the Determination of Magnesium, *Ind. Eng. Chem., Anal. Ed.*, **13**, 499–501 (1941).
125. Gitzen, W. H.: Identification of Free Silica in Dusts and Fumes, *Anal. Chem.*, **20**, 265–267 (1948).
126. Glasser, L. G.: The Ultraviolet Method of Continuous Gas Analysis, *J. Electrochem. Soc.*, **97**, 201c–204c (1950).
127. Goldman, F. H.: "Laboratory Procedures in Industrial Hygiene," prepared for the course in Sanitary Air Analysis at Georgia School of Technology, Atlanta, Ga., 1947.
128. Goldman, F. H., and H. Yagoda: Collection and Estimation of Traces of Formaldehyde in Air, *Ind. Eng. Chem., Anal. Ed.*, **15**, 378 (1943).
129. Goodeve, C. F.: The Removal of Mist by Centrifugal Methods, *Trans. Faraday Soc.*, **32**, 1218–1223 (1936).
130. Goodman, C., J. W. Irvine, Jr., and C. F. Horan: Mercury Vapor Measurement. A Radioactive Method, *J. Ind. Hyg. Toxicol.*, **25**, 275–281 (1943).
131. Gordon, C. L.: Nucleonics (in Analytical Chemistry), *Anal. Chem.*, **21**, 96–101 (1949).
132. Gore, R. C., and E. M. Petersen: Infrared Studies of Polymyxin, *Ann. N.Y. Acad. Sci.*, **51**, 924–934 (1949).
133. Gosline, C. A.: Dispersion from Short Stacks, *Chem. Eng. Progr.*, **48**, 165–172 (1952).
134. Gosline, C. A. (ed.): "Air Pollution Abatement Manual," Manufacturing Chemists' Association, Inc., Washington, D.C., 1951.
135. Graham, R. P., and J. A. Maxwell: Determination of Titanium in Rocks and Minerals, *Anal. Chem.*, **23**, 1123–1126 (1951).
136. Grant, W. M.: Colorimetric Determination of Sulfur Dioxide, *Anal. Chem.*, **19**, 345–346 (1947).
137. Great Britain: Chlorine, *Dept. Sci. Ind. Research (Brit.), Leaflet* 10, 1939.
138. Great Britain: Nitrous Fume, *Dept. Sci. Ind. Research (Brit.), Leaflet* 5, 1939.
139. Green, D. E.: Colorimetric Microdetermination of Zirconium, *Anal. Chem.*, **20**, 370–372 (1948).
140. Green, H. L., and H. H. Watson: Physical Methods for the Estimation of the Dust Hazard in Industry, *Med. Research Council (Brit.), Spec. Rept. Ser. No.* 199, 1935.
141. Grosse, A. V., S. G. Hindin, and A. D. Kirshenbaum: Elementary Isotopic Analysis, *Anal. Chem.*, **21**, 386–390 (1949).
142. Gucker, F. T., Jr.: Determination of Concentration and Size of Particulate Matter by Light Scattering and Sonic Techniques, *Proc. 1st Nat. Air Pollution Symposium*, p. 14, Stanford Research Institute, Los Angeles, Calif., 1949.
143. Gucker, F. T., Jr.: Instrumental Methods for Measuring Mass Concentration and Particulate Concentration in Aerosols, in L. C. McCabe, ed., "Air Pollution," p. 617, McGraw-Hill Book Company, Inc., New York, 1952.
144. Gucker, F. T., Jr., C. T. O'Konski, H. G. Pickard, and J. N. Pitts, Jr.: Photoelectronic Counter for Colloidal Particles, *J. Am. Chem. Soc.*, **69**, 2422–2431 (1947).
145. Gucker, F. T., Jr., H. G. Pickard, and C. T. O'Konski: "Handbook on Aerosols," chap. 10, from Summary Technical Report of Division 10, National Defense Research Committee, U.S. Atomic Energy Commission, Washington, D.C., 1950.
146. Guyton, A. C.: Electronic Counting and Size Determination of Particles in Aerosols, *J. Ind. Hyg. Toxicol.*, **28**, 133–141 (1946).
147. Haagen-Smit, A. J., E. F. Darley, M. Zaitlin, H. Hull, and W. Noble: Investigation on Injury to Plants from Air Pollution in the Los Angeles Area, *Plant Physiol.*, **27**, 18–34 (1952).
148. Haagen-Smit, A. J., and M. M. Fox: Determination of Total Oxidant in Air, Described by McCabe, *Ind. Eng. Chem.*, **45**, 111A (1953).
149. Hall, S. R.: Evaluation of Particulate Concentrations with Collecting Apparatus, *Anal. Chem.*, **24**, 996–1000 (1952).
150. Hamilton, W. F.: Industrial Analysis and Recording of Carbon Dioxide and Oxygen in Air, *Ind. Eng. Chem., Anal. Ed.*, **2**, 233–237 (1930).

151. Hanson, V. F.: Ultraviolet Photometer, Quantitative Measurement of Small Traces of Solvent Vapors in Air, *Ind. Eng. Chem., Anal. Ed.*, **13,** 119–123 (1941).
152. Happ, G. P., D. W. Stewart, and H. F. Brockmyre: The Mass Spectrometer as an Analytical Tool in Industrial Hygiene, *Am. Ind. Hyg. Assoc. Quart.*, **11,** 135–142 (1950).
153. Harrison, G. R., R. C. Lord, and J. R. Loofbourow: "Practical Spectroscopy," Prentice-Hall, Inc., New York, 1948.
154. Harrold, G. C., and L. E. Gordon: Use of the Interferometer for Two-component Mixtures, *J. Ind. Hyg. Toxicol.*, **21,** 491–497 (1939).
155. Hemeon, W. C. L., J. D. Sensenbaugh, and G. F. Haines, Jr.: Measurement of Air Pollution, *Instruments*, **26,** 566–568, 590 (1953).
156. Hemeon, W. C. L.: The Nature of the Air Pollution Problem in the City of New Haven, Industrial Hygiene Foundation of America, Pittsburgh, Pa., 1952.
157. Heyrovsky, J.: The Fundamental Laws of Polarography, *Analyst*, **72,** 229–234 (1946).
158. Horton, A. D., P. F. Thomason, and F. J. Miller: Spectrophotometric Determination of Inorganic Fluoride, *Anal. Chem.*, **24,** 548–550 (1952).
159. Hubbard, D. M.: Determination of Arsenic in Biological Material. A Photometric Method, *Ind. Eng. Chem., Anal. Ed.*, **13,** 915–918 (1941).
160. Hubbard, D. M.: Microdetermination of Bismuth in Biological Material. An Improved Photometric Dithizone Method, *Anal. Chem.*, **20,** 363–364 (1948).
161. Hubbard, D. M.: Microdetermination of Copper in Biological Material. An Improved Dithizone-polarographic Method, personal communication, to be published.
162. Hudgens, J. E., Jr.: Analytical Application of Radiochemical Techniques, *Anal. Chem.*, **24,** 1704–1708 (1952).
163. Hudgens, J. E., Jr., and P. J. Cali: Determination of Antimony by Radioactivation, *Anal. Chem.*, **24,** 171–174 (1952).
164. Huffman, W. T.: Effect on Livestock of Air Contamination Caused by Fluoride Fumes, chap. 5 in L. C. McCabe, ed., "Air Pollution," McGraw-Hill Book Company, Inc., New York, 1952.
165. Hyslop, F., E. D. Palmes, W. C. Alford, A. R. Monaco, and L. T. Fairhall: Toxicity of Beryllium, *Natl. Inst. Health, Bull.* **181,** 1943.
166. Jacobs, M. B.: "The Analytical Chemistry of Industrial Poisons, Hazards and Solvents," 2d ed., Interscience Publishers, Inc., New York, 1949.
167. Jelley, E. E.: "Physical Methods of Organic Chemistry," vol. I, chap. 11, Microscopy, A. Weissberger, ed., Interscience Publishers, Inc., New York, 1945.
168. Jephcott, C. M., J. H. Johnston, and G. R. Finlay: The Fume Exposure in the Manufacture of Alumina Abrasives from Bauxite, *J. Ind. Hyg. Toxicol.*, **30,** 145–159 (1948).
169. Kamen, M. D.: "Radioactive Tracers in Biology," Academic Press, Inc., New York, 1947.
170. Katz, M.: Sulfur Dioxide in the Atmosphere and Its Relation to Plant Life, *Ind. Eng. Chem.*, **41,** 2450–2465 (1949).
171. Katz, M.: Photoelectric Determination of Atmospheric Sulfur Dioxide, *Anal. Chem.*, **22,** 1040–1047 (1950).
172. Katz, M.: Sources of Pollution, *Proc. 2d Nat. Air Pollution Symposium*, p. 95, Stanford Research Institute, Los Angeles, Calif., 1952.
173. Katz, M., and J. Katzman: Rapid Determination of Low Concentrations of Carbon Monoxide in Air, *Can. J. Research*, **26F,** 318–330 (1948).
174. Katz, S. H., D. A. Reynolds, H. W. Frevert, and J. J. Bloomfield: A Carbon Monoxide Recorder and Alarm, *U.S. Bur. Mines, Tech. Paper* 355, 1926.
175. Kaufman, H. S., and I. Fankuchen: X-ray Diffraction, *Anal. Chem.*, **21,** 24–29 (1949); **22,** 16–18 (1950); **24,** 20–22 (1952).
176. Kay, K.: Analytical Methods Used in Air Pollution Study, *Anal. Chem.*, **44,** 1383–1388 (1952).
177. Kay, K.: Rapid Quartz Analysis by X-ray Spectrometry, *Am. Ind. Hyg. Assoc. Quart.*, **11,** 185 (1950).
178. Keenan, R. G., and B. M. Flick: Determination of Cobalt in Atmospheric Samples, *Anal. Chem.*, **20,** 1238–1241 (1948).
179. Keenan, R. G., and D. H. Byers: Rapid Analytical Method for Air Pollution Surveys. The Determination of Total Particulates and the Rapid Semiquantitative Spectrographic Method of Analysis of the Metallic Constituents in High-volume Samples, *Arch. Ind. Hyg. and Occupational Med.*, **6,** 226–230 (1952).
180. Kingsley, G. R., and R. R. Schaffert: Microdetermination of Arsenic and Its Application to Biological Material, *Anal. Chem.*, **23,** 914–919 (1951).
181. Klein, B.: Microdetermination of Sulfate. A Colorimetric Estimation of the Benzidine Sulfate Precipitate, *Ind. Eng. Chem., Anal. Ed.*, **16,** 536–537 (1944).
182. Klotz, I. M., and M. Dole: An Automatic-recording Ultraviolet Photometer for Laboratory and Field Use, *Ind. Eng. Chem., Anal. Ed.*, **18,** 741–745 (1946).

183. Klug, H. P., L. Alexander, and E. Kummer: X-ray Diffraction Analysis of Crystalline Dusts, *J. Ind. Hyg. Toxicol.*, **30**, 166–171 (1948).
184. Klug, H. P., L. Alexander, and E. Kummer: Quantitative Analysis with the X-ray Spectrometer, *Anal. Chem.*, **20**, 607–609 (1948).
185. Kolthoff, I. M., and J. J. Lingane: The Fundamental Principles and Applications of Electrolysis with the Dropping Mercury Electrode and Heyrovsky's Polarographic Method of Chemical Analysis, *Chem. Revs.*, **24**, 1–94 (1939).
186. Kolthoff, I. M., and J. J. Lingane: "Polarography," Interscience Publishers, Inc., New York, 1941.
187. Koppius, O. G.: Detection of Lead in Air with the Aid of a Geiger-Müller Counter, *J. Opt. Soc. Amer.*, **39**, 294–297 (1949).
188. Lacourt, A., G. Sommereyns, and G. Wantier: Quantitative Inorganic Paper Chromatography, *Analyst*, **77**, 943–954 (1952).
189. Lamar, W. L.: Determination of Fluoride in Water. A Modified Zirconium–Alizarin Method, *Ind. Eng. Chem., Anal. Ed.*, **17**, 148–149 (1945).
190. Landahl, H. D., and R. G. Herrmann: Sampling of Liquid Aerosols by Wires, Cylinder and Slides, and the Efficiency of Impaction of the Droplets, *J. Colloid Sci.*, **4**, 103–136 (1949).
191. Landry, A. S.: Simultaneous Determination of Lead and Zinc in Atmospheric Samples. A Polarographic Method Utilizing a Double Internal Standard, *J. Ind. Hyg. Toxicol.*, **29**, 168–174 (1947).
192. Landry, A. S.: Polarographic Determination of Benzene in Presence of Its Homologs, *Anal. Chem.*, **21**, 674–677 (1949).
193. Langer, A.: Macro- and Microvessels for Polarographic Analysis, *Ind. Eng. Chem., Anal. Ed.*, **17**, 454–456 (1945).
194. Largent, E. J.: The Effects of Air-borne Fluorides on Livestock, chap. 6 in L. C. McCabe, ed., "Air Pollution," McGraw-Hill Book Company, Inc., New York, 1952.
195. Larson, G. P.: Second Technical and Administrative Report on Air Pollution Control in Los Angeles County, Los Angeles County Air Pollution Control District, Los Angeles, Calif., 1950–1951.
196. Laug, E. P.: Determination of Bismuth in Biological Material, *Anal. Chem.*, **21**, 188–189 (1949).
197. Lederer, E.: "Progrès récents de la chromatographie," part I, Hermann & Cie, Paris, 1949.
198. Lennox, D., and J. Leroux: Applications of X-ray Diffraction Analysis in the Environmental Field, *Arch. Ind. Hyg. and Occupational Med.*, **8**, 359–370 (1953).
199. Levine, L.: Polarographic Determination of Toxic Metal Fumes in Air, *J. Ind. Hyg. Toxicol.*, **27**, 171–177 (1945).
200. Lingane, J. J.: Systematic Polarographic Metal Analysis Characteristics of Arsenic, Antimony, Bismuth, Tin, Lead, Cadmium, Zinc and Copper in Various Supporting Electrolytes, *Ind. Eng. Chem., Anal. Ed.*, **15**, 583–590 (1943).
201. Lingane, J. J.: Apparatus for Rapid Polarographic Analysis, *Ind. Eng. Chem., Anal. Ed.*, **16**, 329–330 (1944).
202. Lingane, J. J.: Polarographic Theory, Instrumentation, and Methodology, *Anal. Chem.*, **21**, 45–60 (1949); **23**, 86–97 (1951).
203. Lingane, J. J., and H. Kerlinger: Polarographic Determination of Nickel and Cobalt, *Ind. Eng. Chem., Anal. Ed.*, **13**, 77–80 (1941).
204. Littman, F. E., and R. W. Benoliel: Continuous Oxidant Recorder, *Anal. Chem.*, **25**, 1480–1483 (1953).
205. Littman, F. E., D. M. Coulson, and H. J. Eding: Unpublished report to American Petroleum Institute, 1953.
206. Luce, E. N., E. C. Denice, and F. E. Akerlind: Turbidimetric Determination of Small Amounts of Chlorides, *Ind. Eng. Chem., Anal. Ed.*, **15**, 365–366 (1943).
207. Ludwig, E. E., and C. R. Johnson: Spectrophotometric Determination of Magnesium by Titan Yellow, *Ind. Eng. Chem., Anal. Ed.*, **14**, 895–897 (1942).
208. MacDougall, D., and D. A. Biggs: Estimation of Boron in Plant Tissue. Modification of Quinalizarin Method, *Anal. Chem.*, **24**, 566–569 (1952).
209. MacIntire, W. H., L. J. Hardin, and W. Hester: Measurement of Atmospheric Fluorine. Analysis of Rain Water and Spanish Moss Exposures, *Ind. Eng. Chem.*, **44**, 1365–1370 (1952).
210. Machle, W., and K. Kitzmiller: The Effects of the Inhalation of Hydrogen Fluoride. II. The Response Following Exposure to Low Concentration, *J. Ind. Hyg.*, **17**, 223–229 (1935).
211. Machle, W., F. Thamann, K. Kitzmiller, and J. Cholak: The Effects of the Inhalation of Hydrogen Fluoride. I. The Response Following Exposure to High Concentrations, *J. Ind. Hyg.*, **16**, 129–145 (1934).

212. Mader, P. P., W. J. Hamming, and A. Bellin: Determination of Small Amounts of Sulfuric Acid in the Atmosphere, *Anal. Chem.*, **22**, 1181–1183 (1950).
213. Mader, P. P., M. W. Heddon, R. T. Lofberg, and R. H. Koehler: Determination of Small Amounts of Hydrocarbons in the Atmosphere, *Anal. Chem.*, **24**, 1899–1902 (1952).
214. Magill, P. L.: The Los Angeles Smog Problem, *Ind. Eng. Chem.*, **41**, 2476–2486 (1949).
215. Magill, P. L.: Unpublished report to American Petroleum Institute, August, 1953.
216. Maits, C. B., et al.: Carbon Monoxide Survey in Liberty Tubes. *J. Ind. Hyg. Toxicol.*, **14**, 295–300 (1932).
217. Maren, T. H.: The Microdetermination of Antimony, *Bull. Johns Hopkins Univ.*, **77**, 338–344 (1945).
218. Marks, G. W., and E. V. Potter: The Spectrochemical Determination of Thallium in Ores, Concentrates, Dusts and Chemicals, *U.S. Bur. Mines, Rept. Invest.* 4461, 1949.
219. McCabe, L. C., ed.: "Air Pollution," McGraw-Hill Book Company, Inc., New York, 1952.
220. McCord, C. P., and W. N. Witheridge: "Odors, Physiology and Control," McGraw-Hill Book Company, Inc., New York, 1949.
221. McMurdie, H. F.: Application of the Electron Microscope to Air Pollution, chap. 19 in L. C. McCabe, ed., "Air Pollution," McGraw-Hill Book Company, Inc., New York, 1952.
222. McNaught, K. J.: Spectrophotometric Determination of Cobalt in Pastures and Animal Tissues, *New Zealand J. Sci. Technol.*, **30A**, 109–115 (1949).
223. Meetham, A. R.: "Atmospheric Pollution," Pergamon Press Ltd., London, 1952.
224. Meggers, W. F., and B. F. Scribner: Index to the Literature on Spectrochemical Analysis, part I, 1920–1939; part II, 1940–1945, American Society for Testing Materials, Philadelphia, Pa.
225. Mellon, M. G.: "Analytical Absorption Spectroscopy: Absorptometry and Colorimetry," John Wiley & Sons, Inc., New York, 1950.
226. Miller, R. D., and M. B. Russell: Pneumatic Autodetector for Infrared Gas Analysis, *Anal. Chem.*, **21**, 773–777 (1949).
227. Milton, R. F., and W. D. Duffield: Estimation of Mercury Compounds in the Atmosphere, *Analyst*, **72**, 11–13 (1947).
228. Mine Safety Appliance Co.: Explosimeter, Combustible Gas Indicator, Benzol Indicator, Pittsburgh, Pa.
229. Mitchell, H. H., and M. Edman: Fluorine in Soils, Plants and Animals, *Soil Sci.*, **60**, 81–90 (1945).
230. Monnier, D., Y. Ruscone, and P. Wenger: Colorimetric Determinations of Fluoride, *Helv. Chim. Acta*, **29**, 521–525 (1946).
231. Muller, O. H.: Oxidation and Reduction of Organic Compounds at the Dropping-mercury Electrode and the Application of Heyrovsky's Polarographic Method in Organic Chemistry, *Chem. Rev.*, **24**, 95–124 (1939).
232. Munger, H. P.: Meteorological Methods for Studying Air Pollution, presented at the Twelfth International Congress of Pure and Applied Chemistry, New York, Sept. 10–13, 1951, Battelle Memorial Institute, Columbus, Ohio.
233. Nachtrieb, N. H.: "Principles and Practice of Spectrochemical Analysis," McGraw-Hill Book Company, Inc., New York, 1950.
234. National Research Council of Canada: Effect of Sulfur Dioxide on Vegetation (prepared for the Associate Committee on Trail Smelter Smoke), N.R.C. No. 815, Ottawa, Canada, 1939.
235. Nawrocki, C. Z., F. S. Brackett, and H. W. Werner: Determination of the Concentration of Monoalkyl Ethylene Glycol Ethers in Air by Infrared Absorption Spectroscopy, *J. Ind. Hyg. Toxicol.*, **26**, 193–196 (1944).
236. Nichols, M. L., and L. H. Rogers: Determination of Small Amounts of Molybdenum in Plants and Soils, *Ind. Eng. Chem., Anal. Ed.*, **16**, 137–140 (1944).
237. Oshry, H. I., J. W. Ballard, and H. H. Schrenk: Spectrochemical Determination of Lead, Cadmium, and Zinc in Dusts, Fumes and Ores, *J. Opt. Soc. Amer.*, **31**, 627–633 (1941).
238. Owen, C. E., J. C. Delaney, and C. M. Neff: The Spectrochemical Analysis of Air Borne Dusts for Beryllium, *Am. Ind. Hyg. Assoc. Quart.*, **12**, 112–114 (1951).
239. Page, J. E., and F. A. Robinson: Polarographic Studies III. Determination of Vanadium, *Analyst*, **68**, 269–271 (1943).
240. Parker, A., and S. H. Richards: Instruments Used for the Measurement of Atmospheric Pollution in Great Britain, in L. C. McCabe, ed., "Air Pollution," p. 531, McGraw-Hill Book Company, Inc., New York, 1952.

241. Parks, T. D., and L. Lykken: Separation and Microdetermination of Small Amounts of Aluminum, *Anal. Chem.*, **20,** 1102–1106 (1948).
242. Parks, T. D.: Potentiometric, Amperometric and Polarographic Methods for Microanalysis, *Anal. Chem.*, **22,** 1444–1446 (1950).
243. Patterson, G. D., Jr., and M. G. Mellon: Determination of Sulfur Dioxide by Color-changing Gels, *Anal. Chem.*, **24,** 1586–1590 (1952).
244. Patty, F. A.: Calibration and Use of the Gas Interferometer, *J. Ind. Hyg. Toxicol.*, **21,** 469–474 (1939).
245. Patty, F. A., ed.: "Industrial Hygiene and Toxicology," Interscience Publishers, Inc., New York, 1948–1949.
246. Pepkowitz, L. P., and E. L. Shirley: Microdetermination of Sulfur, *Anal. Chem.*, **23,** 1709–1710 (1951).
247. Perley, G. A., and B. F. T. Langsdorf: Problems in the Recording of SO_2 in Polluted Atmospheres, chap. 69 in L. C. McCabe, ed., "Air Pollution," McGraw-Hill Book Company, Inc., New York, 1952.
248. Pollard, F. H., and J. F. W. McOmie: Analysis of Inorganic Compounds by Paper Chromatography, *Endeavor*, **10,** 213–221 (1951), Oct.
249. Prentiss, A. M.: "Civil Defense in Modern War," McGraw-Hill Book Company, Inc., New York, 1951.
250. Price, M. J., and O. J. Walker: Determination of Fluoride in Water by the Aluminum-hemotoxylin Method, *Anal. Chem.*, **24,** 1593–1595 (1952).
251. Pringsheim, P., and M. Vogel: "Luminescence of Liquids and Solids and Its Practical Applications," Interscience Publishers, Inc., New York, 1943.
252. Raben, M. S.: Microdetermination of Iodine Employing Radioactive Iodine, *Anal. Chem.*, **22,** 480–482 (1950).
253. Radley, J. A., and J. Grant: "Fluorescent Analysis in Ultraviolet Light," 3d ed., D. Van Nostrand Company, Inc., New York, 1939.
254. Redmond, J. C.: Quantitative Analysis with the X-ray Spectrometer, *Anal. Chem.*, **19,** 773–777 (1947).
255. Reynold, W. C.: London and Suburban Air, *J. Soc. Chem., Ind.* **49,** 168–72T (1930).
256. Rittenberg, D., and G. L. Foster: A New Procedure for Quantitative Analysis by Isotope Dilution with Application to the Determination of Amino Acids and Fat Acids, *J. Biol. Chem.*, **133,** 737–744 (1940).
257. Robinson, W. O., and G. Edgington: Fluorine in Soils, *Soil Sci.*, **61,** 341–353 (1946).
258. Rock, S. M.: Qualitative Analysis from Mass Spectra, *Anal. Chem.*, **23,** 261–268 (1951).
259. Rooksby, H. P.: Microanalysis Using X-ray Diffraction Technique, *Analyst*, **73,** 326–330 (1948).
260. Russell, J. J.: The Colorimetric Determination of Traces of Boron, National Research Council of Canada, Division of Research, A. E. Project, No. 1596, reissued Chalk River Laboratory, August, 1947.
261. Sandell, E. B.: "Colorimetric Determination of Traces of Metals," 2d ed., Interscience Publishers, Inc., New York, 1950.
262. Sands, A. E., M. A. Grafius, H. W. Wainwright, and M. W. Wilson: The Determination of Low Concentrations of Hydrogen Sulfide in Gas by the Methylene Blue Method, *U.S. Bur. Mines, Rept. Invest.* 4547, 1949.
263. Sartori, M.: "The War Gases," D. Van Nostrand Company, Inc., New York, 1939.
264. Schadt, C., P. L. Magill, R. D. Cadle, and L. Ney: An Electrostatic Precipitator for the Continuous Sampling of Sulfuric Acid Aerosols and Other Air-borne Particulate Electrolytes, *Arch. Ind. Hyg. and Occupational Med.*, **1,** 556–564 (1950).
265. Schrenk, H. H., H. Heimann, G. D. Clayton, W. M. Gafafer, and H. Wexler: Air Pollution in Donora, Pa., *U.S. Public Health Service, Public Health Bull.* 306, 1949.
266. Schmelzer, L. L.: A Rapid X-ray Diffraction Method for the Determination of Quartz in Industrial Dusts, *Arch. Ind. Hyg. and Occupational Med.*, **3,** 121–128 (1951).
267. Schuster, M. C., and E. F. Fullam: Preparation of Powdered Materials for Electron Microscopy, *Ind. Eng. Chem., Anal. Ed.*, **18,** 653–657 (1946).
268. Scott, T. R.: Use of Ether Extraction in the Determination of Uranium, *Analyst*, **74,** 486–491 (1949).
269. Scribner, B. F.: Spectrographic Analysis in Air-pollution Studies, chap. 27 in L. C. McCabe, ed., "Air Pollution," McGraw-Hill Book Company, Inc., New York, 1952.
270. Scribner, B. F., and W. F. Meggers: "Index to the Literature on Spectrochemical Analysis," part III, 1946–1950, American Society for Testing Materials, Philadelphia, Pa., 1954.
271. Shepherd, M.: Rapid Determination of Small Amounts of Carbon Monoxide in Air, *Anal. Chem.*, **19,** 77–81 (1947).
272. Shepherd, M., S. M. Rock, R. Howard, and J. Stormes: Isolation, Identification and

Estimation of Gaseous Pollutants of Air, Examination of Los Angeles Co., Calif., Smog, *Anal. Chem.*, **23**, 1431–1440 (1951).
273. Silverman, L.: Industrial Air Sampling and Analysis, *Ind. Hyg. Foundation, Amer. Chem. and Toxicol. Ser., Bull.* 1, pp. 1–72, 1947.
274. Silverman, L.: Simple Polarographic Procedure for Cadmium Dust and Fumes in Air, *Chemist Analyst*, **35**, 53–55 (1946).
275. Silverman, L., H. F. Schulte, and M. W. First: Further Studies on Sensory Response to Certain Industrial Solvent Vapors, *J. Ind. Hyg. Toxicol.*, **28**, 262–266 (1946).
276. Silverman, L., and F. J. Viles, Jr.: A High-volume Air-sampling and Filter-weighing Method for Certain Aerosols, *J. Ind. Hyg. Toxicol.*, **30**, 124–128 (1948).
277. Silverman, L., J. L. Whittenberger, and J. Muller: Physiological Response of Man to Ammonia in Low Concentrations, *J. Ind. Hyg. Toxicol.*, **31**, 74–78 (1949).
278. Simpson, G. C., et al.: The Investigation of Atmospheric Pollution. Report on Observations in the Year Ended March 31, 1932, *Dept. Sci. Ind. Research* (*Brit.*), 18*th Rept.*, 1933.
279. Singh, A. D.: A Survey of Sulfur Dioxide Pollution in Chicago and Vicinity, *Univ. Illinois Bull.* **36,** No. 37, 1939.
280. Smales, A. A.: The Determination of Small Quantities of Uranium in Rocks and Minerals by Radioactivation, *Analyst*, **77**, 778–789 (1952).
281. Smales, A. A., and B. D. Pate: Determination of Submicrogram Quantities of Arsenic by Radioactivation, *Anal. Chem.*, **24**, 717–721 (1952).
282. Smales, A. A., and B. D. Pate: The Determination of Submicrogram Quantities of Arsenic by Radioactivation. Part II. The Determination of Arsenic in Sea Water, *Analyst*, **77**, 188–195 (1952).
283. Smith, C. S., and R. L. Barrett: Apparatus and Techniques for Practical Chemical Identification by X-ray Diffraction, *J. Appl. Phys.*, **18**, 177–191 (1947).
284. Smith, R. B., and B. S. T. Friis: Portable Motor-driven Impinger Unit for Determination of Sulfur Dioxide, *J. Ind. Hyg.*, **13**, 338–342 (1931).
285. Smith, R. G., A. J. Boyle, W. G. Fredrick, and B. Zak: Spectrographic Determination of Beryllium in Urine and Air, *Anal. Chem.*, **24**, 406–409 (1952).
286. Smith, R. G.: Personal communication.
287. Smith, R. G., and P. Diamond: Microdetermination of Ozone—Preliminary Studies, *Am. Ind. Hyg. Assoc. Quart.*, **13**, 235–238 (1952).
288. Snell, F. D., and C. T. Snell: "Colorimetric Methods of Analysis," D. Van Nostrand Company, Inc., New York, 1936.
289. Sommer, H.: Color Test for Elementary Sulfur, *Ind. Eng. Chem., Anal. Ed.*, **12**, 368–369 (1940).
290. Spector, N. A., and B. F. Dodge: Colorimetric Method for Determination of Traces of Carbon Dioxide in Air, *Anal. Chem.*, **19**, 55–58 (1947).
291. Stamm, R. F., and J. T. Whalen: Calibration of the Rayleigh Refractometers without Recourse to Gas Chambers, *J. Ind. Hyg. Toxicol.*, **29**, 203–217 (1947).
292. Stanford Research Institute: Second Interim Report, The Smog Problem in Los Angeles County, Western Oil and Gas Association, Los Angeles, Calif., 1949.
293. Stanford Research Institute: *Proc.* 1*st Nat. Air Pollution Symposium*, Los Angeles, Calif., 1010.
294. Stang, A. M., J. E. Zatek, and C. D. Robson: A Colorimetric Method for the Determination of Sulfur Dioxide in Air, *Am. Ind. Hyg. Quart.*, **12**, 5–8 (1951).
295. Starr, C. E., Jr., and T. Lane: Accuracy and Precision of Analysis of Light Hydrocarbon Mixtures, *Anal. Chem.*, **21**, 572–582 (1949).
296. Steffens, C.: Measurement of Visibility by Photographic Photometry, *Ind. Eng. Chem.*, **41**, 2396–2399 (1949).
297. Steigmann, A.: Acid-bleached Fuchsin Solution as Analytical Reagent, *Anal. Chem.*, **22**, 492–493 (1950).
298. Stiles, W.: "Trace Elements in Plants and Animals," The Macmillan Co., New York, 1946.
299. Stillman, J. W.: Bibliography of Photoelectric Spectrophotometric Methods of Analysis for Inorganic Ions, *Proc. Am. Soc. Testing Materials*, **44**, 740–748 (1944).
300. Stokinger, H. E., and S. Laskin: Air Pollution and the Particle Size–Toxicity Problem—II, *Nucleonics*, **6**, 15–31 (1950).
301. Strain, H. H.: Chromatography, *Anal. Chem.*, **22**, 41–48 (1950).
302. Strain, H. H.: Chromatographic Systems, *Anal. Chem.*, **23**, 25–38 (1951).
303. Strain, H. H., and G. W. Murphy: Chromatography and Analogous Differential Migration Methods, *Anal. Chem.*, **24**, 50–60 (1952).
304. Strong, J.: A New Method of Measuring the Mean Height of the Ozone in the Atmosphere. I, *J. Franklin Inst.*, **231**, 121–155 (1941).
305. Swain, R. E.: Smoke and Fume Investigations, *Ind. Eng. Chem.*, **41**, 2384–2388 (1949).

306. Talvitie, N. A.: Colorimetric Determination of Fluoride in Natural Waters with Thorium and Alizarin, *Ind. Eng. Chem., Anal. Ed.*, **15**, 620–621 (1943).
307. Taylor, A. E., and R. T. Dillon: Determination of Microgram Quantities of Thorium in Water. Extension of Colorimetric Method, *Anal. Chem.*, **24**, 1624–1625 (1952).
308. Thomas, M. D.: The Present Status of the Development of Instrumentation for the Study of Air Pollution, *Proc. 2d Nat. Air Pollution Symposium*, p. 16, Stanford Research Institute, Los Angeles, Calif., 1952.
309. Thomas, M. D.: Gas Damage to Plants, *Ann. Rev. Plant Physiol.*, **2**, 293–322 (1951).
310. Thomas, M. D., R. H. Hendricks, and G. R. Hill: Sulfur Content of Vegetation, *Soil Sci.*, **70**, 9–18 (1950).
311. Thomas, M. D., J. O. Ivie, J. N. Abersold, and R. H. Hendricks: Automatic Apparatus for Determination of Small Concentrations of Sulfur Dioxide in Air. Application to Hydrogen Sulfide, Mercaptans and Other Sulfur and Chlorine Compounds, *Ind. Eng. Chem., Anal. Ed.*, **15**, 287–290 (1943).
312. Thomas, M. D., J. O. Ivie, and T. C. Fitt: Automatic Apparatus for Determination of Small Concentrations of Sulfur Dioxide in Air, *Ind. Eng. Chem., Anal. Ed.*, **18**, 383–387 (1946).
313. Thomas, M. D., and J. O. Ivie: Automatic Apparatus for the Determination of Small Concentrations of SO_2 and Other Contaminants in the Atmosphere, chap. 70 in L. C. McCabe, ed., "Air Pollution," McGraw-Hill Book Company, Inc., New York, 1952.
314. Thomason, P. F., M. A. Perry, and W. M. Byerly: Determination of Microgram Amounts of Thorium, *Anal. Chem.*, **21**, 1239–1241 (1949).
315. Thorp, C. E.: Starch-iodide Method of Ozone Analysis, *Ind. Eng. Chem., Anal. Ed.*, **12**, 209 (1940).
316. Thrun, W. E.: Rapid Methods for Determining Fluoride in Waters, *Anal. Chem.*, **22**, 918–920 (1950).
317. Toribara, T. Y., and P. S. Chen, Jr.: Separation of Beryllium from Biological Material, *Anal. Chem.*, **24**, 539–542 (1952).
318. Treon, J. F., and W. E. Crutchfield, Jr.: Rapid Turbidimetric Method for Determination of Sulfates, *Ind. Eng. Chem., Anal. Ed.*, **14**, 119–121 (1942).
319. Turk, A., R. M. Smock, and T. I. Taylor: Mass and Infrared Spectra of Apple Vapors, *Food Technol.*, **5**, 58–63 (1951).
320. Turk, A., H. Sleik, and P. J. Messer: Determination of Gaseous Air Pollution by Carbon Adsorption, *Am. Ind. Hyg. Assoc. Quart.*, **13**, 23–28 (1952).
321. Urone, P. F., and H. K. Anders: Determination of Small Amounts of Chromium in Human Blood, Tissues and Urine. Colorimetric Method, *Anal. Chem.*, **22**, 1317–1321 (1950).
322. Urone, P. F., and W. E. Boggs: Acid-bleached Fuchsin in Determination of Sulfur Dioxide in the Atmosphere, *Anal. Chem.*, **23**, 1517–1519 (1951).
323. Urone, P. F., and M. L. Druschel: Infrared Determination of Chlorinated Hydrocarbon Vapors in Air, *Anal. Chem.*, **24**, 626–629 (1952).
324. Urone, P. F., M. L. Druschel, and H. K. Anders: Polarographic Microdetermination of Chromium in Dusts and Mists, *Anal. Chem.*, **22**, 472–476 (1950).
325. U.S. Atomic Energy Commission: "Handbook on Aerosols," Washington, D.C., 1950.
326. Vallee, B. L., and J. G. Gibson, 2d.: An Improved Dithizone Method for the Determination of Small Quantities of Zinc in Blood and Tissue Samples, *J. Biol. Chem.*, **176**, 435–443 (1948).
327. Vedder, E. B.: "The Medical Aspects of Chemical Warfare," The Williams & Wilkins Company, Baltimore, 1925.
328. Vitro Corporation of America: "Vitro Microsensor," New York.
329. Voegtlin, C., and H. C. Hodge, eds.: "Pharmacology and Toxicology of Uranium Compounds," vol. 1, McGraw-Hill Book Company, Inc., New York, 1949.
330. Von Brand, E. K.: Applications of a Portable Continuous Smoke Recorder, *Mech. Eng.*, **72**, 479–481 (1950).
331. Wachtel, C.: "Chemical Warfare," Chemical Publishing Company, Inc., New York, 1941.
332. Wade, H. A., H. B. Elkins, and B. P. W. Ruotolo: Composition of Nitrous Fumes from Industrial Processes, *Arch. Ind. Hyg. and Occupational Med.*, **1**, 81–89 (1950).
333. Wallach, A., and W. A. McQuary: Sampling and Determination of Chlorine in Air, *Am. Ind. Hyg. Assoc. Quart.*, **9**, 63–65 (1948).
334. Waller, R. E.: The Benzpyrene Content of Town Air, *Brit. J. Cancer*, **6**, 8–21 (1952).
335. Washburn, H. W., H. F. Wiley, S. M. Rock, and C. E. Berry: Mass Spectrometry, *Ind. Eng. Chem., Anal. Ed.*, **17**, 74–81 (1945).
336. Washburn, H. W., and R. R. Austin: Some Instrumentation Problems in the Analysis of the Atmosphere, *Proc. 1st Nat. Air Pollution Symposium*, p. 69, Stanford Research Institute, Los Angeles, Calif., 1949.

337. Watkins, J. M., and C. L. Gemmill: Infrared Gas Analyzer for Low Concentrations of Carbon Dioxide, *Anal. Chem.*, **24,** 591 (1952).
338. Watson, J. H. L.: Particle Size Determinations with Electron Microscopes, *Anal. Chem.*, **20,** 576–584 (1948).
339. Wawzonek, S.: Organic Polarography, *Anal. Chem.*, **21,** 61–66 (1949).
340. West Virginia Department of Health: Atmospheric Pollution in the Great Kanawha River Valley Industrial Area. February, 1950–August, 1951, Bureau of Industrial Hygiene, Charleston, W. Va., 1952.
341. White, C. E.: Fluorometric Analysis, *Anal. Chem.*, **21,** 104–108 (1949).
342. Wilkening, M. H.: A Monitor for Natural Atmospheric Radioactivity, *Nucleonics*, **10,** No. 6, 36 (1952).
343. Willard, H. H., and O. B. Winter: Volumetric Method for the Determination of Fluorine, *Ind. Eng. Chem., Anal. Ed.*, **5,** 7–10 (1933).
344. Williams, V. Z.: Infrared Instrumentation and Techniques, *Rev. Sci. Instr.*, **19,** 135–178 (1948).
345. Wilson, W. L., and F. C. Miles: The Use of Exposure Panels in Measuring Air Pollution, paper presented at the meeting of the American Industrial Hygiene Association, Cincinnati, Ohio, April, 1952.
346. Winn, G. S., E. L. Godfrey, and K. W. Nelson: Polarographic Procedure for Urinary Thallium, *Arch. Ind. Hyg. and Occupational Med.*, **6,** 14–19 (1952).
347. Yankwich, P. E.: Radioactive Isotopes as Tracers, *Anal. Chem.*, **21,** 318–321 (1949).
348. Yoe, J. H., and H. Kleinman: "Photometric Chemical Analysis," vol. II, Nephelometry, John Wiley & Sons, Inc., New York, 1929.
349. Zimmerman, P. W.: Impurities in the Air and Their Influence on Plant Life, *Proc. 1st Nat. Air Pollution Symposium*, p. 135, Stanford Research Institute, Los Angeles, Calif., 1949.

Section 12

EXPERIMENTAL TEST METHODS

BY LESLIE SILVERMAN

12.1 INTRODUCTION

Investigators in air pollution work are often called upon to simulate in the laboratory conditions which may exist in the field. It is necessary to produce atmospheres either in a static or stable state or in a dynamic or continuous state. These can be used for exposing subjects to ascertain the effects of various gaseous mixtures or aerosols, for the calibration and testing of instruments, for toxicological investigation of the effects of various contaminants on animals, and for fundamental studies of the stability of mixtures of suspensions in air. Because of the wide variations in the nature of airborne contaminants it is necessary to consider all the various types of gaseous mixtures as well as aerosols which may be in liquid or solid state. Aerosols may vary widely in particle size, ranging from submicron to mesh size, and also in composition, density, shape, etc. Consequently, it is desirable to have work done in the laboratory with fairly flexible test equipment.

Another necessity in the preparation of atmospheres in the laboratory is the testing and calibration of control equipment such as air cleaners. In this case, except with small prepilot scaled-down devices, it is necessary to be able continuously to generate fairly large volumes of contaminated mixtures (hundreds of cubic feet per minute). The generation of large volumes of these mixtures often involves metering of small volumes of gases which are diluted by air or other gases for testing or the metering and dispersion of liquid or solid particles in order to form aerosols. In exposure chambers of small dimensions, fairly simple devices can be used, whereas in the continuous generation of particulate contaminants it is necessary to resort to more complex methods. Each of the contaminants involved, whether gas, vapor, dust, fume, smoke, or mist, requires special treatment. Whereas gas analysts are interested primarily in percentage concentrations, industrial air analysts are interested in amounts ranging downward from hundredths of a per cent or parts per million to faint traces of contaminants (parts per billion) which may create detectable odors or other effects.

It is intended in this section to discuss the design, construction, and operation of exposure chambers of both the static and dynamic types, methods of generating gas and vapor concentrations, useful methods and devices for generating homogeneous and heterogeneous concentrations of solid particles, and methods of generating aerosols containing liquid particles. The problem of metering small liquid and gas flows will be discussed in connection with generation methods, and the problem of testing air- and gas-cleaning devices by means of dynamic concentrations of various contaminants will be covered in some detail.

12.2 DESIGN, CONSTRUCTION, AND OPERATION OF EXPOSURE CHAMBERS

12.2.1 Static Chambers

The general design for an exposure chamber in the analytical laboratory for simulating gaseous mixtures or aerosols is dependent upon the basic applications for such a chamber. In many instances, the chamber is used primarily for analysis or calibration work, for standardizing physical instruments, or for checking analytical, chemical, or sampling procedures. In other cases, the chamber is used to study pathological effects of gases or aerosols on animals or to study sensory (111,147) responses of human beings exposed to subacute concentrations of various contaminants.

The requirements for the exposure chamber vary with the type of problem to be handled and will be described in order of importance.

Volume or Size. In the case of instrument calibration and when studying physical changes in gas or aerosol mixtures, relatively small chambers (1 to 2 cu m) may be employed. In the case of exposing animals or human beings, much larger chambers are necessary, depending upon the size and number of the subjects used for study. In the case of animal and human exposure chambers, it is preferable to have volumes of at least 10 or, if possible, 20 to 30 cu m.

Shape and Surface. The problem of chamber shape is best discussed first from a theoretical standpoint and next from that of physical practicability. The best theoretical chamber would be a sphere, since it encompasses the largest volume with the least wall surface. Wall surface is an important factor since the question of absorption or adsorption of the contaminant by the wall becomes a problem with certain airborne materials. Furthermore, mixing and diffusion in spherical chambers is better, especially if the contaminant is generated or supplied at the center of the sphere and diffusion is outward to the wall. However, spheres are practical spacewise only for small gas volumes of 1 or 2 cu m. They can be made of transparent plastics or spun metal such as aluminum or steel. The latter materials require ports for viewing. In studying the chemistry or physics of aerosol dispersions, the spherical chamber also provides an excellent opportunity for visual observation from all angles. An approximation of the sphere has been used in several laboratories: this is the plastic hemisphere from surplus bomber aircraft. Two of these can easily be cemented together. Metal spheres can be obtained from metal fabricators or spinners.

Nonspherical chambers are, of course, more practical to construct. In shape, the nearest geometric configuration with minimal surface is the cube, and in medium-size (20 to 30 cu m) chambers this dimension is quite practical. Where floor space is not at a premium and height is not critical, the cube is the shape of choice. In many instances, where less obstruction to floor area is desired, rectangular chambers approaching cube shape are used. Cylindrical chambers have been used, as well as combinations of cones and cylinders. In practice it has been found that many laboratories prefer to get maximum ceiling height into their chamber and therefore have been installing chambers with minimum floor space and maximum height. Obviously, in such a chamber it is more difficult to prepare uniform dispersions of

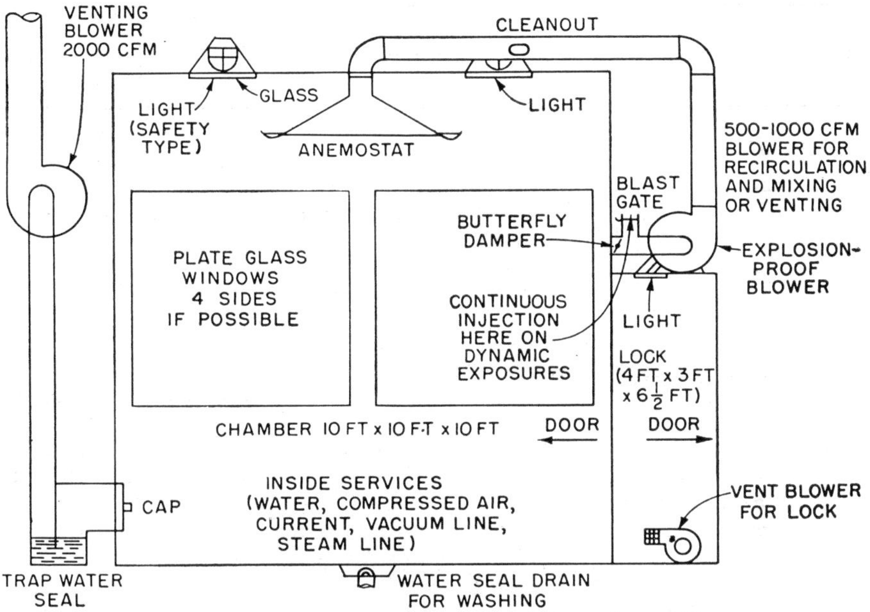

FIG. 12-1. Exposure chamber for static or dynamic conditions.

mixtures; another disadvantage is the larger convection gradient which will exist if quiescent studies are to be undertaken.

Materials. The materials of construction selected for chemical or physical work in static chambers should be ones which will not absorb or react with the various contaminants under consideration. There are certain limits to the availability of such materials, and for practical purposes it is necessary to be concerned only in the case of specific contaminants which are corrosive in low concentrations, such as fluoride gases or some strong acid or alkali mists which may react with metal or glass surfaces. The most economical and successful wall material which has been used in the laboratory is vitreous enameled steel, which appears to be stable to almost all gases and vapors and is not significantly affected by corrosive settled dusts. Glazed white tile gives comparable results but is more expensive. Stainless steel (preferably 316 alloy), aluminum (anodized, if possible) acrylic plastics, and waterproof plywood are excellent materials; the choice depends upon the permanence required and the economics. For most surfaces a coat of corrosion-resistant lacquer is sufficient to prevent appreciable damage and loss of concentrations. Silver (138) reports studies where the actual wall-surface effect was ascertained by direct testing with low gas concentrations. He observed that cellulose acetate surfaces, which should have been reasonably stable, were greatly affected by chlorovinyldichlorarsine, whereas a corrosion-resistant lacquer (Cotoid, available from the Lithgow Corporation, Chicago, Ill.) was only slightly affected. At the present time, flame spraying of polyethylene (Polythene) or tetrafluorethylene (Teflon) will provide a nonreactive surface on many base surfaces such as steel.

The magnitude of surface effects, of course, depends partly on the quantity of surface exposed per unit volume and will be influenced by the shape and size of the chamber as well as by any protuberances or mixing equipment placed inside.

Auxiliary Equipment. Accessories such as lights and fans should be placed outside the chamber wherever possible. Excellent results may be obtained by utilizing external lighting through glass windows and ports and locating the fan or mixing unit at the side of the chamber. Fans or mixing vanes within the chamber are necessary only with static chambers wherein a known volume of contaminant is added for a single mixture. In the dynamic case a continuous throughput of gas mixture or aerosol will maintain good mixing and diffusion within the chamber. Figure 12-1 shows typical construction details for a chamber which may be used for either static or dynamic preparation of concentrations. A number of chambers have been described in the literature by Drinker (40), von Oettingen (112), Irish (73), Miller (102), Silver (138), Cadle and Magill (20), Baurmash et al. (12), and many others.

It is difficult to give optimum proportions, but relative proportions can be maintained for chambers of any size desired. In practice, some very small chambers such as 5-gal glass carboys have been used for instrument calibration as, for example, the device used by Stead and Taylor (157), Setterlind (135), and others.

12.2.2 Continuous or Dynamic Chambers

The essential features in a continuous or dynamic chamber are the provision of a baffle for dispersion of gas and an outlet for the contaminant on the opposite side of the chamber. Such chambers can also be arranged for recirculation if desired, in order to minimize the amount of contaminant necessary for preparing mixtures. It is often desirable to prepare dynamic concentrations which may bypass or avoid the use of a chamber and are therefore fed into some type of testing device or into a helmet or a mask worn by the subject. However, these applications are relatively simple and need no further discussion here except to indicate that the various gas and aerosol generators which will be described in subsequent articles may be applied to them as well as to a chamber.

An important problem in dynamic chamber design is the question of type of aerosol to be used. Gases and vapors offer no serious problem unless they react with the chamber walls or supply and recirculation ducts. If particulate suspensions are made, it is essential that settling and wall losses be avoided by minimizing obstructing surfaces in duct work. Where particulate aerosols are generated in ducts leading to dynamic chambers, it is desirable to have facilities for grounding and for minimizing static electricity and electrostatic effects on the aerosol. In chambers where dynamic suspensions of aerosols are produced, it is impossible in the case of larger particles to avoid some settling and loss; hence services for washing down and cleaning are necessary. For certain physical studies the availability of steam and vacuum lines is also a must.

In the case of gases and vapors it is possible to derive mathematically the equilibrium concentration by means of the usual ventilation equations (135) as shown by Silver (138). These derivations, which will be given below, cannot be applied to particulate aerosols because these are affected by other physical forces in addition to diffusion and because, unless their size approaches molecular dimensions, they are affected by gravity, inertial and thermal forces, Brownian movement, and electrostatic effects. It is therefore apparent that preparation of known concentrations of particulate aerosols is subject to far greater variation than gases.

When a gas or gas mixture is introduced at a uniform rate into a chamber through which a continuous flow of air is provided, the concentration increases until equilibrium is reached. If perfect mixing takes place (in practice a coefficient may be applied to account for departures from the perfect case), the concentration at any time is expressed by

$$C = \frac{M}{Q}(1 - e^{-Qt/V}) \tag{12-1}$$

where C = concentration, mg/liter at time t
M = weight of gas introduced per minute
V = chamber volume, liters
Q = air volume (liters) passing through the chamber per minute

The percentage of the desired concentration M/Q at time t is therefore

$$\text{Per cent} = 100(1 - e^{-Qt/V}) \tag{12-2}$$

Equation (12-1) is a logarithmic increment relationship in which the chamber concentration rises rapidly and then approaches a constant value when time is infinite. For practical purposes, animal or human exposure can begin at 99 per cent of the theoretical value since the rise in concentration is insignificant for any subsequent exposure. By means of Eq. (12-2) it is possible to obtain the 99 per cent value; hence it follows that

$$t_{99} = 4.6\frac{V}{Q} \tag{12-3}$$

From Eq. (12-3) it is apparent that, if the chamber size V and the rate of ventilation air flow Q are known, the time to reach x per cent of equilibrium can be obtained. In general form, Eq. (12-3) may be rewritten as

$$t_x = K\frac{V}{Q} \tag{12-4}$$

where x is the per cent of concentration in time t and K is a constant. The various values of K corresponding to x are as follows:

x	K
99	4.6
95	3.0
90	2.3
85	1.9
80	1.6

Silver (138) has verified the use of this equation in experiments with chlorine in a 629-liter chamber and obtained excellent agreement.

In the case of dynamic chambers he also found that additional mixing fans were unnecessary to obtain good distribution if sufficient air flow were provided. In a 629-liter cubic chamber it was observed that, if the chamber were ventilated at a rate of 500 liters/min, distribution throughout the eight corners was excellent and measured concentrations agreed within 1 per cent of each other.

To promote turbulence and good mixing in a dynamic chamber, the inlet for the gas mixture should be small enough to provide a high velocity. In the large chamber at Harvard (34 cu m, 1,200 cu ft) an anemostat or commercial ceiling diffuser located at the center of the ceiling is used. This device allows large volumes of air to be introduced at high velocity without unnecessary drafts. Recirculation is created purposely by the anemostat, which aids in mixing.

It should be borne in mind that when animal or human subjects are introduced into a chamber they must be previously located there or enter rapidly to avoid dilution unless high flow rates are employed so that a significantly high exit velocity (100 fpm) can be provided. Another factor to consider is loss in concentration due to animal surface or clothing absorption. Silver (138) presents data which show that clothed subjects caused a greater loss than naked ones: losses of 5 to 6 per cent per subject were possible in the first case and 1.5 to 1.6 per cent in the second. As a general rule it is desirable to keep the ratio of volume of the animals to chamber volume at a value less than 5 per cent.

12.3 METHODS OF GENERATING GAS OR VAPOR CONCENTRATIONS IN AIR

The methods of generating gas and vapor concentrations depend upon whether the concentration is to be a static one or a dynamic one where the gas mixture is supplied at a continuous rate to a chamber and exhausted at the same rate. Similarly, mixtures may be supplied at a continuous rate through an extended duct system to devices for testing, cleaning, or altering the composition of the mixture.

12.3.1 Static Concentrations

Preparing known mixtures in static chambers is essentially a problem of supplying a known volume of gas or vapor to the chamber. This may be done by several methods. For gases it is possible to displace a known volume by flushing a gas buret into the chamber with air, as indicated in Fig. 12-2*a*. Booth (15) describes a very useful buret for this purpose, and Moll and Burger (105) have developed a pipet unit for delivering small gas volumes. Displacing the gas from a filled flask, as shown in Fig. 12-2*b*, is also a simple, reliable procedure. The gas may also be admitted to the chamber from a pressurized cylinder, as shown in Fig. 12-2*c*. Methods *a* and *c* are used if the gas would react with mercury or the displacing liquid. In the illustration given, Fig. 12-2*c*, the delivery of the small bomb or pressure cylinder can be calibrated in terms of milliliters per unit of pressure-gauge or manometer reading. Tests have been conducted in the Harvard laboratory for several years with this type of device, and although it was originally provided with a thermometer, it was found that the temperature change on expansion was negligible and did not alter the volumes delivered. A unit of the relative dimensions shown will deliver about 40 to 50 ml per psi gauge reading.

Gaseous concentrations may also be prepared by adding stoichiometrically determined quantities of reacting chemicals; for example, concentrations of hydrofluoric acid may be generated by adding dilute sulfuric acid to weighed amounts of sodium fluoride. The mixture is placed on a hot plate within the chamber to provide complete reaction. Similarly, gas concentrations may be produced in a static chamber by supplying the gas from a plastic or rubber balloon or an inexpensive glass vessel which is then broken within the chamber, the gas being allowed to mix with the chamber air. When it is necessary to produce concentrations of gases or vapors from substances which are ordinarily liquids, it is possible to pipet the liquid from an external buret. Alternatively, the contents of a buret may be transferred to an evaporating dish which may be rapidly placed in the chamber on a hot plate or allowed to evaporate and be dispersed by forced currents from a circulating fan. Where it is desired to do this remotely in known and exact amounts, the syringe-injection device described by the writer (141) several years ago and shown in Fig. 12-3 is a useful unit since it

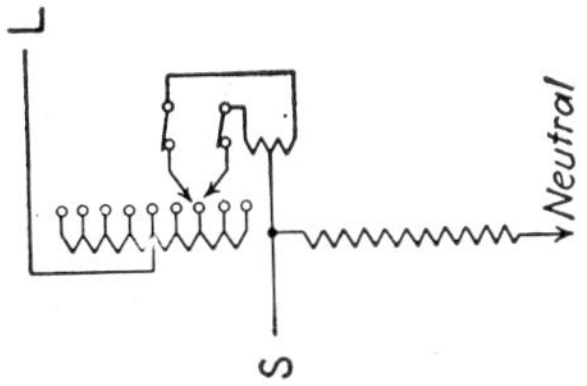

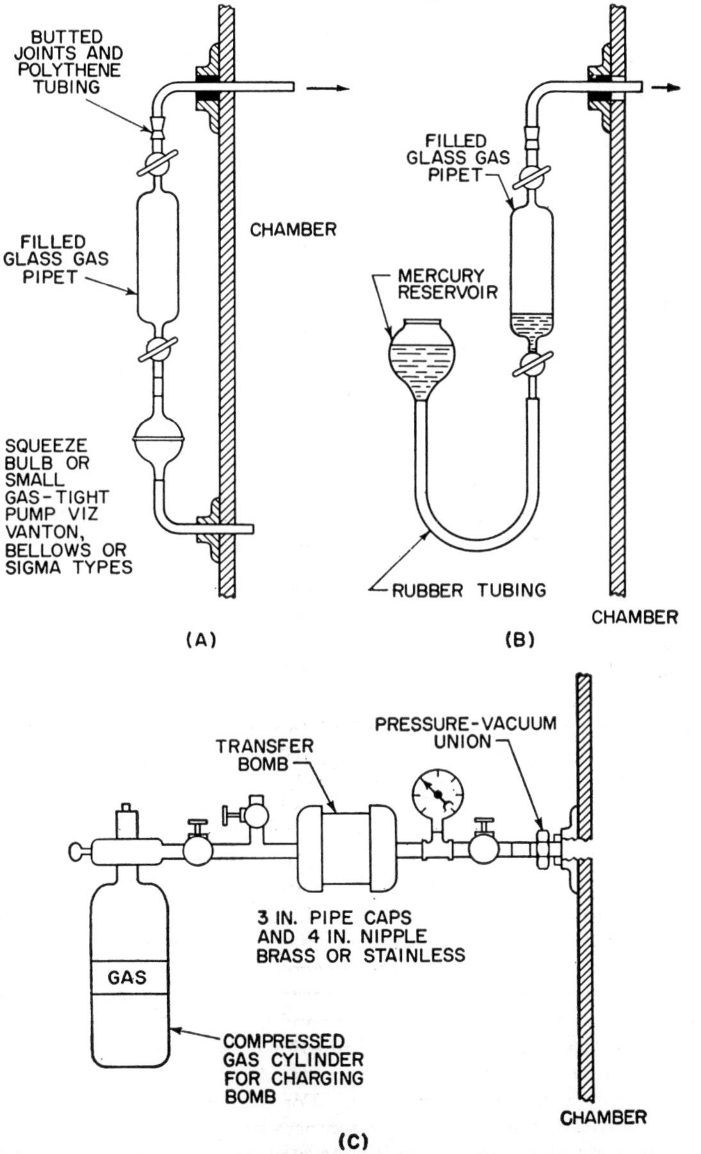

FIG. 12-2. Methods of displacing gas into an exposure chamber for producing static concentrations. (*a*) Flushing from gas pipet or buret, (*b*) displacing with mercury or nonabsorbing liquid, (*c*) use of pressure vessel or bomb for displacement.

permits changes in concentrations by simply rotating the thumbscrew from the outside of the chamber. Heat of the small light bulb furnishes the energy for evaporation.

Vapor concentrations may also be generated by the use of glass ampoules as described by Setterlind (135). An ampoule containing a known weight or volume of the solvent is placed in the chamber and subsequently broken or opened. When the liquid is difficult to vaporize, additional heat supplied by a coil or mantle heater may be necessary.

Another procedure to produce gas or vapor concentrations is to use solids which sublimate, such as carbon dioxide. Here a known weight of material is placed as a solid in the chamber and allowed to sublime, thereby producing a calculated and measurable gas concentration.

Another procedure which will produce desired concentrations in a static chamber is that used in providing constant humidities as described by Walker and Ernst (166). A liquid is utilized which has a vapor-pressure equilibrium corresponding to that of the desired concentration. Saturation concentrations of mercury or low-volatility

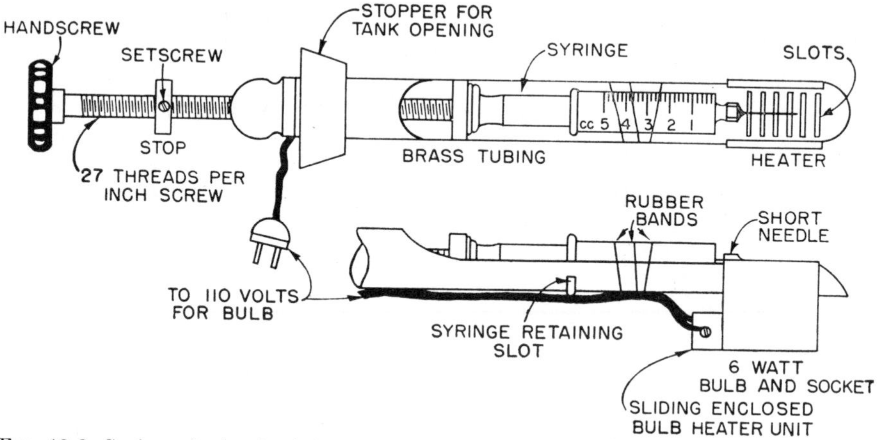

FIG. 12-3. Syringe device for injecting volatile solvents or liquids into closed chambers. (*After Silverman.*) (See Ref. 141.)

solvents are good examples. If it is desired to produce controlled humidities, solutions such as those indicated in Table 12-1, which have definite equilibrium humidity concentrations existing over the liquid, are used. By using these solutions, the vapor phase over the liquid is always at the known partial-pressure condition. Solvent atmospheres may also be produced by the equilibrium concentration over solutions of the solvent in another vehicle; for example, bromobenzene dissolved in paraffin oil can be used to produce small concentrations of methyl bromide for instrument-calibration purposes as shown by Stenger et al. (158).

An important procedure for preparing known stable concentrations of gases in high-pressure cylinders was developed many years ago by the U.S. Bureau of Mines Central Experiment Station. It utilizes a standard compressed-gas cylinder such as the compressed-air type normally used for laboratory and field work (approximately 180 cu ft at atmospheric pressure, from a 1,800 psi pressure). The cylinder is first evacuated and a known volume of gas added. Compressed air or oxygen is admitted from another cylinder or a high-pressure pump. For example, assume that the cylinder after evacuation is filled with gas at room temperature and pressure and then charged with compressed air or oxygen as desired to a pressure of 1,800 to 2,000 lb. The gas mixture thus prepared will be a dilution of 1 atm of gas to 120 atm of air, which represents a volume dilution since the law of partial pressures applies. The resulting preparation must then be mixed. This may be done either by providing a number of marbles or large steel shot in the cylinder, which cause mechanical agitation when the cylinder is tipped several times, or by allowing hot and cold water to flow over each end of the cylinder alternately so that the resulting convection currents will create adequate mixing. These procedures are necessary because of reduced diffusion under conditions of high pressure. If artificial mixing is not provided, the gas concentrations may stay in a quasi-equilibrium or stratified state for several months or perhaps even years if left at reasonably constant temperature conditions. This procedure provides a convenient source of diluted gas concentration, and much higher dilutions can be obtained if smaller amounts of gas are added. Only partial evacuation is necessary in this case, and a reliable pressure gauge will assist in making controlled mixtures. By controlled addition to the evacuated cylinder from a small gasometer, it is possible to prepare concentrations of the order of 1 ppm or less quite accurately and in a volume large enough to supply a calibrating mixture source for several tests of small instruments over an extended period.

Table 12-1. Constant-humidity Atmospheres from Known Solutions

(Percentage humidity at equilibrium within a closed space when in contact with saturated aqueous phase. Temperature at 20°C)

Solid phase or salt	Aqueous tension, mm Hg	Humidity, %
$ZnCl_2 \cdot 1\frac{1}{2}H_2O$[a]	1.74	10
$KC_2H_3O_2$	3.47	20
$CaCl_2 \cdot 6H_2O$	5.61	32.3
$Zn(NO_3)_2 \cdot 6H_2O$	7.29	42
$Na_2Cr_2O_7 \cdot 2H_2O$	9.03	52
$NaNO_3$	11.5	66
$NaClO_3$	13.0	75
$(NH_4)_2SO_4$	14.1	81
$ZnSO_4 \cdot 7H_2O$	15.6	90
$Na_2SO_3 \cdot 7H_2O$	16.5	95
$CaSO_4 \cdot 5H_2O$	17.0	98

[a] Unstable form.

See "International Critical Tables," vol. 1; C. D. Hodgman, "Handbook of Chemistry and Physics," 25th and later editions, Chemical Rubber Publishing Co., Cleveland, Ohio, gives more complete data; D. S. Carr, and B. L. Harris, Solutions for Maintaining Constant Relative Humidity, *Ind. Eng. Chem.*, **41**, 2014 (1949).

12.3.2 Dynamic Concentrations

In preparing dynamic concentrations of gases the important objectives are to provide a known and constant rate of gas for mixing with a measured air-flow rate. For injecting small gas volumes into an air stream, the injection stopcocks described by Brown (17) can be employed. These act as calibrated volumetric bypasses which may be flushed into the air stream. When handling relatively small continuous streams, 50 to 300 liters/min, the compressed-air ejector and small-chamber technique shown in Fig. 12-4 provides continuous dilution with a metered gas stream. This method has been used for studies of the effect of ammonia (149) and other gases on the human respiratory tract. In the case of testing with larger gas volumes or when chambers of large size are to be provided with controlled atmospheres, larger-scale techniques must be employed. Apparatus for this purpose is shown in Fig. 12-5. A primary requirement is that of proper equipment for metering the gas accurately. For this purpose small rotameters or capillary flowmeters may be used to meter gases which are then supplied to the duct system in measured amounts to be mixed with a metered volume of air. A large number of compressed gases are available from commercial suppliers, and suitable reducing and control valves are easily obtained for them. By supplying gases or mixtures on the inlet side of the blower as shown, it is

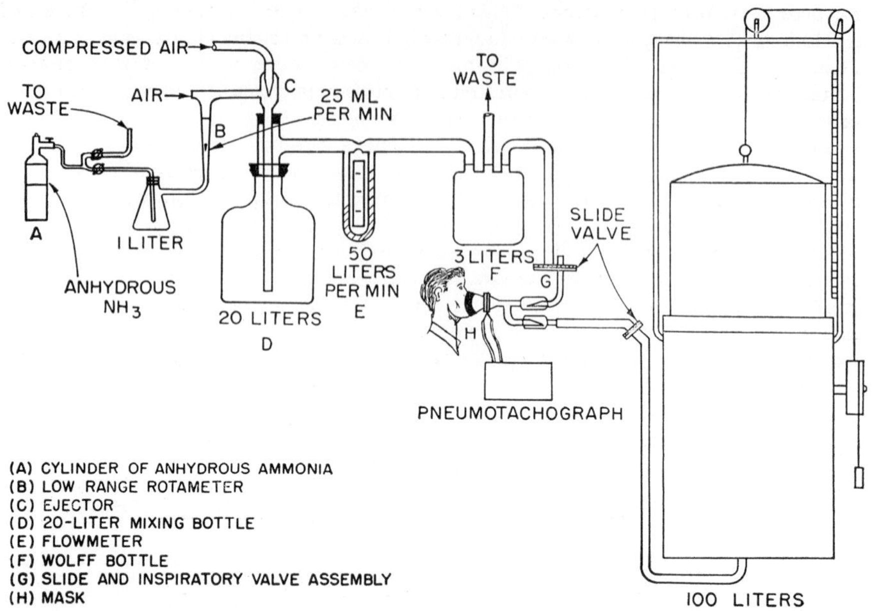

Fig. 12-4. Continuous-flow dilution device utilizing compressed-gas cylinder, rotameter, and flowmeter.

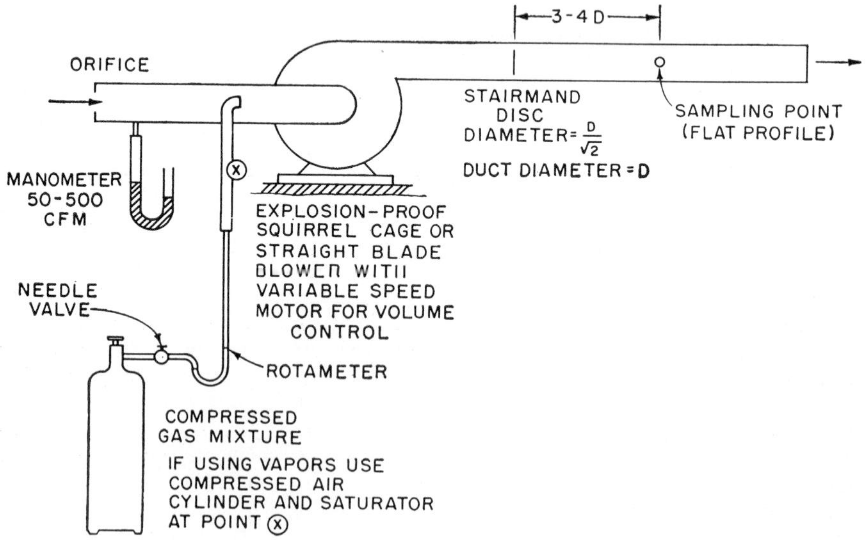

Fig. 12-5. Continuous-flow unit for large-volume dynamic dilution. Total flow 50 to 500 cfm of air.

possible to get a well-mixed sample in the effluent gas stream. This stream may then be continuously sampled or supplied to an exposure chamber if desired. Diffusion baffles such as the Stairmand disk will assist in mixing if only the aspirating or discharge side of a duct is used as shown in Fig. 12-5. If the volume of the rotameter is kept in the range of 1 liter/min and the total-air flowmeter in the range of 1 to 100 cfm, it is possible to provide dilutions of 1 to 1,000 to 1 to 100,000. If lower concentrations are desired, it is possible to obtain them by subsequent dilution while sampling or by means of the primary dilution of mixing the metered gas with a smaller metered volume of air and subsequently diluting this volume in the main stream. An alternative procedure is to use an extremely small-range rotameter in the range of 1 to 10 ml/min. Such gas metering is done best either by a bubble meter, in which the gases bubble through a solution and the number of bubbles of known size are counted, or by means of a pinhole critical orifice from a source of gas at constant pressure. The calibration and metering of small flows will be discussed in a subsequent article at the end of this section; so it will suffice here to describe the general principle of the method.

Small volumes of gas may also be displaced by dropping mercury or another liquid from a source at known head, as indicated in Fig. 12-6. This displacement of the gas can be metered quite accurately by controlling the rate of rise of the mercury in the column. Another mercury pump for gases is described by Irish and Adams (73).

Another procedure for the dynamic generation of gas is to use a chemical reaction at controlled rate. For example, by dropping formic acid into hot sulfuric acid at a controlled rate, carbon monoxide can be obtained as desired. There are other examples of similar rate reactions, such as the method used by Fieldner et al. (53) to generate continuously hydrogen cyanide from sodium cyanide and sulfuric acid. To supply small amounts of gas at a controlled rate, one can use the synchronous-motor screw-driven syringe heads similar to those described by Barrett et al. (11) and Roggoff (122). This type of device can be used for both gases and vapors. In the case of gases it is feasible only where needed for short periods of time, since there will be some gas loss from the syringe with time. It should be borne in mind that the syringe should be maintained at constant temperature so that the gas does not diffuse from the tip. Still another procedure for supplying gas at known rates is to use a motor-operated rising mercury column of the type generally used for producing vapor concentrations. For gases, a continuous rise of the mercury displaces the gas through an orifice or a liquid-sealed valve so that backflow is eliminated.

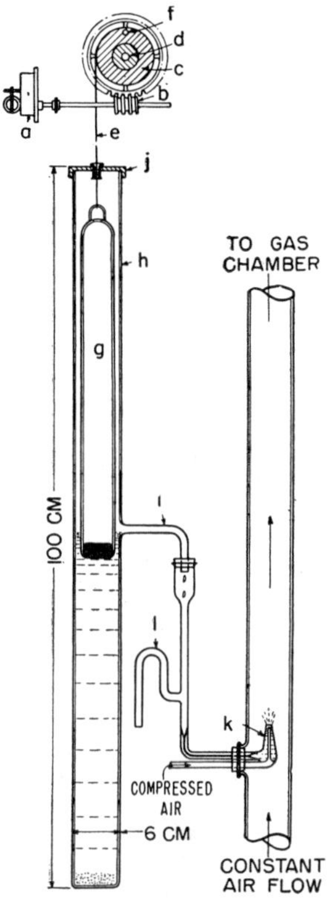

FIG. 12-6. Continuous-displacement device for producing dynamic concentrations of vapors. (*a*) Synchronous clock motor, 1 rpm, (*b*) 100:1 worm gear, (*c*) pulley, (*d*) drive burr, (*e*) piano wire, 26 gauge, (*f*) taper plug, (*g*) glass plunger, (*h*) liquid reservoir, (*i*) side arm, (*j*) cap, (*k*) atomizer, (*l*) overflow to safety trap. (*After U.S. Bureau of Mines, Rept. Invest.* 3323.)

Another method for producing gas concentrations is by the use of controlled electrolysis of solutions. By controlling the current supplied, the evolution of gas may be accurately regulated, and the gas thus generated can then be fed into a diluting air stream for producing known concentrations. The electrolytic system described by Burwell (19) in which mercury is displaced by gas generated by electrolysis of a 30 per cent potassium hydroxide solution can often be utilized. The displaced mercury can force the feed gas (or liquid if desired) continuously into the chamber. King and Davidson (80) describe a similar device.

When it is desired to mix two gases into a reaction chamber at a constant predetermined rate, the constant-pressure flow-ratio regulator described by Heikes (65) will be of interest. These combinations of gases may be led into the systems described previously. The static gaseous concentrations produced from compressed-air–gas mixtures in cylinders, described in Art. 12.3.1, Static Concentrations, can be fed continuously into a diluting air stream to produce dynamic concentrations.

In producing continuous concentrations of gases it is also possible to utilize the vapor-pressure equilibrium of the gas over a liquid by passing a saturated gas stream into the mixing stream continuously. Such a system has been described for mercury concentrations by Shepherd et al. (137) and for halogenated hydrocarbons by Goldman et al. (62). Saturators through which a gas stream may be bubbled usually incorporate a sintered or fritted absorber for producing rapid saturation and a glass-wool filter or trap to remove entrained droplets. The saturator is usually placed in a controlled constant-temperature bath. Further examples of this method of generating concentrations are presented by Jones et al. (76), Heppel et al. (67), and Scott (134). When high concentrations of gases are desired, it is possible to use the type of device described by Lewis and Koepf (90), in which two wet-gas meters are connected by means of a chain. The gas fed to the first meter causes the second meter to turn at a corresponding speed, thus producing a known ratio of gases. This is adequate for producing known percentages of gas mixtures on a continuous basis but is not usually adaptable for the concentration range needed for hygienic studies. However, this automatic-dilution method can be used to control the composition of the diluting gas, which may also require variation with the concentration of the contaminant in some instances.

For preparation of known concentrations of vapors dynamically the problem is in limiting the discussion to a few methods, since many satisfactory devices have been developed for feeding small amounts of liquid continuously into gas streams. These generators may be classified in two groups:

1. Liquid-feed systems in which liquid is fed to a vaporizing device or allowed to evaporate directly into an air stream
2. Vapor-equilibrium devices in which air or gas is saturated with the vapor at a given equilibrium condition and the saturated gas allowed to mix with a diluting air stream

The methods for delivering liquids at constant rates are numerous. Many of the devices described above such as motor-driven syringes, electrolytic cells, and rising mercury columns are better adapted to liquid displacement than to gases since fewer precautions such as precise temperature control and preventing leakage are necessary.

Feed systems using liquids which are dropped through a fine capillary tip or a small nozzle at a constant rate onto a hot plate or heated surface for evaporation have been used by many investigators, for example, see articles by Yant and Frey (177) and Machle et al. (97). It is important to limit the temperature of the evaporating source to a value which will not decompose or ignite the solvent or chemical to be volatilized. Dynamic concentrations for small vapor stream volumes can be prepared by the simple vaporizing techniques described by Olsen et al. (114) and Gisclard (61). In these devices small volumes of volatile solvents may be vaporized into a flowing air stream by placing a known weight or volume into a small vessel and allowing the entering air stream to be directed at the surface of the liquid to cause evaporation. The remaining liquid can be measured by weight or volume and the loss calculated in terms of the air volume which has passed over the solvent.

In the case of testing with larger gas volumes or when chambers of large size are to be given controlled atmospheres, different techniques must be adopted. Small-volume precision-machined gear pumps such as those used to pump viscose solutions

Table 12-2. Gas and Vapor Concentrations Prepared in Dynamic Atmospheres

No.	Gas or vapor	Method of generation	Range of concentrations generated, ppm	Refs.[a]
1	Acetone	Dilution from continuous saturator in constant-temperature bath	100–>5%	76, 111
2	Acetaldehyde	Dilution in air stream after direct feeding of liquid	0–200	147
3	Acrolein	Same as 2	0–5	20, 155
4	Acrylonitrile	Dilution from continuous saturator	0–250	46
5	Ammonia	Dilution from constant feed of gas from cylinder of anhydrous ammonia	100–1,000	149
6	Amyl acetate (iso)	Dilution in air stream after direct feeding of liquid onto hot plate	100–500	111
7	Amyl alcohol (iso)	Same as 6	50–250	111
8	Benzene	Displacement of liquid into moving air stream	50–500	178
9	Butyl acetate	Same as 6	50–300	111
10	Butyl alcohol (normal)	Same as 6	0–100	111
11	Butyl ether (normal)	Same as 6	50–300	147
12	Carbon disulfide	Same as 2	0–320	89, 98
13	Carbon tetrachloride	Same as 1	0–5,000	11, 53, 62
14	Chlorine	Dilution from constant feed of gas from a liquid chlorine cylinder	0–5,000	53
15	Cyclohexanol	Same as 6	0–200	111
16	Cyclohexanone	Same as 6	0–100	111
17	Diacetone alcohol	Same as 6	0–200	147
18	Diallyl phthalate	Same as 6	Saturated vapor	147
19	Diisobutyl carbinol	Same as 6	0–10	147
20	Diisobutyl ketone	Same as 6	0–100	147
21	Dioxane	Same as 6	100–500	147
22	Ethyl acetate	Same as 2	50–500	111
23	Ethylene dichloride	Same as 8, utilizing heated tube	0–1,000	22
24	Ethyl ether	Direct volatilization	0–1%	66
25	Formaldehyde	Direct feeding of known solution into heated crucible in gas chamber	0–20	9
26	Formic acid	Direct dispersion into chamber	0–1	95
27	Gasoline	Direct feeding through steam gasket to air ventilating test chamber	0–2.6%	43
28	Hexane	Same as 6	0–1,000	111
29	Hexylene glycol	Same as 6	0–100	147
30	Hydrogen cyanide	Same as 2	0–2,000	32
31	Hydrogen selenide	Reaction between FeSe and HCl diluted with air	0–12	47
32	Hydrogen sulfide	Cylinder gas diluted with air	0–500	91
33	Isophorone	Same as 6	0–50	147
34	Iso-propyl acetate	Same as 6	0–300	147
35	Iso-propyl ether	Same as 2	0–6%	96, 147
36	Methane	Dilution of compressed gas in air stream	0–10,000	23
37	Methyl alcohol	Same as 2 utilizing atomizer	0–10,000	132
38	Methyl amyl acetate	Same as 6	0–200	147
39	Methyl bromide	Same as 2	0–13,000	74

Table 12-2. Gas and Vapor Concentrations Prepared in Dynamic Atmospheres *(Continued)*

No.	Gas or vapor	Method of generation	Range of concentrations generated, ppm	Refs.[a]
40	Methyl chloride	Continuous dilution of released compressed gas in chamber with air	0–140	170
41	Methyl ethyl ketone	Same as 1	0–500	111
42	Methyl isobutylcarbinol	Same as 6	0–100	147
43	Methyl vinyl carbinol	Same as 6	0–100	147
44	Methylene chloride	Same as 1	0–10,000	67
45	Mercury	Same as 1	0–0.10	137
46	Methyl oxide	Same as 6	0–100	147
47	Mononitroparaffin	Direct feeding of liquid to vapor stream passing through heated tube	0–5%	97, 134
48	Naphtha	Same as 1	0–1%	177
49	Nitric acid	Same as 1	0–1	20, 155
40	Nitric oxide	Treating solution of KNO_3 with H_2SO_4 and sweeping into air stream where it forms NO_2 rapidly	0–1,200	27
51	Nitroethane	Same as 1	0–3%	177
52	Nitrogen dioxide	Glass ampoules with gas broken in chamber	0–100	116
53	Ozone	Passing pure O_2 through electric arc or UV lamp	0–1	31
54	Stoddard solvent	Same as 6	0–1,000	111
55	Sulfur dioxide	Cylinder gas diluted with air	0–10	3
56	Sulfuric acid	Dilution from vaporized acid	0–2	4
57	Toluene	Same as 6	0–500	111
58	Trichloroethylene	Same as 2 or 1	0–1,200	11, 62
59	Trichloropropane	Same as 6	0–200	147
60	Turpentine	Same as 6	0–200	111
61	Xylene	Same as 6	0–300	111

[a] References are given at the end of this section.

through rayon spinnerettes can be used to give a constant liquid flow into an air stream. These pumps are driven by synchronous motors through a gear reduction train, and flow rates of the order of 0.1 ml and less per minute are easily obtained. Chemical proportioning pumps such as those used for adding flotation reagents to large mineral-separation vats are also suitable for this purpose. These may involve vibrating plastic or rubber membranes or tubing constantly milked by an eccentric series of rollers. Precision liquid feeders for small flow rates of liquid which can be adapted to air-concentration generation are described by Calcote (21), who uses a critical-flow orifice, and by Lundsted et al. (94) and Jones et al. (75). A number of the references cited previously on exposure chambers describe methods for continuously adding solvent vapors to air streams, in particular those of von Oettingen (112) and Irish and Adams (73). Table 12-2 presents a number of gases and vapors for which methods have been developed for preparing known static or dynamic concentrations for investigation purposes.

12.4 GENERATION OF PARTICULATE CONCENTRATIONS IN AIR

Basically, the generation of particulate suspensions in gases is one of the most difficult problems facing the experimenter in the field of aerosol technology. Particulate matter, of course, is divided into several categories, which will not be discussed at any length here since they are discussed in Sec. 13. It is important, however, to emphasize, as in Table 12-3, the factors and ranges which are involved. The basic distinctions given will be of importance in discussing methods of preparing aerosols of each type.

Table 12-3. Factors in Particulate Aerosols

Nature of aerosol or descriptive term	Range of particle sizes	Range of specific gravity	Shape	Composition	Method of generation	Remarks
Dust......	Probably 0.05μ to mesh size only. Those below 25μ will stay in suspension in quiescent atmospheres for a significant period. This phenomenon also depends on specific gravity	1.0–10.0 Majority of mineral dusts lie between 2 and 3	Spherical to irregular, including needles and flakes	Mineral, metallic, or organic. Can be salts or oxides, etc.	Crushing, grinding, blasting, pulverizing, decrepitation	Materials considered as solid throughout, although they may be heterogeneous in composition. Composition same as parent material from which size is produced
Fume.....	0.001–1μ. Greater sizes may be produced by flocculation, coagulation, or agglomeration	1.2–10.0	Spherical to irregular. May include chains or clusters	Usually metals or oxides, but several organic materials which sublimate fall in this category	Combustion, sublimation, condensation	These materials all flocculate rapidly
Smoke.....	0.001–0.5μ	1.0–4.0	Spherical to irregular. May include chains	Usually organic, but smokes may be liquid or solid	Burning or chemical reaction	May include particles of low vapor pressure which settle slowly
Mist......	0.01–25μ	1.0–14.0	Spherical	Liquid	Condensation upon nuclei. Atomization or disintegration of liquids	May be homogeneous in composition or heterogeneous if solution contains dissolved salts, etc.

Experimenters working with aerosols in general have two basic requirements. These are generation of aerosols which are homogeneous with respect to size, shape, or specific gravity, so as to minimize the number of variables which must be considered, and of those which are heterogeneous in all respects and simulate the particles generated in nature or by industrial processes. The only group of materials in nature which seem to be fairly uniform in shape and size are microorganisms such as bacteria, spores, fungi, or pollen. Ordinary atmospheric dusts or smokes are widely divergent with respect to size and most of the other physical factors involved. Either homogeneous or heterogeneous aerosols may also be discussed from the point of view of the physical nature of the particles in the suspensoid, that is, whether they are solid or liquid. A discussion will be given first of generating homogeneous aerosols and then of the heterogeneous types. Before discussing them separately, however, a basic discussion of the methods of generation common to all aerosols is presented.

12.4.1 Generation Methods

Concentrations of solid particles may be generated by the following methods:

1. Direct attrition such as grinding, crushing, abrading, or blasting of solid materials
2. Dry dispersal of material previously ground or sized
3. Wet dispersal (atomization of colloidal suspensions) of material previously ground or sized
4. Wet dispersal or atomization of solutions containing a solute of the material which will remain as a particulate after airborne evaporation (spray drying)
5. Vaporization of materials and their subsequent condensation as solid particles
6. Melting and atomization with subsequent solidification of material
7. Burning of combustible materials (liquids or solids whose combustion products are solids)

Concentrations of liquid particles may be generated by the following methods:

1. Direct atomization or disintegration of liquids
2. Vaporization of liquids or solutions and their subsequent condensation as liquid droplets
3. Saturation and entrainment of liquids by gases caused to bubble or form in them
4. Burning of combustible materials in which volatile matter is distilled and subsequently absorbed in droplets formed from the water of combustion

It can be seen from the foregoing listings that many forms of generation are common to both solid and liquid particles. In fact, many solid aerosols are generated by spray or droplet drying of liquids which contain either salts in solution or particles in suspension. Particulate concentrations may be generated for weight, stain, or surface concentration or for enumeration (count) purposes. Requirements for each of these conditions vary.

12.4.2 Generation of Homogeneous Aerosols

The need for generating aerosols of specifically one size, one shape, and one density is fundamental in developing basic laws and information in such fields as aerosol behavior; in gas cleaning, i.e., testing or studying filters, wet collectors, and electrostatic precipitators; and in human lung-retention studies. The availability of a constant or controllable size simplifies the number of variables to be considered and permits a more rapid delineation of the factors involved in any controlled study. Unfortunately, it is difficult to generate these aerosols and still more troublesome to generate them in high enough concentrations that they may be used for large-scale testing.

Homogeneous liquid aerosols can be produced only by two methods at the present time with any outstanding success. These are (1) the use of controlled vaporization, condensation on nuclei, and growth over saturated vapors, and (2) the use of the spinning top or rotating disk to which a liquid is fed and allowed to shear at the edge of the disk in motion. Another possible method is the use of drop generators, but the latter are limited to relatively large sizes and very low concentrations.

The controlled-vaporization apparatus has been described by LaMer and his coworkers in several publications (81,83,151) and has been used successfully to generate aerosols of uniform size from 0.01 to 0.25μ in diameter. A typical unit is shown in Fig. 12-7. Such liquids as sulfuric acid (which is self-nucleating), dioctyl phthalate, dibutyl phthalate, triphenyl phosphate, and certain other oils may be used.

Constant-temperature vaporization of oils such as Diol 55 or dioctyl phthalate (59,93) yields aerosols in the range of 0.3 to 1.0μ, but these are not completely homogeneous.

The spinning-top device, on the other hand, generates a major particle and a satellite as shown by Walton and Prewett (169) and May (100). The latter worker has improved the device, however, and made a satellite extractor leaving only the principal particle in the air stream. Figure 12-8 shows a schematic drawing of a spinning-top generator as used in the Harvard laboratory for methylene blue aerosol

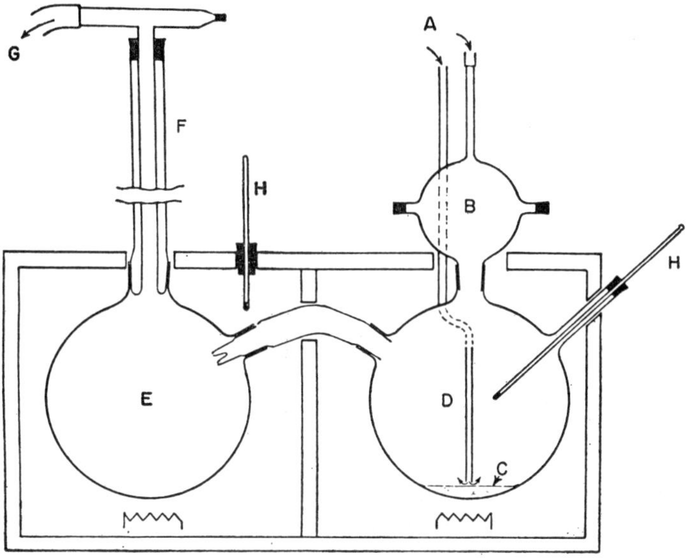

(A) INLETS FOR DUST-FREE AIR
(B) CHAMBER FOR PRODUCTION OF NUCLEI
(C) LIQUID OR SOLID TO BE VAPORIZED
(D) BOILER
(E) REHEATER
(F) DOUBLE WALL GLASS CHIMNEY
(G) OUTLET
(H) THERMOMETERS

FIG. 12-7. Diagrammatic sketch of all-glass monodisperse aerosol generator.

production. In the spinning-top device the aerosol size is predicted from Eq. (12-5), developed by Walton and Prewett.

$$d = \frac{K}{\omega}\sqrt{\frac{T}{D\rho}} \tag{12-5}$$

where d = diameter of drop, cm
ω = angular velocity of top, radians/sec
D = rotor diameter, cm
ρ = density of fluid, g/cu cm
T = surface tension of liquid, dynes/cm
K = constant (average value = 3.8)

Droplets generated ranged from 15μ to 3 mm diameter, depending partly upon the liquid employed. A large number of substances have been studied, including mercury and several phthalates.

The use of vibrating or rotating capillaries as generators has been described by

Vonnegut and Neubauer (164), Cottrell (30), and Rayner and Hurtig (121). While these methods can develop uniform drops, their output is quite limited and not satisfactory except for very low loadings.

Of the methods described above, the only ones satisfactory for solid particles are those which use controlled vaporization of materials which sublime and are then permitted to condense on nuclei and allowed to grow to a definite size. LaMer et al. (82) have described the use of the homogeneous liquid aerosol generator for this purpose, using stearic acid which produces a supercooled viscous liquid droplet which can crystallize on contact. Benzoic acid was also tried in the Harvard laboratory, but it does not remain stable for long enough periods. At the present time microcrystalline waxes and volatile dyes are being used with success by some workers (LaMer's group at Columbia University), but details of these studies have not yet been published.

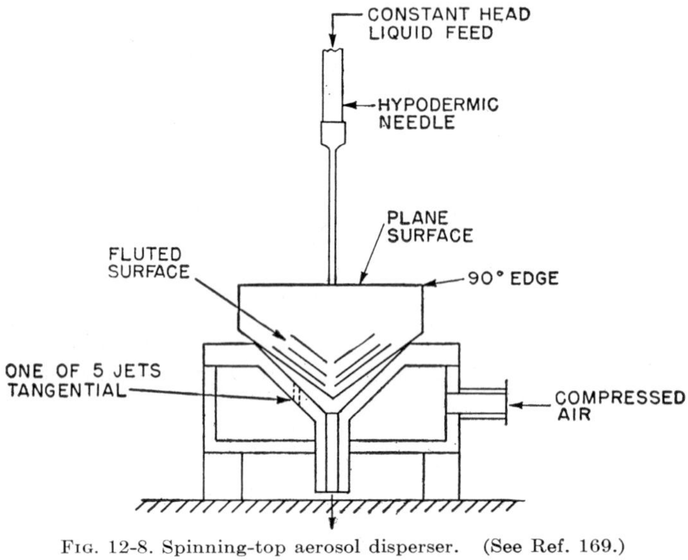

FIG. 12-8. Spinning-top aerosol disperser. (See Ref. 169.)

Another approach to generating a homogeneous solid aerosol is the use of the spinning top as described by Walton and Prewett but using a salt or resin in solution in a solvent. Evaporation of the solvent leaves a homogeneous solid resin particle. This method appears to have some promise. The Harvard group has also done this, as mentioned above (Fig. 12-9), by using solutions of methylene blue in alcohol. Rather than use the major particles, these have been removed by letting them impact on a surrounding surface. The satellites, which are much smaller although not completely homogeneous, are used instead. Obviously other dyes or salts can be used.

The possibility of dry or wet dispersal of uniform-size particles should also be mentioned, although the methods employed are similar to those to be described for heterogeneous aerosols. Some colloidal suspensions of uniform-size latex and resin particles have been prepared and used for electron-microscope standards and similar purposes, but these are too expensive for practical use. A vinyl resin powder of uniform 0.5μ size has also been prepared (see Table 12-4), but the dry dispersion of this as discrete particles has not been successful because of the electrostatic charge developed on the particles. A solution of them or other materials always leaves the distinct possibility that one droplet may contain more than one particle before evaporation. To prevent this, very dilute solutions must be employed, which means very low loadings in air or gases.

The use of bacteria which are fairly uniform in size and in some instances in spherical shape is another possibility in preparing homogeneous aerosols. These are usually sprayed from solutions by a variety of nebulizing devices, as described by Rosebury (123) and Ferry et al. (52).

One major limitation of all the homogeneous generators described is that only small gas streams, not exceeding 1 or 2 cu m/min, can be treated with the types of devices

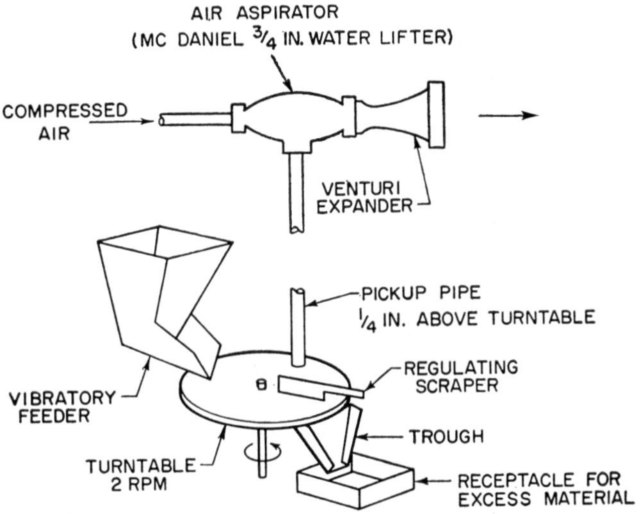

FIG. 12-9. Turntable dust feeder as used at Harvard Air Cleaning Laboratory. (See Ref. 55.)

available because of practical limitations involved in large-scale production of equipment. For this reason also, only low loadings can be used if higher air flows are desired.

12.4.3 Generation of Heterogeneous Aerosols

In discussing the production of heterogeneous aerosols, for which a large use is in existence at the present time, it will be well to discuss the solid-particulate group and follow this with a discussion of procedures for liquids. Again it should be mentioned that those methods used for liquid aerosols also have application if salts or resins in solution can be dried or solidified by evaporation of the vehicle used.

Solid Particles. A number of direct attrition methods have been used to produce dry aerosols. Some have utilized small hammer mills or impact mills for crushing or grinding solid or aggregated materials. Small ball mills or rod mills have also been utilized by passing a continuous air stream through them. A recent method described by Fitzgerald and Detwiler (58) employed a revolving steel-wire brush rubbing against a duraluminum plate to produce a solid aerosol of 2.7 g per cubic centimeter density. An impaction plate was used to capture particles above 5μ, and the final aerosol was in the range of 0.1 to 2.1μ. Scraping or brushing of other solid materials is possible by this procedure. One limitation, however, is that the materials should be of widely differing mechanical resistance to abrasion. Druett and Sowerby (45), by using a rotating drum filled with charcoal, were able to produce controlled dust concentrations by elutriation and varying drum speed. Particle sizes and concentrations corresponding to factory air values could be obtained.

For large-scale testing, energy mills* are capable of producing fine materials down to 1 or 2μ in quantities to be dispersed in as much as 200 to 300 cu ft of air per minute. A number of other companies manufacture laboratory- or portable-size reduction devices which can be adapted to this purpose. The recent article by Smith (152) on the general subject of size reduction gives an extensive list of types, applications, and size production range.

Dry dispersal has had perhaps the widest application in experimental dust studies because finely divided materials of a wide variety of sizes, shapes, specific gravities, and surface characteristics can be readily obtained. These materials can be purchased in grain to ton quantities and in sizes ranging upward from 0.1μ. Table 12-4 lists a number of common materials which have been used for test aerosols, including some "standard test dusts" or mixtures which have been adopted for filter and air-cleaner testing.

One of the chief difficulties in dry dispersion is avoiding disseminating clouds containing agglomerates, or flocculated materials. This is a serious limitation in many instances but can be minimized by the proper dispersion equipment.

Dispersion of materials has been accomplished by hand distribution into a moving air stream, which is quite unreliable, and by a number of more elaborate techniques. The solid feeders using the principles and types discussed by Olive (113) can be applied to small- and large-scale testing by dry dispersion. A large number of devices and modifications of devices have been used for this purpose. Table 12-5 lists several kinds of units, and their characteristics, which have been found satisfactory for various testing purposes. For introducing dusts into an air stream on a mechanically controlled basis, Deichmann (36) has developed a device which is called a "dust-shaker." It consists of a small screen continuously rapped into an air stream which flows under the bottom of the sieve. The device is satisfactory for low concentrations of previously ground materials which should be well dried before dispersal. Dried air is also recommended. Sonkin et al. (154) have modified this device by incorporating a sintered-glass surface, an ejector, and an agitating air stream. Because of their construction, neither of these types is adapted to continuous addition of material but must operate with the initial material charged to the screen surface or the shaker volume.

Ejectors operating from compressed-air streams are usually employed for picking up the dust from a moving trough, belt, or turntable. The simplest of these consists of a small vibratory feeder (Syntron) feeding directly into an ejector. The only limitation to this method is that vibration in the Syntron feeder may cause agglomeration and inadequate dispersion. To minimize this the ejector should be operated at a pressure high enough so that sonic velocity takes place at the primary jet. The dust should therefore be introduced at that point to get maximum dispersion. Some dusts, because of their nature (e.g., charge, surface characteristics, or hygroscopicity), tend to ball or clump to a point where they cannot be fed uniformly to the pickup nozzle, and other procedures must be employed as will be discussed later in this article.

Cadle and Magill (20) describe a feeder which has a moving-trough device which contains the dust to be dispersed. The trough is driven by a motor past the ejector inlet or pickup. An improved form of this device has been used by the American Air Filter Company for several years and is now the recommended feeder for the new Air Filter Institute Code (1). The Farr Company of Los Angeles also has a comparable device using multiple grooves in a tray (Far-Air Model 4). In the Harvard laboratory studies (55), a turntable feeder similar to that shown in Fig. 12-9 is used. This device incorporates an elutriating tower and a method of regulating the amount of dust ribbon on the plate. This is a modification of similar turntable dust feeders

* C. H. Wheeler Company, Philadelphia, Pa.

used by Rowley and Jordan (126), working for the American Society of Heating and Ventilating Engineers. Several others have used this type of feeder.

Another procedure which has been used is the U.S. Bureau of Mines feeder developed by Brown (133) which is shown in Fig. 12-10. This utilizes a rising dust tube fed into a swirling air stream. An improved type in a much more compact feeder has been developed by Wright (176). The Wright feeder, which is available commercially,* utilizes a compacted dust sample such as powdered coal and rotates this sample continuously against a scraper knife. Compressed air is fed through the

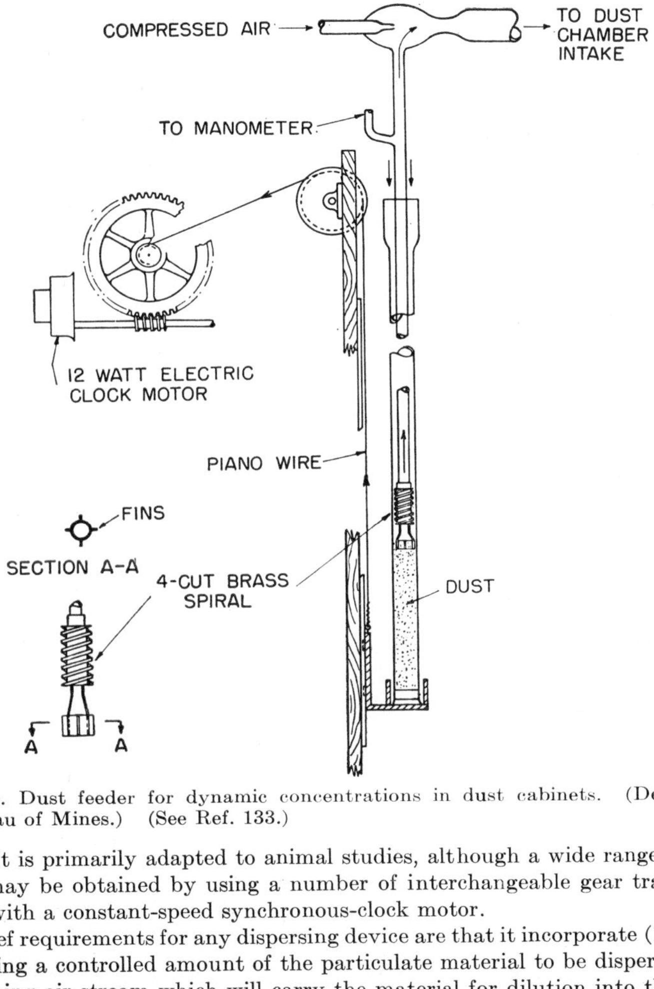

FIG. 12-10. Dust feeder for dynamic concentrations in dust cabinets. (Developed by U.S. Bureau of Mines.) (See Ref. 133.)

device. It is primarily adapted to animal studies, although a wide range of concentrations may be obtained by using a number of interchangeable gear trains in conjunction with a constant-speed synchronous-clock motor.

The chief requirements for any dispersing device are that it incorporate (1) a method of producing a controlled amount of the particulate material to be dispersed and (2) an entraining air stream which will carry the material for dilution into the main air stream. Most of the dust feeders, as mentioned, have employed ejectors or compressed air directed through the unit as a means of aspirating the dust into a jet where it is impinged by the jet air stream and thus dispersed into a conducting air stream or a chamber. Studies have been made in the Harvard laboratory (143) which show that it is necessary to use fairly high velocities (above 100 m/sec) at the

* L. Adams, Ltd., London, England.

Table 12-4. Dusts Used for Test Aerosols

A. Standard Test Dusts

Name or material identification	Characteristics: Mean size		Standard deviation	Specific gravity	Shape	Composition	Remarks	Refs.
Air Filter Institute standard test dust	Army standard dust size given below		....	....	Irregular plus carbon black spheres in chains	72% std. air cleaner dust (Army std. given below) 25% K-I carbon black, 3% cotton linters	Dust available from James H. Herron Co., Cleveland, Ohio	1
	Size, μ	%						
Army standard fine air-cleaner test dust	<2½–3	10	....	2.5	Irregular	Mostly Arizona road-dust fines, silicates, and quartz	Available from A.C. Spark Plug Div., General Motors Corp., Flint, Mich.	50
	0–5	39 ± 2						
	5–10	18 ± 3						
	10–20	16 ± 3						
	20–40	18 ± 3						
	40–80	9 ± 3						
	Through 200 mesh	100						
American Society of Heating & Ventilating Engineers test dust	80% Pocahontas ash through 200 mesh, 20% double-bolted carbon dust through 100 mesh		....	Variable	Irregular	80% Pocahontas ash. 20% double-bolted carbon	No longer in use as standard at present time. Original code used 50% powdered lampblack and 50% Pocahontas ash	128
Atmospheric dust........	0.5μ		1.5–2	Variable 1.2–2.0	Irregular	Variable. Contains silicates, carbon, mixed oxides	Used by several laboratories such as Harvard Air Cleaning Laboratory (14,55), Eastman Kodak Co., and Bell Telephone Laboratories	106
	Size	%						
British standard test dusts	Passing 100 mesh Passing B.S. sieve	100–99	....	2.5	Irregular	Mineral dusts containing quartz, feldspars, limestone, and clay	Designed for testing internal-combustion engines. British Standard 1701:1950, British Standards Institute. Standard now in prepara-	69
	Passing 150 mesh							

	200 / 86–76 $<40\mu$ / 70–60 $<20\mu$ / 46–35 $<10\mu$ / 30–20 (sic: 19–11 for last) 					tion for panel filters using methylene blue aerosol from 1% solution and three test dusts all of aluminum oxide as follows: Aloxite grade / Mass / Mean size, μ 1 / 50 / 5 2 / 175 / 10 3 / 225 / 17.5	
National Bureau of Standards, U.S. Dept. of Commerce, air-filter test dust Cottrell precipitate	0.5μ	2.6	2.0–2.5	Irregular but high percentage of spheres	Carbon, oxides and silicates, Cottrell collected ash from pulverized-fuel plant. Modified dust included lampblack and 96% ash	Sizing done by writer on airborne material collected from NBS-type generator operated at 40 psi. Material available from Potomac Electric Power Co., Buzzard Point Station, Washington, D.C.	38
U.S. Bureau of Mines: Silica dust	0.6μ	1.9	2.65	Irregular	Homogeneous 99% + free silica	Also prepared as water mist having dust in suspension	41, 133, 161
Lead fume	$<0.1\mu$	....	8–9	Irregular	Lead oxide	Prepared by burning lead tetraethyl	
Magnesium oxide	$0.5–0.7\mu$	....	3.6	Irregular	Magnesium oxide	Prepared by burning magnesium ribbon	
Chromic acid mist	$<1–5\mu$	....	2.7	Spherical	Chromic acid	Prepared by electrolyzing acid	
Lead-paint mist	$<1–10\mu$	....	6.6	Irregular	Lead carbonate	Prepared by spraying specified paint	

B. Pulverized or Ground Materials Used as Test Dusts

Name	Characteristics				Remarks	Ref.
	Mean size	Standard deviation	Specific gravity	Shape		
Aluminum oxide	Variable $5–38\mu$	....	4.0	Irregular	A number of sized grits can be obtained from abrasives manufacturers. Reference aerosol. Mean size $<2\mu$ because chamber air floated	37

Table 12-4. Dusts Used for Test Aerosols *(Continued)*

Name	Characteristics				Remarks	Ref.
	Mean size	Standard deviation	Specific gravity	Shape		
Asbestos	$<5\mu$	....	3–3.2	Irregular	Designated as "floats"	14
Bismuth subcarbonate	3.8μ MMD[a]	....	6.9	Irregular	....	86
Calcium carbonate	2.6μ MMD	2.0	2.7	Irregular rhomboidal crystals	Precipitated chalks or ground limestones may be used. Can be analyzed chemically	14
Calcium phosphate	$0.7–4.3\mu$ MMD	....	2.3	Irregular	....	84
Carbon black	0.087μ MMD	1.55	1.77	Discrete particles are spheres	Often called lampblack. Discrete size very small but tends to chain and is difficult to disperse to ultimate size	59
Carbon dust	$<5\mu$	....	2.0	Irregular	Continuously generated by attrition of charcoal in rotating drum	45
Copper sulfate	$<5\mu$	....	3.6	Irregular	Ground chemical dispersed as dust	56
Copper tartrate	$<5\mu$	....	....	Irregular	Standard laboratory reagent. Water elutriated	146
Copper oxide	4.3μ MMD	Max size 9μ	....	Irregular	Cupric oxide powder. Merck or B&A. tech. or ACS grades	153
Dowex resin	$0.05–0.1\mu$	....	....	Spheres	Difficult to disperse to ultimate size. High electrostatic effect. Must be ball-milled	39
Fly ash	16μ MMD	>3	....	Irregular and spheres	Obtained from power plants burning pulverized fuel and using Cottrell precipitators	14
Glass, ground	$<5\mu$	....	2.6	Irregular	Crushed and pulverized cullet. Available from glass companies	99
Glass spheres	7.3μ	>3	2.6	Spheres	Obtained in quantity in larger sizes. Small sizes fines of superbright beads	14, 104
Iron, carbonyl	3μ MMD, 5μ MMD, 8μ MMD	....	7.8	Spheres	Can be obtained in controlled sizes	6
Iron dust	$0.57–1.6\mu$ $1.10–6.60\mu$ MMD	....	7.9	Irregular	Can be obtained in controlled size ranges	37
Ilmenite	4.3μ	2.2	4.5–4.9	Irregular	Obtained by grinding ore	
Kadox	$0.1–0.3\mu$	....	....	Irregular	Zinc oxide pigment	160
Lead dust	2.2μ	2.3	11.3	Irregular	Disintegrated metal product	88
Lithopone	$<5\mu$	....	4.3	Irregular	ZnS, 28–30%; $BaSO_4$, 72–70%	160
Magnesium silicate	2.75μ	2.5	3.3	Irregular	Grade No. 399	171
Metal powders	....	....	1.5–10	Irregular	A wide variety of metals may be disintegrated to suitable sizes	101
Methylene blue	$4.9–11\mu$ MMD	....	1.26	Irregular	Specific gravity corresponds to methyl thionine chloride. Merck No. 73881	86

Mineral dusts	1–25μ	....	2.0–10.0	Irregular	Can be produced by impact pulverizing equipment and micronizers	
Paint pigments (also see Kadox above)	0.59μ MMD (zinc oxide)	1.28	5.5	Irregular	A wide variety of pigments from iron, chromium, zinc, etc. May be difficult to disperse	59, 71, 172
Plastic powders	<5μ	....	16.2	Irregular	Molding powders from polystyrene, polyethylene. Difficult to disperse because of static charge developed	
Talc	2.5μ MMD	1.6	3.3	Irregular	Micronized talc. Available from R. T. Vanderbilt Co., New York, as NYTAL	14
Titanium dioxide	<5μ	....	3.9	Rounded granules	Anatase TiO_2 pigment	160
Silica, amorphous	4–25μμ	....	2.2	Irregular to spherical	Tends to form chains. Four different products. Difficult to disperse	26
Silica, crystalline crushed	1–3μ	....	2.6	Irregular	Dispersed as dry powder. Can be fractionated also by water (elutriation) sedimentation	41, 117, 162
Silica, vaporized	0.56–0.64μ MMD	1.4–1.6	2.2	Spherical	Agglomerated, difficult to disperse. Actual particles in chains, discrete size 0.3μ or less	14
Sodium bicarbonate	0.5μ MMD	....	2.2	Irregular	Reagent-grade crystals	84
Sodium sulfate	4.8μ MMD	....	2.7	Irregular	Reagent-grade crystals	86
Starch	<5μ	....	1.5	Irregular	Dry flow powder	110
Uranium oxides	<5μ	....	....	....	Dispersed as dry powder	87
Vinyl resins	0.5μ	Range 0.5–50μ	1.35	Spheres	Vinylite resin powder VYNV No. 1. Requires ball milling. Difficult to disperse because of charge	159
Zinc cadmium sulfide	2.0μ MMD	Range 0.5–3μ	4.0	Irregular	New Jersey Zinc Co., No. 2266. Fluorescent pigment. Excellent for tracer work. Cannot be completely dispersed	119
Zinc silicate	1.5μ MMD	Range <0.5–5μ	3.5–3.9	Irregular	du Pont Company No. 601 fluorescent pigment. Same use as above but lower fluorescence	119

C. Solids or Liquids Directly Vaporized or Atomized as Aerosols

Name	Characteristics	Remarks	Refs.
Ammonium bifluoride	0.54μ MMD. Standard deviation 2.8	Sublimed from hot plate. Spherical particles	13
Ammonium chloride	0.2–10μ	LaMer-Sinclair generator	160
Aluminum chloride	0.59μ MMD. Standard deviation 2.7	Sublimed from hot plate. Spherical particles	13
Arochlor	0.2–10μ	LaMer-Sinclair generator. Chlorinated naphthalenes and diphenyls	160
Benzoic acid	Variable 0.5–2μ	Use LaMer type generator. Aerosol particles are spherical, but then volatilize after a short period	64

Table 12-4. Dusts Used for Test Aerosols *(Continued)*

Name	Characteristics	Remarks	Refs.
Corn oil	1.4–5μ MMD	Generated by atomizing corn oil	84
Chromic acid	Corrosive acid mist, size <0.5μ to 10μ. Spherical particles	Prepared by atomizing acid in nebulizer followed by elutriation	145
Decalin (decahydro-naphthalene)	30–65μ	Prepared by spinning top	169
Dibutyl phthalate	16–80μ	Prepared by spinning top	83, 169
	0.02–0.18μ	LaMer-Sinclair generator	
Dibutyl phthalate (dyed)	180–600μ	Prepared by spinning disk	169
Dioctyl phthalate	Size 0.3–0.6μ. Spherical particles. Standard deviation variable, depending on control	Prepared by vaporization	83
Diol 55	0.5 by count. Standard deviation 1.5. Spherical	Prepared by vaporization	30, 93, 160
DM and DA smokes	0.2μ median size, max 0.6μ	Prepared by dropping 1% solution in acetone on hot plate. Particles are solid crystals	160
Ethylene dibromide	17–37μ	Prepared by spinning top	169
Glycerol	25–90μ	Prepared by spinning top	169
Methyl salicylate (dyed)	190–2,800μ	Prepared by spinning disk	169
Methyl salicylate plus liquid paraffin (25/75)	12–85μ	Prepared by spinning top	169
Mercury	210–1,020μ	Prepared by spinning disk. Spraying into chamber	109, 169
Metals—aluminum, brass, copper, lead	1–100μ	Vaporized through spray gun using reducing flame	41, 72, 103
Microcrystalline wax	0–1μ	Using LaMer-Sinclair generator	44
Mineral oil (Niyol)	0–1μ	Using LaMer-Sinclair generator	44
Nitric acid	1.0μ MMD	Atomization or heating of nitric acid	8, 57
o-Dichloro-benzene	17–37μ	Prepared by spinning top	169
Oleic acid	<1μ	LaMer-Sinclair generator. Vaporized in boiler may decompose at boiler temperature	160
Orange dye	0.8μ	Vaporized in oil (Calco Oil Orange Y-293)	160
Rosin	0.2–10μ	LaMer-Sinclair generator	160
Stearic acid	0.1–1μ	LaMer-Sinclair generator. Supercooled liquid smoke	160
Sudan red G	<1–10μ	La-Mer Sinclair generator. Supercooled liquid smoke	175
Sulfuric acid	Mean size 0.6 to 4.5μ. Standard deviation 3 (Gibbs reports 0.8 to 5.5μ)	Size depends upon generation conditions. Dilute acid produces coarser aerosol. Droplets can grow in moist air	13, 60
	0.02–0.16μ	LaMer-Sinclair generator	83
Tricresyl phosphate	20–90μ	Prepared by spinning top	169
	<1μ	Prepared by vaporization	160
Triphenyl phosphate	<1μ	Prepared by vaporization. Supercooled liquid smoke may condense on equipment	160

D. Solid Aerosols Produced from Soluble Salts or Suspensions

Name	Characteristics				Remarks	Refs.
	Mean size, μ	Standard deviation	Specific gravity	Shape		
Copper sulfate	0.3μ	....	3.6	Spherical	Fuses into spheres at temperatures above 200°C. Spheres stable for a few days	57, 146
	0.7μ	2.4				
	2.7μ MMD					
Copper tartrate	<1μ	....	......	Irregular	Sprayed as suspension. Suspensions may be prepared 24 hr previous to use and decanted	146
Methylene blue	0.2μ	(Max size 0.4)	1.26	Spherical	1% solution atomized produces spherical particles	34, 124
	2.0μ MMD	1.3		Irregular	0.1% solution (spinning disk)	160, 167
Phenol red	1.6–3.8μ MMD	....	......	Irregular	Nebulized from 0.2% solution	123
Silver iodide	0.3μ max 0.8μ	....	5.7	Irregular	Atomized from 1% aqueous solution plus wetting agent	48
Sodium bicarbonate	0.2μ MMD	....	2.2	Irregular	From 1% solution	84
	0.4μ MMD	....	......	....	From 8% solution. Crystalline particles generated by atomizer and dried in an electrically heated tube	
Sodium chloride	0.4–1.2μ	....	2.2	Irregular (cubic crystals) also spheres (dried at high temp)	Atomized 2% solution or atomized 1% solution. Also made containing Y^{90} as a tracer	34, 48, 168
Sucrose	0.03μ	....	1.58	Irregular	Sprayed 1% solution at 30 psi to give charged spray	160
Uranine	1.3μ MMD	2.0	......	....	Sprayed from broth solution	120
Uranium oxide	0.45μ MMD Range 0.4–0.68μ 90% values 1.71–2.90μ	....	10.9	Irregular	Larger cloud also generated. 1.7μ MMD	87
Uranium trioxide	0.74μ MMD	1.4	7.3	Spheres	Produced by atomizing uranium nitrate and passing through heated tube	14, 57

E. Solid or Liquid Aerosols Produced by Flame Passage, Chemical Reaction, or Burning

Name	Characteristics	Remarks	Refs.
Acetylene	Burned in insufficient air to produce carbon black	Forms chains. Shawinigan black, 0.087μ MMD. Standard deviation 1.55	59
Ammonium chloride	Mean size 0.1μ	Produced by reacting NH_3 and HCl	60

Table 12-4. Dusts Used for Test Aerosols *(Continued)*

Name	Characteristics	Remarks	Refs.
Benzene	Burned in insufficient air	Same as acetylene	41
Butane	Burned in insufficient air	Same as acetylene	70
Fused mineral dusts	Size depends upon ground material passed through flame	Produces spherical particles of most minerals and salts	130, 131
Metal combustion or vaporization	Metallurgical oxides ranging from 0.001 to 0.1μ, mean size $<0.1\mu$. Shape depends upon metal. Magnesium produces cubes, aluminum spheres, zinc irregular, cadmium irregular, etc.	Produced by direct ignition of metal or introducing solid or pulverized material into high-temperature flame with feeder on air side of torch	41
Metallo-organic liquids	Fumes comparable to those above	Obtained by thermal decomposition of liquids; such as lead oxide from burning lead tetraethyl, iron oxide from iron penta carbonyl	161
Phosphorus	Screening smoke $<0.1\mu$	Produced by burning white phosphorus	60
Silver iodide	0.01μ particles produced	Used for ice nucleation	163
Silicon tetrachloride	Produce amorphous oxides $<0.1\mu$ of metals by hydrolysis	Exposing material to moist air	60
Tin tetrachloride	Produce amorphous oxides $<0.1\mu$ of metals by hydrolysis	Exposing material to moist air	60
Titanium tetrachloride	Produce amorphous oxides $<0.1\mu$ of metals by hydrolysis	Exposing material to moist air	60
Zirconium tetrachloride	Produce amorphous oxides $<0.1\mu$ of metals by hydrolysis	Exposing material to moist air	60
Tobacco	0.5–1.0μ MMD	Predominantly liquid for droplets	57

F. Biological Materials Which Can Be Used as Test Dusts[b]

Name	Characteristics	Remarks	Refs.
Albumin, egg	0.5–10μ	Can be dispersed by Sinclair geyser	160
Bacteria:			
E. coli	1 by 7μ cylinders	Dispersed by nebulizers or atomizers	51, 52, **123**
Serratia indica	0.5 by 1.0μ cylinders		
S. marcescens	Cylinders 0.6 by 1.0μ. Sphere of equivalent mass 0.7μ		
B. globigii	Cylinders 0.8 by 1.5μ. Sphere of equivalent mass 0.9μ		
Bacteriophages, *E. coli*-T-3	0.02 to 0.05μ, spherical shape	Used for filter tests	35
Blood cells, dried	7.85μ MMD. Standard deviation 1.08	Dried bovine blood may be employed	59
Lycopodium spores	25 to 40μ. Shape like three-sided pyramid	. . . . . .	5
Pollen: giant ragweed (*Ambrosia trifida*)	Mean size 19.6μ. Standard deviation 1.07	Used for filter testing. Nearly spherical shape with rough surface	37
Orchard grass (*Dactylis glomerata*)	Mean size 31μ. Standard deviation 1.1	Elliptical 4:3. Some irregularities	37
Corn (*Zea mays*)	100μ	. . . . . .	68
Rye (*Secale cereale*)	55μ	. . . . . .	68
Spores: *Pilletia caries*	17μ	. . . . . .	68
Bovista plumbea	5.6μ	. . . . . .	68

[a] MMD = mass median diameter.

[b] A number of pollen, special dusts, furs, etc., used for allergen studies are available from allergist supply houses such as Sharp & Sharp, Everett, Washington, and Stemen Laboratories, Oklahoma City, Okla.

Table 12-5. Characteristics of Dust Feeders Used for Aerosol Generation

Name	General principles	Remarks	Ref.
Air Filter Institute dust feeder	Venturi ejector pickup from level tray driven by rack. Dust height in tray controlled by rotating pinion. Ejector operated at 60 psi with dried air. Feed rate 0.5 to 1 g/min	Complete unit may be obtained from Fortwengler Die and Machine Co., Louisville, Ky. Approximate cost $565. Standard feeder for Air Filter Institute	1
ANP...........	Utilizes dust packed in tube elevated by hydraulic piston into pressurized line using 90 psi dried air fed into dust box for elutriation purposes. Feed rate 1.3 to 73.5 g/min	Developed for feeding copper oxide powder. Batch feeder	153
Atkins..........	Utilizes small screw feeder into recirculating unit provided with fan. Final dilution by ejection with secondary air feed. Feed rate 7.7 g/min or less	Developed for feeding pulverized coal	7
Bureau of Mines (pulverized-coal feeder)	Gravimetric feed to fluidized bed utilizing cyclone for returning material to fluidized section. Can be adapted to pressurized systems. Maximum feed at 26 lb/cu ft	Developed for feeding pulverized coal to synthesis units	2, 8
Bureau of Mines respiratory testing unit	Mechanically elevated column of dust fed by clock feed to spiraling air stream developed by a cut-brass spiral above dust to compressed-air ejector. Used for low feed rates of counting range, viz., mg/min. Fed to cyclone or settling chamber	Used for respirator testing. Requires dust screened through 325 mesh. Batch feeder	133
Bureau of Standards air-filter test unit	Rotating gear with machined-down teeth feeding from small hopper. Air from gear picked up from teeth valleys by diagonally pitched ejector nozzle. Uses dried air at 2 psi and simple ejector fabricated from pipe T. Feed rates 0.25 to 2.5 grains/min	Used for feeding Cottrell ash to air-filter test unit	38
Colorado.......	Modified Bureau of Mines respirator test unit including plenum chamber and bag filter. Dust surface is scraped into spiraling air stream. Column lifted by rotating screw	Used to feed aluminum hydroxide (2μ mean size) to breathing chamber	28
Dustshaker.....	Utilizes oscillating shaking screen 40 to 60 mesh with flushing air stream. Oscillation at 100 impacts per minute. Feed rate in mg/min	Used for toxicity studies. Batch feeder	36
Farr Far-Air No. 4	Utilizes grooved plate (4 grooves) on rack and gear drive. Incorporates rotating leveling gear which slices off thin layer of dust. Dust sliced off is aspirated by 4 ejectors to 4 distributing nozzles. Rapper to keep gears and aspiration tubes clean. Feed rate 0 to 5 g/min	Used for dust-filter testing. Four nozzles to give uniform concentration in test duct	12, 50
Geyser (pneumatic disperser)	Powder chamber connected to air jets dispersing and feeding to aspirating nozzles. Filtered dry air at 60 psi used for dispersing and aspirating. Air flow at 140 liters/min. Loading up to 0.03 mg/liter	Used for lithopone, egg albumin, Kadox. Essentially a batch unit	150
Harvard No. 1..	Pneumatic dispersal from tube filled with test dust rapped by bell clapper. Settling chamber followed by filter used to remove coarse particles. Feed rates, mg/min.	Used for silica dust	42

Table 12-5. Characteristics of Dust Feeders Used for Aerosol Generation (*Continued*)

Name	General principles	Remarks	Ref.
Harvard No. 2. .	Rotating screen fed by Syntron vibratory feeder. Air pickup by blower inlet used for air flow and dispersal. Compressed air fed from below rotating screen. Elutriating tower rapped to prevent agglomeration. Feed rate, g/min	Used for talc, dust, glass spheres, and copper sulfate dust	54, 146
Harvard No. 3. .	Pollen feed unit consisting of rotating wire screen drum 300 mesh, 1½ in. diameter, 2 in. long. End of drum is tube in which charge is placed. Drum is arranged to rotate a variable-speed vertical cylinder. Drum axle extends beyond cylinder and is rapped by eccentric cam and spring-loaded clapper. Pollens, dispersed into vertical cylinder, fall and are aspirated into sampling thief at bottom projecting across cylinder. Loadings 1 to 10 times outdoor pollen levels, i.e., pollen grains/cu yd	Giant ragweed and orchard grass have been dispersed. Essentially an inside batch unit	140
Harvard No. 4. .	Turntable feeder provided with scraper blade and ejector pickup through elutriation tube using 40 psi air. Table is fed by Syntron feeder. Loading to 1,000 cfm air stream, 0.25 to 10 grains/cu ft	Used for talc, calcium carbonate, vaporized silica, copper sulfate, and fly ash	14
Harvard No. 5. .	Screw feeder fabricated from meat grinder and driven by variable-speed motor. Varying rotating orifice plates at screw discharge; can provide limiting feed control as well as speed of screw. Can feed g/min to lb/hr	Feeds flake and other materials which tend to agglomerate in vibrating units	139
Harvard No. 6. .	Combination feeder based on screw feed to moving belt with rotating blade for adjusting height and sonic-velocity ejector pickup. Loading varies from 0 to 300 mg/min	Used to disperse flake materials difficult to feed directly into an air stream	142
Haskell Laboratory	Three elutriating towers containing dust in first tube. Pulsating air from alternately compressed tubing or reciprocating compressor disperses dust through successive tubes which give uniform distribution	Essentially a batch unit	112
Knolls.	Consists of rotating steel-wire brush rubbing against metal surface of aluminum or other metals. Very low loadings in μg/min	Used for tests	58
Minnesota No. 1	Revolving-disk unit on which dust is placed in ribbon and picked off by low-pressure section of venturi tube. Loadings, 0.1 to 2 g/min	Used for ASHAE test dusts	127
Minnesota No. 2	Moving trough unit of aluminum channel moved across venturi aspirator used above. Dust is weighed and is placed uniformly in trough	Used for ASHAE test dusts at low loadings	129
New York State No. 1	Hopper feeds onto slowly moving belt driven by variable-speed motor. Dust falls off belt into hopper through which air is drawn to two parallel impingement nozzles and to four elutriating towers in parallel to control desired size ranges. Feed rate from mg to g/min	Used for determining cloth-filter resistance coefficients with mineral and organic dusts <200 mesh to 1μ	173

Table 12-5. Characteristics of Dust Feeders Used for Aerosol Generation (*Continued*)

Name	General principles	Remarks	Ref.
New York State No. 2	Feeder consists of large hopper, vibrating chute, air ejector, and mechanical drive. Uses 40 psi air and impingement plate if desired for further breakup of particles. Maximum capacity 3 lb/min	Useful for all dry dusts according to authors	174
Research products	Dried compressed air is fed to flask containing test dust. Dust is dispersed by impinging air jet and diluted with moving filter test air stream	Used for testing panel-type air filters since replaced by AFI feeder above	125
Rotating drum..	Uses a drum through which air is passed. Drum is rotated at various speeds. Loadings in mg/min. Produces clouds of 250 mg/cu m in 7.8-cu m chamber	Used for attrition of charcoal to produce carbon dust	45
SRI............	Utilizes moving V trough similar to AFI, Farr, and No. 2 Minnesota units. Uses a glass aspirator to lift dust from moving trough. Used in conjunction with settling chamber to remove aggregates. Used for low loadings, mg/min	Essentially a batch unit used for feeding various powders and carbon black	20
Screw..........	A variety of units may be assembled using small stoker screws or wood augers from ¼ to 2 in. connected to variable-speed motors	Works best on free-flowing dry dusts	
Sonic...........	Utilizes sound projecting cone such as speaker element operated at high-frequency levels and directed at dust deposit to disperse dust into moving gas stream	Experience reported on plaster dust 15μ and others ranging from 2 to 30μ	108
Syntron........	A vibratory feeder consisting of trough or special trays connected to electromagnetic vibrator whose amplitude can be electrically controlled. Smallest unit feeds from g/min to several hundred lb/hr	Available from Syntron Co., Homer City, Pa.	
Texas..........	Modified Bureau of Mines respiratory test dust feeder with air ejector and cyclonic separator	Used for antimony and sulfur dusts	156
University of Chicago No. 1	Agitated sintered-glass filter (40 mesh) dispersed. Air from top of disperser and free flow through bottom of sintered filter aspirates dust. Ejector nozzle utilizes 4 psi air. Resultant feed rate, mg/min	Essentially a batch unit utilizing vibrated filter. Similar to dust-shaker described above	154
University of Chicago No. 2	Multiple impinging nozzles in series fed from dust placed in vessel. Feeds in μg/min	Essentially a batch unit. Successive impinging used for size discrimination	85
Wright.........	Rotating packed drums moved against knife-edge scraper. Dust dispersed into compressed-air stream at low pressure. Air must be cleaned and dried. Variable gears regulate speed of rotating drum. Feed rates from μg to mg/min	Used for pulverized coal, silica dust, lead dust. Available from L. Adams, Ltd., London, England. Essentially a batch feeder but will operate at low loadings for 24-hr periods	176

jet to get complete dispersion. The degree of dispersion obtained depends considerably on the nature of the particulate matter used for test suspension. Dusts which have a tendency to flocculate or which generate a high charge by friction always offer problems, and in many cases these problems are so great that it is impossible to get ultimate dispersion of the dust. The dispersing unit often requires a combination of feeding principles to avoid difficulties. In a recent problem of feeding a flake-iron powder, after several types of units had been tried, including most of those in Table 12-5, a combination feeder (shown in Fig. 12-11) was developed (142) which incorporates a screw feeder plus a belt feeder with an ejector pickup. Further treatment, necessary in the application, involved passing the powder to a flame for burning to oxide so that complete dispersion was obtained by combustion.

To separate particle aggregates, impingement, elutriating towers to settle out coarse sizes, cyclones to cut off aggregated sizes, or even coarse, inefficient filters may be

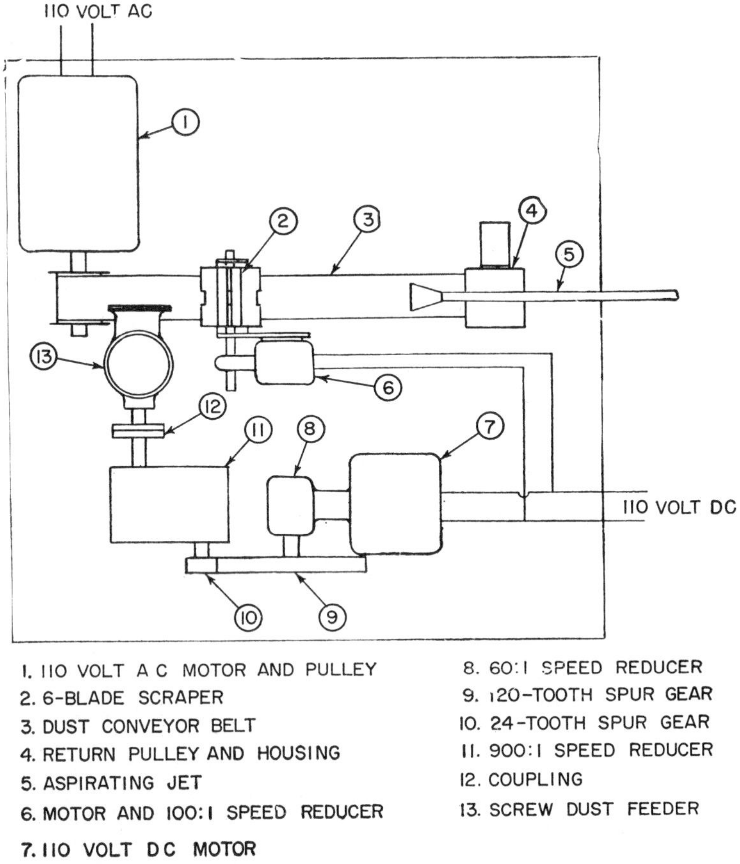

FIG. 12-11. Combination Harvard feeder for handling flake or agglomerating powders. (See Ref. 142.)

used in conjunction with any of the feeders described. Their main drawback is that they reduce dust loadings markedly if weight concentrations are to be utilized. Once the dust has been dispersed, it is often necessary to break it up by further treatment such as that described by First (54) and Atkins (7). In each case the dust is disaggregated by passing through a rotating fan-blower unit. Those particles which aggregate in the blower can be removed by passing the air stream through an elutriating tower. Some experimenters regard the results obtained with redispersed dusts as misleading because the ultimate size of dispersed material is not the same as the originally created material. For example, Billings et al. (14), using vaporized silica (VS) collected from a ferrosilicon alloy electric furnace, found that the ultimate size of the particles which could be generated by redispersion was not so small as the size collected from the furnace stack. However, by continually reworking the material through the generator after collecting in a dust collector, it was found that the size could be reduced to a minimum comparable with the field-collected sample. For example, where the field dust showed a mean size of 0.3μ, the reworked dust in the laboratory could be reduced to a mean size of 0.4μ. Both the originally collected fume and the redispersed material had standard deviations which were comparable. The problem of metering aerosols may arise in heavy loadings, and the techniques described by Farbar (49) may be of assistance.

Any of the dry techniques mentioned above may be employed to obtain desired concentrations within a test chamber. To maintain the concentration in the chamber, the aerosol must be continuously supplied. A static cloud may be produced to

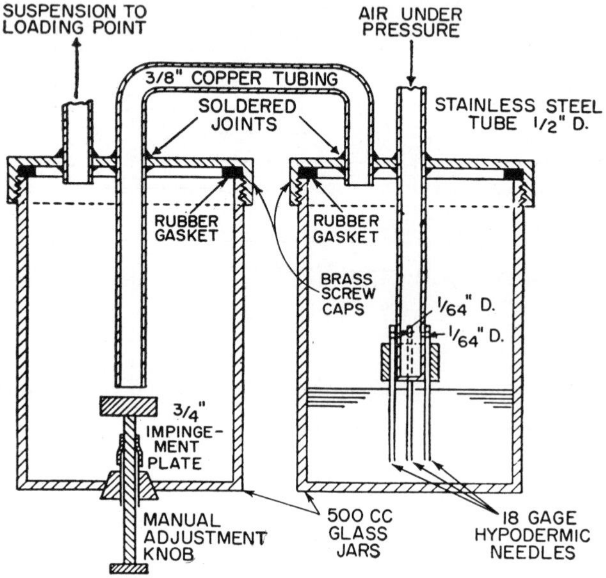

FIG. 12-12. Continuous-nebulizing generator utilizing atomizer and sizing impinger. (*After Fahnoe et al.*) (See Ref. 48.)

observe decay or stirred settling rates. Static concentrations can also be produced by air blasts such as the device developed by the Pennsylvania Salt Manufacturing Company (25) or by use of small CO_2 pressure capsules (soda-water charger size). Small explosive powder charges can also be employed as can sonic dispersing units.

The wet dispersal of colloidal suspensions was mentioned briefly in connection with dispersing microorganisms. The method has merit if low concentrations are desired and very finely ground materials are available. Wetting agents may aid in maintaining the material in suspension. A number of devices of this type have been used, utilizing either a pneumatic atomizing nozzle with air entrainment of the liquid or a hydraulic nozzle operated at high pressure (250 to 500 psi). Laskin (87) has used a multiple hypodermic-needle unit for producing uranium aerosols, and Silverman et al. have produced copper tartrate aerosols by using a pneumatic nozzle for this method (146). Fahnoe and his colleagues (48) utilized a two-stage unit with an impingement plate for controlling particle size, as shown in Fig. 12-12. Any standard nebulizer, atomizer, or pneumatic nozzle can be adapted to this purpose, but it is desirable to provide additional settling or inertial separation to eliminate coarse droplets which may contain several particles each in addition to causing agglomeration. The technique is best adapted to low dust loadings. Glass liquid dispersers for this purpose are also described by Cadle and Magill (20). For higher concentrations, larger-scale mist generators such as those discussed below can be employed. It is essential in this technique to dry the air stream used to entrain the material; in some cases a drying agent or heated air may be necessary.

In the technique of atomizing solutions, the material to be generated is dissolved; for example, sodium chloride or methylene blue dye is dissolved in water or alcohol, and after spraying the resulting mist is dried, leaving the material in a solid form. If the drying is rapid, spheres frequently result. If the material is crystalline in nature, it may appear as regular crystals. Thus, sodium chloride when air-dried at room temperature produces cubes but when dried by passing through a heated zone above the melting point of salt produces spheres. Walton (167,168) describes methods for generating methylene blue and sodium chloride aerosols, utilizing atomization of 1 and 2 per cent solutions of each, respectively, and air-drying. The air-dried methylene blue is spherical, and the sodium chloride is cubical. Davies (34) shows electron photomicrographs of each of the particles formed. Size, of course, can be adjusted within a limited range by varying the concentration of solution sprayed. Since the mass of the particle is a function of the cube of the diameter, the ultimate aerosol size and range can be computed if the size and range of droplets initially produced are known. The solid particle diameter d can be computed from Eq. (12-6).

$$d = D\sqrt[3]{C} \tag{12-6}$$

where D is the droplet diameter and C the decimal concentration in solution. The useful rate of production of aerosol, of course, depends upon the method of atomization employed. For nozzles producing fine aerosols, 1 to 60 ml/min is the ordinary range without using multiple-nozzle systems. For spinning tops, a maximum rate of 1 ml/min to produce 10-μ drops is reported. The number of particles per minute can be computed from

$$\text{Particles per minute} = \frac{\text{volume of feed solution per minute}}{\text{volume per particle}} \tag{12-7}$$

Cottrell (30), using data for 10-μ droplets and Eq. (12-7) for the maximum feed rate of solution, gives 2×10^9 particles per minute. Thus only relatively small volumes of air up to 15 liters/min can be given loadings of 10^5 particles per cubic centimeter of air.

Information on mechanisms producing droplets from atomization has been developed by Castleman (24), and a complete bibliography on spraying has been prepared by Pennsylvania State University (118).

Air atomization or hydraulic pressure may be used to produce sprays. Finer sprays require either high air pressures or high hydraulic pressures. Spheres of copper sulfate and uranium oxide may be prepared by utilizing a simple pneumatic nozzle and tank system, as shown in Fig. 12-13. Additional heat supplied by burners causes both of these aerosols to be produced in spherical shape. The collected copper aerosol on standing over 24 hr gradually hydrates and assumes its crystalline shape.

The spinning top and other droplet-atomizing devices described for homogeneous aerosols can also be used for aerosol generation with salts, resins, or dyes in solution, but if the special precautions to extract satellites are not followed, a heterogeneous cloud will develop. A number of spinning-disk atomizers have been used for spray drying of solids. Special impingement plates and baffles, developed by Dautreband

and his coworkers, are discussed by Cadle and Magill (20). A number of nebulizers for dyes and bacteria are also described in Rosebury's book (123).

The use of liquefied gases for spraying is convenient for static chambers or short experiments with dynamic systems. Details of this method for insecticidal aerosols are given by Goodhue (63). The salts or dyes must be soluble in the liquefied gases. Gases such as methyl chloride or Freon are usually employed, but others may be feasible.

Large volumes of mists containing materials in solution can be produced by venturi atomization, as described by Comings et al. (29).

Heterogeneous aerosols of several materials can be generated merely by heating them to cause evaporation and allowing them to condense in an entraining air stream. Examples are aerosols generated by heating waxes or resins and in some cases metals. In most instances the metals tend to oxidize and produce a change in the aerosol surface characteristics. For example, heated zinc or lead immediately vaporizes and

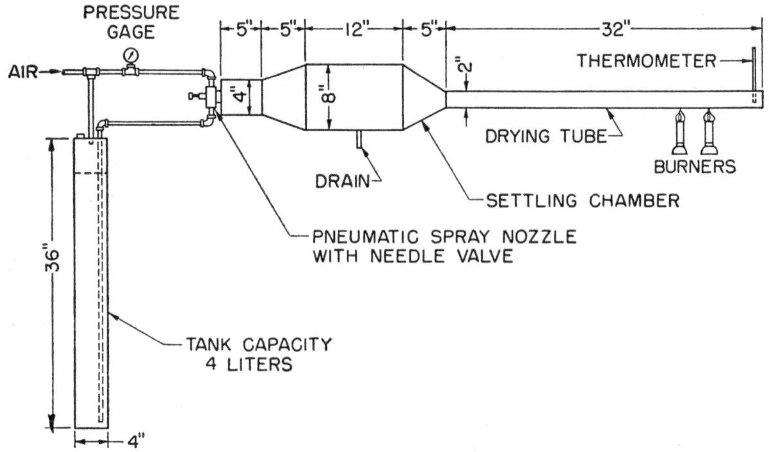

FIG. 12-13. Aerosol generator utilizing soluble salts (e.g., copper sulfate) in solution.

then condenses as the oxide. If prepared in an atmosphere of nitrogen or an inert gas, however, metallic spheroids are produced. Keenan and Fairhall (79) produced a lead aerosol in this manner. Nagel et al. (109) prepared condensation aerosols by spraying mercury and lead. Atomized liquid mercury and tetramethyllead vapor were confined in a chamber and irradiated with a mercury-vapor lamp to produce the aerosol.

Molten metals dispersed from spray guns or atomizers form very stable aerosols. The commercial spray guns have been adapted to spraying of plastic resins and inorganic frits as well as metals. Details of this technique and equipment for this purpose are described in "The Metco Metallizing Handbook" (72). Miller and his colleagues (103) produced aerosols of zinc, lead, iron, brass, and aluminum in sizes predominantly in the 3-μ range.

A large number of the aerosols produced in nature are products of combustion, and the parent materials may be organic (the most common situation) or inorganic (metals). Because aerosols occur so frequently in combustion processes and elsewhere, they have been used quite often for test purposes. Contrary to first impressions or concepts, the aerosols produced by burning are not always solids and may, as in the case of tobacco smoke, consist of droplets containing tarry matter and small particles of carbon. The chief reason for this is the large amount of distillation products present, including water vapor. Condensation of these products by cooling creates the saturated droplets.

A number of techniques have been used for producing aerosols by combustion, and an example of each will be given. The first method is the incomplete combustion of a gas. Acetylene was used by Katz (77); Hill (70) used city gas. Benzene vapor may also be employed. The resulting aerosol is carbon black in each case, and it tends to form chains because of the rapid agglomeration of the carbon particles. Because of the agglomeration and the resulting lack of uniformity, carbon black is not the best aerosol for filter testing. Katz describes a patented washing technique which prevents aggregates from accumulating on the walls.

Metals may be burned by such processes as sparking, electric or gas welding, or metal cutting or by igniting metallo-organic compounds (e.g., lead tetraethyl or zinc ethyl). Silverman and Ege (144) and others have generated lead fume by this method. The lead tetraethyl may also be volatilized into a natural-gas stream and burned continuously, as done by the U.S. Bureau of Mines for respirator-testing purposes (161).

Another approach to generating continuous metal-fume concentrations is to ignite metals such as magnesium which will then continue to burn spontaneously, producing the oxide. Thermit welding or gas welding can also produce stable heterogeneous metal-fume aerosols. When inert-gas welding is employed, the fume may be a mixture of the original metal and oxides.

In a recently developed procedure (88), metal powders of lead or iron are fed by a continuous feeder into the air side of an air-oxygen-acetylene flame, and the resulting fume is a fine metal oxide. Electron photomicrographs in this case show primarily irregular particles, in contrast to other studies where spherical particles are usually seen, as reported by Shekhter et al. (136). Most of the metal smokes are in the size range of 0.001 to 0.5μ.

The formation of smoke by burning organic material can be accomplished with such materials as rosin or tobacco. Tobacco smoke has been used for this purpose, and a generator for static clouds is described by Drinker and Hatch (41), for respirator testing by Katz (78), and for large-scale testing with a continuous dynamic generator by First et al. (57). Tobacco smoke ranges from 0.3 to 0.6μ in diameter, depending upon the dryness of the tobacco, the rate of burning, and other factors.

Carbon arcs, high-voltage condenser sparks, condenser discharge through thin foil, and flash bulbs are all means of producing metallic or organic decomposition-product smokes for study in chambers. Direct simulation of actual processes such as cutting or welding is usually best for producing aerosols which vary widely in composition.

Liquid Particles. The techniques employed for the atomization of liquids are those discussed previously for generating particles from solution. In this case, however, the liquid does not contain solutes or solids, and the mist must either be formed in a saturated atmosphere or have a negligible vapor pressure in order to remain stable.

A number of devices which can be used for this purpose have been mentioned above. These include nebulizers, paint spray guns, atomizers using compressed gases for atomization, and hydraulic nozzles functioning with high liquid pressures to produce fine sprays. The materials which can be used to produce sprays are liquids such as organic solvents, oils, emulsions, or acids. A typical unit for respiration studies is described by von Oettingen (112), and a generator for continuously producing chromic acid mists in the hygienic range was developed by Silverman and Ege (145). In the latter case an impingement surface and small settling chamber serve to elutriate the mist to a fine stable size (below 10μ). Mists of other acids may also be produced by the same method.

For large volumes of mists, the venturi atomizer described by Comings et al. (29) can be applied, as can the Aerojet-venturi described by Boucher (16). Another procedure is jet disintegration of droplets or thin liquid streams. Such approaches can produce only very low loadings.

Typical means of vaporization with subsequent condensation are discussed under Generation of Homogeneous Aerosols (Art. 12.4.2). They involve the heating of the liquid followed by condensation upon artificially produced nuclei or on atmospheric dust. The oil-smoke generator can produce aerosols from Diol 55, engine lubricating oils, dioctyl phthalate, oleic acid, or triphenyl phosphate. Unless condensation and growth are well controlled, a heterogeneous aerosol results. Sulfur or sulfuric acid aerosols are also produced by the same method. The use of liquefied gases by the aerosol-bomb technique has already been discussed.

Two additional procedures are available for producing stable liquid aerosols. When sintered or fritted glass or metal dispersers are submerged below a liquid surface and compressed gases are forced through them, bubbles form which saturate in passing through the liquid. The saturated gases will contain the liquid as mist if it has a high boiling point and low vapor pressure. Liquid aerosols of acids or oils can easily be generated and reduced in concentration by subsequent dilution with air or other gases.

The electrolysis of solutions of acids such as chromic or sulfuric, using inert anodes and cathodes, will produce aerosols by the saturation mechanism as the hydrogen and oxygen gases form and pass to the surface. Rapid agitation of liquids by bubbling compressed gases through them can also produce mists when the bubbles break at the liquid surface. This surface shattering can create droplets at the liquid-gas interface, producing liquid droplets.

Liquid aerosols from combustion are produced as condensation aerosols during the burning. If mists are sprayed into an air stream containing combustion products, the mists will absorb some of the products, forming a mixed liquid aerosol. Desired mixtures can be arrived at by proper design if the combustion products are known or predictable.

12.5 MEASUREMENT OF VERY LOW GAS FLOWS

The importance of measuring very low gas flows in experimental aerosol and gas testing stems from the need for measuring minute quantities of gases to be diluted in large air volumes. It is possible in some instances to employ ordinary volumetric standards, but in other cases special techniques must be considered. For measuring very small gas flows, the primary volumetric standards are the gasometer, Mariotte's bottle, or siphon bottles. Because of the small volumes involved, a very small gasometer, 500 ml or less, of the type shown in Fig. 12-14*a* is used. Because of the desirability of getting greater accuracy for small amounts and greater ease in counterbalancing, the Krogh type of spirometer is preferred. This device, shown in Fig. 12-14*b*, is a pivoted gasometer which can be handled with great ease and reliability, providing a simple means of magnifying readings. These devices have been made for volumes as small as 50 cu cm. Plastic bags or envelopes, weighed before and after use on an extremely sensitive balance, provide approximate results. However, the sensitivity is limited, and there is interference from moisture and temperature changes. Some of the newer plastic films such as Mylar provide impermeable thin membranes. Plastic bags may also develop an electrostatic charge during handling, interfering with weighing.

Another simple volumetric device which simulates the gasometer is the rotating friction syringe developed by Brubach (18). This is a very useful device since it can employ syringes as small as the 1-ml tuberculin type.

Another procedure for measuring very small gas flows involves the use of soap films or soap bubbles. The soap-film procedure as described by Barr (10) has been employed as a means of measuring low air volumes. The characteristics of the soap-

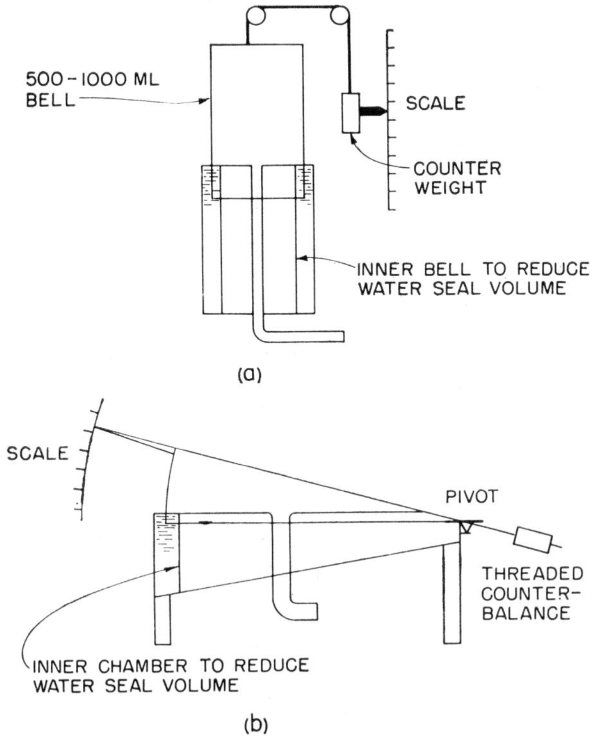

FIG. 12-14. Small gasometers for calibrating devices or for measuring low flows. (*a*) Conventional cylindrical bell type, (*b*) Krogh or tipping bell unit.

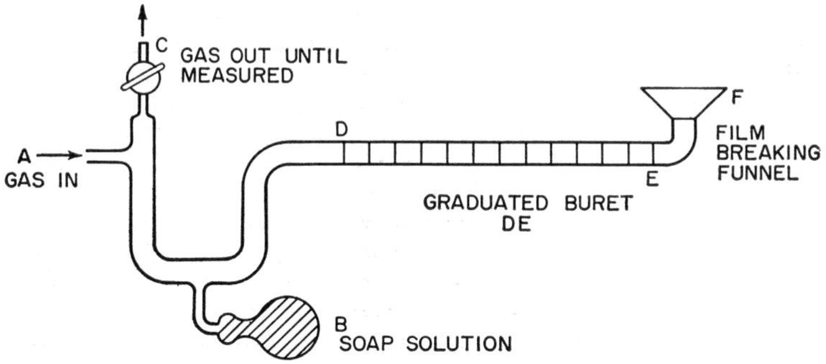

FIG. 12-15. Soap-film gasometer developed by Barr for measuring small gas volumes or flow. (See Ref. 10.)

film device and two linear flowmeters which have application for low air flows are also described by Barr. A sketch of the soap-film gasometer is shown in Fig. 12-15. The resistance of this device can be negligible since the diameter of the soap-film bubble controls the resistance, as indicated in the accompanying summary of the data.

Diameter, mm	*Resistance of Soap Film, mm H_2O*
6	1.52
12	0.25
20	0.13
35	Not measurable

For measuring purposes, a tube of 12- to 20-mm diameter is ordinarily used for low flows. The only limitation to the soap-film technique is the problem of absorption of certain gases by the soap film. It works quite well with air, nitrogen, and other inert gases which are not absorbed in the soap film, but acid gases may offer some problem.

Another application of the soap technique is the bubble device described by Silverman and Thomson (148) which utilizes the increase in diameter of a soap bubble on a capillary tip, thereby offering low resistance measurements at extremely low gas flows.

Another procedure which can be used to measure small gas flows is the submerged-bubble technique. This involves counting the number of bubbles (all of a measured size) which are created when gas is passed through a liquid layer.

The siphon bottle or Mariotte's bottle, arranged with constant head, also gives reliable results. In some instances it is desirable to use mercury, and very small air flows can be measured with very fine capillaries such as those used for the dropping electrode (polarograph). The volume of mercury can be measured by weighing or counting the number of mercury drops per unit time.

In addition to the foregoing methods, there are a number of physical techniques that can be adapted to measuring small gas flows. Thermal meters are available which utilize the rate of cooling of a thermocouple or the rate of gas flow over a heated wire (an electrical-resistance-type flowmeter). The limitation of most thermal devices is that primary calibration must be made with a known standard. They are useful for metering the flow once their calibration is known. Lovelock and Wasilewska (92) have described an ionization-type unit which can easily be adapted to measuring flows, although it was originally developed for measuring very low velocities. It utilizes a collecting plate for collecting ions from a polonium-coated surface. The deflection of the air moving through a tube or between two parallel plates causes ions to miss the collecting surface target, and the number of ions collected therefore is in proportion to the velocity or volume of air flow.

In measuring very low gas flows with secondary devices such as rotameters and resistance flowmeters, it is desirable to use extremely sensitive manometers if small pressure losses are involved. For this purpose the Wahlen gauge (165) or a modified form of it can be employed. In measuring very small gas flows on the order of 1 to 10 cu cm/min, the spiral-tube manometer shown in Fig. 12-15 can be adapted to metering with considerable accuracy because of the extreme sensitivity of the manometer. The use of small rotameters is highly recommended, and units capable of handling flows as small as 0.1 cu cm/min are possible by using small-bore tapered tubes with light hollow bead floats.

Wet meters and dry meters used for higher gas flows are not practical for measuring small gas flows because of slippage. Precision gear or syringe pumps may be used, however, if driven by constant-speed motors. One of the devices which has an advantage for constant metering of small gas flows is the critical-flow orifice. These can be made in sizes small enough to control gas flows in the range of 1 to 100 cu cm/min by using extremely fine pinhole orifices (watch jewels have been used for this purpose). When the velocity through the orifice is sonic, the flow is dependent upon the upstream pressure only. Therefore, if flow is induced by subatmospheric pressures, constant flow is dependent only on maintaining sonic velocity at the orifice. This occurs when the downstream pressure is 53 per cent or less of the upstream absolute pressure. A description of constant-flow orifices and references covering this method are discussed by Page (115).

Another procedure which can be employed for measuring very small gas flows is the use of the dilution gas technique which requires subsequent chemical or physical composition analysis. Morley and Tebbens (107) illustrate this technique for calibrating flow-metering devices. For metering flows it is necessary that a known gas be inserted into a flowing system. If air is added to the gas and its concentration measured continuously after mixing, the original flow can be calculated. If a metered volume of air is added to an unmetered volume of gas and then analyzed, it is possible to calculate the volume of the mixture.

12.6 EXPERIMENTAL TESTING OF DEVICES OR MATERIALS

The use of the tools already discussed may be now briefly considered. In air pollution studies we are fundamentally concerned with producing synthetic atmospheres simulating actual conditions in composition, concentration, and particle size (if particles are involved). Synthetic smogs or gas mixtures have been prepared of many contaminants to study their effect on man, animals, and plants. Similarly, studies of toxicity or sensory response have been conducted in test chambers.

Exposure chambers and dynamic generators can be used to calibrate instruments or check analytical methods for aerosol studies. In working with gases, the efficiency of collecting devices is best determined by the ratio of the quantity retained to the known chamber concentration. The use of similar absorbers or collectors in series is always questionable since the material which passes the first absorber may also pass through the second and any subsequent absorbers. Thus, the chamber or dynamic concentration represents the most reliable standard. In the case of chamber concentrations of gases or vapors which are far below the saturation level, the reliability is well within 1 per cent. If condensation or absorption takes place, then the accuracy will be limited. When working with levels near saturation, dynamic-concentration generators for gases and vapors are essential.

In evaluating sampling or testing devices for particulate concentrations, it is important to note that the chamber concentrations are influenced by gravitation, thermal, electrostatic, and diffusional forces. Particulate concentrations cannot be reproduced with the accuracy obtainable with gases. When working with suspensions of uniform particles in the range less than 0.5μ, it is possible to have fairly stable (within ± 5 per cent) concentrations if the loadings are not so high as to permit coagulation to take place. Coagulation formulas are presented by Drinker and Hatch (41) for homogeneous smoke, and it is possible with these formulas to predict losses in concentration.

In calibrating sampling devices for particles, the foregoing remarks regarding chamber concentrations do not apply. In limited cases, known stable concentrations of fine aerosols can be used with ± 5 per cent accuracy, but if greater precision is desired in determining absolute efficiency of collection, it is suggested that other standards be used. The use of particulate-sampling devices in series as a means of measuring the efficiency of the first is also subject to error if multiples of the same device are used. Slippage of particles through the first unit can continue in the second. It is therefore essential that a reliable absolute device such as an electrostatic or thermal precipitator or ultrafilter be used for standardization even with well-maintained chamber concentrations, since this approach will ensure greater reliability.

The effects of physical factors such as temperature, pressure, and radiation on aerosols can also be studied once a controlled concentration has been prepared. In addition to the kinds of tests mentioned, the efficiency of masks and counteractants in dissipating or removing the contaminants can be measured.

Continuous means of testing masks or protective respiratory devices are described in References 34, 167, and 168, and the following discussion will be limited to the requirements for testing large-scale devices for gas cleaning.

It is important to note that, when testing large equipment such as filters, scrubbers, precipitators, catalytic converters, and other types of air-cleaning devices, it is desirable to treat fairly large volumes of air. This is especially true if full-scale or large components are to be observed. Since it is often impractical to duplicate the machinery producing the dust or fume or to simulate the type of process in the laboratory in many instances, test dusts are used. These dusts should be dispersed under ideal conditions, in a manner which does not introduce any new physical factors such as temperature and electrostatic charge. For this reason, in testing such devices it is sometimes desirable to place settling devices or impingement traps in series with the equipment to remove a portion of the larger sizes of dust. Elutriating towers remove a certain maximum size, but no controlled minimum size can be obtained unless the fines are previously removed. With these techniques it is possible to produce dynamic size variations in the test dust.

Because of the wide range of industrial aerosols encountered in practice, it is usually considered a good idea to test air-cleaning devices with a number of aerosols (5 or 6) with particulate materials differing widely in specific gravity, size, and shape. Atmospheric dust is a convenient aerosol already in suspension, but it varies widely in concentration as well as composition. The size of its particles, however, is small enough to be a severe test for any cleaning equipment. The most important factors to consider in testing air-cleaning equipment have been discussed at length in the "Handbook on Air Cleaning" (59) and by Drinker and Hatch (41).

In air pollution control work the most essential devices are those for control of process effluents. Usually, high loadings of aerosols occur, as well as wide variations in composition and size. Air-conditioning equipment for use in regions of excessive pollution must be studied and evaluated at low resistance and low dust loadings.

A number of tests have been developed, and it is necessary that there be adequate and representative sampling in providing reliable measurement of performance. Isokinetic sampling and proper location of collection points are essential items in experimental testing of devices.

In addition to sampling methods for measuring performance, it is possible to measure the amount of material fed to the device and to collect the total effluent on an absolute filter. Testing of filters for air conditioning on an absolute basis is recommended by the new Air Filter Institute code (1), which uses a spun-glass filter (PF-105, Owens-Corning Glass Company) as a final standard. Flows up to 1,000 cfm are possible with this method.

The efficiency of cleaning devices can be expressed on a volume, mass, or weight basis. In addition, stain production, optical transmission, optical absorption, and particle enumeration may be used as bases. In each of these cases a different particle parameter must be considered.

Testing of gas or vapor collectors is usually done by measuring the volume or the weight of influent and effluent gases, although in some cases odors may be measured as an index of performance. Particulate studies are done on any of the bases mentioned, depending upon the intended application. For pollution control, mass emission per unit time is desired, but in some cases the measurement of visibility is required. Particle enumeration is important only for extremely toxic materials, tracer substances, and where recirculation of fibrosis-producing dusts is intended.

Efficiency of collection can be measured and expressed in several ways, regardless of the actual basis employed for measurements, i.e., weight, stain, or count. One method involves knowing the concentration entering C_E and leaving C_L from which

$$\text{Efficiency} = \frac{(C_E - C_L)}{C_E} 100 \tag{12-8}$$

Another expression is based on measuring the amount entering C_E and the amount collected in the device C_C from which

$$\text{Efficiency} = \frac{C_C}{C_E} 100 \tag{12-9}$$

A third basis of measuring efficiency is to know the effluent concentration C_L and the amount collected C_C from which

$$\text{Efficiency} = \frac{1 - C_L}{C_C + C_L} 100 \tag{12-10}$$

A number of devices are now tested for particle-size discrimination with results expressed as fractional efficiencies based on performance in a given size range. This procedure requires that size measurements be made of either the material leaving the collector or that collected.

REFERENCES

1. Air Filter Institute, Louisville, Ky.: Code for Testing Air Cleaning Devices Used in General Ventilation. Sect. 1. Unit or Panel Type Air Filtering Devices, Aug. 10, 1953.
2. Albright, C. W., et al.: Pneumatic Feeder for Finely Divided Solids, *Chem. Eng.*, **56,** 108–111 (1949).
3. Amdur, M. O., W. W. Melvin, Jr., and Philip Drinker: Effects of Inhalation of Sulfur Dioxide by Man, *Lancet,* **265,** 758–759 (1953).
4. Amdur, M. O., et al.: Inhalation of Sulfuric Acid Mist by Human Subjects, *Arch. Ind. Hyg. and Occupational Med.*, **6,** 305–313 (1952).
5. American Pharmaceutical Association: "National Formulary," 9th ed., 1950.
6. Antara Products, New York, N.Y.
7. Atkins, B. R.: A Technique for Entraining Fine Powders in an Air Stream at a Constant Rate, *J. Sci. Instr.*, **28,** 221 (1951).
8. Barker, K. R., J. S. Sebastian, L. D. Schmidt, and H. P. Simons: Pressure Feeder for Powdered Coal or Other Finely Divided Solids, *Ind. Eng. Chem.*, **43,** 1204–1209 (1951).
9. Barnes, E. C., and H. W. Speicher: Determination of Formaldehyde in Air, *J. Ind. Hyg. Toxicol.*, **24,** 10–17 (1942).
10. Barr, G.: Two Designs of Flow Meter, and a Method of Calibration, *J. Sci. Instr.*, **11,** 321–324 (1934).
11. Barrett, H. M., D. L. McLean, and J. G. Cunningham: A Comparison of the Toxicity of Carbon Tetrachloride and Trichlorethylene, *J. Ind. Hyg. Toxicol.*, **20,** 360–379 (1938).
12. Baurmash, L., F. A. Bryan, R. W. Dickinson, and W. C. Burke, Jr.: A New Exposure Chamber for Inhalation Studies, *Am. Ind. Hyg. Assoc. Quart.*, **14,** 26–30 (1953).
13. Berly, E. M., et al.: Recovery of Soluble Gas and Aerosols from Air Streams, *Ind. Eng. Chem.*, **46,** 1769–1777 (1954).
14. Billings, C. E., R. Dennie, M. W. First, and L. Silverman: Laboratory Performance of Fabric Dust and Fume Collectors. (NYO-1590), Harvard University School of Public Health, Air Cleaning Laboratory, and U.S. Atomic Energy Commission, Aug. 31, 1954.
15. Booth, H. S.: The Baro-buret—A New Accurate Buret, *Ind. Eng. Chem., Anal. Ed.*, **2,** 182–186 (1930).
16. Boucher, R. M. G.: Sur le functionnement de l'epurateur a micro-brouillards "aerojet-Venturi," *Chaleur & ind.*, **33,** 363–377 (1952).
17. Brown, E. H.: An Injection and Sampling Stopcock, *Ind. Eng. Chem., Anal. Ed.*, **14,** 551 (1942).
18. Brubach, H. F.: Some Laboratory Applications of the Low Friction Properties of the Dry Hypodermic Syringe, *Rev. Sci. Instr.*, **18,** 363–366 (1947).
19. Burwell, R. L., Jr.: Precision Feed Device for Catalytic Experiments, *Ind. Eng. Chem. Anal. Ed.*, **12,** 681–682 (1940).
20. Cadle, R. D., and P. L. Magill: Preparation of Solid- and Liquid-in-air Suspensions. *Ind. Eng. Chem.*, **43,** 1331–1335 (1951).
21. Calcote, H. F.: Accurate Control and Vaporizing System for Small Liquid Flows, *Anal. Chem.*, **22,** 1058–1060 (1950).
22. Carpenter, C. P., et al.: The Assay of Acute Vapor Toxicity, and the Grading and Interpretation of Results on 96 Chemical Compounds, *J. Ind. Hyg. Toxicol.*, **31,** 343–346 (1949).
23. Carpenter, T. M., and E. L. Fox: A Gas Analysis Apparatus Modified for the Determination of Methane in Metabolism Experiments, *J. Biol. Chem.*, **70,** 115–121 (1926).
24. Castleman, R. A., Jr.: The Mechanism of the Atomization of Liquids, *Bur. Standards J. Research,* **6,** 369–376 (1931).

25. *Chem. Eng. News*, **25**, 506–507 (1947), New Distributor Improves Dust Testing.
26. *Chem. Eng.*, **60**, No. 12, 122–126 (1953), Surge for Silica.
27. Cholak, J., and R. R. McNary: Determination of the Oxides of Nitrogen in Air, *J. Ind. Hyg. Toxicol.*, **25**, 354–360 (1943).
28. Church, F. W., and F. R. Ingram: Apparatus for Dispensing Aluminum Dust in Treatment of Silicotics, *J. Ind. Hyg. Toxicol.*, **30**, 246–250 (1948).
29. Comings, E. W., C. H. Adams, and E. D. Shippee: High Velocity Vaporizers, *Ind. Eng. Chem.*, **40**, 74–76 (1948).
30. Cottrell, W. D.: Solid Aerosol Generation, ORNL 1666, Oak Ridge National Laboratory, Oak Ridge, Tenn., Feb. 5, 1954.
31. Crabtree, J., and A. R. Kemp: Accelerated Ozone Weathering Test for Rubber, *Ind. Eng. Chem., Anal. Ed.*, **18**, 769–774 (1946).
32. Cupples, H. L.: Equipment for Laboratory Fumigations with Hydrocyanic Acid, *Ind. Eng. Chem., Anal. Ed.*, **5**, 36–38 (1933).
33. Dautrebande, L., B. Highman, and W. C. Alford: Studies on Aerosols. I. Reduction of Dust Deposition in Lungs of Rabbits by Aqueous Aerosols, *J. Ind. Hyg. Toxicol.*, **30**, 103–107 (1948).
34. Davies, C. N.: Fibrous Filters for Dust and Smoke, *Proc. 9th Intern. Congr. Ind. Med.*, September, 1948.
35. Decker, H. M., et al.: Filtration of Microorganisms from Air by Glass Fiber Media, *Heating, Piping Air Conditioning*, **26**, 155–158 (1954).
36. Deichmann, W. B.: A "Dustshaker," *J. Ind. Hyg. Toxicol.*, **26**, 334–335 (1944).
37. Dennis, R., et al.: Particle Size Efficiency Studies on a Design 2 Aerotec Tube, AEC Report (NYO 1583), April, 1952.
38. Dill, R. S.: A Test Method for Air Filters, *Trans. Am. Soc. Heating Ventilating Engrs.*, **44**, 379–386 (1938).
39. Dow Chemical Company, Midland, Mich.
40. Drinker, Philip: Laboratories of Ventilation and Illumination, *J. Ind. Hyg. Toxicol.*, **6**, 57–66 (1924).
41. Drinker, Philip, and T. Hatch: "Industrial Dust," 2d ed., McGraw-Hill Book Company, Inc., New York, 1954.
42. Drinker, Philip, R. M. Thomson, and S. M. Fitchet: Atmospheric Particulate Matter. II. The Use of Electric Precipitation for Quantitative Determinations and Microscopy, *J. Ind. Hyg. Toxicol.*, **5**, 162–185 (1923).
43. Drinker, Philip, C. P. Yaglou, and M. F. Warren: The Threshold Toxicity of Gasoline Vapor, *J. Ind. Hyg. Toxicol.*, **25**, 225–232 (1943).
44. Drozin, V. G., and Hochberg: Columbia University, Chemistry Department, NYO Report 4603, 1954.
45. Druett, H. A., and J. M. Sowerby: An Apparatus for the Maintenance of a Carbon Dust Cloud of Constant Concentration, *Brit. J. Ind. Med.*, **3**, 187–193 (1946).
46. Dudley, H. C., and P. A. Neal: Toxicology of Acrylonitrile (Vinyl Cyanide). I, *J. Ind. Hyg. Toxicol.*, **24**, 27–36 (1942).
47. Dudley, H. C., and J. W. Miller: Toxicology of Selenium. VI. Effect of Subacute Exposure to Hydrogen Selenide, *J. Ind. Hyg. Toxicol.*, **23**, 470–477 (1941).
48. Fahnoe, F., A. E. Lindroos, and R. J. Abelson: Aerosol Build-up Techniques, *Ind. Eng. Chem.*, **43**, 1336–1346 (1951).
49. Farbar, L.: Metering of Powdered Solids in Gas-Solids Mixtures, *Ind. Eng. Chem.*, **44**, 2947–2955 (1952).
50. Farr, R. S., W. N. Pauley, and K. A. Crismon: An Improved Test Method for Rating Air Filters, *Trans. Am. Soc. Heating Ventilating Engrs.*, **54**, 187–200 (1948).
51. Ferry, R. M., et al.: A Study of Freshly Generated Bacterial Aerosols of *Micrococcus candidus*, etc., *J. Infectious Diseases*, **88**, 256–271 (1951).
52. Ferry, R. M., L. E. Farr, Jr., and M. G. Hartmann: The Preparation and Measurement of the Concentration of Dilute Bacterial Aerosols, *Chem. Revs.*, **44**, 389–417 (1949).
53. Fieldner, A. C., G. G. Oberfell, M. C. Teague, and J. N. Lawrence: Methods of Testing Gas Masks and Absorbents, *J. Ind. Eng. Chem.*, **11**, 519–540 (1919).
54. First, M. W.: Cyclone Dust Collector Design, American Society of Mechanical Engineers, ASME Paper 49-A-127, Nov. 27–Dec. 2, New York, 1949.
55. First, M. W., et al.: Air Cleaning Studies, U.S. Atomic Energy Commission, Progress Report (NYO-1581), June 30, 1951.
56. First, M. W., et al.: Air Cleaning Studies. Progress Report for July 1, 1952, to June 30, 1953, U.S. Atomic Energy Commission Report (NYO-1591), August, 1954.
57. First, M. W., R. Moschella, L. Silverman, and E. Berly: Performance of Wet Cell Washers for Aerosols, *Ind. Eng. Chem.*, **43**, 1363–1370 (1951).

58. Fitzgerald, J. J., and C. G. Detwiler: Collection Efficiency of Air Cleaning and Air-sampling Filter Media, Knolls Atomic Power Laboratory, KAPL 1088, Mar. 15, 1954.
59. Friedlander, S. K., L. Silverman, Philip Drinker, and M. W. First: "Handbook on Air Cleaning," Harvard University School of Public Health, Department of Industrial Hygiene, and U.S. Atomic Energy Commission, September, 1952.
60. Gibbs, W. E.: "Clouds and Smokes," P. Blakiston & Son, Philadelphia, 1924.
61. Gisclard, J. B.: Simple Device for Preparing Vapor Air Mixtures, *Ind. Eng. Chem. Anal. Ed.*, **15,** 582 (1943).
62. Goldman, F. H., and C. G. Seegmiller: The Determination of Halogenated Hydrocarbons in the Atmosphere, *J. Ind. Hyg. Toxicol.*, **25,** 181–184 (1943).
63. Goodhue, L. D.: Insecticidal Aerosol Production, *Ind. Eng. Chem.*, **34,** 1456–1459 (1942).
64. Harvard University, School of Public Health, Industrial Hygiene Department: Unpublished data.
65. Heikes, N. L.: A Constant-pressure and Flow-ratio Regulator for Continuously Mixing Two Gases, *Ind. Eng. Chem., Anal. Ed.*, **15,** 133–134 (1943).
66. Henderson, Y., and H. W. Haggard: "Noxious Gases," Reinhold Publishing Corporation, New York, 1943.
67. Heppel, L. A., P. A. Neal, T. L. Perrin, M. L. Orr, and V. T. Porterfield: Toxicology of Dichloromethane. I. Effects of Daily Inhalation, *J. Ind. Hyg. Toxicol.*, **26,** 8–16 (1944).
68. Hewson, E. W.: University of Michigan Engineering Research Project No. 2160 Scientific Report No. 1, October, 1953.
69. Heywood, H.: Filter Efficiency and Standardization of Test Dust, *Inst. Mech. Engrs. Proc.* (B), **1B,** No. 5, 169–174 (1952).
70. Hill, A. S. G.: A Photoelectric Smoke Penetrometer, *J. Sci. Instr.*, **14,** 296–303 (1937).
71. Hodgman, C. D., R. C. Weast, and C. W. Wallace (eds.): "Handbook of Chemistry and Physics," 35th ed., Chemical Rubber Publishing Co., Cleveland, Ohio, 1954.
72. Ingham, H. S., and A. P. Shepard: "The Metco Metallizing Handbook," 5th ed., Metallizing Engineering Co., Long Island City, N.Y., 1951.
73. Irish, D. D., and E. M. Adams: Apparatus and Methods for Testing the Toxicity of Vapors, *Ind. Med., Ind. Hyg. Sec.*, **1,** 1–5 (1940).
74. Irish, D. D., et al.: The Response Attending Exposure of Laboratory Animals to Vapors of Methyl Bromide, *J. Ind. Hyg. Toxicol.*, **22,** 218–230 (1940).
75. Jones, B. W., S. A. Jones, and M. B. Neuworth: Precision Liquid Feeder, *Ind. Eng. Chem.*, **44,** 2233–2234 (1952).
76. Jones, G. W., E. S. Harris, and W. E. Miller: Explosive Properties of Acetone-Air Mixtures, *U.S. Bur. Mines, Tech. Paper* 544, 1933.
77. Katz, S. H.: Generator of Fine Smoke for Direct Impregnation of Filter Paper Used in Gas Masks, U.S. Patent No. 2,528,522, Nov. 7, 1950.
78. Katz, S. H., G. W. Smith, and E. G. Meiter: Dust Respirators. Their Construction and Filtering Efficiency, *U.S. Bur. Mines, Tech. Paper* 394, 1926.
79. Keenan, R. G., and L. T. Fairhall: The Absolute Efficiency of the Impinger and of the Electrostatic Precipitator in the Sampling of Air Containing Metallic Lead Fume, *J. Ind. Hyg. Toxicol.*, **26,** 241–249 (1944).
80. King, R. O., and R. R. Davidson: Liquid Flow at Small Constant Rates, *Can. Mining J.*, **64,** 573–574 (1943).
81. LaMer, V. K.: The Preparation, Collection and Measurement of Aerosols, *Proc. 1st Nat. Air Pollution Symposium*, pp. 5–13, Stanford Research Institute, Los Angeles, Calif., 1949.
82. LaMer, V. K., et al.: Studies on Filtration of Monodisperse Aerosols. Columbia University, Central Aerosol Laboratories (NYO-512), Mar. 31, 1951.
83. LaMer, V. K., E. C. Y. Inn, and I. B. Wilson: The Methods of Forming, Detecting and Measuring the Size and Concentration of Liquid Aerosols in the Size Range of 0.01 to 0.25 Micron Diameter, *J. Colloid Sci.*, **5,** 471–496 (1950).
84. Landahl, H. D., and S. Black: Penetration of Air-borne Particulates through Human Nose, *J. Ind. Hyg. Toxicol.*, **29,** 269–277 (1947).
85. Landahl, H. D., and R. G. Herrmann: On Retention of Air-borne Particulates in Human Lung, *J. Ind. Hyg. Toxicol.*, **30,** 181–188 (1948).
86. Landahl, H. D., and T. Tracewell: Penetration of Air-borne Particulates through Human Nose, *J. Ind. Hyg. Toxicol.*, **31,** 55–59 (1949).
87. Laskin, S.: quoted in H. B. Wilson, G. E. Sylvester, S. Laskin, C. W. LaBelle, and H. E. Stokinger: The Relation of Particle Size of Uranium Dioxide Dust to Toxicity Following Inhalation by Animals, *J. Ind. Hyg. Toxicol.*, **30,** 319–331 (1948).
88. LaTorre, P., and L. Silverman: Collecting Efficiencies of Filter Papers for Sampling Lead Fume, *Arch. Ind. Hyg. and Occupational Med.*, in press, 1955.

89. Lewey, F. H., et al.: Experimental Chronic Carbon Disulfide Poisoning in Dogs; Clinical, Biochemical and Pathological Study, *J. Ind. Hyg. Toxicol.*, **23,** 415–436 (1941).
90. Lewis, R. A., and G. F. Koepf: An Apparatus for Producing Constant Gas Mixtures, *Science*, **93,** 407–408 (1941).
91. Littlefield, J. B., W. P. Yant, and L. B. Berger: A Detector for Low Concentrations of Hydrogen Sulfide, *U.S. Bur. Mines, Rept. Invest.* 3276, 1935.
92. Lovelock, J. E., and E. M. Wasilewska: An Ionization Anemometer, *J. Sci. Instr.*, **26,** 367–370 (1949).
93. Lowry, P. H., D. A. Mazzarella, and M. E. Smith: Ground-level Measurements of Oil-fog Emitted from a Hundred-meter Chimney, *Meteorol. Monographs*, **1,** No. 4, 30–35 (1951).
94. Lundsted, L. G., A. B. Ash, and N. L. Koslin: Constant Rate Feed Device, *Anal. Chem.*, **22,** 626 (1950).
95. McCabe, L. C.: Atmospheric Pollution (I.E.C. Production Forum), reference to a summary work of Mader, Cann, and Palmer of L.A.A.P.C.D. on the Significance of Organic Acids in Air Pollution and Their Effect on Plant Tissues, *Ind. Eng. Chem.*, **46,** No. 12, 93–94A (1954).
96. Machle, W., et al.: The Physiological Response to Isopropyl Ether and to a Mixture of Isopropyl Ether and Gasoline, *J. Ind. Hyg. Toxicol.*, **21,** 72–96 (1939).
97. Machle, W., E. W. Scott, and J. F. Treon: The Physiological Response of Animals to Some Simple Mononitroparaffins and to Certain Derivatives of These Compounds, *J. Ind. Hyg. Toxicol.*, **22,** 315–332 (1940).
98. McKee, R. W., et al.: A Solvent Vapor, Carbon Disulfide. Absorption, Elimination, Metabolism and Mode of Action, *J. Am. Med. Assoc.*, **122,** 217–222 (1943).
99. McKnight, W. H.: Pressed and Sintered Glass Powder Shapes, *Materials & Methods*, **40,** 94–96 (1954).
100. May, K. R.: An Improved Spinning-top Homogeneous Spray Apparatus, *J. Appl. Phys.*, **20,** 932–938 (1949).
101. Metals Disintegrating Co., Elizabeth, N.J., and others.
102. Miller, A. T., Jr.: A Respiratory Chamber for Chronic Exposure to Gases, *J. Lab. Clin. Med.*, **28,** 1854–1858 (1943).
103. Miller, H. I., Jr., G. M. Hama, E. C. J. Urban, and Philip Drinker: Health Hazards in Metal Spraying, *J. Ind. Hyg. Toxicol.*, **20,** 380–387 (1938).
104. Minnesota Mining and Manufacturing Co., St. Paul, Minn.
105. Moll, W. J. H., and H. C. Burger: A Pipet for Measuring a Small Quantity of Gas, *Z. tech. Phys.*, **21,** 203 (1940).
106. Moore, C. E., R. McCarthy, and R. F. Logsdon: A Partial Chemical Analysis of Atmospheric Dirt Collected for Study of Soiling Properties, *Heating, Piping Air Conditioning*, **26,** 145–148 (1954).
107. Morley, M. J., and B. D. Tebbens: The Electrostatic Precipitator-dilution Method of Flow Measurement, *Am. Ind. Hyg. Assoc. Quart.*, **14,** 303–306 (1953).
108. Morse, R. D., and E. F. von Wettberg, Jr.: Solids-Gas Contacting, U.S. Patent 2,667,706 (to du Pont), Feb. 2, 1954.
109. Nagel, P., G. Jander, and G. Scholz: Condensation Aerosols: Mercury Fog and Lead Oxide Smoke, *Kolloid-Z.*, **107,** 194–201 (1944).
110. National Starch Products, New York, N.Y.
111. Nelson, K. W., J. F. Ege, Jr., M. Ross, L. E. Woodman, and L. Silverman: Sensory Response to Certain Industrial Solvent Vapors, *J. Ind. Hyg. Toxicol.*, **25,** 282–285 (1943).
112. Oettingen, W. F. v.: A Laboratory of Industrial Toxicology, *J. Ind. Hyg. Toxicol.*, **18,** 609–622 (1936).
113. Olive, T.: Solids Feeders, *Chem. Eng.*, **59,** No. 11, 163–178 (1952).
114. Olsen, J. C., H. F. Smyth, Jr., G. E. Ferguson, and L. Scheflan: Determination of the Concentration of Vaporized Carbon Tetrachloride, *Ind. Eng. Chem., Anal. Ed.*, **8,** 260–263 (1936).
115. Page, R. T.: Constant-flow Orifice Meters of Low Capacity, *Ind. Eng. Chem., Anal. Ed.*, **7,** 355–358 (1935).
116. Patty, F. A., and G. M. Petty: Nitrite Field Method for the Determination of Oxides of N (except N_2O and N_2O_5), *J. Ind. Hyg. Toxicol.*, **25,** 361–365 (1943).
117. Pennsylvania Quartz Co., Philadelphia, Pa.
118. Pennsylvania State University, Department of Engineering Research: "Bibliography on Sprays," 2d ed., published by the Texas Company Refining Department, Technical and Research Division, New York, December, 1953.
119. Perkins, W. A., et al.: A Fluorescent Atmospheric Tracer Technique for Mesome-

teorological Research, *Proc. 2d Nat. Air Pollution Symposium*, Stanford Research Institute, Los Angeles, Calif., 1952.

120. Phelps, E. B.: Aerobiology, *American Association for the Advancement of Science, Publication* 17, pp. 133–137, 1942.
121. Rayner, A. C., and H. Hurtig: Apparatus for Producing Drops of Uniform Size, *Science*, **120**, 672–673 (1954).
122. Rogoff, J. M.: An Apparatus for Constant Intravascular Injection of Liquids, *J. Lab. Clin. Med.*, **25**, 853–856 (1940).
123. Rosebury, T.: "Experimental Air-borne Infection" (Microbiological Monographs), The Williams & Wilkins Company, Baltimore, 1947.
124. Rossano, A. T., Jr., and L. Silverman: Electrostatic Effects in Fiber Filters for Aerosols, *Heating and Ventilating*, **51**, 101–108 (1954).
125. Rowe, C. B.: Rating, Selection and Use of Panel Type Air Filters, *Heating and Ventilating*, **48**, 58–63 (1951).
126. Rowley, F. B., and R. C. Jordan: Air Filter Performance as Affected by Kind of Dust, Rate of Dust Feed and Air Velocity through Filter, *Heating, Piping Air Conditioning*, **10**, 539–548 (1938).
127. Rowley, F. B., and R. C. Jordan: Air Filter Performance as Affected by Kind of Dust, Rate of Dust Feed and Air Velocity through Filter, *Trans. Am. Soc. Heating Ventilating Engrs.*, **44**, 415–437 (1938).
128. Rowley, F. B., and R. C. Jordan: A Standard Air Filter Test Dust, *Trans. Am. Soc. Heating Ventilating Engrs.*, **45**, 681–695 (1939).
129. Rowley, F. B., and R. C. Jordan: Discoloration Methods of Rating Air Filters, *Trans. Am. Soc. Heating Ventilating Engrs.*, **49**, 487–494 (1943).
130. Salazar, A., and L. Silverman: New Method for Determination of Free Silica in Industrial Dusts, *J. Ind. Hyg. Toxicol.*, **25**, 139–148 (1943).
131. Salazar, A., and L. Silverman: Preparation of Microscopic Spheres from Quartz, *J. Ind. Hyg. Toxicol.*, **27**, 231–233 (1945).
132. Sayers, R. R., et al.: Methanol Poisoning. II. Exposure of Dogs for Brief Periods, etc., *J. Ind. Hyg. Toxicol.*, **26**, 255–259 (1944).
133. Schrenk, H. H.: Testing and Design of Respiratory Protective Devices, *U.S. Bur. Mines, Inform. Circ.* 7086, September, 1939.
134. Scott, E. W.: The Metabolism of Mononitroparaffins. III, *J. Ind. Hyg. Toxicol.*, **25**, 20–25 (1943).
135. Setterlind, A. N.: Preparation of Known Concentrations of Gases and Vapors in Air, *Am. Ind. Hyg. Assoc. Quart.*, **14**, 113–120 (1953).
136. Shekhter, A. B., S. Z. Roginskii, and S. Zakharova: An Electron Microscopic Investigation of Smoke Deposits, *Acta Physicochim. U.S.S.R.*, **21**, 463–468 (1946) (in English).
137. Shepherd, M., S. Schuhmann, R. H. Flinn, J. W. Hough, and P. A. Neal: Hazard of Mercury Vapor in Scientific Laboratories, *J. Research Nat. Bur. Standards*, **26**, 357–375 (1941). (Research Paper 1383.)
138. Silver, S. D.: Constant-flow Gassing Chambers—Principles Influencing Design and Operation, *J. Lab. Clin. Med.*, **31**, 1153–1161 (1946).
139. Silverman, L.: Unpublished data, Harvard University, 1943.
140. Silverman, L., and R. Dennis: Unpublished data, Harvard Air Cleaning Laboratory, 1954.
141. Silverman, L.: A Method of Producing Air-Vapor Mixtures in Gas or Fume Chambers, *Rev. Sci. Instr.*, **11**, 346 (1940).
142. Silverman, L., and I. L. Beauchamp: Unpublished data, Harvard School of Public Health, American Iron and Steel Institute Project, 1954.
143. Silverman, L., and R. Dennis: Unpublished data, Harvard University, 1954.
144. Silverman, L., and J. F. Ege, Jr.: A Filter-paper Method for Lead-fume Collection, *J. Ind. Hyg. Toxicol.*, **25**, 185–188 (1943).
145. Silverman, L., and J. F. Ege, Jr.: A Rapid Method for the Determination of Chromic Acid Mist in Air, *J. Ind. Hyg. Toxicol.*, **29**, 136–139 (1947).
146. Silverman, L., M. W. First, G. S. Reichenbach, Jr., and Philip Drinker: Final Progress Report, Harvard University and U.S. Atomic Energy Commission, (NYO-1527), Feb. 1, 1950.
147. Silverman, L., H. F. Shulte, and M. W. First: Further Studies on Sensory Response to Certain Industrial Solvent Vapors, *J. Ind. Hyg. Toxicol.*, **28**, 262–266 (1946).
148. Silverman, L., and R. M. Thomson: Rapid Determination of Very Small Gas Flows, *Ind. Eng. Chem., Anal. Ed.*, **14**, 928 (1942).
149. Silverman, L., J. L. Whittenberger, and J. Muller: Physiological Response of Man to Ammonia in Low Concentrations, *J. Ind. Hyg. Toxicol.*, **31**, 74–78 (1949).

150. Sinclair, D.: Measurement of Particle Size and Size Distribution, in "Handbook on Aerosols," pp. 79–116, U.S. Atomic Energy Commission, Washington, D.C., 1950.
151. Sinclair, D., and V. K. LaMer: Light Scattering as a Measure of Particle Size in Aerosols. The Production of Monodisperse Aerosols, *Chem. Rev.*, **44,** 245–267 (1949).
152. Smith, J. C.: Size Reduction, *Chem. Eng.*, **59,** No. 8, 151–166 (1952).
153. Smith, W. H., and R. B. O'Brien: Unpublished data, ANP Project Report, unclassified, June, 1953. General Electric Co.
154. Sonkin, L. S., M. A. Lipton, and D. Van Hoesen: An Apparatus for Dispersing Finely Divided Dusts, *J. Ind. Hyg. Toxicol.*, **28,** 273–275 (1946).
155. Stanford Research Institute: The Smog Problem in Los Angeles County, Second Interim Report, Western Oil and Gas Association, Los Angeles, Calif., 1949.
156. Stead, F. M., et al. Dust Feed Apparatus Useful for Exposure of Small Animals to Small and Fixed Concentrations of Dust, *J. Ind. Hyg. Toxicol.*, **26,** 90–93 (1944).
157. Stead, F. M., and G. J. Taylor: Calibration of Field Equipment from Air Vapor Mixtures in a Five Gallon Bottle, *J. Ind. Hyg. Toxicol.*, **29,** 408–412 (1947).
158. Stenger, V. A., S. A. Shrader, and A. W. Beshgetoor: Analytical Methods for Methyl Bromide, *Ind. Eng. Chem., Anal. Ed.*, **11,** 121–124 (1939).
159. Union Carbide and Carbon Co. (Bakelite Division), New York, N.Y.
160. United States Atomic Energy Commission: "Handbook on Aerosols," Washington, D.C., 1950.
161. U.S. Bureau of Mines: Approval Schedule No. 21, Procedure for Testing Filter Type Dust, Fume, and Mist Respirators for Permissibility, approved Aug. 20, 1934.
162. Vapor Blast Co., Milwaukee, Wis.
163. Vonnegut, B.: The Nucleation of Ice Formation by Silver Iodide, *J. Appl. Physics*, **18,** 593–595 (1947).
164. Vonnegut, B., and R. Neubauer: Detection and Measurement of Aerosol Particles by Use of an Electrically Heated Filament, General Electric Research Laboratory Occasional Report No. 29, Project Cirrus Report No. RL-555, 1951.
165. Wahlen, F. G.: The Wahlen Gage, *Univ. Ill. Eng. Exp. Station, Bull.* 120, 1921.
166. Walker, A. C., and E. J. Ernst, Jr.: Preparation of Air of Known Humidity and Its Application to the Calibration of an Absolute-humidity Recorder, *Ind. Eng. Chem., Anal. Ed.*, **2,** 134–138 (1930).
167. Walton, W. H.: The Methylene Blue Particulate Test for Respirator Containers, Porton Spec. 1151, Chemical Defence Center, Porton, England, 1940.
168. Walton, W. H.: The Sodium Flame Particulate Test for Respirator Canisters, Porton Report No. 2161, Porton Spec. 1206, Chemical Defence Center, Porton, England, 1941.
169. Walton, W. H., and W. C. Prewett: Production of Sprays and Mists of Uniform Drop Size by Means of Spinning Disk Type Sprayers, *Proc. Phys. Soc. (London)*, **62B,** 341–350 (1949).
170. White, J. L., and P. P. Somers: Toxicity of Methyl Chloride for Laboratory Animals, *J. Ind. Hyg. Toxicol.*, **13,** 273–275 (1931).
171. Whittaker, Clarke, and Daniels, New York, N.Y.
172. C. K. Williams Co., Easton, Pa.
173. Williams, C. E., et al. Determination of Cloth Area for Industrial Air Filters, *Heating, Piping Air Conditioning*, **12,** 259–263 (1940).
174. Williams, C. E., and W. P. Battista: Constant Dust Feeding Device for Laboratory Use, *J. Ind. Hyg. Toxicol.*, **22,** 152–153 (1940).
175. Witzmann, H.: Elementary Drop and Liquid Holdup in a Packed Tower, *Z. Electrochem.*, **46,** 313–321 (1940).
176. Wright, B. M.: A New Dust Feed Mechanism, *J. Sci. Instr.*, **27,** 12–15 (1950).
177. Yant, W. P., and F. E. Frey: Apparatus for Preparing Vapor-Air Mixtures of Constant Composition, *Ind. Eng. Chem.*, **17,** 692–694 (1925).
178. Yant, W. P., S. J. Pearce, and H. H. Schrenk: A Microcolorimetric Method for the Determination of Toluene, *U.S. Bur. Mines, Rept. Invest.* 3323, December, 1936.

SECTION 13

EQUIPMENT AND PROCESSES FOR ABATING AIR POLLUTION

BY H. F. JOHNSTONE, A. M. CLARK, R. C. COREY, S. K. FRIEDLANDER, K. E. LUNDE, M. S. PETERS, I. G. POPPOFF, H. B. SCHNEIDER, K. T. SEMRAU, AND W. T. SPROULL

13.1 INTRODUCTION

This section describes the equipment and processes used for abating air pollution and the principles to be employed in selecting the proper equipment and processes for different conditions.

Over a period of years it has been found that certain types of processes have been successfully adapted to certain situations, and the experienced users have indicated preferences. Where this has occurred, an effort has been made to point out the reasons for the choice. Because of the wide variation of conditions that are encountered in practice, it is only rarely that direct transfer of equipment or processes can be made from one installation to the next. Thus each case should be considered individually and the controlling factors in each carefully weighed.

13.1.1 Classes of Pollutants

The problems of pollution are usually due to those materials which can be classified as gases or vapors or particulate matter, such as dust or smoke. In this section these terms are used in the following sense:

1. *Gas:* one of the three states of aggregation of matter, having neither independent shape nor volume and tending to expand indefinitely.

2. *Vapor:* the gaseous phase of matter which exists in a liquid or solid state.

3. *Dust:* a loose term applied to solid particles predominantly larger than colloidal and capable of temporary suspension in air or other gases. Formation from larger masses through the application of physical force is usually implied.

4. *Smoke:* small gas-borne particles resulting from incomplete combustion, consisting predominantly of carbon and other combustible material, and present in sufficient quantity to be observable independently of the presence of other solids.

5. *Fume:* very small particles resulting from chemical reaction or from the condensation of vapors produced in combustion, distillation, or sublimation. They are com-

monly metals or metallic oxides, and their composition may be different from that of the parent material from which they originate.

6. *Mist:* droplets of liquids produced by condensation of vapor on suitable nuclei. They are stable if the vapor pressure of the particles is low or if the gas is saturated.

The effects of contaminants can be abated by preventing their entering the atmosphere, by removal processes, or by ejecting them into the atmosphere at such heights that they become sufficiently diluted to be harmless and not a nuisance when they reach the ground.

If contaminant gases are to be eliminated, this may be done by absorption or adsorption processes. The removal of particulate matter is accomplished by a variety of methods, which depend upon the size of the particles, their size distribution, their nature, the concentration of particles, and the degree of elimination desired. The value of the products removed and their ultimate method of disposal also must be considered.

In most cases the products removed at the sources of pollution are, by themselves, the obnoxious materials. But some materials considered relatively innocuous become altered in the atmosphere, either through reaction with materials nearby or merely by the oxidation of air and sunlight, and produce substances more obnoxious than those originally present. These reactive materials may also require removal.

13.1.2 General Engineering Problems

The volume of gas that must be treated varies over a wide range. It may be as little as a few hundred or as much as several million cubic feet per minute. The problem may require removal of both particulate and gaseous materials. For example, the combustion of powdered coal of high sulfur content may require equipment for the removal of both fly ash and sulfur dioxide. In such cases the load on the equipment would be relatively uniform with respect to time, and the equipment could therefore be designed for its continuous and full utilization to capacity. Frequently, however, wide variations in gas-flow rates and materials to be removed are encountered. An example of this is the removal of particulate matter from open-hearth steel-manufacturing operations.

Even if multiple units are used, the size of each unit in large gas-cleaning installations is frequently many times the size of similar equipment employed in gas absorption and of scrubbing units used in chemical processes. This is a result of the large volumes of gas handled and the low concentrations of materials to be removed. This places a high premium on the efficiency of performance and careful selection of processes for gas cleaning.

The concentration of the component to be removed will ordinarily be fairly small, usually a fraction of 1 per cent. In some cases it is desirable to remove components that exist at concentrations of less than 1 ppm.

The temperature of the gases treated may range from 100 to 1000°F or more. Both corrosion and erosion may be encountered. The nature of the substances treated and the conditions will determine the materials of construction, and these often greatly affect the economics of the process and the equipment selected.

Treatment of gases with a scrubbing process using water or aqueous solutions will frequently bring the gases to the wet-bulb temperature saturated with water vapor. Thus when they are discharged into cool air, a condensate plume is apt to form. These plumes of water droplets or fog usually disappear within a few hundred feet of the stack and in fact cause little harm. Nevertheless, their existence sometimes is undesirable from the standpoint of appearance.

Frequently the material recovered has some value as a by-product, but this is by no means the usual situation. Generally, the pollutant either has no value or is in such a form or concentration that its recovery is not economic; it must be abated solely because of its nuisance effects. In such cases, dispersal of the pollutant from high stacks may reduce concentrations at ground level to such a point that collection is unnecessary. The effectiveness of stack disposal is controlled by many factors, including topography, meteorology, and the economics of stack construction.

Where collection of a pollutant is necessary, subsequent disposal of sizable quantities of material may be necessary. If the material is large in quantity and has no value, and no method of disposal is readily available, then disposal itself may constitute the principal cost of abating the pollutant. From some standpoints it is preferable to recover the material as a solid if there is a choice, since it can be more easily stored or dumped in this form. Many of the possible methods of disposal for such materials are obvious and are therefore not treated in detail in this section.

Liquid wastes, such as the effluents from scrubbing devices, have frequently been discharged to streams. When the resulting concentration of pollutant material in the stream is low, this procedure is sometimes permissible, but in general such discharge of untreated effluents is objectionable. Restrictions on pollution of watercourses are frequently much more stringent than existing restrictions on discharge of pollutants to the atmosphere, and no purpose is served by alleviating an air pollution problem by a method which simultaneously creates a water pollution problem. In general, liquid wastes must be treated before discharge to waterways. Solids can be removed by clarifiers, while toxic or otherwise obnoxious materials require chemical treatment.

If their volume is not too great, liquid wastes can, in some cases, be discharged to ponds for disposal by settling, evaporation, or percolation through the soil. Such ponds may sometimes give rise to secondary air pollution problems through release of odors, etc. Thick sludge may be discharged to trenches or area land fills.

Dry solid wastes may be disposed of in dumps or land fills. However, secondary air pollution at the disposal site may be created in this way. Fine dusts normally have low bulk densities and easily become wind-borne and hence easily form dust clouds during handling and in winds. This difficulty may be largely overcome by compacting the material in a land fill, to save space, and covering the fill with a compacted layer of earth to prevent escape of the waste material into the atmosphere.

A problem common to land-disposal methods is the possibility of pollution of ground waters. The geology of a potential disposal area must therefore be considered in selection of the area.

13.2 DISPOSAL OF POLLUTANTS

13.2.1 Dispersal from Stacks

The theoretical aspects of stack dispersal of pollutants are treated in Sec. 5. The following articles give primary attention to engineering problems in the construction of tall stacks.

Effect of Stack Height on Dispersal. Inasmuch as a large percentage of any pollution is introduced into the atmosphere through stacks or chimneys of sizable proportions, their effect on dispersion of pollutants is important.

Under conditions of a light wind, the plume from a stack will gradually rise and flow downwind until dispersion is complete. Strong winds, gusts, and unstable atmospheric conditions such as vertical convection cells will often bring the gases to the ground before the obnoxious concentration has been dissipated. Also during periods of calm or temperature inversion, smoke that is usually dissipated may intersect the ground in high concentrations.

The topography, including man-made structures, in the vicinity of a plant should also receive consideration in a study of air pollution. The turbulent wake on the leeward side of a large building in the vicinity of the stack will bring smoke down to the ground in high concentration if the top of the stack is not above the upper limits of this turbulence.

The exit velocity of the gases from the top of the chimney is also an important factor in dispersion. The kinetic energy of the gas stream gives the gas an upward impetus, and it is important that this impetus carry the gas past the tip vortices at the top of the chimney, if immediate downwash is not to occur in the vicinity of the chimney.

In order to control pollution, therefore, it is necessary to have a proper combination of stack height and exit velocity of the stack gases with consideration given to prevailing atmospheric conditions and surrounding terrain. It is possible to calculate the required height of chimney for each situation, but each location presents a problem of its own, and the best way to ensure the desired results is to use controlled wind-tunnel tests with scale models of the stack, surrounding buildings, and terrain features. The quantitative aspects of such calculations are discussed in more detail in Sec. 5.

Construction of Stack. In addition to the proper height of stack for dispersion of obnoxious gases, the chimney must be designed to provide the draft required for successful and efficient operation. First, the minimum diameter of the chimney is determined by considering the volume of gases flowing. Ordinarily stack velocities vary from 15 to 60 fps, with the tendency toward higher velocities for large chimneys. Inasmuch as friction loss is inversely proportional to the fifth power of the diameter for a given height and given volume of gases, it is readily seen that a high velocity resulting from a small diameter could result in a friction loss exceeding the natural draft.

The foundation for the chimney, as for any other structure, is of primary importance. A careful examination of subsurface conditions should always be made to check the nature and thickness of the strata from the natural surface down to such depths as will leave no doubt regarding the safety of the structure.

Pending actual test data, allowable bearing values for various soils may be selected, using the accompanying tabulation as a guide.

Soil	*Psf*
Clay—moist	2,000
Sand—clean and dry	4,000
Clay—moderately dry	5,000
Sand and gravel—well compacted	6,000
Gravel and coarse sand—well cemented	8,000
Hard pan or stratified clay	10,000
Shale	12,000
Bed rock	60,000

Industrial chimneys are built of steel, brick, or reinforced concrete. Steel stacks exhausting gases whose temperature may fall below the dew point should be lined with brick or gunite to protect the structure against corrosion resulting from condensation. If the stack is lined with brick, there should be a ½- or ¾-in. layer of dense grout between the brick lining and the steel shell.

For ordinary boiler service a radial brick chimney is lined for about one-fifth of its height, the lining starting 2 or 3 ft below the bottom of the flue opening. Where pulverized coal or wood refuse is used as a fuel, the lining should be continued down to the bottom to protect the main walls against damage from possible secondary combustion in the soot deposited in the chimney and falling to the bottom.

Reinforced-concrete chimneys have been built in the United States and Canada since the early 1900s. In the early 1930s, the American Concrete Institute created its original Committee 505, which prepared a tentative specification titled "Proposed

Standard Specification for the Design and Construction of Reinforced Concrete Chimneys." In 1949 they reactivated Committee 505 and revised the tentative standard specifications. Because of the prevailing use of higher-strength concrete, the need for a more accurate approximation of earthquake forces, and the need for a more complete study of the temperature gradient through the chimney walls, a revision was felt necessary. A new "Proposed Standard Specification for the Design and Construction of Reinforced Concrete Chimneys" was prepared by Committee 505 and adopted as an ACI Standard in 1954.

A reinforced-concrete chimney is readily adaptable to space limitations that a brick chimney could not economically meet. For ordinary boiler service, reinforced-concrete chimneys are lined for one-third to one-half the chimney height and are usually equipped with an outside ladder or scaling rungs, a lightning-protection system, a cleanout door, and a cast-iron cap or coping.

With regard to the relative costs of different types of chimneys, steel stacks usually are less expensive in initial construction than brick or reinforced-concrete chimneys. If given proper attention and maintenance, the durability of brick and reinforced-concrete chimneys are about equal. For a chimney having a height less than 150 ft, a perforated radial-brick chimney will generally be less expensive than a reinforced-concrete stack. For chimney heights from 150 to 175 ft, the cost of radial brick and concrete is about the same, and where the chimney height is greater than 175 ft, reinforced-concrete construction is usually less expensive than perforated radial brick.

Chimneys which conduct corrosive gases require special study to determine the type of lining and mortar needed. No definite rules can be given. Process stacks handling sulfur gases at low temperature must be protected against acid attack. The destructive effect of intermittent operation must be considered in many cases.

Occasionally dust-removal apparatus such as a centrifugal device is installed in a chimney, but it is usually more practical and economical to install a removal apparatus in the flue system. Sometimes water sprays are installed in the chimney to scrub out both solid and gaseous products. Unless the lower portion of the chimney is specially lined, the use of a water spray will hasten corrosion and deterioration of the chimney because of the acid nature acquired by the scrubbing water.

It is recommended that industrial chimneys be inspected at least every 2 years, preferably by the builder of the chimney in the case of brick and concrete chimneys, and repairs should be made when indicated. If a chimney is to be used for a service other than that for which it was designed, the builder of the chimney should be consulted as to necessary alterations.

13.2.2 Combustion Processes

Incineration. GENERAL. Incineration has unique advantages over other methods of disposal: the volume of waste is reduced some twentyfold, the residue is noncombustible and contains no organic matter that would ultimately decompose and create a nuisance, and it is inherently hygienic.

The choice of a particular method for municipal- or industrial-waste disposal should be based upon a comprehensive evaluation of such factors as the combustible content of the waste, the total investment and handling costs, salvage returns, convenience, hygiene, and esthetics. A recent publication discusses in detail municipal incineration in relation to other disposal methods (131).

Incineration is any combustion process for disposing of combustible wastes, such as rubbish, refuse, garbage, human and animal remains, and solid, semisolid, liquid, and gaseous by-products. The net heat generated by combustion may or may not be recovered for such useful purposes as preheating the combustion air used for incineration or generating steam for other processes.

Incineration should be conducted under conditions to achieve (1) maximum combustion efficiency, so that minimum quantities of products of incomplete combustion, such as smoke, tars, and malodorous compounds, are discharged to the atmosphere; (2) complete combustion of combustible solids, so that there is the maximum reduction of volume; and (3) maximum retention of solids within the system, so that the least possible quantity of ash and charred residue is discharged to the atmosphere.

Since the incineration of solid wastes can be treated theoretically and practically in the same terms as the combustion of coal or coke in a furnace (20), the principles of the combustion of these fuels are discussed below, especially in relation to the conditions necessary for complete combustion. These principles have been evolved over a span of many years, both from fundamental studies of different types of fuel beds and a wealth of engineering data from furnace-performance tests under a wide variety of conditions.

COMBUSTION THEORY. Combustion is broadly defined as any chemical reaction accompanied by heat and light but is more commonly understood to be the oxidation of substances by the oxygen in air. Substances commonly of interest in connection with incineration, which undergo combustion, are those which consist primarily of compounds of carbon and hydrogen, such as cellulose, fats, oil, plastics, and rubber.

Fuel beds may be classified in terms of the relative directions of flow of the air and fuel (96). Accordingly, there are three elementary types:

1. The overfeed bed, in which the air and fuel move countercurrently. It is characteristic of this type of bed that the ignition plane, which is essentially the area where the fuel reaches its ignition temperature, moves in the same direction as the air and that the hot combustion gases pass through the relatively cold incoming raw fuel. It is best exemplified by hand-fired furnaces, where the fuel is burned on grates through which air is passing.

2. The underfeed bed, in which the air and fuel move concurrently. The ignition plane moves in the opposite direction to the fuel, and the combustion gases pass through ash and ignited fuel. Industrially, pure underfeed burning is rare, but it occurs essentially during the first stage of burning in traveling-grate stokers (12).

3. The cross-feed bed, in which the fuel flows at right angles to the flow of air. Since incinerator performance can be adequately interpreted in terms of overfeed and underfeed burning, this type need not be considered further.

The chemical reactions within these basic fuel beds result in the occurrence of carbon monoxide, hydrogen, smoke, tars, and other products of incomplete combustion in the gases leaving the fuel bed. The relations are most easily explained in terms of air passing through an overfeed bed of carbon, that is, with no volatile matter present. Three reactions are believed to occur:

1. Carbon reacts with at least one-half of the available oxygen according to $C + \frac{1}{2}O_2 \rightarrow CO$; $\Delta H = -52{,}090$ Btu/lb-mol of carbon consumed. ΔH is the enthalpy change in the system, convention requiring that a negative value indicate the evolution of heat (exothermic).

2. The carbon monoxide from 1 reacts in the voids between the fuel particles with the remainder of the oxygen, according to $CO + \frac{1}{2}O_2 \rightarrow CO_2$; $\Delta H = -121{,}630$ Btu/lb-mol CO consumed.

3. The carbon dioxide from 2 reacts with hot carbon according to $CO_2 + C \rightarrow 2CO$; $\Delta H = 69{,}540$ Btu/lb-mol of carbon consumed. Heat is absorbed by this reaction (endothermic).

The reactions in 1 and 2 release 173,720 Btu/lb-mol of carbon consumed and result in an exothermic zone in the fuel bed that may be considered to be bounded by the plane where oxygen first reacts with the carbon and the plane where the oxygen is completely consumed. The thickness of this zone depends upon a number of factors, such as the composition, size, and reactivity of the fuel; the temperature and flow rate of the combustion air; and the heat-transfer losses to the surroundings. For coal and coke, this zone may be no thicker than a few fuel-particle diameters.

An idealized overfeed fuel bed is shown schematically in Fig. 13-1, with the relative distributions of the temperature and composition of the gases at various depths in the bed. It will be noted that oxygen disappears a short distance from the point of entry of the air to the bed; thereafter, both the temperature and the percentage of carbon dioxide decrease, and the percentage of carbon monoxide increases. If high-volatile fuels are burned, the gases leaving the top of the fuel bed would also contain hydrogen, hydrocarbons, tars, tar acids, and smoke, which result from devolatilization of the incoming fuel and thermal cracking of the volatile matter. The product may

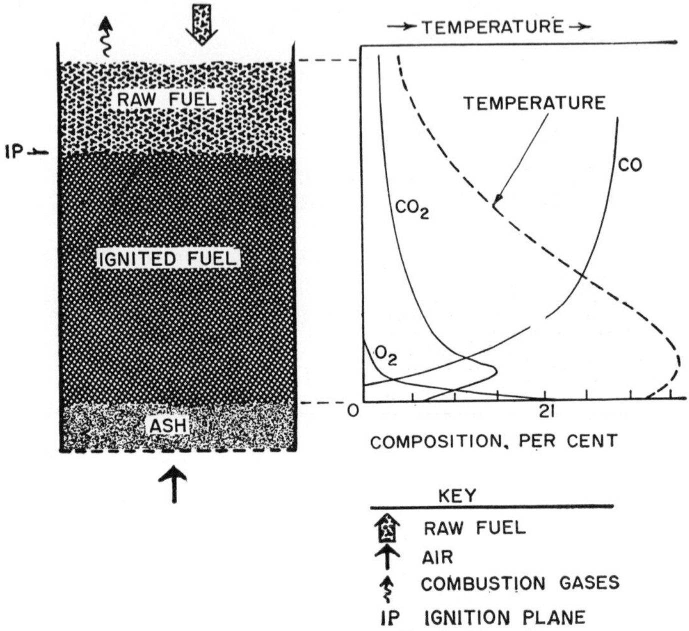

FIG. 13-1. Idealized overfeed fuel bed and relative distribution of temperature and products of combustion. (*Courtesy of U.S. Bureau of Mines.*)

be an acrid white, yellowish, or dense black smoke, depending upon the extent of cracking.

An underfeed fuel bed is shown schematically in Fig. 13-2, with the relative distribution of the temperature and composition of the gases at various depths in the bed. The combustion products pass through ignited fuel and ash, rather than through raw fuel, as in the overfeed; consequently, other things being equal, the exit gases are hotter and contain less smoke, tar, etc.

The important similarity between these two types of fuel beds is that products of incomplete combustion are discharged from the top of the bed, and additional oxygen, as secondary air, must be supplied to the combustion chamber to burn them completely to carbon dioxide and water. The problems involved in securing the most efficient use of secondary air are manifold and will be discussed later.

This analysis of fuel beds deals largely with the over-all results of such combustion processes, and except for certain assumptions that have been made regarding the chemical reactions within the bed, it takes no cognizance of the extremely complex chemical reactions and heat- and mass-transfer processes that govern the rate and the mechanism of combustion. Knowledge of these processes and their interrelationships in the exothermic zone in a fuel bed is as yet imperfect. However, the empirical principles described above are adequate to explain the performance of a fuel bed, whether it is a coal-fired furnace or an incinerator, from the standpoint of securing complete combustion with minimum discharge to the atmosphere of solids, combustible gases and vapors, and odors. For more detailed study of physical and chemical processes within fuel beds, the reader is referred to some recent theoretical studies in this connection (118,125).

Incinerators vary from single-stage hand-fired devices of relatively simple design to multiple-stage automatically stoked devices that vary widely in complexity of design. Whereas there is a wealth of published engineering data for industrial furnaces, from which adequate design parameters may be derived, reliable engineering

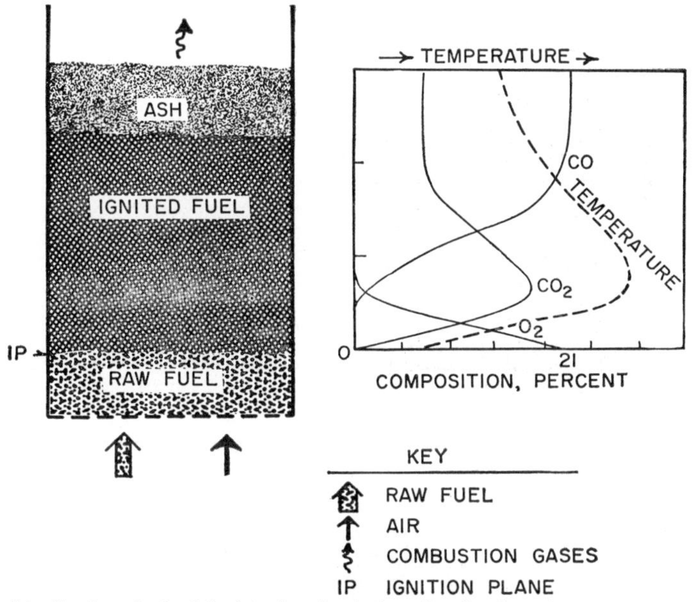

FIG. 13-2. Idealized underfeed fuel bed and relative distribution of temperature and products of combustion. (*Courtesy of U.S. Bureau of Mines.*)

data for incinerators are scarce. Much of this is due to the fact that incinerators, especially batch-fired units, are required to perform satisfactorily over a wide range of operating conditions, which is unusual in the case of industrial furnaces. For example, incinerators generally are charged with heterogeneous materials that contain different kinds and proportions of solid and semisolid wastes. Moreover, when incinerators are charged at random intervals with varying quantities of waste materials, the flow rate and the distribution of air through and above the ignited charge are changed radically, with the result that the temperatures throughout the combustion chamber fluctuate widely. The incinerator must accommodate these operating characteristics, or there will result intermittent discharge to the atmosphere of solids, aerosols, and malodorous constituents, which arise from incomplete combustion of the charge.

These and other adverse factors in the operation of incinerators are illustrated in Fig. 13-3, which shows schematically a conically piled charge of, say, rubbish on a grate in a simple chamber. Undergrate air is passing upward through the grate, and overfire air is entering a port above the grate, both under natural draft. Figure 13-3*A* represents a time after the charge was ignited on the surface and underfeed burning has been established (see Fig. 13-2). The thicknesses of the zones of ash and ignited materials are wholly relative; the zones will be thick for rubbish and other highly combustible materials, for which both the ash content and the bulk density are low as compared to coal or coke.

With reference to the undergrate air, its rate of flow through the bed controls the rate of burning of the charge, other things being constant. If, as often happens in practice, the grate becomes uncovered, much of the air will bypass the charge, and control of the rate of burning will be lost. (Under certain conditions, which will be described later, the air that bypasses the charge may be effectively utilized to burn

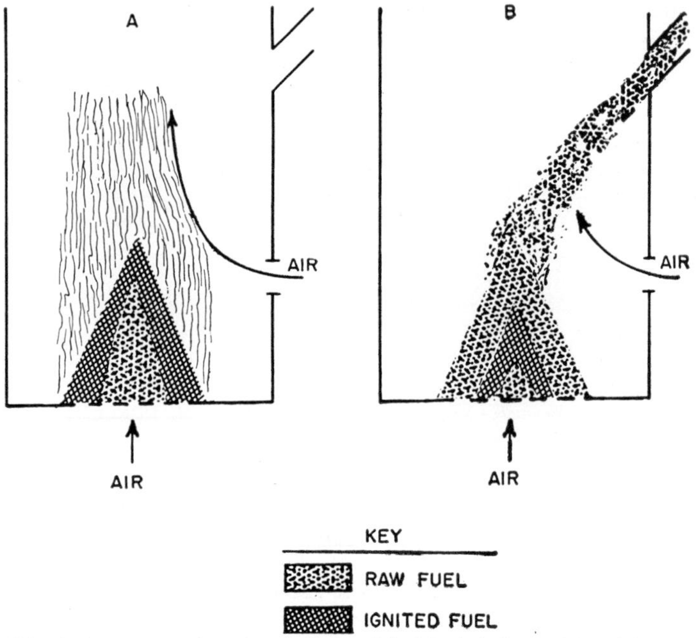

FIG. 13-3. Physical processes in a batch-charged fuel bed (schematic). (*Courtesy of U.S. Bureau of Mines.*)

the charge, but without special provisions for distributing this air, burning conditions are impaired by uncovered areas on the grate.)

With reference to the overgrate air, which is intended to provide oxygen for burning the combustible gases and vapors leaving the bed, unless it has enough velocity and proper direction to achieve thorough mixing with the combustibles, it will be largely ineffective. This is shown schematically in Fig. 13-3*A*, in which the overgrate air has too little momentum to induce turbulent mixing with the products of combustion. The optimum design and placement of overfire air nozzles are highly empirical and frequently are determined by trial and error. The engineering basis of the design of overfire jets has been described (47).

Figure 13-3*B* refers to a later time, when fresh charge is dumped upon the remaining portion of the original burning charge. Unless there is enough heat transfer from the walls of the chamber, or from a flame from a gas or oil burner located in the chamber, to ignite the new surface of the charge rapidly, burning becomes essentially the overfeed type (see Fig. 13-1), and smoke, tars, and combustible gases will be discharged. Accumulation of these materials in the chamber may lead to pulsations, puffs, or even explosions in the chamber.

A recent investigation (21) of the incineration of sawdust and other cellulosic materials has shown that these low-ash high-volatile wastes may be burned with high combustion efficiency in a cylindrical combustion chamber by the use of tangential air above the charge. The vortex flow of gases thus established provides rapid mixing of volatile combustibles with oxygen and has the following advantages with respect to both the combustion characteristics and the retention of particulate matter within the combustion chamber:

1. Both the tangential and the radial velocities of the gas above the bed increase as the gases approach the center of the bed. For a fixed-mass flow rate of air through the tangential ports, which fixes the mean radial velocity, the mean tangential velocity can be increased merely by reducing the diameter of the ports. These phenomena imply high relative velocity of gases with respect to the surface of the charge and thus reduce the diffusional resistance and provide mixing of the air with the products of combustion.

2. For a given net linear gas velocity through the chamber, entrained particles of dust and charred material travel a longer path than if there are no angular flow components and thus have a better chance to be consumed. Moreover, the chamber functions much like a cyclone dust collector and therefore retains a certain amount of entrained solids.

3. Downward flow of air along the walls of the chamber, which is characteristic of cyclone chambers, results in some preheating of the air by convective heat transfer from the walls of the chamber.

Minimum requirements for the design and operation of incinerators include the following factors:

1. The primary and secondary air (undergrate and overgrate) streams should be under complete control. The grate must be covered at all times. Secondary air must mix thoroughly with the gases leaving the charge, which may be achieved most effectively by means of properly designed and operated jets arranged to produce turbulent mixing or to establish vortex motion of the air and gases above the bed.

2. Excess air must be kept to a minimum, consistent with good combustion conditions, to obtain maximum temperatures in the combustion chamber and to secure minimum entrainment of particulate matter in the products of combustion. All openings in the chamber, except those serving a useful purpose, must be closed to prevent induction of unnecessary air.

3. Auxiliary firing should be provided for, especially in the combustion chamber, to maintain adequate temperatures therein when the heat of combustion of the charge is low as a result of excessive moisture and when large quantities of fresh charge are added to the unit.

4. Preheating of combustion air is desirable to increase its heat content and thereby increase the average temperature in the combustion chamber, other things being equal.

5. Continuous recording of smoke density and oxygen or carbon dioxide in the flue gases from industrial, commercial, and municipal incinerators reflects good operating practice.

There is substantial evidence that complete combustion can be accomplished in a single-chamber incinerator and that secondary chambers serve no other purpose than to cool the gases and to remove some of the gas-borne solids (21,131).

Classification of Incinerators. *Domestic incinerators:* Domestic incinerators are intended for the disposal of refuse and garbage (see Table 13-1) from single-family residences. Various types are available for either outdoor or indoor installation and with or without auxiliary firing. The use of auxiliary firing will depend largely on the kind of waste that is to be handled, but owing to the heterogeneity of household wastes, auxiliary firing is preferable in any case. From the standpoint of safety, units for indoor use must be designed to avoid the following hazards: excessive radiation of heat to the surroundings, outward leakage of flue gases, and excessive draft fluctuations. In addition, the auxiliary firing system should be provided with approved safety features.

Table 13-1. Classification of Waste Materials[a]

Type 1 *Waste: Rubbish.* Consists of combustible waste, such as paper, cartons, rags, wood scraps, sawdust, foliage, and floor sweepings from domestic, commercial, and industrial activities. It may contain up to 10 per cent noncombustible solids and up to 25 per cent moisture and have a heating value between about 6500 and 8500 Btu/lb as fired.[b]

Type 2 *Waste: Refuse.* Consists of approximately equal weights of rubbish and garbage and is common to domestic occupancy. Its heating value will be the weighted average of the heating value of the rubbish and garbage.[b]

Type 3 *Waste: Garbage.* Consists of animal and vegetable wastes from restaurants, cafeterias, hotels, hospitals, markets, and similar installations. It may contain up to 85 per cent moisture and up to 5 per cent noncombustible solids and have a heating value as low as 1000 Btu/lb.[b]

Type 4 *Waste: Human and Animal Remains.* Consists of carcasses, organs, and solid organic wastes from hospitals, laboratories, abattoirs, animal pounds, and similar sources.

Type 5 *Waste: Industrial By-products.* Consists of gaseous, liquid, and semiliquid wastes from industrial operations and includes noxious and/or toxic materials. The heating value will depend upon the materials to be disposed of.

Type 6 *Waste: Industrial By-products.* Consists of solid wastes from industrial operations and includes noxious and/or toxic materials. The heating value will depend upon the materials to be disposed of.

[a] This is a tentative classification based upon one under consideration for adoption as a standard by the Air Pollution Control Association.

[b] Refer to Table 13-2 for heating values of various materials that comprise these wastes.

Research is progressing to establish the minimum design requirements for smokeless and odorless operation of domestic incinerators (5), but no experimental data have yet been published in this connection.

Flue-fed incinerators: These are for the purpose of burning rubbish and refuse (see Table 13-1) fed directly into the combustion chamber from one or more floors above the incinerator. They have been widely used in schools, hospitals, hotels, apartment houses, and similar multistory buildings, but, as will be noted later, the trend at present is away from flue-fed incinerators for these applications.

Three basic types are shown schematically in Fig. 13-4. The essential features of each are a combustion chamber, which is usually provided with a horizontal fixed- or dump-type grate and doors above and below the grate level, each with adjustable louvers for admitting air. A gas- or oil-fired burner may be provided above or beneath the grate level to supply auxiliary heat. Overfiring is preferable if adequate heat release is provided to maintain a high temperature in the combustion chamber.

In Fig. 13-4*A* the waste is emptied through a service opening on each floor. The flue thus serves a dual purpose: it is a charging chute and a stack for the products of combustion. Not infrequently, waste will block the flue and impair the draft. Moreover, if this material should ignite, a serious safety hazard would result. For these reasons there is a trend at present to discourage installation of single-flue-fed incinerators.

The National Board of Fire Underwriters has issued standards (95) for this type of incinerator in connection with (1) the provisions for auxiliary firing, (2) the construction of the combustion chamber, (3) the construction of the flue, (4) the design and construction of the service opening, (5) the mounting of the incinerator, and (6) the specifications for the space in which the incinerator is placed. The Incinerator Committee of the Air Pollution Control Association (Mellon Institute, Pittsburgh, Pa.) is working to establish standards for good practice for flue-fed incinerators, as well as all other types.

Some disadvantages of the single-flue-fed incinerator may be mitigated through the use of a double flue, as shown in Fig. 13-4*B*. The recommended procedure for operating such units is to fill the incinerator to the top of the combustion chamber and then to ignite the charge through the firing door. The products of combustion discharge through the smaller flue until the charge burns below the level of the baffle wall; thereafter the gases pass upward through both flues, some regulation of draft being obtained with the damper at the top of the charging flue. If the charging flue should become clogged, the gases are diverted through the openings provided in the division wall.

The design in Fig. 13-4*C* shows certain modifications of the combustion chamber that may be employed in a flue-fed incinerator. For example, some of the waste may be fed manually through a charging door on the front of the unit, and a step grate (as shown) or a sloping hearth may be added to facilitate drying of fresh charges. To supply overfire (secondary) air, ports may be installed above the grate level in the front or side walls or both, unless the secondary air is supplied under forced draft.

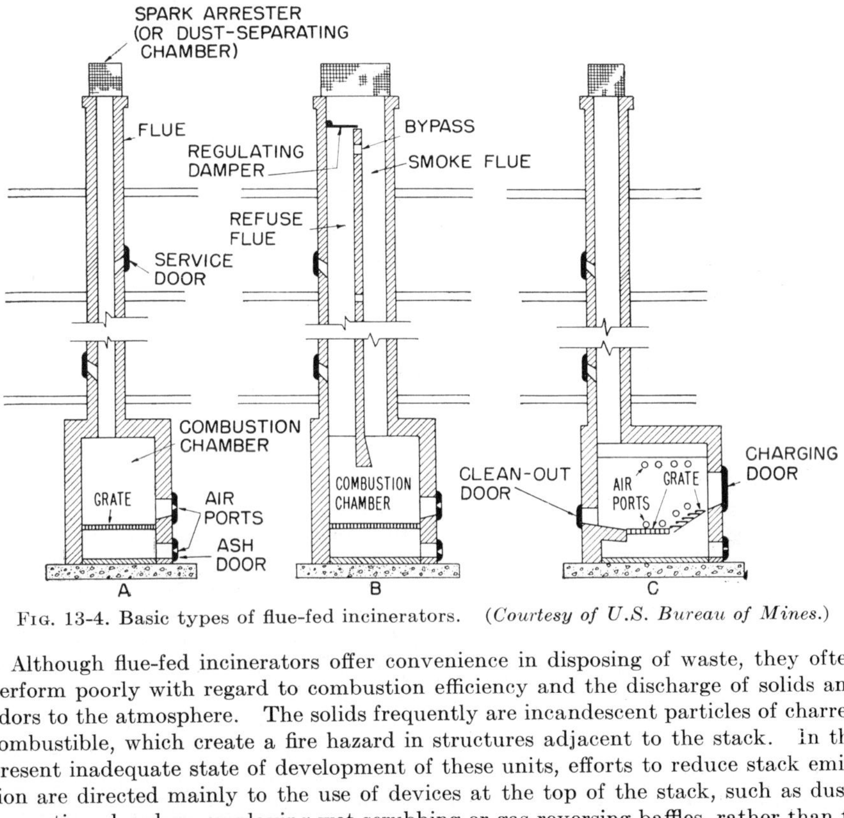

FIG. 13-4. Basic types of flue-fed incinerators. (*Courtesy of U.S. Bureau of Mines.*)

Although flue-fed incinerators offer convenience in disposing of waste, they often perform poorly with regard to combustion efficiency and the discharge of solids and odors to the atmosphere. The solids frequently are incandescent particles of charred combustible, which create a fire hazard in structures adjacent to the stack. In the present inadequate state of development of these units, efforts to reduce stack emission are directed mainly to the use of devices at the top of the stack, such as dust-separating chambers employing wet scrubbing or gas-reversing baffles, rather than to improvements in the combustion chamber. There is a fertile field for engineering research and development in the latter connection.

Commercial- and industrial-type incinerators: This is a broad classification comprising the largest volume of incinerators sold in this country for disposing of wastes from commercial, mercantile, institutional, and industrial activities. Such units should be used instead of flue-fed incinerators* in schools, hospitals, multiple dwellings, etc., when the waste can be collected at one point for disposal. Commercial-

* Flue-fed incinerators may be considered as a special form of this classification. They were treated separately in preceding article for convenience.

type units are designed to dispose of rubbish, garbage, refuse, and wood and textile wastes, as well as other combustible materials from the activities mentioned above. Industrial-type units are usually larger than commercial units and must often be specially designed to dispose of a gas, such as hydrogen sulfide or sewage gas; a semisolid material, such as sewage sludge; or a solid, such as insulation on scrap copper.

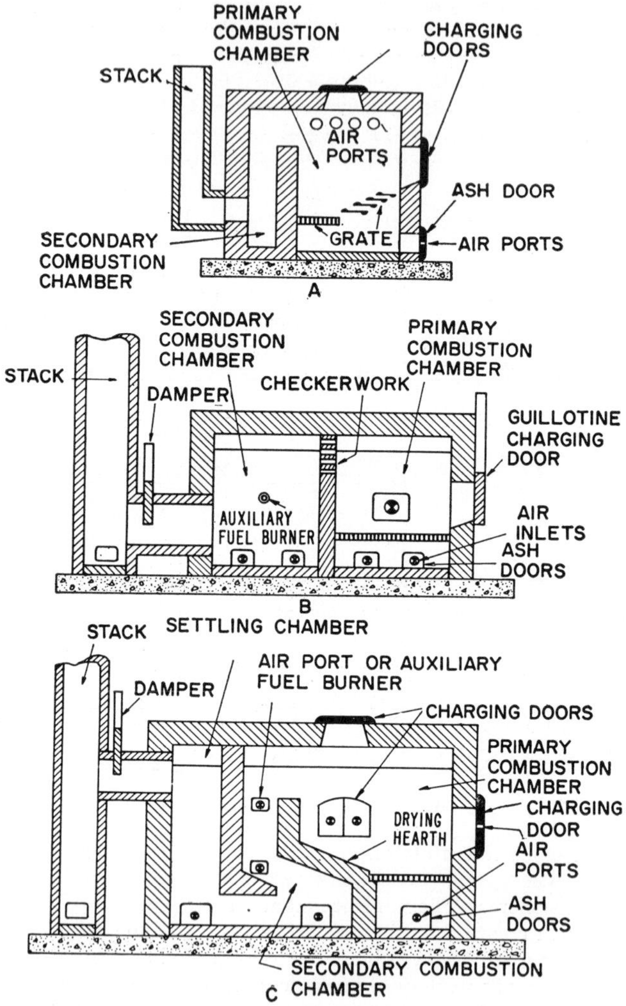

FIG. 13-5. Basic types of small and intermediate commercial and industrial incinerators. (*Courtesy of U.S. Bureau of Mines.*)

Figure 13-5 shows schematically three basic designs of incinerators used for disposing of the more common kinds of commercial or industrial wastes. There are many variations of these types, but the principal features are similar: direct feeding of the material through charging doors at the top, front, or side of the primary combustion chamber, which is provided with some form of grate and may have various ports for introducing primary and secondary air under natural or forced draft; a bridge wall or other means for reversing the direction of the products of combustion to effect mixing of unburned combustibles with secondary air; and one or more chambers thereafter, separated by gas-reversing baffles or partly checkered walls, to effect removal of some particulate matter from the products of combustion.

Depending upon the nature of the charge and the preferences of various manufacturers, the grate may be horizontal and fixed or movable, and it may be supplemented by a step grate or a sloped drying hearth; there may be a mechanical stoking device fed from a hopper to achieve more or less continuous feeding; there may be auxiliary fuel burners in the primary and secondary chambers to obtain complete combustion; and there may be special nozzles in the primary and secondary chambers, or in the bridge wall, to supply secondary air at high velocities for turbulent mixing of the air and the products of combustion.

The design in Fig. 13-5*A* utilizes a bridge wall over which the products of combustion must pass. This accomplishes some mixing of combustibles with excess air, and combustion is supposed to be completed in the downpass chamber, which may be provided with additional secondary air ports and a burner for auxiliary firing.

The design shown in Fig. 13-5*B* also reverses the flow of gases from the primary chamber. The secondary combustion chamber usually is considerably larger than in Fig. 13-5*A* to provide for expansion of the gases and permit some particulate matter to settle out. There are many modifications of the secondary chamber, directed toward preventing reentrainment by the gases of solids that have accumulated on the floor of the chamber and toward securing complete combustion before the gases enter the stack.

The design shown in Fig. 13-5*C* is similar to the others, except for a drying hearth in the primary chamber, over which the products of combustion must pass, and a second baffle to change the direction of the gases again before they enter the exit duct. The practice of permitting the hot gases from the burning charge to pass over and predry fresh wet charges is frequently employed, but careful control of secondary air and gas temperatures is essential to ensure complete combustion of the complex materials that distill from wet charges under such conditions.

Municipal incinerators: Municipal incineration is a broad and complex subject from the standpoint of economics, selection of equipment, handling materials, and air pollution control, and detailed discussion is beyond the scope of this article. It is important to note, however, that there is a growing trend toward the use of incineration, rather than land fill and dumping at sea, for municipal disposal. A survey of collection practices made in 1948 showed that 28 per cent of the group of cities surveyed employed incineration and that the number of cities planning new incinerators almost equaled those using them (35). The most recent informative publications in connection with municipal incineration are those issued by the University of California (131). Discussions of a general nature on this important subject are presented in References 88 and 16.

Incinerators for municipal use are required to handle large quantities of waste material, the composition of which varies locally and nationally with such factors as climate, geography, and the separation requirements of local ordinances and collection agencies. Accordingly, a careful preliminary survey is required to determine the quantity and the quality of material to be destroyed. For preliminary estimates of the quantity to be handled, the accompanying published data are useful (4):

Material	Quantity, lb/capita/year	
	Range	Median
Refuse	587–1,575	794
Rubbish	173– 537	327

Designs for municipal incinerators vary widely. Rectangular batch-fed hand-stoked units similar in design to those shown in Fig. 13-5 are often used, especially for small municipalities. Automatic stoking and step grates or sloped hearths are employed when the charges tend to be rich in wet materials, such as garbage, green foliage, etc. For large capacities, so-called mutually assisting cell-type incinerators, with a burning capacity of 50 to 70 lb/hr-sq ft of grate area, were used frequently in early installations. The present trend is toward designs that provide completely automatic stoking of the charges. Apart from the fact that operating costs are less, this is advantageous when the incinerator is integrated with a boiler to use the hot products of combustion to generate steam for useful purposes, since there is less fluctuation in gas temperatures and gas-flow rates than when hand stoking is used.

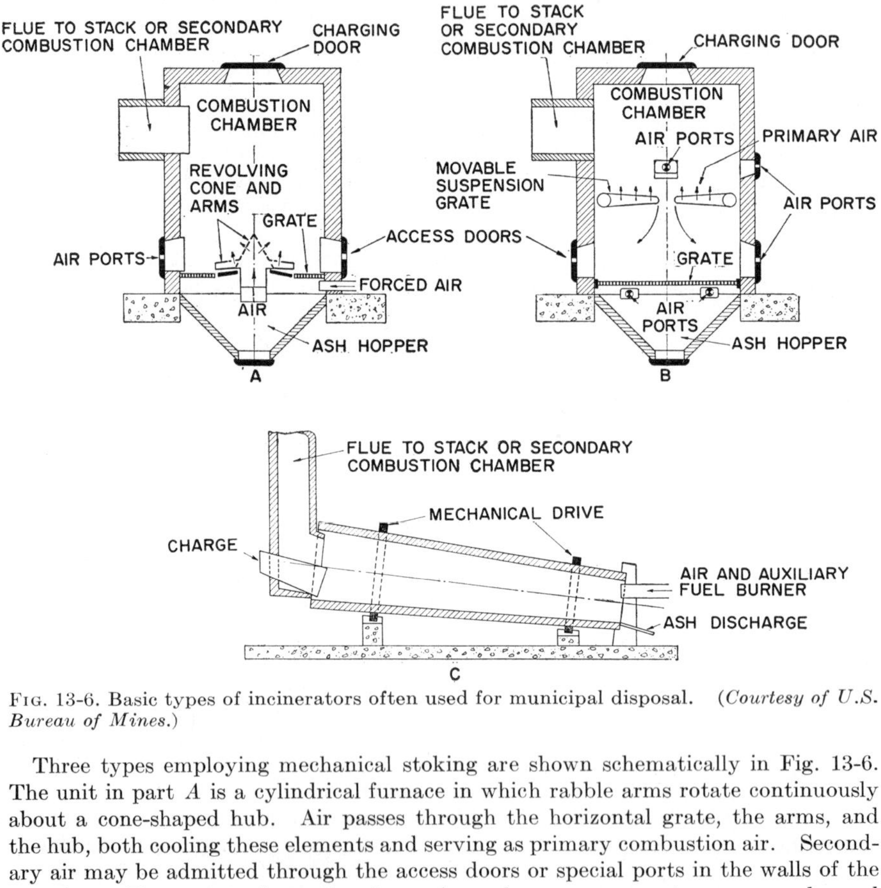

FIG. 13-6. Basic types of incinerators often used for municipal disposal. (*Courtesy of U.S. Bureau of Mines.*)

Three types employing mechanical stoking are shown schematically in Fig. 13-6. The unit in part *A* is a cylindrical furnace in which rabble arms rotate continuously about a cone-shaped hub. Air passes through the horizontal grate, the arms, and the hub, both cooling these elements and serving as primary combustion air. Secondary air may be admitted through the access doors or special ports in the walls of the chamber. The arms agitate the charge in such a manner as to expose unburned material to radiation from the flame and the refractory walls and to move the mass toward the grate. Some possible disadvantages of this system are high maintenance cost of the rotating members and the necessity for some hand stoking. Under normal operating conditions, a unit of this type was reported (131) to burn 102 lb/hr-sq ft, based upon a charge that consisted chiefly of green foliage, wood, and cardboard. The gases from the primary chamber enter the secondary combustion chamber and a settling chamber, provided with water sprays, before discharging to the stack.

The incinerator shown schematically in Fig. 13-6*B*, often referred to as the suspension type or the two-level type, consists of a rectangular primary chamber with two grates, one above the other. The upper grate consists of two longitudinal shafts with fingers attached thereto and extending horizontally toward the center of the chamber. The charge is first dumped on this grate, and heat transfer from the burning charge on the lower grate predries it. Rotation of the shaft moves the fingers downward and drops the dried charge to the lower grate, where it is burned. Primary air is supplied through the fingers, ports in the chamber, and lower grate. Tests under normal operating conditions, with a charge consisting of garbage, paper, wood products, and green foliage, indicated a burning rate of 71.1 lb/hr-sq ft, based upon a grate area of 58.1 sq ft (131). The gases from the primary chamber flowed to a secondary combustion chamber and a large spray chamber before discharging from the stack.

The incinerator shown in Fig. 13-6*C* is a refractory-lined rotary kiln of conventional design. As the shell rotates, the surface of the charge is continuously exposed to air and moves downward toward the outlet end. Owing to counterflow of charge and gases, considerable care is required to achieve complete combustion of the volatiles released from the charge at the upper end. A secondary combustion chamber and a dust-removal system are usually necessary for satisfactory operation. High first costs and maintenance costs must be considered in this type of installation.

Other means of incinerating municipal waste include a traveling or chain grate built into a special combustion chamber or a conventional boiler furnace.

Special-purpose incinerators: Although many systems have been employed for disposing of wood waste from lumbering and finishing mills (6,29,30,137), the silo-type incinerator is largely used when the heat of combustion is not recovered and low first cost is desirable. This type of unit, however, may prove difficult to operate without excessive discharge of smoke and fly ash if it is not designed properly from the standpoint of admitting the air and the raw charge. Recent experimental work with sawdust (21) has shown that high combustion efficiency and negligible discharge of pollutants may be obtained by the use of tangential overfire air, alone, in a cylindrical combustion chamber.

The main design requirements for good combustion characteristics in commercially available silo-type incinerators are shown in Fig. 13-7. Primary air is admitted under forced draft through a small grate, secondary air is admitted under natural or forced draft through tangential ports in the chamber wall above the grate level, and a checkerwork arch is provided as a source of radiant-heat transfer to the charge. The advantages of vortex motion of the gases were discussed previously in connection with the relation of fuel-bed processes to incinerator performance.

The removal of oils, greases, paints, bitumens, latex, and other combustible materials from used cans and drums, so that the containers may be conditioned for resale, occasionally creates a serious air pollution problem. Placing oil in such drums and igniting it in the open, as is frequently done, is not only a community nuisance but expensive. A properly designed tunnel-type furnace can remove combustibles from about 100 large drums an hour at a cost of only a few cents per drum. The basic elements of a suitable furnace for this purpose are shown in Fig. 13-8. The drums, lids removed, are placed upside down on the traveling grate, which may be a conventional traveling grate or chain grate or a suitable substitute, preferably one with a large percentage of open area. Flame from oil or gas burners (oil appears to give better results), located opposite each other in the side walls, impinges on the drums and melts and ignites the combustible residue. Much of the material will drip into the sump, where it is quenched in water, and the remainder will burn off. Air jets beneath the grate and a short distance past the burner zone facilitate combustion of gases that accumulate in the down-ended drums. A burner in the stack is intended

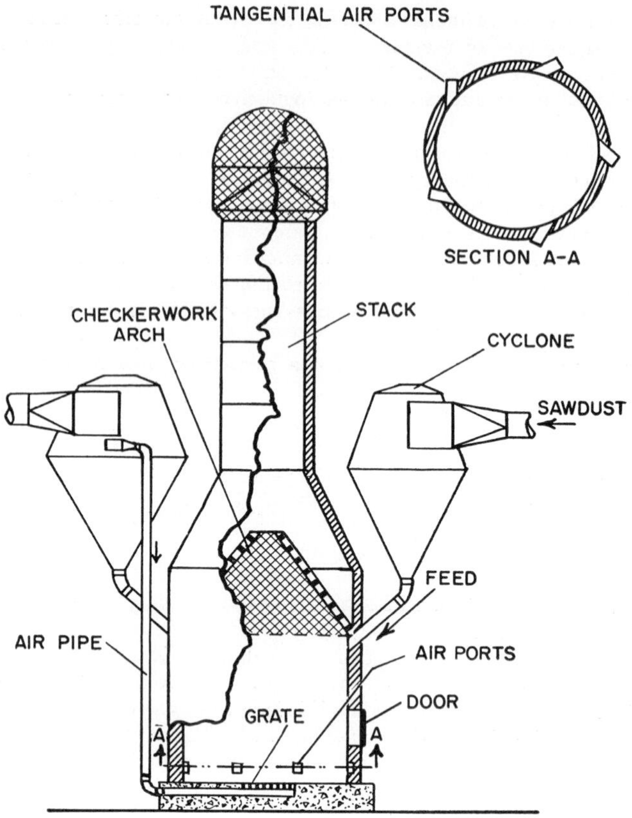

FIG. 13-7. Silo-type incinerator for wood waste. (*Courtesy of U.S. Bureau of Mines.*)

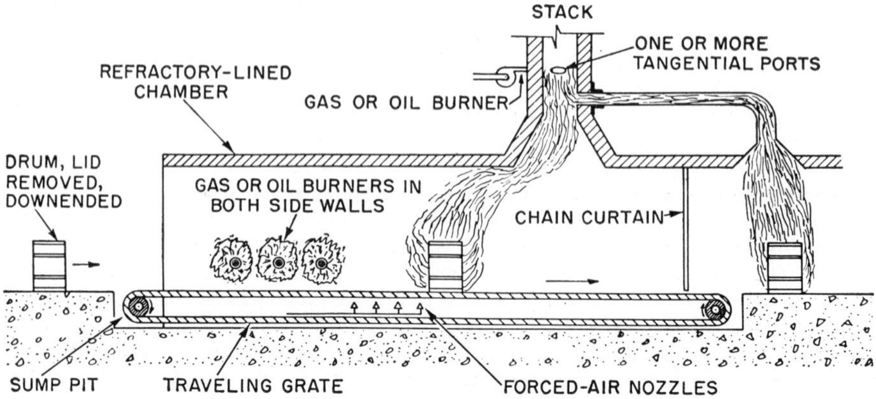

FIG. 13-8. Incinerator for removing organic residues from drums, etc. (*Courtesy of U.S. Bureau of Mines.*)

to complete combustion of organic vapors, etc. The optimum values for the size, location, and adjustment of the fuel burners and the grate speed depend upon the individual requirements. Overheating of drum metal must be avoided, and maximum consumption of residue is desirable to facilitate the usual secondary cleaning operations.

Equipment of this type can be adapted to removing paint, fabrics, rubber, etc., from junked vehicles, preliminary to reclaiming scrap metal.

The burning of insulation and varnish from electrical scrap, preliminary to reclaiming the copper, will create a serious air pollution problem if it is not done under suitably controlled conditions. The incinerator shown schematically in Fig. 13-9 employs the basic elements required. The scrap is placed on a metal cart and wheeled into the first chamber. The second chamber may be used to burn miscellaneous wastes. The last chamber provides expansion for the gases and additional heat and air to complete combustion. Auxiliary fuel burners and air ports must be provided in each

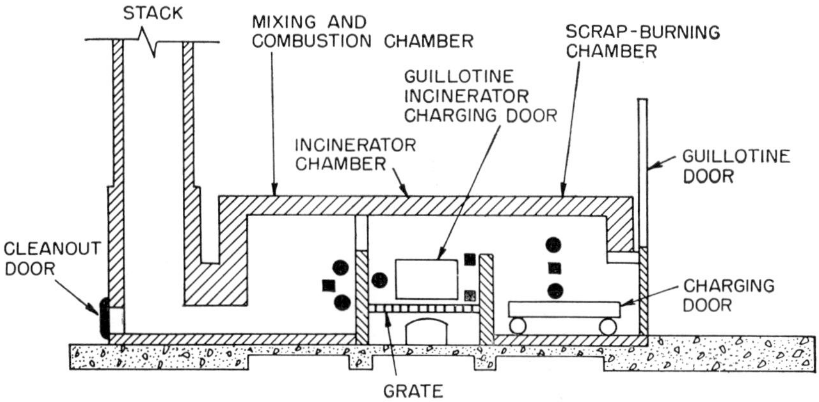

FIG. 13-9. Incinerator for removing organic coatings from scrap metal. (*Courtesy of U.S. Bureau of Mines.*)

chamber; the optimum size, location, and adjustment of these elements is determined largely by trial.

Comments concerning classification of incinerators: The Air Pollution Control Association, Pittsburgh, Pa., through its Incinerator Committee, is presently endeavoring to codify types of wastes, incinerator nomenclature, and the minimum requirements for the design and operation of all the major types of incinerators, which includes units for disposing of wastes that presently are designated as types 4, 5, and 6 (see Table 13-1). It is generally recognized that codes will be subject to constant revision until systematic basic and engineering studies of incineration resolve the many difficulties facing air pollution control bodies in controlling nuisances created by poorly designed or operated incinerators. There is an urgent need for efficient control of odor and particulate discharge from small incinerators at low first cost.

COMBUSTION CALCULATIONS. *Heating value of fuels:* The heating value of a solid fuel is determined by combustion with excess oxygen in a constant-volume bomb calorimeter (ASTM, D271-48). The value reported is the heat of combustion at constant volume and 20°C when the fuel is burned to carbon dioxide (gas), sulfur dioxide (gas), water (liquid), and ash (solid). This value is known as the gross calorific value or the high-heat value (HHV). The net calorific value, or low-heat value (LHV), is calculated by deducting 1030 Btu* from the HHV for each pound of

* These relations refer to a constant-volume process, but they may be used with negligible error for a constant-pressure process, such as a furnace.

water derived from a unit quantity of fuel. The relation is

$$\text{LHV} = \text{HHV} - (9 \times \text{H} \times 1030)$$

where H is the weight fraction of total hydrogen in the fuel. The LHV is usually employed where the temperature of the gases leaving a system is above the dew point, that is, under conditions where no liquid water forms.

A fairly reliable heating value may be calculated from the ultimate analysis of the fuel from the modified Dulong formula:

$$\text{Btu/lb fuel (HHV)} = 145.4\text{C} + 620\left(\text{H} - \frac{\text{O}}{8}\right) + 41\text{S} \tag{13-1}$$

where C, H, O, and S are, respectively, the weight percentages in the fuel of carbon, hydrogen, oxygen, and sulfur.

Weight relationships in combustion: The simple numerical relationships between the reactants and the products of combustion reactions are best illustrated by Eq. (13-2):

$$\text{C} + \text{O}_2 \rightarrow \text{CO}_2 \qquad \Delta H = -173{,}720 \text{ Btu} \tag{13-2}$$

which states that 1 molecule of carbon plus 1 molecule of oxygen forms 1 molecule of carbon dioxide. It also states that 1 lb-mol (lb-mol is the mass in pounds numerically equal to the molecular weight) of carbon (12.01 lb) plus 1 lb-mol of oxygen (32 lb) forms 1 lb-mol of carbon dioxide (44.01 lb). The system releases 173,720 Btu/lb-mol of carbon consumed, as denoted by the enthalpy change ΔH. This energy change often is denoted by $Q_c = 173{,}720$, where Q_c is the heat of combustion.

Most combustion reactions depend upon air for the source of oxygen. For engineering purposes, dry air may be considered to consist of 21 per cent by volume of oxygen of molecular weight 32.00 and 79 per cent by volume of atmospheric nitrogen of molecular weight 28.16. On a weight basis, dry air contains 23.2 per cent oxygen and 76.8 per cent nitrogen. The molecular weight of dry air is 28.97. The volume ratio of nitrogen to oxygen in air is $^{79}/_{21} = 3.762$, which means that each mol of oxygen is accompanied by 3.762 mols of nitrogen or 4.762 mols of air. Equation (13-2) may now be written as follows:

$$\text{C} + \text{O}_2 + 3.762\text{N}_2 \rightarrow \text{CO}_2 + 3.762\text{N}_2 \tag{13-3}$$

The weight of atmospheric nitrogen participating in the reaction remains unchanged.

Volume relationships in combustion: The volumes occupied by 1 lb-mol of every ideal gas are identical for any stated conditions of temperature and pressure. For standard conditions of a gas, defined as 32°F and 1 atm pressure (29.92 in. Hg), the molar volume of a gas is 359 cu ft. At any other temperature or pressure the molar volume is given by Eq. (13-4), which is based upon the ideal gas laws:

$$V_m = 359 \frac{(t + 460)}{492} \frac{29.92}{p} \tag{13-4}$$

where V_m is the molar volume, cu ft; t is temperature, °F; p is absolute pressure, in. Hg. Most combustion gases deviate slightly from the laws for ideal gases, but for engineering calculations no serious errors are caused by assuming them to be ideal.

Referring to Eq. (13-3), the mol quantities of the gases may be expressed as volumes as follows:

$$\text{C} + 359 \text{ cu ft O}_2 + 1350.6 \text{ cu ft N}_2 = 359 \text{ cu ft CO}_2 + 1350.6 \text{ cu ft N}_2 \tag{13-5}$$

that is, each mol of carbon (12.01 lb) combines with 1,710 cu ft of air.

The density of a gas mixture is calculated from the molecular weight of the mixture, which is calculated from the composition by volume of the mixture. Assume, for example, a flue gas with 15 per cent CO_2, 6 per cent O_2, and 79 per cent N_2, and calculate the molecular weight as shown in the accompanying table.

	(a) %/vol	(b) mol weight	(a) × (b) ÷ 100
CO_2	15	44.01	6.60
O_2	6	32.00	1.92
N_2	79	28.16	22.25
Molecular weight of mixture...			30.77

Assume also that the mixture is at 500°F and 1 atm pressure. The density is $30.77/V_m = 30.77/700 = 0.0439$ lb/cu ft at the stated conditions.

The equations given in this article for calculating various quantities are sufficiently accurate for most engineering work.

Air required for combustion: There are two quantities of primary interest in engineering calculations of combustion processes: the theoretical air (which is the exact quantity of air required to burn the fuel completely) and the excess air (which is the quantity of air present beyond that required to burn the fuel completely). A mixture of theoretical air and a given fuel is called a stoichiometric mixture.

Theoretical air required may be calculated either from the ultimate analysis of the fuel,* which is the most accurate method, or from the heating value of the fuel.

The volume of theoretical air based on ultimate analysis of fuel is

$$A_t, \text{ SCF/lb fuel} = 17.10\left(\frac{C}{12} + \frac{H}{4} - \frac{O}{32} + \frac{S}{32}\right) \tag{13-6}$$

where A_t is theoretical air, SCF is standard cubic feet, and C, H, O, and S are the percentages by weight in the fuel of carbon, hydrogen, oxygen, and sulfur. The weight of theoretical air is

$$A_t, \text{ lb/lb fuel} = 1.38\left(\frac{C}{12} + \frac{H}{4} - \frac{O}{32} + \frac{S}{32}\right) \tag{13-7}$$

For engineering estimates, acceptable values for theoretical air based upon the heating value of the fuel may be obtained from the HHV of the fuel:

$$A_t, \text{ lb/lb fuel} = 0.00075 \text{ (HHV)} \tag{13-8}$$

More than the theoretical amount of air is necessary in practice to achieve complete combustion. The excess air is expressed either as a percentage of the theoretical air or as the total air divided by the theoretical air. The former is most frequently used:

$$A_{xs} = \frac{A - A_t}{A_t} 100 = \left(\frac{A}{A_t} - 1\right) 100 \tag{13-9}$$

where A_{xs} is percentage of excess air; A_t is theoretical air, lb/lb fuel; and A is the air actually used, lb/lb.

If it is desired to know the percentage of excess air under operating conditions of a particular combustion process, it may be calculated by means of

$$A_{xs} = \frac{O_2}{0.266N_2 - O_2} 100 \tag{13-10}$$

where O_2 and N_2 are the percentages by volume of these gases in the dry flue gas. This equation is applicable only when the nitrogen in the fuel is negligible and there is

* In all subsequent equations employing the ultimate analysis of the fuel, the percentages by weight refer to the as-fired fuel and H is the total hydrogen.

no carbon monoxide or hydrogen in the flue gases. If these combustible gases are present, the percentage excess air is

$$A_{xs} = \frac{O_2 - 0.5(CO + H_2)}{0.266N_2 - O_2 + 0.5(CO + H_2)} 100 \tag{13-11}$$

where CO and H_2 are the percentages by volume of these gases in the dry flue gas.

The percentage of excess air may also be calculated from the volume percentage of CO_2 in the dry flue gas:

$$A_{xs} = 7{,}900 \frac{(CO_2)_t - CO_2}{CO_2[100 - (CO_2)_t]} \tag{13-12}$$

where $(CO_2)_t$ is the maximum volume percentage of CO_2 obtainable in the dry flue gas for a given fuel and CO_2 is the actual percentage found in the dry flue gas, as determined by the Orsat or other volumetric methods.

If a combustion process is designed for a given percentage of excess air, or this value is known from test data, the actual quantity of air may be obtained by rearrangement of Eq. (13-9).

$$A = \left(\frac{A_{xs}}{100} + 1\right) A_t \tag{13-13}$$

where A and A_t are the actual and theoretical quantities, respectively. Equation (13-6) or (13-7) may be substituted for A_t in Eq. (13-13), depending upon whether the volume or the weight of air is required.

Products of combustion: The following quantities are needed for making heat and material balances over a combustion process and for calculating the quantities and velocities of the gases that must be handled.

1. Quantity of residue formed: If the total residue cannot be weighed, it may be determined from the ash content of representative samples of both the fuel and the residue. The weight of residue is

$$R\text{, lb/lb fuel burned} = \text{ash, lb/lb fuel} \div \text{ash, lb/lb residue} \tag{13-14}$$

2. Quantity of combustible lost in residue: the amount of combustible, which is usually considered to be carbon, lost to the ashpit is

$$C_1\text{, lb/lb fuel} = R \times (\text{percentage of carbon in residue} \div 100) \tag{13-15}$$

If there is considerable residue lost as fly ash, subtract the ash, lb/lb fly ash, from the ash, lb/lb fuel in Eq. (13-14) before calculating C_1.

3. Quantity of dry flue gas: The weight of dry flue gas may be calculated from

$$G_d\text{, lb dry gas/lb fuel} = \frac{(C_f - C_r)}{100} \left[\frac{4(CO_2) + O_2 + 700}{3(CO_2) + CO}\right] \tag{13-16}$$

where C_f and C_r are the percentages by weight of carbon in the fuel and residue, respectively, and (CO_2), (O_2), etc., are the percentages by volume of these gases in the flue gas, as determined by an Orsat analysis or otherwise. This equation assumes hydrogen to be absent or negligible in the dry gas. If the volume of dry gas is required, divide the quantity above by the density of the dry gas at the stated temperature and pressure, which may be calculated by the procedure given previously for gas mixtures.

The volume of dry gas can also be calculated from the ultimate analysis of the fuel and the percentage of excess air, assuming complete combustion and negligible carbon losses in the residue and fly ash, from the following:

$$G_d\text{, SCF dry gas/lb fuel} = 3.59 \left[4.76 \left(\frac{A_{xs}}{100} + 1\right)\left(\frac{C}{12} + \frac{H}{4} - \frac{O}{32}\right) - \frac{H}{4} + \frac{O}{32}\right] \tag{13-17}$$

The quantities C, H, and O are the percentages by weight of these elements in the fuel as fired. The sulfur and nitrogen in the fuel are neglected, since these elements contribute to a negligible extent to the volume of products of combustion of most solid fuels.

Multiplying Eq. (13-17) by the density of the resulting gas at standard conditions will give the weight of gas, lb/lb fuel.

It is evident that the theoretical dry gas, that is, the volume of gas with theoretical air, will be obtained from Eq. (13-17) when A_{xs} is zero.

4. Quantity of wet flue gas: This may be calculated by adding the amount of water equivalent to the total hydrogen in the as-fired fuel to the dry gas calculated by either Eq. (13-16) or (13-17). The relation is

$$G_w = \text{lb wet gas/lb fuel} = G_d, \text{ lb dry gas/lb fuel} + 0.09\text{H} \qquad (13\text{-}18)$$

where H is the total hydrogen, per cent by weight in as-fired fuel. As described previously, dividing this quantity by the density of the gas, lb/cu ft, at any given temperature and pressure will give the volume of gas in cubic feet.

Burning rate: The average rate of fuel consumption is most accurately determined from the weight of fuel charged over a given period of time. Where it is inconvenient to weigh the charges, the burning rate can be calculated from the air rate, the fuel analysis, and the gas analysis. Under fluctuating burning conditions, the air rate and the analysis of the gases should be determined at frequent, regular intervals from the time that ignition of the charge occurs until the last charge burns out, to obtain average values for these quantities. Assuming no combustibles in the flue gases, the following relation, which is based upon carbon and hydrogen balances, will give the weight of fuel burned per unit weight of air supplied:

$$W_f, \text{ lb fuel burned/lb air supplied} = 0.1392 - \frac{13.92 - 0.4142(\text{CO}_2)}{1.264(\text{N}_2)} \qquad (13\text{-}19)$$

The product of W_f and the air rate, lb/hr, gives the pounds of carbon and hydrogen burned per hour.

Combustion temperature: Unless combustion is instantaneous or nearly so, the term combustion temperature or flame temperature is indefinite. In fuel beds and long flames, steep temperature gradients exist and one must decide what is desired, the maximum temperature or the mean temperature.

Often, however, it is desirable to calculate the adiabatic temperature, which is the temperature attained when heat is neither transferred to or from the surroundings; that is, the heat of combustion of the fuel at a standard temperature is used entirely to raise the temperature of the products of combustion. If the fuel and/or the combustion air are preheated to a temperature above the base temperature, the adiabatic flame temperature will be higher than otherwise.

The adiabatic flame temperature is based upon an energy balance, in which the LHV of the fuel and the sensible heat above the base temperature of both the fuel and air are equated to the enthalpy change of the products of combustion. Assuming that the products of combustion are CO_2, H_2O, O_2, and N_2 and that SO_2 is negligible, the energy balance is

$$\begin{aligned}\text{LHV} + FC_1(t_f - t_b) + AC_2(t_a - t_b) \\ = [(\text{CO}_2)C_3 + (\text{H}_2\text{O})C_4 + (\text{O}_2)C_5 + (\text{N}_2)C_6](t_g - t_b) \qquad (13\text{-}20)\end{aligned}$$

where F is 1 lb of as-fired fuel; A is lb air/lb fuel; (CO_2), (H_2O), etc., are the lb/lb fuel of these flue-gas constituents; C_1 to C_6 are the mean specific heats, Btu/lb-°F of the constituents to which each refers; t_f, t_a, and t_g are the temperatures of the fuel, air, and gas, respectively; and t_b is a base temperature, usually 60°F. Calculation of t_g by this relation is complicated by the fact that the specific heats are a function of temperature. Moreover, above about 2500°F, CO_2 and H_2O begin to dissociate appreciably, unless the excess air is of the order of 100 per cent or more, and the actual adiabatic flame temperature will be lower as a result of the energy absorbed in dissociation of the gases. However, dissociation is negligible in most types of combustion equipment operated at atmospheric pressure with some excess air, and Smith and Stinson (121) illustrate the use of a convenient multiparameter chart for calculating the adiabatic flame temperature up to about 3000°F for air-preheat temperatures up to 1200°F. The following data are required: the hydrogen-carbon ratio of the fuel, lb/lb; the fuel-air ratio, lb/lb; the LHV of the fuel; and the moisture content, lb/lb, and temperature, °F, of the combustion air.

It must be remembered that calculated adiabatic temperatures represent the maximum possible value and that heat transfer from the flame or fuel bed to furnace walls, etc., results in appreciably lower temperatures. Direct measurement of flame temperatures under actual operating conditions by means of a high-velocity thermocouple is the best means of determining the mean flame temperature, or the temperature at any point within the flame envelope.

Stack calculations: Stacks serve two purposes: to supply the natural draft necessary to draw the required amount of air for combustion and to convey the products of combustion away from the vicinity of the stack. The principles involved in the dispersion of smoke, fumes, etc., from stacks are discussed elsewhere in this handbook, and only the requirements for combustion are given here. The following equation is useful in this connection:

$$i\text{, draft, in. } H_2O = 0.0192H(d_a - d_g) - \left(1 + 0.09\frac{H}{D}\right)\frac{v_2 d_g}{64.4} \qquad (13\text{-}21)$$

where H is the stack height, ft; d_a is the density of the air, lb/cu ft at the stated conditions; d_g is the density of the flue gas at the mean temperature in the stack; D is the internal diameter of the stack, ft; and v_2 is the velocity of the gases corresponding to d_g, fps. The mean temperature of the gas may be estimated by allowing a drop in temperature of 1°F for each foot of stack height.

Useful tables: Tables 13-1 to 13-6 will be useful in connection with incinerator design

Table 13-2. Approximate Combustion Characteristics of Various Kinds of Waste Materials

Substance	High heat value,[a] Btu/lb maf[b] waste	Theoretical air needed for complete combustion, lb air/lb maf waste[c]
Paper	7,900	5.9
Wood	8,400	6.3
Leaves and grass	8,600	6.5
Rags, wool	8,900	6.7
Rags, cotton	7,200	5.4
Garbage	7,300	5.5
Rubber	12,500	9.4
Suet	16,200	12.1

[a] These values are necessarily approximate, since the ultimate composition of the combustible part of the materials will vary somewhat, depending upon their sources. The heating value of the as-received material is obtained by multiplying the maf Btu by 100-(percentage moisture and ash)/100. For example, garbage containing 35 per cent moisture and 5 per cent ash or other noncombustible material will have an as-fired heating value of 4380 Btu/lb.

[b] Maf means moisture-and-ash-free, where ash refers to total noncombustible material.

[c] These values are also approximate and are based upon 0.75 lb air/1000 Btu for complete combustion. For various percentages of excess air, multiply these values by (100 + per cent excess air)/100. For example, if paper is burned with 100 per cent excess air, (5.9) 200/100 = 11.8 lb air/lb of maf paper will be required.

and performance. Owing to the heterogeneity and the physical structure of most solid wastes, bulk densities, heating values, and other pertinent characteristics vary quite widely. The importance of proximate and ultimate analyses of representative samples of wastes to engineering calculation cannot be overemphasized.

Table 13-3. Approximate Bulk Densities of Various Wastes[a]

	Lb/cu ft	Lb/cu yd	Lb/bu
Paper, loose	5	135	7
Rubbish	10	270	13
Refuse	16	440	21
Wood scraps, air-dried	15	400	20
Wood, shavings or sawdust, air-dried	10	270	13
Garbage	35	945	45

[a] These are average values. Bulk densities will vary with the degree of compactness and the moisture content of the charge. For example, the lowest and highest values found in a survey of garbage were about 30 and 57 lb/cu ft, respectively. For refuse, the lowest and the highest values were about 7 and 30 lb/cu ft, respectively. Satisfactory preliminary engineering estimates for incinerator design may be made with the average values given above.

Table 13-4. Approximate Quantities of Refuse from Various Sources

Source	*Daily Quantity of Refuse*
Apartment buildings having 1- and 2½-room apartments	1 to 1½ lb per person
Apartment buildings having over 4 rooms per apartment	2 lb per person
Hotels	2 lb per guest room plus 2 lb per meal
Cafeterias	½ lb per meal
Restaurants	1 lb per meal
Hospitals	7 lb per bed
Schools	8 lb per classroom

Table 13-5. Combustion Characteristics of Carbon, Hydrogen, and Sulfur for Calculations Based upon Ultimate Analysis of Waste Materials

Substance	Molecular or atomic weight	Heat of combustion		Theoretical air required		Products of combustion	
		Btu/lb	Btu/SCF[a]	SCF/SCF	Lb/lb	SCF/lb	Lb/lb
Carbon	12.01	14,087	. . .	. . . .	11.50	150	12.50
Hydrogen	2.016	60,958[b]	325	2.38	34.34	542	35.38
Sulfur	32.06	3980	. . .	. . . .	4.31	56	5.31

[a] SCF = standard cubic foot (60°F, 30 in. Hg). 1 lb mol = 379 cu ft.
[b] High-heat value (HHV). Often referred to also as gross heating value.

Odor Control by High-temperature Oxidation. GENERAL. The control of objectionable odors, which may be gases, mists, or solids discharged to the atmosphere from commercial and industrial processes, is a matter of growing concern. Hydrogen sulfide, carbon disulfide, mercaptans, products from the decomposition of certain proteins (especially those of animal origin), and certain petroleum hydrocarbons are the commonest malodors. However, there are a great many different kinds of odors that are potentially community nuisances, and no over-all appraisal of air pollution is complete without consideration of all sources of objectionable odors.

The following general procedures may be employed alone or in various combinations to eliminate or diminish odors: (1) modification of a process to prevent formation or evolution of odors; (2) dilution of odorous gases with fresh air; (3) masking, or superimposing a more pleasant odor; (4) oxidation of the gases at elevated temperature, with or without a catalyst; (5) liquid scrubbing of the gases in a suitable absorption unit; (6) application of a dry adsorbent with specific adsorptive characteristics (such as activated carbon in a filter bed); and (7) injection of controlled quantities of chlorine or ozone into the process-gas stream. The economics, characteristics, and applications of these procedures have been discussed elsewhere (87,129). The present discussion deals with procedure 4, high-temperature oxidation of objectionable constituents in process gases.

Table 13-6. Combustion Characteristics of Auxiliary Fuels Used for Firing Incinerators

Substance	Density		Heat of combustion, Btu[a]					Theoretical air required for complete combustion,[c] lb		
			HHV[b]		LHV		HHV per gal			
	Lb/SCF	Lb/gal	Per lb	Per SCF	Per lb	Per SCF		Per SCF	Per lb	Per gal
Natural gas[d]	0.046	...	24,550	1130	22,190	1020	......	0.80	17.2	
Carbureted water gas	0.047	...	11,700	550	10,800	508	......	0.35	7.4	
Propane, commercial	0.114	...	22,550	2570	20,800	2370	......	1.80	15.6	
Butane, commercial	0.146	...	22,320	3260	20,400	2980	......	2.30	15.7	
Fuel oil,[e] Mid-continent	.....	7.4	19,380	....	18,230	....	144,000	....	14.4	107
California	.....	8.0	18,835	....	17,755	....	150,000	....	14.1	113

[a] Owing to variations in the composition of these substances, all values have been rounded off in the last significant place.

[b] HHV = high-heat value, LHV = low-heat value.

[c] For various amounts of excess air, multiply these values by (100 + per cent excess air)/100. For example, natural gas burned with 15 per cent excess air would require 1.15 × 0.8 = 0.92 lb air/SCF of gas.

[d] Based on natural gas in Pittsburgh area.

[e] These are fairly typical values.

OXIDATION PROCESSES. Two systems are employed: (1) direct incineration, in which process gases pass through a combustion chamber where they are raised to a temperature of the order of 1500°F or higher in the presence of excess oxygen; and (2) catalytic incineration, in which the process gases, containing oxygen, pass through specially designed units containing catalyst elements, on the surface of which oxidation occurs.

The simplest equipment for direct incineration is a combustion chamber in which the process gases mix turbulently with a flame produced by an oil- or a gas-fired burner. The principal requirements in the design of such apparatus are to supply sufficient heat of combustion to produce a temperature of the mixed gases of at least 1500°F and to maintain some excess air. For reasons chiefly of fuel economy, the excess air should be the minimum for destruction of odors. The optimum residence time of the gases in the chamber is best determined by trial, since various constituents to be burned react at different rates.

Direct incineration generally is not economical if large volumes of a cold, inert gas, such as air, must be heated to the required temperature. For example, with natural gas at 50 cents per 1,000 cu ft, it would cost roughly 1.4 cents for gas alone to heat 1,000 cu ft of dry air from 60 to 1500°F. Of course, process gases that are already at a high temperature and contain appreciable quantities of combustible vapors, such as benzene, require less auxiliary fuel to attain the desired temperature.

Often, it is expedient to discharge malodorous gases and vapors directly to a boiler furnace, which is normally a complement of the plant operation. This practice has been successfully employed in rendering plants to dispose of the gases from driers and cookers.

In catalytic incineration for odor control, the process gases are passed over special elements coated with a catalytic material, and oxidizable constituents in the gases, such as hydrocarbons and other organic-type malodors mentioned previously, react with excess oxygen to form carbon dioxide and water vapor and to liberate heat. By definition, catalysts are substances which accelerate chemical reactions, but which themselves remain chemically unaffected. There are numerous catalysts for different purposes, but those of interest in oxidation reactions are certain metals and metal oxides. Platinum and platinum-rhodium alloy distributed on a suitable support are highly effective oxidation catalysts. The oxides that constitute furnace refractories are also effective catalysts, but the minimum temperature level for sufficiently rapid oxidation is much higher than for metallic catalysts, and therefore they have little advantage over direct incineration if large quantities of gases must be handled.

Catalytic combustion of most organic constituents that occur in process effluent gases is initiated and self-sustaining at about 500°F. If the gases are colder than this before coming into contact with the catalyst, they must be preheated, and the fuel cost would be approximately 0.4 cent for each 1,000 SCF of air heated from 60 to 500°F by means of natural gas, which is less than one-third the cost noted previously for direct incineration.

If the concentration of oxidizable constituents in the process gases is sufficiently high, enough heat will be liberated by the oxidation reactions at the catalyst surface to make it possible to shut off the auxiliary fuel after combustion has been initiated. If the effluent gas were largely air at 60°F, Table 13-7 shows that a heat liberation of 8.2 Btu/SCF would be required to raise the temperature to 500°F and enable the system to become self-sustaining. For comparative purposes, this corresponds approximately to 2,000 ppm of toluene vapor. Heat losses from the system would increase the required heat liberation.

Table 13-7. Sensible Heat of Air above 60°F

Temp, °F	Enthalpy	
	Btu/lb	Btu/SCF
250	45.7	3.52
500	106.8	8.22
750	169.6	13.06
1000	234.3	18.04
1250	301.0	23.18
1500	369.3	28.44

If the process effluent contains appreciable quantities of combustible gases and vapors, the temperature rise may be high enough to make it economical to employ a heat exchanger to recover the heat for other purposes, such as preheating the effluent gases or generating steam. Comparative analyses (89) of actual installations employing catalytic odor control have shown that substantial savings are possible by the use of heat exchangers and by recycling some of the gases leaving the catalyst system.

Commercial catalyst units* which employ a thin film of platinum alloy supported on a base with suitable physical properties have been reported in successful use for conditioning effluent gases from coffee roasters; paint, enamel, and foundry-core baking ovens; chemical plants discharging maleic and phthalic anhydrides; enameling

* Catalytic Combustion Corp., Detroit, Mich.; Oxy-Catalyst, Inc., Wayne, Pa.

plants discharging xylene and similar hydrocarbon solvents; kettles used for cooking animal, vegetable, and mineral oils; phenolic-resin curing ovens; and refineries burning waste cracking gases. Many other applications of this principle either are being investigated or are under consideration, such as the removal of undesirable constituents from automotive exhaust gases.

The important characteristics for commercial catalyst units are low resistance to the flow of gases, high specific surface, and arrangement in a manner to provide turbulent mixing of process gases that might otherwise be stratified.

There are certain limitations to catalyst elements now available: the optimum temperature range is between 500 and 1800°F, and excessive amounts of particulate matter, such as ash or fumes from zinc, lead, and mercury, rapidly impair the activity of the catalyst. Some types of halogenated hydrocarbons also are harmful to the catalyst.

13.3 COLLECTION OF AEROSOLS

13.3.1 Principles of Gas Cleaning

Introduction. In general, aerosols composed of particles larger than about 50μ are unstable unless the turbulence is extreme. It is doubtful whether there is a lower limit to the size of a particle which can exist in an aerosol. Particles smaller than

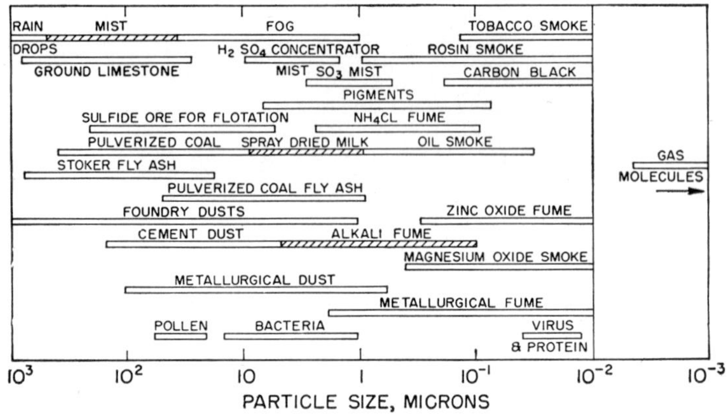

FIG. 13-10. Particle-size ranges for aerosols. (See Ref. 94.)

about 0.01μ have been detected, but their lifetime is probably short because of coagulation with larger particles; relatively little is known about the control and behavior of such particles. Figure 13-10 gives particle-size ranges for aerosols, while Table 13-8 gives mass median diameters and standard geometric deviations for a number of representative dispersions.

Aerosol concentrations vary even more widely than particle sizes. For normal atmospheric air, concentrations of the order of 1 grain/1,000 cu ft are found. In Atomic Energy Commission installations, because of precleaning of the entering air, aerosols are encountered with concentrations as low as 10^{-5} or 10^{-6} grain/cu ft. On the other hand, loadings of some industrial gases may reach several hundred grains per cubic foot, although values of 20 grains/cu ft or less are more common. A generalized chart of aerosol concentrations is shown in Fig. 13-11.

It is evident that, with variations of 10^8 in concentration and 10^3 in particle size, many types of air-cleaning equipment utilizing a variety of mechanisms are required to cope with industrial problems. Before discussing these devices, a brief review of aerosol theory, particularly as applied to gas cleaning, will be helpful.

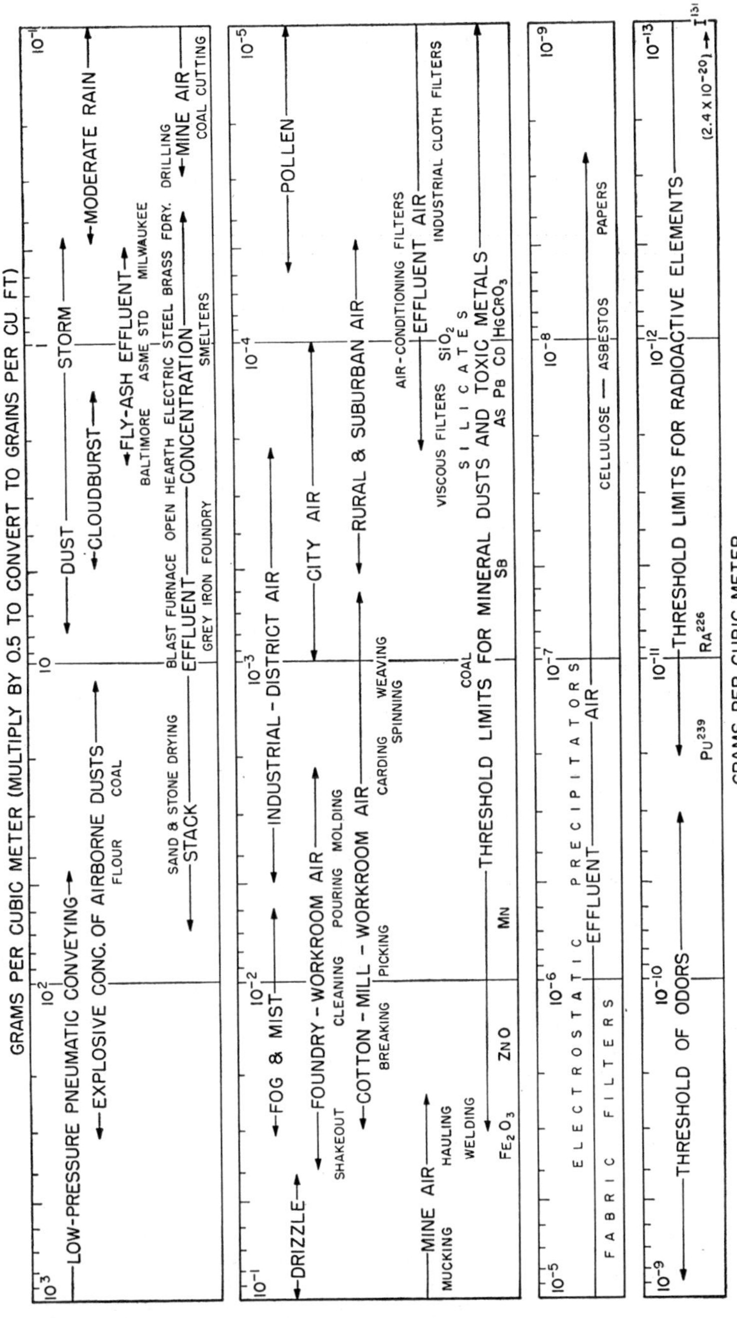

FIG. 13-11. Mass concentrations of various aerosols. (See Ref. 39.)

Table 13-8. Particle Sizes for Some Dispersions of Particles in Air

Aerosol	Diameter mass median, μ	Standard geometric deviation
Water drops from hollow-cone spray nozzle at 100 psi; 0.063 in. orifice diameter	1,220	3.4
Hard, close-grained shale dust containing 29% free silica; produced in a drilling operation in a hard heading in a coal mine	101	3.4
Fly ash	69	2.4
Sea fog	38	1.6
Rye-flour dust in flour mill	16	1.6
Beryllium fluoride (BeF_2 fume from furnace pouring operation; sample taken 10 ft from furnace opening)	2.3	2.2
Atmospheric dust from 14 United States cities (average)	0.97	1.56
Zinc oxide pigment (a fume)	0.59	1.28
Shawinigan acetylene carbon black; longest dimension of individual particles (electron microscope)	0.087	1.55
Red blood cells[a]	7.85	1.08

[a] Red blood cells may be used for calibration in microscopy since they are so homogeneous.
Data from S. K. Friedlander et al., "Handbook on Air Cleaning" (A.E.C.D.-3361; NYO-1572), Harvard University, School of Public Health, Boston, Mass., 1952.

Settling. Particles smaller than about 50μ reach a constant settling velocity within a fraction of a second; the drag on such particles can usually be calculated from Stokes' law, and the terminal settling velocity is given by the equation

$$u_t = \frac{(m_p - m_g)g}{3\pi\mu d_{pe}} \tag{13-22a}$$

$$= \frac{g d_p^2(\rho_p - \rho_g)}{18\mu} \quad \text{for spherical particles} \tag{13-22b}$$

For particles smaller than a few microns, the gas ceases to act as a continuous medium, and the particles fall more rapidly than calculated from Stokes' law, appearing to "slip" between the molecules. As a result, the terminal settling velocity calculated from Stokes' law must be multiplied by the Cunningham correction factor

$$C = 1 + \frac{1.72\lambda}{d_p} \tag{13-23}$$

Stokes' law was derived for spherical particles such as those in mists and fogs. But in dusts and fumes, the particles may be irregularly shaped, they may be crystalline, or they may be composed of aggregates resulting from flocculation. Usually, little error is introduced by using the diameter of an equivalent sphere for particles differing only slightly from spherical, e.g., cubes, octahedrons, or other polyhedral crystals, but a considerable error may result with needles, plates, or chain-shaped flocs. Davies (24) has reviewed the literature on the sedimentation of nonspherical particles. Figure 13-12, taken from his paper, gives values of equivalent Stokes' law diameters for ellipsoidal particles as evaluated from the equations of fluid motion. Theory indicates that an ellipsoid of revolution or any particle with three mutually perpendicular planes of symmetry will not change its initial orientation while falling in the Stokes' law region. Thus, maximum and minimum sedimentation rates can be determined from Fig. 13-12, if the particles are approximately ellipsoidal in shape; unfortunately, the theoretical results have not been checked because of a lack of reliable data.

The theory also indicates that ellipsoids of revolution will fall exactly vertically only when released with the axis of revolution either vertical or horizontal. Otherwise, there is some deviation, particularly for the extreme forms such as a thin flat disk or a long thin needle. For a disk, the maximum deviation from the vertical is about 11° and for a needle shape, about 19°.

For mist and fog particles which coalesce to form spheres, the particle density and the normal density of the original liquid are about the same. For aerosols made up of flocs of fume particles, however, ρ_p may be considerably less than the normal

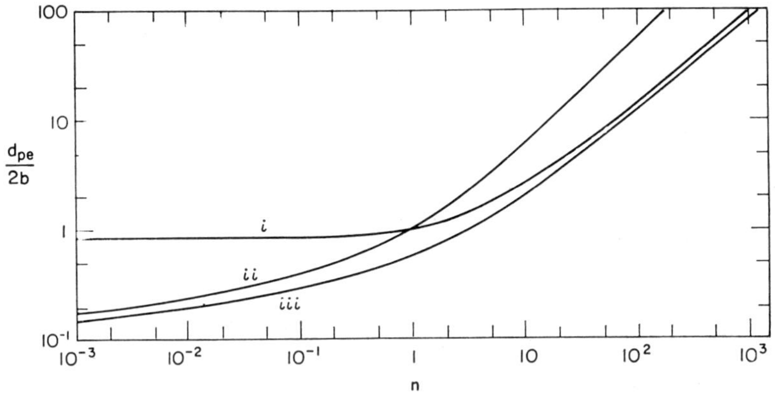

i ELLIPSOID OF REVOLUTION MOVING IN THE DIRECTION OF ITS AXIS OF ROTATION
a = AXIS OF ROTATION = nb, b = c

ii ELLIPSOID OF REVOLUTION MOVING IN A DIRECTION AT RIGHT ANGLES TO ITS AXIS OF ROTATION
b = AXIS OF ROTATION, a = nb = c

iii ELLIPTICAL PLATE MOVING EDGEWAYS
a = nb, c = 0

FIG. 13-12. Sedimentation of ellipsoids in a direction parallel to the axis. (See Ref. 24.)

density. Mean densities for aggregates of several materials are given in Table 13-9. These figures are averages; the densities of individual flocs vary with the packing and with the size of the original particles.

Table 13-9. Particle Densities of Flocs[a]

Material	Floc density, g/cu cm	Normal density, g/cu cm
Silver	0.94	10.5
Mercury	1.70	13.6
Cadmium oxide	0.51	6.5
Magnesium oxide	0.35	3.65
Mercuric chloride	1.27	5.4
Arsenic trioxide	0.91	3.7
Lead monoxide	0.62	9.36
Antimony trioxide	0.63	5.57
Aluminum oxide	0.18	3.70
Stannic oxide	0.25	6.71

[a] Excludes particles of normal density.
Data from R. Whytlaw-Gray and H. S. Patterson, "Smoke," Edward Arnold & Co., London, 1932.

Inertial Deposition. When the path of a moving aerosol changes, the particles, because of their inertia, tend to continue in their original direction and thus may impinge on a collecting surface. This effect is utilized in many air-cleaning devices including cyclones, baffle chambers, fibrous filters, and wet collectors of various kinds.

In the cyclone, the gas makes several complete revolutions before it leaves the apparatus, while in the other equipment mentioned, it flows around baffle plates, fibers, water droplets, or wet surfaces.

In theory, the individual particle trajectories, and therefore the collection efficiency, can be calculated from a force balance and from the velocity field in the neighborhood of the collecting surface. In practice, however, the velocity distribution may be difficult to characterize, and even when this is possible, an analytical solution of the force-balance equations is not always possible. Most of the theoretical and experimental studies of inertial deposition have been concerned with impaction on cylinders and spheres and with centrifugal separation from rotating gas streams.

IMPACTION ON CYLINDERS AND SPHERES. The impaction of aerosol particles on cylinders has been given considerable attention since it provides an insight into the functioning of fibrous filters. For both cylinders and spheres, the impaction efficiency is defined as the fraction of the particles of given size which strike the collecting surface in the volume swept out by the body. For flow around a cylindrical fiber, as

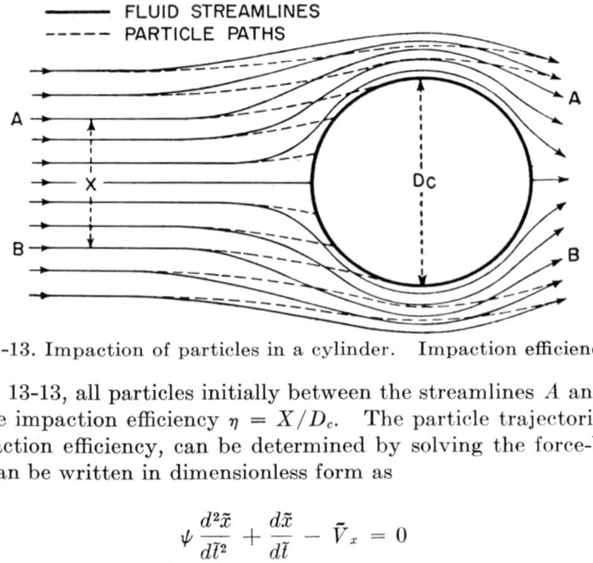

FIG. 13-13. Impaction of particles in a cylinder. Impaction efficiency $= X/D_c$.

shown in Fig. 13-13, all particles initially between the streamlines A and B strike the body, and the impaction efficiency $\eta = X/D_c$. The particle trajectories, and therefore the impaction efficiency, can be determined by solving the force-balance equations which can be written in dimensionless form as

$$\psi \frac{d^2\bar{x}}{d\bar{t}^2} + \frac{d\bar{x}}{d\bar{t}} - \bar{V}_x = 0$$
$$\psi \frac{d^2\bar{y}}{d\bar{t}^2} + \frac{d\bar{y}}{d\bar{t}} - \bar{V}_y = 0 \qquad (13\text{-}24)$$

where ψ is a dimensionless group known as the impaction parameter. The solution of these equations depends on the nature of the velocity distribution around the cylinder; numerical solutions have been obtained for different flow fields, and the results are shown in Fig. 13-14 in which impaction efficiency is plotted as a function of $\sqrt{\psi}$, with $\text{Re} = D_c V_0 \rho_g/\mu$ as the parameter. Wong's (147) experimental results are also shown.

Potential-flow theory and experimental data in the high Re range agree fairly well. It should be noted that theoretical efficiencies for the low Re range are considerably lower than those predicted for potential flow, and it is in the low Re range that most fibrous filters are operated.

The impaction efficiencies shown in Fig. 13-14 were calculated on the assumption that the particles act as point masses. Actually, however, particles with trajectories which are not intercepted by the collecting surface may touch the surface because their radii are greater than the distance from their path to the surface. This effect is known as direct interception. For given values of ψ and Re, i.e., a given set of particle trajectories, direct interception depends on the ratio of the diameter of the particle to that of the obstacle, $\mathcal{R} = d_p/D_c$. Combined impaction and interception efficiencies

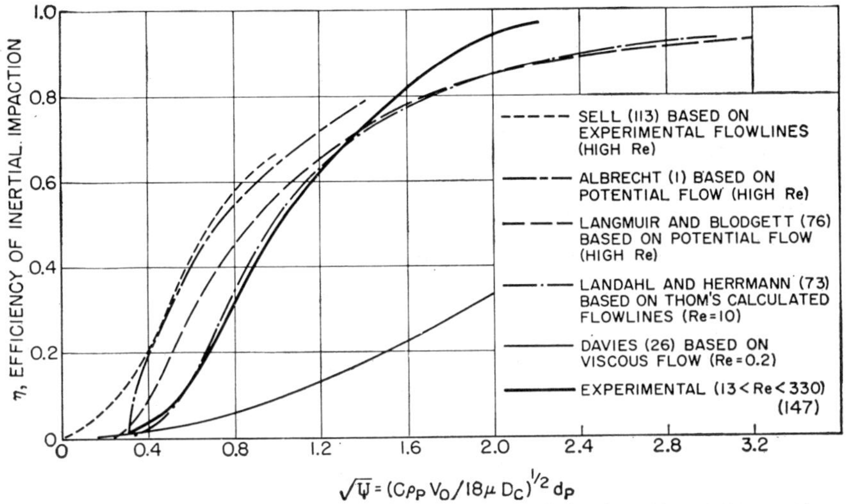

FIG. 13-14. Comparison of theoretical and experimental efficiencies of inertial impaction on circular cylinders. (See Ref. 147.)

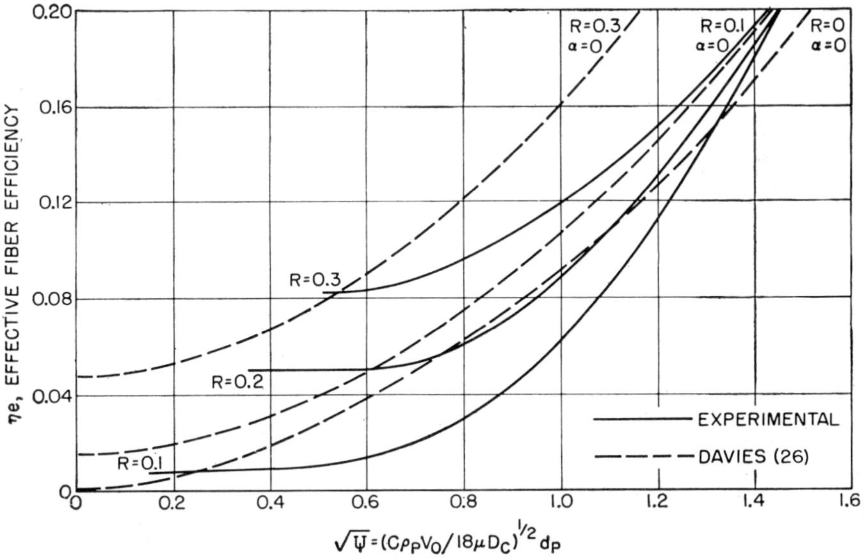

FIG. 13-15. Single-fiber efficiencies for beds of low porosity ($0.04 < \text{Re} < 1.4$). (See Ref. 147.)

are shown in Fig. 13-15 as a function of $\sqrt{\psi}$, with $\mathcal{R}$ as a parameter for the low Re region. The experimental data of Wong (147) for fiber mats of low porosity are compared with the theory of Davies (26) for single cylinders. The fiber efficiencies are designated "effective" since they include orientation and fiber interference effects.

Figure 13-15 indicates that, for values of $\sqrt{\psi}$ below 0.4, the efficiency depends principally on direct interception.

The treatment of impaction on spheres is analogous to the case of cylindrical obstacles. Efficiencies based on potential and laminar flows with interpolated values for the intermediate region have been calculated by Langmuir (76*a*).

CENTRIFUGAL SEPARATION. For the centrifugal separation of particles from a rotating gas stream, the force balance, after several simplifying assumptions, takes the form

$$u_t = \frac{dR}{dt} = \frac{d_p^2(\rho_p - \rho_g)V_t^2}{18\mu R} \tag{13-25}$$

The solution of this equation depends on the variation of the tangential velocity with radius. When a gas enters a cyclone, its velocity undergoes a redistribution, and the tangential component can often be expressed as a function of the radius by an equation of the form $V_t \sim 1/R^n$. The spiral velocity may reach a value several times the average inlet gas velocity. For a freely rotating gas, angular momentum is conserved and n is unity. Because of the wall friction in a cyclone, however, the value of n is less than 1 and, according to Shepherd and Lapple (114,115), varies from 0.5 to 0.7 over a large portion of the cyclone radius. The simplest form of integrated equation is obtained, however, if a constant centrifugal force is assumed, i.e., an n of -0.5:

$$R_1 - R_1 = \frac{Nd_p^2V_c(\rho_p - \rho_g)}{9\mu} \tag{13-26}$$

This equation was originally derived by Rosin, Rammler, and Intelmann (107). Its application to cyclone design is discussed in a later article.

Deposition by Brownian Diffusion. Diffusion due to the Brownian movement plays an important part in the deposition of submicron aerosol particles. The efficiency of collection can, in theory, be calculated from the differential equation of diffusion and the flow distribution in the neighborhood of the collecting surface, just as the efficiency of impaction can be obtained from the equation of particle motion and the flow distribution. However, rigorous analytical solutions for the cases of most interest, i.e., cylindrical fibers and spherical water droplets, have not been obtained; thus, in practice, recourse is had either to an approximate solution or to the semiempirical expressions for heat and mass transfer with single cylinders and spheres. In neither case have the calculations been checked experimentally.

The approximate solution was first obtained by Langmuir (106), who derived expressions for the collection efficiency of a single cylinder by diffusion, by interception, and by diffusion and interception combined. His calculations are based on Lamb's equations for the velocity field around a cylinder and are therefore applicable for Re less than 1. At a given Re, diffusional efficiencies can be expressed as a function of $\mathcal{R}$ and of the diffusion parameter $\mathfrak{D} = D_{BM}/D_cV_0$, where D_{BM} is the particle diffusivity. Figure 13-16, taken from Chen (18), is based on Langmuir's calculations and shows the variation of the combined diffusion and interception efficiency with $\mathfrak{D}$ at a Re of 0.01 with the interception group $\mathcal{R}$ as a parameter. For η above about 0.1, the accuracy of these curves diminishes because of the approximations in the original calculations. The particle diffusivity D_{BM} can be calculated from the Stokes-Einstein equation:

$$D_{BM} = \frac{CkT}{3\pi\mu d_p} \tag{13-27}$$

This expression was derived for particles with diameters of the same order of magnitude as the mean free path of the gas molecules, i.e., 0.1μ for gases at room temperature and atmospheric pressure. Table 13-10 shows values of D_{BM} at 25°C for particles in comparison with the diffusivity of sulfur dioxide molecules.

Table 13-10. Diffusivities of Particles in Air at 25°C

d_p, μ	D_{BM}, (*sq cm/sec*)
0.5	$6.4 \times (10)^{-7}$
0.1	$6.5 \times (10)^{-6}$
SO_2 molecules	$11.8 \times (10)^{-2}$

Data from H. F. Johnstone, Gas Cleaning and Purification, in "Encyclopedia of Chemical Technology," vol. 7, p. 83, R. E. Kirk and D. F. Othmer, eds., Interscience Publishers, Inc., New York, 1951.

Another approach to the problem has been taken by Johnstone and Roberts (62) who suggested that the diffusional efficiency can be estimated from the semiempirical equations for heat and mass transfer to surfaces normal to the flow:

$$\eta = \frac{D_{BM}}{D_c V_0}\left[\frac{1}{\pi} + 0.55\text{Sc}^{1/3}\,\text{Re}^{1/2}\right] \qquad \text{for cylinders} \tag{13-28}$$

$$\eta = \frac{D_{BM}}{D_c V_0}\left[2 + 0.55\text{Sc}^{1/3}\,\text{Re}^{1/2}\right] \qquad \text{for spheres} \tag{13-29}$$

These equations are satisfactory for predicting mass transfer when the Schmidt number $\mu/\rho_g D_{BM}$ is less than about 100, but they have not been checked for the transfer

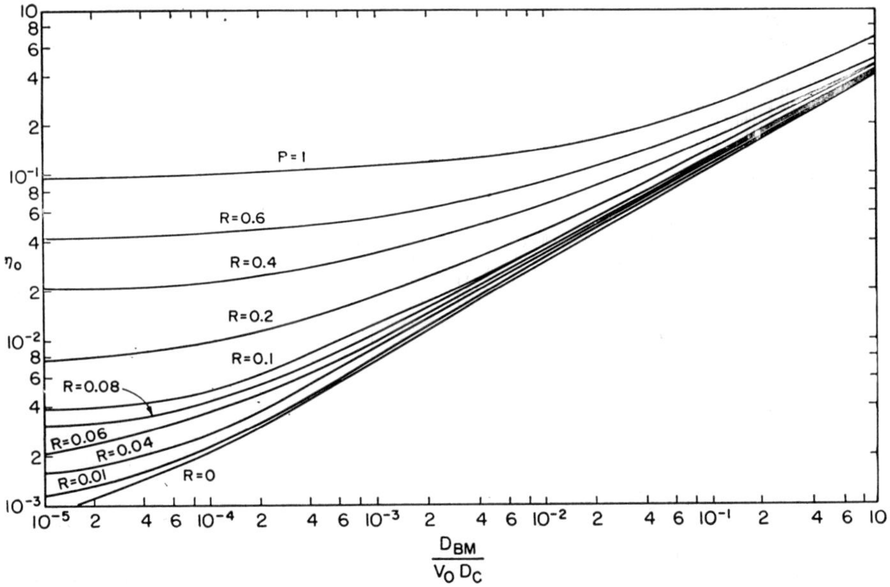

FIG. 13-16. Calculated collection efficiency of an isolated cylinder by diffusion and interception (based on Langmuir's theory, Re $= 10^{-2}$). (See Ref. 18.)

of particles which have considerably larger Schmidt groups. It is interesting to note, however, that efficiencies calculated from Eq. (13-28) compare well with Langmuir's values when $\mathcal{R}$ is zero (no interception).

Effects of Turbulence. The presence of turbulence in an aerosol produces eddy diffusion in the main stream and turbulent deposition at boundary surfaces. In collecting devices such as settling chambers, cyclones, or electrical precipitators, eddy diffusion opposes the forces producing deposition and reduces efficiency. If the turbulence is so great that the particle concentration remains substantially uniform, the removal efficiency for particles of a given size is given by the expressions

$$E = 1 - e^{-u_t A/Q} \qquad \text{for the steady state} \tag{13-30}$$

$$E = 1 - e^{-u_t \Delta t/h} \qquad \text{for the batch case} \tag{13-31}$$

When a concentration gradient exists in the system, the position of the particles can be calculated from the diffusion equations, provided that the eddy diffusivity is known. Sherwood (116), on the basis of experiments performed with gases of varying density, has suggested that the eddy diffusity of very small particles is about the same as that of gases. Longwell and Weiss (85) concluded that the eddy diffusivity of particles is less than that of a carrier gas because of the particle inertia. At present, there are not sufficient experimental data to establish whether there is a difference in eddy diffusivities or at what point the difference becomes significant.

When an aerosol flows in turbulent motion past a surface, some of the particles are deposited even though there is no net velocity in the direction of the surface. This turbulent deposition results from the fluctuating velocity component normal to the collecting area. It occurs in the movement of aerosols through straight ducts or diffuser sections and, in general, on any body whose boundary layer becomes turbulent when passing through a gas containing particles. It undoubtedly contributes to removal of particles in such devices as cyclones and cyclone scrubbers operated at high levels of turbulence.

In essence, turbulent deposition is a form of inertial deposition in which sudden gusts of fluid move toward the surface, change their direction, and thereby cast out the particles which they carry. Recent studies indicate that turbulent deposition increases with the velocity past the collecting surface and with the size of the suspended particles. Both these results would be expected from the inertial mechanism.

Agglomeration. The collection of dispersed matter by gravitational or inertial forces can be facilitated by coagulating the particles. Three mechanisms of coagulation are known: (1) Brownian coagulation, (2) turbulent coagulation, and (3) sonic agglomeration.

BROWNIAN COAGULATION. As a result of their Brownian motion the particles of an aerosol collide with each other. Solid particles form loose clusters or chainlike structures, while liquid droplets usually coalesce. Brownian coagulation is of great importance in the growth of atmospheric dust particles and may be utilized to obtain flocculation of fume particles before collection in settling chambers or by bag filters. The classical studies of the phenomenon were made by Whytlaw-Gray and his collaborators (143). They worked with aerosols of magnesium and cadmium oxides, ammonium chloride, stearic acid, resin, and others. Concentrations were determined by counting with an ultramicroscope the number of particles suspended in a sample of known volume. The coagulation rate was found to be proportional to the square of the particle concentrations:

$$-\frac{dc}{dt} = K_b c^2 \tag{13-32}$$

For heterogeneous aerosols, which have been most commonly studied, the coagulation coefficient is 25 to 50 per cent higher than predicted from the theory of Smoluchowski for a homogeneous dispersion of spherical particles:

$$K_b = \frac{4kT}{3\mu}\left(1 + \frac{1.72\lambda}{d_p}\right) \tag{13-33}$$

The higher coagulation rate for heterogeneous aerosols can be predicted in a semi-theoretical manner.

Table 13-11, calculated from Eqs. (13-32) and (13-33), shows the time necessary to reduce various concentrations of 1-μ particles to one-tenth of their initial value. These figures are applicable to any size above about 0.5μ since the coagulation coefficient is relatively insensitive to diameter. In general, aerosols at concentrations greater than 10^6 particles per cubic centimeter are unstable.

Table 13-11. Coagulation Time as a Function of Particle Concentration

c_0[a]	$t_{0.1}$[b]
10^{10}	3 sec
10^9	0.5 min
10^8	5 min
10^7	50 min
10^6	9 hr

[a] c_0 is the original concentration, particles per cubic foot or particles per cubic centimeter.
[b] $t_{0.1}$ is the time required for c_0 to be reduced to one-tenth its initial value.

At lower pressures, theory indicates that coagulation should take place more rapidly because of the increase in the mean free path of the gas. This has been confirmed experimentally (13). The structure of the agglomerate is also found to affect the coagulation rate. Certain materials such as carbon and zinc and cadmium oxides form chainlike aggregates owing, it is thought, to the dipole nature of the particles. The coagulation rate for such aggregates is considerably greater than would be predicted from the theory.

No practical methods are known for controlling the agglomeration rate of an existing aerosol. The results of investigators who have attempted to influence coagulation by charging of the particles or by the introduction of foreign vapors have often disagreed. DallaValle et al. review this subject and discuss the results of their own research (22).

Turbulent Coagulation. The presence of a velocity gradient in an aerosol results in an increase in the coagulation rate since particles moving at different velocities in adjacent streamlines tend to collide. This mechanism is a significant factor in the coagulation of atmospheric aerosols (128,130) and has been noted in stirred smokes (46). Again a coagulation coefficient K_s can be defined by the equation

$$-\frac{dc}{dt} = K_s c^2 \qquad (13\text{-}34)$$

Teverovskii (128) has reported values for this coefficient ranging from 10 to 4,000 times that due to Brownian coagulation for atmospheric smokes. By means of an ultramicroscope, the smoke was examined at intervals as it moved through the ground layer of the atmosphere. The average radius of the particles increased from 0.2μ initially to 0.4μ at a distance of 1,000 m. Particle concentrations ranged from 10^4 to 10^6 per cubic centimeter. An expression for the coefficient K_s has been derived theoretically by Smoluchowski for a dispersion subjected to a simple shearing stress:

$$K_s = \tfrac{4}{3} d_p{}^3 \omega \qquad (13\text{-}35)$$

The calculation of K_s for an aerosol in a turbulent field is difficult because of the complexity of the flow pattern.

Sonic Agglomeration. Flocculation of particles by high-intensity sound waves has received considerable attention of late. If the particles can be grown to a sufficiently large size in a sound field, the agglomerates can be separated by relatively inexpensive equipment such as a cyclone separator. The speed of agglomeration depends principally on the intensity of the sound field (rate of energy transfer per unit cross-sectional area), on the sound frequency, and on the time of exposure to the sound field. The theory has not as yet been developed sufficiently to allow prediction of coagulation rates. Three mechanisms have been proposed to explain this phenomenon. The most obvious is that of covibration of the particles in the gas. Particles of different sizes vibrate with different amplitudes, thus increasing the probability of collision. Secondly, two particles which have their lines of center perpendicular to the gas vibration move toward each other as a result of the reduced static pressure in the constricted region between them. Finally, sound waves exert a pressure against an obstacle in their path. Small particles in a standing wave tend to concentrate at the position of maximum amplitude where they agglomerate.

From his laboratory studies with ammonium chloride smokes, St. Clair (108) concluded that radiation pressure is the most significant mechanism. However, Stokes' (127) investigations with carbon-black aerosols indicated that covibration of the particles is dominant in commercial equipment where time of exposure to the sound field is only a few seconds.

Electrostatic Attraction. The deposition of aerosols can be effected by electrostatic forces. In some applications, such as electrostatic filters, the existing charges on the particles are utilized. In the more general application to electrostatic precipitators, however, the particles are artificially charged.

Fundamentally, the force $\mathfrak{F}$ causing migration of the particle toward the collecting surface is that caused by the interaction of an electrostatic charge q and an electric field $\mathcal{E}$. This force is often described by the following equation:

$$\mathfrak{F} = q\mathcal{E} \qquad (13\text{-}36)$$

The electric field, which is expressed in terms of volts per unit distance, is a vector quantity representing the potential gradient at a point and is written as follows:

$$\boldsymbol{\mathcal{E}} = -\text{grad } \mathbf{P} \qquad (13\text{-}37)$$

or in terms of components along the cartesian axes

$$\mathcal{E}_x = -\frac{\partial P}{\partial x} \qquad \mathcal{E}_y = -\frac{\partial P}{\partial y} \qquad \mathcal{E}_z = -\frac{\partial P}{\partial z} \qquad (13\text{-}38)$$

A positively charged particle in a field, then, will experience a force in the direction of decreasing potential and a negatively charged particle in the direction of increasing potential.

Electrostatic Precipitators. In the application of electrostatic forces to precipitators, the particulate matter comprising the aerosol is charged by passing through a highly ionized region. The material is then removed from the gas stream by electrostatic forces in an intense electric field. In most designs the migration of particles toward the collection surface is aided by turbulence in the gas stream.

Particles are charged as the result of intense ionization in the corona discharge immediately surrounding an electrode with a small radius of curvature. The electric field strength in this region is very high. If the ionizing electrode is negative with respect to the collecting electrode, any positive ions in this region will be accelerated rapidly toward it and on collision will release electrons which are accelerated toward the collection electrode (83). These in turn may collide with gas molecules, creating more electrons and positive ions. The positive ions are accelerated back to the negative electrode and result in the production of more electrons while the electrons are accelerated toward the positive electrode. The electrons moving toward the positive electrode attach themselves to particles suspended in the gas, thus charging them negatively. The charged particles are in turn accelerated toward the positive (collecting) electrode.

If the ionizing electrode is positive, electrons are accelerated toward it, bombarding molecules on the way. The electrons formed by the bombardment in turn ionize more molecules, resulting in a cascade of electrons toward the positive electrode. The positive ions, meanwhile, are accelerated toward the negative (collecting) electrode. They attach themselves to suspended particles, charging them positively and thus causing them to be accelerated toward the collection surface.

Negative ionization electrodes are usually used since they allow the use of higher electrode voltages, and hence more intense fields, before sparking (breakdown). The ratio of breakdown voltages for negative and positive coronas may be as high as 2:1 (140).

The charge acquired by a particle in the environment described above is a function of time. There are two important mechanisms of particle charging. These are bombardment by ions moving under the force of an electric field and diffusion of ions to the particle. The latter mechanism is important only for particles smaller than 1μ.

Particle charges acquired are given by the following formulas (77,141):

Charging by bombardment:

$$q_t = \left(\frac{\mathcal{E}a^2t}{t + \dfrac{1}{\pi nek}}\right)\left(1 + 2\,\frac{\epsilon - 1}{\epsilon + 2}\right) \tag{13-39}$$

$$q_s = \left(1 + 2\,\frac{\epsilon - 1}{\epsilon + 2}\right)\mathcal{E}a^2 \tag{13-40}$$

Charging by diffusion:

$$q_t = \frac{akt}{e}\ln\left(1 + \frac{\pi a U_{\mathrm{rms}} ne^2}{kT}\,t\right) \tag{13-41}$$

$$q_s = 6.8 \times 10^3 eaT \tag{13-42}$$

The first three formulas may be derived theoretically and agree with experimental measurements. The last formula is empirical.

The charged particles migrate to the collection surface under the influence of the electric field alone or are carried close to the collecting surface by turbulence in the precipitator and removed by the electric field. The turbulence can be the result of turbulent gas flow through the tube combined with the "ionic-wind" velocity (140) or the ionic wind alone. The ionic wind is caused by the transfer of momentum from the ions formed in the corona to the surrounding gas molecules (77,83). The velocities found in practice are of the order of 2 fps (77). The pressure due to this effect is given below (83).

$$p = \mathcal{E}en\delta \tag{13-43}$$

The efficiency of collection for precipitators has been expressed as follows by Deutsch (31,77):

$$E = 1 - e^{-(U_tA/\mathfrak{F})} \tag{13-44}$$

Electrostatic Filters (119). Electrostatic filters usually take advantage of the natural charges on materials to provide the electrostatic forces necessary for particle removal. Most aerosols are charged, as are many fibers. Most synthetic fibers become charged during the process of manufacture. Filters relying on natural charges to produce electric fields lose their efficiency after a short while because of neutralization of fiber charges by the aerosol charges. Three mechanisms have been employed to counteract this effect. One is to polarize the filtering material by placing it in an artificial electric field at 12 to 15 kv. The second mechanism is continually to charge the filter fibers by mechanical carding. The third mechanism involves mechanical-friction charging of fibers at opposite ends of the triboelectric series (120). This is accomplished by continuously wiping a filter at one end of the series with a wiper blade covered with some material from the other end of the series. The generation of surface voltages up to 20 kv have been reported by Silverman et al. (120).

13.3.2 Gas-cleaning Equipment

Introduction. The wide variety of commercially available gas-cleaning equipment can be divided into the following groups: (1) settling chambers, (2) inertial separators, (3) scrubbers and washers, (4) filters, (5) electrical precipitators, and (6) sonic precipitators. This classification is somewhat arbitrary since, for example, the installation of water sprays in a simple settling device or cyclone may change the collection method to scrubbing, while the insertion of baffles changes a settling chamber into an inertial separator.

The choice of an industrial air-cleaning installation depends on the size and size distribution of the particles to be removed, the quantity of gas to be treated, the concentration of the aerosol, the temperature of the gas, and the efficiency of removal that is required. Unfortunately, the design of these systems is still something of an art because of the many variables involved. Chapter 10 of the Manufacturing Chemists' Association "Air Pollution Abatement Manual" is a help in this regard. In the discussion which follows, it will be noted that the design equations developed for simple models of equipment are often difficult to apply to practical systems. For this reason, performance data are included to give an indication of the applications of various devices and the results to be expected from their use.

Settling Chambers. These are the simplest of the collectors. They may consist of nothing more than the enlargement of a flue or duct to reduce the gas velocity and allow the larger particles to settle out. The most common form is a long boxlike structure, set horizontally, often on the ground. The balloon flues leading to baghouses in metallurgical plants serve both for settling and for the cooling of the gases. They may be equipped with sprays for additional cooling in hot weather and with drags, screw conveyors, scrapers, and rapping devices.

The settling rooms or "kitchens" used for collecting arsenic trioxide may run at temperatures of 500 to 600°F and balloon flues and breechings of power plants, at 300 to 400°F. In contrast, the settling chambers in which powdered milk is collected are operated at ordinary room temperature or even less.

In general, it is not practical to catch in settlers particles of subsieve size (less than 40μ). However, fine materials such as carbon black and various metallurgical fumes may form agglomerates which have enough mass to permit collection in settling devices.

The dimensions of a chamber which will provide an efficiency E for particles settling with a velocity u_t can be estimated from the equation

$$E = \frac{N_s u_t BL}{Q} \tag{13-45}$$

In practice, the presence of eddy currents may hinder precipitation, and a more conservative estimate can be obtained by halving the calculated terminal settling velocity. To minimize turbulence and ensure uniform velocity, curtains, rods, or screens may be suspended in the chamber. Another modification is the Howard dust chamber (Fig. 13-17) which was used for many years in the metallurgical industry and in sulfuric acid plants burning pyrites. This settler was fitted with a number of inclined shelves to reduce the distance through which the particles fell.

Gas velocities should be low (less than about 10 fps) to prevent reentrainment. The pressure drop through a settling chamber is usually small and consists mostly of entrance and exit losses. Installation costs are low because of the simple structure.

Centrifugal and Inertial Separators. This group includes cyclones and baffle chambers—devices in which suspended matter is separated from the gas stream by a change in direction. Cyclone collectors are not commonly used when there is an appreciable amount of material smaller than 5μ, but they can be used for hot or cold gases and have a very wide range in capacity. They may be operated at high inlet velocities (several thousand feet per minute) since efficiency increases with the gas rate.

THE CYCLONE SEPARATOR. The cyclone separator is the most widely used gas-cleaning device. It generally consists of a cylinder with a tangential inlet and an inverted cone attached to the base; the cylinder diameter varies from several inches to 15 or 20 ft, depending on the efficiency desired and the amount of gas which must be handled. In the conventional cyclone, the gas enters tangentially either from a horizontal duct or through directing vanes; it then spirals down the annular space into a cylindrical or conical chamber, turns upward, and forms an inner spiral which leaves through the exit duct.

Cyclones are used extensively as primary separators for dusts from rock crushing, ore handling, woodworking, and sand conditioning in foundries. They are not adapted to the collection of fine metallurgical fume particles, dispersed or flocculated. They handle gas volumes ranging from about 30 to 25,000 cfm and at a very wide range in temperatures.

Particle size is fairly critical. If the particles are large (5 to 200μ), a properly designed cyclone will perform adequately with moderate power requirements. For particles larger than 200μ, a settling chamber is suitable and more resistant to abrasion.

In the past, it has been the practice to build large-diameter cyclones, but banks of small collectors in parallel with a single header and a single dust hopper are finding increasing use. Examples of this type are the Multiclone (Western Precipitation

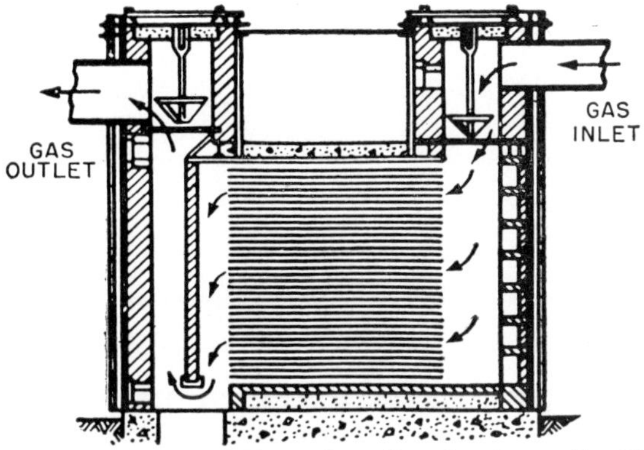

FIG. 13-17. Cross section of Howard settling chamber. (See Ref. 77.)

Corp.), in which the individual cylinder diameter varies from 6 to 24 in.; the Thermix collector (Prat-Daniel Corp.), with a diameter between 6 and 9 in.; and the Aerotec collector (Aerotec Corp.) for free-flowing dusts, in which the tube diameter is only 2 to 3 in. The Sirocco fly-ash collector (American Blower Corp.) consists of multiple horizontal tubes, approximately 15 in. in diameter and 6 ft long, in parallel. The gas is given a spiral motion by vanes located in the inlet, and the dust is concentrated in a small portion of the gas, which is recirculated through a smaller secondary cyclone separator for final collection of the dust. The Multicyclone (Prat-Daniel Corp.) is furnished in multiple units, 2 or 3 ft in diameter, each containing an adjustable inlet damper to compensate for variations in gas throughput. Initial costs for cyclones range from about \$0.04 to \$0.08 per cfm for large units to \$0.25 per cfm and higher for small multitube units. A number of commercial cyclone separators are shown in Fig. 13-18. Performance data for various cyclones are given in Table 13-12.

Conventional cyclone design in this country has become rather standard. Tables of detailed dimensions are available, and many cyclones are constructed by local sheet-metal firms according to specifications furnished them. Cyclones of this type are generally satisfactory for handling very coarse dusts, but more care must be taken in the design of higher-efficiency cyclones for the smaller size range.

There is, however, some disagreement among those who have studied cyclone design (82,107,114,115). One of the most widely used approaches is that of Rosin, Rammler, and Intelmann (107). Assuming a constant centrifugal force to establish the velocity pattern, these investigators derived the following equation for the size of the smallest particle which can be completely removed in a cyclone:

$$d_{pm} = \sqrt{\frac{9\mu(R_2 - R_1)}{\pi N V_c(\rho_p - \rho_g)}} \tag{13-46}$$

Actually, however, in most cyclones the gas path is a double spiral, with the gas rotating downward on the outside and upward on the inside. There may be a secondary gas movement in the form of an eddy at the top of the cyclone, and this eddy

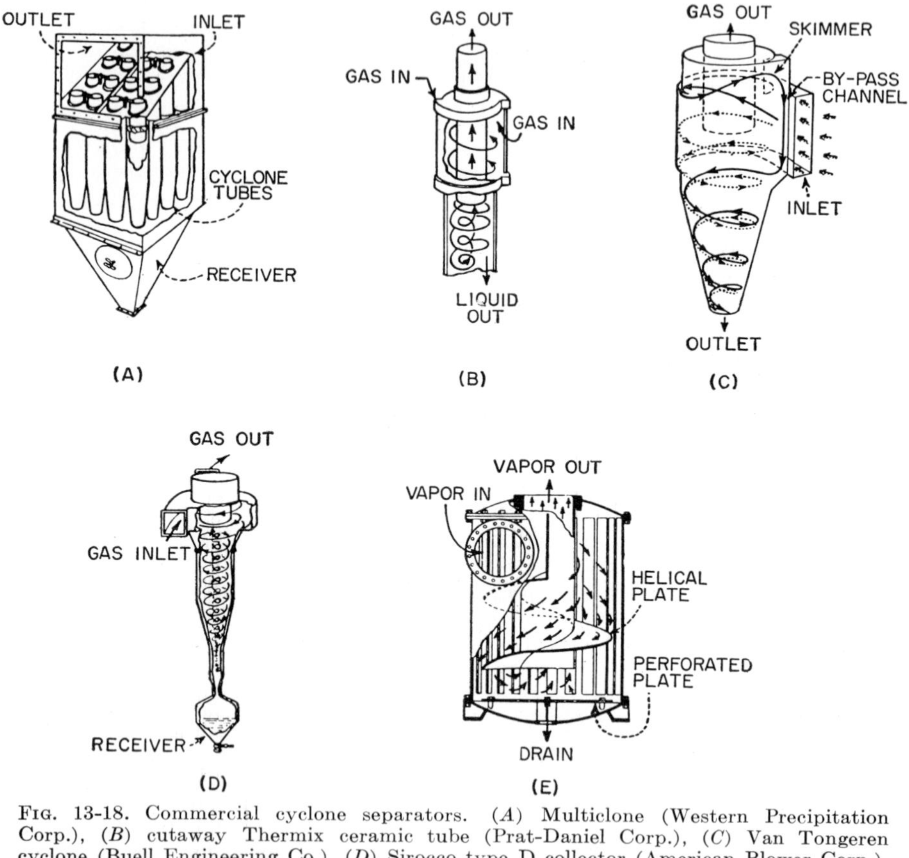

FIG. 13-18. Commercial cyclone separators. (*A*) Multiclone (Western Precipitation Corp.), (*B*) cutaway Thermix ceramic tube (Prat-Daniel Corp.), (*C*) Van Tongeren cyclone (Buell Engineering Co.), (*D*) Sirocco type D collector (American Blower Corp.), (*E*) Flick separator (Wurster & Sanger Inc.). (*From C. F. Montross, Entrainment Separation, Chem. Eng., October*, 1953.)

seems to contribute significantly to removal in some cases. In general, then, the velocity distribution is quite complex and cannot be represented by a simple equation. In addition to the uncertainty concerning the flow pattern, the number of turns N made by the gas is difficult to evaluate. For these reasons Lapple (77) has suggested that N be taken as an empirical coefficient. The value of N should be constant for cyclones of given geometric proportions and can be regarded as an approximate measure of the effectiveness of cyclone design. Lapple reports that N is approximately 5 for cyclones having the proportions shown in Fig. 13-19. The use of vanes to reduce pressure drop results in a decrease in efficiency, and a lower value of N must be employed in the calculations.

Table 13-12. Performance Data for Cyclone Separators

Device	Manufacturer	Aerosol	Inlet concentration, grains/cu ft	Mass median particle size at inlet, μ	Efficiency weight, %	Resistance, in H_2O	Inlet velocity, fpm	Flow rate, cfm	Diam, ft
Cyclone...	American Blower Corp.	Dust from sand and gravel drying	38	8.5	86	1.9	2,000	12,300	9
Sirocco No. 20 cyclone	American Blower Corp.	Dust from stone drying kiln	16.8	5.1	86	....	1,700	4,600	5
Sirocco No. 20 cyclone	American Blower Corp.	Dust from sand drying kiln	18.7	6.3	78	....	1,700	4,800	5
Rotoclone Type D	American Air Filter Co.	Fluffy zinc stearate from rubber dusting operation	0.6	1.75	88	9.0	4,800	3,300	2
Rotoclone Type D	American Air Filter Co.	Fluffy zinc stearate from rubber dusting operation	1.7	1.75	78	....	2,400	750	2
Cyclone....	Plant Construction	Talc dust from abrasive cleaning operation	2.2	128	93	0.33	1,000	2,300	7
Cyclone...	Plant Construction	Wood dust from planing mill operations	0.1	99% >500 (weight)	97	3.7	3,100	3,500	6.5
Cyclone....	Plant Construction	Aluminum dust from grinding operations	0.7	99% >500 (weight)	89	1.2	1,400	4,400	6
Cyclone....	Plant Construction	Cotton flock	0.7	99% >500 (weight)	88	1.6	2,400	2,600	6.5

Data from S. K. Friedlander et al., "Handbook on Air Cleaning" (A.E.C.D.-3361; NYO-1572). Harvard University, School of Public Health, Boston, Mass., 1952.

Ter Linden (82) has also investigated the cyclone separator, and his results disagree in some respects with those of Rosin, Rammler, and Intelmann. Ter Linden concluded that decreasing the diameter of the gas outlet had the biggest effect on increasing efficiency, while reducing the outside diameter actually decreased efficiency to some extent.

The total pressure drop through a cyclone is due to friction along the wall, entry and exit losses, and the kinetic energy of rotation remaining in the gas in the exit duct. The pressure drop is most conveniently expressed in terms of the velocity head based on the inlet area (77). The inlet velocity head Δh, expressed in inches of water, is related to the average inlet gas velocity and density in the following manner:

$$\Delta h = 0.003\rho_g V_c^2 \qquad (13\text{-}47)$$

where ρ_g is in lb/cu ft and V_c in fps. In practice, the friction loss may range from 1 to 20 inlet velocity heads, depending on the geometric proportions. For cyclones of given geometric proportions, the losses are independent of the cyclone size. Attempts to calculate the friction loss or pressure drop from fundamental considerations have not been successful. Shepherd and Lapple (114,115) correlated their data on friction loss in cyclones of the design shown in Fig. 13-19 by means of the equation

$$F = \frac{KB_cH_c}{D_e^2} \tag{13-48}$$

where F is expressed in number of inlet velocity heads. For the conventional arrangement in which the rectangular inlet terminates at the outer wall of the cyclone body, K was found to have a value of 16.0. If the inner wall of the inlet duct is extended into the annular space halfway to the opposite wall to form an inlet vane, the friction loss is reduced by about 50 per cent, and K is found to be 7.5. As indicated above, however, inlet vanes tend to reduce the collection efficiency of the cyclone.

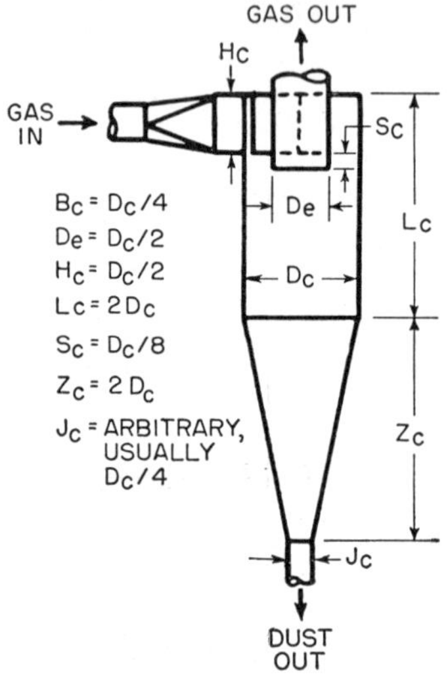

FIG. 13-19. Cyclone design proportions. (See Ref. 77.)

MECHANICAL CENTRIFUGAL SEPARATORS. Instead of the centrifugal force being supplied by the motion of the gas alone, the rotation may be obtained by a fan. In the Type D Rotoclone (American Air Filter Company, Inc.), the fan and dust collector are combined as a single unit. The blades of the fan are specially shaped to direct the separated dust into an annular slot leading to the collection hopper, while the clean gas continues to the scroll. In the Sirocco cinder fan (American Blower Corp.), the gas enters at the periphery of the scroll and passes radially inward through the rotor and out at the center. The dust is thrown to the scroll wall and is concentrated in a small stream of gas which is bypassed to a cyclone collector. The chief advantage of these cleaners is their compactness. Efficiencies are comparable to those of high-velocity cyclones. In some cases, maintenance costs are high because of deposition of collected dust on the fan blades, causing rotor unbalancing.

For the Type D Rotoclone, initial costs, including the fan, range from about $0.17 to $0.22 per cfm for units of 15,000 to 50,000 cfm capacity. These figures do not include charges for freight, duct work, erection, or accessory equipment like motors and drives.

IMPINGEMENT SEPARATORS (Fig. 13-20). Impingement separators depend on the inertial deposition of the particles as the gas passes around an obstruction. The shape of the obstacle may vary from that of simple baffles to more complicated patterns, which give maximum impaction efficiency with minimum pressure loss. In some cases, the surfaces may be wetted with a liquid to facilitate removal of the impacted particles.

Baffle chambers are the simplest type of impingement separators. The tortuous flow is obtained by zigzag plates or shaped obstacles placed in the gas stream. These devices are suitable for removing particles larger than about 20μ and are often employed as cinder traps for cleaning flue gases from stoker-fired furnaces. The Modave dust catcher, widely used in Europe, consists of a series of vertical thimbles of rectangular cross section, wetted by overflow of water from the inside of the tubes. In the Riley flue-gas scrubber (Riley Stoker Corp.), the particles of ash impinge on vertical carbon plates as the gas travels in staggered paths between the elements. The surface of these plates is swept by a flow of water which removes the accumulated particles. The velocity of the gases is approximately 1,000 fpm. The draft loss through the scrubber is about 0.8 in. of water, and about 2½ gal of water is circulated per 1,000 cu ft of flue gases. Such scrubbers are efficient for the removal of particles larger than 20μ.

The Calder-Fox scrubber is an impingement separator developed in England for removing sulfuric acid mist from gases. Two groups of perforated lead plates, ⅛ in. thick, are used in series. Gas passing through small orifices impinges on a plate set at a distance of about ⅛ in. This is followed by a series of collector plates with

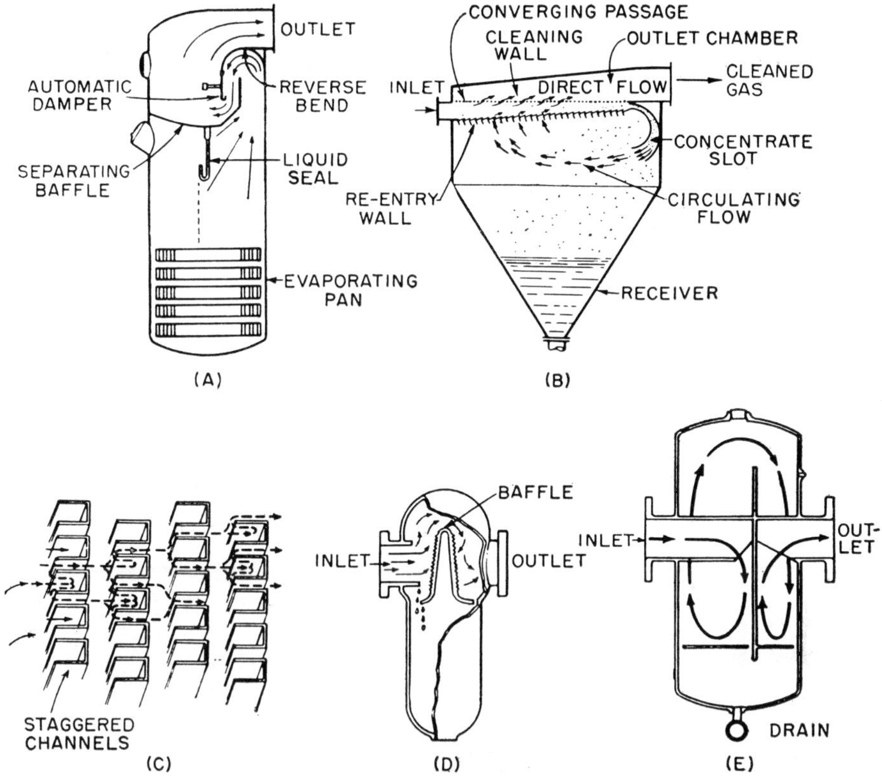

FIG. 13-20. Impingement separators. (*A*) Reverse-flow separator on evaporator (Arthur Harris & Co.), (*B*) impingement separator (mainly for solids, e.g., fly ash) (Buell Engineering Co.), (*C*) Tracyfier (Blaw Knox Co.), (*D*) Type E separator (Wright-Austin Co.), (*E*) PL separator (Ingersoll-Rand Co.). (*From C. F. Montross, Entrainment Separation, Chem. Eng., October*, 1953.)

nonstaggered orifices. The unit is effective in removing mist particles larger than 2μ. The pressure drop is 1 to 5 in. of water for gas velocities (through the orifices) of 50 to 100 fps.

Entrainment separators for the removal of sprays from humidifiers and suspended particles carried over from boiling liquids are usually of the baffle type.

Packed beds of granular solids also serve as impingement separators. In coke boxes used for removing sulfuric acid mist, the size of the coke particles ranges from $\frac{1}{40}$ to ½ in., with thin layers of larger particles acting as supports for the filter bed. Superficial gas velocities range from 2 to 10 fpm, with pressure drops from 1 to 10 in. of water. With the finer sizes of coke, mist particles as small as 0.5μ may be removed at weight-collection efficiencies as high as 99.9 per cent. Granite, quartz, and sand

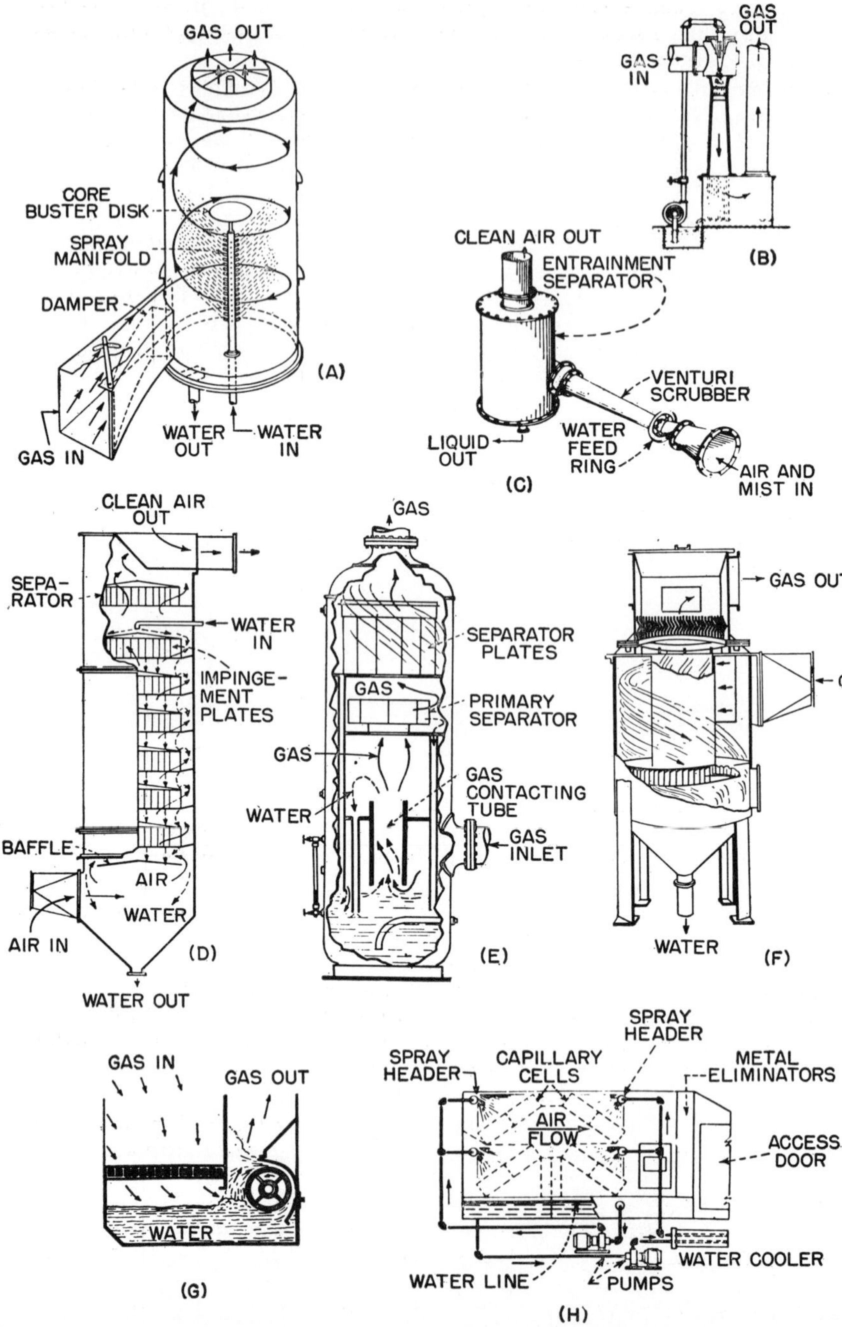

FIG. 13-21. Scrubbers and washers. (*A*) Pease-Anthony cyclone scrubber (Chem. Const. Corp.), (*B*) water-jet scrubber (Schutte and Koerting Co.), (*C*) Pease-Anthony venturi scrubber (Chem. Const. Corp.), (*D*) Multi-wash scrubber (Claude B. Schneible Co.), (*E*) liquid vortex contactor (Blaw Knox Co.), (*F*) Hydroclone type scrubber (Whiting Corp.), (*G*) Centrimerge rotor scrubber (Schmieg Industries), (*H*) wet filter (Air and Refrigeration Corp.).

are also used in dry packed beds with depths of 2 to 6 ft. When high concentrations of dust are present, it is necessary to have a moving bed and to withdraw continuously a portion of the material for cleaning.

Scrubbers and Washers. Scrubbers, washers, and their various modifications are widely used in industry for cleaning and cooling air and other gases. Water is naturally the liquid most often used; caustic may be added for acid mist collection. Wetting agents decrease bubble size, accentuate frothing, increase entrainment, and usually add little to collection efficiency. Oils may be employed for removing substances not readily wetted by water.

In water scrubbers, the gas leaves nearly saturated at a temperature depending on the over-all heat balance. For relatively small ratios of water to gas or when the water is recirculated, both the gas and water leave near the wet-bulb temperature. Dehumidification may be obtained by using large quantities of cold water. Gases cleaned in a water scrubber often show a condensate plume when emitted into cold air. This disappears within a short distance of the stack and is usually not considered objectionable. If there is a danger of stream or sewer pollution, or when there is a shortage of water, thickeners, filters, and storage tanks are necessary auxiliaries to a properly operated plant. The presence of acid constituents frequently poses corrosion problems. Performance data for various gas washers are given in Table 13-13, while some of the devices are shown in Fig. 13-21.

Spray Chambers. The simplest type of gas scrubber is an empty tower into which liquid is introduced at the top through a bank of spray nozzles while the gases pass countercurrent to the falling drops. Spray towers are employed as coolers and as primary cleaners in treating blast-furnace gas and for fly-ash and cinder removal. They are satisfactory for removing coarse dust if high efficiencies are not required. The gas velocity through the tower should be of the order of 3 to 5 fps.

Another type of spray chamber, the commercial air washer, is used for humidity control in air conditioning. In this device, the air flows horizontally through a chamber containing banks of nozzles set at different heights above the level of the sump water. The water may be sprayed with or against the air stream or in both directions by different sets of nozzles. Eliminators, usually zigzag plates, are placed at the exit to remove mist particles, and in some cases wetted baffle plates are installed across the air stream in the chamber itself. Water requirements for air washers range from about 0.5 to 20 gal per 1,000 cu ft of gas.

The Pease-Anthony cyclone spray scrubber (Chemical Construction Corp.) is a modification of the simple spray scrubber. The scrubbing liquid is introduced through nozzles located on a central manifold, and the gases enter through a tangential duct either at the top or bottom of the cylindrical tower. The small droplets are hurled through the gas stream at high velocities by the centrifugal force. An unsprayed section above the nozzles is provided so that the liquid droplets containing the collected particles will have time to reach the walls of the scrubber before the gas stream emerges. These scrubbers have been used for recovering metallurgical fumes and lime-kiln dust and for cleaning blast-furnace gases and flue gases from powdered-coal furnaces. Efficiencies of 97 per cent or better may be obtained when the particle size of the dust is above 1μ. The entrance velocity of the gas into the cyclone may be as high as 200 fps. Centrifugal accelerations are 50 to 300 times that of gravity. Superficial tower velocities are generally in the range of 4 to 8 fps. The pressure drop ranges from 2 to 6 in. of water, and the water rate is 3 to 10 gal per 1,000 cu ft of gas (58). Cyclone scrubbers are not sold as standard units but are especially designed for each cleaning job. Initial costs for mild steel construction range from about $1.50 per cfm for 500-cfm units to $0.40 per cfm for units larger than 25,000 cfm; corresponding figures for stainless steel are $2.25 to $0.80 per cfm. Power requirements usually range from 1 to 3 hp per 1,000 cfm.

Table 13-13 Performance Data for Gas Washers

Device	Manufacturer	Aerosol	Inlet concentration, grains/cu ft	Mass median particle size at inlet, μ	Efficiency, %	Pressure drop,[a] in. H_2O	Flow rate, cfm at STP	Water rate, gpm	Over-all unit dimensions, L × W × H
Type W Rotoclone....	American Air Filter Co.	Dust from castings cleaning by metal shot and sandblast and tumbling	0.75	40	99.3	4.4	4,150	9	8.8 × 8.0 × 15.3
			5.98	47	99.2	3.0	10,000	14	10.2 × 9.6 × 17.8
			7.29	45	99.6	3.3	5,220	7.5	7.0 × 6.6 × 13.2
Type N Rotoclone....	American Air Filter Co.	Dust from foundry tumbling	0.28	140	99.1	2.6	10,100	3.2	8.2 × 8.1 × 16
		Dust from machine-shop grinding	0.0114	37	91.2	2.9	28,000	2.2	19.6 × 8.1 × 18
Air tumbler.........	....	Dust from steam drying alumina	0.525	25	76.0	1.9	3,350	0.3	9 × 3 × 8.5
Multi-wash scrubber..	Schneible Co.	Silicon carbide dust	0.076	20	96.8	0.7	895	4.5	D × H 3 × 13
Ducon..............	Ducon Co.	Dust from stone and sand drying	5.8	4.3	74.0	2.5	9,200	20	6 × 16.7
Spray tower (multiple rows of 300 to 500 psi water sprays)	Buffalo Forge Co.	Fume from ferro-silicon furnace, 90% amorphous silica	0.972	0.89	75.0	1.1	2,800	10	4 × 16

[a] In some cases, the power expended in pumping water to high-pressure sprays may be more significant than power costs for gas flow.

Data from M. W. First et al., Air Cleaning Studies, Progress Report for Feb. 1, 1951, to June 30, 1952, AEC Report No. NYO-1536, Harvard University, School of Public Health, Air Cleaning Laboratory, Boston, Mass., 1952.

The fog filter (R. C. Mahon Co.) is essentially a simple spray scrubber in which special nozzles at very high pressures are used to obtain good disintegration. The recommended water pressure is about 400 psi. The nozzles are sometimes located directly in existing stacks, in cyclones, or in spray-type heat exchangers. The water consumption is 7½ to 10 gal per 1,000 cu ft of gas. Auxiliary collectors are needed to remove the fine water droplets from the gas stream. Pressure loss is relatively low.

Ideally, the efficiency of a spray chamber can be represented (62) by the expression

$$E = 1 - e^{-3/2(\eta/D_c)(W/Q)H} \qquad (13\text{-}49)$$

where η is the individual droplet efficiency. This equation is often difficult to apply because of the heterogeneity of both the aerosol and the spray and because of the complexity of the velocity distribution; it does, however, give a good qualitative picture of scrubber performance under various conditions. For a given chamber with a constant ratio of scrubbing liquid to gas, the efficiency depends on the ratio of the individual drop efficiency to drop diameter η/D_c. For most spray scrubbers, impaction on the droplets is the principal mechanism of removal. The impaction efficiency is greatest for small, fast-moving drops, but since the terminal velocity is proportional to D_c^2, it is the smaller drops which have the lower velocities. Simple spray chambers are ineffective in removing small dust particles since relatively low droplet velocities are acquired in a gravity field. The cyclone scrubber, which utilizes the centrifugal force of the rotating gas to impart high velocities to the drops, is efficient for particles larger than about 2μ, but for smaller particles, the efficiency falls off rapidly unless a high ratio of scrubbing liquid to gas W/Q is used. Calculations by Johnstone and Roberts (62) indicate that reducing drop diameter increases η/D_c, i.e., a considerable improvement in the removal of fine dust by cyclone spray scrubbers can be obtained by using nozzles that provide good atomization of the liquid. Their calculations were made for sprays with mean diameters of 150, 100, and 50μ, using Sell's (113) theoretical impaction efficiencies and assuming a centrifugal field of 100 times gravity. The drops should not be made too small since entrainment may occur, requiring an increase in the height of the tower.

ATOMIZING SCRUBBERS. In the atomizing scrubbers, water is introduced at the throat of a venturi through which the aerosol is passing at a high velocity. During the process of atomization the surface of the liquid is extended in the form of films and filaments. Interfacial velocities are high, and impaction, enmeshment, and electrostatic effects are important. The performance of various atomizing scrubbers is discussed by Boucher and Franck (7) and by Ekman and Johnstone (33).

In the Schutte and Koerting water-jet scrubber, the flow of gas is induced by a high-velocity jet of water passing through the nozzle at 50 to 100 gal per 1,000 cu ft of gas. The intimate mixing in the entrainment section produces high efficiencies for removal of particles larger than 2μ. Scrubbers with capacities up to 50,000 cfm are available.

The Pease-Anthony venturi scrubber (Chemical Construction Corp.) is an atomizing device designed to extend the use of gas scrubbers to removing submicron particles. The scrubbing liquid is introduced at low pressure at the throat of the venturi through which the gas is passed at velocities ranging from 200 to 500 fps. For the removal of submicron particles, the water rate must be in the range of 2 to 6 gal per 1,000 cu ft of gas, and the over-all pressure loss is 13 to 20 in. of water, with efficiencies of 95 to 99 per cent obtained in some instances. For coarser particles, such as those in blast-furnace gases and flue gases from powdered-coal furnaces, lower velocities and water rates may be used, with pressure losses of 6 to 8 in. of water. The venturi scrubber has been used for removing mists and dusts from gases from kraft mill furnaces, open-hearth furnaces, blast furnaces, sulfuric acid converters, sulfuric acid concentrators, and from vaporized organic materials. Venturi scrubbers are especially designed for each cleaning job. Initial costs for mild steel construction range from about $2.00 per cfm for 500-cfm units to $0.50 per cfm for units larger than 25,000 cfm; corresponding figures for stainless steel are $3.00 to $1.00 per cfm. Power requirements range from about 3 to 5 hp per 1,000 cfm.

DEFLECTOR WASHERS. Several types of commercial washers utilize deflector plates to capture and also to help disperse spray droplets. In general, these collectors are designed for removing particles larger than 1 to 5μ. Examples of deflector washers are the Type N Rotoclone (American Air Filter Company), the Multi-wash dust collector (Claude B. Schneible Company), and the Feld scrubber (Bartlett-Hayward Company). The pressure loss across these scrubbers is 2½ to 4 in. of water at rated capacity. Water consumption ranges up to 6 gal per 1,000 cu ft of gas. Initial costs for the Type N Rotoclone vary from about $0.32 to $0.24 per cfm in the 5,000- to 40,000-cfm range. These figures include the exhaust fan but not accessory equipment such as motors and drives.

The Type W Rotoclone (American Air Filter Company) and the Hydroclone (Whiting Corp.) are examples of deflector washers in which there is a continuous change in direction rather than a sudden change of short duration. Initial costs for the Type W Rotoclone vary from about $0.26 to $0.18 per cfm in the 5,000- to 40,000-cfm range; these prices do not include accessories such as motors and drives.

Wetted filters are also used for gas cleaning. The capillary conditioner (Air and Refrigeration Corp.), consisting of a glass-fiber filter on which water is sprayed, serves for both dust removal and humidification. The wetted cells, which are about 3 in. thick, are followed by entrainment eliminators in the form of metal-plate baffles of glass-fiber mats. Water is circulated at a rate of about 3 gal per 1,000 cu ft of gas, while the superficial gas velocity is about 300 fpm. Pressure loss is ½ to 1 in. for each cell. High efficiencies have been obtained with these washers when several cells are used in series. Initial costs for wet filters are approximately $0.10 per cfm. About 2 hr per week of maintenance are required for washing and cleaning the units. The wet cells rarely need to be replaced in less than a year, and the average period of service is 2 years.

MECHANICAL SCRUBBERS. Several designs of gas scrubbers are available in which a liquid is sprayed onto a rotating blade or disk. For dust and fume, the water droplets should not be larger than about 100 times the diameter of the aerosol particles. The Theisen disintegrator (Freyn Engineering Company) has been widely used in the cleaning of blast-furnace gases for many years. The dust-laden gas and the scrubbing liquid are passed outward through a series of rotating and stationary arms. The speed of rotation is 350 to 750 rpm. While the power consumption is about 10 hp per 1,000 cfm very efficient removal of fine dust can be obtained.

The Cycoil gas cleaner (American Air Filter Company) uses oil instead of water. The entrained droplets are removed partly by centrifugal force and partly by a steel-mesh filter. These units can be operated under vacuum, or at pressures up to 50 psi. They have been used to clean air for internal-combustion engines and for the final clean-up of stack gases passing to carbon dioxide recovery units.

Filters. Filters can be divided into two types. In the first, a fibrous medium acts as the separator, and collection takes place in the interstices of the bed. Such collectors usually handle light dust loads, ranging from 0.0002 to 0.01 grain/cu ft. High-efficiency units of this type may be used as aftercleaners, following other collecting devices which separate the greater part, by weight, of the suspended material. In the second type of filter, the particles are removed by a layer of dust which accumulates on the surface of a supporting fabric. Concentrations as high as 100 grains/cu ft are handled by these cloth collectors. The efficiency of separation increases as the dust layer builds up, but the simultaneous rise in resistance limits the permissible dust-layer thickness.

Fibrous Filters. Fibrous filters may be composed of wool, asbestos, cellulose, glass or iron fibers. The most common of these, the air-conditioning filters, handle large volumes of air with low dust loads and can be classed as either viscous or dry. In the viscous type, the filter is coated with a sticky material or "adhesive" to help catch the particles and prevent reentrainment. The adhesive is usually an oil or grease of high flash point and low volatility, and it should be a good wetting agent. The filter medium, generally glass wool, wire screen, animal hair, or hemp fibers, is placed in a metal or cardboard and wire frame approximately 20 in. square and several inches deep. The fibers may be packed with increasing density from front to back so that most of the large particles are removed before the more efficient back part is reached; this prolongs the life of the filter. When resistance to air flow becomes excessive, glass-wool media are generally discarded while the wire mesh is washed with hot water and then reoiled.

Viscous filters are also made up as a belt of metal panels which moves perpendicularly to the air stream (Fig. 13-22). The panels are attached to chains mounted on sprockets at the top and bottom of the filter housing. The belt travels up one side of the sprockets, down the other, and then passes through an oil bath at the bottom of the housing. Here the panels are cleaned and oiled while the dust settles out as a sludge. These units are particularly useful for atmospheric air containing high concentrations of dust, i.e., above 2 grains per 1,000 cu ft.

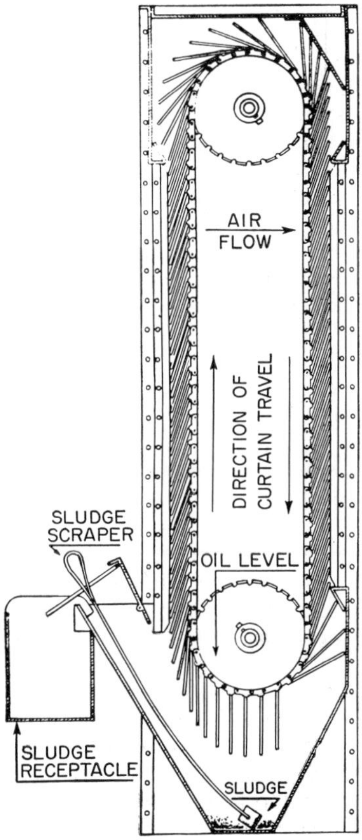

Fig. 13-22. Automatic viscous filter. (*Courtesy of American Air Filter Co.*)

Dry filters for air conditioning are supplied in units similar in size to the viscous type except that the depth of the dry cell is usually greater. The filter material may be paper, glass fibers, or cotton batting. For the most efficient types, air passages are generally smaller than those of the viscous media; consequently, it is necessary to operate at much lower velocities (30 to 60 fpm as contrasted with 300 to 500 fpm for viscous filters) to avoid excessive resistances. Since the velocities are low, filter area may be increased by arranging the surface in an accordion form with pockets and pleats for increased capacity. A cell 2 ft square may have a filter surface of 15 to 30 sq ft. When the pressure drop becomes excessive, the dry-type media are discarded. Performance data for various dry filters are given in Table 13-14 and cost data for viscous and dry filters in Table 13-15.

Dry filters which are automatically vibrated at intervals to dislodge the dust are also available. They handle higher concentrations than the usual dry-cell filters but not the heavy loadings of the cloth collectors.

Several fibrous filters capable of very high efficiencies have been developed for such applications as the removal of radioactive or poisonous particles and the cleaning of air in industrial plants producing photographic film or fine instruments. Composite glass-fiber filters are made by placing coarse, loosely packed pads of glass fibers at the gas inlet and progressively increasing the density of packing while decreasing fiber diameter. Such filters combine high efficiency with long life.

The Atomic Energy Commission makes use of a cellulose-asbestos paper filter originally developed by the Chemical Corps for gas masks. This filter, known as CWS No. 6, consists of fine asbestos fibers mixed with coarser cellulose fibers to give mechanical strength and act as a support for the asbestos. The asbestos mesh does most of the filtering. Efficiency is initially high and increases with use, and for aerosols at low concentrations, replacement may not be necessary for many months. Since these papers cannot withstand prolonged exposure to mists or acid vapors, and since they plug rapidly at high dust concentrations, a glass-fiber prefilter is often used. Efficiency and pressure-drop data are given in Fig. 13-23.

Table 13-14. Performance Data for Dry Filters

Device	Manufacturer	Aerosol	Inlet concentration	Mass median particle size at inlet, μ	Efficiency, wt %	Resistance, in. H_2O	Velocity, fpm	Remarks
Dustop...	Owens-Corning Fiberglas Corp.	City air	0.07 grains/1,000 cu ft	~1	56	0.025	150	A preliminary filter (coarse glass wool) containing a double layer of filters, 4 units high and 5 units wide. Flow rate was 8,300 cfm
Dustop...	Owens-Corning Fiberglas Corp.	35% city air, 65% recirculated air	0.015 grain/1,000 cu ft	~1	42	0.17	167	A preliminary filter (coarse glass wool), containing a double layer of filters, 6 units wide and 6 units high. Flow rate was 24,000 cfm

Data from S. K. Friedlander et al., "Handbook on Air Cleaning" (A.E.C.D-3361; NYO-1572), Harvard University, School of Public Health, Boston, Mass., 1952.

An improved cellulose-asbestos paper has been developed (122) and is now being manufactured by the Hollingsworth and Vose Co. The new paper is made from materials more easily obtainable and provides high efficiency in a thin low-resistance sheet. For individual units, the filters can be made in such sizes as 50, 500, 800, and 1,000 cfm; face velocities are of the order of 5 fpm at a resistance of 1 in. of water. Efficiencies and resistances are very similar to those of the CWS No. 6 medium (Fig. 13-23). The first cost of this paper is about \$0.10 per cfm of gas treated.

All-glass papers, such as those developed by Naval Research Laboratories (10) and made to a limited extent by several paper companies, serve well for cleaning hot or corrosive gases; these papers are the most efficient of the fibrous filters. Efficiency and pressure-drop data for a glass-fiber paper made by the Hurlbut Paper Co. are given in Fig. 13-24. This paper, which contains a resin binder, is composed of 0.5-μ glass fibers.

Another high-efficiency filter, the Hansen resin-wool filter, is made by carding natural or synthetic resins into wool (25,136). The small bits of resin clinging to the wool fibers carry high surface charges which cause deposition of the particles. In filtering dusts or smokes, these charges last for long periods and if dissipated can be restored by recarding the wool. However, mists (particularly those composed of oil droplets) cause rapid breakdown of the filter by spreading over the surfaces of the

Table 13-15. Initial and Operating Costs for Viscous and Dry Filters (1946)

Type of filter	Trade name	Manufacturer	Size of unit	Initial cost	Renewal cost	Rating, cfm	Fixed charges at 1%	Renewal of media	Labor[a] for replacement	Labor for cleaning	Misc. supplies	Operating electric cost[b]	Total annual cost per 1,000 cfm
Throwaway or renewable viscous	Dustop	Owens-Corning Fiberglas Corp.	20 × 20 × 2 in.	$6.00 for 2 7.50 for 4	$1.00	800	$0.72 0.90	$4.00 4.00	$0.40 0.40	$0.80 0.80	....	$2.28 4.55	$10.25 13.31
	Renu-Vent	American Air Filter Co.	20 × 20 × 3½ in.	6.00	1.00	800	0.72	4.00	0.40	0.80	......	1.47	9.24
Automatic viscous	Multi-panel MS	American Air Filter Co.	3 and 4 ft wide; 5–15 ft high	53.00–136.00[c]	.....	3,030 – 23,470	6.39 – 16.38	.....	.....	2.00[c]	0.70[c]	3.90[c]	12.99 – 22.98
	Model A-3	Dollinger Corp.	33 and 51 in. wide; 5–14 ft high	50.00–140.00[c]	.....	3,150 – 21,950	6.00[c] – 16.80	.....	.....	2.00[c]	0.70[c]	3.90[c]	12.60 – 23.40
Cleanable and renewable viscous	M/W 2	American Air Filter Co.	20 × 20 × 2 in.	10.70	.....	800	1.28	.....	.....	2.00	0.24	1.61	6.41
	DPV-2	Dollinger Corp.	20 × 20 × 2 in.	10.70	.....	800	1.28	.....	.....	2.00	0.24	1.07	5.74
Dry............	PL-24	American Air Filter Co.	24 × 24 × 8¾ in.	27.00	0.34	1,000	3.24	1.36	0.67	0.67	......	1.34	7.28
	WKE-4	Dollinger Corp.	24 × 24 × 4 in.	25.00	4.50	1,200	3.00	2.25	......	1.00	......	1.61	6.55
	Coppus	Coppus Eng. Corp.	20 × 20 × 5¾ in.	30.00	7.00	800	3.60	0.88	......	1.00	......	2.68	10.20

[a] Based on $1.00 per hour labor rate.
[b] Basis of calculation: 3,000-hr operation annually, power cost $0.02 per kwhr, 70 per cent fan efficiency, 90 per cent motor efficiency.
[c] Per 1,000 cfm.
Data from W. H. Carrier, R. E. Cherne, and W. A. Grant, "Modern Air Conditioning, Heating, and Ventilating," 2d ed., p. 220, Pitman Publishing Corporation, New York, 1950.

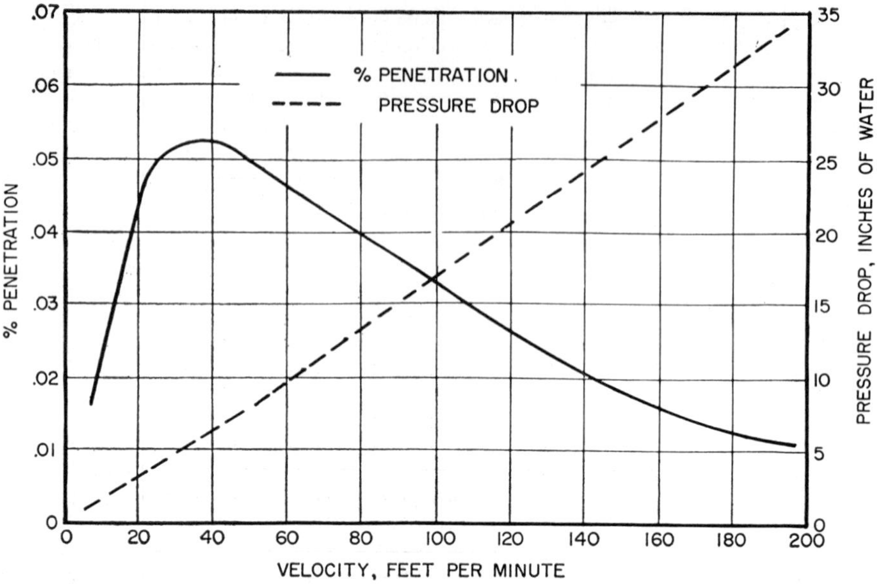

FIG. 13-23. DOP smoke penetration and pressure drop vs. velocity for CWS No. 6 paper (123). (DOP particle size = 0.3μ and concentration = 0.02 grain/cu ft.)

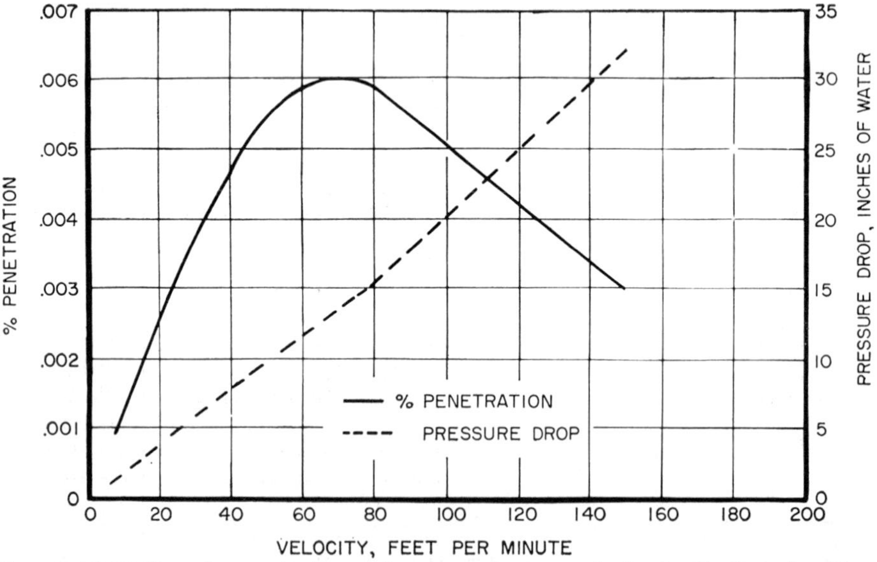

FIG. 13-24. DOP smoke penetration and pressure drop vs. velocity for Hurlbut glass-fiber paper (123). (DOP particle size = 0.3μ and concentration = 0.02 grain/cu ft.)

resin particles. The resin filter has been used in gas masks and, on a limited scale, in bag collectors; as yet, it has found little use in cleaning atmospheric air or process gases. Ideally, the removal efficiency of a filter in which all the fibers are arranged transverse to the flow is given by the expression

$$E = 1 - e^{-\frac{4}{\pi}\frac{\eta_e}{D_c}\left(\frac{1-\alpha}{\alpha}\right)1} \tag{13-50}$$

For a given bed thickness and porosity, the efficiency depends on the ratio of the effective fiber efficiency to fiber diameter, η_e/D_c. The individual fiber efficiency is due to (1) impaction, (2) diffusion, (3) settling, and (4) electrostatic effects. Settling is of importance for particles larger than about 5 or 10μ, impaction for particles larger than 1μ, and diffusion for the submicron range. Both diffusion and impaction increase rapidly as fiber diameter is reduced. Viscous filters are run at high velocities since they depend largely on impaction, but because of the coarseness of their fibers, they are seldom effective for particles smaller than about 5 or 10μ. The dry filters, particularly the glass-fiber type, are somewhat more effective for smaller particles since they are composed of finer fibers. The high-efficiency filters contain many fine fibers, in some cases as small as 0.5μ.

Much progress has been made toward the prediction of filtration efficiency from the properties of the mat and aerosol. The principal difficulty has been the determination of accurate values for η_e. When electrostatic effects and settling can be neglected and when an average fiber diameter can be obtained from microscopic examination, an estimate of filtration efficiency can be made by calculating values for ψ, $\mathfrak{D}$, and $\mathfrak{R}$, checking Re to be sure it is in the viscous range, and then using Figs. 13-15 and 13-16 to obtain impaction and diffusion efficiencies. In general, the parameters will be such that either impaction or diffusion is controlling, but when both are significant, a simple addition of the two efficiencies is probably satisfactory.

A considerable amount of data is available on the pressure drop across fibrous filters; Chen (18) reviews the literature on the subject and from his own research suggests the following expression for mats with porosities below 0.99:

$$\Delta p = \frac{4}{\pi} \frac{K_2}{\ln (K_3\alpha^{-0.5})} \left(\frac{\alpha}{1 - \alpha}\right) \frac{\mu V_s 1}{D_{cs}^2} \tag{13-51}$$

where K_2 and K_3 are constants which depend on the method of formation of the mat. The pressure drop is greatest for densely packed mats composed of small fibers. As the dust load accumulates, pressure drop increases, air throughput decreases, and it becomes necessary to renew the filter medium. For air-conditioning filters, renewal is necessary at intervals ranging from a month to a year, depending on the concentrations. Some operators change filters after the resistance has risen to two or three times its initial value, i.e., a final resistance of about 0.5 in. of water. Another rule of thumb is to allow the pressure drop to increase until the air rate is decreased by 10 per cent.

Cloth Filters. A common cloth filter consists of tubular bags 5 to 18 in. in diameter and 2 to 30 ft in length (Fig. 13-25). Bags for mineral dust are shorter and narrower than those for metallurgical fumes such as lead and zinc oxide. The tubes are suspended with open ends attached to an inlet manifold either at the bottom or top of the housing or both; the lower manifold also serves as a receiving hopper for the dust. As the air enters, it impinges on a baffle plate, causing the larger particles to fall to the hopper. It then passes through the tubes, depositing its particles on the inner surfaces of the cloth. The accumulation of dust on the cloth surface gradually increases the resistance, and it is necessary to vibrate the tube at intervals to detach the layers. For this purpose, the bags are fitted with a shaking device, usually at the top.

Instead of a bag, the filter may be in the form of a cloth envelope pulled over a wire-screen frame like a pillow case (Fig. 13-26). The assembly is commonly 1½ to 3 ft wide, 3 to 4 ft long, and 1 or 2 in. thick. In contrast with the bag filter, air passes from the outside of each envelope into the frame unit and thence out of the collector; if the flow were reversed, the envelopes would be distended and the seams subjected to strain. Screen frames are fitted with a shaking or rapping device for removing accumulated dust.

A more common cleaning arrangement is the intermittent type in which the tubes or screens are vibrated by periodic operation of the shaking device while the exhaust fan is off. Cleaning periods generally last 5 or 10 min and may take place after each 4 or 5 hr of operation. The intermittent type is economical in first cost and maintenance and is to be preferred when short shutdowns are permissible. When these are undesirable or where fluctuations in resistance must be minimized, compartmented continuous-automatic units are employed, in which one group of bags or screens is cleaned while the others are operating. The control of the air flow may be manual or by automatic control equipment. In the continuous equipment, the dampers are often arranged so that air is drawn in the reverse direction through the section being cleaned. Both the initial cost and maintenance for the continuous type are higher than for the intermittent collector, particularly when the unit is highly automatic.

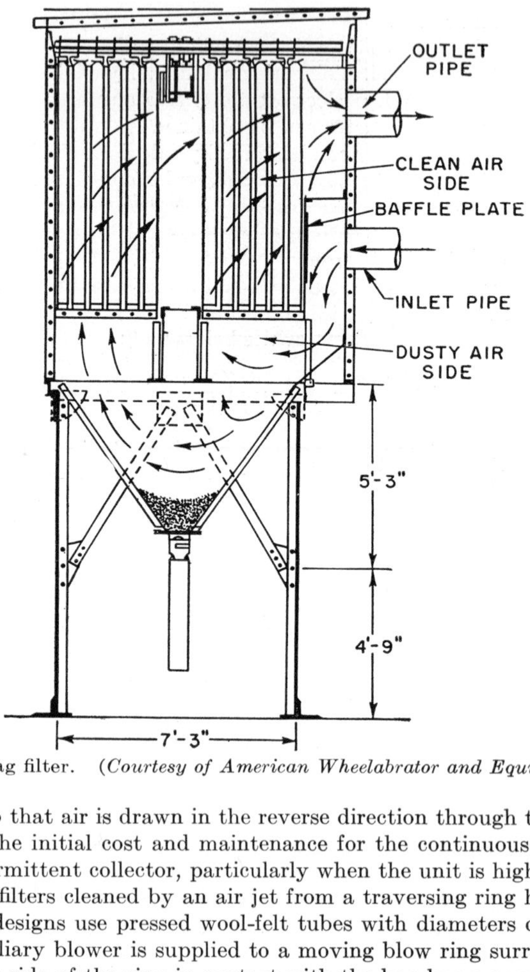

Fig. 13-25. Bag filter. (*Courtesy of American Wheelabrator and Equipment Corp.*)

Recently, bag filters cleaned by an air jet from a traversing ring have been developed. Current designs use pressed wool-felt tubes with diameters of 9 in. or more. Air from an auxiliary blower is supplied to a moving blow ring surrounding the felt tube. The inner side of the ring in contact with the bag has a narrow slot through which the cleaning air passes into the tube, in reverse direction to the normal flow. The traversing ring may work intermittently or continuously and is actuated by the pressure drop across the bag. The reverse-jet filter requires no shutdowns and runs at 10 to 30 fpm as compared with 1 to 6 fpm for the usual cloth filter. In addition, the cloth resistance can be maintained at a nearly constant value in contrast with other cloth filters in which the resistance builds up to the design value.

Initial costs of intermittent cotton-cloth collectors vary from about $1 per sq ft for smaller sizes to $0.70 per sq ft for larger units. Continuous, compartmented collectors vary from about $1.50 per sq ft for smaller sizes to $1 per sq ft for larger units. These figures do not include charges for piping systems, exhauster, drives, motors and controls, wiring, freight, or erection. A cost breakdown for an envelope-screen collector is given in Table 13-16. The initial costs for reverse-jet units are considerably higher than those of the other cloth filters. This higher cost is justified according to their manufacturers by superior performance and by lower power consumption.

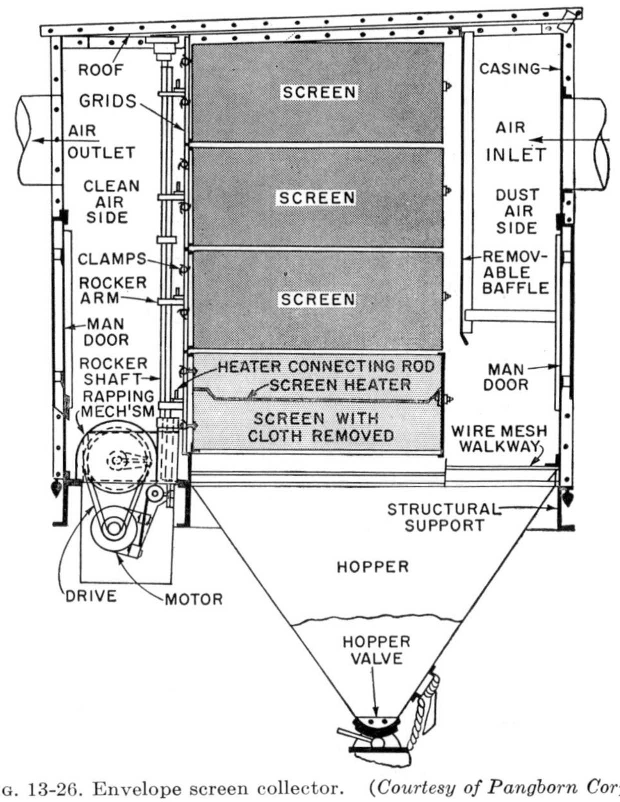

Fig. 13-26. Envelope screen collector. (*Courtesy of Pangborn Corp.*)

Table 13-17 (104) lists the properties of commercially available filter fabrics. Values are given for tensile strength, coefficient of resistance, maximum operating temperature, and relative costs based on a 5- by 70-in. tube. Cotton, the material commonly used, is satisfactory at temperatures below 180°F when acid gases are not present. The more acid-resistant wool fabrics serve for the collection of metallurgical fume and for extremely fine and abrasive dusts such as cement and diatomaceous earth. Recently, several new cloths have been developed, of which the U.S. Rubber Company's Asbeston and du Pont's Orlon are of particular interest as filter fabrics. Asbeston, a silicone-impregnated weave of 15 to 20 per cent cotton fibers and 80 to 85 per cent asbestos fibers, withstands temperatures of 275°F and perhaps higher but is attacked by acid vapors. Orlon can withstand similar temperatures and resists acid. These new fabrics permit the use of cloth filtration for effluents from spray dryers, sintering machines, and blast furnaces.

Table 13-16. Breakdown of Costs for 9,000-sq-ft Envelope-screen Collector

Filter	Cost	Classifier (a baffle chamber), 6 ft long	Shaking mechanisms	Support, 10 ft	Fan	V belt	Motor, 40 hp	Fan pipe	Total	Freight, foundations, erection, piping, wiring	Total	Cost per sq ft
Intermittent..........	$4,398	$493	$134	$809	$1,273	$165	$1,154	$90	$9,056	$6,000	$15,056	$1.68
Automatic continuous..	5,460	493	2,081	904	1,273	165	1,154	90	11,620	7,000	18,620	2.08

Data from S. K. Friedlander et al., "Handbook on Air Cleaning" (A.E.C.D.-3361; NYO-1572), Harvard University, School of Public Health, Boston, Mass., 1952.

The total pressure drop across a cloth filter is composed of the resistances of the filter medium and of the accumulated dust layer. In general, the flow through both the cloth and the dust layer is in the streamline region, and the resistance of each component is therefore proportional to the velocity and to the cloth or layer thickness. The total pressure drop can then be expressed as the sum of the two resistances:

$$\Delta p = K_0 V_s + K_1 V_s w \qquad (13\text{-}52)$$

Values of the cloth resistance coefficient K_0 are given in Table 13-17, while values for K_1 are given in Table 13-18. As shown in Table 13-19, resistances for bag- and

Table 13-17. Properties of Selected Fiber Fabrics—Typical Values for New Cloth

Cloth	Style	Weaver	Thread count		Tensile strength, lb/linear in.	Coefficient of resistance K_0 (in. H_2O/fpm)	Maximum operating temp, °F	Cost factor[a]
			Warp	Fill				
Cotton	F-11	Wellington	46	56	180	0.025	180	1.00
Cotton	F-12	Callaway	104	68	200	0.028	180	1.00
Wool, white	E-22	Albany	36	32	50	0.0027	215	
Wool, black	E-21	Pendleton	28	30	40	0.00455	215	2.43
Wool	19-C	Pendleton	30	26	140	0.0091	215	
Vinyon	D-21	Stevens	37	37	220	0.022	200	2.08
Nylon	C-11	Wellington Sears	37	37	275	0.031	225	2.58
Nylon	C-12	Stevens	...	..	400	0.07	225	2.08
Asbeston	A-11	U.S. Rubber	...	..	100	0.01	275	3.80
Orlon	B-21	Stevens	72	72	170	0.012	275	3.55
Orlon	B-22	Stevens	74	38	270	0.014	275	4.93
Orlon	B-23	Stevens	...	..	200	0.021	275	3.55
Orlon	B-24	Burlington	...	..	90	0.036	275	5.50

[a] Based on a 5- by 70-in. tube.

Data from R. T. Pring, Bag-type Cloth Dust and Fume Collectors, in L. C. McCabe, ed., "Air Pollution," p. 280, McGraw-Hill Book Company, Inc., New York, 1952.

Table 13-18. Filter-resistance Coefficients[a] K_1 for Certain Industrial Dusts on Cloth-type Air Filters

Dust	K_1 for particle size less than						
	20 mesh[b]	140 mesh[b]	375 mesh[b]	90μ[c]	45μ[c]	20μ[c]	2μ[c]
Granite	1.58	2.20	....	....	.....	19.8	
Foundry	0.62	1.58	3.78				
Gypsum	....	....	6.30	....	.....	18.9	
Feldspar	....	....	6.30	....	.....	27.3	
Stone	0.96	....	....	6.30			
Lampblack	....	....	....	....	.....	....	47.2
Zinc oxide	....	....	....	....	.....	....	15.7[d]
Wood	....	....	....	6.30			
Resin (cold)	....	0.62	....	....	.....	25.2	
Oats	1.58	....	....	9.60	11.0		
Corn	0.62	....	1.58	3.78	8.80		

[a] Inches of water per pound of dust per square foot of cloth per foot per minute of filtering velocity.

[b] Coarse.

[c] Less than 90 or 45μ, medium; less than 20 or 2μ, fine; theoretical size of silica, no correction made for materials having other densities.

[d] Flocculated material, not dispersed; size actually larger.

Data from C. E. Williams et al., Determination of Cloth Area for Industrial Air Filters, *Heating, Piping Air Conditioning*, **12,** 259–263 (1940).

Table 13-19. Performance Data for Cloth Filters

Device	Manufacturer	Aerosol	Inlet concentration, grains/cu ft	Mass median particle size at inlet, μ	Effluent concentration, grains/cu ft	Mass median particle size at exit, μ	Efficiency, wt %	Penetration, wt %	Resistance, in. H_2O	Velocity, fpm	Flow rate, cfm	Remarks
Aeroturn	Turner and Haws Eng. Co.	Silicon carbide and aluminum oxide dusts from truing and shaping of abrasive products	0.13	1.40	0.001	0.55	99.2	0.8	4	17.0	18,800	A Hersey type bag filter. 12 bags 1.5 ft diam, 20 ft long, ⅛ in. pressed wool-felt cloth
Aeroturn, *a*	Turner and Haws Eng. Co.	Silicon carbide and aluminum oxide dusts from truing and shaping of abrasive products	2.26	2.50	0.0009	1.25	99.9	0.04	1.35	5.5	930	Aeroturn with interior partition acting as two independent units of equal capacity
b	. . . . . .	. . . . . .	0.66	1.30	0.0008	0.53	99.8	0.121	. . . .	7.2	1,230	4 bags, 1.5 ft diam, 9 ft long, ⅛ in. pressed wool-felt cloth
Aeroturn	Turner and Haws Eng. Co.	Beryllium oxide dust	0.0037	. . . .	0.00000425	. . . .	99.88	0.115	2.8	12.7	7,200	6 bags, 1.5 ft diam, 20 ft long, ⅛ in. pressed wool-felt cloth
Aeroturn	Turner and Haws Eng. Co.	Jewelers' rouge (iron oxide) and lint from buffing and polishing watch cases	0.0048	0.90	0.00012	0.57	97.5	2.5	3.6	27	5,000	4 bags, 1.5 ft diam, 12 ft long, ⅛ in. pressed wool-felt cloth
Aeroturn	Turner and Haws Eng. Co.	Jewelers' rouge (iron oxide) and lint from buffing and polishing watch cases	0.0094	1.21	0.0022	0.74	97.6	2.4	6.4	32	6,000	4 bags, 1.5 ft diam, 12 ft long, ⅛ in. pressed wool-felt cloth
Aeroturn	Turner and Haws Eng. Co.	Tapioca starch granules from drying ovens	3.17	7.7	0.000054	1.1	99.99	0.0015	4.4	11.3	2,100	4 bags, 1.5 ft diam, 11 ft long, ⅛ in. pressed wool-felt cloth
Dustube	American Wheelabrator and Equip. Corp.	Bronze and SiO_2 dusts from casting cleaning operations	0.441	4.5	0.000069	1.7	99.98	0.0156	5.0	0.8	565	94 bags 5 in. diam, 70 in. long, 720 sq, ft cloth area, cotton cloth

Table 13-19. Performance Data for Cloth Filters (*Continued*)

Device	Manufacturer	Aerosol	Inlet concentration, grains/cu ft	Mass median size at inlet, μ	Effluent concentration, grains/cu ft	Mass median particle size at exit, μ	Efficiency, wt %	Penetration, wt %	Resistance, in. H_2O	Velocity, fpm	Flow rate, cfm	Remarks
Dustube........	American Wheelabrator and Equip. Corp.	Iron oxide and SiO_2 dusts from casting cleaning operations	0.68	1.52	0.000015	0.2–0.3 (estimated)	99.99	0.0022	1.5	2.5	7,700	288 bags 12 ft long, 2,820 sq ft cloth area, cotton cloth
Airmat.........	American Air Filter Co.	Fluffy zinc stearate from rubber dusting operation	0.1	1.75	....	......	95	......	......	....	3,300	Pocketed filter cleaned by vibrating at intervals. Paper medium.
Dustube........	American Wheelabrator and Equip. Corp.	Iron scale and sand from casting cleaning operations	0.33	3.45	0.000013	......	99.99	0.0039	2.2	2.2	2,050	96 bags 4 ft long, 940 sq ft cloth area, cotton cloth
Pangborn.......	Pangborn Corp.	Talc dust from rubber dusting	4.3	113	0.002	......	99.85	......	......	....	7,000	
Sly collector.....	Sly Mfg. Co.	Granite dust from chipping	0.032	2.3	0.000028	1.1	99.9	0.088	3.0 (about)	2.3	7,000	3,000 sq ft cloth area, cotton cloth
Sly collector.....	Sly Mfg. Co.	Iron scale and sand from castings cleanings	0.18	1.44	0.000013	......	99.99	0.0072	1.2	1.3	13,900	510 frames, 11,220 sq ft cloth area, cotton cloth
Sly collector.....	Sly Mfg. Co.	Iron scale and sand from castings cleanings	0.39	2.2	0.00063	0.64	99.85	0.16	1.9	1.5	18,200	12,320 sq ft cloth area, cotton cloth

Data from S. K. Friedlander et al., "Handbook on Air Cleaning" (A.E.D.C.-3361; NYO-1572), Harvard University, School of Public Health, Boston, Mass., 1952.

screen-type collectors range from about 2 to 5 in. of water while those of the reverse-jet type are somewhat higher.

The theory of removal of aerosol particles by cloth filters has not been fully developed. The initial deposition on the fabric fibers is probably due to impaction, interception, settling, and perhaps diffusion, but the efficiency at this time is low since the fiber diameters are quite large. Actually, cloth filters depend to a great extent on

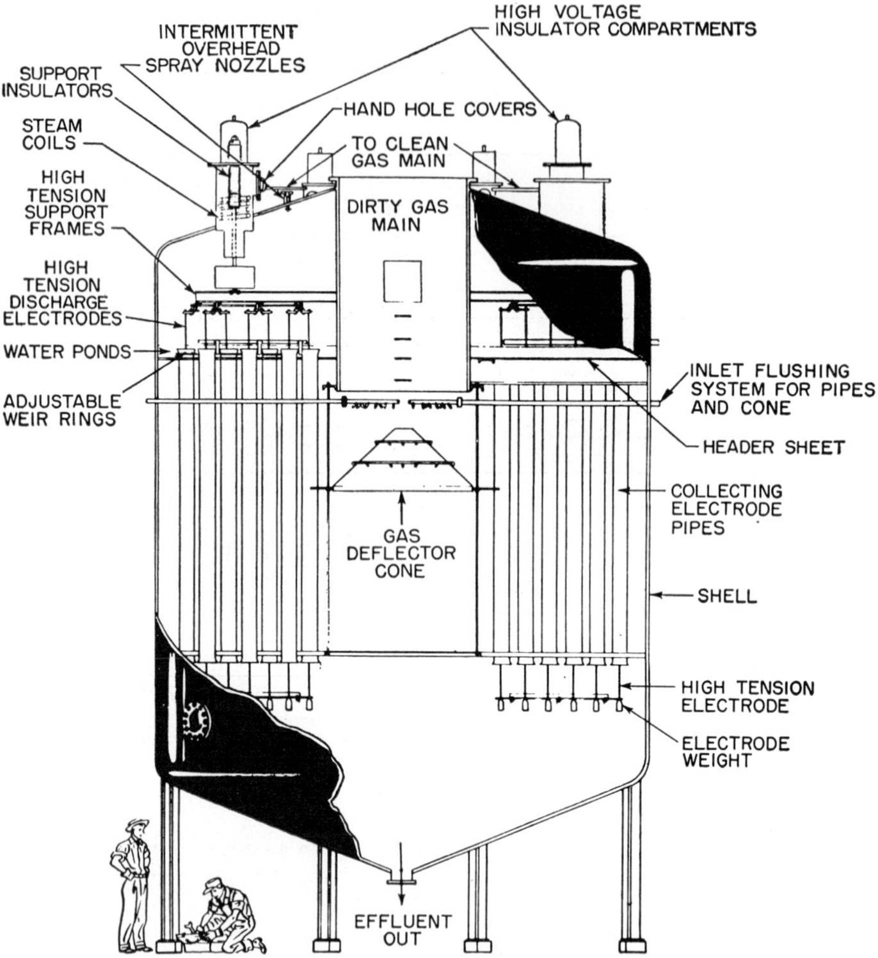

FIG. 13-27. Water-flushed pipe-type precipitator cleaning blast-furnace gases. (*Courtesy of Western Precipitation Corp.*)

the accumulated dust layer for their effectiveness. Once this layer has been built up, efficiencies above 99.9 per cent are not uncommon, and for this reason best results are obtained with high loadings. Separation by the dust layer seems to be due to the mechanisms mentioned above, and in addition sieving or the removal of particles larger than the pores in the layer may also contribute to collection. Table 13-19 gives performance data for various cloth filters.

Attempts have been made to obtain high efficiencies for low-concentration dusts by coating the cloth with a filter aid before starting filtration: over 99.0 per cent removal by weight was reported for atmospheric air with a concentration of 0.025 grain per 1,000 cu ft using a bag filter primed with fine asbestos fibers; filtration velocity was 3 fpm, and the resistance was 0.212 in. of water (37).

Electrostatic Precipitators. TYPES AND FIELDS OF APPLICATION. Electrostatic precipitators may be classified in accordance with the type of use for which they are designed. In general, they include precipitators for air cleaning, industrial purposes, and special uses.

Air-cleaning precipitators for cleaning air in buildings, ships, railway cars, etc., are used to make the air more healthful or pleasant to breathe by removing tobacco smoke, pollen, etc., or to prevent dust from interfering with delicate industrial operations such as manufacture of watches or electronic tubes, or to prevent soiling of draperies, walls, paintings, etc.

Industrial precipitators for collecting dust, smoke, fume, mist, etc., from industrial gases are used to collect fly ash from pulverized-coal-fired boilers, cement-kiln dust, catalyst dust at oil refineries, metallurgical fume, soda fume in pulp mills, sulfuric acid mist, etc. These industrial precipitators may be further classified on the basis of design as pipe-type precipitators and plate-type precipitators.

Pipe-type precipitators, where the gas flow is distributed among numerous vertical pipes usually a foot in diameter or less, have an electrically charged wire suspended along the axis of each pipe. Precipitation of the aerosol particles occurs on the inner pipe walls, from which the material may be dislodged by rapping or flushing with water (Fig. 13-27).

In plate-type precipitators, the gas flow is distributed among several passages or "ducts" between grounded electrodes having a basic structure resembling vertical parallel plates, equally spaced at intervals usually larger than 6 in. and smaller than 14 in. between centers. High-voltage electrodes, such as wires or thin twisted square rods in the mid-plane of each "duct" cause precipitation of the aerosol particles on the grounded electrodes, from which the material is dislodged by rapping, either periodically or continuously. These plate-type precipitators may be further classified on the basis of the direction of gas flow as horizontal-flow (Fig. 13-28) and vertical-flow precipitators.

Special precipitators include small portable precipitators for collecting dust samples and various other experimental or special types. Air-cleaning precipitators are used for the purposes already mentioned, in theaters, public buildings, factories, and in some private homes and clubhouses. Table 13-20 gives the performance characteristics of some precipitators of this type, as tabulated by M. W. First (37).

Industrial precipitators are widely used in the industries and for the applications listed in Table 13-21.

FUNDAMENTAL CONSIDERATIONS AND DESIGN. The electrical precipitator is one of the most versatile devices thus far developed for removing particulate matter from air or other gases. Reduced to its bare essentials, this device consists of (1) a chamber or shell in which the separation of the suspended particles from the gas takes place, (2) electrodes which may be charged to high voltage within the shell, (3) electrical equipment to supply the high voltage required, and (4) means for removing the precipitated material from the electrodes to the desired place without redispersal in the gas.

Air-cleaning precipitators are usually of the two-stage design, in which the first stage consists of a series of fine (0.007 in. diameter) positively charged wires equally spaced at a distance of 1 to 2 in. from a series of parallel grounded tubes or rods and a corona discharge between the wires and the tubes serving to charge the particles suspended in the air as it flows through. The d-c potential applied to the fine wires amounts to 12 or 13 kv, and positive polarity is used rather than negative in order to minimize the formation of ozone.

The second stage consists of parallel metal plates usually less than an inch apart.

In some instances, alternate plates are charged positively and the others negatively to a potential of 6 or 6½ kv direct current, so that the potential difference between adjacent plates is 12 or 13 kv. In other instances, alternate plates are grounded and the others charged to a potential of 12 or 13 kv. These plates may be cleaned periodically by shutting down the unit and flushing with water, or the plates may be automatically dipped in oil and brushed off while submerged, by using an endless-belt or chain principle.

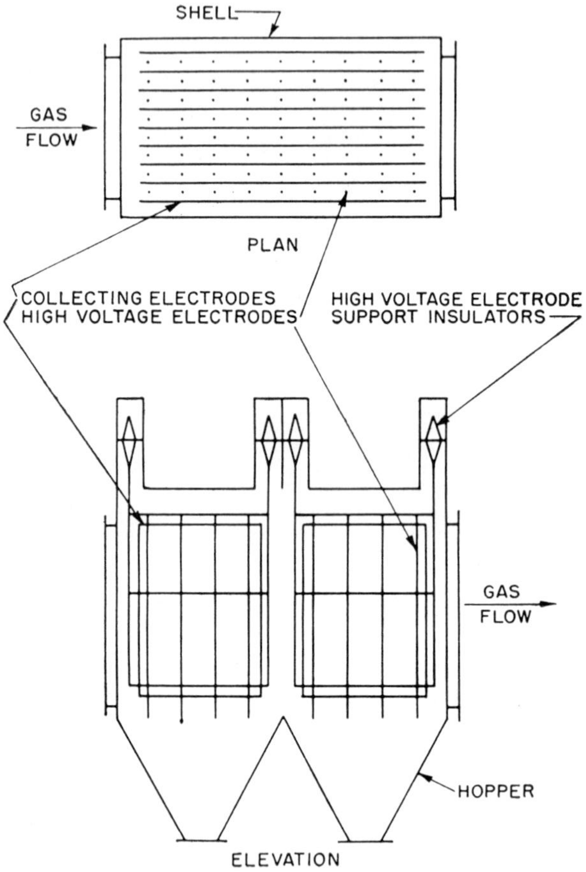

FIG. 13-28. Diagram of horizontal-flow precipitator. (*Courtesy of Western Precipitation Corp.*)

The efficiency of air-cleaning precipitators is usually measured in accordance with a U.S. Bureau of Standards "dust-spot test" at values between 85 and 90 per cent. In such a test, one might observe that 100 cu ft of outlet air drawn through a particular area or spot on a filter paper will darken it just as much as 10 cu ft of inlet air drawn through a second equal area of the same paper. This would indicate a dust-spot efficiency of 90 per cent. Such a 90 per cent stain efficiency may correspond to an efficiency in excess of 99 per cent on a weight basis, because a 10-μ particle does not produce 1,000 times as much blackening as a 1-μ particle, although the weight ratio is 1,000.

In industrial precipitators, the usual procedure is to precipitate the dust upon

Table 13-20. Performance of Air-cleaning Electrostatic Precipitators with Atmospheric Dust

Manufacturer	Test[a] and unit[c]	Method[b] of plate cleaning and coating	Dust description						Air flow rate, STP,[f] cfm	Average face velocity, fpm	Per cent air recirculated (estimated)	Collection efficiency, wt %
			Inlet			Outlet						
			Loading,[d] grains/1,000 cu ft	Median size, μ		Loading,[e] grains/1,000 cu ft	Median size, μ					
				Count	Mass		Count	Mass				
Westinghouse	1(*A*)	Manual washing and water-soluble adhesive[g]	0.0254	0.44	0.90	0.00638	0.39	0.48	21,900	780	50	74.9
	2(*A*)		0.0302	0.44	0.85	0.00803	0.40	0.46	27,150	565	50	76.4
	3(*A*)		0.031	....	....	0.0034	....	....	15,700	154	65	89.4
	4*a*(*A*)		0.0224	....	....	0.011	....	....	15,700	253[h]	50	50.9
	4*b*(*A*)		0.061	0.45	0.74	0.0221	0.39	0.47	15,700	253[h]	50	63.7
AAF	5(*B*)	Automatic cleaning by oil dipping	0.0584	0.44	0.54	0.0171	0.42	0.47	22,000	389	50	70.8
	6(*B*)		0.0632	0.45	0.52	0.0058	0.42	0.48	4,580	255	50	90.9
	7(*C*)		0.0307	0.43	0.58	0.00107	0.39	0.48	7,370	140	50	96.6
Trion	8(*D*)	Built-in water sprays, no adhesive coating	0.0302	0.51	0.63	0.0032	0.46	0.56	1,800	322	45	89.5
	9(*D*)		0.00902	0.42	0.62	0.00228	0.41	0.51	2,200	297[h]	90	74.7

[a] Cell voltages: ionizer 12 to 13.5 kv, plate 6 to 6.5 kv.
[b] Total plate depth: 12 in. (Westinghouse), 10 in. (AAF), and 10.75 in. (Trion).
[c] Letters in parenthesis indicate different designs: (*A,D*) fixed vertical plates; (*B*) moving vertical plates, 2 rev per 24 hr; (*C*) moving horizontal plates, 2 rev per 24 hr.
[d] Outside air in units 1 to 4 precleaned with coarse Fiberglas filters.
[e] Loading refers to concentration downstream of coarse metal-fiber afterfilters (integral parts of units 5 to 9).
[f] STP = 70°F, 760 mm Hg.
[g] No adhesive on plate.
[h] Voltage fluctuation during tests. Arcing in unit 4, low plate current in unit 9.

NOTE: Pressure losses across all units varied from 0.05 to 0.1 in. water.

Data from M. W. First et al., Air Cleaning Studies, Progress Report for Feb. 1, 1951, to June 30, 1952, AEC Report No. NYO 1586, Harvard University, School of Public Health, Air Cleaning Laboratory, Boston, Mass., 1952.

specially constructed "collecting electrodes" which are grounded. The opposing electrodes, usually consisting of slender rods or wires, are electrically insulated and ordinarily charged to a negative potential between 15 and 100 kv. These are called the "discharge electrodes" because they ionize the gas surrounding them and set up

Table 13-21. Common Applications of Industrial Precipitators

Industry	Application	Gas flow range, cfm	Temp range, °F	Dust conc range, grains cu ft	Per cent weight of dust (below 10)	Usual efficiencies, %	Cost range, per cfm
Electric power	Fly ash-pulverized-coal-fired boilers	50,000–750,000	270–600	0.4–5.0	25–75	95–98	$0.50–$0.85
Portland cement	Dust from kilns	50,000–1,000,000	300–750	0.5–15.0	35–75	85–99+	1.00–1.50
	Dust from dryers	30,000–100,000	125–350	1.0–15.0	10–60	95–99	0.60–1.00
	Mill ventilation	2,000–10,000	50–125	5.0–25.0	35–75	95–99	1.50–3.00
Steel........	Cleaning blast-furnace gas for fuel	20,000–100,000	100–150	0.02–0.5	100	95–99	2.00–3.00
	Collecting tars from coke-oven gases	50,000–200,000	100–150	0.1–1.0	100	95–99	0.75–1.75
	Collecting fume from open-hearth and electric furnaces	30,000–75,000	300–700	0.05–3.0	95	90–99	1.00–3.00
Nonferrous metals	Fume from kilns, roasters, sintering machines, aluminum pot lines, etc.,	5,000–1,000,000	150–1100	0.05–50.0	10–100	90–98	2.00–10.00
	Acid mist	See chemical industry					
Pulp and paper	Soda-fume recovery in kraft pulp mills	50,000–200,000	275–350	0.5–2.0	99	90–95	1.00–2.50
	Acid mist	See chemical industry					
Chemical....	Acid mist	2,500–20,000	100–200	0.02–1.0	100	95–99	2.00–3.50
	Cleaning hydrogen, CO_2, SO_2, etc.	5,000–20,000	70–200	0.01–1.0	100	90–99	1.50–5.00
	Separate dust from vaporized phosphorus	2,500–7,500	500–600	0.01–1.0	30–85	99+	5.00–10.00
Petroleum...	Powdered catalyst recovery	50,000–150,000	350–550	0.1–25.0	50–75	90–99.9	1.50–2.50
Rock products	Roofing, magnesite, dolomite, etc.	5,000–200,000	100–700	0.5–25.0	30–45	90–98	1.50–10.00
Gas.........	Tar from gas	2,000–50,000	50–150	0.01–0.2	100	90–98	1.00–1.75
Carbon black	Collecting and agglomerating carbon black	20,000–150,000	300–700	0.03–0.5	100	10–35	0.75–1.50
Gypsum....	Dust from kettles, conveyors, etc.	5,000–20,000	250–350	1.5–5.0	95	90–98	2.10–3.50

an electrical discharge, which charges the particles of the dispersoid. The gas ordinarily is carried to the precipitator through ducts in which the gas velocity is 30 to 60 fps. Care is required in designing the inlet "plenum chambers," distribution plates, etc., in order to distribute this gas so that it will flow at a lower but nearly uniform velocity throughout any transverse cross section of the precipitator (which may be, for example, 50 ft wide and 20 ft high).

Since the gas is passed through the precipitator at velocities of only a few feet per second, the pressure drop is small (¼ to ½ in. of water in plate-type precipitators and ½ to 1 in. in pipe types). If 90 per cent of the suspended material is precipitated by passing the gas through the precipitator, then in general, if the dust-particle size, gas conditions, etc., remain the same, 90 per cent of the remaining 10 per cent can be precipitated by passing the gas through a second identical precipitator, thus raising the efficiency from 90 to 99 per cent. In this case, instead of describing the system as "two precipitators in series," it is usually described as one precipitator having "two sections." By adding a third section in this example, one should expect to attain an efficiency of about 99.9 per cent. Thus the performance (efficiency) of a precipitator is a logarithmic function of its size and cost.

Precipitators have been built in practice for operation at atmospheric pressure and at pressures above and below atmospheric. They have been built for operation at temperatures from about −70 to +1000°F, although operating temperatures above 700°F or below zero are unusual. The material collected has varied in particle size from about 100μ down to 0.01μ or less and includes such diverse categories as those tabulated for 11 different industries in Table 13-21. The concentration of the material entering the precipitator may vary from a small fraction of a grain per cubic foot of gas to 10 grains or more per cubic foot.

Although the versatility of the electrical precipitator has just been emphasized, it does have important limitations, and it should not be pictured as a universal cure-all for every gas-cleaning problem. Four of its most important limitations are as follows:

1. Only particulate matter can be precipitated; one gas cannot be separated by a precipitator from another gas or from a vapor without first condensing the vapor to a fume or mist or introducing some preliminary chemical reaction. A constituent which vaporizes at 500°F, for example, will pass freely through a precipitator in which the gas has a temperature of 600°F and will escape up the stack. Upon reaching the open air, however, this constituent will condense and make it appear as though dense clouds of the material were passing through the precipitator.

2. The physical and electrical characteristics of some materials prevent them from being collected efficiently by a precipitator. One example is carbon black, which has a very low electrical resistivity and an extremely small particle size as it escapes from the burners where it is manufactured from oil or gas. Moreover, it is very light and fluffy. The particles are readily precipitated, but being highly conducting they lose their electrical charge upon touching the collecting electrode and consequently will not adhere to it. Instead, the particles are reentrained in the gas stream, and only a small fraction of the material collects in the hoppers, most of it escaping with the outlet gas which is usually led to large cyclonic collectors. These are able to collect the carbon black efficiently because the particles agglomerate as they pass through the precipitator. Another example is zinc oxide fume, which "quenches" a corona discharge and is moreover very light and fluffy like carbon black. Probably the most important example includes various materials which have such a high electrical resistivity that they coat the collecting electrodes with a highly insulating coating and cause a condition known as "back discharge," which seriously impairs the performance of a conventional precipitator (112). The addition of moisture to the gas will reduce the resistivity and is the usual remedy for back discharge. "Conditioning agents" such as acid mist, ammonia, etc., are also used.

3. The precipitator may cost more than other devices which can be substituted satisfactorily in some cases. For example, if the temperature is low and the material is dry and not corrosive, a bag filter may be cheaper. Or if the material is coarse,

a centrifugal collector may be cheaper. Both of these devices require a much greater pressure difference than a precipitator, however.

4. The cost of a precipitator in terms of dollars per cubic foot of gas treated each minute becomes rapidly greater as the gas volume decreases below, say, 10,000 cfm. Thus, an ordinance requiring every home incinerator and trash burner in a city to be equipped with an electrical precipitator to collect the smoke would be difficult or impossible to enforce because of the prohibitive cost.

Nearly all industrial precipitators built to date have been of the "single-stage" type, in which a single system of electrodes is provided which is required to perform two distinct functions, namely, (1) charge the aerosol particles by bringing them into a corona discharge and (2) precipitate the charged particles by bringing them into a strong electrostatic field which pushes them over against the "collecting electrodes."

"Two-stage" precipitators, first disclosed in a patent by W. A. Schmidt (111) in 1920, have a first stage for charging the particles and a second stage for precipitating them. Very few two-stage precipitators for industrial service have been built to date, but the two-stage design is usual in air-conditioning precipitators.

The mathematical relation between the efficiency of a precipitator, its size and cost, and the rate at which gas flows through it is best expressed by an equation derived by W. Deutsch (31) in 1922. It applies equally well to pipe-type or plate-type precipitators and is

$$E = 1 - e^{-u_t f} \tag{13-53}$$

In this equation, E is the decimal efficiency; that is, if the precipitator catches 9 lb of material out of every 10 lb of suspended material entering, then the efficiency is 90 per cent and $E = 0.90$. The letter f represents the specific collecting area of the precipitator, expressed as square feet of collecting electrode area per cubic foot of gas handled per second, or sq ft/cu ft-sec. The symbol u_t represents the drift velocity, which is the average velocity at which the (charged) dust particles drift laterally toward the collecting electrode under the influence of the strong electric field, expressed in fps, and e is the Napierian log base. Since the equation originated in Europe, u_t is often given in terms of cm/sec. European practice is to express f in sq m/cu m-sec, calculate u_t in m/sec, and then multiply by 100 to convert to cm/sec.

An analysis of the Deutsch equation yields considerable insight into the factors affecting the design of a precipitator. It is seen at once that, in order to attain high efficiency, the value of the product $u_t f$ must be as high as practical. The term f is a rough measure of the expense of a precipitator in terms of cost per cfm of gas treated, and it increases rapidly as the required efficiency approaches 100 per cent, where f theoretically becomes infinite. For a given efficiency, any reduction in the drift velocity u_t must be compensated by a corresponding increase in f. For example, the material being collected may consist of unusually fine particles, or back-discharge conditions may exist; if such circumstances reduce u_t by 50 per cent, then to prevent a decrease in efficiency, f must be doubled, thus approximately doubling the cost of the precipitator.

The drift velocity u_t, on the other hand, can be made greater in at least four different ways: (1) by charging the dust particles with the highest electrical charge possible and keeping them highly charged while they are in the precipitator, the electrostatic force causing the particles to drift toward the collecting electrodes is increased; (2) by maintaining as high an electrostatic field as possible between the electrodes, the force which causes the particles to drift is increased; (3) by making particles agglomerate, since the larger the particles are, the faster they will drift through the gas toward the collecting electrodes, just as coarse dust will settle out by gravity in a quiet chamber faster than fine dust (Stokes' law); and (4) by reducing the viscosity of the gas, which will also increase the drift velocity, just as a pebble will fall faster in water than it will in syrup.

Stokes' law states that the resisting force $\mathfrak{F}_r$ which a spherical particle of radius a encounters in drifting through a gas of viscosity μ at a velocity u_t is

$$\mathfrak{F}_r = 6\pi u_t a\mu \tag{13-54}$$

To propel the particle, therefore, an electrostatic force $\mathcal{E}q$ must be exerted equal to $\mathfrak{F}_r$, where $\mathcal{E}$ is the electric field strength and q the charge on the particle. Consequently,

$$6\pi u_t a\mu = \mathcal{E}q \tag{13-55}$$

and

$$u_t = \frac{\mathcal{E}q}{6\pi a\mu} \tag{13-56}$$

where $\mathcal{E}$ is in cgs electrostatic units, q is in statcoulombs, a is in cm, μ is in poises (dyne-sec/sq cm), and u_t is in cm/sec.

One might infer from this equation that the drift velocity is greater for small particles than for larger ones, but this is untrue because q, the particle charge, is proportional to the square of the radius, when the radius a is greater than 1μ, and hence u_t is proportional to a. For particles having a radius less than 1μ, q is nearly proportional to the radius a, and so the drift velocity of these very fine particles is practically independent of their radius.

Equation (13-56) shows that the gas temperature will affect the operation of a precipitator because the sparking potential decreases in inverse proportion to the absolute temperature. Thus, at 600°F (1060°R),* the sparking voltage in a clean precipitator will be only half as much as it is at 70°F (530°R), which means that $\mathcal{E}$ in Eq. (13-56) will be reduced 50 per cent. Moreover, the viscosity of air at 600°F is about 0.0003 poise as compared to 0.00018 at room temperature, the viscosity of most common flue gases behaving in about the same way. These two factors (sparking voltage and viscosity) therefore reduce the drift velocity at 600°F to a value only about 30 per cent of its value at 70°F. Neglecting other factors, in order to maintain the same efficiency at 600°F as at 70°F, then one must increase the size and cost of the precipitator by about 3⅓ times, assuming the same gas flow rate (cfm) to accomplish the same performance. In an actual case, however, doubling the absolute temperature of the gas would double its volume, thus doubling the flow rate in cfm, so that the area of the collecting electrodes would have to be doubled in order to keep the specific collecting area f from decreasing. Thus, even if one neglects such important considerations as the resistivity of the dust, the electrode area should be increased six or seven times in order to handle the gas at 600°F instead of 70°F, if the efficiency is to be maintained undiminished. Such an increase in the gas temperature, however, introduces other important considerations, which cannot be neglected. For example, the electrical resistivity of the dust will be affected, and this may introduce back-discharge conditions which will require a still further increase in precipitator size if the efficiency is to be maintained undiminished.

Increasing the moisture content of the gas, on the other hand, is helpful, for this can nearly double the sparking potential if as much as 30 per cent moisture is added at 600°F, for example, and the addition of moisture also cools the gas and helps avoid back-discharge troubles, as already mentioned.

The collecting electrodes in industrial precipitators of the plate type may consist of vertical steel rods (rod curtains) to minimize the build-up of a clinging dust coat. Other types intended to minimize reentrainment of light, fluffy dusts include wire screens, perforated or "expanded" sheet steel, specially shaped steel channels having pockets to protect the precipitated dust from the gas stream, screens or expanded-metal covering pockets (screen pockets), metal boxes provided with slots, and others. In cases where corrosion is a severe problem, stainless steel may be used, and in collecting acid mist, a pipe-type precipitator with lead pipes is usual. In plate-type

* °R = Rankine scale or absolute Fahrenheit.

precipitators, concrete-plate electrodes have been used occasionally to facilitate collection of material of high resistance.

Except for these concrete-plate precipitators and water-flushed pipe-type precipitators, it is ordinarily necessary to dislodge the precipitated material from the collecting electrodes and drop it into the hoppers by rapping the collecting electrodes. The concrete-plate electrodes used to date will not withstand rapping shocks, and in some cases the dust is allowed to build up on them until it sloughs off by its own weight. In other cases, mechanical scrapers are used. The high-tension discharge electrodes are commonly rapped also to prevent excessive dust build-up and caking on the discharge rods or wires. Rapping may be accomplished by means of devices which are automatically actuated at suitable intervals by pneumatic or electromagnetic means. If the rapping intervals are shorter than 1 min, the process is usually called "continuous rapping." Otherwise, the rapping interval is usually at least 10 min, and may be as long as 8 hr. In order to avoid a "puff" of dust escaping from the stack during rapping when the interval is long, large precipitators are sometimes subdivided into two or more parallel units so that each one can be isolated by closing dampers while it is rapping, the gas meanwhile passing through the other units.

In general, all air-cleaning precipitators follow the same basic design, with different manufacturers offering innovations in such features as the endless-belt oil cleaning already mentioned, or in "packaged units," or in water-flushing nozzle systems, etc.

Likewise, the various manufacturers of industrial precipitators adhere to similar designs, but there is more variation than in the air-conditioning types. For example, there are occasional instances of two-stage design, and in the conventional single-stage designs there are variations in discharge electrode systems, some using wires and weights, others using twisted square rods in a supporting frame. Other variations include horizontal as opposed to vertical gas flow for the same application, mechanical rectification as opposed to vacuum-tube rectification, electromagnetic rapping as opposed to pneumatic, channel pockets as opposed to screen pockets, etc.

Cost Data and Power Requirements. Air-cleaning precipitators sell in the United States for prices ranging between \$0.14 and \$0.20 per cfm at the present time (1953), the smaller figure applying to very large installations. The power requirement to supply the high voltage is about 50 watts plus 10 watts for each thousand cfm of air handled by the installation. The power is so small that it is usually handled from a single-phase 115-volt line.

Industrial precipitators vary much more widely in cost than air-conditioning types, the range being somewhere between \$0.25 per cfm and \$20.00 per cfm, with \$1.00 per cfm being a rough average figure. The cost per cfm will be high if expensive construction like water-flushed pipes is required, if corrosive materials like sulfuric acid mist require that the precipitator be built of expensive materials like lead, or if the particles are extremely fine, as in fume. In general, if excessive resistivity, or extremely fine particle size, or any other factor makes the drift velocity u_t in Eq. (13-53) as low as 5 cm/sec or less, the specific collecting area f, which represents precipitator cost per cfm, will be high. If the material precipitates easily so that it has a value of 10 cm/sec or more for u_t, then f can be small and the precipitator cost per cfm will be relatively low.

The power requirements for energizing a commercial precipitator may be as high as 50 kw for 50,000 cfm, or as low as 15 kw for 500,000 cfm, in exceptional cases, with 15 kw for 100,000 cfm being a usual figure. Abnormal power consumption may be caused by back-discharge conditions, for example, and subnormal consumption may result from corona quenching by certain fumes.

For applications where the temperatures are below 300°F, where the gas and the particulate matter are both dry and noncorrosive, and where a pressure drop of an inch or two of water can be tolerated, it is usually less expensive to build a bag filter than to build a precipitator for high efficiencies such as 99 per cent or more. However, the maintenance expense for replacement of bags and shaking mechanisms is usually higher for a bag filter than for the equivalent precipitator, and additional power is required to produce the necessary pressure drop.

If the necessary quantities of water are available, and if the collection of the material in the form of a slurry is no disadvantage, then a washer or scrubber may be cheaper than the equivalent precipitator. However, maintenance costs of such equipment are usually higher, and considerable power is required to make the operation effective. Most types of scrubber do not effectively collect material in the size range below 1μ.

Sonic Precipitators (58). The development of an efficient siren-type sound generator (Ultrasonic Corp.) has made possible the application of high-frequency sound waves to industrial gas cleaning (23). These generators consist of a rotor and stator with precision-matched ports around the periphery of each. A compressed gas is passed through the ports of the rotor and then through the stator. As the rotor turns, alternately opening and closing the ports of the stator, the gas flows intermittently through the stator ports. An intense sound wave is thereby created at the interface. This is directed from the generator by an acoustic horn. The intensity of the sound wave is controlled by the pressure of the compressed gas and the frequency by the speed at which the rotor turns. About 225 cfm of gas compressed to 8 psig is required to operate a small-scale generator. The gas flow is obtained from a compressor requiring about 10 hp. Such a unit processes 3,000 cfm of dust-laden gas. Larger models of sound generators have been developed with gas capacities up to 50,000 cfm. The intensity of the acoustic field ranges upward of 150 db, and the frequency is 1,000 to 10,000 cycles/sec. The higher frequencies are required for agglomeration of the smaller particles. The generator efficiency and the effectiveness of the treating chamber are both functions of the frequency, so that a compromise must be made based on development experience. Power consumption for aerosol agglomeration is normally in the range of 2 to 5 kw per 1,000 cfm of gas. The concentration of the aerosol to be treated should exceed 1 grain/cu ft. If the concentration is too low, it may be necessary to add a secondary aerosol such as a water mist.

Sonic agglomeration has been used for the removal of sulfuric acid mists from contact acid gases and for the collection of soda-ash fume from a kraft-paper mill. The acid gases are first humidified and cooled by passing through water sprays. In both installations the agglomeration chamber is of sufficient size to allow an exposure of less than 4 sec to the sound waves before the gases are passed through a cyclone separator for the recovery of the enlarged particles. Collection efficiencies of about 90 per cent are obtained.

13.3.3 Nomenclature

A	area of collecting surface, sq ft
a	particle radius, cm or μ
B	width of settling chamber, ft
B_c	width of rectangular cyclone inlet duct, ft
C	Cunningham correction factor $= 1 + 1.72\ (\lambda/d_p)$, dimensionless
c	concentration, particles/cu ft or particles/cu cm
c_0	initial concentration, particles/cu ft or particles/cu cm
$\mathfrak{D}$	diffusion parameter $= D_{BM}/D_c V_0$, dimensionless
D_{BM}	diffusivity, sq cm/sec
D_c	diameter of cylinder or sphere, cm
D_{cs}	surface average fiber diameter, cm or μ
D_e	diameter of cyclone exit duct, ft

d_p	particle diameter, cm or μ
d_{pe}	equivalent diameter of ellipsoid to be used in denominator of Eq. (13-22*a*), cm or μ
d_{pm}	diameter of smallest particle completely removed in cyclone, cm or μ
$\mathcal{E}$	electric field strength
E	efficiency of collector as measured by fraction of dispersoid collected by weight, dimensionless
e	electronic charge or Napierian log base
$\mathfrak{F}$	force
$\mathfrak{F}_r$	resisting force
F	friction loss in number of inlet velocity heads based on inlet duct area H_cB_c, dimensionless
f	specific collecting area
g	local acceleration of gravity, ft/sec/sec or cm/sec/sec
H	distance through which droplets travel, ft
H_c	height of rectangular cyclone inlet duct, ft
h	distance to collecting surface, ft
K, K_2, K_3	constants
K_b	coefficient of Brownian coagulation, cu cm/sec
K_1	coefficient of resistance for dust layer, in. H_2O/(lb dust/sq ft)(fpm)
K_0	coefficient of resistance for fabric, in. H_2O/fpm
K_s	coefficient of shearing coagulation, cu cm/sec
k	Boltzmann constant
L	length of settling chamber in direction of gas flow, ft
l	depth of filter bed, ft
M	ion mobility
m	mass of particle, g
N	number of spiral turns made by gas
N_s	number of shelves in settling chamber
n	ion concentration
P	potential
Q	volumetric flow rate, cfs
q	charge
q_s	saturation charge
q_t	charge at time t
$\mathcal{R}$	interception parameter $= d_p/D_c$, dimensionless
R_1	any radial distance in cyclone, ft
R_2	outer radius of cyclone, ft
Re	Reynolds number $= D_cV_0\rho_g/\mu$, dimensionless
Sc	Schmidt group $= \mu/\rho_gD_{BM}$, dimensionless
T	temperature, °K or °R
t	time, sec
$\tilde{t}$	reduced time $= t/(D_c/V_0)$, dimensionless
U_{rms}	root-mean-square velocity of ions
u_t	terminal particle velocity, fps or cm/sec
V_c	gas velocity at cyclone inlet, fps
V_0	initial, undisturbed velocity of aerosol stream in direction of flow, fps or cm/sec
V_s	superficial gas velocity, fps
V_t	tangential gas velocity, fps
V_x	velocity in x direction, fps or cm/sec
V_y	velocity in y direction, fps or cm/sec
$\tilde{V}_x$	reduced velocity of fluid in x direction $= V_x/V_0$, dimensionless
$\tilde{V}_y$	reduced velocity of fluid in y direction $= V_y/V_0$, dimensionless

W	volumetric water rate, cfs
w	weight of dust collected per unit area of filter, psf
x, y	position coordinates, cm
$\tilde{x}$	reduced position coordinate $= x/D_c$, dimensionless
$\tilde{y}$	reduced position coordinate $= y/D_c$, dimensionless
α	fiber volume per unit volume of mat, dimensionless
ΔH	enthalpy change
Δh	inlet velocity head, in. H_2O
Δp	pressure drop, in. H_2O or psi
Δt	time of contact, sec
δ	depth of gas
ϵ	dielectric constant
η	efficiency of individual fiber or droplet
ηe	effective fiber efficiency
λ	mean free path of gas, cm or μ
μ	gas viscosity or micron, viscosity in gm/cm sec
ρ_g	gas density, lb/cu ft or g/cc
ρ_p	particle density, lb/cu ft or g/cc
ψ	impaction parameter $= C\rho_p V_0 d_p^2/18\mu D_c$, dimensionless
ω	velocity gradient (cm/sec)/cm

13.4 CONTROL OF GASEOUS POLLUTANTS

13.4.1 Principles of Gas Purification

Removal of gases or vapors from gas streams may be accomplished by absorption in liquids or, occasionally, in solids, and by adsorption on solids. In absorption, the gas being absorbed passes through the phase boundary to become distributed through the body of the solid or liquid. The absorption process may or may not be accompanied by chemical reaction. In adsorption, the adsorbed gas is retained on the surface of the solid as a result of surface forces existing in the latter. The surfaces of the solid may be largely internal, as in the case of highly porous structures. The adsorption process may be primarily physical in nature, or may be akin to chemical reaction.

The nature of absorption and adsorption phenomena and the principles and methods of design and selection of equipment are treated in detail in References 81, 90, 100, and 117. Therefore, only a summary of the main principles of gas absorption and adsorption processes will be given here and will follow the treatment of Sherwood and Pigford (117), and the bulk of this article will be devoted to discussion of processes for collection of materials which frequently appear as atmospheric contaminants.

Diffusion and Transfer of Material between Phases. Both absorption and adsorption are diffusional operations, which involve transfer of gaseous material between phases as the result of molecular or eddy diffusion. Within a single fluid phase also, movement of a substance through the phase may occur as the result of molecular or eddy diffusion. Molecular diffusion depends primarily upon the nature of the fluids involved. Eddy diffusion, on the other hand, depends principally upon the nature of the turbulent fluid and the conditions of flow and is much more rapid than molecular diffusion. The process of interphase transfer is much more complex than the movement or diffusion of one substance through a single phase.

In order to obtain practical rates of mass transfer in gas absorbers or adsorbers, it is necessary to bring about the highest possible degree of turbulence in the fluid phase or phases. Transfer through the main body of a fluid is by eddy diffusion, the molecular diffusivity being relatively small. But near the phase boundary the fluid turbulence is less, and transfer is mostly by the slower process of molecular diffusion. The rate of transfer for both types of diffusion is proportional to the concentration of diffusing substance.

In practice, the relative importance of eddy diffusion and molecular diffusion is not known. Hence, the effective value of the diffusivity and the average length of the diffusion path are unknown. Therefore, the latter factors are lumped together in the form of a "mass-transfer coefficient" defined by the equations

$$N_A = k_G(p - p_i) \qquad \text{(gas streams)} \qquad (13\text{-}57)$$

$$N_A = k_L(c_i - c) \qquad \text{(liquid streams)} \qquad (13\text{-}58)$$

The subscript i refers to the partial pressure or concentration in equilibrium with the phase boundary.

The greater proportion of the resistance to mass transfer normally lies in the relatively nonturbulent region near the phase boundary where the eddy diffusivity is low. From this fact was developed a "film concept" which assumed that the entire resistance to diffusion lay in a film of stagnant fluid at the phase boundary, through which material passed by molecular diffusion alone.

If this concept were correct, the rate of transfer into a fluid stream would be given by the equations for steady-state molecular diffusion (117):

$$N_A = \frac{D_v P}{RTB_f p_{BM}} (p - p_i) \qquad \text{(gas film)} \qquad (13\text{-}59)$$

$$N_A = \frac{D_v}{B_f} (c_i - c) \qquad \text{(liquid film)} \qquad (13\text{-}60)$$

B_f represents the thickness of the hypothetical "film."

From a comparison of Eqs. (13-57), (13-58), (13-59), and (13-60) it is seen that

$$k_G = \frac{D_v P}{RTB_f p_{BM}} \qquad (13\text{-}61)$$

$$k_L = \frac{D_v}{B_f} \qquad (13\text{-}62)$$

The fictitious film thickness B_f is that of a stationary fluid layer which would present the same resistance to molecular diffusion as is encountered in the actual process of transfer from the turbulent main fluid stream to the phase boundary. In fact, while most of the resistance is in a region near the phase boundary, the resistance to transfer by eddy diffusion is a significant part of the total. The value of the coefficient k_G or k_L is not proportional to the molecular diffusivity, as suggested by the film theory; it is actually dependent on the degree of fluid turbulence and is proportional to the molecular diffusivity raised to a power less than unity (117). Coefficients k_G and k_L may be considered as empirical coefficients and be evaluated by experiment.

Gas Absorption. TWO-FILM CONCEPT IN ABSORPTION. In the absorption of a gas by a liquid, the solute gas must diffuse out of the gas phase and into the liquid phase. The gas and liquid are in motion relative to the interface between them, and a film may be considered to exist in each phase adjacent to the interface (see Fig. 13-29). The transfer process may then be visualized as one in which the solute gas passes successively through the two films. It is assumed that the two phases are in equilibrium at all points on the surface of contact (117). The concentration c_i in the liquid at the interface and the partial pressure p_i in the gas at the interface are related by the equilibrium conditions.

For steady-state transfer of a solute from a gas stream to a liquid stream, the rate of diffusion of solute from the main body of the gas stream to the interface must be the same as the rate of diffusion from the interface to the main body of the liquid. Hence:

$$N_A = k_G(p - p_i) = k_L(c_i - c) \qquad (13\text{-}63)$$

In order to calculate N_A from this equation, it is necessary to know both k_G and k_L as well as p_i and c_i. However, it is difficult to obtain experimentally the individual film coefficients. Therefore, to calculate N_A without knowledge of p_i or c_i, it is convenient to use "over-all coefficients," defined by the equations

$$N_A = K_G(p - p^*) = K_L(c^* - c) \tag{13-64}$$

In these equations, p^* is the equilibrium partial pressure of solute over a solution having the same concentration c as that in the main liquid stream, and c^* is the concentration of a solution which would be in equilibrium with the solute partial pressure p existing in the main gas stream. The over-all coefficients K_G and K_L can

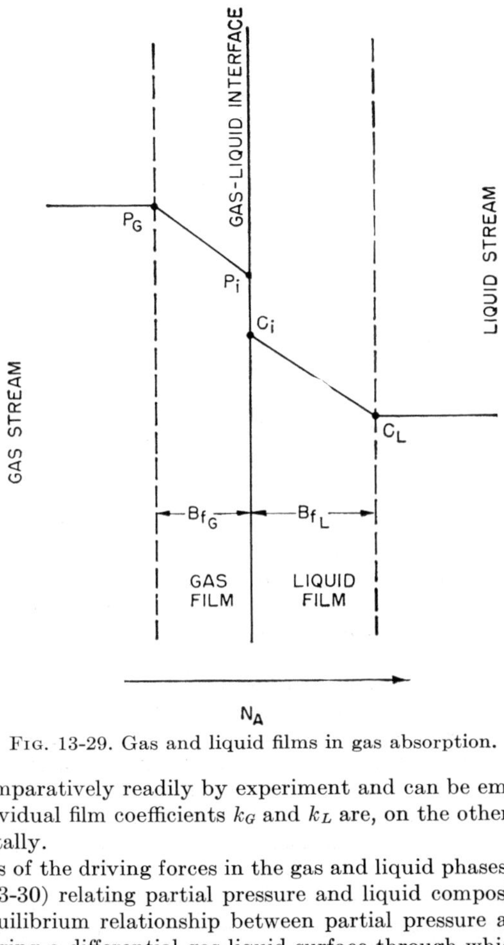

FIG. 13-29. Gas and liquid films in gas absorption.

be determined comparatively readily by experiment and can be employed directly in design. The individual film coefficients k_G and k_L are, on the other hand, difficult to obtain experimentally.

The magnitudes of the driving forces in the gas and liquid phases are illustrated by the graph (Fig. 13-30) relating partial pressure and liquid composition. Curve *OD* represents the equilibrium relationship between partial pressure and liquid concentration. Considering a differential gas-liquid surface through which solute is diffusing, point *A* represents the compositions of the main bodies of gas and liquid, while point *B* on the equilibrium curve represents the compositions of the two phases at the interface. The average partial pressure of the diffusing gas in the main gas stream is given by the ordinate p and the average solute concentration in the main liquid stream by the abscissa c. The liquid concentration c^* and pressure p^* are the equilibrium values corresponding, respectively, to the partial pressure p and concentration c. The driving force in terms of pressures $p - p^*$ of Eq. (13-64) is represented by the vertical distance AE, while the driving force in terms of concentrations $c^* - c$ is represented by the horizontal distance FA.

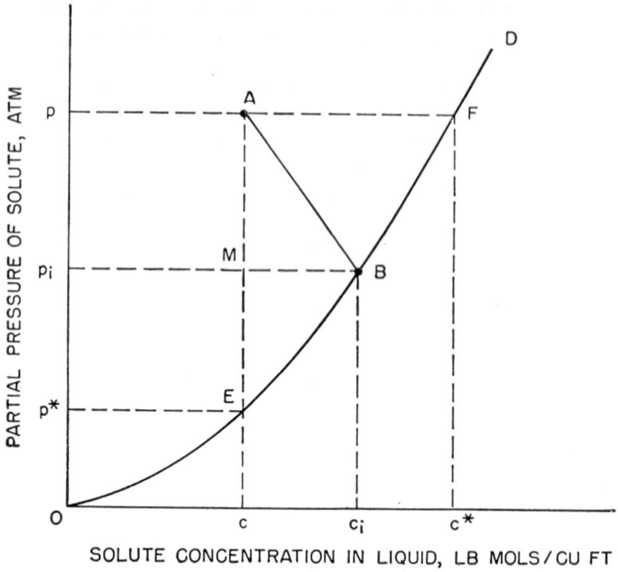

FIG. 13-30. Driving forces for gas absorption.

By reference to Eq. (13-63), it will be seen that the driving forces $p - p_i$ and $c_i - c$ are represented, respectively, by the vertical distance AM and the horizontal distance BM. Therefore,

$$\frac{p - p_i}{c_i - c} = \frac{k_L}{k_G} \tag{13-65}$$

and the line AB has a negative slope k_L/k_G.

In a case where the equilibrium relation obeys Henry's law, the equilibrium curve is a straight line represented by the equation

$$p^* = Hc \tag{13-66}$$

in which H is the Henry's law constant.

By elimination of c_i and p_i from Eqs. (13-63), (13-64), and (13-66), the following relations between individual film and over-all coefficients are obtained:

$$\frac{1}{K_G} = \frac{1}{k_G} + \frac{H}{k_L} \tag{13-67}$$

$$\frac{1}{K_L} = \frac{1}{k_L} + \frac{1}{Hk_G} \tag{13-68}$$

The coefficients k_L and k_G are expected to depend on the nature of the solute and solvent and on the conditions of fluid turbulence, but should be independent of the concentration of solute in the gas or liquid phases. However, Eqs. (13-67) and (13-68) show that K_G and K_L will vary with concentrations unless the equilibrium curve is straight. Strictly, the use of over-all coefficients is not valid unless the equilibrium line is straight over the region which includes the interfacial and bulk compositions. Over-all coefficients should be employed only for conditions similar to those under which they were measured and cannot be employed for other concentration ranges unless the equilibrium curve is straight (117).

The reciprocals of the mass-transfer coefficients may be regarded as resistances to mass transfer by diffusion. The over-all resistances $1/K_G$ and $1/K_L$ are equal to the sum of the resistances of the individual films, which are represented by the terms on the right-hand sides of Eqs. (13-67) and (13-68). The solubility of the solute gas in the liquid is given by the coefficient H. From Eqs. (13-67) and (13-68), it is evident that if H is sufficiently small the liquid-film resistance may be negligible compared to the gas-film resistance. Hence, for absorption of very soluble gases the gas-film resistance is controlling, and, conversely, the liquid-film resistance is controlling in the absorption of relatively insoluble gases.

Design Methods. Apparatus for gas absorption has as its objective the promotion of intimate contact between gas and liquid over a large interphase surface. While many varieties of equipment have been used, most of these may be classified under one of four types (117):

1. Spray towers, in which liquid sprays are injected into an empty tower through which the gas stream passes.
2. Units in which the gas is dispersed in the form of fine bubbles through pools of the liquid. Dispersion of the gas may be accomplished by passage through a porous plate or by mechanical agitation.
3. Bubble-plate and sieve-plate absorbers. In these, bubbles are formed under the surface of a shallow pool of liquid on the plate by passage of the gas through small holes in the plate or from under the edge of bubble caps.
4. Packed towers, in which the liquid stream is subdivided to provide a large interfacial area as it flows by gravity over the surface of a packing material.

The general approach to gas-absorber design will be illustrated using the countercurrent packed tower as an example. In a packed tower, it is difficult if not impossible to evaluate the effective interfacial area. The effective area is somewhat smaller than the area of the dry packing, which may not all be wet because of channeling of the liquid or of the tendency of the liquid to collect at the points of contact of packing pieces (117). Hence, it is convenient to employ a variable a, which represents the effective interfacial area per unit of tower volume. While K_G and a cannot be determined individually, they both depend on the nature of the packing and the liquid and gas flow rates. Hence they may be lumped together as a product K_Ga, which represents the over-all mass transfer, or capacity, coefficient on a volume basis and is defined by the equation

$$N_Aa\,dV = K_Ga(p - p^*)\,dV \tag{13-69}$$

Here N_Aa is the rate of transfer in moles per unit volume of equipment and V is the volume of packing. Similarly, K_La is defined by the equation

$$N_Aa\,dV = K_La(c^* - c)\,dV \tag{13-70}$$

Equations (13-69) and (13-70) may also be written in a form to express driving forces in mol fractions:

$$N_Aa\,dV = K_Ga(p - p^*)\,dV = K_GaP(y - y^*)\,dV$$

and

$$N_Aa\,dV = K_La(c^* - c)\,dV = \frac{K_La}{\rho_M}(x^* - x)\,dV \tag{13-71}$$

Countercurrent contacting of gas and liquid, as in a packed tower, is illustrated diagrammatically in Fig. 13-31. The absorbent enters the top at the rate of L'_M lb moles per hour of solute-free liquid per square foot of tower cross section, containing X_2 lb moles of solute per mole of solvent. The gas enters the bottom at the rate of

G'_M lb moles per hour of solute-free gas per square foot of tower cross section, with a partial pressure of solute gas p_1 atm. At the top of the tower the partial pressure of the solute in the exit gas has been reduced to p_2, while at the bottom of the tower the solute concentration in the exit liquid has increased to X_1. If the pressure drop across the tower is negligible relative to the total pressure P, the over-all material balance may be written

$$L'_M(X_1 - X_2) = G'_M(Y_1 - Y_2) = G'_M\left(\frac{p_1}{P - p_1} - \frac{p_2}{P - p_2}\right) \tag{13-72}$$

In general, if the concentrations at any point in the tower are X and p, then

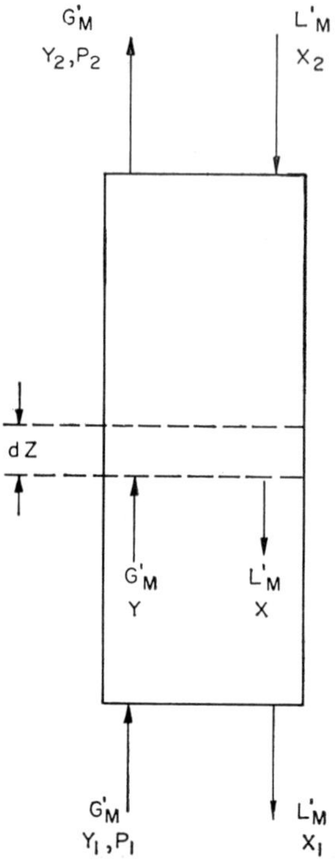

FIG. 13-31. Countercurrent contacting of gas and liquid.

$$L'_M(X_1 - X) = G'_M\left(\frac{p_1}{P - p_1} - \frac{p}{P - p}\right) \tag{13-73}$$

This equation, which is based on a material balance only, expresses the relation between gas and liquid concentrations at any point in the apparatus.

If gas and liquid concentrations are low, as is frequently the case in absorption of atmospheric pollutants, a simplified calculation procedure may be used. The mass flow rates of solute-free gas and liquid are approximately equal to the total flows, and the concentrations in stoichiometric units (moles of solute per mole of solute-free gas or liquid) may be taken as approximately equal to the concentrations expressed in mol fractions:

$$G'_M \doteq G_M \quad \text{and} \quad L'_M \doteq L_M \tag{13-74}$$

$$Y \doteq y = \frac{p}{P} \quad \text{and} \quad X \doteq x = \frac{c}{\rho_M} \tag{13-75}$$

The material balance in stoichiometric units

$$L'_M(X_1 - X) = G'_M(Y_1 - Y) \tag{13-76}$$

may be written approximately as

$$L_M(x_1 - x) = G_M(y_1 - y)$$

or

$$y_1 - y = \frac{L_M}{G_M}(x_1 - x) \tag{13-77}$$

In Fig. 13-32, the line AB represents graphically the material balance equation, Eq. (13-77), and is termed the "operating line." Point B represents the conditions at the gas inlet and liquid outlet, while point A represents those at the gas outlet and liquid inlet. The curve OC represents the equilibrium relationship. The over-all driving force based on the gas phase at a point in the tower is represented by the vertical distance between the operating line and equilibrium curve.

For a differential element of tower height dZ (Fig. 13-31), the rate equation may be written

$$L_M\,dx = G_M\,dy = K_Ga P(y - y^*)\,dZ = \frac{K_La}{\rho_M}(x^* - x)\,dZ \tag{13-78}$$

The required tower height may be calculated in terms of either the gas-phase or liquid-phase composition:

$$Z = \frac{G_M}{K_G a P} \int_{y_2}^{y_1} \frac{dy}{y - y^*} \tag{13-79}$$

$$Z = \frac{L_M}{\rho_M K_L a} \int_{x_2}^{x_1} \frac{dx}{x - x^*} \tag{13-80}$$

The height of a tower may be computed from Eq. (13-79) or (13-80), using values of $K_G a$ or $K_L a$ derived from experiments or generalized correlations. The value of the integral may be obtained by graphical or analytical methods (117).

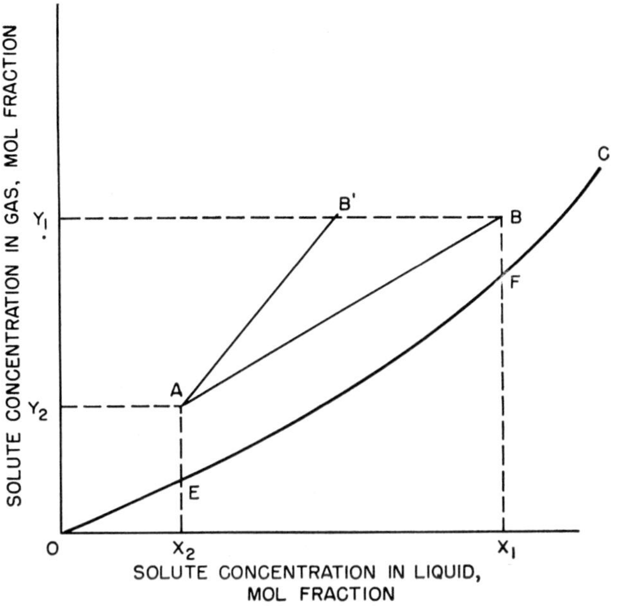

FIG. 13-32. Operating lines and equilibrium line for countercurrent column.

Alternatively, the "transfer-unit" concept (117) may be employed. The number of over-all gas-phase transfer units is defined by

$$N_{OG} = \int_{y_2}^{y_1} \frac{dy}{y - y^*} \tag{13-81}$$

N_{OG} is a dimensionless quantity which expresses the difficulty of absorbing the solute from the gas. The smaller the mean driving force and the larger the required change in gas composition, the greater the number of transfer units.

The "height of a transfer unit," or HTU, is defined by

$$H_{OG} = \frac{G_M}{K_G a P} \tag{13-82}$$

and has the dimension of length only. The height of the tower may then be expressed as

$$A = H_{OG} N_{OG} \tag{13-83}$$

The transfer-unit concept may also be applied on the basis of the liquid phase:

$$N_{OL} = \int_{x_2}^{x_1} \frac{dx}{x^* - x} \tag{13-84}$$

$$H_{OL} = \frac{L_M}{\rho_M K_L a} \tag{13-85}$$

H_{OG} or H_{OL} may be determined experimentally and applied as an alternative to determining the capacity coefficient K_Ga or K_La.

The evaluation of the integral representing N_{OG} can be carried out in general by graphical or analytical methods, as noted above. However, in some cases a mean value of driving force can be used, avoiding the necessity of carrying out the integration. Where the equilibrium curve and operating line are linear over the range in which they are used, it has been shown (117) that the correct mean driving force is the logarithmic mean of the terminal values:

$$\int_{y_2}^{y_1} \frac{dy}{y - y^*} = \frac{y_1 - y_2}{(y - y^*)_{Av}} \tag{13-86}$$

where

$$(y - y^*)_{Av} = \frac{(y - y^*)_1 - (y - y^*)_2}{\ln \dfrac{(y - y^*)_1}{(y - y^*)_2}} \tag{13-87}$$

From Eq. (13-58) it is seen that the slope of the operating line (Fig. 13-32) is L_M/G_M and that a large value of L_M/G_M corresponds to a steep operating line and large values of the driving force. In Fig. 13-32 the lines AB and AB' represent operation over the same range of gas composition with different liquid-to-gas ratios. As L_M/G_M is decreased, the operating line approaches the equilibrium curve, and the driving force is decreased, so that larger equipment is required to accomplish the required degree of solute recovery. Touching of the operating line and equilibrium curves represents equilibrium in the apparatus; this cannot be attained in an actual apparatus of finite length. Below this minimum value of L_M/G_M the absorber could not operate, regardless of its size.

The maximum gas and liquid rates in the countercurrent tower are determined by the tendency to flooding, entrainment, and carry-over, which are discussed in References 81, 100, and 117.

The same concepts and methods of design can be applied to other types of absorption equipment. In plate towers, gas-liquid contacting takes place in a series of discrete steps rather than continuously, as in the packed tower. Design is frequently based upon the concept of the "number of theoretical plates" rather than upon that of the height of a transfer unit (117). The transfer-unit concept is also applied to spray towers (61,63,102,117).

Cocurrent operation, or parallel flow, is illustrated in Fig. 13-33. Point B represents the inlet conditions and point A the outlet conditions, as in Fig. 13-32. The extent of solute recovery is limited by the approach to equilibrium at the outlet. However, it is possible to use liquid-to-gas ratios in a cocurrent contactor which would produce flooding in countercurrent contactors, and a large value of average driving force can be attained, making possible the use of smaller equipment. Cocurrent contacting may be used if the desired exit-solute concentration in the gas phase can be attained with a practical liquid rate.

Crossflow represents a case intermediate between cocurrent and countercurrent contacting. It may be illustrated by the passage of liquid droplets across a gas stream at right angles to the direction of gas flow. Along the line of travel of the droplet the gas composition remains constant, but the solute concentration in the droplet increases, so that the driving force producing diffusion decreases the farther the droplet travels.

ABSORPTION AND CHEMICAL REACTION. In the selection of an absorbent for scrubbing a gas, the usual object is to find a liquid which has the capacity to absorb a large quantity of the solute without building up an appreciable equilibrium back pressure. This may be accomplished by using a solution of a chemical with which the solute reacts irreversibly. Simultaneous absorption and chemical reaction is treated in Reference 117.

In most cases of combined absorption and chemical reaction the rate of absorption is limited by both the resistance to diffusion and the velocity of the reaction. At the present time, the theory applicable to these cases is not sufficiently advanced to afford prediction of whether the rate in a given case will be limited by diffusional or chemical resistance. Each case must be studied experimentally (117). A limiting case is that in which a very rapid, irreversible reaction takes place so that there is essentially no back pressure of solute over the solution. In such a case (see Figs. 13-32 and 13-33), $y^* = 0$, and there is no difference between countercurrent and cocurrent contacting with regard to driving force or the possible degree of solute recovery.

Gas-absorption Equipment. Absorption of gaseous components from gas streams for control of atmospheric pollution frequently involves treatment of very large gas volumes containing only low concentrations of pollutant material. Because of the

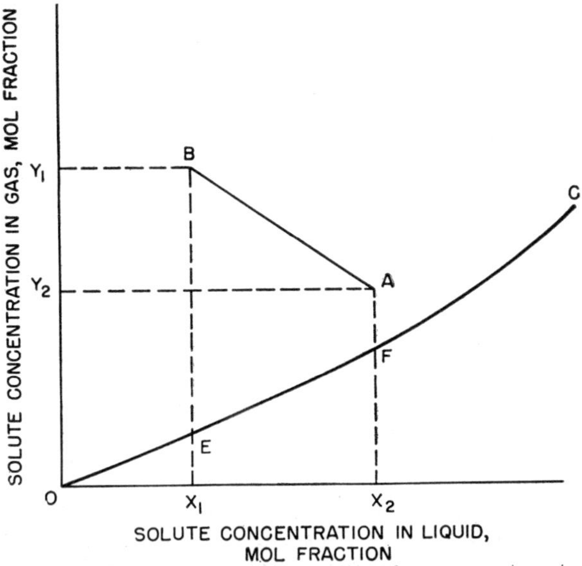

Fig. 13-33. Operating and equilibrium lines for cocurrent contacting.

large gas volume and the low driving force for diffusion, the size of the equipment may be very large relative to that of gas-absorption equipment used in chemical processing. The pressure drop through the equipment may be critical. In addition, the gas stream may contain dust, and the gas absorber may be required to act simultaneously as a wet dust collector.

Packed towers have a tendency to become plugged with solids. Hence, grid packings have frequently been favored over random-dumped packings such as Raschig rings because of a lesser tendency to plug as well as because of lower pressure drop (65,81,117).

While they may involve higher pressure drops than packed towers, plate towers are more readily cleaned. Certain types of countercurrent plate towers are used as dust collectors, and may be used for simultaneous gas absorption (see Art. 13.3).

Simple spray towers (66,102), which may be either countercurrent or cocurrent in operation, have a low pressure drop and are not subject to plugging. They may also be used as dust collectors. While it had been thought that the interior of the drops were stagnant, and that absorption of relatively insoluble gases should therefore be slow, recent studies (102,117) have indicated that this is not the case, that there is circulation within the drops. Spray chambers may therefore be fairly effective in the absorption of even relatively insoluble gases. Some have been found to compare favorably with other types of equipment for certain applications (102,117). In the countercurrent towers, however, mixing of spray and gas is such that appreciable countercurrent action is apparently not obtained. Hence, the application of spray towers is probably best suited to operations in which only a few transfer units are required.

A tendency to entrainment of droplets in the exit-gas streams frequently requires use of entrainment eliminators.

The problem of entrainment may be met by the use of cyclonic spray towers, such as the Pease-Anthony (61,63) or Fog-Filter (71). Relatively high gas rates may be employed without excessive entrainment (61). However, the gas-liquid contact is of a crossflow type and is equivalent only to about one theoretical stage. These devices are frequently employed primarily as dust collectors, in which service they are also more effective than simple spray chambers.

The venturi scrubber (59,64) may also be classified with the spray towers. The gas stream passes at high velocity through a venturi, and the scrubbing liquid is introduced under low pressure at the venturi throat, where it is atomized by the gas stream. Contacting is cocurrent. The entrained liquid is removed from the gas stream by a cyclone or other type of entrainment separator. The venturi scrubber is commonly used for dust or mist collection but may also be effective for gas absorption.

Porous-plate absorbers and agitated vessels are normally used for absorption of relatively insoluble gases in which the liquid-phase resistance is controlling.

Regenerative Processes. In absorption and adsorption processes it is frequently desirable or necessary for technical or economic reasons to recover or dispose of the solute gas and recycle the absorbing or adsorbing medium. In a regenerative process, the absorbing or adsorbing medium is treated by either chemical or physical means to remove the solute gas and thus provide a fresh medium for recycle. The method used for a regeneration process is dependent upon the chemical or physical characteristics of the solute, the gas-solute concentration, the absorbing or adsorbing medium, and economic factors. The process may be either physical or chemical. The physical methods of regeneration employed are distillation, heating, stripping with an inert gas, steam stripping, or vacuum stripping. In the processes which employ chemical means for regeneration of the absorbing solution, the absorbed solute is reacted with other chemicals which precipitate or volatilize the solute gas in an altered chemical form.

Nonregenerative Processes. In some cases the absorbing or adsorbing medium may not be regenerated but may be discarded or converted into useful products. Many factors will affect the general choice of the process to be employed for specific situations. The concentration of the gas to be removed in the gas phase is the most important single factor which must be considered in this choice. Obviously, the higher the initial concentration to be removed, the greater the likelihood that economical and profitable recovery of this gas can be accomplished. As a broad generalization, processes which are regenerative by physical means are best adapted to situations with a high initial concentration of gas to be removed, except in those situations where the final concentration of the solute gas in the exit gas stream must be held to low values.

Another factor, of course, is the total quantity of gas which is to be removed as well as its unit value. In some instances, this quantity is small enough so that it could not be marketed economically. In such cases, nonregenerative processes often offer the most economic solution to the problem of removal.

In many cases, stack gases must be purified, and these usually contain some proportion of oxygen. When organic absorbing solutions are used, the oxidation of the solution may be great enough to render its use economically prohibitive.

The choice of the process may also be affected by the presence of contaminants such as dust particles (if an efficient dust collector does not precede the absorption equipment) or products of incomplete combustion. Dust particles may tend to clog certain types of absorption and adsorption equipment and may influence the marketability of the product of the process. Products of incomplete combustion may seriously affect oxidation rates or otherwise contaminate the absorbing medium.

In processes where absorption is adversely affected by high temperatures, the wet-bulb temperature of the gases to be treated may be important. High-temperature stack gases can be cooled to the wet-bulb temperature readily by evaporation of relatively small quantities of water preceding the absorption or adsorption step. Further cooling, however, requires the removal of water vapor which, in turn, requires quantities of cooling water which may be excessive.

Gas Adsorption. Adsorption processes involve contacting a gas mixture with a solid under such conditions that some of the fluid is adsorbed on the surface of the solid, with a resulting change in composition of the unadsorbed fluid. The phenomenon of adsorption is difficult to define or limit because the mechanism is relatively complex; several different types of adsorption are recognized. These are physical adsorption or condensation of gases on solids at temperatures considerably above the dew point; chemical or activated adsorption, in which definite chemical bonds are produced between the atoms or molecules on the surface of the solid and the adsorbed atoms or molecules; and ion exchange. This article is concerned primarily with physical adsorption and desorption.

For the adsorption of gases, solids of an essentially porous nature, each with a decided affinity for the adsorption of certain substances, have been developed for industrial use in the recovery of solvents, in fractionation of mixed gases, and in other applications. The materials which were developed for commercial uses include a variety of clays, chars, activated carbons, gels, aluminas, and silicates. They are more or less granular in form and are supported in beds or columns of suitable thickness through which the gas passes. Inasmuch as adsorption may be practically complete even with very low vapor content, the procedure lends itself readily to recovery and purification operations. An example is the odor- and taste-removal properties of gas-adsorbent carbon. Bulkeley (9) has reviewed the use of adsorbents.

Adsorbents can collect 8 to 25 per cent of their weight in vapors. The power consumption of each system is primarily a function of the depth of the adsorbing bed, which may vary from a few inches to several feet. The adsorbent produces condensation of vapors at temperatures above the saturation temperature. In this process, the latent heat of vaporization is released to the bed and to the unadsorbed gas, making it necessary in some cases to precool the gas. It is important to note that capacity and efficiency of adsorption decrease with rising temperature of the adsorbent.

Carbon does not selectively adsorb water vapor, and hence moist gases do not raise the temperature of the carbon bed and the gas stream as would occur with some other adsorbents, e.g., silica gel. If carbon beds are regenerated with steam, the resulting moisture is rapidly removed from the bed owing to selective displacement by solvent vapor when the adsorption process is resumed. The stripping of the moisture cools the bed, making it unnecessary to precool the gases.

Vapors are usually recovered from the adsorbent by heating the bed above the boiling temperature of the solvent at atmospheric pressure. The heating may be done by submerged heating elements or by circulating hot gases or steam through the bed. The volatilized solvent is condensed, and the remaining gases are sent through another adsorber for stripping. The regenerated adsorber is then cooled and returned to service.

In some cases the adsorbers may be regenerated by vacuum stripping, eliminating the heating and cooling cycle. The volatilized solvent is condensed in vacuo in the latter process. Occasionally, adsorption processes may operate at pressures greater than atmospheric.

There are three common types of arrangement of adsorbent units, the choice depending primarily on the requirements of the process. The types of arrangement are a single unit for discontinuous operations; a double unit for alternate on and off use; and a three-unit setup, interconnected so that two units may be in operation while the third unit is being regenerated.

Adsorption units have high installation costs, but operation and maintenance costs are low. Small units may be operated as efficiently as large systems. The value of the recovered solvent occasionally reduces the cost of the investment.

The major factors to be considered in the design of adsorbers of the "fixed-bed" type are (8):

1. Quantity of fluid handled per unit time
2. Amount of material to be adsorbed
3. Allowable pressure drop through bed
4. Duration of adsorption portion of cycle
5. Time required for reactivation, purging, etc.

The mass or volume of adsorbent is primarily determined by factors 2, 3, and 4, whereas the diameter and depth of adsorbent are dependent upon factors 1 and 5. The duration of the various portions of the adsorbing-desorbing cycle and also the number of adsorbers on stream at one time can be determined by an economic balance based on experience and judgment. The effect of process variables on the pressure drop through the adsorbing bed can be obtained from test data or estimated by the usual methods (90,100). The time required for reactivation (or dumping and refilling) is best obtained from experimental data or experience. If the adsorbing capacity of the bed is known for the conditions in question, the optimum design for the adsorber system can be obtained directly from material balances, pressure-drop relationships, and economic considerations.

13.4.2 Removal of Sulfur Dioxide

The principal potential sources of sulfur dioxide as an atmospheric pollutant are metallurgical processes handling sulfur-containing ores, sulfuric acid plants, and combustion processes which use high-sulfur coal or fuel oil. The sulfur dioxide concentration in waste gases from metallurgical operations is normally relatively high, and many smelters have installed processes which recover many hundreds of tons of valuable sulfur products per day. The concentration of sulfur dioxide in sulfuric acid plant waste gases and in the stack gases from the combustion of fuels is quite low. In such instances the economic removal and recovery of sulfur dioxide is difficult, and only a few installations for the removal of sulfur dioxide at these low concentrations exist. Probably the major and certainly the most widespread source of air pollution from sulfur dioxide is derived from the combustion of high-sulfur fuels, and in the future more stringent laws may require removal of sulfur dioxide from these gases to a far greater extent than in the past.

Many of the processes described below can be used to recover sulfur dioxide in one or more marketable forms, such as sulfuric acid, elemental sulfur, ammonium sulfate, or gypsum. Generally, when pure or concentrated sulfur dioxide is recovered, it is converted to sulfuric acid, although small quantities of liquid sulfur dioxide are also produced.

Processes Regenerative by Physical Means. DIMETHYLANILINE PROCESS. In this process (see Fig. 13-34) the gases are first treated to remove particulate matter. They are then passed to an absorbing tower in which they are contacted countercurrently with dimethylaniline which absorbs the sulfur dioxide. The purified gases are then contacted with a solution of sodium carbonate which removes some or all of the sulfur dioxide escaping the absorbing tower and which further collects any entrained dimethylaniline carried over from the absorbing tower. The waste gases are finally contacted with a dilute sulfuric acid solution which reacts with any dimethylaniline vapor in the waste gases and absorbs it in the acid as the sulfate.

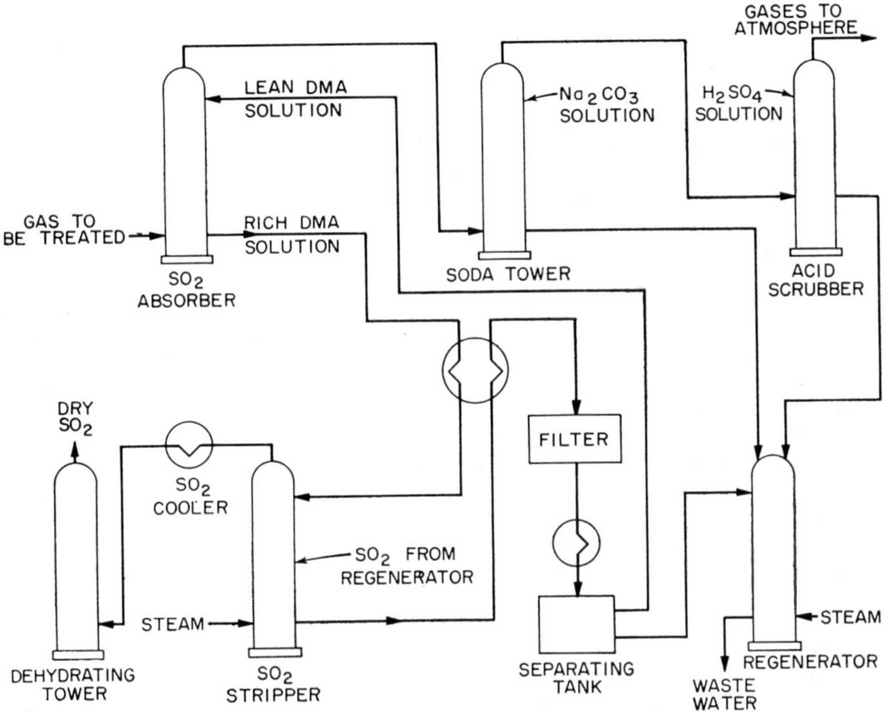

FIG. 13-34. Dimethylaniline process.

The rich dimethylaniline solution from the absorber passes through heat exchangers and then to a stripping tower where the sulfur dioxide is stripped out of the dimethylaniline by contact with steam. The sulfur dioxide driven off from the dimethylaniline is cooled to condense some water and dimethylaniline vapors which go overhead from the stripping tower. After the cooling step, the sulfur dioxide passes to the dehydrating tower and then to its ultimate use.

The lean solution from the stripping tower is passed through heat exchangers and a filter and finally returned to a decanting tank to separate the dimethylaniline from entrained water. The decanted dimethylaniline is then returned to the absorber for further pickup of sulfur dioxide. Effluent solutions from the sodium carbonate tower, the acid towers, and the water-decanting tank are treated in a dimethylaniline regenerator with steam. In this tower the dimethylaniline is distilled off the above-mentioned solutions and returned to the sulfur dioxide stripping tower (40).

This process has been used on Dwight-Lloyd sintering-machine gases containing

5.5 per cent sulfur dioxide, with the removal of 99 per cent of the gas. Lead-coated steel bubble towers were employed both for absorption and desorption (17).

The vapor pressure of sulfur dioxide over dimethylaniline has been reported (50). The data indicate that dimethylaniline is suitable only for the removal of relatively high concentrations of sulfur dioxide, such as those derived from smelting operations.

Sulfidine Process. This process is similar to the dimethylaniline process but uses a mixture of xylidine and water as an absorbing solution, normally in a 1:1 ratio. The water and xylidine enter the absorber in two phases. When the sulfur dioxide concentration reaches about 100 g/liter, however, a homogeneous liquid is formed. As with dimethylaniline, some entrained xylidine liquid and vapors escape the absorber with the waste gases, and these are recovered by contact with alkaline bisulfite solutions or mineral acids such as sulfuric acid. The latter solutions are reactivated by heating with steam to a temperature of 80 to 100°C.

Sulfur dioxide that has been oxidized to sulfur trioxide is removed from the system by treating a small portion of the circulating stream with sodium carbonate. The xylidine associated with the oxidized sulfur dioxide is then freed and separated from the water solution containing sodium sulfate. The xylidine is then returned to the system and the sodium sulfate solution wasted (138). Raschig-ring packed towers are used for both absorption and desorption.

Plants have been built using this process on oil-refinery gases containing 0.5 to 8 per cent sulfur dioxide and on smelter gases containing about 4 per cent sulfur dioxide. It is said to be adaptable to gases ranging in sulfur dioxide concentration from 1 to 16 per cent (43). Operating on gases containing 7 per cent sulfur dioxide, the steam requirements amount to 1 to 1.2 tons per ton of sulfur dioxide produced (138).

Ammonia Process. In this process a solution of ammonium sulfite is contacted countercurrently with the sulfur dioxide-containing gases, forming a solution of ammonium sulfite and bisulfite. The effluent solution from this tower is then passed to a reactivator where the sulfur dioxide is driven off with steam. The regenerated ammonium sulfite solution is returned after heat exchange to the absorption tower. In the original application of this process, the rich solution was simply heated and then flashed to drive off the sulfur dioxide. Considerable study has been given to more efficient means of regeneration, and vapor recompression has been suggested for this purpose (54,67,100).

This process has been considered for use with relatively low-concentration gases. Under such conditions, however, the proportion of ammonium sulfite oxidized to ammonium sulfate has been very large.

Comprehensive vapor-pressure data on the sulfur dioxide–ammonia–water system as well as distillation studies have been reported (55,56,60).

Basic Aluminum Sulfate Process. In this process, gases are contacted countercurrently with a solution of basic aluminum sulfate which is prepared by treating a solution of aluminum sulfate with lime. Entrained droplets from the absorption towers are removed in a mechanical separator.

The effluent liquid from the absorption towers is heated by steam to drive off essentially pure sulfur dioxide and to regenerate the basic aluminum sulfate solution. The sulfur dioxide from the regenerating tower is cooled to condense moisture, and the sulfur dioxide is finally dehydrated by contact with sulfuric acid.

A wood grid tower has been used for absorption, and the regeneration tower contained ceramic rings. A special film-type heater must be used in the regeneration tower in order to avoid precipitation of an insoluble salt.

The vapor pressure of sulfur dioxide above basic aluminum sulfate solutions as well as process-performance data have been reported (2).

This process has been successfully operated on smelter gases containing 5 per cent sulfur dioxide and may be useful with sulfur dioxide concentrations greater than 1 per cent. Substantial oxidation of the absorbed sulfur dioxide in the solution occurs. This is removed from the circulating solution by precipitation of calcium sulfate with limestone. If hydrogen sulfide in appreciable concentrations is present in the gas to be treated, another side stream of liquor must be reacted with copper sulfate in order to prevent fouling the solution with elemental sulfur.

MISCELLANEOUS. A number of solutions have been investigated as possible absorbing media for sulfur dioxide, including the ethanolamines. For the most part, these solutions have not offered any substantial advantages over those solutions described above (27).

A mixture of sodium sulfite and boric acid has also been investigated as an absorbing solution which would be regenerated by heating with steam (19).

Processes Regenerative by Chemical Means. SODIUM SULFITE–ZINC SULFITE PROCESS. In this process sulfur dioxide is removed from the gas stream by contact with a solution of sodium sulfite, forming sodium bisulfite. The rich solution from the tower is reacted with zinc oxide which precipitates zinc sulfite and re-forms a caustic solution for further use in the absorber. The precipitated zinc sulfite is then calcined to drive off pure sulfur dioxide and re-form zinc oxide.

Sulfates formed as a result of oxidation in the absorber are removed from the circulating system in a somewhat indirect manner. This involves precipitation of calcium sulfite followed by the reaction of sodium sulfate in solution with the precipitated calcium sulfite under a high partial pressure of sulfur dioxide. Under such conditions the precipitated calcium sulfite goes into solution and reacts with the sodium sulfate to form calcium sulfate, which then precipitates. The calcium sulfate is filtered and discarded.

In the absorption step, either countercurrent or parallel flow may be employed because of the negligible vapor pressure of sulfur dioxide above the sodium carbonate–sodium sulfite solution. Counterflow in wood grid towers and crossflow in a wet cyclone have both been employed for absorption of sulfur dioxide with these solutions, and each has performed very well. A specially designed radiant-heated flash calciner has been used for the decomposition of the zinc sulfite.

Comprehensive design data have been published on this process (61,63,64), but it has not been operated on a large enough scale to establish the magnitude of chemical losses. This process may be operated to yield nearly complete removal of sulfur dioxide, but it may not be economical with initial sulfur dioxide concentrations much below 0.5 per cent, owing to the proportionately increased effect of oxidation to sulfate (see Oxidation, page 13–89).

WET THIOGEN PROCESS. In the wet thiogen process, sulfur dioxide is absorbed countercurrently in water or a recirculated solution containing small quantities of barium thionates, sulfites, etc. The effluent solution from the tower is reacted with powdered barium sulfite. This forms a precipitate composed primarily of elemental sulfur, barium sulfite, and barium thiosulfate. After settling and filtration, the mother liquor is returned to the absorption tower and the filtered precipitate is heated to distill off elemental sulfur, the major product of the process. The residue after distillation, composed primarily of barium sulfite and barium sulfate, is mixed with coke or other reducing agents and reduced to barium sulfide for recycle in the process. The presence of calcium is undesirable since it forms soluble thiosulfates, thionates, and sulfites, which raise the specific gravity of the solution to such a point that settling becomes difficult.

This was one of the earliest processes for the removal of sulfur dioxide from smelter gases and was abandoned after pilot-plant operation. Possibly the major objection to the process is the necessity of using water (which has little capacity for sulfur dioxide when the initial concentration is low) as the absorbing agent. Owing to the relatively high vapor pressure of sulfur dioxide above water, the process is limited to relatively high concentrations of sulfur dioxide (139).

Nonregenerative Processes. AMMONIA–SULFURIC ACID PROCESS. As in the regenerative ammonia process, sulfur dioxide is removed by contacting countercurrently with the solution of ammonium sulfite and ammonium bisulfite. The composition of the solution entering the tower is adjusted so as to provide a good removal of sulfur dioxide without an excessive loss of ammonia in the waste gases. If the proportion of ammonium bisulfite in the incoming solution were too high, the maximum removal of sulfur dioxide would be limited. If the solution were too high in proportion to ammonium sulfite, the vapor pressure of ammonia above the solution would be so large as to cause excessive ammonia losses. Both sulfur dioxide removal and ammonia loss are adversely affected by high temperatures, and consequently the initial gases to be treated must be cooled as much as possible.

The effluent solution from the absorption tower is reacted with sulfuric acid, which forms ammonium sulfate and sulfur dioxide. One mol of sulfuric acid produces between one and two mols of sulfur dioxide gas. The exact amount of sulfur dioxide varies with the ratio of ammonium sulfite to ammonium bisulfite in the effluent solution from the tower. The pure sulfur dioxide generated is dried with sulfuric acid. The ammonium sulfate from the reaction is stripped with steam in order to eliminate any residual sulfur dioxide in the ammonium sulfate solution. The ammonium sulfate is then crystallized from the stripped solution, centrifuged, and finally dried.

Raschig-ring packed towers and wood grid towers have both been used for absorption. Nearly saturated solutions are used in the absorbing towers, and in one installation four towers are employed in series with ammonia make-up between towers (11,28,32,67).

One modification of the process described above is the use of dilute solutions of ammonium sulfite–bisulfite in the absorber. After acidification of the effluent solution from the absorber, the dilute solution of ammonium sulfate is contacted by hot stack gases. The partial evaporation of the ammonium sulfate solution by hot stack gases reduces the steam requirements in subsequent crystallization. The advantage of this modification is that ammonia losses in the waste gases from the absorption tower can be reduced considerably below that occurring when nearly saturated solutions are used in the absorber.

A proposed modification to this process is to treat the effluent solutions from the absorber with a small quantity of air in order nearly or completely to oxidize the sulfite and bisulfite solutions to ammonium sulfate. In this process little or no sulfur dioxide would be formed, but also little or no sulfuric acid would be required (57).

This process is adaptable to operation on stack gases containing low concentrations of sulfur dioxide and has been commercially used with gases containing initially only 0.1 per cent sulfur dioxide. Nearly complete removal can be accomplished, and although oxygen present in the gas stream will oxidize the sulfur dioxide absorbed in the solution, the sulfate produced is in a marketable form.

LIME-NEUTRALIZATION PROCESS. In this process, sulfur dioxide is absorbed by water containing suspended calcium hydroxide, calcium sulfite, and calcium sulfate. Initially the sulfur dioxide dissolves in the water and then reacts with calcium hydroxide to form calcium sulfite, which is comparatively soluble in the tower effluent. The oxygen contained in the stack gases, however, also dissolves and rapidly reacts with the calcium sulfite to form calcium sulfate. The calcium sulfate becomes supersaturated in the water to a marked extent. It will, however, precipitate readily on walls, on the tower packing, and in subsequent lines and vessels. The presence of suspended calcium sulfate in the water provides feeding surfaces on which the calcium sulfate can deposit rather than on the packing and other equipment. The effluent from the absorption tower passes into vessels where desupersaturation of the calcium sulfate is completed. After desupersaturation the slurry is passed into settling cones where the heavy particles settle out and are filtered. The overflow from the tower containing small seed crystals of calcium sulfate and other small solids is returned to the absorption tower after make-up lime has been added.

Wood grid towers have been employed with this process, although parallel or cross-flow equipment could be used equally successfully. A cyclonic type of contactor might in fact be less susceptible to calcium sulfate scaling than a wood grid tower. The desupersaturation vessels must be large enough to give substantial residence time, but turbulent flow conditions in the desupersaturation vessel must be maintained in order to minimize precipitation of calcium sulfate on the walls and to hasten the desupersaturation. Normally, the calcium sulfite is not completely oxidized in the absorption tower, and if a sulfite-free calcium sulfate is desired, or if it is desired to maintain the maximum quantity of calcium sulfate in the circulating stream, a small oxidizing tower can be introduced to the system. In the oxidizing tower the circulating stream is contacted with a small flow of air (43,80,99).

This process will remove sulfur dioxide from the stack gas almost completely and is best adapted to stack gases having low initial sulfur dioxide concentrations. The calcium sulfate produced by this process may be marketable as gypsum. However, if any large proportion of dust is present in the stack-gas stream, the calcium sulfate may be contaminated by dust collected in the absorber. Also, the particle size of the calcium sulfate is smaller than natural gypsum.

Absorption by Alkaline Water. In this process a hard or alkaline water is used to absorb the sulfur dioxide from the stack-gas stream. The sulfur dioxide reacts with the sodium, magnesium, calcium, etc., compounds in the water and is thereby neutralized. If counterflow is used, some unreacted sulfur dioxide may be dissolved in the tower effluent, and this may be neutralized by the addition of lime after passage of the water through the tower.

Obviously, very large quantities of water must pass through the absorber, and normally the process is feasible only when the plant is located adjacent to large supplies of water. Usually the liquid effluent from the tower must be aerated before discharge into rivers or lakes. This is necessary in order to convert any sulfite to sulfate and thereby minimize the effect of the waste on the biological oxygen demand.

If enough water is employed, nearly complete removal of sulfur dioxide can be attained, and as with the lime neutralization process, the absorption in water is best adapted to those situations where the initial sulfur dioxide concentration is low (34,43,98).

Catalytic Oxidation to Sulfuric Acid. In this process, sulfur dioxide is absorbed by countercurrent contact with plain water. Simultaneously, oxygen in the gases to be treated is also absorbed. This reacts with the sulfur dioxide in solution to form sulfuric acid. The oxidation is promoted by the presence of small quantities of iron or manganese in the solution.

Sulfuric acid concentrations as high as 30 per cent have been prepared on a laboratory scale. The rate of absorption falls off rapidly, however, as the concentration of acid in the solution increases, and it is difficult to attain acid concentrations much above 10 per cent. The catalytic activity of the iron and manganese is also extremely susceptible to poisoning by a number of materials.

This process is adaptable to relatively low concentrations of sulfur dioxide and can effect nearly complete removal. Except under special circumstances the acid produced is dilute and cannot be economically concentrated to commercial strength.

Factors Affecting Choice of Process. For the most part, sulfur dioxide removal processes have been developed, designed, and operated by individual companies to meet specific situations. Some performance data have been published on these processes from which general design information can be derived. This information is adequate for preliminary evaluation of the feasibility of the various processes under conditions other than those under which the process was developed.

Some of the most important factors to be considered in the selection of a process are as follows:

Initial and final sulfur dioxide concentrations in stack gases: As noted above, the processes regenerative by physical means are adaptable only to conditions where the initial sulfur dioxide concentration in the stack gases is relatively high, at least greater than 1 per cent. Also, these processes cannot practically accomplish essentially complete removal of sulfur dioxide.

The processes regenerative by chemical means have not been proved to be economically feasible. In comparison with other processes, however, their best range of applicability would seem to be between about 0.5 and 1 per cent sulfur dioxide initially in the gas stream. They are adapted to essentially complete removal of sulfur dioxide.

The nonregenerative processes are best adapted to initial sulfur dioxide concentrations below 1 per cent. Where no marketable products are recovered, obviously the lower the initial concentration (for a given volume of total stack gases), the less costly the process is. The nonregenerative processes are adapted to essentially complete removal of sulfur dioxide.

Recovered form of sulfur: Pure sulfur dioxide is generally the most desirable form in which to recover the sulfur. A limited market for liquid sulfur dioxide exists, but the major usage would be as sulfuric acid or elemental sulfur. Sulfur dioxide may be converted into sulfuric acid in standard sulfuric acid plants, and it may be converted to elemental sulfur by reduction with coke or by reaction with hydrogen sulfide.

Ammonium sulfate is a valuable sulfur product, but the market for this material is relatively limited.

The calcium sulfate produced by the lime-neutralization process might conceivably be used in the manufacture of wallboard, provided that it were not contaminated by foreign substances. Its particle size is, however, too small to be used directly as an agricultural gypsum. Under special circumstances it might be feasible to convert the calcium sulfate to ammonium sulfate. The ammonium carbonate may be reacted with calcium sulfate in an autoclave to produce ammonium sulfate, and the lime is thereby regenerated for further use in the absorption of sulfur dioxide.

Oxidation: It is predictable on theoretical grounds and confirmed by experience that the amount of sulfur dioxide oxidized in solution to sulfate is essentially independent of the initial sulfur dioxide concentration in the gas. Thus as the initial concentration of sulfur dioxide in the gas becomes smaller, the proportion of sulfur dioxide oxidized to sulfate becomes greater. This is one of the major reasons why processes designed to recover sulfur dioxide as such fail when treating gases of low initial sulfur dioxide content.

Obviously, oxidation is greatly dependent on the concentration of oxygen in the gases to be treated. Processes for recovering sulfur dioxide may therefore be adaptable to relatively low sulfur dioxide concentration gases if the oxygen content is also low.

A number of materials have been tried as oxidation inhibitors. Some of these have been partially successful, but none has been completely so.

There is some evidence to indicate that the type of equipment used for absorption of sulfur dioxide affects the oxidation of sulfur dioxide in the solution. Wet cyclone absorbers, for example, have been found to produce less oxidation of sodium sulfite solutions than grid towers.

Liquid flows: The cost of the sulfur dioxide removal process other than that of the absorption equipment may be very greatly influenced by the volume of liquid that must be circulated. In general, with processes such as lime neutralization where the solubility of the active absorbing substance in water is low, the volumes of solutions circulated are very large. This not only requires a relatively large capital investment in liquid handling equipment but may entail large power requirements.

Temperature of absorption: In those processes where sulfur dioxide has an appreciable partial vapor pressure above the absorbing solution, the temperature at which absorption takes place becomes critical. Normally the gases to be treated are at elevated temperatures, and even though they may be cooled by humidification with water prior to absorption, their wet-bulb temperatures are usually not below 120 to 140°F. The latter temperatures may still be too high for efficient absorption where the absorbing solution has a significant partial vapor pressure of sulfur dioxide. Further cooling below the wet-bulb temperature requires large quantities of cooling water.

Materials of construction: In those processes where sulfur dioxide is recovered, acidproof construction materials must be employed in the liquid-handling system. In those processes where sulfur dioxide is neutralized, mild steel can normally be employed.

Steam and power requirements: The steam consumption of the processes regenerative by physical means is an important factor in determining their over-all economics. The dimethylaniline process, when operated under essentially anhydrous conditions, and the sulfidine process may both exhibit particularly low steam requirements.

When dealing with low-concentration gases in particular, the pressure drop through the coolers and absorbers becomes more important. For this reason, grid-type towers are frequently used in handling low-concentration gases.

13.4.3 Removal of Hydrogen Sulfide

Hydrogen sulfide is principally produced in the coking of coals and in petroleum-refining operations. It also occurs in natural gas, and some is formed in the production of manufactured gas. There are numerous other sources, such as the evaporation of black liquor in the kraft pulping process and some smelter gases. The majority of hydrogen sulfide removal and recovery plants have been installed for the purpose of purifying a useful product (i.e., natural gas). In recent years, however, many plants have been installed for the special purpose of eliminating atmospheric pollution.

Many hydrogen sulfide removal processes produce elemental sulfur directly. In the majority of the remaining processes, hydrogen sulfide is converted to elemental sulfur but may be burned to sulfur dioxide for subsequent sulfuric acid manufacture.

Processes Regenerative by Physical Means. PHENOLATE PROCESS. In this process the sour gas is contacted countercurrently with a solution which contains approximately 3 mols of sodium hydroxide and 2 mols of phenol per liter. The rich solution from the bottom of the absorbing tower passes through heat exchangers and then to the top of a regeneration tower where the hydrogen sulfide is stripped out of the absorbing solution by contact with steam from the solution in the tower reboiler. The stripping steam is condensed from the overhead product from the regeneration column and returned to the system. The lean stripped solution from the bottom of the regenerating tower passes through heat exchangers and is returned to the absorber. A simplified flow diagram which is characteristic of this and other processes regenerative by physical means is illustrated in Fig. 13-35.

In gases containing both carbon dioxide and hydrogen sulfide, it has been found that the phenolate solution is partially selective in the absorption of hydrogen sulfide because it is absorbed faster than carbon dioxide. This solution has a quite high absorption capacity for hydrogen sulfide. Normally, however, the efficiency of hydrogen sulfide removal by this process is quite low, and the consumption of steam in regeneration is rather high (14,68,103).

TRIPOTASSIUM PHOSPHATE PROCESS. In this process a water solution containing approximately 32 per cent tripotassium phosphate is used as the absorbing solution.

Recently, modifications to the flow diagram shown in Fig. 13-35 have been made. High efficiency in the removal of hydrogen sulfide has been accomplished by the process, as depicted in Fig. 13-36. In this process a lean solution nearly free of hydrogen sulfide enters the top of the tower; part-way down the column a semilean solution containing considerably more hydrogen sulfide is also introduced to the absorber. The two streams mix in the lower part of the column and emerge as a single rich solution. This technique permits a nearly complete removal of hydrogen

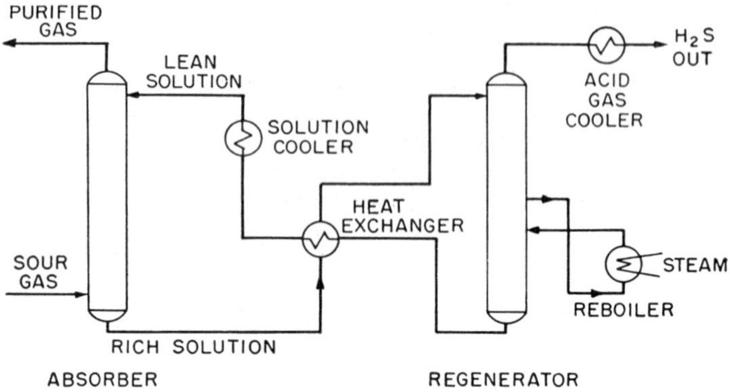

FIG. 13-35. Regenerative process for hydrogen sulfide removal.

sulfide from the gas stream without the necessity of completely regenerating the entire circulating stream. The rich solution from the bottom of the absorber, after passage through a heat exchanger, is introduced to the top of the regenerator. Midway down the regenerating tower, when the solution is partially stripped of hydrogen sulfide, a side stream is bled off the tower and fed to a reconcentrator. The vapors from the reconcentrator are fed back to the top part of the regenerator, whereas the bottoms are split. Part of the bottoms, after passage through heat exchangers, is reintroduced

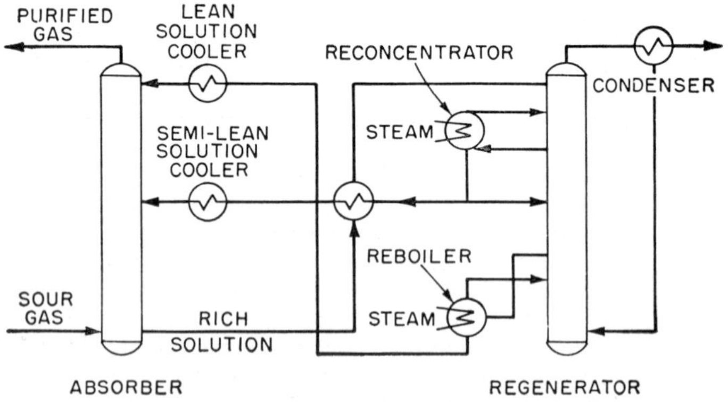

FIG. 13-36. Modified regenerative process for hydrogen sulfide removal.

to the absorption tower as the semilean solution. The other portion of the bottoms from the reconcentrator enters the lower half of the regenerator and is completely stripped of hydrogen sulfide. The stripped solution passes through solution coolers and is introduced to the top of the absorption tower as the lean solution. Steam in the overhead from the regeneration tower is condensed and returned to the system.

Tripotassium phosphate solutions are quite selective for hydrogen sulfide in the presence of carbon dioxide, and the process may be operated at relatively high temperatures and still obtain good hydrogen sulfide removal. The process is nonvolatile and can tolerate contaminants in the gas stream. It does require a relatively high steam consumption (68,105,134).

Alkazid Process. In this process, sodium alanine or potassium dimethyl glycine is used as the absorbing solution in a process similar to that shown in Fig. 13-35.

The potassium dimethyl glycine has a very high absorption capacity for hydrogen sulfide, but since the selectivity in gases also containing carbon dioxide is dependent on the more rapid rate of absorption of hydrogen sulfide, contact times are critical. For this reason, disintegrators rather than packed towers are sometimes used as absorbers (105,134).

Amines. In this process, monoethanol amine, diethanol amine, methyl diethanol amine, or triethanol amine is used as the absorbing solution. The basic regenerative process is used with these solutions, although a split-flow process also has been developed. This differs from the split-flow process shown in Fig. 13-36 principally in that no reconcentrator is used on the semilean solution and that part of the rich solution from the bottom of the absorber tower is fed to the top of the regenerator and part to the lower part of the regenerator after the semilean side stream has been taken off.

Monoethanol amine has a high absorption capacity for hydrogen sulfide and absorbs it rapidly. It is, however, comparatively volatile, sensitive to contaminants such as carbonyl sulfide and carbon disulfide, and essentially nonselective when treating gas streams containing both carbon dioxide and hydrogen sulfide. Diethanol amine absorbs hydrogen sulfide much less rapidly than monoethanol amine but is comparatively nonvolatile, less sensitive to contaminants, and partially selective in treating gases containing carbon dioxide as well as hydrogen sulfide. Triethanol amine has a lower absorption rate and a lower capacity for absorbing hydrogen sulfide than mono- or diethanol amine. It is, however, quite selective for the absorption of hydrogen sulfide from gases containing both hydrogen sulfide and carbon dioxide. Methyl ethanol amine has a higher absorption capacity for hydrogen sulfide than triethanol amine and is somewhat more selective in hydrogen sulfide removal from mixed carbon dioxide–hydrogen sulfide streams. The amine solutions generally require a comparatively small amount of steam for regeneration (44,68,78,93,97,105,135).

Processes Regenerative by Chemical Means. Iron Oxide–Dry-box Process. In this process, the gas to be treated passes through a bed of iron oxide sponge which usually consists of wood shavings impregnated with iron oxide. The hydrogen sulfide reacts with the iron oxide to form an iron sulfide. After the bed is exhausted, the sponge is treated with air to form elemental sulfur and regenerate the iron oxide.

This process will effect almost complete removal of hydrogen sulfide. It will not absorb carbon dioxide and remains efficient even in the presence of tars, hydrogen cyanide, ammonia, carbonyl sulfide, carbon disulfide, and other contaminants. Being a batch operation, it is not well adapted to conditions where relatively high concentrations of hydrogen sulfide are encountered. A number of attempts have been made to improve its operation in this respect by carrying out oxidation in place, rather than removing the sponge from the equipment, followed by extraction of the elemental sulfur. More recently pilot-plant studies have been carried out on a fluidized-bed–iron oxide process. In this case continuous operation is effected, which may overcome many of the objections to the use of iron oxide as a means of removing hydrogen sulfide from gases containing high concentrations of hydrogen sulfide (41,52,68,91, 132,133,146).

Ferrox Process. In this process, hydrogen sulfide is absorbed by a solution of soda ash containing suspended iron oxide or hydroxide. Optimum concentrations in the liquid are about 0.4 per cent ferric hydroxide and 3 per cent total alkali. In the absorption the hydrogen sulfide reacts with the sodium carbonate and iron oxide to produce elemental sulfur and iron sulfide. The solution from the bottom of the tower is aerated in long troughs, and additional free sulfur is formed from the conversion of the iron sulfide back to iron oxide. The regenerated solution is then returned to the absorber, and the free sulfur close to the top of the solution in the aerating trough is skimmed off.

This process can accomplish high removal of hydrogen sulfide and is selective for hydrogen sulfide when treating mixed gas streams also containing carbon dioxide. The sulfur produced is in the form of fine particles and contains considerable occluded iron oxide (68,105,109).

SEABOARD AND VACUUM CARBONATE PROCESSES. In the Seaboard process, hydrogen sulfide is absorbed by a solution of about 3 to 3½ per cent sodium carbonate. The rich solution from the bottom of the absorbing tower passes to a reactivating tower in which compressed air is introduced. The air strips the hydrogen sulfide, and the sodium carbonate solution is regenerated for further use.

This process is partially selective for hydrogen sulfide in the presence of carbon dioxide, but contaminants such as hydrogen cyanide react with this solution. The power requirements for this process are high, and it suffers from the major disadvantage that it releases hydrogen sulfide to the atmosphere.

A modification of the Seaboard process, known as the vacuum carbonate process, supplants air in the reactivation step with steam. The steam stripping is carried out at a vacuum of about 25 in. of mercury. The reactivation of the absorbing solution is more complete in this process than in the Seaboard, and lower solution circulation rates are needed for the removal of a given quantity of hydrogen sulfide (51, 68,105).

THYLOX PROCESS. In this process, hydrogen sulfide is absorbed in a neutral solution of sodium thioarsenate. As in the Seaboard process, regeneration of the solution is accomplished by bubbling compressed air through the regeneration tower, forming fine particles of elemental sulfur which float on the top of the tower and which are skimmed and filtered.

High degrees of hydrogen sulfide removal have been obtained with this process, and the solution is selective for hydrogen sulfide in the presence of carbon dioxide. Hydrogen cyanide is absorbed by the solution. The finely divided sulfur produced by the process is contaminated by arsenic and other impurities.

Nonregenerative Processes. ALKALINE LIQUORS. In some instances, such as coal-coking plants, a continuous supply of ammonia liquor is available. This has been used to remove hydrogen sulfide from gas. Some selectivity can be obtained in the presence of carbon dioxide by reducing the contact time to take advantage of the greater absorption of hydrogen sulfide (36,148).

Solutions of sodium hydroxide or soda ash also may be used to absorb hydrogen sulfide from gas streams without regeneration of the solution. In one installation the gas is contacted with atomized alkaline solutions in a specially designed scrubber. Effective removal of both hydrogen sulfide and sulfur dioxide has been obtained with this equipment (69,70).

CATALYTIC CONVERSION TO SULFUR. In this process, sulfur dioxide is added to the gas stream containing hydrogen sulfide. The mixture is then passed through a bed of alumina at a temperature of 60 to 90°C, and the hydrogen sulfide and sulfur dioxide react according to the Simon-Carves reaction to form water and elemental sulfur. A moving bed of alumina is employed, and the spent alumina withdrawn from the reaction vessel is transferred to a regeneration vessel which is heated to a temperature of 500°C. This vaporizes the sulfur formed and regenerates the alumina for reuse. A slight excess of sulfur dioxide over that necessary for the reaction is used (98).

Factors Affecting Choice of Process. *Initial hydrogen sulfide concentration and degree of removal:* In general, the regenerative processes, particularly those employing liquid absorbents, are best adapted to operation on gases containing relatively high concentrations of hydrogen sulfide, perhaps on the order of 20 grains/SCF or greater. Under favorable conditions some of the liquid-absorption processes can effect quite complete removal of hydrogen sulfide.

In situations where the initial concentration of hydrogen sulfide is low or the total quantity of sulfur that could be recovered is small, regenerative batch processes such as the iron oxide box process, or nonregenerative processes, may be the most economical.

Operating pressure: Many of the hydrogen sulfide removal processes have been employed on gases under pressures of several hundred psi. With many of the solutions, absorption will not be so complete when it is necessary to operate at atmospheric pressure as when the gas is under high pressure. A process suitable for high-pressure operation may therefore not be feasible for atmospheric-pressure operation.

Operating temperature: The majority of the regenerative processes are operated best at temperatures at or slightly above ambient temperature. If the gas to be treated is at a relatively high temperature and cannot be technically or economically cooled, solutions such as tripotassium phosphate are more desirable absorbents than those wherein the vapor pressure of hydrogen sulfide above the solution is sensitive to high temperatures.

Considerable loss of absorbent may also be experienced by vaporization of the absorbent when high-temperature absorption must be employed.

Contaminants: Many processes which would otherwise be suitable cannot be used when small quantities of certain contaminants are present. The organic absorbents are particularly sensitive in this respect. Tars, cyanogen, organic acids, oxygen, and organic sulfides may all seriously hamper the operation of amine solutions. Also, small quantities of contaminants such as valve grease may cause difficulties by foaming in amine-recovery plants. The iron oxide–dry-box process and the tripotassium phosphate process are particularly suitable for gases which contain contaminants that would react with many of the other hydrogen sulfide removal solutions.

Selectivity: As pointed out above, many of the absorbents for hydrogen sulfide will also partially absorb carbon dioxide. The absorption of carbon dioxide along with the hydrogen sulfide not only dilutes the hydrogen sulfide stripped from the regenerating tower but also requires the employment of much higher liquid-circulation rates in order to obtain the same degree of removal.

Many of the processes which are selective for hydrogen sulfide depend on the more rapid rate of solution of hydrogen sulfide than of carbon dioxide, and if too-prolonged contact times are permitted in the absorber, the carbon dioxide may actually displace the hydrogen sulfide from the solution. With selective removal it is therefore important to choose absorption equipment which will yield adequate contact time to obtain the desired hydrogen sulfide removal but not enough contact time to remove an appreciable proportion of the carbon dioxide.

Steam and power requirements: Steam requirements for regeneration have been an important factor in selection of hydrogen sulfide removal plants, and in many instances one process has been chosen over another primarily on this account. Of those processes which employ steam for reactivation, the amine solutions generally require the least.

Power requirements may become important in cases where large volumes of low-pressure gases are to be treated. In this case it is important to choose absorption equipment which will yield the desired removal of hydrogen sulfide at a minimum pressure drop through the absorber. Power requirements may also be an important factor in processes such as the thylox process where reactivation is accomplished by blowing compressed air through a tall column of liquid.

13.4.4 Removal of Fluorides

The most common fluoride found in industrial waste gases is hydrogen fluoride (HF). Gaseous silicon tetrafluoride is evolved when materials containing calcium fluoride or calcium fluoapatite ($Ca_{10}F_2(PO_4)_6$) are calcined or treated with acid in the presence of sand. Silicon tetrafluoride is readily hydrolized by the moisture in air to give hydrogen fluoride. Consequently, the stack gases from phosphate fertilizer plants contain considerable quantities of fluorides. Aluminum plants, fused tricalcium phosphate furnaces, nodulizing kilns, and calcium metaphosphate furnaces evolve fluorides. Substantially all these fluorides are in the form of hydrogen fluoride which can readily be removed by scrubbing processes.

The maximum permissible concentration of hydrogen fluoride for industrial operations is 3 ppm (74); however, this value is much too high for prolonged vegetation exposure. Concentrations as low as 0.01 ppm may cause damage to sensitive vegetation such as sweet potato plants, white pine trees, peach trees, or gladioli (86). (Also see Effect of Air Pollution on Plants, Sec. 9 of this handbook.) Therefore, to avoid local atmospheric pollution, it is necessary to reduce the fluoride content of stack gases to a very low value.

Absorption in Water-spray Towers. Water has a very high affinity for hydrogen fluoride even at liquid concentrations as high as 2 per cent by weight. Consequently, water is often used for removing hydrogen fluoride from gases.* The necessary contact between liquid water and the gases can be obtained by spraying water into an unpacked tower at several points along the tower length, with the gases rising through the tower countercurrent to the descending water droplets. The effluent liquor may be treated with lime slurry to remove the fluoride ion as insoluble calcium fluoride.

Water-spray towers are capable of handling large volumes of stack gases and can remove more than 97 per cent of the entering fluorides. Scrubbing towers of this type have been used commercially on nodulizing kilns in which the evolved gases contain some sodium fluoride dust as well as hydrogen fluoride. In one industrial application, a water-spray tower consistently removed more than 97 per cent of the total fluorides in the entering gases while handling approximately 50,000 cu ft of gas per minute and removing about 4,000 lb of total fluorine per day. Table 13-22 presents representative data obtained with this tower.

Table 13-22. Data from Water-spray Tower for Removing Fluorides from Gases

Tower diameter = 18 in.
Tower height = 80 ft
Number of spray injection points = 6
Gas inlet temperature = 300°C
Gas exit temperature = 72°C
Gas inlet rate = 52,000 cfm
Gas exit rate = 26,500 cfm
Rate of inlet hydrogen fluoride = 4,000 lb fluorine per day
Rate of exit hydrogen fluoride = 97 lb fluorine per day
Percentage fluorine removed (as hydrogen fluoride) = 97.6 per cent
Rate of inlet sodium fluoride as dust = 340 lb fluorine per day
Rate of exit sodium fluoride as dust = 14 lb fluorine per day
Percentage removal of fluorine in dust = 95.9 per cent
Water flow = 700 gpm
Hydrated lime consumption = 24,000 lb per day

Lump-limestone Bed. Hignett and Siegel (49) have presented a method for removing hydrogen fluoride from gases which involves passing the gases through a bed of lump limestone at temperatures higher than the dew point of the stack gas. The reaction product is calcium fluoride, and this can be removed from the lump-limestone bed in the form of fine particles. Portions of the bed are withdrawn from the tower from time to time, and the fines are removed by screening. The oversize material from the screening together with fresh limestone for make-up are fed back into the

* If elemental fluorine is involved, water should not be used since, under some conditions, the danger of explosion exists.

tower. If a sufficiently large bed is used, this process permits removal of more than 99 per cent of the entering hydrogen fluoride.

Hignett and Siegel conducted tests on a lump-limestone bed 4 ft deep and slightly over 3 ft in diameter, using gas rates of 275 cfm. They were able to remove 95 per cent of the net fluorine from entering gases containing 0.02 to 0.33 lb of net fluorine per 1,000 cu ft of gas. They estimated that 99.9 per cent of the net fluorine would be removed if the bed height were increased to 9 ft. The calcium fluoride product was found to contain 80 to 90 per cent calcium fluoride, which is in the range of commercial grades of fluorspar.

Wet-cell Washers. Wet cells containing coarse Saran fibers in the form of mats have shown excellent efficiencies for removing hydrogen fluoride from gases (38). The mats may be arranged in series to permit virtually 100 per cent removal of hydrogen fluoride from the entering gases. The individual mats are wetted by water delivered

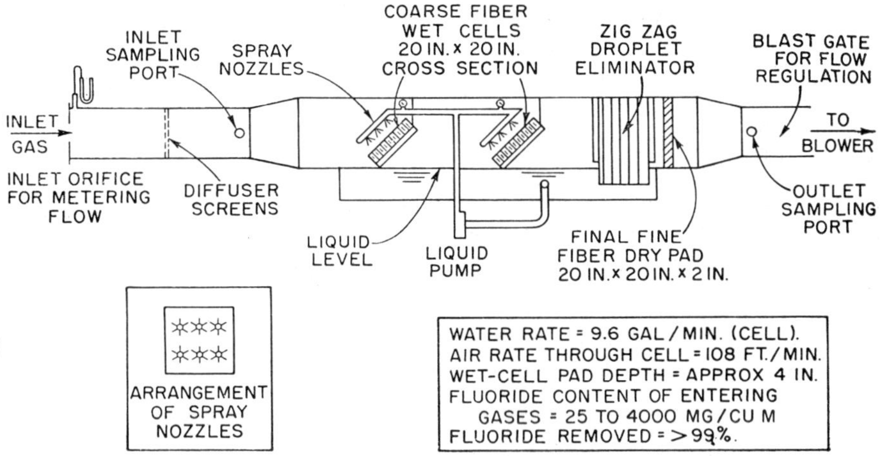

FIG. 13-37. Wet-cell washer for removing hydrogen fluoride from gases. (*Courtesy of Air Cleaning Laboratory, Harvard University.*)

from coarse flooding nozzles or other devices which do not produce fine droplets. When the water is delivered from fine-mist nozzles, an aqueous hydrogen fluoride mist is formed which tends to penetrate the wet cells and reduce the removal efficiency.

Figure 13-37 shows equipment used for testing the performance of wet cells. Two wet cells in series consistently removed more than 99 per cent of the net fluorine from gases containing 25 to 4,000 mg of net fluorine per cubic meter of air.

Fluoride-mist aerosols may also be removed by wet-cell washers. However, in order to obtain good removal efficiencies, it is necessary to have an effective final dry pad in the system.

Wet-cell washers have the advantage of being able to handle large volumes of gases efficiently. These washers are easily constructed and do not require a large amount of space. Since Saran shows good resistance to hydrogen fluoride, the pads do not have to be replaced frequently. The main disadvantage is the tendency for the pads to become clogged with solid material removed from the gases.

Absorption in Sodium Hydroxide Solution. Landau and Rosen (75) have described a continuous commercial process for removing fluorine and hydrogen fluoride from large quantities of gases. In this process, fluorine and hydrogen fluoride are continuously absorbed in a 5 to 10 per cent sodium hydroxide solution. The gases flow countercurrent to the sodium hydroxide solution in a packed tower at a temperature of about 100°F. The effluent liquid is treated with a calcium hydroxide slurry which regenerates the sodium hydroxide and precipitates calcium fluoride. The following chemical reactions are involved:

$$F_2 + 2NaOH \rightarrow \tfrac{1}{2}O_2 + 2NaF + H_2O \qquad (13\text{-}88)$$
$$HF + NaOH \rightarrow NaF + H_2O \qquad (13\text{-}89)$$
$$2NaF + CaO + H_2O \rightarrow CaF_2 + 2NaOH \qquad (13\text{-}90)$$

It is necessary to maintain the caustic concentration higher than 2 per cent to destroy any intermediate toxic fluorine oxide (OF_2) which might be formed from the fluorine reaction. The precipitated calcium fluoride is removed continuously from a settling tank, and the concentration of caustic is maintained constant. Results obtained with this process show that a fluoride content of less than 3 ppm can be obtained in the exit gases from a tower handling up to 500 lb of hydrogen fluoride per hour.

Miscellaneous Processes. A bed composed of porous sodium fluoride pellets prepared by the thermal decomposition of sodium acid fluoride may be used to remove hydrogen fluoride from petroleum gases, hydrocarbon liquids, air, or nitrogen (84). The bed is operated at a temperature of about 100°C, and the pellets are regenerated by heating in a stream of gas at 300 to 500°C. By the use of a series of beds, it is possible to remove essentially 100 per cent of the hydrogen fluoride in the entering gases.

Volatile inorganic fluorides can be removed from gaseous mixtures by scrubbing the gases with a countercurrent stream of dihydroxyfluoboric acid (53). The fluorides may be desorbed from the absorbing acid by the application of heat.

Hydrogen fluoride may be removed from light gases, such as hydrogen, methane, ethane, and propane which are vented from an alkylation process, by countercurrent scrubbing in a packed absorption tower. A liquid containing olefin fractions is used as the absorbing medium (79). The hydrogen fluoride reacts with the olefins to form alkyl fluorides which may be returned to the reaction zone as alkylating agents.

Economics. Water has a high affinity for hydrogen fluoride and is also very cheap. Processes which remove hydrogen fluoride from gases by water absorption are generally more economical than processes involving direct chemical reactions. Water-spray towers show excellent removal efficiencies, and, from an economic viewpoint, they are superior to the other methods when large quantities of gases must be handled. The equipment for the wet-cell process does not require so much space as the water-spray towers and also requires a relatively small investment.

If elemental fluorine is present in the gases, pure water should not be used as the absorbing agent because fluorine does not always react with water and explosions may occur under some circumstances. Therefore, more expensive processes such as the sodium hydroxide process must be used when elemental fluorine is involved.

13.4.5 Removal of Nitrogen Oxides

Nitric oxide (NO) and nitrogen dioxide (NO_2) are the common forms of nitrogen oxides evolved in industrial processes. Waste gases from nitric acid plants, chamber sulfuric acid plants, and units for the regeneration of cracking catalysts contain small amounts of nitrogen oxides. Certain metal-pickling processes also evolve nitrogen oxides.

Concentrations as high as 0.4 ppm have been reported in the Los Angeles atmosphere; however, these oxides are seldom present to a sufficient extent to cause noticeable physiological effects. Since nitrogen dioxide can react with the water vapor in air or with raindrops to produce nitric acid, small concentrations in the atmosphere can cause undue corrosion to metal surfaces in the immediate vicinity of the releasing stacks. Plant life may also be damaged if the concentrations become excessive (see

Sec. 9). The oxides of nitrogen may affect atmospheric pollution in an indirect manner since they can act as catalysts for certain reactions such as the oxidation of sulfur dioxide.

Nitrogen dioxide and nitrogen tetroxide (N_2O_4) react with water to form nitric acid and nitric oxide. The nitric oxide formed in the process can be oxidized to nitrogen dioxide which, in turn, can react with more water. The following chemical reactions are involved:

$$2NO_2 \text{ (or } N_2O_4) + H_2O \rightarrow HNO_3 + HNO_2 \quad (13\text{-}91)$$
$$2HNO_2 \rightarrow NO + NO_2 \text{ (or } \tfrac{1}{2}N_2O_4) + H_2O \quad (13\text{-}92)$$
$$NO + \tfrac{1}{2}O_2 \rightarrow NO_2 \quad (13\text{-}93)$$
$$2NO_2 \rightarrow N_2O_4 \quad (13\text{-}94)$$

Experimental tests indicate that the rate of the removal of nitrogen oxides from gases is controlled by the speed of the chemical reactions rather than by diffusional resistances (101). Reaction (13-91) occurs in the films at the interface between the gas and the liquid. The oxidation of nitric oxide is relatively slow but proceeds essentially to completion. Reactions (13-91) and (13-92) do not go to completion in the presence of concentrated nitric acid, but the reactions do approach completion in the presence of pure water.

Bubble-cap Plate Columns. Nitrogen oxides may be removed from gaseous mixtures by passing the gases through a series of bubble-cap plates countercurrent to water or aqueous nitric acid. The nitric oxide evolved in the chemical reactions is primarily oxidized to nitrogen dioxide in the vapor space between the plates, although some oxidation by absorbed oxygen may occur in the liquid phase. The gases entering the tower must contain a large amount of oxygen in order to permit high rates of nitric oxide oxidation.

The efficiency of bubble-cap plate columns decreases as the concentration of the gaseous oxides decreases. At low concentration a large fraction of the oxides is in the form of nitrogen dioxide, while at high concentrations the fraction of oxides in the form of nitrogen tetraoxide is high. This offers an explanation for the low efficiency of bubble-cap columns at low concentrations since the rate of reaction between nitrogen tetraoxide and water is undoubtedly much higher than the rate of reaction between nitrogen dioxide and water.

Table 13-23 presents data for nitrogen oxide removal efficiencies obtained in a conventional bubble-cap plate column. At the low concentrations ordinarily encountered in stack gases, conventional bubble-cap columns are very inefficient.

Venturi Injector. Experimental tests using a venturi injector for removing nitrogen oxides from gases have shown promising results. Water is sprayed axially into the gas flowing at high velocities through the venturi throat. The large amount of interfacial area between the gas and atomized liquid gives high rates of absorption of the oxides. The rate of oxygen absorption into the liquid is also very high with this type of equipment, and tests indicate that a considerable portion of the nitric oxide oxidation is accomplished in the liquid phase.

It is possible to obtain even greater rates of removal of nitrogen oxides by adding steam to the entering gases. The increased water-vapor pressure in the gases tends to promote the gas-phase reactions and increase the efficiency of removal of nitrogen oxides. The gases leaving the venturi contain a considerable amount of nitric acid mist, and this mist should be removed by a cyclone or other separating equipment before the cleaned gas is exhausted to the atmosphere.

Packed Towers and Spray Towers. Packed towers and spray towers operated with countercurrent water and gas flow are capable of removing nitrogen oxides from gases. However, these types of equipment give poor efficiencies at low gas concentrations.

Adsorption on Silica Gel. The nitric oxide in flue gases containing 1 to 1.5 per cent nitric oxide may be removed from the gases by oxidizing the nitric oxide to nitrogen dioxide and then adsorbing the latter (42). Silica gel containing adsorbed nitrogen dioxide is used to catalyze the nitric oxide oxidation. The nitrogen dioxide formed in the oxidation reaction is removed from the gases by adsorption on silica gel. The adsorbed nitrogen dioxide may be recovered from the silica gel by heating.

Table 13-23. Efficiency Data for Bubble-cap Plate Column

(Removal of nitrogen oxides from gases by reaction with pure water)

Efficiency = per cent of entering oxides (as NO_2 and N_2O_4) removed by one plate.
Operating pressure = 1 atm
Operating temperature = 25°C

Entering gas concentration ($NO_2 + 2N_2O_4$), mol %	Fraction of entering $NO_2 + 2N_2O_4$ in form of N_2O_4	Slot gas velocity, fps	Efficiency, %
10	0.428	1.18	33
8	0.390	1.18	32
4	0.274	1.18	25
1	0.104	1.18	10
10	0.428	1.65	31
8	0.390	1.65	29
4	0.274	1.65	23
1	0.104	1.65	9
10	0.428	2.35	25
8	0.390	2.35	24
4	0.274	2.35	19
1	0.104	2.35	6

Data from M. S. Peters, C. P. Ross, and L. B. Andersen, Removal of Nitrogen Oxides from Gaseous Mixtures, AEC Technical Report No. 10 (COO-1011), University of Illinois, Engineering Experiment Station, 1953.

Economics. The commercial processes for removing nitrogen oxides from gases all involve equipment requiring large investments. When dealing with gases containing less than about 1.0 per cent nitrogen oxides, the removal efficiencies are very poor in bubble-cap, packed, or spray towers. Of the three types of tower, the bubble-cap tower is the most economical if an appreciable fraction of the oxides must be removed from the gases.

The silica-gel process requires a large investment and needs special attention to handle the silica-gel absorption-desorption cycles. Ordinary stack gases would foul the silica gel because of carry-over of dust, moisture, and other materials. The stack gases should be given a cleaning treatment before entering the silica-gel beds, and this alone would make the silica-gel process impractical for most industrial purposes. The advantage of this process is that the nitrogen oxides can be recovered in a concentrated form and can be used immediately for making nitric acid or for other purposes.

The use of the venturi injector appears to be the most economical method for removing nitrogen oxides from stack gases. The required investment is not prohibitive, and any reasonable degree of oxide removal can be obtained by using a number of venturi injectors in series.

13.4.6 Nomenclature

- a area of interphase contact, sq ft/cu ft
- B_f thickness of stagnant film, ft
- c solute concentration in the main body of the liquid, lb moles/cu ft
- c^* concentration of solute in liquid phase corresponding to equilibrium with gas, lb moles/cu ft

c_i solute concentration in the liquid at the interphase boundary, lb moles/cu ft
Δ difference, as in concentrations
D_v molecular diffusivity, sq ft/hr
G_M superficial molar mass velocity of gas, lb moles/hr-sq ft
G_M' superficial molar mass velocity of inert gas, lb moles/hr-sq ft
H Henry's law constant, p^*/c
H_{OG} height of an over-all gas-phase transfer unit, ft
H_{OL} height of an over-all liquid-phase transfer unit, ft
k_G gas-film coefficient, lb moles/hr-sq ft (atm)
K_G over-all gas-film coefficient, lb moles/hr-sq ft
K_Ga over-all gas-film coefficient on a volume basis, lb moles/hr-cu ft (atm)
K_L over-all liquid-film coefficient, lb moles/hr-sq ft (unit Δc)
K_La over-all liquid-film coefficient on a volume basis, lb moles/hr-cu ft (unit Δc)
L_M superficial molar mass velocity of liquid, lb moles/hr-sq ft
L_M' superficial molar mass velocity of solute-free liquid, lb moles/hr-sq ft
N molal density of liquid, lb mole/cu ft
N_{OG} number of over-all gas-phase transfer units
N_{OL} number of over-all liquid-phase transfer units
p partial pressure of solute in main gas stream, atm
p_{BM} log mean of inert-gas pressures at film boundaries, atm
p_i partial pressure of solute at liquid-gas interface, atm
p^* pressure of solute in equilibrium with concentration of main body of liquid, atm
p_1 partial pressure of solute in gas entering, atm
p_2 partial pressure of solute in gas leaving, atm
P total pressure on system, atm
R gas-law constant = 1,544 ft-lb/lb-mole (°R)
T absolute temperature, °R
V volume of absorption apparatus, cu ft
x mol fraction of solute in main liquid stream
x^* mol fraction of solute in liquid at equilibrium with main gas stream
x_i mol fraction of solute in liquid at interface
x_1 mol fraction of solute in liquid stream at concentrated end of apparatus
x_2 mol fraction of solute in liquid stream at dilute end of apparatus
X solute concentration in liquid, lb mole solute/lb mole solvent
X_1 solute concentration in liquid at concentrated end of apparatus, lb mole solute/lb mole solvent
X_2 solute concentration in liquid at dilute end of apparatus, lb mole solute/lb mole solvent
y mol fraction of solute in main gas stream
y^* mol fraction of solute in equilibrium with main body of liquid
y_i mol fraction of solute in gas at interface
y_1 mol fraction of solute in gas stream at concentrated end of apparatus
y_2 mol fraction of solute in gas stream at dilute end of apparatus
Y solute concentration in gas stream, lb mole solute/lb mole inert gas
Y_1 solute concentration in gas stream at concentrated end of apparatus, lb mole solute/lb mole inert gas
Y_2 solute concentration in gas stream at dilute end of apparatus, lb mole solute/lb mole inert gas
Z height of packed section of tower, ft

REFERENCES

1. Albrecht, F.: Theoretical Research Concerning the Settling of Dust from Streaming Air and Its Application to the Theory of the Dust Filter, *Physik. Z.*, **32**, 48–56 (1931).

2. Appleby, M. P.: The Recovery of Sulphur from Smelter Gases, *J. Soc. Chem. Ind.*, **56**, 139–46T (1937).
3. Audas, F. G.: A Continuous Dry Process for the Removal of Hydrogen Sulfide from Industrial Gases, *Coke and Gas*, **13**, 229–234 (1951).
4. American Public Works Association: Refuse Collection Practice, Committee on Refuse Collection and Disposal, Chicago, Ill., 1941.
5. Ballman, Harry, Executive Secretary, Air Pollution Control Association: Written communication.
6. Barkley, J. F., and R. E. Morgan: Burning Wood Waste for Commercial Heat and Power, *U.S. Bur. Mines, Inform. Circ.* 7580, 1950.
7. Boucher, R. M. G., and M. L. Franck: Venturi Washers for the Cleaning of Gases, *Ind. Chemist*, **29**, 51–55 (1953).
8. Brown, G. G.: "Unit Operations," John Wiley & Sons, Inc., New York, 1950.
9. Bulkeley, C. A.: Adsorption, Absorption and Condensation in Recovery of Solvents, *Chem. Met. Eng.*, **45**, No. 6, 300–305 (1938).
10. Callinan, T. D., and R. T. Lucas: Manufacture and Properties of Paper Made from Ceramic Fibers, Naval Research Laboratory, Washington, D.C., NRL Report 4044, Oct. 20, 1952.
11. *Can. Chem. Process Inds.*, **26**, 138–139 (1942), Elemental Sulfur; Extraction and Reduction of SO_2 from Roaster Gases at Trail.
12. Carman, E. P., and W. T. Reid: Ignition through Fuel Beds on Traveling- or Chain-grate Stokers, *Trans. Am. Soc. Mech. Engrs.*, **67**, 425–437 (1945).
13. Carrier, W. H., R. E. Cherne, and W. A. Grant: "Modern Air Conditioning, Heating, and Ventilating," 2d ed., p. 220, Pitman Publishing Corporation, New York, 1950.
14. Carvlin, G. M.: The Use of Sodium Phenolate for Hydrogen Sulfide Removal, *Proc. Am. Petroleum Inst., 8th Mid-year Meeting, Sec.* III, **19**, 23–33 (1938).
15. Cawood, W., and R. Whytlaw-Gray: The Influence of Pressure on the Coagulation of Ferric Oxide Smokes, *Trans. Faraday Soc.*, **32**, 1059–1068 (1936).
16. Chass, R. L., and A. H. Rose, Jr.: Discharge from Municipal Incinerators. W. P. Fannon, Municipal Incineration in Baltimore. W. S. Foster, Some Principles in the Design and Operation of Municipal Incinerators, *Air Repair*, **3**, No. 2, 119–129 (1953).
17. *Chem. Eng. (Pacific Ed.)*, **56**, PPI-5 (1949), Sulfur Dioxide Recovery Unit.
18. Chen, C. Y.: Filtration of Aerosols by Fibrous Media, University of Illinois, Engineering Experiment Station, Annual Report, Jan. 30, 1954.
19. Cole, C. S.: The Effect of Boric Acid on the Regeneration of Sodium Sulfite Solutions in the Recovery of Sulfur Dioxide from Waste Gases, B.S. thesis, University of Illinois, 1939.
20. Corey, R. C.: Some Fundamental Considerations in the Design and Use of Incinerators in Controlling Atmospheric Contamination, in L. C. McCabe, ed., "Air Pollution," p. 394, McGraw-Hill Book Company, Inc., New York, 1952.
21. Corey, R. C., L. A. Spano, C. H. Schwartz, and H. Perry: Experimental Study of Effects of Tangential Overfire Air on the Incineration of Combustible Wastes, *Air Repair*, **3**, No. 2, 109–116 (1953).
22. DallaValle, J. M., Clyde Orr, Jr., and B. L. Hinkle: The Aggregation of Aerosols, p. 92, Report of Symposium V, Aerosols, Chemical Corps, Chemical and Radiation Laboratories, Army Chemical Center, Md., June 22, 1953. (PB 111-411.)
23. Danser, H. W., and E. P. Neumann: Industrial Sonic Agglomeration and Collection Systems, *Ind. Eng. Chem.*, **41**, 2439–2442 (1949).
24. Davies, C. N.: The Sedimentation of Small Suspended Particles, Symposium on Particle Size Analysis, Supplement to *Trans. Inst. Chem. Eng. & Soc. Chem. Ind.*, **25**, 25 (1947).
25. Davies, C. N.: Fibrous Filters for Dust and Smoke, *Proc. 9th Intern. Congr. Ind. Med., London*, 1948, pp. 162–196 (1949).
26. Davies, C. N.: The Separation of Airborne Dust and Particles, *Proc. Inst. Mech. Engrs. (London)*, **B1**, 185–198 (1952).
27. Dean, R. S., H. W. St. Clair, P. M. Ambrose, G. W. Marks, and F. S. Wartman: Fixation of Sulphur from Smelter Smoke, *U.S. Bur. Mines, Rept. Invest.* 3339, May, 1937.
28. Dean, R. S., and R. E. Swain: Report Submitted to the Trail Smelter Arbitral Tribunal, *U.S. Bur. Mines, Bull.* 453, 1944.
29. DeLorenzi, O.: Recent Development in Burning Wet Wood, *Combustion*, **21**, 39–43 (1949).
30. DeLorenzi, O.: Furnaces for Wood Refuse and Bagasse, "Combustion Engineering," chap. 12, Combustion Engineering Co., Inc., New York 1947.
31. Deutsch, W.: Motion and Charge of a Charged Particle in the Cylindrical Condenser, *Ann. Physik (4th series)*, **68**, 335–344 (1922).
Deutsch, W.: Townsend Discharges in Dense Clouds in the Presence of Space Charges, *Ann. Physik (5th series)*, **10**, 847–867 (1931).

Deutsch, W.: Is the Action of Electrical Gas Purification Due to "Electric Wind?" *Ann. Physik* (*5th series*). **9,** 249–264 (1931).
32. Diamond, R. W.: The Trail Heavy-chemical Plants, *Trans. Can. Inst. Mining Met.*, **37,** 442–460 (1934).
33. Ekman, F. O., and H. F. Johnstone: Collection of Aerosols in a Venturi Scrubber, *Ind. Eng. Chem.*, **43,** 1358–1363 (1951).
34. *Engineering*, **136,** 459–461, 556–558 (1933), Flue Gas Washing Plant at the Battersea Generating Station of the London Power Company. (From G. W. Hewson, S. L. Pearce, A. Pollitt, and R. L. Rees, The Application to the Battersea Power Station of Researches into the Elimination of Noxious Constituents from Flue Gases, and the Treatment of Resulting Effluents, paper before Chemical Engineering Group, Annual Meeting of the Society of Chemical Industry, July 11, 1933).
35. *Engineering News Record*, **140,** 712–715 (1948), Refuse Disposal Practice and Cost.
36. Eymann, C.: Removal of Ammonia and Hydrogen Sulfide from Gases by Water or Ammonia Liquor, *Gas- u. Wasserfach*, **90,** 505–512, 534–538, 568–570, 577–581 (1949).
37. First, M. W., et al.: Air Cleaning Studies, Progress Report for Feb. 1, 1951, to June 30, 1952, AEC Report No. NYO-1586, Harvard University, School of Public Health, Air Cleaning Laboratory, Boston, Mass., 1952.
38. First, M. W., R. Moschella, L. Silverman, and E. Berly: Performance of Wet Cell Washers for Aerosols, *Ind. Eng. Chem.*, **43,** 1363–1370 (1951).
39. First, M. W., and P. Drinker: Concentrations of Particulates Found in Air, *Arch. Ind. Hyg. and Occupational Med.*, **5,** 387 (1952).
40. Fleming, E. P., and T. C. Fitt: Recovery of SO_2 from Gas Mixtures, U.S. Patent 2,295,587, Sept. 15, 1942.
41. Förch, J. H.: The Mechanism of Hydrogen Sulfide Removal from a Gas by a Mass of Iron Oxide, *Chem. Weekblad*, **49,** 149–151 (1953).
42. Foster, E. G., and F. Daniels: Recovery of Nitrogen Oxides by Silica Gel, *Ind. Eng. Chem.*, **43,** 986–992 (1951).
43. Francis, W.: Flue Gas Washing Processes, *Power & Works Engr.*, **41,** 17–21, 25; 37–40; 75–77; 102–105 (1946).
44. Frazier, H. D., and A. L. Kohn: Selective Absorption of Hydrogen Sulfide from Gas Streams, *Ind. Eng. Chem.*, **42,** 2288–2292 (1950).
45. Friedlander, S. K., et al.: "Handbook on Air Cleaning" (A.E.C.D.-3361; NYO-1572), Harvard University, School of Public Health, Boston, Mass., 1952.
46. Gillespie, T., and G. O. Langstroth: Aging of Ammonium Chloride Smokes, *Can. J. Research*, **29,** 201–216 (1951).
47. Hammond, E.: The Effects of Turbulence on the Space Required for the Combustion of Coal on a Grate, *Sheffield Univ. Fuel Soc. J.*, **3,** 39–44 (1952).
48. Heinrich, R.: The Present Status of Electrical Gas Purification, *Electrotechnik. Z.*, **60,** 7–10, 43–46 (1939).
49. Hignett, T. P., and M. R. Siegel: Recovery of Fluorine from Stack Gases, *Ind. Eng. Chem.* **41,** 2493–2498 (1949).
50. Hill, A. E., and T. B. Fitzgerald: The Compounds of Sulfur Dioxide with Various Amines, *J. Am. Chem. Soc.*, **57,** 250–254 (1935).
51. Hollings, H.: Progress in Gas Purification, Institute of Gas Engineers, copyright Pub. No. 407, 1952; *Gas World*, **135,** 585–589, 594, 694–696; *Gas J.*, **270,** 696–700.
52. Hopton, G. U.: Removal of Hydrogen Sulfide from Coal Gas by Means of Iron Oxide, *Gas World*, **128,** 538–544 (1948).
53. Hughes, E. C.: Volatile-fluoride Recovery (to Standard Oil of Ohio), U.S. Patent 2,440,542, 1948.
54. Johnstone, H. F.: Sulfur Dioxide, *Ind. Eng. Chem.*, **34,** 1017–1028 (1942).
55. Johnstone, H. F.: Recovery of Sulfur Dioxide from Waste Gases (Equilibrium Partial Vapor Pressures over Solutions of the Ammonia–Sulfur Dioxide–Water System), *Ind. Eng. Chem.*, **27,** 587–593 (1935).
56. Johnstone, H. F.: Recovery of SO_2 from Waste Gases, *Ind. Eng. Chem.*, **29,** 1396–1398 (1937).
57. Johnstone, H. F.: Recovery of SO_2 from Dilute Gases, *Pulp & Paper Mag. Can.*, **53,** 105–112 (1952).
58. Johnstone, H. F.: Gas Cleaning and Purification, in R. E. Kirk and D. F. Othmer, eds., "Encyclopedia of Chemical Technology," vol. 7, p. 83, Interscience Publishers, Inc., New York, 1951.
59. Johnstone, H. F., R. B. Field, and M. C. Tassler: Gas Absorption and Aerosol Collection in a Venturi Atomizer, *Ind. Eng. Chem.*, **46,** 1601–1608 (1954).
60. Johnstone, H. F., and D. B. Keyes: Recovery of Sulfur Dioxide from Waste Gases, *Ind. Eng. Chem.*, **27,** 659–665 (1935).
61. Johnstone, H. F., and R. V. Kleinschmidt: The Absorption of Gases in Wet Cyclone Scrubbers, *Trans. Am. Inst. Chem. Engrs.*, **34,** 181–198 (1938).

62. Johnstone, H. F., and M. H. Roberts: Deposition of Aerosol Particles from Moving Gas Streams, *Ind. Eng. Chem.*, **41,** 2417–2423 (1949).
63. Johnstone, H. F., and H. E. Silcox: Gas Absorption and Humidification in Cyclone Spray Towers, *Ind. Eng. Chem.*, **39,** 808–817 (1947).
64. Johnstone, H. F., and A. D. Singh: The Recovery of Sulfur Dioxide from Dilute Waste Gases by Chemical Regeneration of the Absorbent, *Univ. Illinois Bull., Eng. Expt. Sta., Bull. Ser.* 324, (1940).
65. Johnstone, H. F., and A. D. Singh: Recovery of Sulfur Dioxide from Waste Gases. Design of Scrubbers for Large Quantities of Gases, *Ind. Eng. Chem.*, **29,** 286–297 (1937).
66. Johnstone, H. F., and G. C. Williams: Absorption of Gases by Liquid Droplets. Design of Simple Spray Scrubbers, *Ind. Eng. Chem.*, **31,** 993–1001 (1939).
67. Kirkpatrick, S. D.: Trail Solves Its Sulfur Problem, *Chem. Met. Eng.*, **45,** No. 9, 483–485 (1938).
68. Kohl, A. L.: Selective Hydrogen Sulfide Absorption—A Review of Available Processes, *Petroleum Processing*, **6,** 26–31 (1951).
69. van Krevelen, D. W., P. J. Hoftijzer, and C. J. van Hooren: Gas Absorption, IV. Simultaneous Gas Absorption and Chemical Reaction, *Rec. trav. chim.*, **67,** 133–152 (1948) (in English).
70. Kropp, E. P.: Scrubbing Devices Successfully Control Air Pollution from Refinery Processes, *Petroleum Processing*, **5,** 627–629 (1950).
71. Kropp, E. P., and R. N. Simonsen: Scrubbing Devices for Air Pollution Control, *Proc. Air Pollution and Smoke Prevention Assoc. Amer.*, **45,** 48–53 (1952).
72. Labbe, A. L.: Acid Conditioning of Metallurgical Smoke for Cottrell Precipitation, *J. Metals*, **188,** *Trans.* 692–693 (1950).
73. Landahl, H. D., and R. G. Herrmann: Sampling of Liquid Aerosols by Wires, Cylinders, and Slides, and the Efficiency of Impaction of the Droplets, *J. Colloid Sci.*, **4,** 103–136 (1949).
74. Landau, R., and R. Rosen: Industrial Handling of Fluorine, *Ind. Eng. Chem.*, **39,** 281–286 (1947).
75. Landau, R., and R. Rosen: Fluorine Disposal, *Ind. Eng. Chem.*, **40,** 1389–1393 (1948).
76. Langmuir, I., and K. B. Blodgett: A Mathematical Investigation of Water Droplet Trajectories, General Electric Research Laboratory, Schenectady, N.Y., Report RL 225, 1944–1945.
76*a*. Langmuir, I.: The Production of Rain by a Chain Reaction in Cumulus Clouds at Temperatures above Freezing, *J. Meteor.*, **5,** 175–192 (1948).
77. Lapple, C. E.: Dust and Mist Collection, in J. H. Perry, ed., "Chemical Engineers' Handbook," 3d ed., p. 1039, McGraw-Hill Book Company, Inc., New York, 1950.
78. Leĭbush, A. G., and A. L. Shneerson: Absorption of Hydrogen Sulfide and of Its Mixtures with Carbon Dioxide by Ethanolamines, *J. Appl. Chem.* (*U.S.S.R.*), **23,** 149–157 (1950) (English translation); *Zhur. Priklad. Khim.*, **23,** 145–152.
79. Leonard, A. B., and G. R. Hettick: Recovery of Hydrogen Fluoride from Gases (to Phillips Petroleum Co.), U.S. Patent 2,425,745, 1947.
80. Lessing, R.: The Development of a Process of Flue-gas Washing without Effluent, *J. Soc. Chem. Ind.*, **57,** 373–388T (1938).
81. Leva, Max.: "Tower Packings and Packed Tower Design," United States Stoneware Co., Akron, Ohio, 1952.
82. Linden, A. J. ter: Investigations into Cyclone Dust Collectors, *Proc. Inst. Mech. Engrs.*, **160,** 233–251 (1949).
83. Loeb, L. B.: "Fundamental Processes of Electrical Discharge in Gases," John Wiley & Sons, Inc., New York, 1939.
84. Long, K. E., and H. W. Cromer: Removal of Hydrogen Fluoride from Nonaqueous Fluids (to Harshaw Chemical Co.), U.S. Patent 2,426,558, 1947.
85. Longwell, J. P., and M. A. Weiss: Mixing and Distribution of Liquids in High Velocity Air Streams, *Ind. Eng. Chem.*, **45,** 667–677 (1953).
86. McCabe, L. C., ed.: "Air Pollution," p. 100, McGraw-Hill Book Company, Inc., New York, 1952.
87. McCabe, L. C., ed.: "Air Pollution," pp. 248–263, McGraw-Hill Book Company, Inc., New York, 1952.
88. McCabe, L. C., ed.: "Air Pollution," pp. 404–407, discussion by W. Xanten and H. K. Kugel of chap. 46, McGraw-Hill Book Company, Inc., New York, 1952.
89. McCabe, L. C.: Atmospheric Pollution, *Ind. Eng. Chem.*, **45,** 109A (1953); **45,** 115–117A (1953). P. H. Goodell, Catalytic Energy Recuperation for Industrial Process Heating, *Ind. Heating*, **20,** 1560–1570, 1617 (1953).
90. Mantell, C. L.: "Adsorption," 2d ed., McGraw-Hill Book Company, Inc., New York, 1951.

91. Marshall, J. R.: Gas Purification—Wet and Dry, *Gas World*, **129**, No. 3338, Coking Sec., 105–113, 114 (1948).
92. Mierdel, B., and R. Seeliger: Investigation upon the Physical Process in Electrical Filtration, *Arch. Elektrotech.*, **29**, 149–172 (1935).
93. Miller, F. E., and A. L. Kohl: Here's a New Solution for Selective Absorption of Hydrogen Sulfide, *Oil Gas J.*, **51**, No. 51, 175–176, 178, 180, 183 (1953).
94. Munger, H. P.: The Spectrum of Particle Size and Its Relation to Air Pollution, in L. C. McCabe, ed., "Air Pollution," p. 159, McGraw-Hill Book Company, Inc., New York, 1952.
95. National Board of Fire Underwriters: Standard for Incinerators, Pamphlet 82, October, 1953.
96. Nicholls, P.: Underfeed Combustion, Effect of Preheat and Distribution of Ash in Fuel Beds, *U.S. Bur. Mines, Bull.* 378, 1934.
97. Norris, W. E., and F. R. Clegg: Investigation of a Girbotol Unit Charging Cracked Refinery Gases Containing Organic Acids, *Petroleum Refiner*, **26**, No. 11, 107–109 (1947).
98. Pearce, S. L., G. H. Hewson, A. Pollitt, and R. L. Rees: The Application to the Battersea Power Station of Researches into the Elimination of Noxious Constituents from Fuel Gases and Treatment of Resulting Effluents, *J. Soc. Chem. Ind.*, **52**, 593–594 (1933).
99. Pearson, J. L., G. Nonhebel, and P. H. N. Ulander: The Removal of Smoke and Acid Constituents from Flue Gases by a Non-effluent Water Process, *J. Inst. Elec. Engrs.*, **77**, No. 463, 1–48 (1935).
100. Perry, J. H., Ed.: "Chemical Engineers' Handbook," 3d ed., McGraw-Hill Book Company, Inc., New York, 1950.
101. Peters, M. S., C. P. Ross, and L. B. Andersen: Removal of Nitrogen Oxides from Gaseous Mixtures, University of Illinois, Engineering Experimental Station, AEC Technical Report No. 10 (COO-1011), 1953.
102. Pigford, R. L., and C. Pyle: Performance Characteristics of Spray-type Absorption Equipment, *Ind. Eng. Chem.*, **43**, 1649–1662 (1951).
103. Powell, A. R.: Recovery of Sulfur from Fuel Gases, *Ind. Eng. Chem.*, **31**, 789–796 (1939).
104. Pring, R. T.: Bag-type Cloth Dust and Fume Collectors, in L. C. McCabe, ed., "Air Pollution," p. 280, McGraw-Hill Book Company, Inc., New York, 1952.
105. Reed, R. M., and N. C. Updegraff: Removal of Hydrogen Sulfide from Industrial Gases, *Ind. Eng. Chem.*, **42**, 2269–2277 (1950).
106. Rodebush, W. H., I. Langmuir, and V. K. LaMer: Filtration of Aerosols and the Development of Filter Materials, O.S.R.D. Report No. 865, Sept. 4, 1942 (PB 99669).
107. Rosin, P., E. Rammler, and W. Intelmann: Principle and Limits of Cyclone-dust Removal, *Z. Ver. deut. Ing.*, **76**, 433–437 (1932).
108. St. Clair, H. W.: Agglomeration of Smoke, Fog, or Dust Particles by Sonic Waves, *Ind. Eng. Chem.*, **41**, 2434–2438 (1949).
109. Saunders, J. J.: The Composition and Activity of Manchester Process Plant Liquor, *Gas World*, **135**, 200–201 (1952).
110. Schmidt, W. A., and E. Anderson: Electrical Precipitation, *Elec. Eng.*, **57**, 332–338 (1938).
111. Schmidt, W. A.: Electrical Precipitation and Mechanical Dust Collection, *Ind. Eng. Chem.*, **41**, 2428–2434 (1949).
112. Schmidt, W. A.: Apparatus for Electrical Separation of Suspended Particles from Gases, U.S. Patent 1,343,285, 1920.
113. Sell, W.: Dust Precipitation on Simple Bodies and in Air Filters, *Forsch. Gebiete Ingenieurw.*, **2**, Forschungsheft, 347 (1931).
114. Shepherd, C. B., and C. E. Lapple: Flow Pattern and Pressure Drop in Cyclone Dust Collectors, *Ind. Eng. Chem.*, **31**, 972–984 (1939).
115. Shepherd, C. B., and C. E. Lapple: Flow Pattern and Pressure Drop in Cyclone Dust Collectors. Cyclone without Inlet Vane, *Ind. Eng. Chem.*, **32**, 1246–1248 (1940).
116. Sherwood, T. K.: Mass Transfer and Friction in Turbulent Flow, in "Fluid Mechanics and Statistical Methods in Engineering," p. 55, University of Pennsylvania Press, Philadelphia, Pa., 1941; Also *Trans. Am. Inst. Chem. Engrs.*, **36**, 817–840 (1940).
117. Sherwood, T. K., and R. L. Pigford: "Absorption and Extraction," 2d ed., McGraw-Hill Book Company, Inc., New York, 1952.
118. Silver, R. S.: Theoretical Treatment of Combustion in Fuel Beds. I. Gas Composition and Heat Release. II. Temperature Attained in Combustion, *Fuel*, **32**, 121–150 (1953).
119. Silverman, L.: New Developments in Air Cleaning, *Am. Ind. Hyg. Assoc. Quart.*, **15**, 183–192 (1954).

120. Silverman, L., E. W. Conners, and D. Anderson: Mechanical Electrostatic Charging of Fabrics for Air Filters (in preparation, 1954).
121. Smith, M. L., and K. W. Stinson: "Fuels and Combustion," pp. 102–113, McGraw-Hill Book Company, Inc., New York, 1952.
122. Smith, W. J., and E. Stafford: Development of a Dry Fibrous Filter, in L. C. McCabe, ed., "Air Pollution," p. 64, McGraw-Hill Book Company, Inc., New York, 1952.
123. Smith, W. J., and N. F. Surprenant: Properties of Various Filtering Media for Atmospheric Dust Sampling, preprint before presentation to American Society for Testing Materials, July 1, 1953.
124. Spaeth, W.: The Stuttgart Coke-oven-gas Plant, *Gas- u. Wasserfach.*, **94,** 230–235 (1953).
125. Spalding, D. B.: Conditions for the Stable Burning of Carbon in an Air Stream, *J. Inst. Fuel*, **26,** 289–294 (1953).
126. Sproull, W. T., and Y. Nakada: Operation of Cottrell Precipitators, *Ind. Eng. Chem.*, **43,** 1350–1358 (1951).
127. Stokes, C. A.: Sonic Agglomeration of Carbon Black Aerosols, *Chem. Eng. Prog.*, **46,** 423–432 (1950).
128. Teverovskiĭ, N.: Coagulation of Aerosol Particles in a Turbulent Atmosphere, 12, *Akad. Nauk, Ser. Geogr. Geofiz.* No. 1, 1948.
129. Tremaine, B. K.: Masking Industrial Malodors, *Air Repair*, **3,** No. 2, 59–63 (1953); J. von Bergen, The Counteraction of Industrial Odors, *Air Repair*, **3,** No. 2, 64–67 (1953).
130. Tunitskiĭ, N. N.: Diffusion Processes under Conditions of Natural Turbulence, *J. Phys. Chem.* (*U.S.S.R.*), **20,** 1137–1141 (1946).
131. University of California, Sanitary Engineering Research Project: Municipal Incineration, *Tech. Bull.* 5, June, 1951; *Tech. Bull.* 6, November, 1951.
132. Vajna, S.: Removal of Hydrogen Sulfide from Industrial Gases by a Purifying Mass Containing Iron Hydroxide. I. *Acta Chim. Acad. Sci. Hung.*, **2,** 163–174 (1952) (in English).
133. dé Voogd, J. G.: Determination of the Value of Fresh Gas-purifying Material, *Het Gas*, **64,** 73–74 (1944); *Chem. Zentr.*, **1944,** II, 1343–1344.
134. Wainwright, H. W., G. C. Egleson, C. M. Brock, J. Fisher, and A. E. Sands: Selective Absorption of Hydrogen Sulfide from Synthesis Gas, *Ind. Eng. Chem.*, **45,** 1378–1384 (1953).
135. Wainwright, H. W., G. C. Egleson, C. M. Brock, J. Fisher, and A. E. Sands: Removal of Hydrogen Sulfide and Carbon Dioxide from Synthesis Gas Using Di- and Triethanolamine, *U.S. Bur. Mines, Rept. Invest.* 4891, 1952.
136. Walton, W. H.: The Electrical Characteristics of Resin Impregnated Filters, Porton Report No. 2465, Chemical Defense Experimental Station, Porton, England, Dec. 15, 1942. (Declassified.)
137. Webber, L. E.: Furnaces for Wood Burning, *Power*, **85,** 471–473 (1941).
138. Weidmann, H., and G. Roesner: Process for the Manufacture of Pure Sulfur Dioxide, *Metallges. Periodic Rev.* **11,** 7–13, (1936).
139. Wells, A. E.: The Wet Thiogen Process for Recovering Sulfur from Sulfur Dioxide in Smelter Gases, *U.S. Bur. Mines, Bull.* 133, 1917.
140. White, H. J.: Role of Corona Discharge in Electrical Precipitation Process, *Elec. Eng.*, **71,** 67–73 (1952).
141. White, H. J.: Particle Charging in Electrostatic Precipitation, *Trans. Am. Inst. Elec. Engrs.*, **70** (Pt. II), 1186–1191 (1951).
142. White, H. J., L. M. Roberts, and C. W. Hadberg: Electrostatic Collection of Fly Ash, *Mech. Eng.*, **72,** 873–880 (1950).
143. Whytlaw-Gray, R., and H. S. Patterson: "Smoke," Edward Arnold & Co., London, 1932.
144. Williams, C. E., et al.: Determination of Cloth Area for Industrial Air Filters, *Heating, Piping Air Conditioning*, **12,** 259–263 (1940).
145. Williamson, R. H., and J. E. Garside: Application of Fluidized-solids Technique to Coal-gas Purification, *Inst. Gas Engrs., Commun.* 345; *Gas World*, **129,** 809–811, 948–952 (1948).
146. Williamson, R. H., and J. E. Garside: Application of Fluidized-solids Technique to Coal-gas Purification. II. *Inst. Gas Engrs., Commun.* 357 (1949); *Gas World*, **130,** 1999–2001 (1949); **131,** 119–125 (1950).
147. Wong, J. B., and H. F. Johnstone: Collection of Aerosols by Fiber Mats, *Univ. Illinois, Eng. Expt. Sta., Tech. Rept.* II (COO-1012), Oct. 31, 1953.
148. Zankl: Possibilities and Limits for Hydrogen Sulfide Removal by Ammonia Scrubbing, *Gas- u. Wasserfach*, **94,** 178–180 (1953).

SECTION 14

AIR POLLUTION LEGISLATION

BY FRANK L. SEAMANS

14.1 INTRODUCTION

Whether air pollution should be controlled by legislation and, if so, by what kind of legislation are questions which immediately cause many thinking people to line up in two opposing camps. This section is not intended to argue either side of these questions or to take up the fight for any specific approach to the problem.

The purpose here is to (1) review some of the basic legal considerations involved, (2) point out some of the issues with which pollution legislation will need to deal, and (3) report what has been done and is proposed by way of air pollution legislation.

One salient factor stands out clearly in this field, as in all subjects: good decisions must be made as to whether any legislation is needed, and, if so, the type of legislation to be adopted must be based on facts and not hysteria, on scientific investigation and not emotion. If the facts about the problem to be attacked are discovered, it is much easier to find the solution most likely to achieve the desired result.

Factually speaking, there can be no doubt that the public is demanding action in the field of air pollution abatement. Public consciousness of air pollution came later than awareness of many other imperfect conditions in our environment about which society has concerned itself, but no one can now doubt that the man on the street prefers cleaner air and that the public, as a mass, is exerting constant pressure to get something done about it.

What is to be done? It is hoped that this section may be of some aid to those who are giving intelligent consideration to that question.

14.1.1 Definitions

When one speaks of air pollution, he is using a term that does not define itself and which does not mean the same thing to all people and to all sections of the country. Air pollution means principally getting rid of the dirt by smoke control in Pittsburgh, while in Los Angeles one of the chief concerns is to eliminate the chemical combination causing eye irritation and reducing visibility.

Is it then to include in-plant exposures to employees, a subject now generally covered by state health department regulations and workmen's compensation laws, or shall we talk about the problem of the general public and not the relationship of employer-employee?

It must be realized, of course, that everyone contributes to air pollution every day by smoking cigarettes, driving automobiles, burning trash, and so on. All these sources send unnatural substances into the air. Determining a definition is always a good way to start consideration of a subject with sharp thinking. The kind of air pollution dealt with in this section can be defined as the presence in the atmosphere of substances, resulting from acts of man, in quantities which are or may become injurious to human, plant, or animal life or to property; all aspects of employer-employee relationship as to health and safety hazards are excluded.

14.1.2 Enforcement and Performance

Whatever the type of air pollution legislation considered, the aim is to achieve some degree of abatement where needed by empowering some person or board to "enforce" the requirement of abatement. And here is what seems to the author to be the nub of the problem. Only the agency representing the public can enforce; yet only the offender (be it industrial plant, hospital, or household incinerator) can abate. The enforcement agency cannot install abatement equipment and probably will not require a plant to shut down for lack of proper abatement measures. The latter merely avoids, not solves, the problem and in addition puts people out of work, thus causing economic hardship to the community.

The public, industry, and government all are now convinced that air pollution abatement is important and that some improvement must be effected. Gradually, it is being realized that the improvement which must be made is almost entirely a matter of economics because the technical "know-how" is available for the most part.

The alleged offender (for example, an industrial plant) is the only one that understands its particular problem well enough and has the necessary engineers and technical staff (together with available outside consultants) to design and install abatement devices which will really work. Consequently, blind enforcement cannot answer the problem—the public's desire for both jobs and cleaner air will not be realized. The offender (still an industrial plant) must be induced to attack the problem itself, wholeheartedly, for it is the only one, in the final analysis, that can produce abatement.

Therefore, in the judgment of the writer, any air pollution legislation will have missed the target badly if it does not create a framework which will convince the offending industries that:

1. The abatement effort is a bona fide, good-faith attempt to do a real job—not a witch hunt or a temporary political expediency.

2. The public agency will be manned by conscientious, competent experts who are capable of understanding what the project really involves and who will be reasonably patient with an industrial plant which tries to do a good job.

3. The offender will be allowed to work out the solution that best fits its own conditions and it will not have to follow arbitrary, blanket requirements prescribed for all sources of emission, regardless of kind or character.

4. No rules will be forced upon one offender more than another; sources must abate; one firm won't be at a competitive disadvantage if it invests money while others ignore the problem.

If such a framework is presented, industry will have the inducement and the encouragement to do a real cleanup job itself.

14.2 FUNDAMENTAL CONCEPTS

Since the beginning of life on this planet, the atmosphere has functioned as an acceptor of emitted wastes. Breathing creatures inhale oxygen and exhale carbon dioxide, which is transferred through the air to green vegetation where the oxygen is released to complete the cycle. There was no reason to question releasing waste products into the air—not until the amount became "too much."

The objective of air pollution control should be to limit the amount of foreign material in the air so that there will not be too much, but at the same time allow the atmosphere to function usefully to its fullest capacity. This principle was recognized in a recent Oregon bill (6) which provided for "reasonable or natural use" of the air. It is logical to regard the atmosphere as one of many natural resources which are being and should be used to technical and economic advantage. Today modern industry is paying more and more attention to atmospheric conditions in considering new plant sites and seeking the services of meteorologists in addition to chemists and engineers to aid in planning for satisfactory operation.

Each localized area is unique in its air pollution problem. Atmospheric dispersion of airborne wastes depends upon air movements and topography. The waste-load capacity of a given area is influenced also by the height and relative location of the various points of emission. Obviously a heavy concentration of low stacks in a narrow valley would be more likely to create an undesirable situation than would scattered tall stacks in open country, assuming the same rate and kinds of emissions and comparable surroundings as to habitation, use of property, etc.

Superficial consideration of the air pollution problem occasionally leads people to the fallacious view that the remedy is to establish concentration limits for each air contaminant at the source of emission and apply them uniformly throughout the state or nation. This is undesirable, both technically and economically, for it would require everyone to meet the restrictions required for the worst situation. Further, concentration values alone are almost meaningless. They become significant only when applied in conjunction with such factors as rate of emission, stack height, and wind velocity.

It is proper to judge air pollution only on the basis of relating air conditions at the point of contact (e.g., ground level) to effects produced by them. As technical knowledge about air pollution increases, it may become established that a certain limit of dustfall represents the tolerable level in city areas or that a certain limit of sulfur dioxide is all that specific crops can stand, etc. Information of this sort emphasizes the fundamental validity of the concept that air pollution problems vary for each area.

The report of a recent survey of stream pollution control activities contains the following (2):

"The principle was expressed by practically all governmental authorities that where a stream is polluted by an industry upon which the life of a community depends, it is better to permit some degree of pollution and maintain the community, than to abate the pollution and destroy the community. This principle was more picturesquely expressed by one of the general public . . . when . . . asked if the deathly stench from a nearby pulpmill was not unbearable. 'Yes' was the reply, 'but it's less unbearable than the deathly stench from a dead town.'"

This principle is equally appropriate in the realm of air pollution. If it can be demonstrated that reduction of pollution to an innocuous level is not economically feasible, the people whose livelihood would be affected should have a voice in deciding the outcome.

Responsibility for enforcement of an air pollution law must rest in the municipal, county, state, or Federal government, or a subdivision or combination of one or more of these entities, either preexisting or specially established for this purpose. The emphasis to date has been on the enactment of city ordinances, though in some states it is first necessary for the state legislature to pass an enabling act to authorize local municipalities to enact ordinances in this field. The emphasis on municipal control of air pollution results from the fact that air pollution to many means "smoke control," and "smoke" is principally found in cities. Also, municipal ordinances have been found easier to pass than legislation covering a wider unit or a combination of both.

Air pollution control limited to the boundaries of a municipality is insufficient in many cases and, to be effective, is more costly than many cities can afford. In some instances one air pollution law covers a city and another the whole county in which that city is located. This causes some duplication of effort and conflict in methods of achieving results. On the other hand, clean air cannot be obtained in the city without restrictions on the surrounding industry in the entire county, and vice versa.

Also, in some areas city limits arbitrarily divide population centers; thus, one metropolitan area may include portions of several cities, two or more counties, and even two or more states.

For these reasons, many authorities advocate the creation of "control zones," where there is a sameness of the problem even though the "zone" cuts across existing, but artificial, political boundaries.

The placement of responsibility in a particular bureau or commission of the city, county, or state is a matter that must be determined by local and political considerations. Frequently, city ordinances have created new "smoke bureaus" or air pollution departments. In other cities this job has been assigned to the public health department or to the city engineer.

Of paramount importance is the placement of responsibility in a way that will permit the establishment of a competent staff, both in number and in specialized training. Problems of air pollution will be solved only through cooperation and technical skills. Any worthwhile legislation must emphasize both these cardinal points, and the department to which responsibility is assigned must be free in every sense of the word to gather a staff of competent, trained people. The problem of air pollution must not become a political football.

So far, only the state of Oregon has passed a state air pollution law (Oregon Laws, 1951, c.425). This statute creates an additional division within the State Board of Health to be known as the Air Pollution Authority of the State of Oregon.

14.3 LEGAL BASES FOR CONTROL

14.3.1 Common Law

At common law a person aggrieved by the pollution emission of another might sue either as a "trespass" or "nuisance" (trespass on the case according to old pleading forms): "trespass" if the result was direct and immediate and the action intentional and "nuisance" if the action was unintentional and the result or damage indirect and consequential.

As is plain from such a brief statement of the common-law rules, the distinctions are highly elusive, and the application of the rules always debatable. Error in choosing the wrong form of action at common law carries severe consequences.

In later years, however, reformation in pleading practices has abolished the niceties of such distinctions. It became, and is today, adequate common-law pleading merely to assert the facts of the defendant's action and of the plaintiff's resulting damage. Whether the cause of action so alleged is in reality one of "trespass" or of "nuisance" loses all importance generally except where a different statute of limitations applies to the two types of action. Gradually, it has become customary to speak of all actions involving alleged damage due to airborne emanations as "nuisance" cases.

The common law, however, makes one further distinction—that between a "private" and a "public" nuisance. If John Doe thinks his gladioli were ruined by the fumes from X plant, John Doe declares upon a "private" nuisance. The conduct of the X Company, against which complaint is brought, has produced special and personal damage to Mr. Doe. Mr. Doe can sue for an injunction or damages or both.

If, however, the Rosewood section of Urbanville has been bothered by fumes that stain curtains and make life miserable for all the citizens in that section alike, then a "public" nuisance exists, and an injunction suit (but not damages) may be brought by a competent public representative of the whole group.

It can therefore be flatly asserted that there now exists at common law (and without special statutes) an *available* legal remedy for all the John Does and for all citizens of Rosewood. In other words, John Doe does not need any special legislation to enable him to sue for his gladioli damages, nor do the citizens of Rosewood need a new law to enable them to stop the offender from fumigating their end of town. Yes, the legal remedy for air pollution is *available* now without special legislation—but is it *adequate?*

An aggrieved person bringing a common-law private-nuisance action will be awarded proper damages if he proves that the defendant's emissions caused his damage, regardless of what effort the defendant has made to control the fumes. The plaintiff can also get an injunction to stop the emission of such fumes, assuming that the cause-and-effect relationship has been established *only,* however, by meeting certain other tests. These are variously stated in different jurisdictions but generally include consideration of whether the plaintiff knowingly moved into an industrial area, whether the fumes or dust now are worse than when the plaintiff moved in, and the doctrine of balancing the equities. By the latter term is meant the consideration most equity courts give to the hardship to the plaintiff if the fumes are not prohibited and to the hardship to the defendant, the community, and even the nation if an injunction is granted.

Frequently when a court feels a complete shutdown of the plant to be out of reason, some lesser degree of control regulation may be ordered, with operation of the plant to continue under the eye of the court for a stated period.

In view of the fact that there is a legal remedy for a nuisance without special legislation, one may ask why should air pollution legislation ever be considered (beyond meeting the popular demand expressed by the phrase "There ought to be a law")? The following would appear to be reasonable grounds for considering additional legislation:

1. Bringing and successfully concluding a nuisance action is an expensive undertaking for one person.

2. While a nuisance action may produce an award of damages, that may not solve the continuing problem, and it may be impossible to obtain an injunction because of balancing the equities.

3. The solution that will best serve the whole community is abatement on the part of the offender. A court order will not put an affirmative duty on the guilty plant to install a type A scrubber, and the court (or plaintiff) would not know what remedy to prescribe even if such an order were to be entered.

4. The plant sued is likely to be reluctant to expend vast sums for abatement unless its competitors must do so likewise (and quite often its competitors have not been sued for one reason or another).

5. Perhaps a good law can bring scientific help, through trained experts, into the study to help the offending source find the cheapest and most effective means of abatement.

14.3.2 Statute

When a statute is enacted, no doubt exists, and the decisions are uniform. The state may, by the proper exercise of its police power to protect the public health, safety, and morals, pass effective air pollution control legislation. Naturally, the different state constitutions contain different requirements and limitations regarding the form and content of such laws, but the power to pass such statutes is clear and indubitable.

14.3.3 Legal Power of the Community

We spoke above of the power of the state to pass air pollution control legislation. Whether and how that can be done by lesser political entities, such as cities, counties, etc., is a different question which must be separately examined in each state.

Suffice it to say that the state, which is the residuum of the police power, may under proper circumstances delegate the exercise of that power for specific purposes to cities, counties, etc. In some states, such delegation is found in the state constitution or in old laws creating those cities or counties. In other states, a new and specific enabling act must first be passed by the state legislature.

14.3.4 Reports and Licenses

Effective aid to industry with respect to air pollution, as well as control to be exercised for the public good, cannot intelligently be given or exercised unless the government agency charged with responsibility has knowledge of the facts. The fee and permit system, in which the operator must submit his plans for approval, is an attempt to supply those facts.

Another method is to require registration of points of emission with the bureau in charge, whether municipal, county, or state. Ilion and Herkimer, N.Y., have incorporated this concept in their new smoke-abatement ordinances. The data to be given in either system would include:

1. Location of outlet
2. Size of outlet
3. Height of outlet
4. Rate of emission from outlet
5. Composition of effluent

The usual exemptions from registration would apply, such as single noncontinuous emissions, accidental discharges, small units, etc. Other exemptions can be provided as needed in that area.

When the agency charged with responsibility receives data respecting the points of emission, a pattern is readily established locating the probable trouble spots and permitting intelligent expenditure of time for constructive improvement.

14.3.5 Cause and Effect

No law will gain the confidence of the sources of emission nor achieve its goal of obtaining cleaner air unless it is based on a study of cause-and-effect relationships. The effect which is found to be undesirable and which is to be minimized cannot be successfully reached until the causes are determined. Because of the complexity of air pollution, the cause is often extremely difficult to determine. Nevertheless, to avoid arbitrary blanket regulations, concentrations, or permissible levels, the law should require the agency establishing those standards to do so after a cause-and-effect study; or, stating it another way, after it is known what effects are desired, the law should provide that any regulations issued must be determined on the basis of eliminating the cause of these undesirable effects.

14.3.6 Use of Technical Information

Recognizing that advancement of technical knowledge is vital to solving the overall problem of air pollution, the law should authorize the agency selected to (1) conduct investigations on both the control and effects of air pollution and assist in solving the technical problems of air pollution control within the control unit; (2) consult and cooperate with other branches of the state, local, and Federal governments and with other groups in furthering the objective of the air pollution control law; (3) assemble and distribute information about air pollution and its control; (4) represent the control unit in dealing with interstate air pollution problems or in cooperating with international commissions.

14.3.7 Legal Safeguards

For the purpose of examining sources of emission, personnel of the agency selected should have the right to enter private or public property, subject to reasonable limitations so as not to infringe unduly on the historic sanctity of private property. The conduct of investigating personnel should conform to established regulations (e.g., where special safety or fire precautions may be required) and cause a minimum of interference with operations.

Secret process information and confidential competitive data should, of course, be protected from exposure by the industry.

14.3.8 Statement of Law in Principle

In brief, the fundamental basis for an air pollution law might be characterized as follows (3):

"Pollution of the atmosphere to the detriment of health or property shall be unlawful. This shall not be construed as contrary to the reasonable or natural use of air for dispersing waste products within the proper capacity to do so. Each localized area, i.e., affected by the same sources of pollution, shall be considered unique, and pollution within the area judged on the basis of its specific effects. The requirements for pollution reduction and control shall remain within the bounds of scientific knowledge and economic feasibility. The responsibility for proper control of emissions shall lie with the potential offender. Where community livelihood would be adversely affected by drastic reduction of air pollution, the right of local option is to be exercised in determining the extent to which reduction of pollution shall be required."

14.4 PRESENT TYPES OF LEGISLATION

Most of the existing air pollution control legislation is on a city or municipal basis, with counties second in amount of legislation. Legislation is generally aimed at eliminating smoke and dirt, with attempts to establish limits on fly-ash and smoke emission from stacks.

Generally speaking, existing legislation is of one or more of the following three types:

14.4.1 Survey Authority

In this general group we find enactments declaring the public interest in air pollution and authorizing the expenditure of money for a survey and investigation with the proviso that a subsequent report be filed.

14.4.2 Enabling Legislation

In this grouping we have enactments authorizing cities and counties and other lesser political entities to provide the mechanics for limiting emissions and controlling air pollution, subject to the broad general terms of the enabling statute.

14.4.3 Control Legislation

This includes the laws creating a real control system which involves the formation of a new, or the designation of an existing, agency to act as the control bureau or commission; authority in that agency to promulgate rules and regulations regarding emission; requirements regarding licensing or registration of information; the authority to make investigations; the mechanics of enforcement; and penalties for violation.

In general, it can be said that existing legislation covers a great variety of viewpoints in approach and varies greatly in scope and detail. Much of what has been enacted is not much different from existing common law on the subject of nuisances, as explained above. However, it has no doubt been felt in these instances that codification on this subject achieves a reawakening of interest and perhaps an ease of procedure.

14.5 EXISTING STATE AND MUNICIPAL LEGISLATION

The only statewide air pollution enactment to date is the Oregon Air Pollution Act (Oregon Laws, 1951, c.425). It creates an additional division within the State Board of Health to be known as the Air Pollution Authority of the State of Oregon. This Authority consists of five members, the State Health Officer and four appointed members, who serve staggered terms. The members receive no compensation but are reimbursed for expenses.

The Air Pollution Authority is charged with the development of a comprehensive program for the prevention and control of all sources of pollution in the air in the state, cooperating with other agencies, encouraging air pollution studies, collecting and disseminating information, making rules and regulations, considering complaints, making investigations, and holding hearings. The Authority may employ specialists and enter premises for inspections.

The Act provides:

"The discharge into the air of solids, liquids or gases so as to cause such injury to human, plant or animal life or to property as constitutes a public nuisance is contrary to the public authority of the State of Oregon."

Through the Attorney General, the Authority may bring proceedings in the name of the state. To abate a nuisance, a specific procedure is provided wherein, after the Authority determines that the Act or an order of the Authority has been violated, notice is given specifying the facts and requiring correction within a reasonable time period. If the person complained against does not choose to comply, he is entitled to a hearing before the Authority, following which the Authority enters an order. The order of the Authority becomes final unless appeal to the courts is taken within 20 days. A record of the proceedings before the Authority is kept, and this record constitutes the record on appeal.

In addition to the power given the Authority to enforce its orders, if sustained by the courts, violations of its regulations are made misdemeanors subject to the usual penalties.

There have been no Federal enactments.

Most air pollution legislation has been centered in municipalities. The Air Pollution Control Association, Pittsburgh, Pa., has completed a very interesting and significant statistical tabulation of air pollution control bureaus in the United States and Canada (1). From another recent study of municipal ordinances (5), it appeared that approximately 80 per cent of them have been adopted within the past 10 years. A breakdown of the ordinances studied reveals the following:

14.5.1 Smoke Density

The allowable smoke emission for stationary stacks was examined for 82 city, county, or state smoke-abatement ordinances. In most instances the information was taken directly from the ordinances. Additional information was obtained from a comparison of ordinances by E. B. Brundage (4).

Two principal methods for measuring smoke density are used in ordinances. Both are optical methods, are somewhat subjective, and at best are only an index to the amount being discharged by the stack. Variations in results are due to differences in the light intensity, opaqueness of the atmosphere, and individual differences in making the estimates. The methods follow:

Ringelmann Chart. This is the most widely used method and was sponsored by the U.S. Bureau of Mines. The smoke density is expressed on a scale of 0, 1, 2, 3, 4, and 5. No. 1 smoke allows 80 per cent of the transmitted light to pass through, No. 2 allows 60 per cent, No. 3 allows 40 per cent, No. 4 allows 20 per cent, and No. 5 allows none of the transmitted light to pass through. In actual use, the smoke density at the stack top is visually compared with a chart depicting the Ringelmann scale. Its use is applicable only to black smoke such as emanates from powerhouses; it cannot be applied to white smokes like those from cement kilns.

Umbrascope. The darkness of smoke is determined as follows: one thickness of gray glass of 60 per cent opacity is No. 1; two thicknesses is No. 2; three thicknesses, No. 3; and four thicknesses, No. 4. Ringelmann No. 3 is equivalent to No. 1 on the Umbrascope scale. Thus, any ordinances which continuously allow any smoke measured on the Umbrascope scale are no more restrictive than ordinances which permit a continuous No. 3 Ringelmann discharge.

Smokescope. An instrument developed by Mine Safety Appliance Company has created considerable interest. The M.S.A. Smokescope is an optical instrument for use in estimating the smoke in stack effluent. It embodies several features which maintain the accuracy and reproducibility of readings even though the stack is viewed from different angles or under different conditions of lighting and background. The stack is viewed through one barrel of the instrument, and the smoke is seen through an aperture in the center of a screen. Light from the area adjacent to the stack enters a second barrel where it illuminates standard-density film which is used for comparison with the smoke. An image of the reference film appears on the screen surrounding the aperture and so permits ease of comparison. By means of a lens, the image of the reference is made to appear to be at a distance equivalent to that of the stack so that refocusing of the eye is not needed while making a reading. Because illuminations of the smoke and the reference are simultaneously influenced by the same factors, automatic compensation is achieved for varying conditions. Accurate estimations of smoke density will always require the skill and accuracy of a trained observer. However, the Smokescope should supplement his skill and improve his accuracy.

A considerable variation exists in the provisions of the local ordinances on the Ringelmann smoke-density emission. Ordinances generally state the permissible smoke density, how long and under what circumstances denser smoke may be emitted. The more poorly written or older ordinances prohibit but do not define "dense smoke."

In order to rate the various ordinances, a classification system was devised and is defined in Table 14-1. The rating is generally of decreasing severity of requirements from Class I to VII, but this relation does not always hold. For example, in Class IB, Ringelmann No. 3 smoke may be exceeded for fire cleaning or building but may not in Class IIA.

Table 14-1. Definition of Ordinance Classifications Based on the Measurement of Smoke Density

Class I: Ordinances which allow only Ringelmann No. 1 smoke:
- A. With no exceptions stated
- B. Except for stated periods when cleaning or building fires or for other reason

Class II: Ordinances which allow short periods of No. 2 smoke:
- A. But may not exceed No. 3 for stated periods for fire cleaning or building
- B. And may exceed No. 3 for stated periods for fire cleaning or building

Class III: Ordinances which allow No. 2 smoke at all times:
- A. But may not exceed No. 2 at any time
- B. But may not exceed No. 3 for fire cleaning or building
- C. And may exceed No. 3 for fire cleaning or building

Class IV: Ordinances which allow short periods of No. 3 smoke:
- A. But may not exceed No. 3 at any time
- B. And may exceed No. 3 for fire cleaning and building

Class V: Ordinances which allow No. 3 smoke at all times:
- A. But may not exceed No. 3 at any time
- B. And allow periods in excess of No. 3 without specifying or in addition to fire cleaning or building
- C. And may exceed No. 3 for fire cleaning or building

Class VI: Ordinances which define smoke density using the Umbrascope and not in terms of the Ringelmann chart.

Class VII: Ordinances which do not, or only loosely, define the smoke density prohibited.

The 82 ordinances studied for which information was available have been classified and are shown in Table 14-2. Where possible, the year of adoption or year of latest revision is shown. The breakdown of the number of cities in each class is shown in Table 14-3, excluding the model ordinances such as suggested by the American Society of Mechanical Engineers (ASME). With the exception of the periods for cleaning or building fires, the ordinances of Classes I and II are roughly equivalent to, or more strict than, the model ASME ordinance. These account for about 40 per cent of the ordinances studies. About 7 per cent of the ordinances (Class VII) do not provide for a means of measuring smoke density.

Table 14-4 indicates the provisions of the four cities which utilize the Umbrascope method of scaling. All allow continuous discharge of Umbrascope No. 1, which is equivalent to Ringelmann No. 3. For this reason the ordinances of Class VI cannot be more strict than Class V, if the former are defined in equivalent Ringelmann chart terms.

14.5.2 Dust-loading Requirements

Sources of Information. Copies of ordinances of 70 cities were examined with respect to dust-loading requirements. The requirements of two, Toledo and Cleveland, were available from other sources.

Table 14-2. Ordinances on Basis of Smoke-density Classifications

Class I

Subclass A		Subclass B	
Year	Location	Year	Location
	Des Moines, Iowa	1928	Cedar Rapids, Iowa
1949	Utica, N.Y.	1948	East Orange, N.J.
		1949	Indianapolis, Ind.
		1945	Jefferson City, Mo.
		1948	Kingsport, Tenn.
		1945	Knoxville, Tenn.
		1949	Louisville, Ky.
		1941	Madison, Wis.
		1950	Montreal, Canada
		1947	Nashville, Tenn.
		1949	New Rochelle, N.Y.
		1947	NIMLO[a]
		1946	Pittsburgh, Pa.
		1948	St. Louis, Mo.
		1948	Syracuse, N.Y.
		1950	Yonkers, N.Y.

Class II

Subclass A		Subclass B	
Year	Location	Year	Location
1948	ASME[b]	1946	Baltimore, Md.
1947	CPCSA[c]	1949	Columbus, Ohio
1949	Charlotte, N.C.	1947	Cumberland, Md.
	Cleveland, Ohio	1948	Milwaukee Co., Wis.
	Davenport, Iowa[d]	1946	Monroe, Mich.
1947	Detroit, Mich.[d]	1947	Omaha, Nebr.
1950	Green Bay, Wis.	1947	Providence, R.I.
	Greenville, N.C.		
1947	Niagara Falls, N.Y.[d]		
	Ottawa, Canada[d]		
1949	St. Paul, Minn.[d]		

Class III

Subclass A		Subclass B	
Year	Location	Year	Location
1940	Alexandria, Va.	1939	Chicago, Ill.
	Washington, D.C.	1948	Kansas City, Mo.
		1948	Philadelphia, Pa.
		1950	Tacoma, Wash.

Subclass C	
Year	Location
1949	Allegheny County, Pa.
1947	Buffalo, N.Y.
1949	East Chicago, Ind.
1949	Evansville, Ind.
	Memphis, Tenn.
1949	Roanoke, Va.
1937	Rockford, Ill.
1948	Tonawanda, N.Y.

Table 14-2. Ordinances on Basis of Smoke-density Classifications (*Continued*)

Class IV

Subclass A		Subclass B	
Year	Location	Year	Location
	Atlanta, Ga.	1947	Akron, Ohio
		1949	Chattanooga, Tenn.
		1931	Duluth, Minn.
		1947	Richmond, Va.
			Toledo, Ohio
		1924	Wheeling, W.Va.
		1946	Wilmington, Del.

Class V

Subclass A		Subclass B	
Year	Location	Year	Location
1922	Sioux City, Iowa		Boston, Mass.
		1947	State of California
		1942	Charleston, W.Va.
		1948	Erie, Pa.
		1949	Fort Wayne, Ind.
		1921	Grand Rapids, Mich.
		1948	Highland Park, Mich.
		1931	Hudson Co., N.J.
		1947	Los Angeles, Calif.
			Peoria, Ill.
			Toronto, Canada
		1924	Wheeling, W.Va.

Subclass C

Year	Location
1913	Albany County, N.Y.
	Birmingham, Ala.
1914	Hartford, Conn.

Class VI

Year	Location
1947	Cincinnati, Ohio
	Hamilton, Ohio
1928	Lansing, Mich.
1948	Wyandotte, Mich.

Class VII

Year	Location
1917	Flint, Mich.
	Minneapolis, Minn.
1949	New York, N.Y.
	Windsor, Canada
	Winnipeg, Canada
1949	Winston Salem, N.C.

[a] Model Law of National Institute of Municipal Law Officers, Report No. 120, 1947.
[b] Model Law of American Society of Mechanical Engineers.
[c] Model Law of Coal Producers Committee for Smoke Abatement (same as ASME for smoke emission).
[d] Exactly the same provisions as ASME model ordinance concerning smoke-density emission.

Dust-loading Requirements of Ordinances. Of the 72 ordinances on which dust-loading requirements were available, 34 provided specific requirements with respect to the allowable dust loadings. The remaining ordinances either do not mention the subject or merely state that discharges should not be sufficient to cause nuisance conditions.

Table 14-3. Distribution of Localities According to Smoke-density Classifications[a]

Class	Subclass	Localities: Number	Localities: % of total	Class	Subclass	Localities: Number	Localities: % of total
I	A	2		IV	A	1	
	B	15			B	7	
	Total	17	21		Total	8	10
II	A	9		V	A	1	
	B	7			B	12	
					C	3	
	Total	16	20				
					Total	16	20
III	A	2					
	B	4		VI	Total	4	5
	C	8		VII	Total	6	7
	Total	14	17		Grand total	81	100

[a] Model ordinances excluded.

The classification of the 34 ordinances is given in Tables 14-5 and 14-6. In Table 14-5 those designated as Class I are shown; they are the same as, or similar to, the ASME or Coal Producers' Committee for Smoke Abatement (CPCSA) model ordinances in the dust-loading requirements. There are 21 such ordinances, about 30 per cent of those studied and 62 per cent of those with definite dust-loading limitations. The ordinances of Class I, Table 14-5, have been divided into four subclasses, depending on collection-efficiency requirements.

Table 14-4. Provisions of Ordinances in Class VI

Umbrascope scale: The scale number is equal to the number of thicknesses of gray glass of 60 per cent opacity. Number 1 is equivalent to Ringelmann No. 3.

Cincinnati, Ohio: Umbrascope scale No. 4 (dense smoke) prohibited except for one 6-min period/8 hr for fire cleaning; one 6-min period/24 hr for flue blows; and one 6-min period/24 hr for building a new fire.

Wyandotte, Mich.: Umbrascope scale No. 2 allowed 10 min/hr; No. 3 is allowed for three 2-min periods/hr; No. 4 is a violation.

Hamilton, Ohio: Umbrascope scale No. 4 (dense smoke) prohibited except for 10 min/hr.

Lansing, Mich.: Umbrascope scale No. 4 (dense smoke) prohibited except for 9 min/hr (not exceeding 2 min of continuous emission at any one time). One 20-min period/hr once every 24 hr permitted for building a new fire.

Subclass D has exactly the same provisions as the ASME model ordinance discussed in Art. 14.7.2. Subclass C, represented by Roanoke, Va., provides only for 85 per cent collection efficiency and does not differentiate between old and new plants as does the ASME ordinance. Subclass B ordinances have the provisions of the CPCSA model, which stipulated that no more than 15 per cent of the ash entering a collector can pass out through the stack. This may be very restrictive in actual enforcement of the ordinance, since dust loadings in certain instances may have to be well below 0.85 lb of dust per 1,000 lb of gas to obtain the 85 per cent efficiency required. Subclass A makes no requirements of collection efficiency but specifies the allowable dust loading. In certain instances this means that over 85 per cent efficiency is required to attain 0.85 lb dust per 1,000 lb gas. Subclasses A and B

Table 14-5. Dust-loading Requirements of Ordinances

Class I[a]

Subclass A		Subclass B		Subclass C		Subclass D	
Year	Location	Year	Location	Year	Location	Year	Location
1953	Akron, Ohio[b]	1947	CPCSA Model	1949	Roanoke, Va.	1948	ASME Model
1949	Columbus, Ohio	1947	Cincinnati, Ohio			1948	Allegheny Co., Pa.[c]
1950	Green Bay, Wis.	1946	Monroe, Mich.			1949	Charlotte, N.C.
1948	Milwaukee Co., Wis.	1950	Montreal, Canada				Davenport, Iowa
1950	New York, N.Y.[d]	1947	Niagara Falls, N.Y.			1949	East Chicago, Ind.
1947	Omaha, Nebr.	1948	Syracuse, N.Y.			1950	Tacoma, Wash.
1948	Philadelphia, Pa.						
1947	Providence, R.I.						
1948	St. Louis, Mo.[e]						
1949	St. Paul, Minn.						
1950	Toledo, Ohio[f]						

[a] Class I: Those ordinances which require 0.85 lb fly ash per 1,000 lb flue gas at 12 per cent CO_2 or 50 per cent excess air (0.25 gr/cu ft at 500°F and 50 per cent excess air).

Subclass A: No other provisions.

Subclass B: Not more than 15 per cent of ash entering a collector can leave the stack, i.e., 85 per cent collection efficiency is required.

Subclass C: No more than 85 per cent collection efficiency need be attained.

Subclass D: Same as Subclass C for collectors installed after the ordinance is in force, but no more than 75 per cent for collectors installed before (ASME provision).

[b] Effective date for this provision in ordinance adopted in 1949.

[c] Allegheny County requirements for collection on installations before the ordinance is not to exceed 2.50 lb per 1,000 lb at 50 per cent excess air or greater than 70 per cent collection.

[d] Except for steam-generating equipment.

[e] Not over 0.5 lb fly ash per 1,000 lb flue gas shall be retained on a 325-mesh screen.

[f] Exact subclass not known, since ordinance was not available.

Table 14-6. Dust-loading Requirements of Ordinances

Class II[a]

Location	Date	Ordinance requirements
Kingsport, Tenn.	1948	0.35 grain/cu ft at 350°F and 50% excess air (0.295 grains/cu ft at 500° F or 1.0 lb per 1,000 lb gas)
Detroit, Mich.	1947	0.30 grain/cu ft at 500°F and 50% excess air (1.0 lb per 1,000 lb gas)
Highland Park, Mich.	1948	Same as Detroit
Los Angeles, Calif.	1947	0.40 grain/cu ft at 12% CO_2, no temperature specified[b]
Cleveland, Ohio	....	0.425 grain/cu ft at 500°F and 50% excess air (1.45 lb per 1,000 lb)
Richmond, Va.	1947	2.0 lb per 1,000 lb gas at 12% CO_2 (0.60 grain/cu ft at 500°F)[c]
Baltimore, Md.	1946	Same as Baltimore[d]
Cumberland, Md.	1947	0.75 grain/cu ft at 500°F and 50% excess air (2.5 lb per 1,000 lb)[d]
Indianapolis, Ind.	1949	Same as Baltimore[d]
Nashville, Tenn.	1947	Same as Baltimore[d]
Pittsburgh, Pa.	1946	Same as Baltimore[d]
Pittsburgh proposed	....	0.30 grain/cu ft at 500°F and 50% excess air (1.0 lb/1,000 lb)
Tonawanda, N.Y.	....	0.75 grain/cu ft at stack temperature and 50% excess air
Utica, N.Y.	1949	Same as Baltimore[d]

[a] Class II: Those ordinances which permit specific limits that are greater than 0.85 lb fly ash per 1,000 lb of flue gas at 12 per cent CO_2 (0.25 grain/cu ft at 500°F and 50 per cent excess air).

[b] Special table of allowable discharge rate per hour depending on process weight per hour.

[c] Not more than 25 per cent of the dust entering a collector may leave it.

[d] Not over 0.2 grain/cu ft shall be retained on a 325-mesh screen.

may be highly restrictive under certain circumstances, while C and D may be more lenient.

In Table 14-6 are the 13 Class II ordinances that permit greater dust loadings than required by Class I laws. The requirements range from 1.0 to 2.5 lb dust per 1,000 lb gas at 12 per cent CO_2. Almost half of these (six) permit 2.5 lb dust per 1,000 lb gas (0.75 grain/cu ft at 500°F).

Effective Oct. 1, 1950, New York City adopted the Air Pollution Control Association (APCA) recommendations for solids emission from a stack; i.e., the steam load connected to the stack determines the allowable solids rate of emission. For fuel-burning or refuse-burning equipment not used for steam generation, New York City permits 0.85 lb dust per 1,000 lb of gas at 50 per cent excess air for combustion products with not over 0.40 lb of that dust larger than 325 mesh (0.0017 in. diameter).

In addition to the provision for solid products of combustion in Table 14-6, the Los Angeles ordinance also limits the rate of emission of dust* or fumes according to the rate of use of process weight. Process weight is the "total weight of all materials introduced into any specific process, which process may cause any discharge into the atmosphere. Solid fuels charged will be considered as part of the process weight, but liquid and gaseous fuels and combustion air will not."

Fuel-restriction Requirements. Table 14-7 lists those ordinances of which copies were obtained. Fourteen of these (20 per cent) place a restriction on the sale and/or use of high-volatile coal for hand-fired fuel-burning equipment. Mechanically fired equipment is generally exempt from these restrictions. In fact, if it can be shown that hand-fired burners can utilize higher-volatile coal without violating other ordinance provisions, then the limit on coal type can usually be exceeded.

Annual or Periodic Inspections. In Table 14-7 are shown the 23 ordinances (32 per cent) which require annual or periodic inspections.

Penalties. Penalties listed in Table 14-7 show considerable variation. Several ordinances do not specify the penalties for noncompliance. Many only fine the offender. In some cases either fines or jail sentences, or both, may be meted out; in others, jail sentences may be resorted to when the offender cannot pay the imposed fine.

Prohibition of Other Pollutants. The ASME model ordinance defines fumes as "gases or vapors that are of such character as to create an uncleanly, destructive, offensive or unhealthful condition." Cities with ordinances which use this definition word for word or with some slight variations are:

Akron, Ohio
Buffalo, N.Y.
Chattanooga, Tenn.
Cincinnati, Ohio
Green Bay, Wis.
Highland Park, Mich.
Kingsport, Tenn.
Knoxville, Tenn.
Milwaukee County, Wis.
Davenport, Iowa
Columbus, Ohio
Detroit, Mich.
East Chicago, Ind.
New York, N.Y.
Providence, R.I.
Omaha, Nebr.
St. Paul, Minn.
Richmond, Va.

Of somewhat more elaborate nature are other definitions of fumes similar to the following: "escape of noxious acids, fumes or gases in such place or manner as to cause injury, detriment, nuisance or annoyance, to any person or persons or to the public or to endanger the comfort or repose, health or safety of any such person or persons or the public or in such manner as to cause or have natural tendency to cause

* Dusts are defined as "minute solid particles released into the air by natural forces or by mechanical processes such as crushing, grinding, milling, drilling, demolishing, shoveling, conveying, bagging, sweeping, etc."

Table 14-7. Ordinance Fuel Restrictions, Inspection Requirements, and Penalties

Location	Date[a]	Max vol content of coal for hand firing, %	Inspection requirement	Max fines per offense	
				Dollars	Days sentence
Akron, Ohio	1949	26	Annual	100	
Albany, N.Y.	1913	. .	....	50[b] 500[c]	None
Alexandria, Va.	1940	. .	....	300	3 mo.
Allegheny Co., Pa.	1949	23	....	100	30
Baltimore, Md.	1939	23	....	100	None
Bessemer, Ala.	1942				
Birmingham, Ala.					
Buffalo, N.Y.	1947	. .	Annual	250	None
California, state of					
Cedar Rapids, Iowa	1928	. .	....	100	30
Charleston, W.Va.					
Charlotte, N.C.	1950	. .	Periodic		
Chattanooga, Tenn.	1949	. .	Annual	50	None
Chicago, Ill.	1947	. .	Annual	200	None
Cincinnati, Ohio	1947	26	Annual		
Columbus, Ohio	1949	. .	Annual	50[b] 100[c] 200[d]	None
Cumberland, Md.	1947	. .	....	100	None
Davenport, Iowa	....	. .	....	100	None
Des Moines, Iowa	....	. .	....	100	30
Detroit, Mich.	1947	23	Annual	100	30
Duluth, Minn.	1931	. .	....	100	85
East Chicago, Ind.	1949	. .	Annual	300	180
East Orange, N.J.	1948	. .	Annual	200	30
Erie, Pa.	1948	. .	....	25[b] 50[c] 100[d]	None
Evansville, Ind.	1949	. .	....	300	180
Flint, Mich.	1917	. .	....	100	90
Fort Wayne, Ind.	1949	. .	....	300	None
Grand Rapids, Mich.	1921	. .	....	100	90
Green Bay, Wis.	1950	. .	....	50[b] 100[c]	30
Hamilton, Ohio					
Hartford, Conn.	1914	. .	....	50[b] 100[c]	None
Highland Park, Ill.	1948	23	....	100	30
Hudson Co., N.J.	1931	. .	....	50	30
Indianapolis, Ind.	1948	. .	Annual	300	None
Jefferson City, Mo.	1945	. .	Annual	100	None
Kansas City, Mo.	1948	. .	Periodic		
Kingsport, Tenn.	1948	25	....	.....	None
Knoxville, Tenn.	1945	. .	Annual		
Lansing, Mich.	1928	. .	....	50	30
Los Angeles, Calif.	1950				
Louisville, Ky.	1949	. .	Annual	100	None
Madison, Wis.	1941	. .	....	100	None
Massachusetts, state of	1934	. .	....	100	None
Memphis, Tenn.	....	. .	....	50	None
Milwaukee Co., Wis.	1948	23	....	100	30
Monroe, Mich.	1946	. .	....	100	90
Montreal, Canada	1950	25	....	40	60
Nashville, Tenn.	1947	23	Annual	50	30
New Orleans, La.	....	. .	Annual		
New Rochelle, N.Y.	1949				
New York, N.Y.	1949	24	....	100[b] 500[c]	3 mo. 6 mo.
Niagara Falls, N.Y.	1947	. .	....	100[b] 500[c]	None None

Table 14-7. Ordinance Fuel Restrictions, Inspection Requirements, and Penalties *(Continued)*

Location	Date[a]	Max vol content of coal for hand firing, %	Inspection requirement	Max fines per offense	
				Dollars	Days sentence
Omaha, Neb	1947	..	Annual	50[b]	None
				100[c]	None
				200[d]	None
Philadelphia, Pa	1948	..	Periodic	100	30
Pittsburgh, Pa	1941	20	Annual	100	30
Providence, R.I	1947	..	Annual	100	None
Richmond, Va	1947	..	....	100	None
Roanoke, Va	1949	..	....	300	None
Rochester, N.Y	1914	..	....	150	150
Rockford, Ill	1937	..	....	100	None
St. Louis, Mo	1948	23	....	100	None
St. Paul, Minn	1949	..	....	100	30
Sioux City, Iowa	1922	..	....	100	None
Syracuse, N.Y	1948	23	Annual	100	30
Tacoma, Wash	1950	..	....	300	90
Tonawanda, N.Y	1948	..	Annual	100	30
Utica, N.Y	....	..	....	100	30
Wheeling, W.Va	1924	..	....	100	1 yr.
Wilmington, Del	1946	..	....	50	None
Winston-Salem, N.C	1949	..	....	100[b]	None
				500[c]	None
Wyandotte, Mich	1944	..	....	100	90
Yonkers, N.Y	1950	..	....	150	100

[a] Date of passage or latest revision.
[b] First offense.
[c] Second offense.
[d] Third offense.

injury or damage to business or property." This or similar definitions are found in the ordinances of the following cities:

Charlotte, N.C.	New Rochelle, N.Y.
Fort Wayne, Ind.	Philadelphia, Pa.
Indianapolis, Ind.	Roanoke, Va.
Kansas City, Mo.	Utica, N.Y.
Louisville, Ky.	Pittsburgh, Pa.
Nashville, Tenn.	Tonawanda, N.Y.

Certain ordinances specifically include the obstruction of visibility in the definition or prohibition of fumes. These cities include:

Chattanooga, Tenn.	Roanoke, Va.
Kingsport, Tenn.	Niagara Falls, N.Y.
Knoxville, Tenn.	

Roanoke prohibits "visible fumes other than smoke regardless of color or shade of a density equal to or greater than No. 1 of the Ringelmann Chart" even though the Ringelmann chart cannot be applied to a white smoke or to a colored gas such as nitrogen tetroxide.

Detroit specifically includes metallic dusts in its definition of fumes.

Some ordinances are fairly specific in definitions of terms used in their ordinances. Niagara Falls, for example, uses the following definitions:

"Fumes—Colloidal systems which are formed from chemical reactions or by processes like combustion, distillation, sublimation, calcination or condensation.

"Mists—Disperse systems formed by the condensation of liquid vapor on nuclei such as submicroscopic particles of dust or gaseous ions, or by the atomization of liquids.

"Vapors—Any material in a gaseous state which is formed from a substance, usually a liquid, by increase in temperature."

Niagara Falls also has the only ordinance that includes "deleterious" effects of fumes on trees, plants, or other vegetation in its prohibitions.

The only ordinance that specifically limits the discharge of a particular material (except fly ash) is that of Los Angeles. Sulfur compounds expressed as SO_2 could not be discharged if they were present in a gas in excess of 0.2 per cent by volume. The discharge of material on the basis of process weight utilization rate has already been discussed. This applies to processes other than stationary boiler operation.

14.6 FUTURE LEGISLATION

It is probable that future legislation will be more restrictive, whether the enactments be by state or local governments. This is the consequence of greater demands for control of pollution since the Donora smog incident and the dissemination of the stricter model ordinances, such as those of the ASME or APCA.

Of cardinal importance is a determination of objectives based on facts before a law is enacted. Consequently, the law must encourage the maximum cooperation by industry. Lasting results cannot be achieved overnight. Everyone is persuaded that a job must be done. Intelligence dictates that the job can best be done with cooperation and understanding.

The 1953 sessions of state legislatures and city councils were bombarded with proposed bills on the subject of air pollution. The most significant trend would seem to be the popularity of proposing legislation at the state level either for the purpose of authorizing a survey and study, or for the purpose of establishing a state-wide air pollution control agency.

In these legislative bills and proposed ordinances we find a great variety of definitions of air pollution and a great difference of viewpoint in the kind of control bureau or agency which is to be established. Similarly, there is no unanimity of thought on the mechanics of enforcement. Some bills provide for the bureau to act as the first step in the judicial processes with appeals to the courts, and others provide, in essence, that if the alleged violator does not comply with a directive of the bureau after receiving due notice from the bureau then the bureau is to certify the matter to a district attorney or other public law authority who will prosecute the alleged offender through existing court procedures. In such arrangements it is frequently not clear what effect, if any, the prior findings of the bureau are to have in the evidence presented to the court.

One current bill, interestingly, would, by state enactment, enlarge the authority of each municipal air pollution bureau to cover an area within 3 miles of the territorial limits of the municipality involved.

Another aspect receiving a great deal of current attention in bills offered to the various legislatures is that of correlating and combining laws pertaining to water pollution, air pollution, and in-plant exposure to employees.

One potential danger in some of the current bills offered to the legislatures, which bills would add additional provisions to and supplement existing laws, is the possibility that such a state may end up with far too many cooks supervising the pollution stew, with state boards, regional boards, special bureaus, district attorneys, the attorney general, and others all vested with authority to investigate and prosecute violations. Such a hodgepodge can never be constructive and will inevitably fail to do an effective job.

14.7 RECOMMENDED LEGISLATION

14.7.1 Manufacturing Chemists' Association

The Air Pollution Abatement Committee of the Manufacturing Chemists' Association, after considerable study of this matter, has published a pamphlet entitled "A Rational Approach to Air Pollution Legislation." This emphasizes the need for determining objectives before legislation is enacted and presents many helpful considerations that should be fully explored before any legislation is attempted.

14.7.2 Model Ordinances

Several organizations have suggested model ordinances containing specific provisions regulating smoke density and dust loading; these are referred to below.

Smoke Density. The model smoke ordinance section as finally approved by the ASME in 1948 has been approved by the Stoker Manufacturers' Association and has been discussed jointly at meetings of the ASME and APCA.

The smoke-emission provisions of the ordinance prohibit the emission of No. 2 smoke or darker on the Ringelmann chart except (1) for a period or periods aggregating 4 min in any 30 min when No. 2, but not darker, is allowed; (2) for a period or periods aggregating 3 min in any 15 min of No. 3, but not darker, when building a new fire or when breakdown of equipment occurs.

The National Institute of Municipal Law Officers (NIMLO) is an organization of municipal legal officials, such as city attorney, city council, etc. The NIMLO ordinance was published in 1947.

The ordinance prohibits smoke of No. 2 Ringelmann or darker. Greater than No. 2 is permitted for a period of not over 6 min in any hour when a fire is being cleaned or a new fire is being built.

The smoke-emission allowances of the model ordinance of the Coal Producers' Committee for Smoke Abatement (CPCSA) are the same as those in the ASME ordinance.

Dust-loading Requirements. The ASME model ordinance provides that the dust loading of stack gases shall not exceed 0.85 lb dust per 1,000 lb of gases adjusted to 50 per cent excess air for fuel-burning equipment. Exceptions to this are: an efficiency of more than 85 per cent is not required for collectors installed subsequent to the effective date of the ordinance, and not more than 75 per cent efficiency is required for collectors before the effective date.

The subcommittee on Emission of Solids from Chimneys of the APCA proposed a standard for solids emission in the spring of 1950. At this writing it still has not been officially endorsed by the Society. The solids-emission rate (lb dust/hr) allowable from a stack is based on the total average steam-generation output (lb steam/hr) produced by the boilers connected to that stack. Below 100,000 and above 1,000,000 lb steam per hour, the allowable solids emission is directly proportional to the steam load; i.e., a tenfold increase in steam generation permits a tenfold increase in solids emission. Between 100,000 and 1,000,000 lb steam per hour, the increase in allowable solids emission is no longer linear with respect to increases in the steam-generation rate. A tenfold increase in steam load between 100,000 and 1,000,000 lb steam per hour results in less than a fourfold increase in allowable solids-emission rate.

The increased efficiency required of the higher-capacity plants penalizes the larger plants. The APCA committee justified this on the basis that the cost of air pollution prevention should be weighed against the cost of the plant and each size plant be required "to conform to the highest standards practical at the present state of the art so that each plant will then be doing the best that can be reasonably expected of it."

While the solids-emission provisions of the CPCSA ordinance were based on the ASME model, the following change was made. In place of the ASME provision that not more than 85 per cent of the solids entering the collection device need be retained, CPCSA provides: "In no case shall more than 15 per cent of the total dust measured before entering the collection device be emitted from the stack." The allowable 0.85 lb dust per 1,000 lb of stack gas is the same as the ASME model.

Obviously, the 15 per cent allowance makes this ordinance more strict than the ASME, since in some cases, the dust loading may have to be reduced below 0.85 lb per 1,000 lb to meet this other provision.

14.8 ACKNOWLEDGMENTS

Grateful acknowledgment is made of the generosity of the following people and organizations who have both permitted the use of their material and given freely of their time and comments regarding the preparation of this section: Manufacturing Chemists' Association and the Air Pollution Abatement Committee of that Association; Air Pollution Control Association; L. L. Falk; E. B. Brundage; and Thomas C. Wurts, Director, Bureau of Smoke Control of Allegheny County, Pa.

REFERENCES

1. Air Pollution Control Association: Statistical Tabulation of Air Pollution Control Bureaus in the United States and Canada, Pittsburgh, Pa., 1952.
2. Babbit, H. E.: The Administration of Stream Pollution Prevention in Some States, presented at the Sixth Industrial Waste Conference at Purdue University on Feb. 21, 1951.
3. Best, B. E.: A Rational Approach to Air Pollution Legislation, *Arch. Ind. Hyg. and Occupational Med.*, **5**, 517–526 (1952).
4. Brundage, E. B.: Comparison of ASME Code for Smoke Emission with Codes of Cities, *Proc.* 42d Annual Meeting, Smoke Prevention Association of America, Birmingham, Ala., May 23–27, 1949.
5. Falk, L. L.: Survey of Air Pollution Ordinances, E. I. du Pont de Nemours & Co., Inc. (extracted from Report ED 2143, to Manufacturing Chemists' Association, Inc., committee members), Apr. 26, 1941.
6. Oregon, State of: Senate Bill No. 9, 46th Legislative Assembly, Regular Session, 1951.

TABLE OF CONVERSION FACTORS

To Convert from	*To*	*Multiply by*
Angstrom units	Centimeters	1×10^{-8}
Angstrom units	Microns	1×10^{-4}
Atmospheres	Centimeters of mercury at 0°C	76
Atmospheres	Dynes per square centimeter	1.01325×10^{6}
Atmospheres	Feet of water at 39.1°F	33.899
Atmospheres	Grams per square centimeter	1,033.3
Atmospheres	Inches of mercury at 32°F	29.921
Atmospheres	Millimeters of mercury	760
Atmospheres	Pounds per square inch	14.696
Atmospheres	Bars	1.0133
Barrels, oil	Gallons (U.S.)	42
Boiler horsepower	Btu per hour	33,479
Btu	Calories (gram)	252
Btu	Foot-pounds	777.98
Btu	Joules (absolute)	1,054.8
Btu per minute	Foot-pounds per second	12.96
Calories, gram	Btu	3.968×10^{-3}
Calories, gram	Foot-pounds	3.087
Calories, gram	Joules	4.185
Candles (International)	Lumens (International) per steradian	1.0000
Candlepower (spherical)	Lumens	12.566
Candles per square inch	Lamberts	0.48695
Centimeters	Angstrom units	1×10^{8}
Centimeters	Feet (British or U.S.)	0.032808
Centimeters	Inches (U.S.)	0.393700
Centimeters of mercury at 0°C	Atmospheres	0.013158
Centimeters of mercury at 0°C	Dynes per square centimeter	1.33322×10^{4}
Centimeters of mercury at 0°C	Kilograms per square meter	135.95
Centimeters of mercury at 0°C	Pounds per square foot	27.845
Centimeters per second	Feet per minute	1.9685
Centimeters per second	Miles per hour	0.02237
Centipoises	Pound mass per foot-second	0.672×10^{-3}
Centipoises	Kilogram mass per meter-second	1×10^{-3}
Cubic feet (U.S.)	Gallons (U.S.)	7.481
Cubic feet (U.S.)	Liters	28.316
Dynes per square centimeter	Millimeters of mercury	7.5006×10^{-4}
Dynes per square centimeter	Pounds per square foot	0.0020886
Dynes per square centimeter	Pounds per square inch	1.4504×10^{-5}
Ergs	Btu (mean)	9.4805×10^{-11}

To Convert from	*To*	*Multiply by*
Ergs	Calories, gram (mean)	2.3889×10^{-8}
Ergs	Joules	1×10^{-7}
Ergs per square centimeter	Dynes per centimeter	1
Feet (U.S.)	Centimeters	30.48006096
Feet per minute	Centimeters per second	0.508001
Feet per minute	Meters per second	0.00508001
Feet per minute	Miles per hour	0.011364
Foot-candles	Lumens incident per square foot	1
Foot-candles	Lumens per square meter	10.764
Gallons (U.S.)	Barrels (U.S., liquid)	0.031746
Gallons (U.S.)	Liters	3.78533
Gallons (U.S.)	Cubic feet	0.13368
Grains	Grams	0.064798918
Grains	Milligrams	64.798918
Grains	Pounds (avoirdupois)	1/7,000
Grains per cubic foot	Milligrams per cubic meter	2.796×10^{-3}
Grains per gallon (U.S.)	Parts per million	17.118
Grams	Grains	15.4324
Gravity	Centimeters per square second	980.665
Gravity	Feet per square second	32.174
Kilometers	Centimeters	1×10^{5}
Kilometers	Feet	3280.83
Kilometers	Miles (U.S.)	0.6213699495
Liters	Cubic feet	0.035316
Liters	Gallons (U.S.)	0.26417762
Liters	Ounces (U.S., fluid)	33.8143
Liters	Quarts (U.S., liquid)	1.056681869
Lumens per square foot	Foot-candles	1
Meters	Feet (U.S.)	3.280833333
Meters	Inches (U.S.)	39.3700
Micrograms per cubic meter	Grains per cubic foot	0.35757
Microns	Inches	3.937×10^{-5}
Microns	Centimeters	1×10^{-4}
Miles (U.S., statute)	Feet	5,280
Miles (U.S., statute)	Yards	1,760
Miles per hour	Centimeters per second	44.7041
Miles per hour	Feet per minute	88
Miles per hour	Feet per second	1.4667
Miles per hour	Meters per minute	26.82
Millibars	Atmospheres	98.692
Millibars	Millimeters of mercury	1.329×10^{-5}
Millibars	Dynes per square centimeter	1.000×10^{9}
Milligrams	Grains	0.01543236
Milligrams	Ounces (avoirdupois)	3.52739×10^{-5}
Milligrams per cubic meter	Grains per cubic foot	357.57
Milliliters	Cubic centimeters	1.000027
Milliliters	Cubic inches	0.061025
Millimeters	Inches (U.S.)	0.0393700
Millimeters of mercury at 0°C	Atmospheres	0.00131579
Millimeters of mercury at 0°C	Dynes per square centimeter	1,333.22
Millimeters of mercury at 0°C	Grams per square centimeter	1.3595

To Convert from	*To*	*Multiply by*
Millimicrons or micromillimeters	Angstrom units	10
Millimicrons	Meters	1×10^{-9}
Ounces (avoirdupois)	Grams	28.349527
Ounces (avoirdupois)	Grains	437.5
Ounces (U.S., fluid)	Milliliters	29.5729
Parts per million	Per cent	1×10^{4}
Per cent	Parts per million	1×10^{-4}
Pints (U.S., dry)	Liters	0.550599
Pints (U.S., liquid)	Cubic inches	28.875
Pints (U.S., liquid)	Liters	0.473168
Pounds (avoirdupois)	Grains	7,000
Pounds (avoirdupois)	Grams	453.5924277
Pounds (avoirdupois)	Tons (long)	4.464286×10^{-4}
Pounds (avoirdupois)	Tons (metric)	4.5359243×10^{-4}
Pounds (avoirdupois)	Tons (short)	5×10^{-4}
Pounds per hour	Grams per second	0.12599
Pounds per hour	Tons per stack day	1.2×10^{-2}
Quarts (U.S., dry)	Cubic centimeters	1,101.23
Quarts (U.S., dry)	Cubic inches	67.2006
Quarts (U.S., dry)	Cubic feet	0.038889
Quarts (U.S., dry)	Liters	1.10120
Quarts (U.S., liquid)	Milliliters	946.358
Quarts (U.S., liquid)	Cubic inches	57.749
Quarts (U.S., liquid)	Liters	0.946333
Square centimeters	Square feet	0.0010764
Square centimeter	Square inch	0.15500
Square feet (U.S.)	Square meters	0.09290341
Square feet (U.S.)	Acres	2.29568×10^{-5}
Square inches (U.S.)	Square yards	1/1,296
Square kilometers	Square miles (U.S.)	0.3861006
Square meters	Square yards (U.S.)	1.195985
Square miles	Acres	640
Square miles	Square feet	2.78784×10^{7}
Square miles	Square yards	3.0976×10^{6}
Square miles	Square meters	2,589,998
Square yards (U.S.)	Square meters	0.83613
Tons per stack day	Pounds per hour	83.3
Yards (U.S.)	Meters	0.91440183
Yards (U.S.)	Miles	5.68182×10^{-4}

INDEX